A Second Course in Calculus

Harley Flanders Tel Aviv University
Robert R. Korfhage Southern Methodist University
Justin J. Price Purdue University

ACADEMIC PRESS **New York and London**
A Subsidiary of Harcourt Brace Jovanovich, Publishers

ACADEMIC PRESS, INC.
111 Fifth Avenue, New York, New York 10003

United Kingdom Edition published by
ACADEMIC PRESS, INC. (LONDON) LTD.
24/28 Oval Road, London NW1

Library of Congress Cataloging in Publication Data

Flanders, Harley.
A second course in calculus.

1. Calculus. I. Korfhage, Robert R., joint author. II. Price, Justin J., joint author. III. Title.
QA303.F5823 515 73-18943
ISBN 0-12-259662-5

PRINTED IN THE UNITED STATES OF AMERICA

Differentiation Rules

$f(t)$	$\dfrac{df}{dt} = \dot{f}(t)$
cu	$c\dot{u}$
$u + v$	$\dot{u} + \dot{v}$
uv	$u\dot{v} + v\dot{u}$
$\dfrac{u}{v}$	$\dfrac{v\dot{u} - u\dot{v}}{v^2}$
$u[v(t)]$	$\dot{u}[v(t)]\dot{v}(t)$

$$\frac{d}{dt} f[x(t), y(t)]$$
$$= \frac{\partial f}{\partial x} \dot{x} + \frac{\partial f}{\partial y} \dot{y}.$$

Differentiation Formulas

$f(t)$	$\dfrac{df}{dt} = \dot{f}(t)$
c	0
t^α	$\alpha t^{\alpha-1}$
e^t	e^t
a^t	$(\ln a)a^t$
$\ln t$	$\dfrac{1}{t}$
$\sin t$	$\cos t$
$\cos t$	$-\sin t$
$\tan t$	$\sec^2 t$
$\cot t$	$-\csc^2 t$
$\sec t$	$\tan t \sec t$
$\csc t$	$-\cot t \csc t$
$\arcsin t$	$\dfrac{1}{\sqrt{1 - t^2}}$
$\arctan t$	$\dfrac{1}{1 + t^2}$
$\sinh t$	$\cosh t$
$\cosh t$	$\sinh t$
$\tanh t$	$\operatorname{sech}^2 t$

Polar Coordinates

$$\begin{cases} x = r \cos \theta \\ y = r \sin \theta \end{cases}$$

Spherical Coordinates

$$\begin{cases} x = \rho \sin \phi \cos \theta \\ y = \rho \sin \phi \sin \theta \\ z = \rho \cos \phi \end{cases}$$

Integrals (constant of integration omitted)

1. $$\int u\, dv = uv - \int v\, du$$

2. $$\int \frac{dx}{x} = \ln |x|$$

3. $$\int \frac{dx}{a^2 + x^2} = \frac{1}{a} \arctan \left(\frac{x}{a}\right)$$

4. $$\int \frac{dx}{a^2 - x^2} = \frac{1}{2a} \ln \left| \frac{x + a}{x - a} \right|$$

5. $$\int \frac{dx}{(a^2 + x^2)^n} = \frac{x}{(2n - 2)a^2(a^2 + x^2)^{n-1}} + \frac{2n - 3}{(2n - 2)a^2} \int \frac{dx}{(a^2 + x^2)^{n-1}} \qquad (n > 1)$$

6. $$\int \frac{dx}{\sqrt{a^2 - x^2}} = \arcsin \left(\frac{x}{a}\right)$$

7. $$\int \sqrt{a^2 - x^2}\, dx = \frac{x}{2} \sqrt{a^2 - x^2} + \frac{a^2}{2} \arcsin \left(\frac{x}{a}\right)$$

8. $$\int \frac{dx}{\sqrt{x^2 \pm a^2}} = \ln |x + \sqrt{x^2 \pm a^2}|$$

9. $$\int \sqrt{x^2 \pm a^2}\, dx = \frac{x}{2} \sqrt{x^2 \pm a^2} \pm \frac{a^2}{2} \ln |x + \sqrt{x^2 \pm a^2}|$$

(continued inside back cover)

Vectors

$$\mathbf{a} \cdot \mathbf{b} = a_1 b_1 + a_2 b_2 + a_3 b_3$$

$$\mathbf{a} \times \mathbf{b} = \begin{vmatrix} \mathbf{i} & \mathbf{j} & \mathbf{k} \\ a_1 & a_2 & a_3 \\ b_1 & b_2 & b_3 \end{vmatrix}$$

$$\nabla = \mathbf{i} \frac{\partial}{\partial x} + \mathbf{j} \frac{\partial}{\partial y} + \mathbf{k} \frac{\partial}{\partial z}$$

$$\nabla f = \operatorname{grad} f = (f_x, f_y, f_z)$$

$$\nabla \cdot \mathbf{v} = \operatorname{div} \mathbf{v} = u_x + v_y + w_z$$

$$\nabla \times \mathbf{v} = \operatorname{curl} \mathbf{v} = \begin{vmatrix} \mathbf{i} & \mathbf{j} & \mathbf{k} \\ \partial/\partial x & \partial/\partial y & \partial/\partial z \\ u & v & w \end{vmatrix}$$

Contents

Preface xi

1. INFINITE SERIES AND INTEGRALS

1. Infinite Series 1
2. Convergence and Divergence 4
3. Tests for Convergence 8
4. Series with Positive and Negative Terms 12
5. Improper Integrals 16
6. Convergence and Divergence 23
7. Relation to Infinite Series 30
8. Other Improper Integrals 35
9. Some Definite Integrals [optional] 40
10. Stirling's Formula [optional] 43

2. TAYLOR APPROXIMATIONS

1. Introduction 52
2. Polynomials 52
3. Taylor Polynomials 56
4. Applications 63
5. Taylor Series 67
6. Derivation of Taylor's Formula 69

3. POWER SERIES

1. Introduction 72
2. Ratio Test 78
3. Expansions of Functions 82
4. Further Techniques 90

5. Binomial Series 97
6. Alternating Series 102
7. Applications to Definite Integrals [optional] 105
8. Uniform Convergence [optional] 109

4. SOLID ANALYTIC GEOMETRY

1. Coordinates and Vectors 116
2. Length and Dot Product 123
3. Lines and Planes 130
4. Linear Systems and Intersections 133
5. Cross Product 141
6. Applications 147

5. VECTOR CALCULUS

1. Vector Functions 157
2. Space Curves 161
3. Curvature 168
4. Velocity and Acceleration 174
5. Integrals 181
6. Polar Coordinates 188
7. Polar Velocity and Acceleration [optional] 193

6. FUNCTIONS OF SEVERAL VARIABLES

1. Introduction 197
2. Domains 199
3. Continuity 205
4. Graphs 209
5. Partial Derivatives 213
6. Maxima and Minima 217

7. LINEAR FUNCTIONS AND MATRICES

1. Introduction 224
2. Linear Transformations 227
3. Matrix Calculations 232

4. Applications 239
5. Quadratic Forms 244
6. Quadric Surfaces 253
7. Inverses 261
8. Characteristic Roots [optional] 269

8. SEVERAL VARIABLE DIFFERENTIAL CALCULUS

1. Differentiable Functions 281
2. Chain Rule 287
3. Tangent Plane 292
4. Gradient 297
5. Directional Derivative 302
6. Applications 306
7. Implicit Functions 310
8. Differentials 316
9. Proof of the Chain Rule [optional] 319

9. HIGHER PARTIAL DERIVATIVES

1. Mixed Partials 326
2. Higher Partials 328
3. Taylor Polynomials 333
4. Maxima and Minima 339
5. Applications 345
6. Three Variables 349
7. Maxima with Constraints [optional] 352
8. Further Constraint Problems [optional] 360

10. DOUBLE INTEGRALS

1. Introduction 369
2. Special Cases 372
3. Iterated Integrals 377
4. Applications 382
5. General Domains 387
6. Polar Coordinates 399

11. MULTIPLE INTEGRALS

1. Triple Integrals 409
2. Cylindrical Coordinates 419

3. Spherical Coordinates 427
4. Center of Gravity 440
5. Moments of Inertia 448

12. INTEGRATION THEORY

1. Introduction 457
2. Step Functions 458
3. The Riemann Integral 463
4. Iteration 472
5. Change of Variables 482
6. Applications of Integration 489
7. Improper Integrals [optional] 502
8. Numerical Integration [optional] 511

13. DIFFERENTIAL EQUATIONS

1. Introduction 517
2. Separation of Variables 520
3. Linear Differential Equations 526
4. Homogeneous Equations 530
5. Non-Homogeneous Equations 532
6. Applications 540
7. Approximate Solutions 549

14. SECOND ORDER EQUATIONS AND SYSTEMS

1. Linear Equations 556
2. Homogeneous Equations 557
3. Particular Solutions 562
4. Applications 565
5. Power Series Solutions 576
6. Matrix Power Series 580
7. Systems 585
8. Uniqueness of Solutions [optional] 591

15. COMPLEX ANALYSIS

1. Introduction 596
2. Complex Arithmetic 598
3. Polar Form 602

4. Complex Exponentials 610
5. Integration and Differentiation 615
6. Applications to Differential Equations 619
7. Applications to Power Series 622

Mathematical Tables

1. Trigonometric Functions 629
2. Trigonometric Functions for Angles in Radians 630
3. Four-Place Mantissas for Common Logarithms 632
4. Antilogarithms 634
5. Exponential Functions 636

Answers to Selected Exercises 643

Index 683

Preface

This text, designed for a second year calculus course, can follow any standard first year course in one-variable calculus. Its purpose is to cover the material most useful at this level, to maintain a balance between theory and practice, and to develop techniques and problem solving skills.

The topics fall into several categories:

Infinite series and integrals

Chapter 1 covers convergence and divergence of series and integrals. It contains proofs of basic convergence tests, relations between series and integrals, and manipulation with geometric, exponential, and related series. Chapter 2 covers approximation of functions by Taylor polynomials, with emphasis on numerical approximations and estimates of remainders. Chapter 3 deals with power series, including intervals of convergence, expansions of functions, and uniform convergence. It features calculations with series by algebraic operations, substitution, and term-by-term differentiation and integration.

Vector methods

Vector algebra is introduced in Chapter 4 and applied to solid analytic geometry. The calculus of one-variable vector functions and its applications to space curves and particle mechanics comprise Chapter 5.

Linear algebra

Chapter 7 contains a practical introduction to linear algebra in two and three dimensions. We do not attempt a complete treatment of foundations, but rather limit ourselves to those topics that have immediate application to calculus. The main topics are linear transformations in $\mathbf{R}^2$ and $\mathbf{R}^3$, their matrix representations, manipulation with matrices, linear systems, quadratic forms, and quadric surfaces.

Differential calculus of several variables

Chapter 6 contains preliminary material on sets in the plane and space, and the definition and basic properties of continuous functions. This is followed by partial derivatives with applications to maxima and minima. Chapter 8 continues with a careful treatment of differentiability and applications to tangent planes, gradients, directional derivatives, and differentials. Here ideas from linear algebra are used judiciously. Chapter 9 covers higher order partial derivatives, Taylor polynomials, and second derivative tests for extrema.

Multiple integrals

In Chapters 10 and 11 we treat double and triple integrals intuitively, with emphasis on iteration, geometric and physical applications, and coordinate changes. In Chapter 12 we develop the theory of the Riemann integral starting with step functions. We continue with Jacobians and the change of variable formula, surface area, and Green's Theorem.

Differential equations

Chapter 13 contains an elementary treatment of first order equations, with emphasis on linear equations, approximate solutions, and applications. Chapter 14 covers second order linear equations and first order linear systems, including matrix series solutions. These chapters can be taken up any time after Chapter 7.

Complex analysis

The final chapter moves quickly through basic complex algebra to complex power series, shortcuts using the complex exponential function, and applications to integration and differential equations.

Features

The key points of one-variable calculus are reviewed briefly as needed.

Optional topics are scattered throughout, for example Stirling's Formula, characteristic roots and vectors, Lagrange multipliers, and Simpson's Rule for double integrals.

Numerous worked examples teach practical skills and demonstrate the utility of the theory.

We emphasize simple line drawings that a student can learn to do himself.

Acknowledgments

We appreciate the invaluable assistance of our typists, Sara Marcus and Elizabeth Young, the high quality graphics of Vantage Art Inc., and the outstanding production job of the Academic Press staff.

1. Infinite Series and Integrals

1. INFINITE SERIES

One of the most important topics in mathematical analysis, both in theory and applications, is infinite series. The basic problem is how to add up a sum with infinitely many terms. At first that seems impossible; life is too short. However, suppose we look at the sum

$$1 + \frac{1}{2} + \frac{1}{4} + \cdots + \frac{1}{2^n} + \cdots$$

and start adding up terms. We find $1, \frac{3}{2}, \frac{7}{4}, \frac{15}{8}, \frac{31}{16}, \cdots$, numbers getting closer and closer to 2. The message is clear: in some limit sense the total of all the terms is 2.

If we try to add up terms of the sum

$$1 + 1 + 1 + \cdots,$$

we find $1, 2, 3, 4, \cdots$, numbers becoming larger and larger. The situation is hopeless; there is no reasonable total.

Let us now consider in some detail two important infinite sums.

Geometric Series

A **geometric series** is an infinite sum in which the ratio of any two consecutive terms is always the same:

$$a + ar + ar^2 + \cdots + ar^n + \cdots \qquad (a \neq 0, \quad r \neq 0).$$

Let s_n denote the sum of all terms up to ar^n,

$$s_n = a + ar + ar^2 + \cdots + ar^n.$$

If $r = 1$, then $s_n = a + a + \cdots + a = (n+1)a$, so $s_n \longrightarrow \pm\infty$. If $r \neq 1$, there is a simple formula for s_n:

$$s_n = a(1 + r + r^2 + \cdots + r^n) = a\left(\frac{1 - r^{n+1}}{1 - r}\right).$$

(To check, multiply both sides by $1 - r$.) If $|r| < 1$, then $r^{n+1} \longrightarrow 0$ as n increases. Hence a logical choice for the "sum" of the geometric series is

$a/(1-r)$. But if $|r| > 1$, then r^{n+1} grows beyond all bound, and the situation is hopeless. If $r = -1$, then s_n is alternately a and 0. There is no reasonable sum in this case either.

> An infinite geometric series
> $$a + ar + ar^2 + \cdots + ar^n + \cdots$$
> has the sum $a/(1-r)$ if $|r| < 1$, but no sum if $|r| \geq 1$.

Harmonic Series

The series

$$1 + \frac{1}{2} + \frac{1}{3} + \cdots + \frac{1}{n} + \cdots$$

is known as the **harmonic series.** It is not at all obvious, but the sums $s_n = 1 + \frac{1}{2} + \frac{1}{3} + \cdots + n^{-1}$ increase beyond all bound, so the series has no sum. To see why, we observe that

$$s_1 = 1 > \frac{1}{2},$$

$$s_2 = s_1 + \frac{1}{2} > \frac{1}{2} + \frac{1}{2} = \frac{2}{2},$$

$$s_4 = s_2 + \left(\frac{1}{3} + \frac{1}{4}\right) > s_2 + \left(\frac{1}{4} + \frac{1}{4}\right) > \frac{2}{2} + \frac{1}{2} = \frac{3}{2},$$

$$s_8 = s_4 + \left(\frac{1}{5} + \frac{1}{6} + \frac{1}{7} + \frac{1}{8}\right) > s_4 + \left(\frac{1}{8} + \frac{1}{8} + \frac{1}{8} + \frac{1}{8}\right) > \frac{3}{2} + \frac{1}{2} = \frac{4}{2}.$$

Similarly, $s_{16} > 5/2$, $s_{32} > 6/2$, $\cdots$, $s_{2^n} > (n+1)/2$. Now the sequence of sums s_n increases, and our estimates show s_n eventually passes any given positive number. (This happens very slowly it is true; around 2^{15} terms are needed before s_n exceeds 10 and around 2^{29} terms before it exceeds 20.)

REMARK: Both the geometric series for $0 < r < 1$ and the harmonic series have positive terms that decrease toward zero, yet one series has a sum and the other does not. This indicates the subtlety we must expect in our further study of infinite series.

EXERCISES

Find the sum:

1. $1 + \frac{1}{3} + \frac{1}{3^2} + \cdots + \frac{1}{3^9}$

2. $1 - \frac{1}{3} + \frac{1}{3^2} - + \cdots - \frac{1}{3^9}$

3. $\frac{1}{2}+\frac{1}{4}+\cdots+\frac{1}{256}$

4. $\left(\frac{2}{3}\right)^2+\left(\frac{2}{3}\right)^3+\cdots+\left(\frac{2}{3}\right)^6$

5. $3+\frac{3^2}{x}+\frac{3^3}{x^2}+\cdots+\frac{3^{n+1}}{x^n}$

6. $1-y^2+y^4-+\cdots+y^{20}$

7. $r^{1/2}+r+r^{3/2}+\cdots+r^4$

8. $(x+1)+(x+1)^2+\cdots+(x+1)^5$.

Find the sum of the series:

9. $1-\frac{2}{5}+\left(\frac{2}{5}\right)^2-\left(\frac{3}{5}\right)^3+-\cdots$

10. $\frac{1}{2}-\frac{1}{4}+\frac{1}{8}-\frac{1}{16}+-\cdots$

11. $\frac{1}{2^{10}}+\frac{1}{2^{11}}+\frac{1}{2^{12}}+\cdots$

12. $\frac{1}{3}+\frac{1}{27}+\frac{1}{243}+\cdots$

13. $\frac{1}{2+x^2}+\frac{1}{(2+x^2)^2}+\frac{1}{(2+x^2)^3}+\cdots$

14. $\frac{\cos\theta}{2}+\frac{\cos^2\theta}{4}+\frac{\cos^3\theta}{8}+\cdots$.

15. A certain rubber ball when dropped will bounce back to half the height from which it is released. If the ball is dropped from 3 ft and continues to bounce indefinitely, find the total distance through which it moves.
16. Trains A and B are 60 miles apart on the same track and start moving toward each other at the rate of 30 mph. At the same time, a fly starts at train A and flies to train B at 60 mph. Then it returns to train A, then to B, etc. Use a geometric series to compute the total distance it flies until the trains meet.
17. (cont.) Do Ex. 16 without geometric series.
18. A line segment of length L is drawn and its middle third is erased. Then (step 2) the middle third of each of the two remaining segments is erased. Then (step 3) the middle third of each of the four remaining segments is erased, etc. After step n, what is the total length of all the segments deleted?

Interpret the repeating decimals as geometric series and find their sums:

19. $0.11111\cdots$

20. $0.101010\cdots$

21. $0.434343\cdots$

22. $0.185185185\cdots$.

Show that the series have no sums:

23. $\frac{1}{2}+\frac{1}{4}+\frac{1}{6}+\frac{1}{8}+\cdots$

24. $1+\frac{1}{3}+\frac{1}{5}+\frac{1}{7}+\cdots$.

25. Find n so large that

$$\frac{1}{101}+\frac{1}{102}+\cdots+\frac{1}{n}>2.$$

26. Aristotle summarized Zeno's paradoxes as follows:

> I can't go from here to the wall. For to do so, I must first cover half the distance, then half the remaining distance, then again half of what still remains. This process can always be continued and can never be completed.

Explain what is going on here.

2. CONVERGENCE AND DIVERGENCE

It is time to formulate the ideas of Section 1 more precisely.

An **infinite series** is a formal sum

$$a_1 + a_2 + a_3 + \cdots.$$

Associated with each infinite series is its sequence $\{s_n\}$ of **partial sums** defined by

$$s_1 = a_1, \qquad s_2 = a_1 + a_2, \qquad \cdots, \qquad s_n = a_1 + a_2 + \cdots + a_n.$$

A series **converges** to the number S, or has **sum** S, if $\lim_{n\to\infty} s_n = S$. A series **diverges,** or has no sum, if $\lim_{n\to\infty} s_n$ does not exist.

A series that converges is called **convergent;** a series that diverges is called **divergent.**

Let us recall the meaning of the statement $\lim_{n\to\infty} s_n = S$. Intuitively, it means that as N grows larger and larger, the greatest distance $|s_n - S|$, for *all* $n \geq N$, becomes smaller and smaller. Precisely, for each $\epsilon > 0$, there is a positive integer N such that $|s_n - S| < \epsilon$ for all $n \geq N$. Let us rephrase the definition of convergence accordingly.

The infinite series $a_1 + a_2 + a_3 + \cdots$ converges to S if for each $\epsilon > 0$, there is a positive integer N such that

$$|(a_1 + a_2 + \cdots + a_n) - S| < \epsilon$$

whenever $n \geq N$.

Thus, no matter how small ϵ, you will get within ϵ of S by adding up enough terms. For each ϵ, the N tells how many terms are "enough". Naturally the smaller ϵ is, the larger N will be. From the way convergence is defined, the study of infinite series is really the study of *sequences* of partial sums. Hence we may apply everything we know about sequences.

We know that inserting, deleting, or altering any finite number of elements of a sequence does not affect its convergence or divergence. The same holds for series. For instance, if we delete the first 10 terms of the series $a_1 + a_2 + a_3 + \cdots$, then we decrease each partial sum s_n (for $n > 10$) by the amount $a_1 + a_2 + \cdots + a_{10}$. If the original series diverges, then so does the modified series. If it converges to S, then the modified series converges to $S - (a_1 + a_2 + \cdots + a_{10})$.

WARNING: In problems where we must decide whether a given infinite series converges or diverges, we shall often, without prior notice, ignore or change a (finite) batch of terms at the beginning. This, we now know, does not affect convergence.

Notation

The first term of a series need not be a_1. Often it is convenient to start with a_0 or with some other a_k.

It is also convenient to use summation notation and abbreviate $a_1 + a_2 + a_3 + \cdots$ by $\sum_{n=1}^{\infty} a_n$, and even simply $\sum a_n$. In summation notation, the partial sums s_n of an infinite series $\sum_{n=1}^{\infty} a_n$ are given by

$$s_n = \sum_{k=1}^{n} a_k.$$

Cauchy Criterion

Recall the Cauchy criterion for convergence of sequences:

> A sequence $\{s_n\}$ converges if and only if for each $\epsilon > 0$, there is a positive integer N such that
>
> $$|s_m - s_n| < \epsilon$$
>
> whenever $m, n \geq N$.

Thus all elements of the sequence beyond a certain point must be within ϵ of each other. The advantage of the Cauchy criterion is that it depends only on the elements of the sequence itself; you don't have to know the limit of a sequence in order to show convergence. That's a great help; sometimes it is very hard to find the exact limit of a sequence, whereas you may only need to know that the sequence does indeed converge to some limit.

Let us apply the Cauchy criterion to the partial sums of a series. We simply observe (for $m > n$) that

$$\begin{aligned} s_m - s_n &= (a_1 + a_2 + \cdots + a_n + a_{n+1} + \cdots + a_m) - (a_1 + a_2 + \cdots + a_n) \\ &= a_{n+1} + a_{n+2} + \cdots + a_m. \end{aligned}$$

> **Cauchy Test** An infinite series $\sum a_n$ converges if and only if for each $\epsilon > 0$, there is a positive integer N such that
>
> $$|a_{n+1} + a_{n+2} + \cdots + a_m| < \epsilon$$
>
> whenever $m > n \geq N$.

Thus beyond a certain point in the series, any block of consecutive terms, *no matter how long*, must have a very small sum.

In the last section we proved the harmonic series diverges by producing blocks of terms arbitrarily far out in the series whose sum exceeds $\frac{1}{2}$. In other words, we showed that the Cauchy test fails for $\epsilon = \frac{1}{2}$.

Suppose the Cauchy test is satisfied, and take $m = n + 1$. Then the block consists of just one term a_m, so $|a_m| < \epsilon$ when $m \geq N$. In other words, $a_m \longrightarrow 0$.

Necessary Condition for Convergence If the series $\sum a_n$ converges, then $\lim_{n\to\infty} a_n = 0$.

WARNING: This condition is not sufficient for convergence. The harmonic series $1 + \frac{1}{2} + \frac{1}{3} + \cdots$ diverges even though $1/n \longrightarrow 0$.

Positive Terms

Suppose an infinite series has only non-negative terms. Then its partial sums form an increasing sequence, $s_1 \leq s_2 \leq s_3 \leq s_4 \leq \cdots$. Recall that an increasing sequence must be one of two types: Either (a) the sequence is bounded above, in which case it converges; or (b) it is not bounded above, and it marches off the map to $+\infty$.

We deduce corresponding statements about series:

A series $a_1 + a_2 + a_3 + \cdots$ with $a_n \geq 0$ converges if and only if there exists a positive number M such that

$$a_1 + a_2 + \cdots + a_n \leq M \qquad \text{for all} \quad n \geq 1.$$

Using this fact, we can often establish the convergence or divergence of a given series by comparing it with a familiar series.

Comparison Test Suppose $\sum a_n$ and $\sum b_n$ are series with non-negative terms.

(1) If $\sum a_n$ converges and if $b_n \leq a_n$ for all $n \geq 1$, then $\sum b_n$ also converges.

(2) If $\sum a_n$ diverges and if $b_n \geq a_n$ for all $n \geq 1$, then $\sum b_n$ also diverges.

Proof: Let s_n and t_n denote the partial sums of $\sum a_n$ and $\sum b_n$ respectively. Then $\{s_n\}$ and $\{t_n\}$ are increasing sequences.

(1) Since $\sum a_n$ converges, $s_n \leq \sum_1^\infty a_n = M$ for all $n \geq 1$. Since $b_k \leq a_k$ for all k, we have $t_n \leq s_n$ for all n. Hence $t_n \leq s_n \leq M$ for all $n \geq 1$, so $\sum b_n$ converges.

(2) Since $\sum a_n$ diverges, the sequence $\{s_n\}$ is unbounded. Since $b_k \geq a_k$, we have $t_n \geq s_n$. Hence $\{t_n\}$ is also unbounded, so $\sum b_n$ diverges.

NOTE: It is important to apply the Comparison Test correctly. Roughly speaking, (1) says that "smaller than small is small" and (2) says that "bigger than big is big". However the phrases "smaller than big" and "bigger than small" contain little useful information.

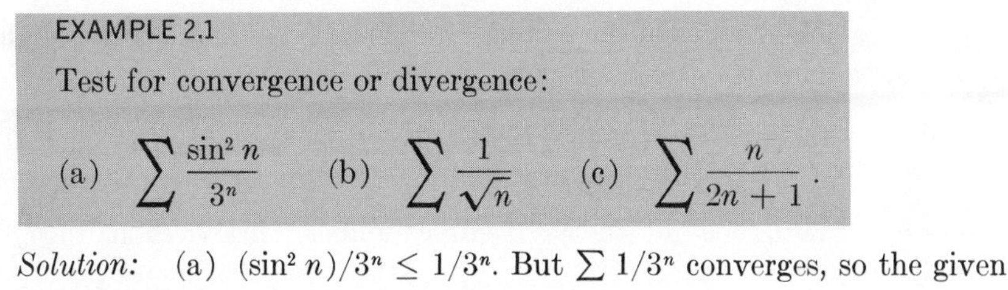

EXAMPLE 2.1

Test for convergence or divergence:

$$\text{(a)} \sum \frac{\sin^2 n}{3^n} \qquad \text{(b)} \sum \frac{1}{\sqrt{n}} \qquad \text{(c)} \sum \frac{n}{2n+1}.$$

Solution: (a) $(\sin^2 n)/3^n \leq 1/3^n$. But $\sum 1/3^n$ converges, so the given series converges.

(b) $1/\sqrt{n} \geq 1/n$. But $\sum 1/n$ diverges, so the given series diverges.

(c) Diverges because $a_n = n/(2n+1) \longrightarrow \frac{1}{2} \neq 0$.

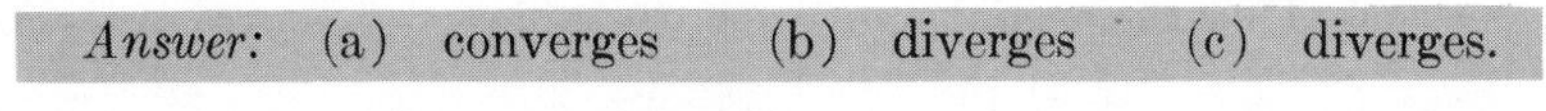

Answer: (a) converges (b) diverges (c) diverges.

p-Series

The comparison test is useful provided you have a good supply of known series. An excellent class of series for comparisons are those of the form $\sum 1/n^p$.

> The series $\sum 1/n^p$ diverges if $p \leq 1$ and converges if $p > 1$.

Proof: If $0 < p \leq 1$, then $1/n^p \geq 1/n$ and the series diverges by comparison with the divergent series $\sum 1/n$.

If $p > 1$, we shall show that the partial sums of the series are bounded. We use an important trick: we interpret s_n as an area and compare it with a region below the curve $y = 1/x^p$. See Fig. 2.1.

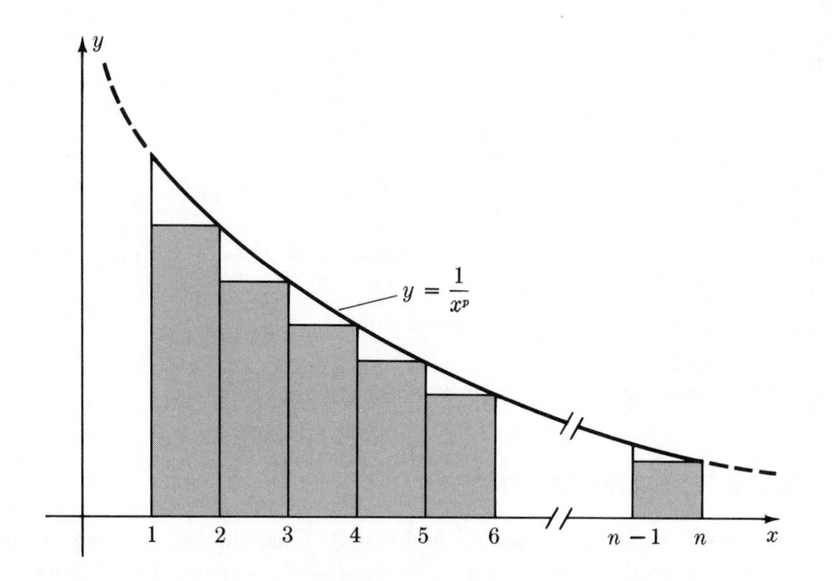

FIG. 2.1 Area under the curve exceeds the rectangular sum.

The combined areas of the rectangles shown is less than the area under the decreasing curve between $x = 1$ and $x = n$. Therefore

$$\frac{1}{2^p} + \frac{1}{3^p} + \cdots + \frac{1}{n^p} < \int_1^n \frac{dx}{x^p} = \frac{-1}{p-1}\frac{1}{x^{p-1}}\bigg|_1^n = \frac{1}{p-1}\left(1 - \frac{1}{n^{p-1}}\right).$$

Since $p - 1 > 0$, the right side is a positive number, a little less than $1/(p-1)$ for all values of n. Hence

$$s_n = 1 + \left(\frac{1}{2^p} + \frac{1}{3^p} + \cdots + \frac{1}{n^p}\right) < 1 + \frac{1}{p-1}$$

for $n \geq 1$. Thus the partial sums are bounded if $p > 1$, so the series converges.

EXERCISES

Determine whether the series converges or diverges:

1. $\sum \frac{1}{n^2+1}$
2. $\sum \frac{1}{2^n\sqrt{n}}$
3. $\sum \frac{n}{4n+3}$
4. $\sum \frac{1}{(2n-1)^2}$
5. $\sum \frac{1}{n\sqrt{n+3}}$
6. $\sum \frac{1+\sqrt[3]{n}}{n}$
7. $\sum \frac{n^2}{2n^4+7}$
8. $\sum \frac{1}{\ln n}$
9. $\sum \frac{1}{n^n}$
10. $\sum \frac{n}{(n+1)(n+3)(n+5)}$.
11. Show that $\sum_{n=1}^{\infty} 1/n^2 \leq 2$. [*Hint:* See text.]
12. Show that $\sum_{n=0}^{\infty} 1/n! < 3$. [*Hint:* Compare $n!$ with 2^n.]
13. Prove that if $\sum a_n$ and $\sum b_n$ converge, then so does $\sum (a_n + b_n)$, and find the sum.
14. If $\sum a_n$ and $\sum b_n$ diverge, show by examples that $\sum (a_n + b_n)$ may either converge or diverge.

Let $\sum a_n$ be a convergent series of positive terms:

15. Prove that $\sum a_n^2$ converges.
16. Show by examples that $\sum \sqrt{a_n}$ may either converge or diverge.

3. TESTS FOR CONVERGENCE

Suppose $\sum a_n$ is a given series, and $c \neq 0$. Then the two series $\sum a_n$ and $\sum ca_n$ either both converge or both diverge. For the partial sums of the series are $\{s_n\}$ and $\{cs_n\}$. Clearly these sequences converge or diverge together.

We can extend these remarks to a pair of series $\sum a_n$ and $\sum b_n$ where the ratios b_n/a_n are not constant, but restricted to a suitable range. Throughout the rest of this section *all series will have only positive terms.*

Let $\sum a_n$ and $\sum b_n$ be given series with positive terms. Suppose there exist positive numbers c and d such that

$$c \le \frac{b_n}{a_n} \le d$$

for all sufficiently large n. Then the series both converge or both diverge.

Proof: If $\sum a_n$ converges, then so does $\sum da_n$. But $b_n \le da_n$, so $\sum b_n$ converges.

If $\sum a_n$ diverges, then so does $\sum ca_n$. But $b_n \ge ca_n$, so $\sum b_n$ diverges. Done.

The conditions of the preceding test are automatically satisfied if the ratios b_n/a_n actually approach a positive limit L. Then, by the definition of limit with $\epsilon = \frac{1}{2}L$, all ratios satisfy $\frac{1}{2}L < b_n/a_n < \frac{3}{2}L$, except perhaps for a finite number of them.

Let $\sum a_n$ and $\sum b_n$ have positive terms. If $\lim b_n/a_n = L$ exists and if $L > 0$, then either both series converge or both series diverge.

EXAMPLE 3.1

Test for convergence or divergence:

(a) $\displaystyle\sum \frac{1}{n + \sqrt{n}}$ (b) $\displaystyle\sum \frac{4n+1}{3n^3 - n^2 - 1}$.

Solution: (a) When n is very large, n is much larger than $\sqrt{n}$. This suggests that the terms behave roughly like $1/n$, so we apply the test with $a_n = 1/n$ and $b_n = 1/(n + \sqrt{n})$:

$$\frac{b_n}{a_n} = \frac{n}{n + \sqrt{n}} = \frac{1}{1 + 1/\sqrt{n}} \longrightarrow \frac{1}{1+0} = 1, \qquad \text{as} \quad n \longrightarrow \infty.$$

The ratios have a positive limit. Therefore $\sum b_n$ diverges since $\sum 1/n$ diverges.

(b) When n is very large, the terms appear to behave like $4n/3n^3 = 4/3n^2$. This suggests comparison with the convergent series $\sum 1/n^2$. Let $a_n = 1/n^2$ and $b_n = (4n+1)/(3n^3 - n^2 - 1)$. Then

$$\frac{b_n}{a_n} = \frac{(4n+1)n^2}{3n^3 - n^2 - 1} = \frac{4 + 1/n}{3 - 1/n - 1/n^3} \longrightarrow \frac{4}{3}.$$

The ratios have a positive limit. Therefore $\sum b_n$ converges because $\sum 1/n^2$ converges.

Answer: (a) diverges (b) converges.

The Ratio Test

In a geometric series, the ratio a_{n+1}/a_n is a constant, r. If $|r| < 1$, the series converges, basically because its terms decrease rapidly. By analogy, we should expect convergence in general if the ratios are small, not necessarily constant.

Let $\sum a_n$ be a series of positive terms.

(1) The series converges if

$$\frac{a_{n+1}}{a_n} \leq r < 1$$

from some point on, that is for $n \geq N$.

(2) The series diverges if

$$\frac{a_{n+1}}{a_n} \geq 1$$

from some point on.

Proof: (1) Suppose $a_{n+1}/a_n \leq r < 1$ starting with $n = N$. Then

$$a_{N+1} \leq a_N r, \qquad a_{N+2} \leq a_{N+1} r \leq a_N r^2,$$

and by induction, $a_{N+k} \leq a_N r^k$, that is, $a_n \leq a_N r^{n-N} = (a_N r^{-N}) r^n$ for all $n \geq N$. It follows that the series $\sum a_n$ converges by comparison with the convergent geometric series $\sum r^n$.

(2) From some point on, $a_{n+1} \geq a_n$. The terms increase, hence the series diverges.

WARNING: Note that the test for convergence requires $a_{n+1}/a_n \leq r < 1$, not just $a_{n+1}/a_n < 1$. The ratios must stay away from 1. If $a_{n+1}/a_n < 1$ but $a_{n+1}/a_n \longrightarrow 1$, we may have divergence. For example, take $a_n = 1/n$. Then $a_{n+1}/a_n = n/(n+1) = 1 - 1/(n+1) < 1$, but $\sum 1/n$ diverges.

It often happens that the ratios a_{n+1}/a_n approach a limit.

Ratio Test Let $\sum a_n$ be a series of positive terms. Suppose $a_{n+1}/a_n \longrightarrow r$.

(1) The series converges if $r < 1$.

(2) The series diverges if $r > 1$.

(3) If $r = 1$, the test is inconclusive; the series may either converge or diverge.

Proof: (1) If $r < 1$, choose ϵ so small that $r + \epsilon < 1$. By definition of the statement $a_{n+1}/a_n \longrightarrow r$, there is a positive integer N such that $a_{n+1}/a_n < r + \epsilon < 1$ for all $n \geq N$. Therefore the series converges by the preceding test.

(2) Similarly, if $r > 1$, then $a_{n+1}/a_n \geq 1$ from some point on. The series diverges.

(3) If $r = 1$, this test cannot distinguish between convergent and divergent series. For example, take $a_n = 1/n^p$. The series converges for $p > 1$, diverges for $p \leq 1$. But for all values of p,

$$\frac{a_{n+1}}{a_n} = \frac{n^p}{(n+1)^p} = \left(\frac{n}{n+1}\right)^p = \left(1 - \frac{1}{n+1}\right)^p \longrightarrow (1-0)^p = 1.$$

EXAMPLE 3.2

Test for convergence or divergence:

(a) $\sum \frac{n}{2^n}$ (b) $\sum \frac{10^n}{n!}$.

Solution: (a) Set $a_n = n/2^n$. Then

$$\frac{a_{n+1}}{a_n} = \frac{n+1}{2^{n+1}} \bigg/ \frac{n}{2^n} = \frac{n+1}{2n} = \frac{1}{2}\left(1 + \frac{1}{n}\right) \longrightarrow \frac{1}{2}.$$

Since $\frac{1}{2} < 1$, the series converges by the ratio test.

(b) Set $a_n = 10^n/n!$. Then

$$\frac{a_{n+1}}{a_n} = \frac{10^{n+1}}{1\cdot2\cdots n(n+1)} \bigg/ \frac{10^n}{1\cdot2\cdots n} = \frac{10}{n} \longrightarrow 0.$$

Since $0 < 1$, the series converges by the ratio test.

Answer: (a) converges (b) converges.

EXERCISES

Test for convergence or divergence:

1. $\sum \frac{1}{n^2 - 3}$
2. $\sum \frac{1}{\sqrt{2n^3 - n}}$
3. $\sum \frac{1}{4n - 1}$
4. $\sum \frac{5 + \sqrt{n}}{1 + n}$
5. $\sum \frac{n^3}{n!}$
6. $\sum n^3 \left(\frac{3}{4}\right)^n$
7. $\sum ne^{-n}$
8. $\sum \frac{3^n + 1}{5e^n + n}$

9. $\sum \dfrac{2^n + n}{3^n - n}$

10. $\sum \dfrac{1}{(\ln n)^n}$

11. $\sum \dfrac{n!}{1 \cdot 3 \cdot 5 \cdots (2n-1)}$

12. $\sum \dfrac{(n!)^2}{(2n)!}$.

Find all real numbers x for which the series converges:

13. $\sum \dfrac{x^{2n}}{n!}$

14. $\sum \dfrac{\sin^2 nx}{n^2}$

15. $\sum (3x)^{2n}$

16. $\sum nx^{2n}$.

17. **(Root Test)** If $a_n > 0$ and if $\sqrt[n]{a_n} \leq r < 1$ for $n \geq 1$, show that $\sum a_n$ converges.

18*. Let $\sum a_n$ and $\sum b_n$ be series with positive terms. If $\sum a_n$ converges, and if $b_{n+1}/b_n \leq a_{n+1}/a_n$, show that $\sum b_n$ also converges.

19. Let $\sum a_n$ and $\sum b_n$ be series with positive terms. Suppose $b_n/a_n \longrightarrow 0$. Find an example where $\sum b_n$ converges while $\sum a_n$ diverges. Does this contradict the text?

4. SERIES WITH POSITIVE AND NEGATIVE TERMS

Infinite series with both positive and negative terms are generally more complicated than series with terms all of the same sign. In this section, we discuss two common types of mixed series that are manageable.

Alternating Series

An **alternating series** is one whose terms are alternately positive and negative. Examples:

$$1 - \frac{1}{2} + \frac{1}{3} - \frac{1}{4} + - + - \cdots,$$

$$1 - x^2 + x^4 - x^6 + - + - \cdots \qquad \text{(alternating for all } x \neq 0\text{)},$$

$$x - \frac{x^2}{4} + \frac{x^3}{9} - \frac{x^4}{16} + - + - \cdots \qquad \text{(alternating only for } x > 0\text{)}.$$

Such series have some extremely useful properties, two of which we now state.

(1) If the terms of an alternating series decrease in absolute value to zero, then the series converges.

(2) If such a series is broken off at the n-th term, then the remainder (in absolute value) is less than the absolute value of the $(n + 1)$-th term.

These assertions provide a very simple convergence criterion and an immediate remainder estimate for *alternating* series. We shall not give a formal proof; rather we shall show geometrically that they make good sense. However, a proof is outlined in Exs. 15 and 16.

Suppose $\sum a_n$ is an alternating series whose terms decrease in absolute value to zero. (To be definite, assume $a_1 > 0$.) Let $s_n = a_1 + a_2 + \cdots + a_n$. Plot these partial sums (Fig. 4.1). The partial sums oscillate back and forth as shown. But since the terms decrease to zero, the oscillations become shorter and shorter. The odd partial sums decrease and the even ones increase, squeezing down on some number S. Thus, the series converges to S.

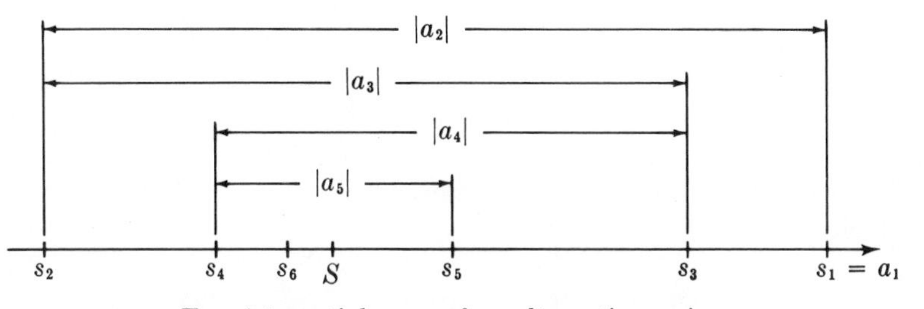

FIG. 4.1 partial sums of an alternating series

If the series is broken off after n terms, the remainder is $|S - s_n|$. But from Fig. 4.1,

$$|S - s_n| < |s_{n+1} - s_n| = |a_{n+1}|.$$

Thus, the remainder is less than the absolute value of the $(n + 1)$-th term.

EXAMPLE 4.1

Find all values of x for which the series

$$1 - x^2 + \frac{x^4}{2} - \frac{x^6}{3} + \frac{x^8}{4} - + \cdots$$

converges.

Solution: From the ratio test, it is easily seen that the series converges for $|x| < 1$ and diverges for $|x| > 1$. But what happens if $|x| = 1$? At $x = \pm 1$, the series is

$$1 - 1 + \frac{1}{2} - \frac{1}{3} + \frac{1}{4} - + \cdots,$$

an alternating series whose terms decrease in absolute value to zero. Such a series is guaranteed to converge by the statement above.

Answer: Converges for $|x| \leq 1$.

NOTATION: A useful device for abbreviating alternating series is use of the factor $(-1)^n$, an automatic sign reverser. For example, the alternating harmonic series can be written $\sum (-1)^{n-1}/n$.

EXAMPLE 4.2

It is known that the series $\sum_{n=0}^{\infty} x^n/n!$ converges to e^x for all real x. Use this series to estimate $1/e$ to 3-place accuracy.

Solution: Set $x = -1$. Then $e^{-1} = \sum_{n=0}^{\infty} (-1)^n/n!$. The signs alternate and the terms decrease in absolute value to 0. Therefore,

$$e^{-1} = 1 - \frac{1}{1!} + \frac{1}{2!} - + \cdots + \frac{(-1)^n}{n!} + \text{remainder},$$

where

$$|\text{remainder}| < \frac{1}{(n+1)!}.$$

For 3-place accuracy, we need $|\text{remainder}| < 5 \times 10^{-4}$, so we want an n for which

$$\frac{1}{(n+1)!} < 5 \times 10^{-4}, \qquad (n+1)! > \frac{1}{5} \times 10^4 = 2000.$$

Now $6! = 720$ and $7! = 5040$. So we choose $n + 1 = 7$, that is, $n = 6$.

Answer: $1 - 1 + \frac{1}{2} - \frac{1}{6} + \frac{1}{24} - \frac{1}{120} + \frac{1}{720} \approx 0.368$.

Absolute Convergence

How is it that the harmonic series $\sum 1/n$ diverges but the alternating harmonic series $\sum (-1)^{n-1}/n$ converges? Essentially the harmonic series diverges because its terms don't decrease quite fast enough, like $1/n^2$ or $1/2^n$ for example. Its partial sums consist of a lot of small terms which have a large total. The terms of $\sum 1/2^n$, however, decrease so fast that the total of any large number of them is bounded.

The alternating harmonic series converges, not by smallness of its terms alone, but also because strategically placed minus signs cause lots of cancellation. Just look at two consecutive terms:

$$+\frac{1}{n} - \frac{1}{n+1} = \frac{1}{n(n+1)}.$$

Cancellation produces a term of a convergent series! Thus $\sum (-1)^{n-1}/n$ converges because its terms get small *and* because a delicate balance of positive and negative terms produces important cancellations.

Some series with mixed terms converge by the smallness of their terms alone; they would converge even if all the signs were $+$. We say that a series $\sum a_n$ **converges absolutely** if $\sum |a_n|$ converges. As we might expect, absolute convergence implies (is even stronger than) convergence.

> If a series $\sum a_n$ converges absolutely, then it converges.

Proof: Suppose $\sum |a_n|$ converges. By the Cauchy test, for each $\epsilon > 0$ there is an N such that

$$|a_{n+1}| + |a_{n+2}| + \cdots + |a_m| < \epsilon, \qquad m > n \geq N.$$

But

$$|a_{n+1} + a_{n+2} + \cdots + a_m| \leq |a_{n+1}| + \cdots + |a_m| < \epsilon$$

by the triangle inequality. Therefore $\sum a_n$ converges by the Cauchy test.

REMARK: In studying series with mixed terms, it is a good idea to check first for absolute convergence. Just change all signs to $+$, then use any test for convergence of positive series.

EXAMPLE 4.3

Test for convergence and absolute convergence:

(a) $1 + \dfrac{1}{2^2} - \dfrac{1}{3^2} + \dfrac{1}{4^2} + \dfrac{1}{5^2} - \dfrac{1}{6^2} + + - \cdots,$

(b) $1 - \dfrac{1}{\sqrt{2}} + \dfrac{1}{\sqrt{3}} - \dfrac{1}{\sqrt{4}} + - \cdots.$

Solution: (a) The series of absolute values is $\sum 1/n^2$, which converges. The series is absolutely convergent, hence convergent.

(b) The series of absolute values is $\sum 1/\sqrt{n}$ which diverges. Hence the given series does not converge absolutely. It does converge, nevertheless, because it satisfies the test for alternating series: the terms decrease in absolute value to zero.

Answer: (a) converges and converges absolutely,
(b) converges but not absolutely.

EXERCISES

Test the series for convergence and for absolute convergence:

1. $\sum (-1)^n \dfrac{1}{\ln n}$

2. $\sum n e^{-n}$

3. $\sum (-1)^n \dfrac{n}{3n + 1}$

4. $\sum (-1)^n \dfrac{\ln n}{n}$

5. $\sum (-1)^n \dfrac{n^2}{(1.01)^n}$

6. $\sum (-1)^n \dfrac{5n^2}{2n^3 - 1}$

7. $\sum (-1)^n \dfrac{1}{n + \ln n}$ 8. $\sum (-1)^n \sin \dfrac{2\pi}{n}$

9. $\sum \dfrac{\sin n}{n^2}$ 10. $\sum (-1)^{n(n+1)/2} \dfrac{1}{\sqrt{e^{-n} + n^3}}$.

Test for convergence or divergence:

11. $1 + \frac{1}{2} - \frac{1}{3} + \frac{1}{4} + \frac{1}{5} - \frac{1}{6} + + - \cdots$.

12. $1 + \frac{1}{2} - \frac{1}{3} - \frac{1}{4} + + - - \cdots$.

13. Estimate $1/\sqrt{e}$ to 4-place accuracy by using the technique of Example 4.2.

14*. Suppose the series $\sum a_n x^n$ converges for the positive value x_0. Show that the series converges absolutely for $|x| < x_0$.

15*. Suppose $\sum a_n$ is an alternating series whose terms decrease in absolute value toward 0. Suppose the first term is $a_1 > 0$. If $\{s_n\}$ denotes the sequence of partial sums, show that the subsequence $\{s_{2n}\}$ is increasing and bounded above and the subsequence $\{s_{2n-1}\}$ is decreasing and bounded below.

16*. (cont.) Conclude that the two sequences coverge and have the same limit S. Show that $\sum a_n = S$.

17*. Suppose $\sum a_n^2$ and $\sum b_n^2$ both converge. Prove that $\sum a_n b_n$ converges absolutely.

18*. Suppose $\sum a_n$ converges, but not absolutely. Let $\sum b_j$ and $\sum c_k$ be the series made of the positive a_n's and negative a_n's respectively. Prove that both $\sum b_j$ and $\sum c_k$ diverge.

5. IMPROPER INTEGRALS

In scientific problems, one frequently meets definite integrals in which one (or both) of the limits is infinite. Here is an example.

Imagine a particle P of mass m at the origin. Consider the **gravitational potential** at a point $x = a$ due to P. This potential is the work required to move a unit mass from the point $x = a$ to infinity, against the force exerted by P. According to Newton's Law of Gravitation, the force is km/d^2, where d is the distance between the two masses and k is a proportionality constant. The work done in moving the unit mass from $x = a$ to $x = b$ is

$$\int_a^b (\text{force})\, dx = \int_a^b \frac{km}{x^2}\, dx = km\left(-\frac{1}{x}\right)\Bigg|_a^b = km\left(\frac{1}{a} - \frac{1}{b}\right).$$

Let $b \longrightarrow \infty$. Then $1/b \longrightarrow 0$, hence

$$\int_a^b \frac{km}{x^2}\, dx \longrightarrow km\left(\frac{1}{a} - 0\right) = \frac{km}{a}.$$

Thus km/a is the work required to move the mass from a to ∞. It is convenient to set

$$\int_a^\infty \frac{km}{x^2}\,dx = \lim_{b\to\infty}\int_a^b \frac{km}{x^2}\,dx = \frac{km}{a}.$$

A definite integral whose upper limit is ∞, whose lower limit is $-\infty$, or both, is called an **improper integral**

Limits

In order to give a precise definition of infinite integrals, we must recall what is meant by an expression such as $\lim_{x\to\infty} F(x) = L$. The definition is patterned after the definition of the limit of a sequence.

> Let F be defined for $x \geq a$, where a is some real number. Then
>
> $$\lim_{x\to\infty} F(x) = L$$
>
> if for each $\epsilon > 0$, there is a number b such that
>
> $$|F(x) - L| < \epsilon \qquad \text{for all} \quad x \geq b.$$

For increasing and decreasing functions the basic fact is analogous to the one for sequences. (We say F is **increasing** provided $F(x_1) \leq F(x_2)$ whenever $x_1 \leq x_2$.)

> Let F be an increasing function. Then $\lim_{x\to\infty} F(x)$ exists if and only if $F(x)$ is bounded above, i.e., if and only if there exists a number M such that
>
> $$F(x) \leq M$$
>
> for all $x \geq a$, that is, on the domain of F.
>
> Similarly if $F(x)$ is a decreasing function, then $\lim_{x\to\infty} F(x)$ exists if and only if $F(x)$ is bounded below.

We shall not give the proof; with some modification it is the same proof as for sequences.

Another important fact, also similar in spirit to one for sequences, is the Cauchy criterion:

> **Cauchy Criterion** $\lim_{x\to\infty} F(x)$ exists if and only if for each $\epsilon > 0$, there exists b such that
>
> $$|F(x) - F(z)| < \epsilon$$
>
> whenever $x \geq b$ and $z \geq b$.

There is a similar discussion for limits of the form $\lim_{x\to-\infty} F(x)$ which is hardly worth writing down. Obviously, if we set $G(x) = F(-x)$, then $\lim_{x\to-\infty} F(x)$ is equal to $\lim_{x\to\infty} G(x)$, so limits at $-\infty$ involve nothing new.

Definition of Improper Integrals

Suppose $f(x)$ is defined for $x \geq a$ and is integrable on each interval $a \leq x \leq b$ for $b \geq a$. Set

$$F(b) = \int_a^b f(x)\,dx.$$

Now $\lim_{b\to\infty} F(b)$ may or may not exist.

Define

$$\int_a^\infty f(x)\,dx = \lim_{b\to\infty} \int_a^b f(x)\,dx,$$

provided the limit exists. If it does, the integral is said to **converge**, otherwise to **diverge.**

Similarly, define

$$\int_{-\infty}^b f(x)\,dx = \lim_{a\to-\infty} \int_a^b f(x)\,dx,$$

provided the limit exists.

Finally, define

$$\int_{-\infty}^\infty f(x)\,dx = \int_{-\infty}^0 f(x)\,dx + \int_0^\infty f(x)\,dx,$$

provided both integrals on the right converge.

REMARK: An integral from $-\infty$ to ∞ may be split at any convenient finite point just as well as at 0.

An improper integral need not converge. As an example, take

$$\int_1^\infty \frac{dx}{x}.$$

Since

$$\int_1^b \frac{dx}{x} = \ln b,$$

and $\ln b \longrightarrow \infty$ as $b \longrightarrow \infty$, the limit

$$\lim_{b\to\infty} \int_1^b \frac{dx}{x} = \lim_{b\to\infty} \ln b$$

does not exist; the integral diverges.

Remember that a definite integral of a positive function represents the area under a curve. We interpret the improper integral

$$\int_a^\infty f(x)\,dx, \qquad f(x) \geq 0,$$

as the area of the infinite region in Fig. 5.1. If the integral converges, the area is finite; if the integral diverges, the area is infinite.

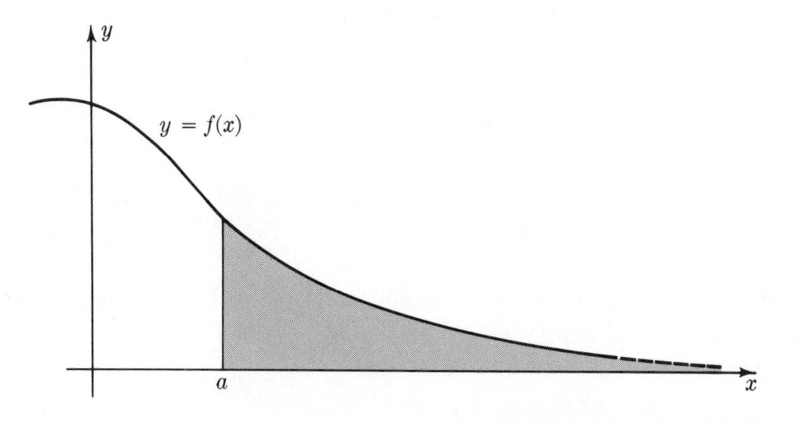

FIG. 5.1 area of infinite region

At first it may seem unbelievable that a region of infinite extent can have finite area. But it can, and here is an example. Take the region under the curve $y = 2^{-x}$ to the right of the y-axis (Fig. 5.2). The rectangles shown in Fig. 5.2 have base 1 and heights 1, $\frac{1}{2}$, $\frac{1}{4}$, $\frac{1}{8}$, $\cdots$. Their total area is

$$1 + \frac{1}{2} + \frac{1}{4} + \frac{1}{8} + \cdots = 2.$$

Therefore, the shaded infinite region has finite area less than 2.

EXAMPLE 5.1

Compute the exact area of the shaded region in Fig. 5.2.

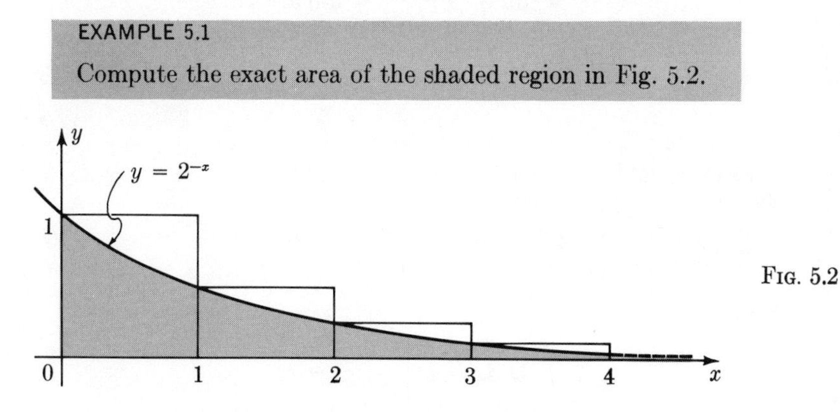

FIG. 5.2

Solution: The area is given by the improper integral

$$\int_0^\infty 2^{-x}\,dx = \lim_{b\to\infty} \int_0^b 2^{-x}\,dx.$$

An antiderivative of 2^{-x} is $-2^{-x}/\ln 2$. (Use $2^{-x} = e^{-x \ln 2}$.) Hence

$$\int_0^b 2^{-x}\,dx = \left(-\frac{1}{\ln 2}\,2^{-x}\right)\Bigg|_0^b = \frac{1}{\ln 2}\,(1 - 2^{-b}),$$

$$\int_0^\infty 2^{-x}\,dx = \lim_{b\to\infty} \frac{1}{\ln 2}\,(1 - 2^{-b}) = \frac{1}{\ln 2}\,.$$

Answer: $\dfrac{1}{\ln 2} \approx 1.4427.$

REMARK: This answer is reasonable. A look at Fig. 5.2 shows that the area of the shaded region is between 1 and 2. A closer look shows that the area is slightly less than 1.5. Why?

EXAMPLE 5.2

Evaluate $\displaystyle\int_0^\infty \frac{dx}{1 + x^2}\,.$

Solution:

$$\int_0^b \frac{dx}{1 + x^2} = \text{arc tan}\, x \Bigg|_0^b = \text{arc tan}\, b.$$

Let $b \longrightarrow \infty$. Then arc tan $b \longrightarrow \pi/2$. Hence

$$\int_0^\infty \frac{dx}{1 + x^2} = \lim_{b\to\infty} \int_0^b \frac{dx}{1 + x^2} = \lim_{b\to\infty} \text{arc tan}\, b = \frac{\pi}{2}\,.$$

Answer: $\dfrac{\pi}{2}\,.$

EXAMPLE 5.3

Evaluate $\displaystyle\int_{-\infty}^3 e^x\,dx.$

Solution:

$$\int_a^3 e^x\,dx = e^x \Bigg|_a^3 = e^3 - e^a.$$

Let $a \longrightarrow -\infty$. Then $e^a \longrightarrow 0$. Hence

$$\int_{-\infty}^{3} e^x\,dx = \lim_{a\to-\infty} (e^3 - e^a) = e^3.$$

Answer: e^3.

The following integral arises in various applications such as electrical circuits, heat conduction, and vibrating membranes:

$$\int_0^\infty e^{-sx} f(x)\,dx.$$

It is called the **Laplace Transform** of $f(x)$.

EXAMPLE 5.4

Evaluate $\displaystyle\int_0^\infty e^{-sx}\cos x\,dx, \qquad s > 0.$

Solution: From integral tables or integration by parts,

$$\int_0^b e^{-sx}\cos x\,dx = \frac{e^{-sx}}{s^2+1}(-s\cos x + \sin x)\Big|_0^b$$

$$= \frac{e^{-sb}}{s^2+1}(-s\cos b + \sin b) + \frac{s}{s^2+1}.$$

Now let $b \longrightarrow \infty$:

$$\int_0^\infty e^{-sx}\cos x\,dx = \lim_{b\to\infty}\int_0^b e^{-sx}\cos x\,dx = 0 + \frac{s}{s^2+1}.$$

Answer: $\dfrac{s}{s^2+1}$.

EXAMPLE 5.5

Evaluate $\displaystyle\int_{-\infty}^{\infty} \frac{dx}{3e^{2x}+e^{-2x}}.$

Solution: By definition, the value of this integral is

$$\int_0^\infty \frac{dx}{3e^{2x}+e^{-2x}} + \int_{-\infty}^{0} \frac{dx}{3e^{2x}+e^{-2x}},$$

provided *both* improper integrals converge. From integral tables,

$$\int_0^b \frac{dx}{3e^{2x}+e^{-2x}} = \frac{1}{2\sqrt{3}}\arctan(e^{2x}\sqrt{3})\Big|_0^b$$

$$= \frac{1}{2\sqrt{3}}\left[\arctan\left(e^{2b}\sqrt{3}\right) - \arctan\sqrt{3}\right].$$

Now $\arctan(e^{2b}\sqrt{3}) \longrightarrow \pi/2$ as $b \longrightarrow \infty$. Note that $\arctan\sqrt{3} = \pi/3$. Hence

$$\int_0^\infty \frac{dx}{3e^{2x} + e^{-2x}} = \lim_{b\to\infty} \int_0^b \frac{dx}{3e^{2x} + e^{-2x}} = \frac{1}{2\sqrt{3}}\left[\frac{\pi}{2} - \frac{\pi}{3}\right].$$

Similarly,

$$\int_{-\infty}^0 \frac{dx}{3e^{2x} + e^{-2x}} = \lim_{a\to-\infty} \int_a^0 \frac{dx}{3e^{2x} + e^{-2x}}$$

$$= \frac{1}{2\sqrt{3}}\left[\arctan\sqrt{3} - \arctan 0\right] = \frac{1}{2\sqrt{3}}\left[\frac{\pi}{3} - 0\right].$$

Thus both improper integrals converge. The answer is the sum of their values.

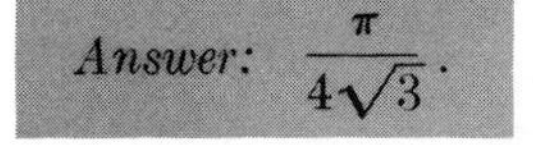

REMARK: Do you prefer this snappy calculation?

$$\int_{-\infty}^\infty \frac{dx}{3e^{2x} + e^{-2x}} = \frac{1}{2\sqrt{3}} \arctan\left(\sqrt{3}e^{2x}\right)\Big|_{-\infty}^{\infty}$$

$$= \frac{1}{2\sqrt{3}}(\arctan\infty - \arctan 0) = \frac{1}{2\sqrt{3}}\frac{\pi}{2} = \frac{\pi}{4\sqrt{3}}.$$

Warning. Try the same slick method on

$$\int_{-\infty}^\infty \frac{dx}{x^2}.$$

It fails. Why?

EXERCISES

Evaluate:

1. $\displaystyle\int_2^\infty \frac{dx}{x^3}$
2. $\displaystyle\int_5^\infty e^{-x}\,dx$
3. $\displaystyle\int_0^\infty xe^{-x}\,dx$
4. $\displaystyle\int_{-\infty}^{-1} \frac{dx}{x^2}$
5. $\displaystyle\int_{-\infty}^{-1} \frac{dx}{1+x^2}$
6. $\displaystyle\int_4^\infty \frac{dx}{x\sqrt{x}}$

7. $\int_{-\infty}^{\infty} e^{-|x|}\,dx$

8. $\int_{-\infty}^{\infty} xe^{-x^2}\,dx$

9. $\int_1^{\infty} \frac{dx}{x\sqrt{9+x^2}}$

10. $\int_0^{\infty} e^{-sx}\sin x\,dx \quad (s>0)$

11. $\int_0^{\infty} \frac{x\,dx}{x^4+1}$ (let $u = x^2$)

12. $\int_0^{\infty} xe^{-sx}\,dx \quad (s>0)$

13. $\int_0^{\infty} x^2e^{-sx}\,dx \quad (s>0)$

14. $\int_0^{\infty} x^ne^{-sx}\,dx \quad (s>0)$

15. $\int_0^{\infty} e^{ax}e^{-sx}\,dx \quad (s>a)$

16. $\int_0^{\infty} xe^{ax}e^{-sx}\,dx \quad (s>a)$

17. $\int_0^{\infty} e^{-sx}\cosh x\,dx \quad (s>1)$

18. $\int_0^{\infty} xe^{-sx}\sin x\,dx \quad (s>0)$.

Is the area under the curve finite or infinite?

19. $y = 1/x$; from $x = 5$ to $x = \infty$
20. $y = 1/x^2$; from $x = 1$ to $x = \infty$
21. $y = \sin^2 x$; from $x = 0$ to $x = \infty$
22. $y = (1.001)^{-x}$; from $x = 0$ to $x = \infty$.

Solve for b:

23. $\int_0^b e^{-x}\,dx = \int_b^{\infty} e^{-x}\,dx$

24. $\int_0^b \frac{dx}{1+x^2} = \int_b^{\infty} \frac{dx}{1+x^2}$.

25. Find b such that 99% of the area under $y = e^{-x}$ between $x = 0$ and $x = \infty$ is contained between $x = 0$ and $x = b$.

Denote the Laplace Transform of $f(x)$ by $L(f)(s)$, so

$$L(f)(s) = \int_0^{\infty} e^{-sx}f(x)\,dx.$$

26*. Suppose $f(x)$ is continuous for $x \geq 0$, and for some n and some constant c we have $|f(x)| \leq cx^n$ for all x sufficiently large. Prove that $L(f)(s)$ exists for all $s > 0$.

27*. Suppose f has a continuous derivative f' for $x \geq 0$ and $|f'(x)| \leq cx^n$ for x sufficiently large. Prove for $s > 0$ that

$$L(f')(s) = -f(0) + sL(f)(s).$$

28*. For the f in Ex. 26, set $g(x) = \int_0^{\infty} f(t)\,dt$. Prove for $s > 0$ that

$$L(g)(s) = \frac{1}{s}L(f)(s).$$

6. CONVERGENCE AND DIVERGENCE

Whether an improper integral converges or diverges may be a subtle matter. The following example illustrates this.

EXAMPLE 6.1

For which positive numbers p does the integral $\int_1^\infty \frac{dx}{x^p}$ converge? diverge?

Solution: Suppose $p \neq 1$. Then

$$\int_1^b \frac{dx}{x^p} = -\frac{1}{p-1} \cdot \frac{1}{x^{p-1}}\bigg|_1^b = \frac{1}{p-1}\left(1 - \frac{1}{b^{p-1}}\right).$$

As $b \longrightarrow \infty$,

$$\frac{1}{b^{p-1}} \longrightarrow 0 \qquad \text{if} \quad p - 1 > 0,$$

and

$$\frac{1}{b^{p-1}} \longrightarrow \infty \qquad \text{if} \quad p - 1 < 0.$$

Hence

$$\lim_{b\to\infty} \int_1^b \frac{dx}{x^p}$$

exists if $p > 1$, does not exist if $p < 1$. That means the given integral converges if $p > 1$, diverges if $p < 1$.

If $p = 1$,

$$\int_1^b \frac{dx}{x} = \ln b \longrightarrow \infty \qquad \text{as} \quad b \longrightarrow \infty;$$

the integral diverges.

Answer: Converges if $p > 1$, diverges if $p \leq 1$.

REMARK 1. Obviously the same is true of the integral

$$\int_a^\infty \frac{dx}{x^p}$$

for any positive number a.

REMARK 2: Now, a subtle question. Why should this integral converge if $p > 1$ but diverge if $p \leq 1$? (See Fig. 6.1.) The curves $y = 1/x^p$ all decrease as x increases. The key is in their *rate* of decrease. If $p \leq 1$, the curve decreases slowly enough that the shaded area (Fig. 6.1a) increases without bound as $b \longrightarrow \infty$. If $p > 1$, the curve decreases fast enough that the shaded area (Fig. 6.1b) is bounded by a fixed number, no matter how large b is.

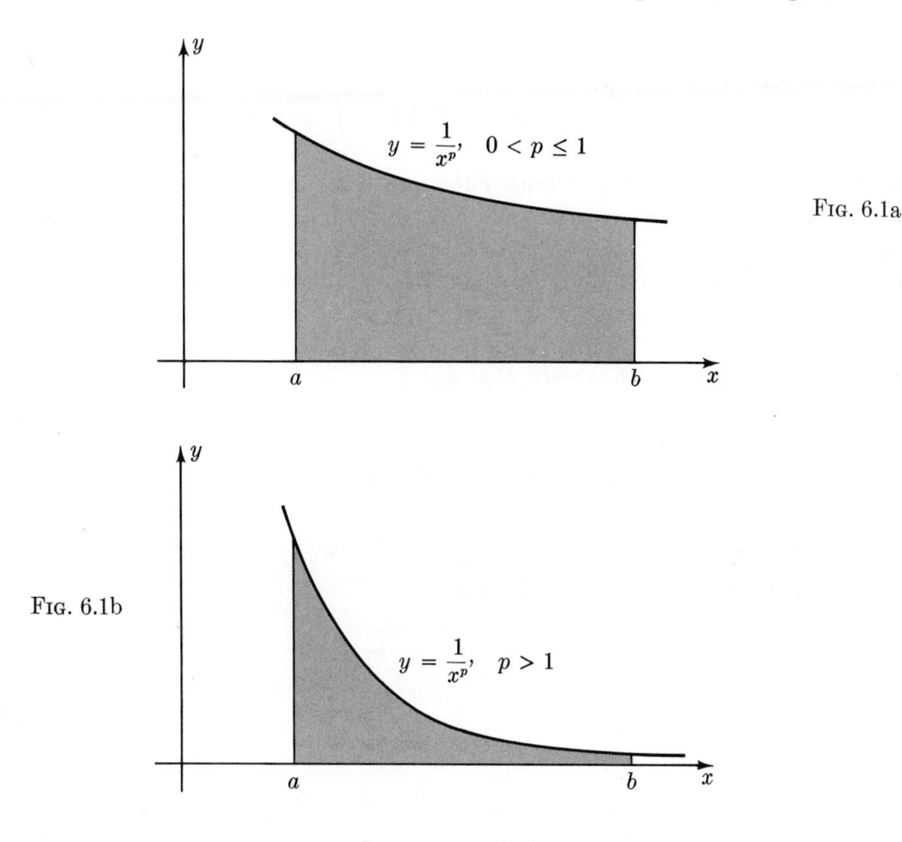

FIG. 6.1a

FIG. 6.1b

Convergence Criteria

We know that the integral

$$\int_a^\infty \frac{dx}{x^2} \qquad (a > 0)$$

converges. Suppose that $0 \leq g(x) \leq 1/x^2$. Then

$$\int_a^\infty g(x)\, dx$$

also converges because the area under the curve $y = g(x)$ is even smaller than the area under $y = 1/x^2$. This illustrates a general principle:

> If $f(x) \geq 0$ and if $0 \leq g(x) \leq f(x)$ for $a \leq x < \infty$, then the convergence of
>
> $$\int_a^\infty f(x)\, dx \quad \text{implies the convergence of} \quad \int_a^\infty g(x)\, dx.$$

Proof: We imitate the proof of the comparison test for positive series.

Let

$$F(b) = \int_a^b f(x)\,dx, \qquad G(b) = \int_a^b g(x)\,dx.$$

Since $f(x) \geq 0$ and $g(x) \geq 0$, both $F(b)$ and $G(b)$ are increasing functions of b. Also $g(x) \leq f(x)$, so $G(b) \leq F(b)$. By hypothesis $\lim_{b\to\infty} F(b) = \int_a^\infty f(x)\,dx$ exists, hence $F(b) \leq \int_a^\infty f(x)\,dx$. It follows that $G(b) \leq F(b) \leq \int_a^\infty f(x)\,dx$, so $\lim_{b\to\infty} G(b)$ exists, that is,

$$\int_a^\infty g(x)\,dx \quad \text{converges.}$$

EXAMPLE 6.2

Show that the integrals converge:

$$\text{(a)} \int_1^\infty \frac{dx}{x^2 + \sqrt{x}}; \qquad \text{(b)} \int_0^\infty \frac{e^{-x}}{x+1}\,dx; \qquad \text{(c)} \int_3^\infty \frac{\sin^2 x}{x^3}\,dx.$$

Solution: Note that

$$\frac{1}{x^2 + \sqrt{x}} < \frac{1}{x^2}, \qquad 1 \leq x < \infty,$$

$$\frac{e^{-x}}{x+1} \leq e^{-x}, \qquad 0 \leq x < \infty,$$

$$\frac{\sin^2 x}{x^3} \leq \frac{1}{x^3}, \qquad 3 \leq x < \infty.$$

Since the integrals

$$\int_1^\infty \frac{dx}{x^2}, \qquad \int_0^\infty e^{-x}\,dx, \qquad \int_3^\infty \frac{dx}{x^3}$$

all converge, the given integrals converge by the preceding test.

A second important convergence criterion is this:

Suppose $f(x) \geq 0$ and $g(x)$ is bounded, i.e., $|g(x)| \leq M$ for some constant M. Then the convergence of

$$\int_a^\infty f(x)\,dx \quad \text{implies the convergence of} \quad \int_a^\infty f(x)g(x)\,dx.$$

Proof: Let

$$F(b) = \int_a^b f(x)\,dx, \qquad H(b) = \int_a^b f(x)\,g(x)\,dx.$$

We are given that $\lim_{b\to\infty} F(b)$ exists, and we must show that $\lim_{b\to\infty} H(b)$ exists.

By the Cauchy criterion, given $\epsilon > 0$, there exists B such that

$$|F(c) - F(b)| < \frac{\epsilon}{M} \qquad \text{whenever} \quad c > b \geq B.$$

It follows that

$$|H(c) - H(b)| = \left|\int_b^c f(x)\,g(x)\,dx\right| \leq \int_b^c |f(x)|\,|g(x)|\,dx$$

$$\leq M \int_b^c f(x)\,dx = M[F(c) - F(b)] < M \cdot \frac{\epsilon}{M} = \epsilon$$

whenever $c > b \geq B$. Therefore by the Cauchy criterion, $\lim_{b\to\infty} H(b)$ exists, that is,

$$\int_a^\infty f(x)\,g(x)\,dx \quad \text{converges.}$$

EXAMPLE 6.3

Show that the integrals converge:

$$\text{(a)} \quad \int_0^\infty e^{-x} \sin^3 x\,dx, \qquad \text{(b)} \quad \int_1^\infty \frac{\ln x}{x^3}\,dx.$$

Solution: Apply the above criterion.

(a) Since

$$\int_0^\infty e^{-x}\,dx$$

converges and $|\sin^3 x| \leq 1$, the given integral converges.

(b) Write

$$\frac{\ln x}{x^3} = \frac{1}{x^2} \cdot \frac{\ln x}{x}.$$

The integral

$$\int_1^\infty \frac{dx}{x^2}$$

converges and $(\ln x)/x$ is bounded. [The maximum of $(\ln x)/x$ is $1/e$.] Hence the given integral converges.

REMARK: Both convergence criteria apply also to improper integrals of the forms

$$\int_{-\infty}^{b} f(x)\,dx \qquad \text{and} \qquad \int_{-\infty}^{\infty} f(x)\,dx.$$

A Divergence Criterion

Here is a simple criterion for divergence of an improper integral.

> If $f(x) \geq 0$ and if $g(x) \geq f(x)$ for $a \leq x < \infty$, then the divergence of
>
> $$\int_{a}^{\infty} f(x)\,dx$$
>
> implies the divergence of
>
> $$\int_{a}^{\infty} g(x)\,dx.$$

Proof: This time

$$G(b) = \int_{a}^{b} g(x)\,dx \geq \int_{a}^{b} f(x)\,dx = F(b)$$

and $F(b)$ is unbounded, so $G(b)$ is unbounded. Hence

$$\int_{a}^{\infty} g(x)\,dx \quad \text{diverges.}$$

This criterion is obvious geometrically; since $g(x) \geq f(x)$, the region under $y = g(x)$ contains the region under $y = f(x)$. If the second region has infinite area, so does the first.

EXAMPLE 6.4

Show that the integrals diverge:

(a) $\displaystyle\int_{1}^{\infty} \frac{\sqrt{x}}{1+x}\,dx$ (b) $\displaystyle\int_{2}^{\infty} \frac{dx}{\sqrt{x} - \sqrt[3]{x}}$

(c) $\displaystyle\int_{3}^{\infty} \frac{\ln x}{x}\,dx.$

Solution: Note that

$$\frac{\sqrt{x}}{1+x} \geq \frac{1}{1+x}, \qquad 1 \leq x < \infty,$$

$$\frac{1}{\sqrt{x} - \sqrt[3]{x}} > \frac{1}{\sqrt{x}}, \qquad 2 \leq x < \infty,$$

$$\frac{\ln x}{x} \geq \frac{\ln 3}{x} > \frac{1}{x}, \qquad 3 \leq x < \infty.$$

Since the integrals

$$\int_1^\infty \frac{dx}{1+x}, \quad \int_2^\infty \frac{dx}{\sqrt{x}}, \quad \int_3^\infty \frac{dx}{x}$$

all diverge, the given integrals diverge by the preceding criterion.

EXERCISES

Does the integral converge or diverge?

1. $\displaystyle\int_0^\infty \frac{dx}{x+1}$

2. $\displaystyle\int_1^\infty \frac{dx}{x^2+x}$

3. $\displaystyle\int_0^\infty \frac{x^2 e^{-x}}{1+x^2}\,dx$

4. $\displaystyle\int_2^\infty e^{-x^3}\,dx$

5. $\displaystyle\int_0^\infty \cosh x\,dx$

6. $\displaystyle\int_0^\infty \frac{x\,dx}{\sqrt{x^2+3}}$

7. $\displaystyle\int_{-\infty}^\infty \frac{\sin x}{1+x^2}\,dx$

8. $\displaystyle\int_0^\infty \sin x\,dx$

9. $\displaystyle\int_1^\infty \frac{\cos x}{\sqrt{x}\,(x+4)}$

10. $\displaystyle\int_2^\infty \frac{dx}{\ln x}$

11. $\displaystyle\int_3^\infty \frac{x^3}{x^4-1}\,dx$

12. $\displaystyle\int_0^\infty \frac{dx}{1+x+e^x}.$

13. Show that $\displaystyle\int_2^\infty \frac{dx}{x(\ln x)^p}$ converges if $p > 1$, diverges if $p \leq 1$.

    [*Hint:* Use the substitution $u = \ln x$.]

14. Show that $\displaystyle\int_3^\infty \frac{dx}{x \ln x[\ln(\ln x)]^p}$ converges if $p > 1$, diverges if $p \leq 1$.

15. Denote by R the infinite region under $y = 1/x$ to the right of $x = 1$. Suppose R is rotated around the x-axis, forming an infinitely long horn. Show that the volume

of this horn is finite. Its surface area, however, is infinite (the surface area is certainly larger than the area of R). Here is an apparent paradox: You can fill the horn with paint, but you cannot paint it. Where is the fallacy?

Find all values of s for which the integral converges:

16. $\displaystyle\int_0^\infty e^{-sx}e^{x}\,dx$

17. $\displaystyle\int_0^\infty \frac{e^{-sx}}{1+x^2}\,dx$

18. $\displaystyle\int_0^\infty e^{-sx}e^{-x^2}\,dx$

19. $\displaystyle\int_1^\infty \frac{x^s}{(1+x^3)^s}\,dx.$

7. RELATION TO INFINITE SERIES

We have already seen a number of similarities between infinite series and infinite integrals. In this section we discuss a very useful connection between them which often enables us to establish the convergence or divergence of a series by studying a related integral. This is important, for usually it is easier to find the value of an integral than the sum of a series.

Consider the relation between the series

$$\frac{1}{2^2}+\frac{1}{3^2}+\cdots+\frac{1}{n^2}+\cdots$$

and the convergent integral

$$\int_1^\infty \frac{dx}{x^2}.$$

(See Fig. 7.1.) The rectangles shown in Fig. 7.1 have areas $1/2^2$, $1/3^2$, $\cdots$. Obviously the sum of these areas is finite, being less than the finite area under the curve. Hence, the series converges. This illustrates a general principle:

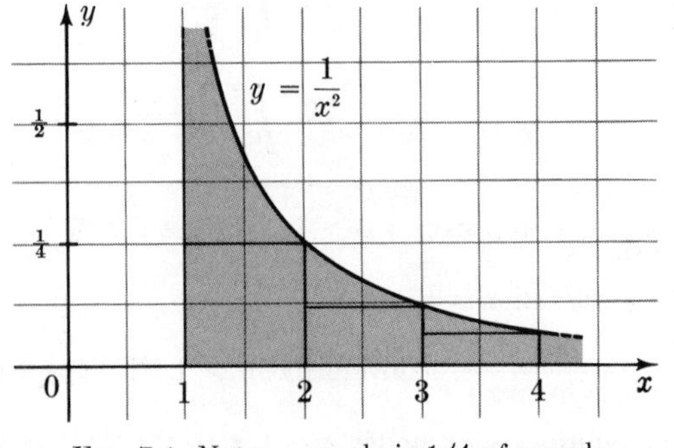

FIG. 7.1 *Note:* x scale is 1/4 of y scale.

> Suppose $f(x)$ is a positive decreasing function. Then the series
> $$f(1) + f(2) + \cdots + f(n) + \cdots$$
> converges if the integral
> $$\int_1^\infty f(x)\,dx$$
> converges, and diverges if the integral diverges.

Proof: The argument given above for $f(x) = 1/x^2$ holds for any positive decreasing function $f(x)$. Figure 7.1 indicates that

$$f(2) + f(3) + \cdots + f(n) \le \int_1^n f(x)\,dx.$$

If the infinite integral converges, then

$$s_n = f(1) + f(2) + \cdots + f(n) \le f(1) + \int_1^n f(x)\,dx \le f(1) + \int_1^\infty f(x)\,dx.$$

Hence the increasing partial sums are bounded; the series converges.

If the infinite integral diverges, the rectangles are drawn above the curve (Fig. 7.2). Their areas are $f(1), f(2), \cdots$. This time

$$s_n = f(1) + f(2) + \cdots + f(n) \ge \int_1^{n-1} f(x)\,dx.$$

But the integrals on the right are unbounded. Hence the increasing sequence $\{s_n\}$ is unbounded; the series diverges.

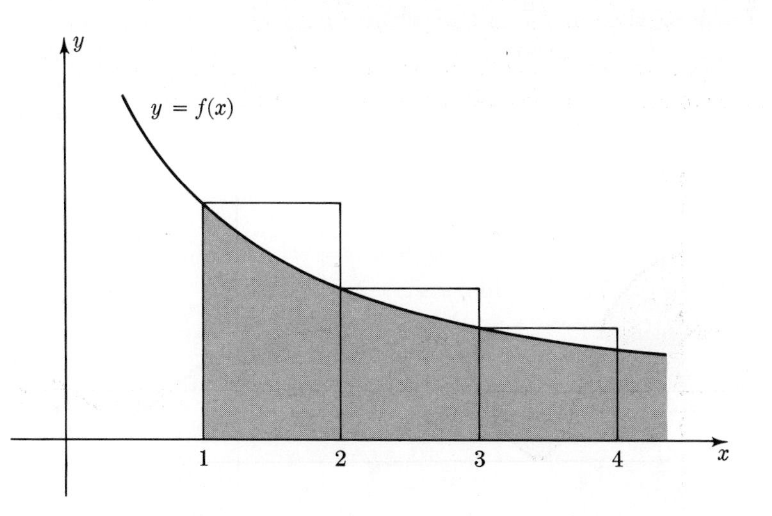

FIG. 7.2 The rectangular sum exceeds the integral.

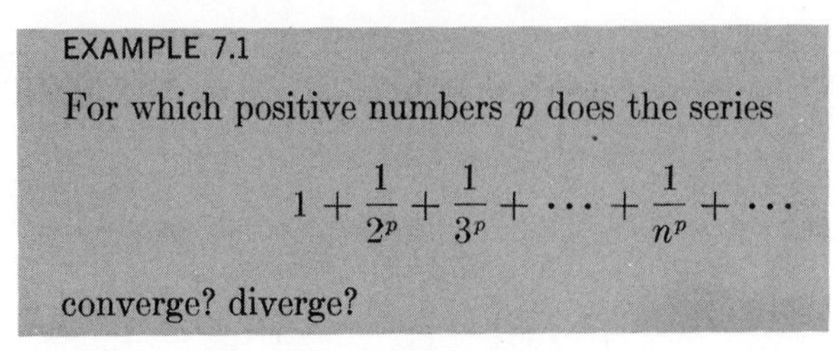

EXAMPLE 7.1

For which positive numbers p does the series

$$1 + \frac{1}{2^p} + \frac{1}{3^p} + \cdots + \frac{1}{n^p} + \cdots$$

converge? diverge?

Solution: The series can be written

$$f(1) + f(2) + \cdots + f(n) + \cdots,$$

where $f(x) = 1/x^p$, a positive decreasing function. By the preceding principle, the given series converges or diverges as

$$\int_1^\infty \frac{dx}{x^p}$$

converges or diverges.

Answer: Converges if $p > 1$, diverges if $p \leq 1$.

Convergence of Integrals

Sometimes we can turn the tables and use the convergence of a series to establish the convergence of an integral. If the integrand changes sign regularly, we may be able to compare the integral with an alternating series.

EXAMPLE 7.2

Prove the convergence of $\displaystyle\int_0^\infty \frac{\sin x}{x}\, dx.$

Solution: First sketch the graph of $y = (\sin x)/x$. See Fig. 7.3. There is no trouble at $x = 0$ because $(\sin x)/x \longrightarrow 1$ as $x \longrightarrow 0$.

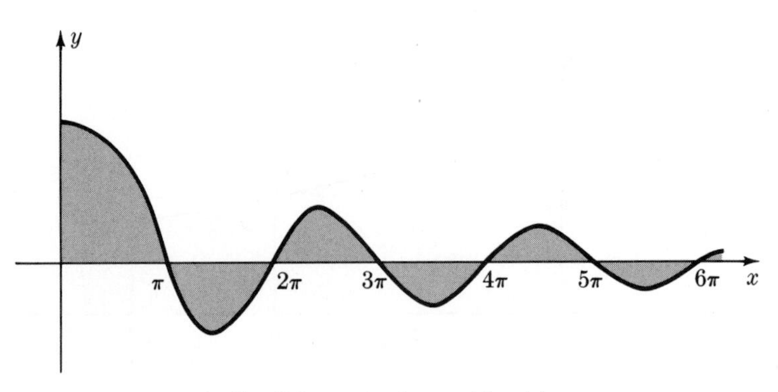

FIG 7.3 graph of $y = (\sin x)/x$

The figure suggests that the integral is given by an alternating series. To be precise, let

$$a_n = (-1)^n \int_{n\pi}^{(n+1)\pi} \frac{\sin x}{x}\, dx.$$

Then $a_n > 0$; in fact a_n is the area of the n-th shaded region in Fig. 7.3. Now

$$a_n = \int_{n\pi}^{(n+1)\pi} \frac{|\sin x|}{x}\, dx < \int_{n\pi}^{(n+1)\pi} \frac{1}{x}\, dx < \frac{\pi}{n\pi} = \frac{1}{n},$$

so $a_n \longrightarrow 0$. Furthermore

$$a_n = (-1)^n \int_0^{\pi} \frac{\sin(x + n\pi)}{x + n\pi}\, dx = \int_0^{\pi} \frac{\sin x}{x + n\pi}\, dx,$$

so $a_n > a_{n+1}$ because

$$\frac{\sin x}{x + n\pi} > \frac{\sin x}{x + (n+1)\pi}, \qquad 0 < x < \pi.$$

Therefore the alternating series

$$a_0 - a_1 + a_2 - a_3 + \cdots$$

converges. But

$$a_0 - a_1 + a_2 - \cdots + (-1)^n a_n = \int_0^{(n+1)\pi} \frac{\sin x}{x}\, dx,$$

so we have the existence of

$$\lim_{n\to\infty} \int_0^{(n+1)\pi} \frac{\sin x}{x}\, dx,$$

where n can take only *integer* values.

If b is any positive *real* number, there is an integer n such that $n\pi \le b < (n+1)\pi$. Then

$$\int_0^b \frac{\sin x}{x}\, dx = \int_0^{(n+1)\pi} \frac{\sin x}{x}\, dx - \int_b^{(n+1)\pi} \frac{\sin x}{x}\, dx.$$

If $b \longrightarrow \infty$, then $n \longrightarrow \infty$, so the first term on the right converges to a limit. The second term approaches 0 because

$$\left| \int_b^{(n+1)\pi} \frac{\sin x}{x}\, dx \right| \le \left| \int_{n\pi}^{(n+1)\pi} \frac{\sin x}{x}\, dx \right| = a_n \longrightarrow 0.$$

Therefore

$$\int_0^\infty \frac{\sin x}{x}\,dx = \lim_{b\to\infty}\int_0^b \frac{\sin x}{x}\,dx$$

converges.

EXERCISES

Does the series converge or diverge?

1. $1 + \dfrac{1}{\sqrt{2}} + \dfrac{1}{\sqrt{3}} + \dfrac{1}{\sqrt{4}} + \cdots$
2. $\dfrac{1}{e} + \dfrac{2}{e^2} + \dfrac{3}{e^3} + \dfrac{4}{e^4} + \cdots$
3. $1 + \dfrac{1}{2^3} + \dfrac{1}{3^3} + \dfrac{1}{4^3} + \cdots$
4. $\dfrac{1}{1+\sqrt{1}} + \dfrac{1}{2+\sqrt{2}} + \dfrac{1}{3+\sqrt{3}} + \dfrac{1}{4+\sqrt{4}} + \cdots$
5. $\dfrac{1}{2\ln 2} + \dfrac{1}{3\ln 3} + \dfrac{1}{4\ln 4} + \cdots$
6. $\dfrac{1}{2(\ln 2)^p} + \dfrac{1}{3(\ln 3)^p} + \dfrac{1}{4(\ln 4)^p} + \cdots$
7. $\dfrac{1}{1+1^2} + \dfrac{1}{1+2^2} + \dfrac{1}{1+3^2} + \cdots$
8. $\dfrac{1}{1^2} + \dfrac{3}{2^2} + \dfrac{5}{3^2} + \dfrac{7}{4^2} + \cdots.$
9. Show geometrically that the sum of $1 + \dfrac{1}{2^2} + \dfrac{1}{3^2} + \dfrac{1}{4^2} + \cdots$ is less than 2. See Fig. 7.1. (It is known that the exact sum is $\pi^2/6$, a startling fact.)
10. Use the method of inscribing and circumscribing rectangles to show that

$$\ln(n+1) < 1 + \frac{1}{2} + \frac{1}{3} + \cdots + \frac{1}{n} < 1 + \ln n.$$

Is $1 + \dfrac{1}{2} + \dfrac{1}{3} + \cdots + \dfrac{1}{1000}$ more or less than 10?

11. Estimate how many terms of the series $1 + \dfrac{1}{2} + \dfrac{1}{3} + \dfrac{1}{4} + \cdots$ must be added before the sum exceeds 1000.

Does the series converge or diverge?

12. $\displaystyle\sum_{n=1}^{\infty} \frac{n}{4+n^3}$

13. $\displaystyle\sum_{n=2}^{\infty} \frac{\ln n}{1+n^2}$

14. $\displaystyle\sum_{n=1}^{\infty} \frac{n^2 e^{-n}}{1+n^2}$

15. $\displaystyle\sum_{n=2}^{\infty} \frac{1}{\sqrt{n}\ln n}.$

16*. Prove convergent:

$$\int_0^{\infty} \frac{\sin x}{\sqrt{x}}\, dx$$

17*. (cont.) Prove convergent:

$$\int_0^{\infty} \sin x^2\, dx.$$

[*Hint:* Set $x^2 = u$.]

18*. Let $0 < a < b$. Prove convergent:

$$\int_0^{\infty} \frac{\cos ax - \cos bx}{x}\, dx.$$

[*Hint:* Separate the difficulties at 0 and ∞.]

19*. Let $0 < a < b$. Prove convergent:

$$\int_0^{\infty} \frac{\arctan bx - \arctan ax}{x}\, dx.$$

8. OTHER IMPROPER INTEGRALS

A definite integral

$$\int_a^b f(x)\, dx, \qquad a \text{ and } b \text{ finite},$$

is called **improper** if $f(x)$ "blows up" at one or more points in the interval $a \leq x \leq b$. Examples are

$$\int_0^3 \frac{dx}{x}, \qquad \int_1^5 \frac{dx}{x^2 - 4}, \qquad \int_6^{10} \frac{dx}{\ln(x-5)}.$$

The first integrand "blows up" at $x = 0$, the second at $x = 2$, the third at $x = 6$. Such bad points are called **singularities** of the integrand.

We shall discuss integrals

$$\int_a^b f(x)\, dx$$

where $f(x)$ has exactly one singularity which occurs either at $x = a$ or at $x = b$. This is the most common case.

Consider the integral

$$\int_0^3 \frac{dx}{\sqrt{x}}$$

whose integrand has a singularity at $x = 0$. What meaning can we give to this integral?

Except at $x = 0$, the integrand is well-behaved. Hence if h is any positive number, no matter how small, the integral

$$\int_h^3 \frac{dx}{\sqrt{x}}$$

makes sense. Its value is easily computed:

$$\int_h^3 \frac{dx}{\sqrt{x}} = 2\sqrt{x}\Big|_h^3 = 2(\sqrt{3} - \sqrt{h}).$$

It is reasonable to *define*

$$\int_0^3 \frac{dx}{\sqrt{x}} = \lim_{h \to 0} \int_h^3 \frac{dx}{\sqrt{x}} = 2\sqrt{3}.$$

Next, consider the integral

$$\int_0^3 \frac{dx}{x}.$$

We try to "sneak up" on the integral as before by computing

$$\int_h^3 \frac{dx}{x} = \ln 3 - \ln h,$$

then letting $h \longrightarrow 0$. But $\ln h \longrightarrow -\infty$ as $h \longrightarrow 0$. Hence

$$\int_h^3 \frac{dx}{x} \longrightarrow \infty \qquad \text{as} \quad h \longrightarrow 0.$$

There is no reasonable value for this integral.

Motivated by these examples, we make the following definitions:

Suppose $f(x)$ has one singularity, at $x = a$, and that $a < b$. Define

$$\int_a^b f(x)\,dx = \lim_{h \to 0} \int_{a+h}^b f(x)\,dx, \qquad (h > 0)$$

provided the limit exists. If it does, the improper integral **converges,** otherwise, it **diverges.**

Similarly, if $f(x)$ has one singularity, at $x = b$, define

$$\int_a^b f(x)\,dx = \lim_{h \to 0} \int_a^{b-h} f(x)\,dx, \qquad (h > 0)$$

provided the limit exists.

EXAMPLE 8.1

For which positive numbers p does the improper integral $\int_0^3 \frac{dx}{x^p}$ converge? diverge?

Solution: The case $p = 1$ was just discussed; the integral diverges. Now assume $p \neq 1$. By definition, the value of the integral is

$$\lim_{h \to 0} \int_h^3 \frac{dx}{x^p},$$

provided the limit exists. Now

$$\int_h^3 \frac{dx}{x^p} = -\frac{1}{p-1} \frac{1}{x^{p-1}} \bigg|_h^3 = \frac{1}{p-1} \left(\frac{1}{h^{p-1}} - \frac{1}{3^{p-1}} \right).$$

But, as $h \longrightarrow 0$,

$$\frac{1}{h^{p-1}} \longrightarrow 0 \qquad \text{if} \quad p - 1 < 0,$$

and

$$\frac{1}{h^{p-1}} \longrightarrow \infty \qquad \text{if} \quad p - 1 > 0.$$

Hence the limit exists only if $p < 1$. In that case

$$\int_0^3 \frac{dx}{x^p} = \lim_{h \to 0} \int_h^3 \frac{dx}{x^p} = -\frac{1}{(p-1)3^{p-1}}.$$

Answer: Converges if $p < 1$, diverges if $p \geq 1$.

REMARK 1: The answer applies as well to

$$\int_0^b \frac{dx}{x^p}$$

for each positive number b since the upper limit plays no essential part in the discussion. Only the behavior of $1/x^p$ in the immediate neighborhood of $x = 0$ counts.

REMARK 2: If $p \geq 1$, the curve $y = 1/x^p$ increases so fast as $x \longrightarrow 0$ that the area of the shaded region (Fig. 8.1a) tends to infinity. If $p < 1$, the curve rises so slowly that the area of the shaded region (Fig. 8.1b) is bounded.

CAUTION: Do not confuse these results with those of Example 6.1 concerning

$$\int_1^\infty \frac{dx}{x^p}.$$

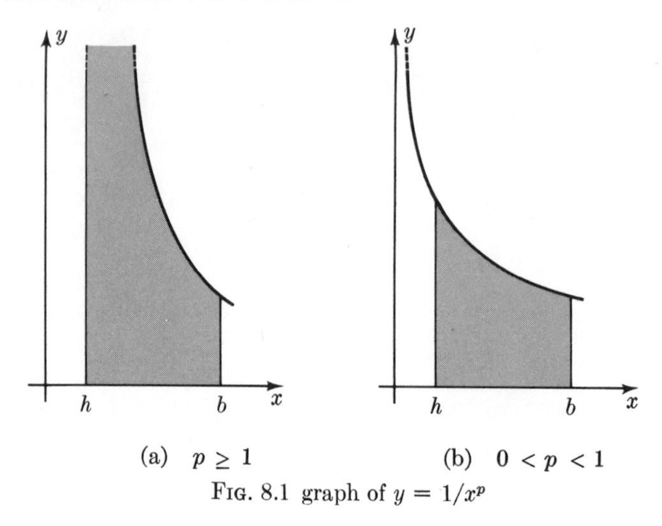

(a) $p \geq 1$ (b) $0 < p < 1$

FIG. 8.1 graph of $y = 1/x^p$

In fact,

$$\int_0^1 \frac{dx}{x^p} \quad \begin{cases} \text{converges} & \text{if} \quad p < 1, \\ \text{diverges} & \text{if} \quad p \geq 1, \end{cases} \qquad \int_1^\infty \frac{dx}{x^p} \quad \begin{cases} \text{diverges} & \text{if} \quad p \leq 1, \\ \text{converges} & \text{if} \quad p > 1. \end{cases}$$

EXAMPLE 8.2

For which positive numbers p does the improper integral

$$\int_3^4 \frac{dx}{(x-3)^p}$$

converge? diverge?

Solution: Change variable. Let $u = x - 3$. Then

$$\int_3^4 \frac{dx}{(x-3)^p} = \int_0^1 \frac{du}{u^p}.$$

But this integral was discussed above.

Answer: Converges if $p < 1$, diverges if $p \geq 1$.

The techniques of the two preceding examples yield a general fact (for a and b finite):

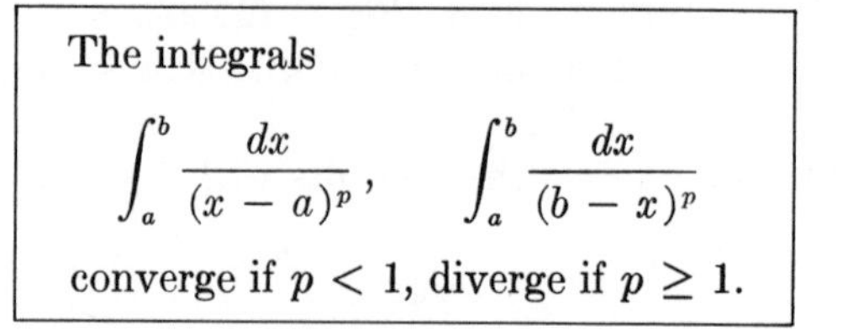
The integrals

$$\int_a^b \frac{dx}{(x-a)^p}, \qquad \int_a^b \frac{dx}{(b-x)^p}$$

converge if $p < 1$, diverge if $p \geq 1$.

A Convergence Criterion

The convergence criteria given in Section 6 have analogues for improper integrals with finite limits. We state (without proof) just one of these.

> Suppose that $f(x) \geq 0$, and that $f(x)$ has a singularity at $x = a$ or at $x = b$. Suppose $g(x)$ is a bounded function. Then the convergence of
>
> $$\int_a^b f(x)\,dx \quad \text{implies the convergence of} \quad \int_a^b f(x)g(x)\,dx.$$

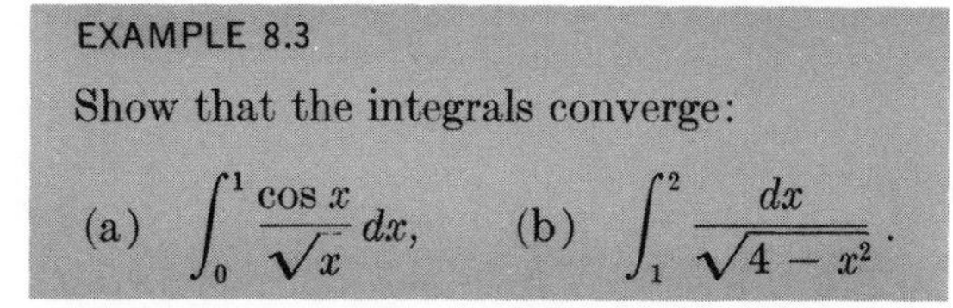

EXAMPLE 8.3

Show that the integrals converge:

(a) $\displaystyle\int_0^1 \frac{\cos x}{\sqrt{x}}\,dx$, (b) $\displaystyle\int_1^2 \frac{dx}{\sqrt{4-x^2}}$.

Solution: Use the preceding criterion.

(a) Let $f(x) = 1/\sqrt{x}$ and $g(x) = \cos x$.

(b) Let $f(x) = 1/\sqrt{2-x}$ and $g(x) = 1/\sqrt{2+x}$.

EXERCISES

Does the integral converge or diverge?

1. $\displaystyle\int_0^1 \frac{dx}{\sqrt[3]{x}}$

2. $\displaystyle\int_0^{\pi/4} \cot x\,dx$

3. $\displaystyle\int_0^2 \frac{e^x}{x}\,dx$

4. $\displaystyle\int_0^1 \frac{dx}{x^3-1}$

5. $\displaystyle\int_0^1 \frac{dx}{(1-x^2)^2}$

6. $\displaystyle\int_1^3 \frac{\sin x}{\sqrt{3-x}}\,dx$

7. $\displaystyle\int_3^5 \frac{dx}{\sqrt{x^2-9}}$

8. $\displaystyle\int_0^5 \frac{\sin^3 2x}{\sqrt{x}}\,dx$

9. $\displaystyle\int_0^1 \ln x\,dx$

10. $\displaystyle\int_0^3 \frac{dx}{\sqrt{x}\,(3+x)}$

11. $\displaystyle\int_0^{1/2} \frac{dx}{x\ln x}$

12. $\displaystyle\int_2^4 \frac{dx}{\sqrt{-x^2+6x-8}}$

13. $\displaystyle\int_1^2 \frac{dx}{\sqrt[3]{x^3-4x^2+4x}}$

14. $\displaystyle\int_0^{\pi/2} \sec x\,dx$

15. $\displaystyle\int_0^1 \sqrt{\frac{1+x}{1-x}}\,dx$ 16. $\displaystyle\int_0^{\pi/2} \frac{\cos x}{\sqrt[3]{x}}\,dx.$

9. SOME DEFINITE INTEGRALS [optional]

Sometimes it is important to know the exact value of an improper integral, not just that it converges. Certain (improper) definite integrals can be found exactly by tricks, even though the corresponding indefinite integrals are difficult, or even impossible to find. The most common type of trick involves a change of variable.

EXAMPLE 9.1

Prove $\displaystyle\int_0^{\pi/2} \ln \sin\theta\,d\theta = -\frac{\pi}{2}\ln 2.$

Solution: Since $0 < \sin\theta \le 1$ for $0 < \theta \le \frac{1}{2}\pi$, we have $-\infty < \ln\sin\theta \le 0$. The integrand approaches $-\infty$ as $x \longrightarrow 0+$. To prove convergence, note that $(2/\pi)\theta < \sin\theta < 1$ for $0 < \theta < \frac{1}{2}\pi$, hence $\ln\theta + \ln(2/\pi) < \ln\sin\theta < 0$. Therefore the integrand does not change sign and the integral converges by comparison with the convergent integral

$$\int_0^1 \ln\theta\,d\theta.$$

Now set

$$I = \int_0^{\pi/2} \ln\sin\theta\,d\theta.$$

Make the change of variable $\theta = \frac{1}{2}\pi - \alpha$. Then

$$I = -\int_{\pi/2}^{0} \ln\cos\alpha\,d\alpha = \int_0^{\pi/2} \ln\cos\theta\,d\theta.$$

We now have two expressions for I; average them:

$$\begin{aligned}
I &= \frac{1}{2}\int_0^{\pi/2} (\ln\sin\theta + \ln\cos\theta)\,d\theta \\
&= \frac{1}{2}\int_0^{\pi/2} \ln(\sin\theta\cos\theta)\,d\theta = \frac{1}{2}\int_0^{\pi/2} \ln\left(\frac{\sin 2\theta}{2}\right)d\theta \\
&= \frac{1}{2}\int_0^{\pi/2} (\ln\sin 2\theta - \ln 2)\,d\theta \\
&= \frac{1}{2}\int_0^{\pi/2} \ln\sin 2\theta\,d\theta - \frac{\pi\ln 2}{4}.
\end{aligned}$$

But

$$\int_0^{\pi/2} \ln \sin 2\theta \, d\theta = \frac{1}{2} \int_0^{\pi} \ln \sin \theta \, d\theta$$

$$= \frac{1}{2} \int_0^{\pi/2} \ln \sin \theta \, d\theta + \frac{1}{2} \int_{\pi/2}^{\pi} \ln \sin \theta \, d\theta$$

$$= \frac{1}{2} I + \frac{1}{2} \int_{\pi/2}^{\pi} \ln \sin \theta \, d\theta.$$

Therefore

$$I = \frac{1}{4} I + \frac{1}{4} \int_{\pi/2}^{\pi} \ln \sin \theta \, d\theta - \frac{\pi \ln 2}{4}.$$

It is obvious by symmetry (or by the transformation $\theta = \pi - \alpha$) that

$$\int_{\pi/2}^{\pi} \ln \sin \theta \, d\theta = \int_0^{\pi/2} \ln \sin \theta \, d\theta = I,$$

hence we have

$$I = \frac{1}{4} I + \frac{1}{4} I - \frac{\pi \ln 2}{4},$$

and the formula follows.

REMARK: It is known that the indefinite integral

$$\int \ln \sin \theta \, d\theta$$

cannot be expressed in terms of elementary functions (composite functions built with rational functions, radicals, exponentials, logs, and trigonometric functions).

In the next two examples we compute by tricks an improper integral whose corresponding indefinite integral can be worked out, but is complicated. Compare Exs. 9 and 10.

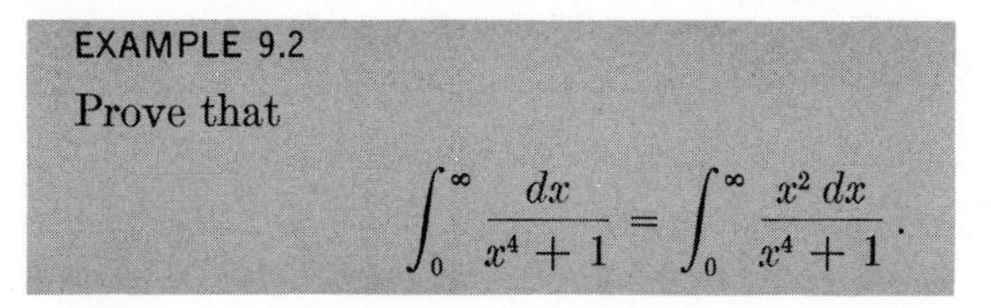

EXAMPLE 9.2

Prove that

$$\int_0^{\infty} \frac{dx}{x^4 + 1} = \int_0^{\infty} \frac{x^2 \, dx}{x^4 + 1}.$$

Solution: Convergence is obvious. The problem suggests a change of variables. Try $x = 1/u$. Then

$$\frac{dx}{x^4 + 1} = \frac{-du/u^2}{(1/u^4) + 1} = \frac{-u^2 \, du}{u^4 + 1}.$$

Hence, if $0 < a < b$,

$$\int_a^b \frac{dx}{x^4+1} = -\int_{1/a}^{1/b} \frac{u^2\,du}{u^4+1} = \int_{1/b}^{1/a} \frac{u^2\,du}{u^4+1}.$$

Let $a \longrightarrow 0$ and then let $b \longrightarrow \infty$. Then $1/a \longrightarrow \infty$ and $1/b \longrightarrow 0$, so the stated formula follows.

EXAMPLE 9.3

Prove that

$$\int_0^\infty \frac{dx}{x^4+1} = \int_0^\infty \frac{x^2\,dx}{x^4+1} = \frac{\pi\sqrt{2}}{4}.$$

Solution: The two integrals are equal by the previous example. This suggests averaging to increase the symmetry:

$$\int_0^\infty \frac{dx}{x^4+1} = \frac{1}{2}\left(\int_0^\infty \frac{dx}{x^4+1} + \int_0^\infty \frac{x^2\,dx}{x^4+1}\right) = \frac{1}{2}\int_0^\infty \frac{x^2+1}{x^4+1}\,dx.$$

Now we need a really clever change of variable. Consider

$$u = x - \frac{1}{x}.$$

Clearly $u \longrightarrow -\infty$ as $x \longrightarrow 0+$ and $u \longrightarrow \infty$ as $x \longrightarrow \infty$. Also

$$\frac{du}{dx} = 1 + \frac{1}{x^2} = \frac{x^2+1}{x^2} > 0,$$

so $u = u(x)$ is a strictly increasing function taking the interval $(0, \infty)$ onto $(-\infty, \infty)$. Furthermore,

$$u^2 = x^2 - 2 + \frac{1}{x^2}, \qquad u^2 + 2 = x^2 + \frac{1}{x^2} = \frac{x^4+1}{x^2},$$

$$\frac{du}{u^2+2} = \frac{(x^2+1)/x^2}{(x^4+1)/x^2}\,dx = \frac{x^2+1}{x^4+1}\,dx;$$

hence

$$\frac{1}{2}\int_0^\infty \frac{x^2+1}{x^4+1}\,dx = \int_{-\infty}^\infty \frac{du}{u^2+2} = \frac{\sqrt{2}}{2}\int_{-\infty}^\infty \frac{dt}{t^2+1},$$

$$= \frac{\sqrt{2}}{2} \arctan t \Big|_{-\infty}^{\infty} = \frac{\pi\sqrt{2}}{2}.$$

EXERCISES

Prove:

1. $\displaystyle\int_0^\pi x \ln \sin x \, dx = -\frac{\pi^2}{2} \ln 2$ [*Hint:* $\displaystyle\int_0^\pi = \int_0^{\pi/2} + \int_{\pi/2}^\pi = \int_0^{\pi/2} + \int_0^{\pi/2}$ (?).]

2. $\displaystyle\int_0^{\pi/2} \ln \tan x \, dx = 0$

3. $\displaystyle\int_0^{\pi/2} \sin x \,(\ln \sin x)\, dx = \ln 2 - 1$ [*Hint:* Integrate by parts.]

4. $\displaystyle\int_0^\infty \frac{dx}{(1+x)\sqrt{x}} = \pi$

5. $\displaystyle\int_0^1 (\ln x)^n \, dx = (-1)^n n!$ for $n = 1, 2, 3, \cdots$

6. $\displaystyle\int_0^1 x \ln(1-x)\, dx = -\frac{3}{4}$

7. $\displaystyle\int_0^\infty \frac{dx}{\cosh x} = \frac{\pi}{2}$

8. $\displaystyle\int_0^\infty \frac{dx}{x^3+1} = \int_0^\infty \frac{x\, dx}{x^3+1} = \frac{2\pi\sqrt{3}}{9}.$

9. Start with $x^4 + 1 = (x^2 + \sqrt{2}x + 1)(x^2 - \sqrt{2}x + 1)$ and obtain the partial fraction decomposition

$$\frac{1}{x^4+1} = \frac{\frac{1}{4}\sqrt{2}x + \frac{1}{2}}{x^2 + \sqrt{2}x + 1} + \frac{-\frac{1}{4}\sqrt{2}x + \frac{1}{2}}{x^2 - \sqrt{2}x + 1}.$$

10*. (cont.) Use this to obtain

$$\int_{-\infty}^\infty \frac{dx}{x^4+1} = \lim_{b\to\infty} \int_{-b}^b \frac{dx}{x^4+1} = \frac{\pi\sqrt{2}}{2}.$$

(Compare Example 9.3.)

10. STIRLING'S FORMULA [optional]

In this section we obtain several useful and important results in analysis by exploiting the method of approximating integrals by sums.

Euler's Constant

We know that the harmonic series $\sum 1/n$ diverges. Now we show much more: that the partial sum s_n is approximately $\ln n$. Consider Fig. 10.1a. Comparing the area under the curve $y = 1/x$ between 1 and n with a sum of rectangles, we see that

$$1 + \frac{1}{2} + \cdots + \frac{1}{n-1} = \int_1^n \frac{dx}{x} + c_n = \ln n + c_n,$$

where c_n is the area of the shaded regions.

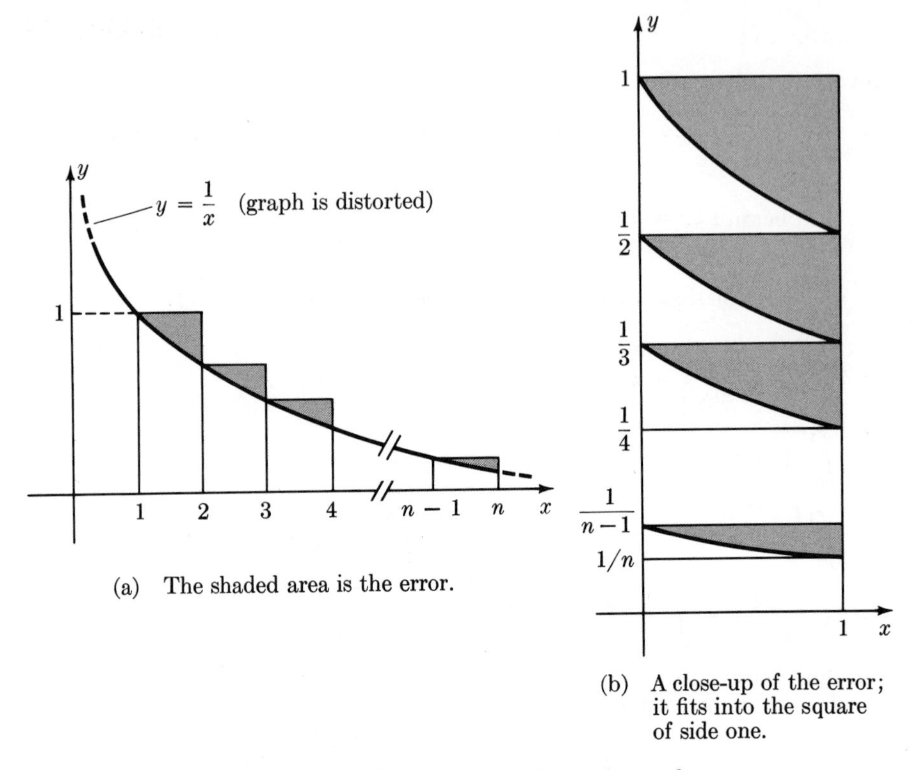

(a) The shaded area is the error.

(b) A close-up of the error; it fits into the square of side one.

FIG. 10.1 approximation of s_n by an integral

Now shift the shaded areas to the left as in Fig. 10.1b. They fit inside a square of side one! It follows that $\{c_n\}$ is an increasing sequence bounded above by 1. Hence $\lim_{n\to\infty} c_n = \gamma$ exists, and we have

$$1 + \frac{1}{2} + \cdots + \frac{1}{n} - \ln n = \frac{1}{n} + c_n \longrightarrow \gamma.$$

> There is a positive constant γ such that
>
> $$\lim_{n\to\infty}\left(\sum_{k=1}^{n}\frac{1}{k} - \ln n\right) = \gamma.$$

The number γ is called **Euler's constant.** Its value is approximately 0.57721566. In view of Fig. 10.1b, this value seems reasonable. For the curve $y = 1/x$ is convex, so the combined shaded areas fill out a bit more than half of the square.

REMARK: In many computations, the exact value of γ is not needed. What counts is that the difference between $\sum 1/n$ and $\ln n$ is bounded, so for large values of n, the approximation $\sum_{k=1}^{n} 1/k \approx \ln n$ is often accurate enough.

Wallis's Product

This is an old and remarkable formula:

Wallis's Product

$$\lim_{n\to\infty} \frac{1}{2n+1}\left[\frac{2\cdot4\cdot6\cdots(2n)}{1\cdot3\cdot5\cdots(2n-1)}\right]^2 = \frac{\pi}{2}.$$

Its derivation is an exercise in integration. Set

$$J_n = \int_0^{\pi/2} \cos^n\theta\, d\theta = \int_0^{\pi/2} \cos^{n-1}\theta \cos\theta\, d\theta.$$

Integrate by parts with $u = \cos^{n-1}\theta$ and $v = \cos\theta$:

$$J_n = 0 + \int_0^{\pi/2} (n-1)\cos^{n-2}\theta \sin^2\theta\, d\theta.$$

But

$$\cos^{n-2}\theta \sin^2\theta = \cos^{n-2}\theta(1-\cos^2\theta) = \cos^{n-2}\theta - \cos^n\theta,$$

hence

$$J_n = (n-1)(J_{n-2} - J_n).$$

Solve for J_n:

$$J_n = \frac{n-1}{n} J_{n-2}.$$

Now apply this reduction formula repeatedly and eventually reach either

$$J_0 = \int_0^{\pi/2} d\theta = \frac{\pi}{2} \qquad \text{or} \qquad J_1 = \int_0^{\pi/2} \cos\theta\, d\theta = 1.$$

Clearly there are two cases, n even and n odd. The results are

$$\begin{cases} J_{2n} = \dfrac{1\cdot3\cdot5\cdots(2n-1)}{2\cdot4\cdot6\cdots(2n)}\,\dfrac{\pi}{2}, \\[2ex] J_{2n+1} = \dfrac{2\cdot4\cdot6\cdots(2n)}{1\cdot3\cdot5\cdots(2n+1)}. \end{cases}$$

Since $0 < \cos\theta < 1$ for $0 < \theta < \frac{1}{2}\pi$, we have $\cos^{2n-1}\theta > \cos^{2n}\theta > \cos^{2n+1}\theta$, hence $J_{2n-1} > J_{2n} > J_{2n+1}$. Substitute, then rearrange:

$$\frac{2\cdot4\cdots(2n-2)}{1\cdot3\cdots(2n-1)} > \frac{1\cdot3\cdots(2n-1)}{2\cdot4\cdots(2n)}\frac{\pi}{2} > \frac{2\cdot4\cdots(2n)}{1\cdot3\cdots(2n+1)},$$

$$\left[\frac{2\cdot4\cdot6\cdots(2n-2)}{1\cdot3\cdot5\cdots(2n-1)}\right]^2(2n) > \frac{\pi}{2} > \left[\frac{2\cdot4\cdot6\cdots(2n)}{1\cdot3\cdot5\cdots(2n-1)}\right]^2\frac{1}{2n+1}.$$

For simplicity, introduce the quantities

$$H_n = \frac{1}{2n+1}\left[\frac{2\cdot4\cdot6\cdots(2n)}{1\cdot3\cdot5\cdots(2n-1)}\right]^2.$$

Now divide the last inequality by H_n:

$$\frac{2n+1}{2n} > \frac{\pi}{2H_n} > 1.$$

It follows easily that $\pi/2H_n \longrightarrow 1$, hence $H_n \longrightarrow \frac{1}{2}\pi$ as $n \longrightarrow \infty$. Done.

Let us express H_n in terms of factorials. The product of evens is easy:

$$2\cdot4\cdot6\cdots(2n) = 2^n(1\cdot2\cdot3\cdots n) = 2^n(n!).$$

To get the product of odds, throw in the missing evens and compensate by dividing them right out again:

$$1\cdot3\cdot5\cdots(2n-1) = \frac{1\cdot2\cdot3\cdot4\cdots(2n-1)(2n)}{2\cdot4\cdot6\cdots(2n)} = \frac{(2n)!}{2^n n!}.$$

It follows that

$$H_n = \frac{1}{2n+1}\left[\frac{2^{2n}(n!)^2}{(2n)!}\right]^2$$

and that Wallis's formula can be expressed in the form:

$$\boxed{\lim_{n\to\infty}\frac{1}{2n+1}\left[\frac{2^{2n}(n!)^2}{(2n)!}\right]^2 = \frac{\pi}{2}.}$$

There are a number of interesting applications of this formula. Here is one concerning probability.

Suppose a coin is tossed $2n$ times, where n is large. Then the number of heads that can be expected is about n. Yet it seems unlikely that *exactly* n heads will appear. Just what is the probability of this event?

Let p_n be the probability of n heads in $2n$ tosses. Then

$$p_n = \frac{1}{2^{2n}}\binom{2n}{n} = \frac{1}{2^{2n}}\frac{(2n)!}{(n!)^2}.$$

Why? Because the probability that a given sequence of $2n$ heads and tails occurs is 2^{-2n}, and there are $\binom{2n}{n}$ such sequences that contain n heads and n tails. (This is the number of ways to choose n positions for the heads from among $2n$ possible positions.)

Now, Wallis's formula may be rewritten by taking reciprocals:

$$\lim_{n\to\infty} (2n+1)\left[\frac{(2n)!}{2^{2n}(n!)^2}\right]^2 = \frac{2}{\pi}.$$

The quantity in brackets is p_n. Hence, for large values of n,

$$(2n+1)p_n^2 \approx \frac{2}{\pi},$$

that is,

$$p_n \approx \sqrt{\frac{2}{(2n+1)\pi}} \approx \frac{1}{\sqrt{n\pi}}.$$

Thus, for example, the probability of exactly 10,000 heads in 20,000 tosses of a coin is

$$p_{10000} \approx \frac{1}{\sqrt{10{,}000\pi}} \approx 0.00564.$$

This is fairly small, roughly 1 in 180. Yet when you realize that there are 20,001 possibilities (no heads, one head, etc.), it is relatively quite large.

Stirling's Formula

The numbers $n!$ occur frequently in applications of mathematics. They grow rapidly as n increases, and are tedious to compute. Still in many problems we need at least an estimate of their size. Such an estimate is provided by the remarkable formula of Stirling:

Stirling's Formula

$$n! \approx \sqrt{2\pi n}\, n^n e^{-n}.$$

More precisely,

$$\lim_{n\to\infty}\left(\frac{n!}{\sqrt{2\pi n}\, n^n e^{-n}}\right) = 1.$$

Note that Stirling's formula does not give a close estimate in the usual sense, but rather an "order of magnitude" estimate. For n large, the difference between $n!$ and $\sqrt{2\pi n}\, n^n e^{-n}$ is also large. However, this difference is small relative to $n!$ In other words, for n large enough, $\sqrt{2\pi n}\, n^n e^{-n}$ approximates $n!$ to within, say, 1%. But 1% of 100! is a huge number.

Proof: It is easier to work with $\ln n!$ than $n!$ itself. Accordingly, let

$$S_n = \ln n! = \ln 1 + \ln 2 + \ln 3 + \cdots + \ln n.$$

This is just the kind of sum we expect to be related to an integral. Let us consider a trapezoidal approximation to $\int_1^n \ln x\, dx$. See Fig. 10.2. Comparing the area of the trapezoids shown to the area under the curve, we see that

$$\int_1^n \ln x\, dx \approx \tfrac{1}{2}\ln 1 + \ln 2 + \ln 3 + \cdots + \ln(n-1) + \tfrac{1}{2}\ln n$$
$$= S_n - \tfrac{1}{2}\ln n.$$

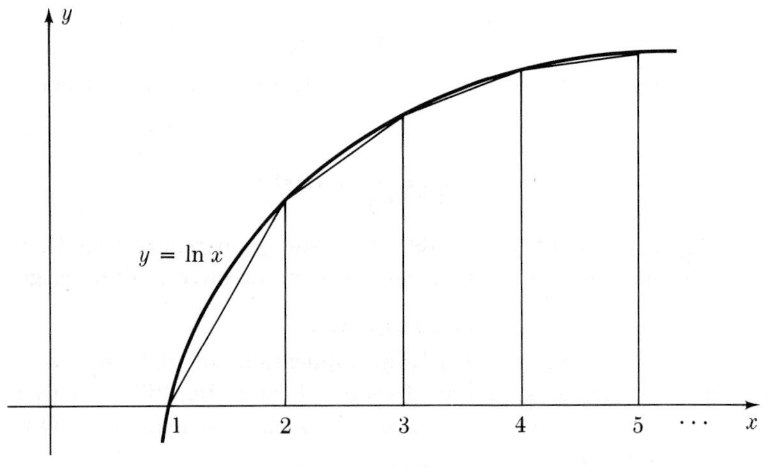

FIG. 10.2 trapezoidal approximation

Set

$$T_n = \int_1^n \ln x\, dx.$$

Then we have $T_n \approx S_n - \frac{1}{2}\ln n$, that is,

$$S_n \approx T_n + \tfrac{1}{2}\ln n.$$

We shall show that this approximation is close, more precisely, that the difference of the two quantities approaches a constant. Therefore, we study the difference

$$A_n = T_n + \tfrac{1}{2}\ln n - S_n.$$

Our strategy is to prove the existence of $\lim A_n$ by showing that $\{A_n\}$ is increasing and bounded. This we do in two steps. Then we complete the proof of Stirling's formula in two further steps.

Step 1: To prove $A_1 < A_2 < A_3 < \cdots$.
We note that

$$A_k - A_{k-1} = (T_k - T_{k-1}) + [\tfrac{1}{2}\ln k - \tfrac{1}{2}\ln(k-1)] - (S_k - S_{k-1}).$$

But

$$T_k - T_{k-1} = \int_{k-1}^{k} \ln x\, dx \qquad \text{and} \qquad S_k - S_{k-1} = \ln k,$$

so

$$A_k - A_{k-1} = \int_{k-1}^{k} \ln x\, dx - \tfrac{1}{2}[\ln k + \ln(k-1)].$$

From the trapezoidal approximation in Fig. 10.3a, we see that

$$\int_{k-1}^{k} \ln x\, dx > \tfrac{1}{2}[\ln k + \ln(k-1)].$$

Therefore $A_k - A_{k-1} > 0$; the sequence increases.

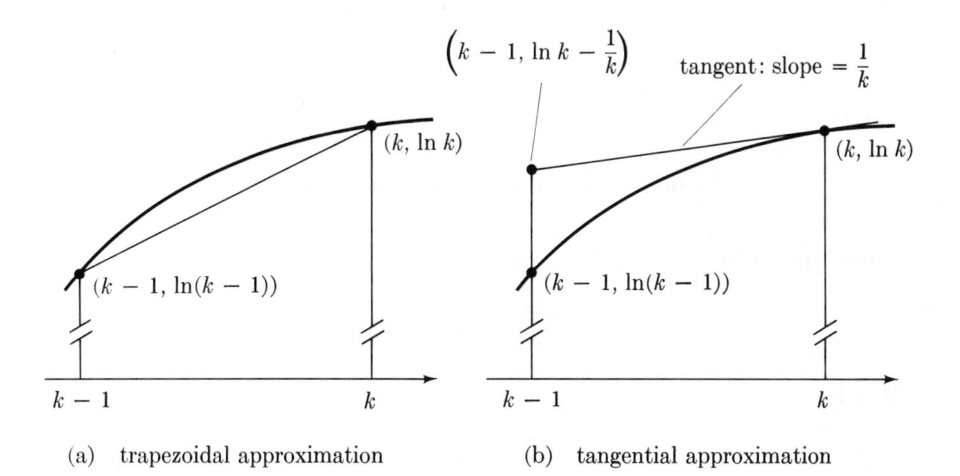

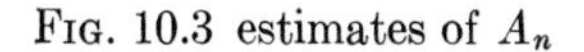
FIG. 10.3 estimates of A_n

Step 2: To find an upper bound for $\{A_n\}$.
Consider Fig. 10.3b. The tangent at $(k, \ln k)$ lies above the convex curve, hence

$$\int_{k-1}^{k} \ln x\, dx < \frac{1}{2}\left[\ln k + \left(\ln k - \frac{1}{k}\right)\right] = \ln k - \frac{1}{2k}$$

for $k \geq 2$. By summing, we conclude that

$$\int_1^n \ln x\, dx < (\ln 2 + \ln 3 + \cdots + \ln n) - \frac{1}{2}\left(\frac{1}{2} + \frac{1}{3} + \cdots + \frac{1}{n}\right),$$

$$T_n < S_n - \frac{1}{2}\left(\frac{1}{2} + \frac{1}{3} + \cdots + \frac{1}{n}\right).$$

Therefore

$$A_n = T_n + \frac{1}{2}\ln n - S_n < \frac{1}{2}\left[\ln n - \left(\frac{1}{2} + \frac{1}{3} + \cdots + \frac{1}{n}\right)\right].$$

But the quantity on the right is bounded, as we have seen in studying Euler's constant. Therefore $\{A_n\}$ is bounded above.

Step 3: The sequence $\{A_n\}$ is increasing and bounded, hence there is a number C such that $A_n \longrightarrow C$, that is,

$$T_n + \frac{1}{2}\ln n - S_n \longrightarrow C.$$

But

$$T_n = \int_1^n \ln x\, dx = n \ln n - n + 1,$$

so

$$n \ln n - n + 1 + \frac{1}{2}\ln n - \ln n! \longrightarrow C.$$

Take exponentials:

$$\frac{n^n e^{-n} e \sqrt{n}}{n!} \longrightarrow e^C,$$

that is,

(*) $$\frac{n!}{n^n e^{-n}\sqrt{n}} \longrightarrow K,$$

where $K = e^{1-C}$.

Step 4: Complete the proof by showing $K = \sqrt{2\pi}$.

We exploit Wallis's product, which provides a relation between $(2n)!$ and $(n!)^2$. From relation (*) above:

$$\left[\frac{n!}{n^n e^{-n}\sqrt{n}}\right]^2 \longrightarrow K^2 \quad \text{and} \quad \frac{(2n)!}{(2n)^{2n} e^{-2n}\sqrt{2n}} \longrightarrow K.$$

The quotient approaches $K^2/K = K$. Divide and simplify:

$$\sqrt{2}\left[\frac{2^{2n}(n!)^2}{(2n)!\sqrt{n}}\right] \longrightarrow K.$$

But the quantity in parentheses has limit $\sqrt{\pi}$ as is seen from Wallis's product. Therefore $K = \sqrt{2\pi}$; the proof of Stirling's formula is complete.

EXERCISES

1. Estimate the number of digits in 100!.
2. Estimate the number of bridge hands (13 cards) that can be formed from a 52-card deck.

Find the limit:

3. $\lim_{n\to\infty}\left(\frac{1}{n+1}+\frac{1}{n+2}+\cdots+\frac{1}{2n}\right)$

4. $\lim_{n\to\infty}\left(\frac{1}{n+1}+\frac{1}{n+2}+\cdots+\frac{1}{n^2}\right)$

5. $\lim_{n\to\infty}\left(\frac{1}{n^2+1}+\frac{1}{n^2+2}+\cdots+\frac{1}{3n^2+n}\right)$

6. $\lim_{n\to\infty}\frac{1}{n}(n!)^{1/n}$.

7*. Use the method of the text to prove the existence of

$$\lim_{n\to\infty}\left(\sum_{k=1}^{n}\frac{1}{\sqrt{k}} - 2\sqrt{n}\right).$$

2. Taylor Approximations

1. INTRODUCTION

In this chapter we shall study approximations of functions by polynomials. Why approximate functions by polynomials? Because values of polynomials can be computed by addition and multiplication, simple operations well-suited for hand or machine computations.

Suppose, for example, you need a 6-place table of

$$f(x) = e^{3x^2}$$

at 1000 equally spaced values of x between -1 and 1. If possible, find a polynomial $p(x)$ such that

$$e^{3x^2} = p(x) + \epsilon(x),$$

where $|\epsilon(x)|$ is less than 5×10^{-7} for $-1 \leq x \leq 1$. Then program a computer to tabulate the corresponding values of $p(x)$.

We shall discuss methods for finding polynomial approximations and obtain estimates for the errors in such approximations.

2. POLYNOMIALS

We begin with a basic algebraic property of polynomials: every polynomial can be expressed not only in powers of x, but also in powers of $(x - a)$, where a is any number. This form of the polynomial is convenient for computations near $x = a$.

EXAMPLE 2.1

Express $x^2 + x + 2$ in powers of $x - 1$.

Solution: Set $u = x - 1$. Then $x = u + 1$, and

$$\begin{aligned} x^2 + x + 2 &= (u + 1)^2 + (u + 1) + 2 \\ &= (u^2 + 2u + 1) + (u + 1) + 2 = u^2 + 3u + 4. \end{aligned}$$

Answer: $(x - 1)^2 + 3(x - 1) + 4.$

EXAMPLE 2.2

Express $x^3 - 6x^2 + 11x - 6$ in powers of $x - 2$.

Solution: Set $u = x - 2$. Then $x = u + 2$, and

$$x^3 - 6x^2 + 11x - 6 = (u + 2)^3 - 6(u + 2)^2 + 11(u + 2) - 6$$
$$= u^3 - u.$$

Answer: $(x - 2)^3 - (x - 2)$.

REMARK: The answer reveals a symmetry about the point $(2, 0)$ not evident in the original expression for the polynomial.

EXAMPLE 2.3

Express x^4 in powers of $x + 1$.

Solution: $x^4 = [(x + 1) - 1]^4$.

Answer: $(x + 1)^4 - 4(x + 1)^3 + 6(x + 1)^2 - 4(x + 1) + 1$.

The methods used in these examples is simple. To express

$$p(x) = A_0 + A_1x + A_2x^2 + A_3x^3 + \cdots + A_nx^n$$

in powers of $x - a$, write $u = x - a$. Then substitute $u + a$ for x:

$$p(x) = A_0 + A_1(u + a) + A_2(u + a)^2 + \cdots + A_n(u + a)^n.$$

Expand each of the powers by the Binomial Formula and collect like powers of u. The result is a polynomial in $u = x - a$, as desired.

This method is laborious when the degree of $p(x)$ exceeds three or four. We now discuss a simpler, more systematic method.

Suppose $p(x)$ is a polynomial expressed in powers of $x - a$:

$$p(x) = A_0 + A_1(x - a) + A_2(x - a)^2 + \cdots + A_n(x - a)^n.$$

What are the coefficients $A_0, A_1, A_2, \cdots$?

There is an easy way to compute A_0. Just replace x by a. Then all terms on the right vanish except the first:

$$p(a) = A_0.$$

Now modify this trick to compute A_1. Differentiate $p(x)$:

$$p'(x) = A_1 + 2A_2(x - a) + \cdots + nA_n(x - a)^{n-1}.$$

Substitute $x = a$; again all terms vanish except the first:

$$p'(a) = A_1.$$

Differentiate again to find A_2:

$$p''(x) = 2A_2 + 3\cdot 2A_3(x - a) + \cdots + n(n - 1)A_n(x - a)^{n-2}.$$

Substitute $x = a$:

$$p''(a) = 2A_2.$$

Once again:

$$p'''(x) = 3\cdot 2A_3 + \cdots + n(n-1)(n-2)A_n(x-a)^{n-3},$$

$$p'''(a) = 3\cdot 2A_3 = 3!\,A_3.$$

Continuing in this way yields

$$p^{(4)}(a) = 4!\,A_4, \quad p^{(5)}(a) = 5!\,A_5, \cdots, p^{(n)}(a) = n!\,A_n.$$

(Here $p^{(k)}$ is the k-th derivative.)

If $p(x)$ is a polynomial of degree n and if a is a number, then

$$p(x) = p(a) + p'(a)(x-a) + \frac{1}{2!}p''(a)(x-a)^2$$

$$+ \frac{1}{3!}p'''(a)(x-a)^3 + \cdots + \frac{1}{n!}p^{(n)}(a)(x-a)^n.$$

EXAMPLE 2.4

Express $p(x) = x^3 - x^2 + 1$

(a) in powers of $x - \dfrac{1}{2}$, (b) in powers of $x - 10$.

Solution: Use the preceding formula with $n = 3$. Compute three derivatives:

$$p'(x) = 3x^2 - 2x, \qquad p''(x) = 6x - 2, \qquad p'''(x) = 6.$$

For (a), evaluate at $x = \frac{1}{2}$:

$$p\left(\frac{1}{2}\right) = \frac{7}{8}, \qquad p'\left(\frac{1}{2}\right) = -\frac{1}{4}, \qquad p''\left(\frac{1}{2}\right) = 1, \qquad p'''\left(\frac{1}{2}\right) = 6.$$

By the formula,

$$p(x) = \frac{7}{8} - \frac{1}{4}\left(x - \frac{1}{2}\right) + \frac{1}{2!}\cdot 1\left(x - \frac{1}{2}\right)^2 + \frac{1}{3!}\cdot 6\left(x - \frac{1}{2}\right)^3.$$

For (b), evaluate at $x = 10$:

$$p(10) = 901, \qquad p'(10) = 280, \qquad p''(10) = 58, \qquad p'''(10) = 6.$$

By the formula,

$$p(x) = 901 + 280(x-10) + \frac{1}{2!}\cdot 58(x-10)^2 + \frac{1}{3!}\cdot 6(x-10)^3.$$

Answer:

(a) $$\frac{7}{8} - \frac{1}{4}\left(x - \frac{1}{2}\right) + \frac{1}{2}\left(x - \frac{1}{2}\right)^2 + \left(x - \frac{1}{2}\right)^3,$$

(b) $$901 + 280(x - 10) + 29(x - 10)^2 + (x - 10)^3.$$

The next example illustrates the computational advantages gained by expanding polynomials in powers of $x - a$.

EXAMPLE 2.5

Let $p(x) = x^3 - x^2 + 1$. Compute $p(0.50028)$ to 5 places.

Solution: Use answer (a) of the preceding example:

$$p(0.50028) = p\left(\frac{1}{2} + 0.00028\right)$$

$$= \frac{7}{8} - \frac{1}{4}(0.00028) + \frac{1}{2}(0.00028)^2 + (0.00028)^3.$$

The last two terms on the right are smaller than 10^{-7}. Therefore to 5 places, $p(0.50028)$ agrees with

$$\frac{7}{8} - \frac{1}{4}(0.00028) = 0.87500 - 0.00007.$$

Answer: 0.87493.

Because we shall write polynomials frequently, it is convenient to use summation notation:

$$\sum_{i=0}^{n} A_i x^i = A_0 + A_1 x + A_2 x^2 + \cdots + A_n x^n.$$

The formula for an n-th degree polynomial in powers of $x - a$ can be abbreviated:

$$\boxed{p(x) = \sum_{i=0}^{n} \frac{p^{(i)}(a)}{i!}(x - a)^i.}$$

Here $p^{(i)}(a)$ denotes the i-th derivative of $p(x)$ evaluated at $x = a$, with the special convention $p^{(0)}(a) = p(a)$. (Also recall the convention $0! = 1$.)

EXERCISES

Expand in powers of $x - a$:

1. $x^2 + 5x + 2$; $a = 1$

2. $x^3 - 3x^2 + 4x;\quad a = 2$
3. $2x^3 + 5x^2 + 13x + 10;\quad a = -1$
4. $x^4 - 5x^2 + x + 2;\quad a = 2$
5. $2x^4 + 5x^3 + 4x + 16;\quad a = -2$
6. $3x^3 - 2x^2 - 2x + 1;\quad a = 1$
7. $5x^5 + 4x^4 - 3x^3 - 2x^2 + x + 1;\quad a = -1$
8. $x^5 + 2x^4 + 3x^2 + 4x + 5;\quad a = -2$
9. $x^4 - 7x^3 + 5x^2 + 3x - 6;\quad a = 0.$

Evaluate to 4 significant digits:

10. $x^3 - 3x^2 + 2x + 1;\quad x = 1.004$
11. $x^5 + x^4 + x^3 + x^2 + x + 1;\quad x = 1.994$
12. $4x^4 - 3x^2 + 10x + 12;\quad x = -0.9890$
13. $10x^3 + 12x^2 - 6x - 5;\quad x = -3.042.$

3. TAYLOR POLYNOMIALS

Consider this problem: Given a function $f(x)$ and a number a, find a polynomial $p(x)$ which approximates $f(x)$ for values of x near a.

One approach that seems reasonable is to construct a polynomial $p_n(x)$ of degree n so that

$$p_n(a) = f(a),\quad p_n'(a) = f'(a),\quad p_n''(a) = f''(a),\quad \cdots,\quad p_n^{(n)}(a) = f^{(n)}(a).$$

Thus $p_n(x)$ mimics $f(x)$ and its first n derivatives at $x = a$.

Let us find $p_n(x)$ explicitly. We write

$$p_n(x) = A_0 + A_1(x - a) + A_2(x - a)^2 + \cdots + A_n(x - a)^n$$

and choose the coefficients A_k appropriately. But in the last section, we saw that $A_k = p_n^{(k)}(a)/k!$. Since we want $p_n^{(k)}(a) = f^{(k)}(a)$, we must choose $A_k = f^{(k)}(a)/k!$

The n-th **degree Taylor polynomial** of $f(x)$ at $x = a$ is

$$p_n(x) = f(a) + f'(a)(x - a) + \frac{1}{2!}f''(a)(x - a)^2 + \cdots + \frac{1}{n!}f^{(n)}(a)(x - a)^n.$$

When $f(x)$ is itself a polynomial of degree n, then

$$p_n(x) = f(x).$$

This was shown in the last section; $p_n(x)$ is precisely the expression for $f(x)$

in powers of $x - a$. Furthermore, in this case,

$$p_n(x) = p_{n+1}(x) = p_{n+2}(x) = \cdots.$$

(Why?) Thus for an n-th degree polynomial $f(x)$, the n-th degree and all higher Taylor polynomials *equal* $f(x)$.

Here are the first three Taylor polynomials explicitly:

$$p_1(x) = f(a) + f'(a)(x - a),$$

$$p_2(x) = f(a) + f'(a)(x - a) + \frac{1}{2}f''(a)(x - a)^2,$$

$$p_3(x) = f(a) + f'(a)(x - a) + \frac{1}{2}f''(a)(x - a)^2 + \frac{1}{6}f'''(a)(x - a)^3.$$

The graph of the linear function $y = p_1(x)$ is the tangent to the graph of $y = f(x)$ at $(a, f(a))$. The graph of the quadratic function $y = p_2(x)$ is a parabola through $(a, f(a))$, also with tangent $y = p_1(x)$, and curved in the same direction as $y = f(x)$. See Fig. 3.1.

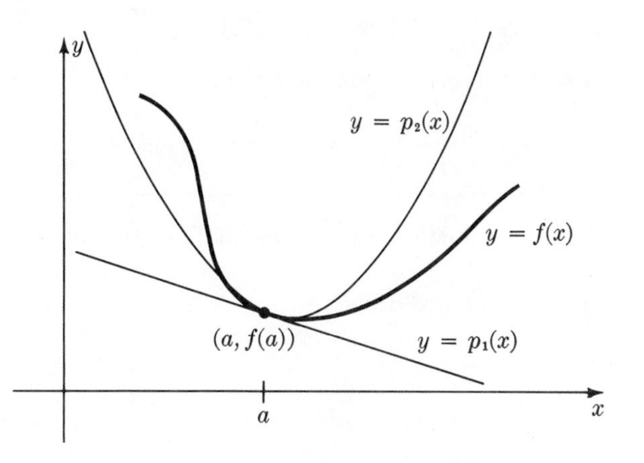

FIG. 3.1

In general, each Taylor polynomial is derived from the preceding one by the addition of a single term:

$$p_{n+1}(x) = p_n(x) + \frac{1}{(n + 1)!}f^{(n+1)}(a)(x - a)^{n+1}.$$

We anticipate that $p_n(x)$ is a good approximation to $f(x)$; the error is $f(x) - p_n(x)$. We try to reduce this error by adding an additional term $f^{(n+1)}(a)(x - a)^{n+1}/(n + 1)!$ to $p_n(x)$, thereby obtaining $p_{n+1}(x)$, an even better approximation (we hope). In Section 6 we shall justify the following formula for the error:

Taylor's Formula with Remainder Suppose $f(x)$ has derivatives up to and including $f^{(n+1)}(x)$ near $x = a$. Write

$$f(x) = p_n(x) + r_n(x),$$

where $p_n(x)$ is the n-th degree Taylor polynomial at $x = a$ and $r_n(x)$ is the remainder (or error). Then

$$r_n(x) = \frac{1}{n!}\int_a^x (x-t)^n f^{(n+1)}(t)\,dt.$$

Usually the integral expressing $r_n(x)$ cannot be computed exactly. Nevertheless, the integral can be estimated. Here is one important estimate:

Estimate of Remainder Suppose

$$f(x) = p_n(x) + r_n(x),$$

where $p_n(x)$ is the n-th Taylor polynomial at $x = a$. If

$$|f^{(n+1)}(x)| \leq M$$

in some interval including $x = a$, say $b \leq x \leq c$, then

$$|r_n(x)| \leq \frac{M}{(n+1)!}|x-a|^{n+1} \qquad \text{for} \quad b \leq x \leq c.$$

This assertion is verified by a direct estimate of the integral:

$$|r_n(x)| = \frac{1}{n!}\left|\int_a^x (x-t)^n f^{(n+1)}(t)\,dt\right|$$

$$\leq \frac{M}{n!}\left|\int_a^x (x-t)^n\,dt\right| = \frac{M}{(n+1)!}|x-a|^{n+1}.$$

EXAMPLE 3.1

Find the n-th degree Taylor polynomial of $f(x) = e^x$ at $x = 0$. Estimate the error.

Solution:

$$f(x) = e^x, \qquad f'(x) = e^x, \qquad f''(x) = e^x, \qquad \cdots;$$

$$f(0) = 1, \qquad f'(0) = 1, \qquad f''(0) = 1, \qquad \cdots.$$

Hence

$$p_n(x) = \sum_{i=0}^{n} \frac{f^{(i)}(0)}{i!}x^i = \sum_{i=0}^{n} \frac{x^i}{i!}.$$

The $(n + 1)$-th derivative is $f^{(n+1)}(t) = e^t$. If $x \geq 0$, the largest value of $f^{(n+1)}(t)$ for t between 0 and x is e^x. By the remainder estimate with $M = e^x$,

$$|r_n(x)| \leq \frac{e^x}{(n+1)!} x^{n+1} \qquad \text{for} \quad x \geq 0.$$

If $x \leq 0$, the largest value of $f^{(n+1)}(t)$ between 0 and x is $e^0 = 1$. By the remainder estimate with $M = 1$,

$$|r_n(x)| \leq \frac{|x|^{n+1}}{(n+1)!}.$$

Answer: $p_n(x) = 1 + x + \dfrac{x^2}{2!} + \dfrac{x^3}{3!} + \cdots + \dfrac{x^n}{n!}$,

$$|r_n(x)| \leq \frac{e^x}{(n+1)!} x^{n+1} \qquad \text{for} \quad x \geq 0,$$

$$|r_n(x)| \leq \frac{|x|^{n+1}}{(n+1)!} \qquad \text{for} \quad x \leq 0.$$

REMARK: At first sight, the answer to Example 3.1 seems circular: the error estimate in approximating e^x involves e^x itself. However, if the e^x in the remainder is replaced by something a little larger, say 3^x, we still get a useful estimate of the remainder:

$$|r_n(x)| \leq \frac{3^x}{(n+1)!} x^{n+1} \qquad \text{for} \quad x \geq 0.$$

EXAMPLE 3.2

Find the Taylor polynomials for $\sin x$ at $x = 0$. Estimate the remainders.

Solution: Compute derivatives:

$$f(x) = \sin x, \qquad f'(x) = \cos x, \qquad f''(x) = -\sin x,$$

$$f'''(x) = -\cos x, \qquad f^{(4)}(x) = \sin x, \quad \cdots,$$

repeating in cycles of four. At $x = 0$, the values are

$$0, \ 1, \ 0, \ -1, \ 0, \ 1, \ 0, \ -1, \ 0, \ \cdots.$$

Hence the n-th degree Taylor polynomial of $\sin x$ is

$$p_n(x) = x - \frac{x^3}{3!} + \frac{x^5}{5!} - \frac{x^7}{7!} + \frac{x^9}{9!} - \cdots,$$

where the last term is $\pm x^n/n!$ if n is odd and $\pm x^{n-1}/(n-1)!$ if n is even.

For example,

$$p_3(x) = p_4(x) = x - \frac{x^3}{3!},$$

$$p_5(x) = p_6(x) = x - \frac{x^3}{3!} + \frac{x^5}{5!},$$

$$p_7(x) = p_8(x) = x - \frac{x^3}{3!} + \frac{x^5}{5!} - \frac{x^7}{7!}.$$

Thus $p_{2m-1}(x) = p_{2m}(x)$ and the sign of the last term is plus if m is odd, minus if m is even:

$$p_{2m-1}(x) = p_{2m}(x) = x - \frac{x^3}{3!} + \frac{x^5}{5!} - \cdots + (-1)^{m-1}\frac{x^{2m-1}}{(2m-1)!}$$

$$= \sum_{i=1}^{m} (-1)^{i-1}\frac{x^{2i-1}}{(2i-1)!}.$$

The remainder estimate is easy: $|f^{(n+1)}(x)| \leq 1$ because $f^{(n+1)}(x) = \pm \sin x$ or $\pm \cos x$. Hence

$$|r_n(x)| \leq \frac{|x|^{n+1}}{(n+1)!}.$$

Since $p_{2m-1}(x) = p_{2m}(x)$, it follows that

$$|r_{2m-1}(x)| = |r_{2m}(x)| \leq \frac{|x|^{2m+1}}{(2m+1)!}.$$

Answer: $\sin x = p_{2m}(x) + r_{2m}(x)$,

$$p_{2m}(x) = \sum_{i=1}^{m} (-1)^{i-1}\frac{x^{2i-1}}{(2i-1)!}, \quad |r_{2m}(x)| \leq \frac{|x|^{2m+1}}{(2m+1)!}.$$

[Note: $p_{2m-1}(x) = p_{2m}(x)$.]

REMARK: For the cosine, a similar argument shows

$$\cos x = 1 - \frac{x^2}{2!} + \frac{x^4}{4!} - \frac{x^6}{6!} + \cdots + (-1)^m\frac{x^{2m}}{(2m)!} + r_{2m}(x),$$

$$|r_{2m}(x)| = |r_{2m+1}(x)| \leq \frac{|x|^{2m+2}}{(2m+2)!}.$$

EXAMPLE 3.3

Find the n-th degree Taylor polynomial of $\ln x$ at $x = 5$. Estimate the remainder if $4 \leq x \leq 6$.

Solution: Let $f(x) = \ln x$. Then

$$f'(x) = \frac{1}{x}, \qquad f''(x) = -\frac{1}{x^2}, \qquad f'''(x) = \frac{1\cdot 2}{x^3},$$

$$f^{(4)}(x) = -\frac{3!}{x^4}, \quad \cdots, \quad f^{(i)}(x) = (-1)^{i-1}\frac{(i-1)!}{x^i}.$$

The coefficient of $(x-5)^i$ in the Taylor polynomial is

$$\frac{1}{i!}f^{(i)}(5) = (-1)^{i-1}\cdot\frac{(i-1)!}{i!}\cdot\frac{1}{5^i} = (-1)^{i-1}\frac{1}{i\cdot 5^i}.$$

Therefore,

$$p_n(x) = \ln 5 + \frac{1}{1\cdot 5}(x-5) - \frac{1}{2\cdot 5^2}(x-5)^2$$

$$+ \frac{1}{3\cdot 5^3}(x-5)^3 - \cdots + (-1)^{n-1}\frac{1}{n\cdot 5^n}(x-5)^n.$$

An error estimate in the range $4 \le x \le 6$ is

$$|r_n(x)| \le \frac{M}{(n+1)!}|x-5|^{n+1} \le \frac{M}{(n+1)!}(1)^{n+1},$$

where M is a bound for $|f^{(n+1)}(x)|$. Now

$$|f^{(n+1)}(x)| = \left|\pm\frac{n!}{x^{n+1}}\right| \le \frac{n!}{4^{n+1}} \qquad \text{for} \quad 4 \le x \le 6.$$

Take this number as M. The resulting estimate is

$$|r_n(x)| \le \frac{1}{(n+1)!}\frac{n!}{4^{n+1}} = \frac{1}{(n+1)4^{n+1}}.$$

Answer: For $4 \le x \le 6$,

$$p_n(x) = \ln 5 + \sum_{i=1}^{n}(-1)^{i-1}\frac{1}{i\cdot 5^i}(x-5)^i,$$

$$|r_n(x)| \le \frac{1}{(n+1)4^{n+1}}.$$

Summary

Taylor's Formula with Remainder:

$$f(x) = p_n(x) + r_n(x),$$

$$p_n(x) = \sum_{i=0}^{n}\frac{f^{(i)}(a)}{i!}(x-a)^i, \qquad r_n(x) = \frac{1}{n!}\int_a^x (x-t)^n f^{(n+1)}(t)\,dt.$$

ESTIMATE OF THE REMAINDER:

If $|f^{(n+1)}(t)| \le M$ for all t between a and x, then

$$|r_n(x)| \le \frac{M}{(n+1)!}|x-a|^{n+1}.$$

SPECIAL FUNCTIONS:

$$e^x = \sum_{i=0}^{n} \frac{x^i}{i!} + r_n(x), \qquad |r_n(x)| \le \begin{cases} \dfrac{e^x x^{n+1}}{(n+1)!} & \text{if } x \ge 0 \\[2ex] \dfrac{|x|^{n+1}}{(n+1)!} & \text{if } x \le 0. \end{cases}$$

$$\frac{1}{1-x} = \sum_{i=0}^{n} x^i + \frac{x^{n+1}}{1-x} \qquad \text{if } x \ne 1.$$

$$\sin x = \sum_{i=1}^{m} \frac{(-1)^{i-1}}{(2i-1)!} x^{2i-1} + r_{2m-1}(x), \qquad |r_{2m-1}(x)| \le \frac{|x|^{2m+1}}{(2m+1)!}.$$

$$\cos x = \sum_{i=0}^{m} \frac{(-1)^i}{(2i)!} x^{2i} + r_{2m}(x), \qquad |r_{2m}(x)| \le \frac{|x|^{2m+2}}{(2m+2)!}.$$

EXERCISES

Find the Taylor polynomials at the given point; estimate the remainder:

1. $f(x) = \sin 2x;\quad x = 0$
2. $f(x) = \sin 2x;\quad x = \pi/2$
3. $f(x) = xe^x;\quad x = 0$
4. $f(x) = xe^x;\quad x = 1$
5. $f(x) = x^2 \ln x;\quad x = 1$
6. $f(x) = x^2 \ln x;\quad x = e$
7. $f(x) = x^2e^{-x};\quad x = 0$
8. $f(x) = x^2e^{-x};\quad x = 1$
9. $f(x) = x \sin x;\quad x = 0$
10. $f(x) = x \sin x;\quad x = \pi/2$
11. $f(x) = \sin x + \cos x;\quad x = 0$
12. $f(x) = \cosh x;\quad x = 0$
13. $f(x) = \sinh x + \sin x;\quad x = 0$
14. $f(x) = 1 + e^x + e^{2x};\quad x = 0.$
15. Let $p(x) = x^4 - 4x^3 + 6x^2 - 3x + 2$. Estimate the error in the range $\frac{3}{4} \le x \le \frac{5}{4}$ if this polynomial is approximated by its linear Taylor polynomial about $x = 1$.
16. Compare the Taylor polynomials for $\sin x$ at $x = \pi/2$ with those for $\cos x$ at $x = 0$.
17. Let $p_n(x)$ be the n-th degree Taylor polynomial of $f(x)$ at $x = a$. Verify that $p_n^{(i)}(a) = f^{(i)}(a)$, for $i = 0, 1, 2, \cdots, n$.

4. APPLICATIONS

Taylor's Formula with Remainder provides a practical method for approximating functions. First, it gives a simple procedure for obtaining a polynomial approximation. Second, it supplies an estimate of the error.

EXAMPLE 4.1

Find a polynomial $p(x)$ such that $|e^x - p(x)| < 0.001$

(a) for all x in the interval $-\frac{1}{2} \le x \le \frac{1}{2}$,
(b) for all x in the interval $-2 \le x \le 2$.

Solution: By the answer to Example 3.1, a logical choice for $p(x)$ is one of the Taylor polynomials

$$p_n(x) = 1 + x + \frac{x^2}{2!} + \frac{x^3}{3!} + \cdots + \frac{x^n}{n!}.$$

We want to choose n so that

$$|e^x - p_n(x)| < 10^{-3}.$$

To minimize computation and round-off error, we prefer n as small as possible.

(a) If $-\frac{1}{2} \le x \le \frac{1}{2}$, then

$$|r_n(x)| \le \frac{e^{1/2}}{(n+1)!}\left(\frac{1}{2}\right)^{n+1}.$$

We choose n so that

$$\frac{e^{1/2}}{(n+1)!}\left(\frac{1}{2}\right)^{n+1} < \frac{1}{1000},$$

that is,

$$\frac{1}{(n+1)!2^{n+1}} < \frac{1}{1000e^{1/2}} < \frac{1}{1648}.$$

A few trials show

$$\frac{1}{4!2^4} = \frac{1}{384}, \qquad \frac{1}{5!2^5} = \frac{1}{3840}.$$

Hence we take $n + 1 = 5$, that is, $n = 4$. The desired polynomial is

$$p_4(x) = 1 + x + \frac{x^2}{2!} + \frac{x^3}{3!} + \frac{x^4}{4!}.$$

(b) If $-2 \le x \le 2$, then

$$|r_n(x)| \le \frac{e^2}{(n+1)!}\cdot 2^{n+1}.$$

This time we choose n so that

$$\frac{e^2}{(n+1)!}\cdot 2^{n+1} < \frac{1}{1000},$$

that is,

$$\frac{2^{n+1}}{(n+1)!} < \frac{1}{7389}.$$

A few trials show

$$\frac{2^{10}}{10!} = \frac{1024}{3{,}628{,}800} = \frac{4}{14{,}175} \approx \frac{1}{3544},$$

$$\frac{2^{11}}{11!} = \frac{2^{10}}{10!}\cdot\frac{2}{11} \approx \frac{2}{38{,}981} \approx \frac{1}{19{,}500}.$$

Hence we take $n + 1 = 11$, that is, $n = 10$. The desired polynomial is

$$p_{10}(x) = 1 + x + \frac{x^2}{2!} + \frac{x^3}{3!} + \cdots + \frac{x^{10}}{10!}.$$

Answer: (a) $p(x) = \sum_{i=0}^{4} \frac{x^i}{i!}$ (b) $p(x) = \sum_{i=0}^{10} \frac{x^i}{i!}$.

Tables of Sines and Cosines

Suppose we want to construct 5-place tables of $\sin x$ and $\cos x$. A logical method is approximation by Taylor polynomials. These should be of low degree (few terms) to limit the number of arithmetic operations necessary and to prevent accumulation of round-off errors. Note that we need tabulate $\sin x$ and $\cos x$ only for $0 \leq x \leq \pi/4$. (Why?)

Let us concentrate on $\sin x$; a similar discussion applies to $\cos x$. First we consider the third degree Taylor polynomial at $x = 0$,

$$p_3(x) = x - \frac{x^3}{6}.$$

From Section 3,

$$|\sin x - p_3(x)| \leq \frac{x^5}{5!} = \frac{x^5}{120}.$$

For 5-place accuracy, the error must be less than 5×10^{-6}. Therefore we want

$$\frac{x^5}{120} < 5 \times 10^{-6}, \qquad x^5 < 6 \times 10^{-4}.$$

An easy computation with slide rule (or 4-place log tables) shows this is the case if $x < 0.22$ rad $\approx 12.5°$. Thus up to about 12°, the third degree Taylor polynomial yields 5-place accuracy.

Next we try

$$p_5(x) = x - \frac{x^3}{3!} + \frac{x^5}{5!}.$$

Since

$$|\sin x - p_5(x)| \leq \frac{x^7}{7!},$$

5-place accuracy is obtained if

$$\frac{x^7}{7!} < 5 \times 10^{-6}, \qquad x^7 < 0.0252.$$

Using the C.R.C. table of seventh powers, we find this is the case provided $x < 0.59$ rad $\approx 34°$.

Next we try

$$p_7(x) = x - \frac{x^3}{3!} + \frac{x^5}{5!} - \frac{x^7}{7!}.$$

The error is

$$|\sin x - p_7(x)| \leq \frac{x^9}{9!}.$$

For angles up to 45°, this error is at most

$$\frac{1}{9!}\left(\frac{\pi}{4}\right)^9 \approx \frac{1}{9!}(0.7854)^9 < \frac{1}{9!}(0.79)^9.$$

The C.R.C. table of ninth powers shows $(79)^9$ is slightly less than 12×10^{16}. Hence $(0.79)^9$ is slightly less than 12×10^{-2}. The C.R.C. tables show also that $1/9! \approx 0.276 \times 10^{-5}$. Therefore

$$|\text{error}| < (12 \times 10^{-2})(0.28 \times 10^{-5}) < 4 \times 10^{-7}.$$

Conclusion: For $0 \leq x \leq \pi/4$, the Taylor polynomial $p_7(x)$ yields 6-place accuracy in approximating $\sin x$. Furthermore, since $p_7(x)$ involves only four terms, the probability of large accumulated round-off error is low. Thus $p_7(x)$ provides a practical way of constructing a table of sines.

It is not necessary to limit ourselves to Taylor polynomials at $x = 0$. For example, if we are interested only in angles close to 45°, it is better to use Taylor polynomials at $\pi/4$. They provide greater accuracy for the same amount of computation.

The third degree Taylor polynomial of $\sin x$ at $\pi/4$ is

$$p_3(x) = \left(\frac{\sqrt{2}}{2}\right)\left[1 + \left(x - \frac{\pi}{4}\right) - \frac{1}{2}\left(x - \frac{\pi}{4}\right)^2 - \frac{1}{6}\left(x - \frac{\pi}{4}\right)^3\right]$$

and

$$|\sin x - p_3(x)| \le \frac{1}{4!}\left(x - \frac{\pi}{4}\right)^4.$$

If x differs from 45° by at most 0.1 rad $\approx$ 5.7°, then the error is bounded by

$$\frac{1}{4!}(0.1)^4 < 4.2 \times 10^{-6}.$$

Hence for x between 39.3° and 50.7°, $p_3(x)$ yields 5-place accuracy. Between 44° and 46°, the error is bounded by

$$\frac{1}{4!}(0.0175)^4 < 4 \times 10^{-9}, \qquad (1° \approx 0.01745 \text{ rad})$$

and so $p_3(x)$ yields 8-place accuracy.

We see that near $\pi/4$, the Taylor polynomial $p_3(x)$ about $x = \pi/4$ gives the same accuracy as does $p_7(x)$ about $x = 0$. This is typical of Taylor polynomials: for values near $x = a$, the accuracy achieved by an n-th degree Taylor polynomial about $x = 0$ is matched by a lower degree Taylor polynomial about $x = a$. Generally the lower degree polynomial means less computation and less round-off error.

What is not typical is that every other coefficient is zero in the Taylor polynomials of $\sin x$ and $\cos x$ about $x = 0$. For computation, this is excellent; it means relatively little computation yields extraordinary accuracy. For example, the polynomial $p_7(x)$ of $\sin x$ actually involves only four terms, yet provides an approximation to within $|x|^9/9!$. This is why the Taylor polynomials about points other than $x = 0$ are rarely used for $\sin x$ and $\cos x$.

EXERCISES

1. Find a polynomial $p(x)$ such that $|e^{-x^2} - p(x)| < 0.001$ for all x in the interval $-1 \le x \le 1$.
2. What degree Taylor polynomial about $x = 0$ is needed to approximate $f(x) = \cos x$ for $-\pi/4 \le x \le \pi/4$ to 5 decimal places?
3. Estimate the error in approximating $f(x) = \ln(1 + x)$ for $-\frac{1}{3} \le x \le \frac{1}{3}$ by its 10-th degree Taylor polynomial about $x = 0$.
4. Approximate $f(x) = 1/(1 - x)^2$ for $-\frac{1}{4} \le x \le \frac{1}{4}$ to 3 decimal places by a Taylor polynomial about $x = 0$.
5. Approximate $\sin^2 x$ by its 4-th degree Taylor polynomial about $x = 0$. Estimate the error if $|x| \le 0.1$. [*Hint:* $\sin^2 x = \frac{1}{2}(1 - \cos 2x)$.]
6. Show that for $100 \le x \le 101$, the approximation $\sqrt{x} \approx 10 + \frac{1}{20}(x - 100)$ is correct to within 0.0002.

7. Show that $|\sin x - p_9(x)| < 5 \times 10^{-6}$ for $0 \le x \le \pi/2$, where $p_9(x)$ is the 9-th degree Taylor polynomial for $\sin x$ at $x = 0$.
8. Find the smallest positive integer n such that for $0 \le x \le \frac{1}{3}$,

$$\left| \frac{1}{1-x} - (1 + x + x^2 + \cdots + x^n) \right| < 5 \times 10^{-6}.$$

9. Compute $\sin(5\pi/8)$ to 5-place accuracy. [Use Taylor polynomials of $\sin x$, but not at $x = 0$.]
10. If $f'(a) = f''(a) = \cdots = f^{(n-1)}(a) = 0$, but $f^{(n)}(a) \neq 0$, show that a reasonable approximation to $f(x)$ near a is $f(x) \approx f(a) + \dfrac{f^{(n)}(a)}{n!}(x-a)^n$.
11. (cont.) Suppose n is even; show that $f(x)$ has a maximum at $x = a$ if $f^{(n)}(a) < 0$, and a minimum at $x = a$ if $f^{(n)}(a) > 0$. Suppose n is odd; show that $f(x)$ has neither a maximum nor a minimum at $x = a$.

5. TAYLOR SERIES

Consider once again the Taylor polynomials and remainders of the exponential function at $x = 0$:

$$e^x = \sum_{i=0}^{n} \frac{x^i}{i!} + r_n(x),$$

where

$$|r_n(x)| \le \frac{g(x)\,|x|^{n+1}}{(n+1)!}; \qquad \begin{cases} g(x) = 1 & \text{for} \quad x < 0 \\ g(x) = e^x & \text{for} \quad x \ge 0. \end{cases}$$

This estimate shows that no matter what x is, the error is very small if n is large enough. In other words, for fixed x,

$$\frac{g(x)\,|x|^{n+1}}{(n+1)!} \longrightarrow 0 \quad \text{as} \quad n \longrightarrow \infty.$$

Here is the reason. The number x is fixed; set $A = |x|$. Pick m so that $m > 10A$. Then

$$\frac{A}{m+1} < \frac{1}{10}, \quad \frac{A}{m+2} < \frac{1}{10}, \quad \frac{A}{m+3} < \frac{1}{10}, \quad \cdots, \quad \frac{A}{m+k} < \frac{1}{10},$$

for any k. If $n = m + k$, then

$$\frac{|x|^n}{n!} = \frac{A^n}{n!} = \frac{A^m}{m!} \cdot \frac{A}{m+1} \cdot \frac{A}{m+2} \cdots \frac{A}{m+k}$$

$$< \frac{A^m}{m!} \cdot \frac{1}{10} \cdot \frac{1}{10} \cdots = \frac{A^m}{m!} \cdot \frac{1}{10^k}.$$

Now $k \longrightarrow \infty$ as $n \longrightarrow \infty$. Since $g(x)A^m/m!$ is fixed, the right-hand term $\longrightarrow 0$ as $n \longrightarrow \infty$. This means that $r_n(x) \longrightarrow 0$ as $n \longrightarrow \infty$ for each x, hence the infinite series $\sum_{i=0}^{\infty} x^i/i!$ converges to e^x:

$$e^x = \sum_{i=0}^{\infty} \frac{x^i}{i!} \qquad \text{for all } x.$$

Similarly

$$\left.\begin{aligned} \sin x &= \sum_{i=1}^{\infty} \frac{(-1)^{i-1}}{(2i-1)!} x^{2i-1} \\ \cos x &= \sum_{i=0}^{\infty} \frac{(-1)^i}{(2i)!} x^{2i} \end{aligned}\right\} \qquad \text{for all } x.$$

Such representations of functions by what look like polynomials of infinite degree, are called **Taylor series.** A familiar one is the infinite geometric series:

$$\frac{1}{1-x} = \sum_{i=0}^{\infty} x^i \qquad \text{for} \quad |x| < 1.$$

In general, suppose $f(x)$ is a function that has derivatives of all orders. (We say that $f(x)$ is **infinitely differentiable.**) Then for each n,

$$f(x) = \sum_{i=0}^{n} \frac{f^{(i)}(a)}{i!} (x-a)^i + r_n(x).$$

If we can show that $r_n(x) \longrightarrow 0$ as $n \longrightarrow \infty$ for certain values of x, then we can write

$$f(x) = \sum_{i=0}^{\infty} \frac{f^{(i)}(a)}{i!} (x-a)^i.$$

This formula is called the **expansion** of $f(x)$ in a **Taylor series** at $x = a$.

In practice, it may be tedious to compute successive derivatives of $f(x)$ and difficult to determine whether $r_n(x) \longrightarrow 0$. In Chapter 3, we shall discuss the theory of representing functions by series and shall present various practical techniques for doing so.

EXERCISES

Find the Taylor series about the given point:

1. $f(x) = \sin 3x; \quad x = 0$
2. $f(x) = \cos \frac{1}{2}x; \quad x = 0$

3. $f(x) = \sin^2 x; \quad x = 0$
4. $f(x) = \cos^2(2x - 1); \quad x = \frac{1}{2}$
5. $f(x) = e^{-2x}; \quad x = 0$
6. $f(x) = \cosh x; \quad x = 0$
7. $f(x) = e^{3x+2}; \quad x = -\frac{2}{3}$
8. $f(x) = \sin x + 2\cos x; \quad x = 0$
9. $f(x) = \dfrac{1}{1 - 3x}; \quad x = 0$
10. $f(x) = \dfrac{1}{2 + x}; \quad x = 0$
11. $f(x) = \dfrac{1}{x}; \quad x = 2.$
12. An early model desk computer had an exponential pack, but no trigonometric one. Its program to compute the sine used the approximation

$$\sin x \approx \frac{1}{2}(e^x - e^{-x}) - \frac{1}{3}x^3\left[1 + \frac{1}{840}\left(x^4 + \frac{1}{7920}x^8\right)\right].$$

Express the exact error as a power series.

13*. (cont.) Prove $|\text{error}| < 10^{-8}$ for $|x| < \frac{1}{2}\pi$. You may assume $15! > 1.3 \times 10^{12}$ and $(\frac{1}{2}\pi)^{15} < 10^3$.

6. DERIVATION OF TAYLOR'S FORMULA

We shall derive Taylor's Formula as stated in Section 3:

$$f(x) = p_n(x) + r_n(x),$$

where

$$p_n(x) = \sum_{i=0}^{n} \frac{f^{(i)}(a)}{i!}(x - a)^i$$

and

$$r_n(x) = \frac{1}{n!}\int_a^x (x - t)^n f^{(n+1)}(t)\, dt.$$

This is a consequence of the following assertion (actually a special case of Taylor's Formula):

> If
>
> $$g(a) = g'(a) = g''(a) = \cdots = g^{(n)}(a) = 0,$$
>
> then
>
> $$g(x) = \frac{1}{n!}\int_a^x (x - t)^n g^{(n+1)}(t)\, dt.$$

If this assertion is correct, how does Taylor's Formula follow? Suppose $f(x)$ is any function and $p_n(x)$ is its n-th degree Taylor polynomial at $x = a$. Set

$$g(x) = f(x) - p_n(x).$$

Now $p_n(x)$ agrees with $f(x)$, and its first n derivatives agree with those of

$f(x)$ at $x = a$. Therefore

$$g(a) = g'(a) = g''(a) = \cdots = g^{(n)}(a) = 0.$$

Thus $g(x)$ satisfies the conditions of the assertion, so

$$g(x) = \frac{1}{n!}\int_a^x (x - t)^n g^{(n+1)}(t)\,dt.$$

But $g(x) = f(x) - p_n(x)$ and $g^{(n+1)}(t) = f^{(n+1)}(t)$ because the $(n + 1)$-th derivative of the polynomial $p_n(x)$ is zero. Therefore,

$$f(x) - p_n(x) = \frac{1}{n!}\int_a^x (x - t)^n f^{(n+1)}(t)\,dt,$$

which is precisely Taylor's Formula.

Let us return to the assertion and verify it for low values of n. We integrate by parts, noting that a and x are fixed.

Case $n = 0$:

$$\frac{1}{0!}\int_a^x (x - t)^0 g'(t)\,dt = \int_a^x g'(t)\,dt = g(t)\Big|_a^x = g(x).$$

(Do not forget $g(a) = 0$.)

Case $n=1$: To evaluate

$$\frac{1}{1!}\int_a^x (x - t)g''(t)\,dt,$$

set $u(t) = (x - t)$ and $v(t) = g'(t)$. Then

$$\int_a^x (x - t)g''(t)\,dt = \int_a^x u\,dv = u(t)v(t)\Big|_a^x - \int_a^x v(t)u'(t)\,dt.$$

Therefore

$$\int_a^x (x - t)g''(t)\,dt = (x - t)g'(t)\Big|_a^x + \int_a^x g'(t)\,dt = \int_a^x g'(t)\,dt = g(x),$$

by the previous case. (Do not forget $g'(a) = 0$.)

Case $n = 2$: Set $u(t) = \frac{1}{2}(x - t)^2$ and $v(t) = g''(t)$. Then

$$\frac{1}{2!}\int_a^x (x - t)^2 g'''(t)\,dt = \int_a^x u\,dv$$

$$= u(t)v(t)\Big|_a^x - \int_a^x v(t)u'(t)\,dt = 0 + \int_a^x (x - t)g''(t)\,dt = g(x),$$

by the previous case. Note that $v(a) = 0$ by the hypothesis $g''(a) = 0$.

The general case is handled the same way. One integration by parts, with

$$u(t) = \frac{(x - t)^n}{n!} \qquad \text{and} \qquad v(t) = g^{(n)}(t),$$

reduces

$$\frac{1}{n!} \int_a^x (x - t)^n g^{(n+1)}(t)\, dt \qquad \text{to} \qquad \frac{1}{(n - 1)!} \int_a^x (x - t)^{n-1} g^{(n)}(t)\, dt.$$

The latter is $g(x)$ by the previous case.

3. Power Series

1. INTRODUCTION

In this chapter we study power series

$$a_0 + a_1(x - c) + a_2(x - c)^2 + \cdots + a_n(x - c)^n + \cdots$$

and their applications. In most of our examples $c = 0$, but the discussion applies to $c \neq 0$ as well.

Power series serve two important purposes. First, they express known functions in a form particularly suitable for computation. Second, they define functions which are not simple to specify otherwise. Certainly nobody objects to defining a function by a polynomial,

$$f(x) = a_0 + a_1x + a_2x^2 + \cdots + a_nx^n.$$

Then why not define a function by a power series,

$$f(x) = a_0 + a_1x + a_2x^2 + \cdots,$$

provided, of course, that the series converges? We shall see, in fact, that in many ways power series resemble polynomials.

One useful power series is the geometric series

$$1 + x + x^2 + \cdots + x^n + \cdots,$$

which converges to $1/(1 - x)$ for $|x| < 1$ (but diverges for $|x| \geq 1$).

Other examples of power series were discussed in the preceding chapter:

$$e^x = 1 + x + \frac{x^2}{2!} + \frac{x^3}{3!} + \cdots = \sum_{n=0}^{\infty} \frac{x^n}{n!},$$

$$\sin x = x - \frac{x^3}{3!} + \frac{x^5}{5!} - \frac{x^7}{7!} + \cdots = \sum_{n=1}^{\infty} (-1)^{n-1} \frac{x^{2n-1}}{(2n-1)!},$$

$$\cos x = 1 - \frac{x^2}{2!} + \frac{x^4}{4!} - \frac{x^6}{6!} + \cdots = \sum_{n=0}^{\infty} (-1)^n \frac{x^{2n}}{(2n)!}.$$

These series were derived from Taylor's Formula with Remainder. Each converges for all values of x, whereas the geometric series converges only for $|x| < 1$.

Convergence and Divergence

For each fixed value of x, a power series is an infinite series of numbers. Therefore convergence is defined just as it was in Chapter 1.

A power series $\sum_{k=0}^{\infty} a_k(x - c)^k$ **converges** at a point x if the sequence of partial sums

$$s_n(x) = \sum_{k=0}^{n} a_k(x - c)^k$$

converges. The power series **diverges** at x if the sequence $\{s_n(x)\}$ diverges.

You can think of a power series as infinitely many numerical series at once, one for each x. The series may converge at some points x and diverge at others. If it converges on the set of points **D**, then the sum of the series will generally vary as x varies in **D**; in other words, the sum is a *function* $F(x)$ with domain **D**. We say that the power series converges to $F(x)$ on **D**. Here is the precise definition:

The power series $\sum_0^{\infty} a_k(x - c)^k$ converges to the function $F(x)$ on a set **D**, if given x in **D** and any $\epsilon > 0$, there exists a positive integer N such that

$$\left| \sum_{k=0}^{n} a_k(x - c)^k - F(x) \right| < \epsilon \qquad \text{for all} \quad n \geq N.$$

The number N tells how soon the partial sums are within ϵ of $F(x)$, thus it measures "rapidity of convergence". The series may converge at different rates for different points x; so for ϵ given, N may depend on x.

For example, take the geometric series $\sum_0^{\infty} x^n$, which converges to $F(x) = 1/(1 - x)$ in the interval $|x| < 1$. Suppose $\epsilon = 0.01$. We ask how large n must be so that

$$\left| F(x) - \sum_{k=0}^{n} x^k \right| < \epsilon, \quad \text{that is,} \quad \left| \frac{1}{1 - x} - \frac{1 - x^{n+1}}{1 - x} \right| = \left| \frac{x^{n+1}}{1 - x} \right| < 0.01.$$

For $x = \frac{1}{5}$, we require

$$\left(\frac{1}{5}\right)^{n+1} \Big/ \frac{4}{5} < \frac{1}{100}, \qquad \frac{1}{5^n} < \frac{1}{400}, \qquad 5^n > 400.$$

Clearly $n \geq 4$ will do, hence $N = 4$.

For $x = \frac{1}{2}$, we require

$$\left(\frac{1}{2}\right)^{n+1} \Big/ \frac{1}{2} < \frac{1}{100}, \qquad 2^n > 100.$$

This time we need $n \geq 7$, so $N = 7$.

For $n = \frac{9}{10}$, we require

$$\left(\frac{9}{10}\right)^{n+1} \Big/ \frac{1}{10} < \frac{1}{100}, \qquad \left(\frac{9}{10}\right)^{n+1} > \frac{1}{1000}.$$

Using logarithms, we find that we need $n \geq 65$; in this case $N = 65$.

These results show that the geometric series converges less and less rapidly (N increases) as x increases towards 1. Nevertheless it does converge for each x in the interval $-1 < x < 1$.

Radius of Convergence

The set on which a power series converges turns out to be simple: it is either a single point, the whole line or an interval. The proof of this statement is based on the following fact.

Lemma If a power series $\sum a_n x^n$ converges at $x = x_1$, where $x_1 \neq 0$, then it converges absolutely in the interval $|x| < |x_1|$.

Proof: If $\sum a_n x_1{}^n$ converges, its terms approach 0. Hence the terms are bounded, that is, there exists a positive number M such that $|a_n x_1{}^n| \leq M$. Now

$$|a_n x^n| = \left| a_n x_1{}^n \left(\frac{x}{x_1}\right)^n \right| \leq M \left|\frac{x}{x_1}\right|^n.$$

If $|x| < |x_1|$, then $|x/x_1| < 1$ and the series converges absolutely by comparison with a convergent geometric series.

The proof of the next result will require a form of the basic completeness property of the field **R** of real numbers: *Each set of real numbers that has an upper bound has a unique least upper bound (supremum).* This means that if **S** is a set of real numbers such that $x \leq B$ for some B and *all* x in **S**, then there is a number L such that $x \leq L$ for all x in **S** and no smaller number has this property.

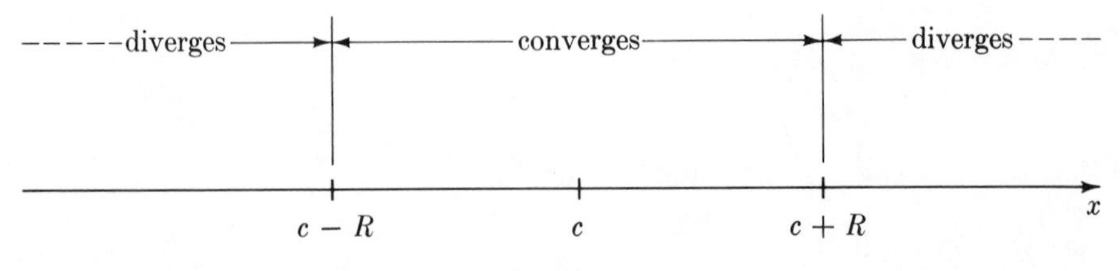

Fig. 1.1

Theorem Given a power series $\sum a_n(x-c)^n$, precisely one of the following three cases holds:

(i) The series converges only for $x = c$.

(ii) The series converges for all values of x.

(iii) There is a positive number R such that the series converges for each x satisfying $|x - c| < R$ and diverges for each x satisfying $|x - c| > R$. See Fig. 1.1.

Proof: We may take $c = 0$ to simplify the notation, that is, we simply replace $x - c$ by x, a translation along the x-axis that takes c to 0. Now let **D** be the domain of convergence of $\sum a_n x^n$. Certainly 0 is in **D** because $\sum a_n 0^n$ converges no matter what the a_n are. We distinguish two cases.

Case 1: **D** is unbounded. Then there are numbers x_1 with $\sum a_n x_1^n$ convergent and $|x_1|$ arbitrarily large. By the Lemma, **D** contains the interval $|x| < |x_1|$ in each such case. Since $|x_1|$ can be taken arbitrarily large, this means that **D** contains all real numbers, Case (ii) in the Theorem.

Case 2: **D** is bounded. Then the set **S** of all numbers $|x_1|$ where $\sum a_n x_1^n$ converges is bounded above. Let R be the supremum. Thus, if $\sum a_n x_1^n$ converges, then $|x_1| \leq R$, and R is the smallest number with this property.

It follows that $\sum a_n x^n$ converges for *some* points x with $|x| < R$, and diverges for *all* points x with $|x| > R$. It remains to show that the series converges for *all* x with $|x| < R$. Suppose $\sum a_n x_2^n$ diverges; it is enough to show that $R \leq |x_2|$. By the Lemma, the series must diverge for every x with $|x_2| < |x|$ otherwise $\sum a_n x_2^n$ would converge. Therefore $|x_2|$ is an upper bound for **S**, so $R \leq |x_2|$ since R is the least upper bound.

If $R > 0$ we have (iii) of the Theorem. Otherwise $R = 0$ and we have (i). The proof is complete.

Case (i) is an extreme case, unimportant and uninteresting. It occurs when the coefficients a_n grow so fast that the power series can converge only if all terms after the first vanish. An example is

$$1 + x + 2^2x^2 + 3^3x^3 + 4^4x^4 + \cdots.$$

For each non-zero x, note that $|n^n x^n| = |nx|^n \longrightarrow \infty$ as $n \longrightarrow \infty$.

Case (ii) is the opposite extreme and occurs when the coefficients a_0, a_1, a_2, $\cdots$ become small very rapidly. The series for e^x is an example. Here the general term is $x^n/n!$, which for each x tends to zero very quickly since the coefficients $a_n = 1/n!$ become small so fast.

Case (iii) lies in between. The coefficients do not increase so rapidly that the series never converges (except for $x = c$), nor do they decrease so rapidly that the series always converges. A typical example is the geometric series $\sum x^n$, where each $a_n = 1$. This series converges for $|x| < 1$ and diverges for $|x| > 1$, hence $R = 1$.

Note that in Case (iii) nothing is said about the endpoints $x = c + R$ and $x = c - R$. The series may or may not converge at either point.

The number R in Case (iii) is called the **radius of convergence.** By convention, $R = 0$ in Case (i), convergence for $x = c$ only, and $R = \infty$ in Case (ii), convergence for all x. (The word "radius" will be clarified in Chapter 16, Section 7.)

The interval $c - R < x < c + R$ in Case (iii) is called the **interval of convergence.** (Refer to Fig. 1.1.) By convention, the interval of convergence in Case (i) is the single point c; in Case (ii) it is the entire x-axis.

EXAMPLE 1.1

Find the sum of the power series and its radius of convergence R:

(a) $1 + \dfrac{x}{3} + \dfrac{x^2}{3^2} + \cdots + \dfrac{x^n}{3^n} + \cdots,$

(b) $1 + 5x + 5^2x^2 + \cdots + 5^nx^n + \cdots.$

Solution: Each is a geometric series

$$\sum_{n=0}^{\infty} y^n = \frac{1}{1-y}, \qquad |y| < 1.$$

In (a), $y = x/3$; hence the series converges for $|x/3| < 1$ and diverges for $|x/3| \geq 1$, that is, converges for $|x| < 3$ and diverges for $|x| \geq 3$. Thus $R = 3$. In (b), $y = 5x$, which implies convergence for $|x| < \frac{1}{5}$ and divergence for $|x| \geq \frac{1}{5}$.

Answer: (a) $\dfrac{1}{1 - \dfrac{x}{3}} = \dfrac{3}{3 - x}, \quad R = 3;$

(b) $\dfrac{1}{1 - 5x}, \quad R = \dfrac{1}{5}.$

Note that the series with smaller coefficients has the larger radius of convergence.

EXAMPLE 1.2

Find the radius of convergence and the sum of the power series

$$1 - x^2 + \frac{x^4}{2!} - \frac{x^6}{3!} + \frac{x^8}{4!} - \cdots.$$

Solution: The series has the form

$$1 + y + \frac{y^2}{2!} + \frac{y^3}{3!} + \cdots, \qquad \text{where} \quad y = -x^2.$$

Since the series converges to e^y for all values of y, we may replace y by $-x^2$ for any value of x.

Answer: $R = \infty$; the series converges to e^{-x^2} for all values of x.

EXERCISES

Find the sum of the series and its radius of convergence:

1. $1 + (x-3) + (x-3)^2 + (x-3)^3 + \cdots + (x-3)^n + \cdots$

2. $1 + \left(\dfrac{x+10}{2}\right) + \left(\dfrac{x+10}{2}\right)^2 + \left(\dfrac{x+10}{2}\right)^3 + \cdots + \left(\dfrac{x+10}{2}\right)^n + \cdots$

3. $1 - x^2 + x^4 - x^6 + \cdots + (-1)^n x^{2n} + \cdots$

4. $\left(\dfrac{x}{3}\right)^2 + \left(\dfrac{x}{3}\right)^3 + \left(\dfrac{x}{3}\right)^4 + \cdots + \left(\dfrac{x}{3}\right)^n + \cdots$

5. $1 + \dfrac{(x+1)}{1!} + \dfrac{(x+1)^2}{2!} + \dfrac{(x+1)^3}{3!} + \cdots + \dfrac{(x+1)^n}{n!} + \cdots$

6. $1 - \dfrac{2x}{1!} + \dfrac{4x^2}{2!} - \dfrac{8x^3}{3!} + \cdots + \dfrac{(-2x)^n}{n!} + \cdots$

7. $1 + (5x)^3 + (5x)^6 + (5x)^9 + \cdots + (5x)^{3n} + \cdots$

8. $2x - \dfrac{8x^3}{3!} + \dfrac{32x^5}{5!} - \dfrac{128x^7}{7!} + \cdots + (-1)^{n-1}\dfrac{(2x)^{2n-1}}{(2n-1)!} + \cdots$

9. $1 - \dfrac{(x-1)^4}{1!} + \dfrac{(x-1)^8}{2!} - \dfrac{(x-1)^{12}}{3!} + \cdots + (-1)^n \dfrac{(x-1)^{4n}}{n!} + \cdots.$

Find the sum of the series and all values of x for which the series converges:

10. $\dfrac{1}{x} + \dfrac{1}{x^2} + \dfrac{1}{x^3} + \cdots + \dfrac{1}{x^n} + \cdots$

11. $1 + e^x + e^{2x} + e^{3x} + \cdots + e^{nx} + \cdots$

12. $\cos^2 x + \cos^4 x + \cos^6 x + \cdots + \cos^{2n} x + \cdots$

13. $1 - \dfrac{\sin 3x}{1!} + \dfrac{\sin^2 3x}{2!} - \dfrac{\sin^3 3x}{3!} + \cdots + (-1)^n \dfrac{\sin^n 3x}{n!} + \cdots$

14. $1 + \dfrac{\ln x}{1!} + \dfrac{\ln^2 x}{2!} + \dfrac{\ln^3 x}{3!} + \cdots + \dfrac{\ln^n x}{n!} + \cdots$

15. $1 + 2\sqrt{x} + 4x + 8x\sqrt{x} + 16x^2 + \cdots + (2\sqrt{x})^n + \cdots$

16. $\ln x + \ln(x^{1/2}) + \ln(x^{1/4}) + \cdots + \ln(x^{1/2^n}) + \cdots$

17. $1 + (x^2 + a^2) + (x^2 + a^2)^2 + (x^2 + a^2)^3 + \cdots$

18. $(x^2-1)+\dfrac{(x^2-1)^2}{2}+\dfrac{(x^2-1)^3}{4}+\cdots+\dfrac{(x^2-1)^n}{2^{n-1}}+\cdots$

19. $1-\dfrac{x^{2/3}}{2!}+\dfrac{x^{4/3}}{4!}-\dfrac{x^{6/3}}{6!}+\cdots+(-1)^n\dfrac{x^{2n/3}}{(2n)!}+\cdots.$

2. RATIO TEST

For most power series that arise in practice, the radius of convergence can be found by the following criterion. Roughly speaking, if the coefficients of a power series behave very much like those of a geometric series with radius of convergence R, then the power series also has radius of convergence R. Such a geometric series is $\sum a_n x^n = \sum (1/R)^n x^n$. Note that for this series $a_n/a_{n+1} = R$.

Ratio Test Suppose the power series

$$a_0 + a_1(x-c) + a_2(x-c)^2 + \cdots + a_n(x-c)^n + \cdots$$

has non-zero coefficients. If

$$\left|\frac{a_n}{a_{n+1}}\right| \longrightarrow R \quad \text{as} \quad n \longrightarrow \infty,$$

where R is 0, positive, or ∞, then R is the radius of convergence.

Proof: For simplicity in notation, let us assume $c = 0$. Suppose $|x| < R$. Then

$$\frac{|a_{n+1}x^{n+1}|}{|a_n x^n|} = \left|\frac{a_{n+1}}{a_n}\right| |x| \longrightarrow \frac{|x|}{R} < 1,$$

hence the series $\sum a_n x^n$ converges (absolutely) by the ratio test for a series of constants (Chapter 1, Section 3).

Suppose $0 < R < \infty$ and $|x| > R$. Then

$$\frac{|a_{n+1}x^{n+1}|}{|a_n x^n|} \longrightarrow \frac{|x|}{R} > 1,$$

so for n sufficiently large, $\{|a_n x^n|\}$ is an increasing sequence. Therefore $\sum a_n x^n$ diverges because it terms do not approach 0. A slight modification of this argument applies if $R = 0$.

Thus the series converges for $|x| < R$ and diverges for $|x| > R$, that is, its radius of convergence is R.

EXAMPLE 2.1

Find the radius of convergence in each case:

(a) $1 + \frac{x}{1} + \frac{x^2}{2} + \frac{x^3}{3} + \cdots + \frac{x^n}{n} + \cdots,$

(b) $(x-5) - 4(x-5)^2 + 9(x-5)^3 - \cdots + (-1)^{n-1}n^2(x-5)^n + \cdots,$

(c) $1 + \frac{x}{2+1} + \frac{x^2}{2^2+2} + \frac{x^3}{2^3+3} + \cdots + \frac{x^n}{2^n+n} + \cdots,$

(d) $1 - x + \frac{x^2}{2^2} - \frac{x^3}{3^3} + \cdots + (-1)^n \frac{x^n}{n^n} + \cdots,$

(e) $\frac{x^3}{\sqrt{3}} + \frac{x^6}{\sqrt{6}} + \frac{x^9}{\sqrt{9}} + \cdots + \frac{x^{3n}}{\sqrt{3n}} + \cdots.$

Solution: In each case apply the Ratio Test.

(a) Here $a_n = 1/n$ and

$$\left|\frac{a_n}{a_{n+1}}\right| = \frac{1}{n} \Big/ \frac{1}{n+1} = \frac{n+1}{n}.$$

Hence

$$\left|\frac{a_n}{a_{n+1}}\right| \longrightarrow 1 \quad \text{as} \quad n \longrightarrow \infty,$$

so $R = 1$.

(b) $a_n = (-1)^{n-1}n^2$,

$$\left|\frac{a_n}{a_{n+1}}\right| = \frac{n^2}{(n+1)^2} = \left(\frac{n}{n+1}\right)^2 \longrightarrow 1 \quad \text{as} \quad n \longrightarrow \infty,$$

so $R = 1$.

(c)

$$\left|\frac{a_n}{a_{n+1}}\right| = \frac{1}{2^n+n} \Big/ \frac{1}{2^{n+1}+n+1} = \frac{2^{n+1}+n+1}{2^n+n}$$

$$= \frac{2 + (n+1)\cdot 2^{-n}}{1 + n\cdot 2^{-n}} \longrightarrow \frac{2+0}{1+0} = 2$$

as $n \longrightarrow \infty$. Hence $R = 2$.

(d)

$$\left|\frac{a_n}{a_{n+1}}\right| = \frac{1}{n^n} \bigg/ \frac{1}{(n+1)^{n+1}}$$

$$= \frac{(n+1)^{n+1}}{n^n} = (n+1)\left(\frac{n+1}{n}\right)^n > n+1.$$

Hence $|a_n/a_{n+1}| \longrightarrow \infty$ as $n \longrightarrow \infty$, so $R = \infty$; the series converges for all values of x.

(e) The Ratio Test does not apply directly because "two-thirds" of the coefficients in this power series are zero. Nevertheless, the series may be written

$$\frac{y}{\sqrt{3}} + \frac{y^2}{\sqrt{6}} + \frac{y^3}{\sqrt{9}} + \cdots + \frac{y^n}{\sqrt{3n}} + \cdots,$$

where $y = x^3$. The Ratio Test does apply to the series in this form:

$$\frac{\sqrt{3(n+1)}}{\sqrt{3n}} = \sqrt{\frac{n+1}{n}} \longrightarrow 1.$$

Hence the y-series converges for $|y| < 1$ and diverges for $|y| > 1$. Therefore, the original series converges for $|x^3| < 1$ and diverges for $|x^3| > 1$, i.e., for $|x| < 1$ and $|x| > 1$, respectively. Hence $R = 1$.

Answer: (a) 1; (b) 1; (c) 2; (d) ∞; (e) 1.

REMARK: The ratio test does not apply to all series. An example is

$$1 + 2x + x^2 + 2x^3 + x^4 + 2x^5 + \cdots.$$

For this series

$$\frac{a_0}{a_1} = \frac{1}{2}, \quad \frac{a_1}{a_2} = 2, \quad \frac{a_2}{a_3} = \frac{1}{2}, \quad \frac{a_3}{a_4} = 2, \quad \text{etc.}$$

The ratios are alternately $\frac{1}{2}$ and 2; they do not have a limit and so the ratio test does not apply. Nevertheless, the given series is a perfectly decent one. It is in fact the sum of two geometric series,

$$1 + x^2 + x^4 + \cdots = \frac{1}{1 - x^2},$$

and

$$2x + 2x^3 + 2x^5 + \cdots = \frac{2x}{1 - x^2},$$

both of which converge for $|x| < 1$ and diverge for $|x| \geq 1$. Hence

$$1 + 2x + x^2 + 2x^3 + \cdots = \frac{1}{1-x^2} + \frac{2x}{1-x^2} = \frac{1+2x}{1-x^2}, \qquad R = 1.$$

EXERCISES

Find the radius of convergence:

1. $1 + x + 2x^2 + 3x^3 + \cdots + nx^n + \cdots$
2. $(x-1) + \dfrac{(x-1)^2}{2} + \dfrac{(x-1)^3}{3} + \cdots + \dfrac{(x-1)^n}{n} + \cdots$
3. $x + \dfrac{x^2}{3} + \dfrac{x^3}{5} + \dfrac{x^4}{7} + \cdots + \dfrac{x^n}{2n-1} + \cdots$
4. $x - \dfrac{x^2}{5} + \dfrac{x^3}{9} - \dfrac{x^4}{17} + \cdots + (-1)^{n-1}\dfrac{x^n}{2^n+1} + \cdots$
5. $1 + \dfrac{x}{1\cdot 2} + \dfrac{x^2}{2\cdot 4} + \dfrac{x^3}{3\cdot 8} + \cdots + \dfrac{x^n}{n\cdot 2^n} + \cdots$
6. $\dfrac{(x+1)}{1\cdot 2\cdot 3} + \dfrac{(x+1)^2}{2\cdot 3\cdot 4} + \dfrac{(x+1)^3}{3\cdot 4\cdot 5} + \cdots + \dfrac{(x+1)^n}{n(n+1)(n+2)} + \cdots$
7. $x + \sqrt{2}x^2 + \sqrt{3}x^3 + \cdots + \sqrt{n}x^n + \cdots$
8. $1 - \dfrac{x}{2} + \dfrac{x^2}{2^4} - \dfrac{x^3}{2^9} + \cdots + \dfrac{(-x)^n}{2^{n^2}} + \cdots$
9. $(e-1)x + (e^2-1)x^2 + (e^3-1)x^3 + \cdots + (e^n-1)x^n + \cdots$
10. $-\dfrac{2x}{1^3} + \dfrac{3x^2}{2^3} - \dfrac{4x^3}{3^3} + \cdots + (-1)^n\dfrac{(n+1)x^n}{n^3} + \cdots$
11. $\dfrac{1}{4} + \dfrac{3(x-5)}{4^2} + \dfrac{3^2(x-5)^2}{4^3} + \cdots + \dfrac{3^n(x-5)^n}{4^{n+1}} + \cdots$
12. $1 + x + 2!x^2 + 3!x^3 + \cdots + n!x^n + \cdots$
13. $\dfrac{x^2}{2+\ln 2} + \dfrac{x^3}{3+\ln 3} + \dfrac{x^4}{4+\ln 4} + \cdots + \dfrac{x^n}{n+\ln n} + \cdots$
14. $1 + \dfrac{1}{2}x + \dfrac{1\cdot 3}{2\cdot 4}x^2 + \dfrac{1\cdot 3\cdot 5}{2\cdot 4\cdot 6}x^3 + \cdots + \dfrac{1\cdot 3\cdot 5\cdots(2n-1)}{2\cdot 4\cdot 6\cdots(2n)}x^n + \cdots$
15. $\dfrac{1}{30} + \left(\dfrac{1}{5^2} - \dfrac{1}{6^2}\right)x + \left(\dfrac{1}{5^3} - \dfrac{1}{6^3}\right)x^2 + \cdots + \left(\dfrac{1}{5^{n+1}} - \dfrac{1}{6^{n+1}}\right)x^n + \cdots$
16. $1 + x^2 + x^{10} + x^{12} + x^{20} + x^{22} + \cdots + x^{10n} + x^{10n+2} + \cdots$
17. $4\cdot 5x^4 + 8\cdot 9x^8 + \cdots + 4n(4n+1)x^{4n} + \cdots$
18. $1 + (1+2)x + (1+2+4)x^2 + (1+2+4+8)x^3 + \cdots$
    $+ (1 + 2 + 2^2 + \cdots + 2^n)x^n + \cdots$

19. $x + \frac{x^2}{2^2} + \frac{x^3}{3^3} + \cdots + \frac{x^n}{n^n} + \cdots$

20. $x + \frac{x^2}{(2!)^2} + \frac{x^3}{(3!)^2} + \cdots + \frac{x^n}{(n!)^2} + \cdots.$

3. EXPANSIONS OF FUNCTIONS

For many applications, it is convenient to expand functions in power series. We have already done this for e^x, $\sin x$, and $\cos x$ by computing the coefficient of x^n from the formula

$$a_n = \frac{f^{(n)}(0)}{n!}.$$

Generally, however, computation of higher derivatives is extremely laborious. Try to find the seventh derivative of $\tan x$ or of $1/(1 + x^4)$ and you will soon agree.

In this section we describe several techniques for deriving power series without tedious differentiation. They all depend on one basic principle:

Uniqueness of Power Series If $f(x)$ has a power series expansion

$$f(x) = \sum_{n=0}^{\infty} a_n (x - c)^n,$$

convergent in some interval $|x - c| < R$, then

$$a_n = \frac{f^{(n)}(c)}{n!}.$$

Thus there is only one possible choice for the coefficients; the series is unique.

Once you find a power series for $f(x)$ at c by any method, fair or foul, then you have it! There is no other series for $f(x)$.

Proof: The proof is based on a property of power series that will be discussed in Section 4: within its interval of convergence, a power series can be differentiated term-by-term, infinitely often. Having this, the rest is easy. If $f(x) = \sum a_n (x - c)^n$, we differentiate n times, then set $x = c$:

$$f^{(n)}(c) = n!a_n, \qquad a_n = \frac{f^n(c)}{n!}.$$

This is exactly the same method we used in Chapter 2, Section 2 to derive the coefficients in the Taylor polynomials of $f(x)$.

The first new technique we consider is addition and subtraction of power series.

Two power series may be added or subtracted term-by-term within their common interval of convergence.

Proof: Suppose $f(x) = \sum_0^\infty a_n(x - x_0)^n$ and $g(x) = \sum_0^\infty b_n(x - x_0)^n$. If $s_n(x)$ and $t_n(x)$ denote the partial sums of these series then $s_n(x) \longrightarrow f(x)$ and $t_n(x) \longrightarrow g(x)$ for each x in the common interval of convergence. But by a basic theorem on limits, $s_n(x) \pm t_n(x) \longrightarrow f(x) \pm g(x)$. That means

$$\sum_0^n (a_k \pm b_k)(x - x_0)^k \longrightarrow f(x) \pm g(x);$$

in other words, $\sum_0^\infty (a_n \pm b_n)(x - x_0)^n = f(x) \pm g(x)$.

EXAMPLE 3.1

Express $\cosh x$ in a power series at $x = 0$.

Solution: $\cosh x = \frac{1}{2}(e^x + e^{-x})$. But

$$e^x = 1 + x + \frac{x^2}{2!} + \frac{x^3}{3!} + \cdots \qquad \text{for all } x,$$

$$e^{-x} = 1 - x + \frac{x^2}{2!} - \frac{x^3}{3!} + \cdots \qquad \text{for all } x.$$

Add these series and divide by 2:

$$\frac{1}{2}(e^x + e^{-x}) = 1 + \frac{x^2}{2!} + \frac{x^4}{4!} + \cdots.$$

Answer: $\cosh x = 1 + \frac{x^2}{2!} + \frac{x^4}{4!} + \cdots$ for all x.

The next technique is formal multiplication of series. To simplify notation we shall stick to $c = 0$.

Theorem Let

$$f(x) = \sum_{n=0}^\infty a_n x^n \qquad \text{and} \qquad g(x) = \sum_{n=0}^\infty b_n x^n$$

for $|x| < R$. Then the product $f(x)g(x)$ has the power series expansion

$$f(x)g(x) = \sum_{n=0}^\infty \Big(\sum_{i+j=n} a_i b_j \Big) x^n,$$

also valid for $|x| < R$.

The theorem means that the power series for $f(x)g(x)$ is obtained by multiplying each term $a_i x^i$ of $f(x)$ by each term $b_j x^j$ of $g(x)$ and collecting terms, just as in multiplying polynomials. To remember the rule, start with the lowest terms and work up:

$$\begin{aligned} f(x)g(x) &= (a_0 + a_1 x + a_2 x^2 + \cdots)(b_0 + b_1 x + b_2 x^2 + \cdots) \\ &= a_0 b_0 + (a_0 b_1 + a_1 b_0)x + (a_0 b_2 + a_1 b_1 + a_2 b_0)x^2 + \cdots. \end{aligned}$$

The proof of this theorem for products is difficult and best postponed to an advanced course.

EXAMPLE 3.2

Compute the terms up to x^6 in the power series of $x^2 e^x \sin 2x$ at $x = 0$.

Solution:

$$x^2 e^x \sin 2x = x^2\left(1 + x + \frac{x^2}{2!} + \frac{x^3}{3!} + \cdots\right)\left(2x - \frac{(2x)^3}{3!} + \frac{(2x)^5}{5!} - + \cdots\right).$$

Since only terms involving x^6 and lower powers are required, it suffices to compute the product

$$x^2\left(1 + x + \frac{x^2}{2!} + \frac{x^2}{3!}\right)\left(2x - \frac{(2x)^3}{3!}\right) = x^2\left(1 + x + \frac{x^2}{2} + \frac{x^3}{6}\right)\left(2x - \frac{4x^3}{3}\right)$$

$$= x^2\left[2x + 2x^2 + \left(1 - \frac{4}{3}\right)x^3 + \left(\frac{1}{3} - \frac{4}{3}\right)x^4 + \cdots\right].$$

Answer: For all x,

$$x^2 e^x \sin 2x = 2x^3 + 2x^4 - \frac{1}{3}x^5 - x^6 + \cdots.$$

The next technique is substitution; it is used to find the power series for a composite function from the series for the component functions.

Theorem Let

$$f(z) = a_0 + a_1 z + a_2 z^2 + \cdots$$

converge in the interval $|z| < R$. Suppose $g(x)$ is a function with domain **D** and such that $|g(x)| < R$ for all x in **D**. Then

$$(*) \qquad f[g(x)] = a_0 + a_1 g(x) + a_2[g(x)]^2 + a_3[g(x)]^3 + \cdots.$$

If in addition $g(x)$ is a power series with $g(0) = 0$, that is,

$$g(x) = b_1 x + b_2 x^2 + b_3 x^3 + \cdots$$

for $|x| < r$, then the series on the right side of (*) can be converted into a power series by formally squaring, cubing, etc., and collecting terms. The resulting power series is the valid expansion of $f[g(x)]$ for $|x| < r$.

The proof of this result is definitely beyond the scope of this course.

REMARK: Note that $g(x)$ lacks a constant term, that is, $b_0 = 0$. Without this restriction, there are infinitely many terms contributing to each each coefficient of $f[g(x)]$, a tricky situation to handle.

The theorem often allows us to find series for various functions by simple modifications of known series. Here are a few everyday examples:

$$\frac{1}{1-x^2} = 1 + x^2 + x^4 + x^6 + \cdots \qquad (|x| < 1).$$

$$\frac{1}{1+2x^3} = 1 + (-2x^3) + (-2x^3)^2 + (-2x^3)^3 + \cdots$$

$$= 1 - 2x^3 + 4x^6 - 8x^9 + - \cdots \qquad (|x| < 1/\sqrt[3]{2}).$$

$$e^{-x^2/2} = 1 + \left(-\frac{x^2}{2}\right) + \frac{1}{2!}\left(-\frac{x^2}{2}\right)^2 + \frac{1}{3!}\left(-\frac{x^2}{2}\right)^3 + \cdots$$

$$= 1 - \frac{x^2}{2} + \frac{x^4}{2^2 \cdot 2!} - \frac{x^6}{2^3 \cdot 3!} + - \cdots \qquad (\text{all } x).$$

Let us now consider a less simple example.

EXAMPLE 3.3

Find all terms up to the term in x^8 in the power series at $x = 0$ of the function

$$f(x) = \frac{1}{1 - \sin(x^2)}.$$

Assume $|x| < 1$.

Solution: If $|x| < 1$, then $|\sin(x^2)| < 1$; hence by (*),

$$f(x) = 1 + \sin(x^2) + [\sin(x^2)]^2 + [\sin(x^2)]^3 + \cdots.$$

Now, from the series for $\sin x$, we have

$$\sin(x^2) = x^2 - \frac{x^6}{6} + \frac{x^{10}}{120} - + \cdots.$$

We substitute this series into the expression for $f(x)$. We square, cube, etc., collecting powers of x up to the eighth power:

$$f(x) = 1 + \left(x^2 - \frac{x^6}{6} + \cdots\right) + \left(x^2 - \frac{x^6}{6} + \cdots\right)^2 + \left(x^2 - \frac{x^6}{6} + \cdots\right)^3$$

$$+ \left(x^2 - \frac{x^6}{6} + \cdots\right)^4 + \cdots$$

$$= 1 + \left(x^2 - \frac{x^6}{6} + \cdots\right) + \left(x^4 - \frac{x^8}{3} + \cdots\right)$$

$$+ (x^6 + \cdots) + (x^8 + \cdots) + \cdots$$

$$= 1 + x^2 + x^4 + \left(1 - \frac{1}{6}\right)x^6 + \left(1 - \frac{1}{3}\right)x^8 + \cdots.$$

Answer: $f(x) = 1 + x^2 + x^4 + \frac{5}{6}x^6 + \frac{2}{3}x^8 + \cdots.$

Odd and Even Functions

The power series at $x = 0$ for the odd function $\sin x$ lacks x^2, x^4, x^6, $\cdots$. Likewise the series at $x = 0$ for the even function $\cos x$ lacks x, x^3, x^5, $\cdots$. These examples illustrate a useful principle:

If $f(x)$ is an odd function, $f(-x) = -f(x)$, then its power series at $x = 0$ has the form

$$f(x) = a_1x + a_3x^3 + a_5x^5 + \cdots.$$

If $g(x)$ is an even function, $g(-x) = g(x)$, then its power series at $x = 0$ has the form

$$g(x) = a_0 + a_2x^2 + a_4x^4 + \cdots.$$

These statements are easy to remember: an odd (even) function involves only odd (even) powers of x at $x = 0$.

The proofs of the principles depend in turn on another pair of important facts:

The derivative of an odd function is an even function; the derivative of an even function is an odd function.

Reason: Suppose $f(x)$ is odd, $f(-x) = -f(x)$. Differentiate, using the Chain Rule on the left-hand side: $-f'(-x) = -f'(x)$. Hence $f'(-x) = f'(x)$, so $f'(x)$ is even. Likewise, if $g(-x) = g(x)$, then $g'(-x) = -g'(x)$.

Now return to power series of odd and even functions. Suppose $f(x)$ is odd and at $x = 0$ has the power series

$$f(x) = a_0 + a_1x + a_2x^2 + \cdots.$$

Since $f(x)$ is odd, $f(0) = 0$. Hence $a_0 = 0$. The derivative $f'(x)$ is even and its derivative $f''(x)$ is again odd. Hence $f''(0) = 0$. But $a_2 = f''(0)/2!$, so $a_2 = 0$. Likewise $a_4 = 0$, $a_6 = 0$, $\cdots$, so

$$f(x) = a_1x + a_3x^3 + a_5x^5 + \cdots.$$

The statement about even functions can be proved similarly.

EXAMPLE 3.4

Find the power series for $\tan x$ at $x = 0$ up to terms in x^7.

Solution: The power series for $\tan x$ is obtained by long division: the series for $\sin x$ is divided by the series for $\cos x$. Here is a systematic way to carry out the long division. Set

$$\tan x = a_1 x + a_3 x^3 + a_5 x^5 + a_7 x^7 + \cdots,$$

valid because $\tan x$ is an odd function. Write the identity $\tan x \cos x = \sin x$ in terms of power series:

$$(a_1 x + a_3 x^3 + a_5 x^5 + a_7 x^7 + \cdots)\left(1 - \frac{x^2}{2} + \frac{x^4}{24} + \frac{x^6}{720} \cdots\right)$$
$$= x - \frac{x^3}{6} + \frac{x^5}{120} - \frac{x^7}{5040} + \cdots.$$

Multiply the two series on the left, then equate coefficients.

$$x: \qquad a_1 = 1;$$

$$x^3: \qquad a_3 - \frac{1}{2}a_1 = -\frac{1}{6};$$

$$x^5: \qquad a_5 - \frac{1}{2}a_3 + \frac{1}{24}a_1 = \frac{1}{120};$$

$$x^7: \qquad a_7 - \frac{1}{2}a_5 + \frac{1}{24}a_3 - \frac{1}{720}a_1 = -\frac{1}{5040}.$$

Solve these equations successively for a_1, a_3, a_5, a_7:

$$a_1 = 1, \qquad a_3 = \frac{1}{3}, \qquad a_5 = \frac{2}{15}, \qquad a_7 = \frac{17}{315}.$$

Alternate Solution: Write

$$\tan x = \sin x \cdot \frac{1}{\cos x}.$$

Now

$$\cos x = 1 - \left(\frac{x^2}{2} - \frac{x^4}{4!} + \frac{x^6}{6!} - + \cdots\right) = 1 - z,$$

$$\frac{1}{\cos x} = \frac{1}{1 - z} = 1 + z + z^2 + z^3 + \cdots.$$

Compute up to x^6 only:

$$z = \frac{x^2}{2} - \frac{x^4}{24} + \frac{x^6}{720} - \cdots,$$

$$z^2 = \left(\frac{x^2}{2} - \frac{x^4}{24} + \cdots\right)^2 = \left(\frac{x^2}{2}\right)^2\left(1 - \frac{x^2}{12} + \cdots\right)^2 = \frac{x^4}{4} - \frac{x^6}{24} + \cdots,$$

$$z^3 = \left(\frac{x^2}{2} - \cdots\right)^3 = \frac{x^6}{8} - \cdots.$$

The quantities z^4, z^5, $\cdots$ can be ignored; they do not contain x^6 or lower powers of x. Collect terms:

$$\frac{1}{\cos x} = 1 + \frac{1}{2}x^2 + \left(-\frac{1}{24} + \frac{1}{4}\right)x^4 + \left(\frac{1}{720} - \frac{1}{24} + \frac{1}{8}\right)x^6 + \cdots$$

$$= 1 + \frac{1}{2}x^2 + \frac{5}{24}x^4 + \frac{61}{720}x^6 + \cdots.$$

Therefore

$$\tan x = \sin x \cdot \frac{1}{\cos x}$$

$$= \left(x - \frac{x^3}{6} + \frac{x^5}{120} - \frac{x^7}{5040} + \cdots\right)\left(1 + \frac{x^2}{2} + \frac{5}{24}x^4 + \frac{61}{720}x^6 + \cdots\right)$$

$$= x + \left(-\frac{1}{6} + \frac{1}{2}\right)x^3 + \left(\frac{1}{120} - \frac{1}{12} + \frac{5}{24}\right)x^5$$

$$+ \left(-\frac{1}{5040} + \frac{1}{240} - \frac{5}{144} + \frac{61}{720}\right)x^7 + \cdots$$

$$= x + \frac{1}{3}x^3 + \frac{2}{15}x^5 + \frac{17}{315}x^7 + \cdots.$$

Answer:

$$\tan x = x + \frac{1}{3}x^3 + \frac{2}{15}x^5 + \frac{17}{315}x^7 + \cdots.$$

REMARK: The power series for $\tan x$ converges to $\tan x$ for $|x| < \pi/2$. This is the largest interval about $x = 0$ in which the denominator $\cos x$ is non-zero.

EXAMPLE 3.5

What is the value of the seventh derivative of $\tan x$ at $x = 0$?

Solution: Denote $\tan x$ by $f(x)$. The coefficient of x^7 in its power series is $f^{(7)}(0)/7!$ But that coefficient is $17/315$ from the answer to the preceding problem. Hence

$$\frac{f^{(7)}(0)}{7!} = \frac{17}{315}, \qquad f^{(7)}(0) = \frac{(17)(5040)}{315}.$$

Answer: 272.

EXERCISES

Find the power series at $x = 0$:

1. $\dfrac{1}{1-5x}$
2. $\dfrac{x^2}{1+x^3}$
3. e^{x^3}
4. $\cos 2x$
5. $\cosh\sqrt{x}$
6. $x(\sin x + \sin 3x)$
7. $(x-1)e^x$
8. $\dfrac{1-\cos x}{x^2}$
9. $\dfrac{x^2}{1-x}$
10. $\dfrac{\sin x - x}{x^3}$
11. $\sin^2 x$
[*Hint:* Use a trigonometric identity.]
12. $\sin(x^2)$.

Compute the terms up to and including x^6 in the power series at $x = 0$:

13. $\dfrac{1}{(1-2x)(1-3x^2)}$
14. $e^x \sin(x^2)$
15. $\dfrac{1}{1-x^2e^x}$
16. $\dfrac{1}{1+x^2+x^4}$
17. $\sin^3 x$
18. $\ln\cos x$.

Find $f^{(8)}(0)$:

19. $f(x) = \dfrac{1}{1-x^4}$
20. $f(x) = x\cos x$
21. $f(x) = e^{-x^2}$
22. $f(x) = \arctan(x^3)$.
23. Show that the series $1 - \dfrac{x}{2!} + \dfrac{x^2}{4!} - \dfrac{x^6}{6!} + \cdots$ converges to $\cos\sqrt{x}$ for $x \geq 0$.
24. (cont.) Show that it converges to $\cosh\sqrt{-x}$ for $x \leq 0$.

4. FURTHER TECHNIQUES

The methods of the last section enable us to derive new power series from known ones. Here is another important method for doing so. We postpone the proof until Section 7.

> A power series may be differentiated or integrated term-by-term within its interval of convergence.

EXAMPLE 4.1

Find power series at $x = 0$ for $(1 - x)^{-2}$ and $(1 - x)^{-3}$.

Solution: For $|x| < 1$,

$$(1 - x)^{-2} = \frac{d}{dx}(1 - x)^{-1}$$

$$= \frac{d}{dx}(1 + x + x^2 + \cdots + x^n + \cdots)$$

$$= \frac{d}{dx}(1) + \frac{d}{dx}(x) + \frac{d}{dx}(x^2) + \cdots + \frac{d}{dx}(x^n) + \cdots$$

$$= 1 + 2x + 3x^2 + \cdots + nx^{n-1} + \cdots.$$

Differentiate again:

$$(1 - x)^{-3} = \frac{1}{2}\frac{d}{dx}(1 - x)^{-2} = \frac{1}{2}\frac{d}{dx}(1 + 2x + 3x^2 + \cdots + nx^{n-1} + \cdots)$$

$$= \frac{1}{2}[2 + 6x^2 + 12x^3 + \cdots + n(n - 1)x^{n-2} + \cdots].$$

Answer: For $|x| < 1$,

$$(1 - x)^{-2} = \sum_{n=1}^{\infty} nx^{n-1}, \quad (1 - x)^{-3} = \sum_{n=2}^{\infty} \frac{n(n - 1)}{2} x^{n-2}.$$

EXAMPLE 4.2

Find the power series at $x = 0$ for $\ln(1 + x)$.

Solution: Notice that $\ln(1 + x)$ is an antiderivative of $(1 + x)^{-1} = 1 - x + x^2 - x^3 + \cdots$. Therefore

$$\begin{aligned}
\ln(1 + x) &= \int_0^x \frac{dt}{1+t} \\
&= \int_0^x [1 - t + t^2 - \cdots + (-1)^n t^n + \cdots]\, dt \\
&= \int_0^x dt - \int_0^x t\, dt + \int_0^x t^2\, dt - \cdots + (-1)^n \int_0^x t^n\, dt + \cdots \\
&= x - \frac{x^2}{2} + \frac{x^3}{3} - \cdots + (-1)^n \frac{x^{n+1}}{n+1} + \cdots.
\end{aligned}$$

Answer: For $|x| < 1$, $\ln(1 + x)$

$$= x - \frac{x^2}{2} + \frac{x^3}{3} - \frac{x^4}{4} + \cdots + (-1)^n \frac{x^{n+1}}{n+1} + \cdots.$$

EXAMPLE 4.3

Sum the series $x + \dfrac{x^3}{3} + \dfrac{x^5}{5} + \cdots + \dfrac{x^{2n-1}}{2n-1} + \cdots.$

Solution: By the ratio test, the series converges for $|x| < 1$ to some function $f(x)$. Write

$$f(x) = x + \frac{x^3}{3} + \frac{x^5}{5} + \cdots + \frac{x^{2n-1}}{2n-1} + \cdots.$$

Each term is the integral of a power of x. This suggests that $f(x)$ is the integral of some simple function. Differentiate term-by-term:

$$f'(x) = 1 + x^2 + x^4 + \cdots + x^{2n-2} + \cdots = \frac{1}{1 - x^2}.$$

Therefore, $f(x)$ is an antiderivative of $1/(1 - x^2)$. Since $f(0) = 0$, it follows that

$$f(x) = \int_0^x \frac{dt}{1 - t^2} = \frac{1}{2} \ln \frac{1+t}{1-t}\bigg|_0^x = \frac{1}{2} \ln \frac{1+x}{1-x}.$$

Answer: $\dfrac{1}{2} \ln \dfrac{1+x}{1-x}.$

EXAMPLE 4.4

Sum the series $x + 4x^2 + 9x^3 + \cdots + n^2x^n + \cdots$.

Solution: Write

$$f(x) = x + 4x^2 + 9x^3 + \cdots + n^2x^n + \cdots = xg(x),$$

where

$$g(x) = 1 + 2^2x + 3^2x^2 + \cdots + n^2x^{n-1} + \cdots.$$

Now

$$g(x) = \frac{d}{dx}(x + 2x^2 + 3x^3 + 4x^4 + \cdots + nx^n + \cdots)$$

$$= \frac{d}{dx}[x(1 + 2x + 3x^2 + 4x^3 + \cdots + nx^{n-1} + \cdots)].$$

By Example 4.1,

$$1 + 2x + 3x^2 + \cdots + nx^{n-1} + \cdots = (1 - x)^{-2}.$$

Hence

$$g(x) = \frac{d}{dx}\left[\frac{x}{(1-x)^2}\right] = \frac{1+x}{(1-x)^3},$$

so

$$f(x) = xg(x) = \frac{x + x^2}{(1-x)^3}.$$

Answer: $\dfrac{x + x^2}{(1-x)^3}$.

CHECK: Evaluate the series at $x = 0.1$:

$$f(0.1) = \frac{1}{10} + \frac{4}{10^2} + \frac{9}{10^3} + \frac{16}{10^4} + \frac{25}{10^5} + \cdots.$$

According to the answer, the sum is

$$\frac{(0.1) + (0.1)^2}{(1 - 0.1)^3} = \frac{0.11}{(0.9)^3} \approx 0.15089\ 16324.$$

It is easy to make a convincing numeric check. Start with the first term and add successive terms: 0.1, 0.14, 0.149, 0.1506, 0.15085, 0.15088 6, 0.15089 09, 0.15089 154, 0.15089 1621, 0.15089 16310.

Partial Fractions

The power series for $1/(x-5)$ can be obtained from the geometric series. Just write

$$\frac{1}{x-5} = -\frac{1}{5-x} = -\frac{1}{5}\frac{1}{1-\dfrac{x}{5}},$$

and expand in a geometric series:

$$\frac{1}{x-5} = -\frac{1}{5}\left[1 + \left(\frac{x}{5}\right) + \left(\frac{x}{5}\right)^2 + \cdots\right],$$

which converges if $|x/5| < 1$, that is, $|x| < 5$.

Combined with this trick, the method of partial fractions is useful in finding power series for rational functions (quotients of polynomials).

EXAMPLE 4.5

Find a power series at $x = 0$ for $f(x) = \dfrac{1}{(x-2)(x-5)}$.

Solution: By partial fractions

$$\frac{1}{(x-2)(x-5)} = \frac{1}{3}\left(\frac{1}{x-5} - \frac{1}{x-2}\right).$$

Expand each fraction by the preceding trick:

$$\frac{1}{x-2} = -\frac{1}{2}\sum_{0}^{\infty}\left(\frac{x}{2}\right)^n = -\sum_{0}^{\infty}\frac{x^n}{2^{n+1}} \qquad \text{for} \quad |x| < 2,$$

$$\frac{1}{x-5} = -\frac{1}{5}\sum_{0}^{\infty}\left(\frac{x}{5}\right)^n = -\sum_{0}^{\infty}\frac{x^n}{5^{n+1}} \qquad \text{for} \quad |x| < 5.$$

Therefore if $|x| < 2$, both series converge and

$$\frac{1}{3}\left(\frac{1}{x-5} - \frac{1}{x-2}\right) = \frac{1}{3}\sum_{0}^{\infty}\left(\frac{1}{2^{n+1}} - \frac{1}{5^{n+1}}\right)x^n.$$

Answer: For $|x| < 2$,

$$\frac{1}{(x-2)(x-5)} = \frac{1}{3}\sum_{n=0}^{\infty}\left(\frac{1}{2^{n+1}} - \frac{1}{5^{n+1}}\right)x^n.$$

EXAMPLE 4.6

Find the power series at $x = 0$ for $\dfrac{-1}{x^2 + x - 1}$.

Solution: The denominator can be factored:

$$x^2 + x - 1 = (x - a)(x - b),$$

where a and b are the roots of the equation

$$x^2 + x - 1 = 0.$$

By the quadratic formula,

$$a = \frac{-1 + \sqrt{5}}{2}, \qquad b = \frac{-1 - \sqrt{5}}{2}.$$

Notice that $ab = -1$ because the product of the roots of $x^2 + px + q = 0$ is q. (Why?)

By partial fractions,

$$\frac{-1}{x^2 + x - 1} = \frac{-1}{(x - a)(x - b)} = \frac{1}{a - b}\left(\frac{1}{x - b} - \frac{1}{x - a}\right).$$

Suppose $|x| < \frac{1}{2}(-1 + \sqrt{5})$, the smaller of the numbers $|a|$ and $|b|$. Now expand:

$$\frac{-1}{x^2 + x - 1} = \frac{1}{a - b}\left(\sum_{n=0}^{\infty} \frac{x^n}{a^{n+1}} - \sum_{n=0}^{\infty} \frac{x^n}{b^{n+1}}\right) = \sum_{n=0}^{\infty} \frac{1}{a - b}\left(\frac{1}{a^{n+1}} - \frac{1}{b^{n+1}}\right) x^n.$$

Note that

$$a - b = \frac{-1 + \sqrt{5}}{2} - \frac{-1 - \sqrt{5}}{2} = \sqrt{5},$$

and (since $ab = -1$)

$$\frac{1}{a} = -b = \frac{1 + \sqrt{5}}{2}, \qquad \frac{1}{b} = -a = \frac{1 - \sqrt{5}}{2}.$$

Answer: For $|x| < \dfrac{1}{2}(\sqrt{5} - 1)$,

$$\frac{-1}{x^2 + x - 1} = \sum_{n=0}^{\infty} c_n x^n, \text{ where}$$

$$c_n = \frac{1}{\sqrt{5}}\left[\left(\frac{1 + \sqrt{5}}{2}\right)^{n+1} - \left(\frac{1 - \sqrt{5}}{2}\right)^{n+1}\right].$$

REMARK: It may not seem so, but the numbers c_n are actually integers! The first few are 1, 1, 2, 3, 5, 8, 13, $\cdots$; each is the sum of the previous two. See Ex. 17 below.

EXAMPLE 4.7

Find a power series for $\dfrac{1}{(1-x)(1+x^2)}$.

Solution: By partial fractions,

$$\frac{1}{(1-x)(1+x^2)} = \frac{1}{2}\left(\frac{1}{1-x} + \frac{1+x}{1+x^2}\right) = \frac{1}{2}\left(\frac{1}{1-x} + \frac{1}{1+x^2} + \frac{x}{1+x^2}\right)$$

$$= \frac{1}{2}[(1 + x + x^2 + x^3 + \cdots) + (1 - x^2 + x^4 - x^6 + - \cdots) + (x - x^3 + x^5 - x^7 + - \cdots)]$$

$$= 1 + x + x^4 + x^5 + x^8 + x^9 + \cdots.$$

Answer: For $|x| < 1$,

$$\frac{1}{(1-x)(1+x^2)} = \sum_{n=0}^{\infty} (x^{4n} + x^{4n+1}).$$

REMARK: This example also can be done by multiplying together the series for $(1-x)^{-1}$ and $(1+x^2)^{-1}$, also by multiplying the series for $(1-x^4)^{-1}$ by $1+x$.

EXERCISES

Obtain the power series at $x = 0$:

1. $\dfrac{x+1}{x^2-5x+6}$
2. $\dfrac{1}{(2x+1)(3x+4)}$
3. $\dfrac{1}{(x-1)(x-2)(x-3)}$
4. $\dfrac{1}{(x-3)(x^2+1)}$
5. $\dfrac{1}{1-x+x^2}$
6. $\dfrac{1+x^2}{1-x^5}$.

Find the sum of the series:

7. $4 + 5x + 6x^2 + \cdots + (n+4)x^n + \cdots$
8. $1 + 4x + 9x^2 + \cdots + (n+1)^2x^n + \cdots$
9. $\dfrac{x^4}{4} + \dfrac{x^8}{8} + \dfrac{x^{12}}{12} + \cdots + \dfrac{x^{4n}}{4n} + \cdots$

   [*Hint:* Differentiate.]

10. $\dfrac{x^2}{1\cdot 2} - \dfrac{x^3}{2\cdot 3} + \dfrac{x^4}{3\cdot 4} - \cdots + \dfrac{(-1)^n x^n}{(n-1)n} + \cdots$

11. $2 + 3x + 4x^2 + \cdots + (n+2)x^n + \cdots$
[*Hint:* Multiply by x and integrate.]

12. Evaluate $1 + \dfrac{2}{5} + \dfrac{3}{25} + \dfrac{4}{125} + \cdots + \dfrac{n}{5^{n-1}} + \cdots$

[*Hint:* Consider $\displaystyle\sum_{n=1}^{\infty} nx^{n-1}$.]

Verify by expressing both sides in power series:

13. $\dfrac{d}{dx}(\sin x) = \cos x$

14. $\dfrac{d}{dx}(e^{-x^2}) = -2xe^{-x^2}$

15. $\displaystyle\int x^2 \cos x\, dx = 2x \cos x + (x^2 - 2) \sin x + C$

16. $\displaystyle\int \arctan x\, dx = x \arctan x - \frac{1}{2} \ln(1 + x^2) + C.$

17. Consider Example 4.6 from a different viewpoint. Suppose, a_0, a_1, a_2, $\cdots$ is a sequence of integers satisfying $a_0 = a_1 = 1$ and $a_{n+2} = a_n + a_{n+1}$ for $n \geq 1$. Set $f(x) = a_0 + a_1 x + a_2 x^2 + \cdots$. Show that $xf(x) + x^2 f(x) = f(x) - 1$, and conclude that $f(x) = (1 - x - x^2)^{-1}$. Now obtain a formula for a_n. (This is the method of **generating functions.**)
18. Which power series is more efficient for computing $\ln(\frac{3}{2})$, the one for $\ln(1 + x)$, or the one for $\ln[(1 + x)/(1 - x)]$? See Examples 4.2 and 4.3. Compute $\ln(\frac{3}{2})$ to 5 places.
19. Expand $\arctan x$ in a Taylor series at $x = 0$.
[*Hint:* Integrate its derivative.] Conclude that

$$\arctan x = x - \frac{x^3}{3} + \frac{x^5}{5} - \frac{x^7}{7} + \cdots \qquad \text{for} \quad |x| < 1.$$

It is known that the series in Ex. 19 is valid for $x = 1$, hence that

$$1 - \frac{1}{3} + \frac{1}{5} - \frac{1}{7} + \cdots = \frac{\pi}{4}.$$

This is an interesting but poor formula for computing π, since its terms decrease very slowly. Exercises 20–23 develop an efficient way to compute π.

20. Prove the formula $\arctan x + \arctan y = \arctan\left(\dfrac{x+y}{1-xy}\right)$ for $|x| < 1$ and $|y| < 1$. Conclude that $\dfrac{\pi}{4} = \arctan \dfrac{1}{2} + \arctan \dfrac{1}{3}$.

21. (cont.) Use this expression for $\pi/4$ and the power series in Ex. 19 to compute π to 4 places.

22. (cont.) Show that the expression in Ex. 20 for $\pi/4$ can be modified to

$\frac{\pi}{4} = 2 \arctan \frac{1}{3} + \arctan \frac{1}{7}$. Why is this even better for computing π?

23. (cont.) Finally modify the expression in Ex. 22 to

$\frac{\pi}{4} = 2 \arctan \frac{1}{5} + \arctan \frac{1}{7} + 2 \arctan \frac{1}{8}$. Now compute π to 7 places.

5. BINOMIAL SERIES

The Binomial Theorem asserts that for each positive integer p,

$$(1+x)^p = 1 + px + \frac{p(p-1)}{2!}x^2 + \frac{p(p-1)(p-2)}{3!}x^3 + \cdots$$
$$+ \frac{p(p-1)(p-2)\cdots(p-n+1)}{n!}x^n + \cdots + \frac{p!}{p!}x^p.$$

Standard notation for the coefficients in this identity is

$$\binom{p}{0} = 1, \qquad \binom{p}{n} = \frac{p(p-1)(p-2)\cdots(p-n+1)}{n!}, \qquad 1 \le n \le p.$$

With this notation the expansion of $(1+x)^p$ can be abbreviated:

$$(1+x)^p = \sum_{n=0}^{p} \binom{p}{n} x^n.$$

A generalization of the Binomial Theorem is the **binomial series** for $(1+x)^p$, where p is not necessarily a positive integer.

Binomial Series Suppose p is any number. Then

$$(1+x)^p = \sum_{n=0}^{\infty} \binom{p}{n} x^n, \qquad -1 < x < 1,$$

where the coefficients in this series are

$$\binom{p}{0} = 1, \qquad \binom{p}{n} = \frac{p(p-1)(p-2)\cdots(p-n+1)}{n!}, \qquad n \ge 1.$$

REMARK: In case p happens to be a positive integer and $n > p$, then the coefficient

$$\binom{p}{n}$$

equals 0 because it has a factor $(p - p)$. In this case, the series breaks off after the term in x^p. The resulting formula,

$$(1 + x)^p = \sum_{n=0}^{p} \binom{p}{n} x^n,$$

is the old Binomial Theorem again. But if p is not a positive integer or zero, then each coefficient is non-zero, so the series has infinitely many terms.

The binomial series is just the Taylor series for $y(x) = (1 + x)^p$ at $x = 0$. Notice that

$$y'(x) = p(1 + x)^{p-1},$$

$$y''(x) = p(p - 1)(1 + x)^{p-2},$$

$$\cdots\cdots\cdots\cdots\cdots$$

$$y^{(n)}(x) = p(p - 1)(p - 2)\cdots(p - n + 1)(1 + x)^{p-n}.$$

Therefore the coefficient of x^n in the Taylor series is

$$\frac{y^{(n)}(0)}{n!} = \frac{p(p - 1)(p - 2)\cdots(p - n + 1)}{n!} = \binom{p}{n}.$$

The binomial series converges for $|x| < 1$. When p is an integer this is obvious because the series terminates. When p is not an integer the ratio test applies:

$$\left|\frac{a_n}{a_{n+1}}\right| = \left|\binom{p}{n} \Big/ \binom{p}{n + 1}\right| = \left|\frac{n + 1}{p - n}\right| \longrightarrow 1,$$

so the radius of convergence is $R = 1$.

This, however, does not prove that the sum of the series *is* $(1 + x)^p$. That requires a delicate piece of analysis beyond the scope of this course.

EXAMPLE 5.1

Find the power series for $\dfrac{1}{(1 + x)^2}$.

Solution: Use the binomial series with $p = -2$. The coefficient of x^n is

$$\binom{-2}{n} = \frac{(-2)(-3)(-4)\cdots(-2 - n + 1)}{n!}$$

$$= (-1)^n \frac{2\cdot 3\cdot 4\cdots n\cdot (n + 1)}{n!} = (-1)^n(n + 1).$$

Therefore

$$(1+x)^{-2} = \sum_{n=0}^{\infty} \binom{-2}{n} x^n = \sum_{n=0}^{\infty} (-1)^n (n+1) x^n$$

$$= 1 - 2x + 3x^2 - 4x^3 + \cdots.$$

Answer: For $|x| < 1$, $\dfrac{1}{(1+x)^2} = \displaystyle\sum_{n=0}^{\infty} (-1)^n (n+1) x^n.$

CHECK:

$$\frac{1}{(1+x)^2} = -\frac{d}{dx}\left(\frac{1}{1+x}\right) = -\frac{d}{dx}(1 - x + x^2 - x^3 + - \cdots)$$

$$= -(-1 + 2x - 3x^2 + - \cdots).$$

EXAMPLE 5.2

Express $\dfrac{1}{\sqrt{1+x}}$ as a power series.

Solution: Use the binomial series with $p = -\frac{1}{2}$. The coefficient of x^n is

$$\binom{-\frac{1}{2}}{n} = \frac{\left(-\frac{1}{2}\right)\left(-\frac{3}{2}\right)\left(-\frac{5}{2}\right)\cdots\left(-\frac{2n-1}{2}\right)}{n!}$$

$$= (-1)^n \frac{1 \cdot 3 \cdot 5 \cdots (2n-1)}{2^n \cdot n!}.$$

But

$$1 \cdot 3 \cdot 5 \cdots (2n-1) = \frac{1 \cdot 2 \cdot 3 \cdot 4 \cdots (2n)}{2 \cdot 4 \cdot 6 \cdot 8 \cdots (2n)} = \frac{(2n)!}{2^n \cdot n!}.$$

Hence the coefficient of x^n is

$$(-1)^n \frac{(2n)!}{2^n \cdot n!} \cdot \frac{1}{2^n \cdot n!} = (-1)^n \frac{(2n)!}{(2^n \cdot n!)^2}.$$

Answer: For $|x| < 1$, $\dfrac{1}{\sqrt{1+x}} = \displaystyle\sum_{n=0}^{\infty} (-1)^n \frac{(2n)!}{(2^n \cdot n!)^2} x^n.$

The following example will be used in Section 7.

EXAMPLE 5.3

Express $\sqrt{1-x}$ as a power series.

Solution: Use the binomial series with $p = \frac{1}{2}$ and replace x by $-x$. The constant term is

$$\binom{\frac{1}{2}}{0} = 1.$$

The term in x^n is

$$\binom{\frac{1}{2}}{n}(-x)^n = \frac{\left(\frac{1}{2}\right)\left(-\frac{1}{2}\right)\left(-\frac{3}{2}\right)\cdots\left(-\frac{2n-3}{2}\right)}{n!}(-x)^n$$

$$= (-1)^{n-1}\frac{1\cdot3\cdot5\cdots(2n-3)}{2^n\cdot n!}(-1)^n x^n.$$

But

$$1\cdot3\cdot5\cdots(2n-3) = \frac{1\cdot2\cdot3\cdot4\cdots(2n-2)}{2\cdot4\cdot6\cdot8\cdots(2n-2)}$$

$$= \frac{(2n-2)!}{2^{n-1}(n-1)!} = \frac{(2n)!}{(2^n\cdot n!)(2n-1)}.$$

Therefore the term in x^n is

$$-\frac{(2n)!}{(2^n\cdot n!)(2n-1)}\cdot\frac{1}{2^n\cdot n!}x^n = -\frac{(2n)!}{(2^n\cdot n!)^2(2n-1)}x^n.$$

Answer: For $|x| < 1$,

$$\sqrt{1-x} = 1 - \sum_{n=1}^{\infty}\frac{(2n)!}{(2^n\cdot n!)^2(2n-1)}x^n.$$

EXAMPLE 5.4

Estimate $\sqrt[3]{1001}$ to 7 places.

Solution: Write

$$\sqrt[3]{1001} = [1000(1+10^{-3})]^{1/3} = 10(1+10^{-3})^{1/3}.$$

Use the binomial series with $x = 10^{-3}$ and $p = \frac{1}{3}$:

$$10(1+10^{-3})^{1/3} = 10\left[1 + \binom{\frac{1}{3}}{1}(10^{-3}) + \binom{\frac{1}{3}}{2}(10^{-3})^2 + \cdots\right]$$

$$= 10\left[1 + \frac{1}{3}\cdot10^{-3} - \frac{1}{9}\cdot10^{-6} + \cdots\right].$$

The first three terms yield the estimate

$$\sqrt[3]{1001} \approx 10.0033322222\cdots.$$

The error in this estimate is precisely the remainder

$$10\left[\binom{\frac{1}{3}}{3}(10^{-3})^3 + \binom{\frac{1}{3}}{4}(10^{-3})^4 + \cdots\right].$$

Now each binomial coefficient above is less than 1 in absolute value:

$$\left|\binom{\frac{1}{3}}{n}\right| = \left|\frac{\left(\frac{1}{3}\right)\left(-\frac{2}{3}\right)\left(-\frac{5}{3}\right)\cdots\left(-\frac{3n-4}{3}\right)}{1\cdot 2\cdot 3\cdots n}\right| = \frac{1}{3}\cdot\frac{2}{6}\cdot\frac{5}{9}\cdots\frac{3n-4}{3n} < 1.$$

Therefore, the error (in absolute value) is less than

$$10[(10^{-3})^3 + (10^{-3})^4 + \cdots] = 10^{-8}[1 + 10^{-3} + 10^{-6} + 10^{-9} + \cdots]$$

$$= 10^{-8}\frac{1}{1 - 10^{-3}} \approx 10^{-8}.$$

Answer: To 7 places, $\sqrt[3]{1001} \approx 10.0033322.$

REMARK: The estimate is actually accurate to at least 8 places. When we estimated each binomial coefficient by 1, we were too generous since

$$\binom{\frac{1}{3}}{n} < \frac{1\cdot 2\cdot 5}{3\cdot 6\cdot 9} = \frac{10}{162}, \qquad n \geq 3.$$

(Why?) Therefore, the error estimate can be reduced by a factor of $\frac{10}{162}$, which guarantees at least 1 more place accuracy.

EXERCISES

Expand in power series at $x = 0$:

1. $\dfrac{1}{(1+x)^3}$

2. $\dfrac{1}{(1-x)^{1/3}}$

3. $\dfrac{1}{(1-4x^2)^2}$

4. $\sqrt{2-x}$

   [*Hint:* $2 - x = 2\left(1 - \dfrac{x}{2}\right)$.]

5. $(1+2x^3)^{1/4}$

6. $\dfrac{1}{(3-4x^2)^2}$.

Expand in power series at $x = 1$:

7. $\sqrt{1+x}$ [*Hint:* Write $1 + x = 2 + (x - 1) = 2\left(1 + \dfrac{x-1}{2}\right)$.]

8. $\dfrac{1}{(3+x)^2}$.

Compute the power series at $x = 0$ up to and including the term in x^4:

9. $\sqrt{1 + x^2 e^x}$

10. $\dfrac{1}{(1 + \sin x)^2}$

11. $(\sin 2x)\sqrt{3 + x}$

12. $\dfrac{1}{\sqrt{1 + x + x^2}}$ [*Hint:* Write $x + x^2 = u$.]

13. $\dfrac{\cos 2x}{\left(1 + \dfrac{x}{3}\right)^4}$.

Compute to 4-place accuracy using the Binomial Theorem:

14. $\sqrt{16.1}$

15. $\sqrt[4]{82}$

16. $\dfrac{1}{(1.03)^5}$.

17. Expand arc sin x in a Taylor series at $x = 0$.
[*Hint:* Integrate its derivative.]

6. ALTERNATING SERIES

In Chapter 1, Section 4, we saw that alternating series whose terms approach 0 in absolute value are convenient to work with. In particular, remainder estimates are very simple and quite accurate: if you break off such a series, then the remainder is less than the absolute value of the first term omitted.

EXAMPLE 6.1

The power series at $x = 0$ for $\ln(1 + x^2)$ is broken off after n terms. Estimate the remainder.

Solution: The series is obtained by integrating from 0 to x the series for the derivative of $\ln(1 + t^2)$.

$$\frac{d}{dt}[\ln(1 + t^2)] = \frac{2t}{1 + t^2} = 2t(1 - t^2 + t^4 - t^6 + - \cdots)$$

$$= 2(t - t^3 + t^5 - t^7 + - \cdots), \qquad |t| < 1.$$

It follows that

$$\ln(1+x^2) = \int_0^x \frac{2t\,dt}{1+t^2} = 2\left(\frac{x^2}{2} - \frac{x^4}{4} + \frac{x^6}{6} - \frac{x^8}{8} + - \cdots\right)$$

$$= x^2 - \frac{x^4}{2} + \frac{x^6}{3} - \frac{x^8}{4} + - \cdots, \qquad |x| < 1.$$

This series alternates; since $|x| < 1$, its terms decrease in absolute value to zero. Therefore, the remainder after n terms is less than the absolute value of the $(n+1)$-th term.

Answer: $|\text{remainder}| < \dfrac{x^{2n+2}}{n+1}$.

Occasionally we encounter an alternating series whose terms ultimately decrease in magnitude, but whose first few terms do not. If the successive terms decrease starting at the k-th term, the series will still converge, and the remainder estimate is still valid—beyond the k-th term. The front end of the series, up to the $(k-1)$-th term, is a finite sum; it causes no trouble. The important part is the tail end, that is, the series starting with the k-th term. It is this series that we test for convergence.

EXAMPLE 6.2

The power series for e^{-x} at $x = 0$ is broken off after n terms. Estimate the remainder for positive values of x.

Solution: If $x > 0$ the power series

$$e^{-x} = 1 - x + \frac{x^2}{2!} - \frac{x^3}{3!} + - \cdots$$

alternates. If $0 < x \leq 1$, the terms decrease to 0, and the above remainder estimate for alternating series applies. If $x > 1$, however, the first few terms may not decrease. (Take $x = 6$ for example:

$$e^{-6} = 1 - 6 + \frac{6^2}{2!} - \frac{6^3}{3!} + - \cdots = 1 - 6 + 18 - 36 + - \cdots.)$$

Nevertheless, for any fixed x, the terms do decrease ultimately.

To see why, note that the ratio of successive terms is

$$\frac{x^{n+1}}{(n+1)!} \Big/ \frac{x^n}{n!} = \frac{x}{n+1}.$$

Take a fixed value of x. Then

$$\frac{x}{n+1} < 1$$

as soon as $n+1>x$; from then on the terms decrease. Furthermore

$$\frac{x}{n+1}<\frac{1}{2}$$

as soon as $n+1>2x$. From then on each term is less than one-half the preceding term. Hence, the terms decrease to 0.

Result: if the series is broken off at the n-th term, the remainder estimate for an alternating series applies, provided $n+1>x$.

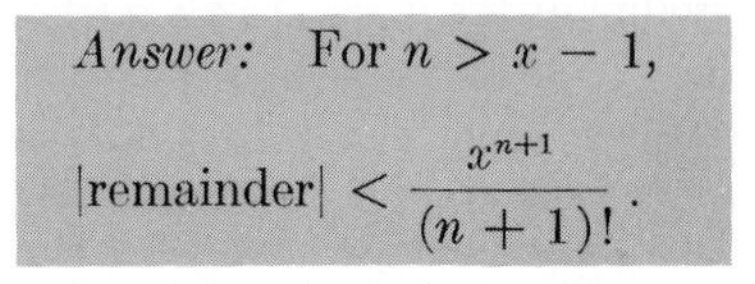
Answer: For $n > x-1$,

$$|\text{remainder}| < \frac{x^{n+1}}{(n+1)!}.$$

As another application of this method, recall Example 5.4. There it was shown that three terms of the binomial series for

$$10(1+10^{-3})^{1/3}$$

provide 7-place accuracy in computing $\sqrt[3]{1001}$. But this series alternates and its terms decrease towards 0. Therefore, one deduces that the remainder after three terms is less than the fourth term, which is

$$10\binom{\frac{1}{3}}{3}(10^{-3})^3=\frac{\left(\frac{1}{3}\right)\left(-\frac{2}{3}\right)\left(-\frac{5}{3}\right)}{3!}10^{-8}=\frac{10}{162}\cdot 10^{-8}<10^{-9},$$

so 8-place accuracy is assured. This estimate is both easier and more precise than the one in Example 5.4.

EXERCISES

1. The power series $\sum \frac{x^n}{n}$ converges *inside* the interval $-1<x<1$. Show that it converges also at $x=-1$.
2. Give an example of an alternating power series that does not converge.
3. Compute $e^{-1/5}$ to 5-place accuracy.
4. Find values of x for which the formula $\cos x \approx 1-\frac{x^2}{2!}+\frac{x^4}{4!}$ yields 5-place accuracy.
5. How many terms of the power series for $\ln x$ at $x=10$ are needed to compute $\ln(10.5)$ with 5-place accuracy? (Assume the value of ln 10 is known.)
6. How many terms of the binomial series for $\sqrt{1+x}$ will yield 5-place accuracy for $0<x\leq 0.1$?

7. APPLICATIONS TO DEFINITE INTEGRALS [optional]

Power series are used in approximating definite integrals which cannot be computed exactly.

EXAMPLE 7.1

Estimate $\int_0^x e^{-t^2}\,dt$ to 6 places for $|x| \le \frac{1}{2}$.

Solution: A numerical integration formula such as Simpson's Rule can be used, but a power series method is simpler. Expand the integrand in a power series:

$$e^{-t^2} = 1 + (-t^2) + \frac{(-t^2)^2}{2!} + \frac{(-t^2)^3}{3!} + \cdots = 1 - t^2 + \frac{t^4}{2!} - \frac{t^6}{3!} + \cdots.$$

Since this series converges for all x, it can be integrated term-by-term:

$$\int_0^x e^{-t^2}\,dt = \int_0^x \left(1 - t^2 + \frac{t^4}{2!} - \frac{t^6}{3!} + \cdots\right) dt$$

$$= x - \frac{x^3}{3} + \frac{x^5}{5\cdot 2!} - \frac{x^7}{7\cdot 3!} + \frac{x^9}{9\cdot 4!} - \cdots.$$

Because of the large denominators, this series converges rapidly if x is fairly small. For $|x| \le \frac{1}{2}$, the sixth term is at most

$$\left(\frac{1}{2}\right)^{11} \frac{1}{11\cdot 5!} < 4 \times 10^{-7}.$$

Since the series alternates, it follows that five terms provide 6-place accuracy.

Answer: For all x,

$$\int_0^x e^{-t^2}\,dt = x - \frac{x^3}{3} + \frac{x^5}{5\cdot 2!} - \frac{x^7}{7\cdot 3!} + \cdots.$$

For $|x| \le \frac{1}{2}$, five terms yield 6-place accuracy.

REMARK: The series converges for all values of x, but for large x it converges slowly. For example, it would be ridiculous to compute

$$\int_0^{10} e^{-t^2}\,dt$$

by this method, since more than 100 terms at the beginning are greater than 1.

Elliptic Integrals

EXAMPLE 7.2

Express the integral $\int_0^{\pi/2} \sqrt{1 - k^2 \sin^2 t}\, dt$, where $k^2 < 1$, as a power series in k.

Solution: By Example 5.3,

$$\sqrt{1-x} = (1-x)^{1/2} = 1 - \sum_{n=1}^{\infty} a_n x^n, \qquad |x| < 1,$$

where

$$a_n = \frac{(2n)!}{(2^n \cdot n!)^2 (2n-1)}.$$

Substitute $k^2 \sin^2 t$ for x. This is permissible because $k^2 \sin^2 t \le k^2 < 1$.

$$(1 - k^2 \sin^2 t)^{1/2} = 1 - \sum_{n=1}^{\infty} a_n k^{2n} \sin^{2n} t.$$

Now integrate term-by-term:

$$\int_0^{\pi/2} (1 - k^2 \sin^2 t)^{1/2}\, dt = \frac{\pi}{2} - \sum_{n=1}^{\infty} a_n k^{2n} \int_0^{\pi/2} \sin^{2n} t\, dt.$$

The integrals on the right are evaluated by a reduction formula (see p. 45), and their values are listed inside the front cover:

$$\int_0^{\pi/2} (\sin x)^{2n}\, dx = \frac{(2n)!}{(2^n \cdot n!)^2} \cdot \frac{\pi}{2}.$$

Therefore

$$a_n \int_0^{\pi/2} \sin^{2n} t\, dt = \left[\frac{(2n)!}{(2^n \cdot n!)^2 (2n-1)}\right]\left[\frac{(2n)!}{(2^n \cdot n!)^2} \cdot \frac{\pi}{2}\right]$$

$$= \left[\frac{(2n)!}{(2^n \cdot n!)^2}\right]^2 \frac{1}{(2n-1)} \cdot \frac{\pi}{2}.$$

Substitute this expression to obtain the answer.

Answer: For $k^2 < 1$, $\int_0^{\pi/2} \sqrt{1 - k^2 \sin^2 t}\, dt$

$$= \frac{\pi}{2}\left\{1 - \sum_{n=1}^{\infty} \left[\frac{(2n)!}{(2^n \cdot n!)^2}\right]^2 \frac{k^{2n}}{2n-1}\right\}.$$

REMARK: This integral arises in computing the arc length of an ellipse. Suppose an ellipse is given in the parametric form

$$x = a\cos\theta, \qquad y = b\sin\theta,$$

where $b > a$. If s denotes arc length,

$$\left(\frac{ds}{d\theta}\right)^2 = \left(\frac{dx}{d\theta}\right)^2 + \left(\frac{dy}{d\theta}\right)^2 = (-a\sin\theta)^2 + (b\cos\theta)^2$$

$$= b^2(\sin^2\theta + \cos^2\theta) - (b^2 - a^2)\sin^2\theta = b^2(1 - k^2\sin^2\theta),$$

where $k^2 = (b^2 - a^2)/b^2$. The length of the ellipse is

$$4\int_0^{\pi/2}\left(\frac{ds}{d\theta}\right)d\theta = 4b\int_0^{\pi/2}\sqrt{1 - k^2\sin^2\theta}\,d\theta.$$

EXAMPLE 7.3

Estimate $\dfrac{2}{\pi}\displaystyle\int_0^{\pi/2}\sqrt{1 - \frac{1}{5}\sin^2 x}\,dx$ to 4-place accuracy.

Solution: By the last example with $k^2 = \frac{1}{5}$,

$$\frac{2}{\pi}\int_0^{\pi/2}\sqrt{1 - \frac{1}{5}\sin^2 x}\,dx = 1 - \sum_{n=1}^{\infty} b_n\left(\frac{1}{2n-1}\right)\left(\frac{1}{5}\right)^n,$$

where

$$b_n = \left[\frac{(2n)!}{2^{2n}(n!)^2}\right]^2 = \left[\frac{1\cdot3\cdot5\cdots(2n-1)}{2\cdot4\cdot6\cdots(2n)}\right]^2.$$

Break off the series after the p-th term. The remainder (error) is

$$\epsilon_p = -\sum_{n=p+1}^{\infty} b_n\left(\frac{1}{2n-1}\right)\left(\frac{1}{5}\right)^n.$$

The problem: choose p large enough that $|\epsilon_p| < 5\times10^{-5}$.

The error is a complicated expression. It can be enormously simplified in two steps, each of which causes a certain amount of overestimation.

First, each coefficient b_n is less than 1:

$$b_n = \left[\frac{1\cdot3\cdot5\cdots(2n-1)}{2\cdot4\cdot6\cdots(2n)}\right]^2 = \left[\frac{1}{2}\cdot\frac{3}{4}\cdot\frac{5}{6}\cdots\frac{2n-1}{2n}\right]^2 < 1;$$

hence

$$|\epsilon_p| = \sum_{n=p+1}^{\infty} b_n\left(\frac{1}{2n-1}\right)\left(\frac{1}{5}\right)^n < \sum_{n=p+1}^{\infty}\left(\frac{1}{2n-1}\right)\left(\frac{1}{5}\right)^n.$$

Second, for $n \geq p + 1$,

$$\frac{1}{2n-1} \leq \frac{1}{2(p+1)-1} = \frac{1}{2p+1};$$

hence replace $1/(2n-1)$ by $1/(2p+1)$:

$$|\epsilon_p| < \sum_{n=p+1}^{\infty} \left(\frac{1}{2p+1}\right)\left(\frac{1}{5}\right)^n = \frac{1}{2p+1} \sum_{n=p+1}^{\infty} \left(\frac{1}{5}\right)^n.$$

On the right is a geometric series; consequently

$$|\epsilon_p| < \frac{1}{2p+1}\left(\frac{1}{5}\right)^{p+1}\left[1 + \left(\frac{1}{5}\right) + \left(\frac{1}{5}\right)^2 + \cdots\right]$$

$$= \frac{1}{2p+1}\left(\frac{1}{5}\right)^{p+1} \frac{1}{1-\frac{1}{5}} = \frac{1}{4(2p+1)5^p}.$$

To obtain 4-place accuracy, choose p so that

$$\frac{1}{4(2p+1)5^p} < 5 \times 10^{-5}.$$

To find a suitable value of p, take reciprocals:

$$4(2p+1)5^p > 20000, \qquad (2p+1)5^p > 5000.$$

Trial and error shows

$$(2\cdot 3 + 1)5^3 = 875, \qquad (2\cdot 4 + 1)5^4 = 5625.$$

Accordingly, choose $p = 4$ and estimate the integral by

$$1 - \left(\frac{1}{2}\right)^2\left(\frac{1}{5}\right) - \left(\frac{1\cdot 3}{2\cdot 4}\right)^2\left(\frac{1}{5}\right)^2\left(\frac{1}{3}\right) - \left(\frac{1\cdot 3\cdot 5}{2\cdot 4\cdot 6}\right)^2\left(\frac{1}{5}\right)^3\left(\frac{1}{5}\right) - \left(\frac{1\cdot 3\cdot 5\cdot 7}{2\cdot 4\cdot 6\cdot 8}\right)^2\left(\frac{1}{5}\right)^4\left(\frac{1}{7}\right)$$

$$= 1 - \frac{1}{20} - \frac{3}{1600} - \frac{1}{6400} - \frac{7}{409600} \approx 0.9480.$$

Answer: 0.9480.

EXERCISES

Express as a power series in x:

1. $\displaystyle\int_0^x \frac{\sin t}{t}\,dt$

2. $\displaystyle\int_0^x \frac{1-\cos t}{t}\,dt$

3. $\displaystyle\int_0^x \frac{t}{1+t^4}\,dt$

4. $\displaystyle\int_0^x \sin(t^2)\,dt.$

Compute to 5-place accuracy:

5. $\displaystyle\int_0^{0.1} e^{x^3}\,dx$

6. $\displaystyle\int_0^{0.2} e^{-x^2}\,dx$

7. $\displaystyle\int_0^{1/4} \sqrt{1+x^3}\,dx$

8. $\displaystyle\int_{3.00}^{3.01} \frac{e^x}{1+x}\,dx$ [Expand at $x = 3$.]

9. Compute to 4-place accuracy the arc length of an ellipse with semi-axes 40 and 41.

10. Estimate the value of x for which

$$\int_0^x \frac{t}{1+t^4}\,dt = 0.1.$$

[*Hint:* Approximate the integral by the first significant term of its power series; use Ex. 3.]

11. (cont.) Refine your estimate of x to 4-place accuracy by taking the first two significant terms of the power series. Use Newton's Method to solve approximately the resulting equation.

8. Uniform Convergence [optional]

Any basic information we need about the convergence of an infinite series $\sum f_n(x)$ of functions is carried in the corresponding *sequence* $\{s_n(x)\}$ of partial sums. In this section we shall study some important properties of sequences of functions, then apply the results to power series.

Definition Let $\{u_n(x)\}$ be a sequence of functions, all with the same domain $\mathbf{D}$. The sequence is said to **converge uniformly** on $\mathbf{D}$ to $u(x)$ if given any $\epsilon > 0$, there exists an N such that

$$|u_n(x) - u(x)| < \epsilon$$

for all $n \geq N$ and *all* x in $\mathbf{D}$.

The last four words are the key to this concept. We can control the degree of approximation of $u_n(x)$ to $u(x)$ *independent of* x. The next three results show the usefulness of uniform convergence.

Theorem 8.1 Let $\{u_n(x)\}$ be a sequence of continuous functions on $\mathbf{D}$ and let $u_n(x) \longrightarrow u(x)$ uniformly on $\mathbf{D}$. Then $u(x)$ is continuous on $\mathbf{D}$.

Proof: Let $\epsilon > 0$. Then there is an N such that $|u_N(x) - u(x)| < \frac{1}{3}\epsilon$ for *all* x in $\mathbf{D}$. Take any point c of $\mathbf{D}$. Since u_N is continuous at c, there is $\delta > 0$ such that $|u_N(x) - u_N(c)| < \frac{1}{3}\epsilon$ for all x in $\mathbf{D}$ such that $|x - c| < \delta$.

If x is in **D** and $|x - c| < \delta$, then

$$\begin{aligned}|u(x) - u(c)| &= |u(x) - u_N(x) + u_N(x) - u_N(c) + u_N(c) - u(c)| \\ &\leq |u(x) - u_N(x)| + |u_N(x) - u_N(c)| + |u_N(c) - u(c)| \\ &< \tfrac{1}{3}\epsilon + \tfrac{1}{3}\epsilon + \tfrac{1}{3}\epsilon = \epsilon.\end{aligned}$$

This proves the continuity of u at c.

Theorem 8.2 Let $\{u_n(x)\}$ be a sequence of continuous functions on a closed interval $[a, b]$, and suppose $u_n(x) \longrightarrow u(x)$ uniformly on $[a, b]$. Then

$$\int_a^b u(x)\,dx = \lim_{n\to\infty} \int_a^b u_n(x)\,dx.$$

Proof: The function u is continuous by the previous theorem, hence integrable. Let $\epsilon > 0$. Then there is an N such that $|u_n(x) - u(x)| < \epsilon/(b - a)$ for all $n \geq N$ and for *all* x in $[a, b]$.

If $n \geq N$, then

$$\begin{aligned}\left|\int_a^b u_n(x)\,dx - \int_a^b u(x)\,dx\right| &= \left|\int_a^b [u_n(x) - u(x)]\,dx\right| \\ &\leq \int_a^b |u_n(x) - u(x)|\,dx < \int_a^b \frac{\epsilon}{b - a}\,dx = \epsilon.\end{aligned}$$

Hence

$$\int_a^b u_n(x)\,dx \longrightarrow \int_a^b u(x)\,dx.$$

Theorem 8.3 Let $\{u_n(x)\}$ be a sequence of continuous functions on an open interval (a, b) and suppose $u_n(x) \longrightarrow u(x)$ for each x in (a, b). Assume also that each $u_n(x)$ is differentiable, that the derivatives $u_n'(x)$ are continuous on (a, b), and that $u_n'(x) \longrightarrow v(x)$ uniformly on (a, b). Then $u(x)$ is differentiable and $u'(x) = v(x)$.

Proof: Fix c in (a, b) and let x be any point of this interval. Then

$$\int_c^x u_n'(t)\,dt \longrightarrow \int_c^x v(t)\,dt$$

by the previous theorem. Hence

$$u_n(x) - u_n(c) \longrightarrow \int_c^x v(t)\,dt.$$

But $u_n(x) \longrightarrow u(x)$ and $u_n(c) \longrightarrow u(c)$, so

$$u(x) - u(c) = \int_c^x v(t)\, dt.$$

By the Fundamental Theorem of Calculus, u is differentiable and $u' = v$.

REMARK: The last two results can be interpreted in terms of interchanging operations. Theorem 8.2 says that $\lim_{n\to\infty}$ and $\int_a^b$ can be interchanged (under suitable hypotheses) and Theorem 8.3 says that $\lim_{n\to\infty}$ and d/dx can be interchanged. Note the key role played by uniform convergence in these results.

Infinite Series

It is easy to translate the results above into statements about infinite series of functions. The starting point is the elementary fact that everything reasonable holds for *finite* sums—that is what the partial sums of a series are.

Theorem 8.4 Let $\{f_n(x)\}$ be a sequence of continuous functions on a domain **D** and assume $\sum_{n=1}^{\infty} f_n(x)$ converges uniformly on **D** to $F(x)$. Then

(1) $F(x)$ is continuous on **D**.

(2) If **D** $= [a, b]$, then $F(x)$ is integrable on $[a, b]$ and

$$\int_a^b F(x)\, dx = \sum_{n=1}^{\infty} \int_a^b f_n(x)\, dx.$$

(3) Suppose the functions $f_n(x)$ have continuous derivatives and $\sum f_n'(x)$ converges uniformly to $G(x)$ on an open interval **D** $= (a, b)$. Then $F(x)$ is differentiable and

$$F'(x) = G(x) = \sum_{n=1}^{\infty} f_n'(x)$$

for each x in **D**.

Proof: Let $s_n(x) = \sum_{k=1}^{n} f_k(x)$. Then $s_n(x) \longrightarrow F(x)$ uniformly on **D**. But $s_n(x)$ is continuous, being a finite sum of continuous functions; hence $F(x)$ is continuous by Theorem 8.1. This proves (1).

Now $F(x)$ is continuous, hence integrable on $[a, b]$. By Theorem 8.2,

$$\int_a^b F(x)\,dx = \lim_{n\to\infty} \int_a^b s_n(x)\,dx = \lim_{n\to\infty} \int_a^b \left[\sum_{k=1}^n f_k(x)\right] dx$$

$$= \lim_{n\to\infty} \sum_{k=1}^n \int_a^b f_k(x)\,dx = \sum_{k=1}^\infty \int_a^b f_k(x)\,dx.$$

This proves (2).

By Theorem 8.3, the function $F(x)$ is differentiable and

$$F'(x) = G(x) = \lim_{n\to\infty} s_n'(x) = \lim_{n\to\infty} \frac{d}{dx}\left[\sum_{k=1}^n f_k(x)\right]$$

$$= \lim_{n\to\infty} \sum_{k=1}^n f_k'(x) = \sum_{k=1}^\infty f_k'(x).$$

This proves (3).

Remark: Parts (2) and (3) of Theorem 8.4 are again results about interchanging the order of operations. They may be stated formally as follows:

(2) $$\int_a^b \sum_{n=1}^\infty f_n(x)\,dx = \sum_{n=1}^\infty \int_a^b f_n(x)\,dx,$$

(3) $$\frac{d}{dx} \sum_{n=1}^\infty f_n(x) = \sum_{n=1}^\infty \frac{df_n(x)}{dx}.$$

The M-test

In applying these results, the first step is always proving the uniform convergence of some series. Often we make use of the following criterion for uniform convergence, called the **Weierstrass *M*-test.**

> **Theorem 8.5 (*M*-test)** Suppose $f_n(x)$ are given on a domain **D** and there are constants M_n such that
>
> (1) $\sum M_n$ converges,
>
> (2) $|f_n(x)| \le M_n$ for all x in **D** and all n.
>
> Then $\sum f_n(x)$ converges uniformly on **D**.

Before reading the proof, it is advisable to review the Cauchy Test in Chapter 1, Section 2.

Proof: Let $\epsilon > 0$. By the Cauchy Test, there is an N such that

$$M_{n+1} + M_{n+2} + \cdots + M_m < \tfrac{1}{2}\epsilon$$

whenever $m > n \geq N$. Therefore

$$\left| \sum_{k=n+1}^{m} f_k(x) \right| \leq \sum_{k=n+1}^{m} |f_k(x)| \leq \sum_{k=n+1}^{m} M_k < \tfrac{1}{2}\epsilon$$

for *all* x in **D** whenever $m > n \geq N$.

By the Cauchy Test again ("if" part), for *each* x in **D**, the series $\sum f_k(x)$ converges to a number $F(x)$. Now let $m \longrightarrow \infty$ in the last displayed inequality:

$$\left| \sum_{k=n+1}^{\infty} f_k(x) \right| \leq \tfrac{1}{2}\epsilon < \epsilon$$

for all $n \geq N$ and *all* x in **D**. An equivalent statement is

$$\left| F(x) - \sum_{k=1}^{n} f_k(x) \right| < \epsilon,$$

that is,

$$|F(x) - s_n(x)| < \epsilon$$

for all $n \geq N$ and *all* x in **D**, where $s_n(x)$ is the n-th partial sum of $\sum f_k(x)$. This says that the series converges uniformly to $F(x)$ on **D**.

Power Series

Let us derive the main facts about integration and differentiation of power series. First we need a result on the uniform convergence of power series.

Theorem 8.6 If a power series $\sum a_n(x - c)^n$ converges in the interval $|x - c| < R$, then it converges uniformly in each smaller interval $|x - c| \leq r$, where $0 < r < R$.

Proof: If $|x - c| \leq r$, then

$$|a_n(x - c)^n| \leq |a_n r^n|.$$

Since $r < R$, the series $\sum a_n r^n$ converges absolutely. Therefore the M-test applies with $M_n = |a_n r^n|$. Done.

Next we need uniform convergence of the series $\sum n a_n(x - c)^{n-1}$ of term-by-term derivatives.

Theorem 8.7 If a power series $\sum a_n(x - c)^n$ converges in the interval $|x - c| < R$, then the series $\sum n a_n(x - c)^{n-1}$ converges uniformly in each smaller interval $|x - c| \leq r$, where $0 < r < R$.

Proof: By Theorem 8.6, it suffices to prove that the power series $\sum na_n(x-c)^{n-1}$ converges in $|x-c|<R$.

Let $|x-c|<R$. Choose r so that $|x-c|<r<R$. Set $b=|x-c|/r$, so $0<b<1$. Then $nb^{n-1}\longrightarrow 0$ (by Lhospital's rule for instance); hence $nb^{n-1}<r$ for n sufficiently large. This implies

$$|n(x-c)^{n-1}|<r^n, \qquad |na_n(x-c)^{n-1}|<|a_nr^n|$$

for n sufficiently large. But $\sum a_nr^n$ converges absolutely since $r<R$. Therefore $\sum na_n(x-c)^{n-1}$ converges by the M-test.

Now we can state the main results of this section.

Theorem 8.8 Let $F(x)=\sum a_n(x-c)^n$, where the power series converges on the interval $|x-c|<R$. Then

(1) $F(x)$ is continuous on $|x-c|<R$.

(2) $F(x)$ is integrable on $|x-c|<R$ and

$$\int_c^x F(t)\,dt=\sum_{n=0}^{\infty}\frac{a_n}{n+1}(x-c)^{n+1}$$

for $|x-c|<R$.

(3) $F(x)$ is differentiable on $|x-c|<R$ and

$$F'(x)=\sum_{n=1}^{\infty}na_n(x-c)^{n-1}.$$

Proof: Except for a small technical detail, everything follows routinely from the previous results. We cannot conclude right off the bat that $F(x)$ is continuous on $|x-c|<R$ because we do not know that $\sum a_n(x-c)^n$ converges uniformly on *all* of the interval $|x-c|<R$. We know only that it converges uniformly on *each* closed interval $|x-c|\leq r$, where $r<R$. But that is enough, for if x_0 is any point such that $|x_0-c|<R$, we can choose r so that $|x_0-c|<r<R$. Hence x_0 belongs to the open interval $|x-c|<r$. But Theorem 8.6 shows that the power series converges uniformly on $|x-c|\leq r$, hence $F(x)$ is continuous on $|x-c|\leq r$. In particular $F(x)$ is continuous at x_0. (Of course we have used the obvious continuity of the summands—polynomials—$a_n(x-c)^n$.) The rest of the theorem is proved similarly.

The last statement in the theorem can be applied to F', then to F'', etc. Thus F can be differentiated repeatedly, and

$$\frac{d^kF}{dx^k}=\sum_{n=k}^{\infty}n(n-1)\cdots(n-k+1)a_nx^{n-k} \qquad \text{for} \quad |x-c|<R.$$

REMARK: We have carefully avoided the endpoints of the interval of convergence. If $F(x) = \sum a_n x^n$ has radius of convergence R, then the power series converges for $|x| < R$, and we may not know what happens at $x = \pm R$. But *suppose* the series converges at $x = R$. Then it can be proved that $\sum a_n x^n$ converges *uniformly* on $0 \leq x \leq R$, and hence that $F(x)$ is continuous and can be integrated on $0 \leq x \leq R$. This is a fairly deep result, and its proof is beyond our scope. It is needed to prove such statements as

$$\ln 2 = \int_0^1 \frac{dx}{1+x} = 1 - \frac{1}{2} + \frac{1}{3} - \frac{1}{4} - \cdots,$$

$$\frac{\pi}{4} = \arctan 1 = \int_0^1 \frac{dx}{1+x^2} = 1 - \frac{1}{3} + \frac{1}{5} - \frac{1}{7} + \cdots.$$

EXERCISES

1. Let $f_n(x) = x^n$. Show that the sequence $\{f_n(x)\}$ converges uniformly on $[0, \frac{9}{10}]$ but not on $[0, 1]$.

2. Let $f_n(x) = \begin{cases} nx, & 0 \leq x \leq 1/n \\ 2 - nx, & 1/n \leq x \leq 2/n \\ 0 & 2/n < x. \end{cases}$

   Prove that $f_n(x) \longrightarrow 0$ for all $x \geq 0$, but the convergence is not uniform on $[0, \infty)$.

3. Prove that $xe^{-nx} \longrightarrow 0$ uniformly on $[0, \infty)$.
   [*Hint:* Find the maximum value of xe^{-nx}.]

4. Determine whether $x^2e^{-nx} \longrightarrow 0$ uniformly on $[0, \infty)$.

5. Prove that $\sum_1^\infty (\sin nx)/n^2$ is continuous on R.

6. Prove that $\sum_1^\infty 1/(1 + x^n)$ is continuous for $x > 1$.

7. Prove that $f(x) = \sum_{n=1}^{\infty} e^{-nx} \sin nx$ is continuous for $x > 0$.

8. (cont.) Justify the formula
$$\int_1^2 f(x)\,dx = \sum_{n=1}^{\infty} \int_1^2 e^{-nx} \sin nx\,dx.$$

9. Justify the formula
$$\frac{d}{dx}\left[\sum_{n=1}^{\infty} \frac{\sin nx}{n^3}\right] = \sum_{n=1}^{\infty} \frac{\cos nx}{n^2}.$$

10*. In Theorem 8.3, suppose the hypothesis $u_n(x) \longrightarrow u(x)$ for *each* x in (a, b) is replaced by $\{u_n(c)\}$ converges for *some* c in (a, b). Prove that $\{u_n(x)\}$ converges for each x follows anyhow.

11. If $\sum a_n$ is an absolutely convergent series, prove that $\sum a_n \sin nx$ converges uniformly.

12. Suppose $f_n(x) \longrightarrow F(x)$ uniformly on $[a, b]$. Does it follow that $f_n'(x) \longrightarrow F'(x)$ on $[a, b]$?

4. Solid Analytic Geometry

1. COORDINATES AND VECTORS

In this chapter we shall develop tools useful for geometric applications of calculus and for the study of functions of several variables. Although we shall work in the euclidean three-space $\mathbf{R}^3$, we note that most of what we do applies as well to the euclidean plane $\mathbf{R}^2$.

First we introduce a rectangular coordinate system in $\mathbf{R}^3$. We select an origin $\mathbf{0}$ and three mutually perpendicular real axes through the origin (Fig. 1.1a).

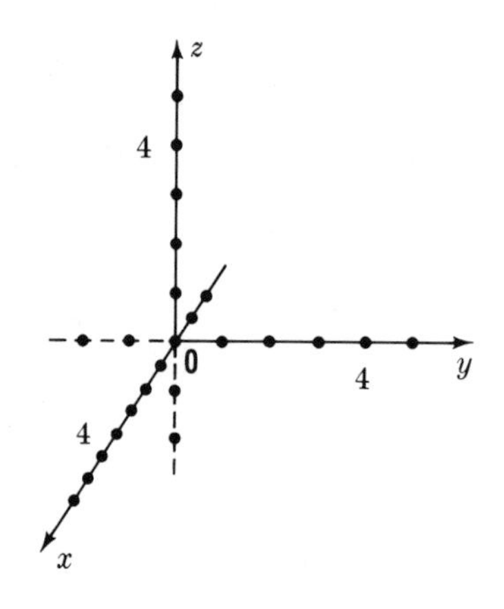

(a) rectangular coordinates in $\mathbf{R}^3$

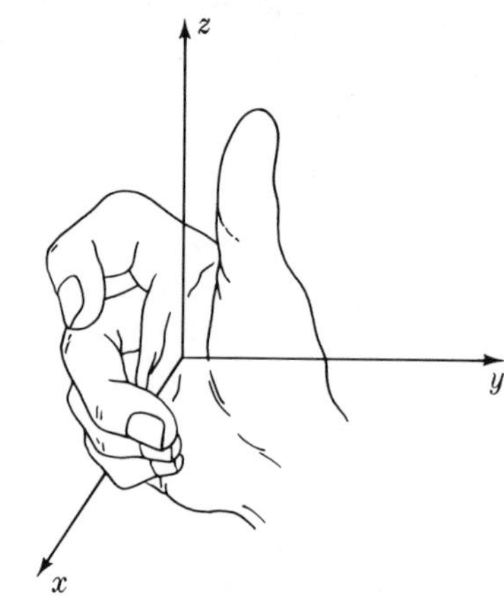

(b) right-hand rule

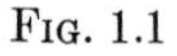

Fig. 1.1

Once directions are fixed on the x- and y-axes, the direction on the z-axis is determined by the **right-hand rule:** curl the fingers of your right hand from the positive x-direction towards the positive y-direction; the thumb will point in the positive z-direction (Fig. 1.1b). (In drawings of 3-space, it is convenient to think of the x-axis as pointing straight up from the paper).

We refer to the plane of the x-axis and y-axis as the **x, y-cordinate plane** or simply the **x, y-plane,** etc. (Fig. 1.2a).

Now take any point $\mathbf{x}$ in space. Pass planes through $\mathbf{x}$ parallel to each coordinate plane. Their intersections with the coordinate axes determine three

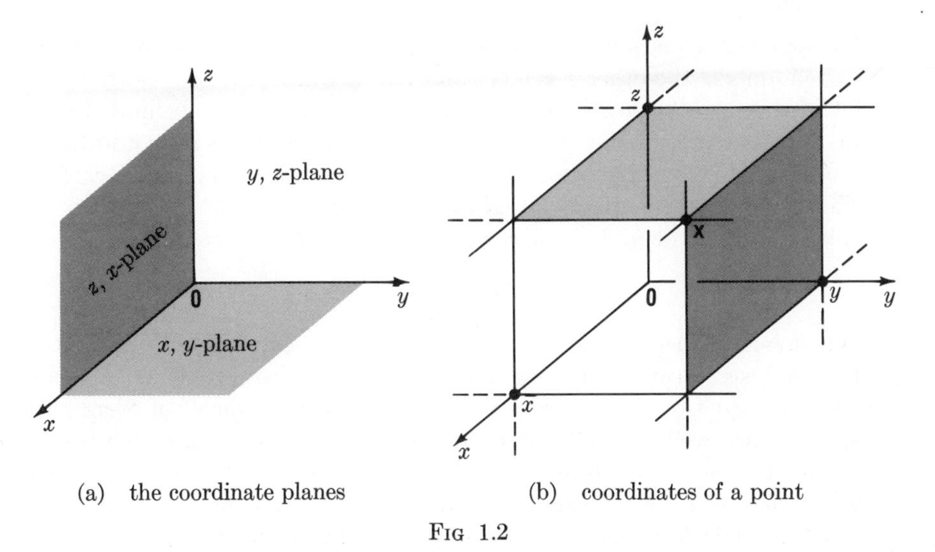

(a) the coordinate planes (b) coordinates of a point

Fig 1.2

numbers x, y, z, called the **coordinates** of $\mathbf{x}$. See Fig. 1.2b. Conversely, each triple (x, y, z) of real numbers determines a unique point $\mathbf{x}$ in space. We shall write

$$\mathbf{x} = (x, y, z).$$

A point (x, y, z) is located by marking its projection $(x, y, 0)$ in the x, y-plane and going up or down the corresponding amount z. (From the habit of living in the x, y-plane for so long, we think of the z-direction as "up".) See Fig. 1.3a for some examples.

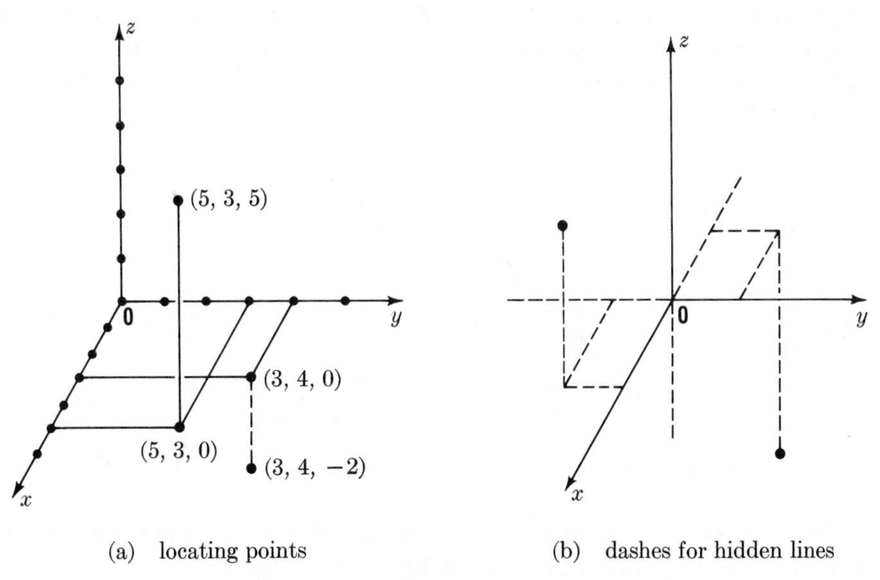

(a) locating points (b) dashes for hidden lines

Fig. 1.3

The portion of space where x, y, and z are positive is called the first **octant.** (No one numbers the other seven octants.) Sometimes part of a figure which is not in the first octant is shown; dotted lines indicate it is behind the coordinate planes (Fig. 1.3b). The angle at which the x-axis is drawn in the y, z-plane is up to you. Choose it so that your drawing is as uncluttered as possible. Actually it is perfectly alright to take a projection into other than the y, z-plane, so that the y- and z-axes are not drawn perpendicular.

Vectors

We now introduce the concept of a vector, then vector algebra and, later, vector analysis. Vectors are most useful for handling problems in space because (1) equations in vector form are independent of choice of coordinate axes, hence are well suited to describe physical situations; (2) each vector equation replaces three ordinary equations; and (3) several frequently occurring procedures can be summarized neatly in vector form.

Let the origin **0** be fixed once and for all. A **vector** in space is a directed line segment that begins at **0**; it is completely determined by its terminal point. Denote vectors by bold-faced letters **x**, **v**, **F**, **r**, etc. (In written work use $\underline{x}$ or $\vec{x}$ instead of **x**.)

A point (x, y, z) in space is often identified with the vector **x** from the origin to the point.

A vector is determined by two quantities, *length* (or *magnitude*) and *direction*. Many physical quantities are vectors: force, velocity, acceleration, electric field intensity, etc.

Remember that the origin **0** is fixed, and that each vector starts at **0**. We often draw vectors starting at other points, but in computations they all originate at **0**. For example, if a force **F** is applied at a point **x**, we may draw Fig. 1.4a because it is suggestive. But the correct figure is Fig. 1.4b. One must specify both the force vector **F** (magnitude and direction) and its point of application **x**.

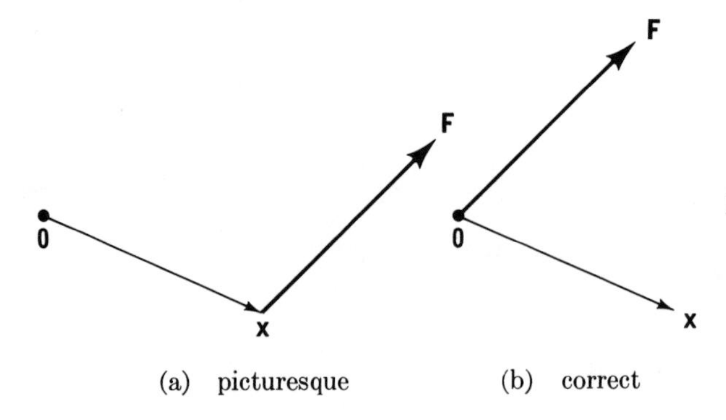

FIG. 1.4 drawing vectors

With respect to coordinate axes, each vector **x** has three *components* (coordinates) x, y, and z, which we indicate by the notation

$$\mathbf{x} = (x, y, z).$$

See Fig. 1.5. Sometimes it is convenient to index the components, writing $\mathbf{x} = (x_1, x_2, x_3)$ instead of $\mathbf{x} = (x, y, z)$.

The **zero vector** (origin) will be written $\mathbf{0} = (0, 0, 0)$. For this vector only, direction is undefined.

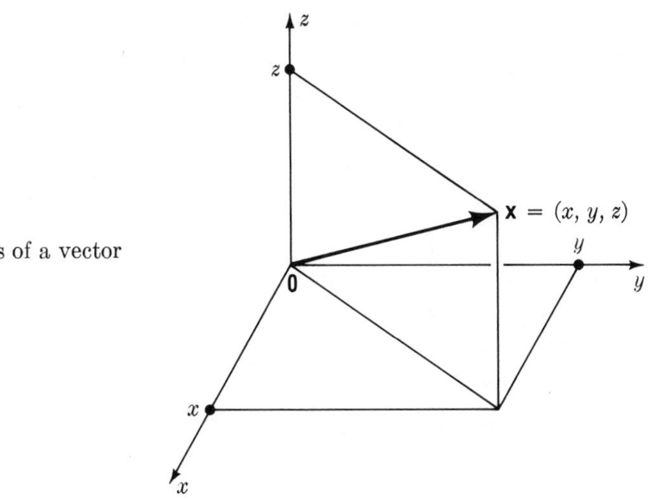

FIG. 1.5 components of a vector

Addition of Vectors

The **sum** $\mathbf{u} + \mathbf{v}$ of two vectors is defined by the parallelogram law (Fig. 1.6). The points $\mathbf{0}$, $\mathbf{u}$, $\mathbf{v}$, $\mathbf{u} + \mathbf{v}$ are vertices of a parallelogram with $\mathbf{u} + \mathbf{v}$ opposite to $\mathbf{0}$.

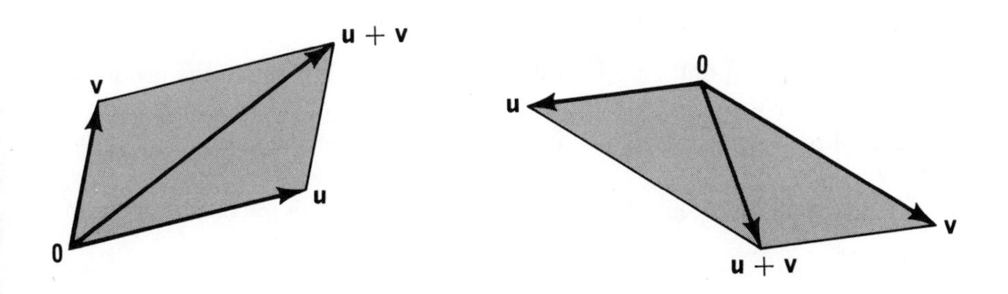

FIG. 1.6 parallelogram law of vector addition

Vectors are added numerically by adding their components:

$$(u_1, u_2, u_3) + (v_1, v_2, v_3) = (u_1 + v_1, u_2 + v_2, u_3 + v_3).$$

For example,

$$(-1, 3, 2) + (1, 1, 4) = (0, 4, 6), \quad (0, 0, 1) + (-1, 0, 1) = (-1, 0, 2).$$

Let us prove that the sum of vectors, defined *geometrically* by the parallelogram law, can be computed *algebraically* by adding corresponding components. We pass planes P, Q, R through $\mathbf{u}$, $\mathbf{v}$, and $\mathbf{w} = \mathbf{u} + \mathbf{v}$ parallel to the x_3, x_1-plane (Fig. 1.7). They meet the x_2-axis at u_2, v_2, and w_2. Because $\mathbf{vw}$ and $\mathbf{0u}$ are parallel, the directed distance from Q to R equals the directed distance from the x_3, x_1-plane to P. Hence $w_2 - v_2 = u_2$, that is, $w_2 = u_2 + v_2$. Similarly $w_1 = u_1 + v_1$ and $w_3 = u_3 + v_3$.

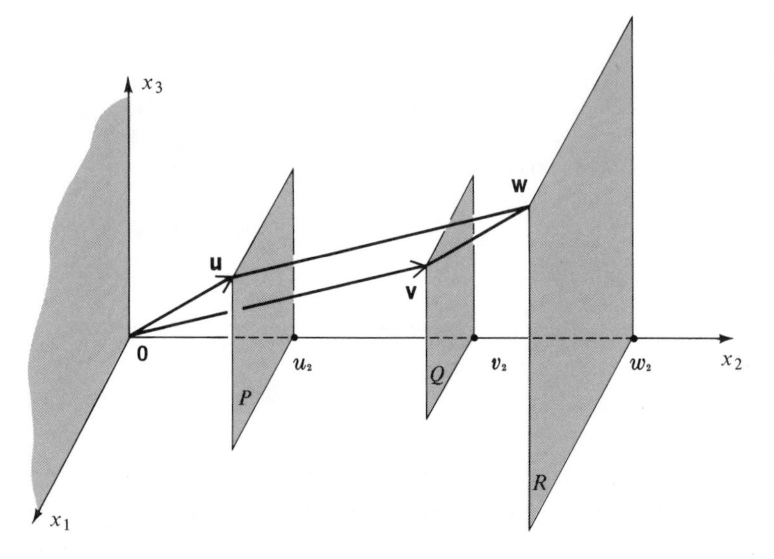

FIG. 1.7 proof of componentwise addition

Multiplication by a Scalar

Let $\mathbf{v}$ be a vector and let a be a number (scalar). We define the product $a\mathbf{v}$ to be the vector whose length is $|a|$ times the length of $\mathbf{v}$ and which points in the same direction as $\mathbf{v}$ if $a > 0$, in the opposite direction if $a < 0$. If $a = 0$, then $a\mathbf{v} = \mathbf{0}$.

The physical idea behind this definition is simple. If a particle moving in a

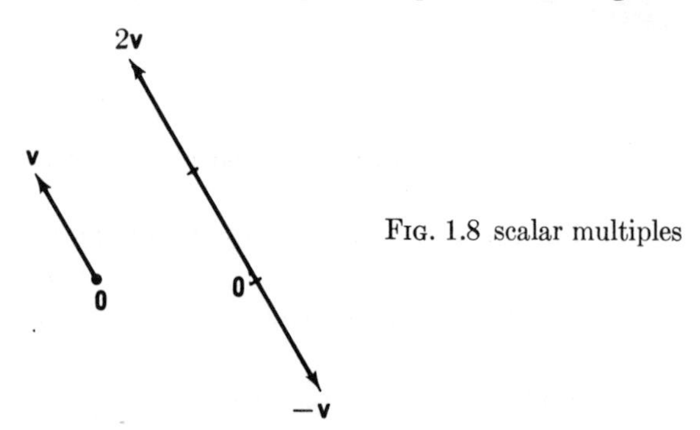

FIG. 1.8 scalar multiples

certain direction doubles its speed, its velocity vector is doubled; if a horse pulling a cart in a certain direction triples its effort, the force vector triples. Figure 1.8 illustrates multiples of a vector.

Scalar multiples are computed in components by the following rule.

$$a(v_1, v_2, v_3) = (av_1, av_2, av_3).$$

This rule is proved by similar triangles (Fig. 1.9). The triangle $\mathbf{0}v_2\mathbf{v}$ is similar to $\mathbf{0}w_2\mathbf{w}$, hence $w_2 = av_2$, etc.

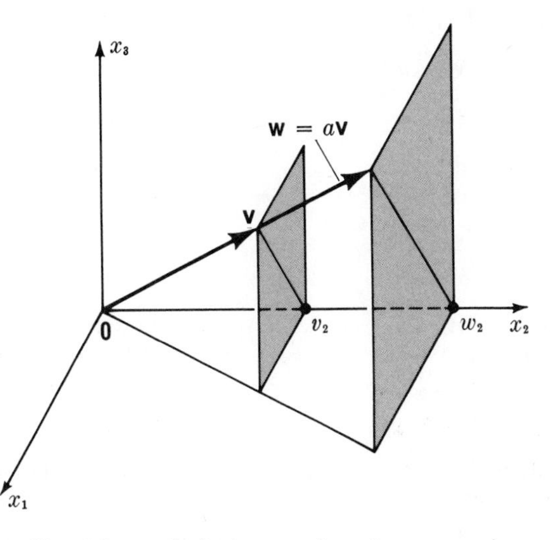

FIG. 1.9 proof of $a(v_1, v_2, v_3) = (av_1, av_2, av_3)$

The difference $\mathbf{v} - \mathbf{w}$ of two vectors is defined by

$$\mathbf{v} - \mathbf{w} = \mathbf{v} + (-\mathbf{w}).$$

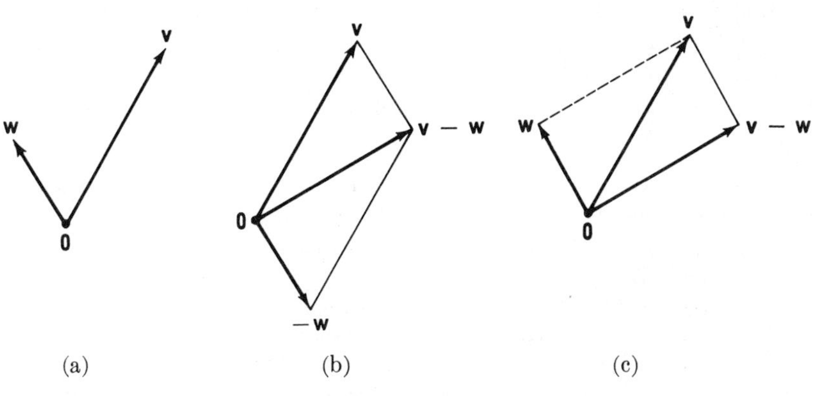

FIG. 1.10 difference of vectors

See Fig. 1.10. (The vector $-\mathbf{w}$ has the same length as $\mathbf{w}$ but points in the opposite direction.)

The segment from the tip of $\mathbf{w}$ to the tip of $\mathbf{v}$ (the dashed line in Fig. 1.10c) has the same length and direction as $\mathbf{v} - \mathbf{w}$. Hence if two points are represented by vectors $\mathbf{v}$ and $\mathbf{w}$, the distance between them is the length of $\mathbf{v} - \mathbf{w}$.

The basic rules of vector algebra follow directly from the coordinate formulas for addition and multiplication by a scalar.

Rules of Vector Algebra

$\mathbf{v} + \mathbf{0} = \mathbf{0} + \mathbf{v} = \mathbf{v}$	$\mathbf{v} + (-\mathbf{v}) = (-\mathbf{v}) + \mathbf{v} = \mathbf{0}$
$\mathbf{u} + \mathbf{v} = \mathbf{v} + \mathbf{u}$	$\mathbf{u} + (\mathbf{v} + \mathbf{w}) = (\mathbf{u} + \mathbf{v}) + \mathbf{w}$
$0\mathbf{v} = \mathbf{0} \quad 1\mathbf{v} = \mathbf{v}$	$a(b\mathbf{v}) = (ab)\mathbf{v}$
$(a + b)\mathbf{v} = a\mathbf{v} + b\mathbf{v}$	$a(\mathbf{v} + \mathbf{w}) = a\mathbf{v} + a\mathbf{w}$

EXERCISES

Draw axes as in Fig. 1.1a and locate each point accurately:

1. $(1, 2, 3)$, $(1, 3, 4)$
2. $(2, 4, 1)$, $(2, -4, 1)$
3. $(-1, 2, 1)$, $(2, 2, -1)$
4. $(1, -3, 3)$, $(3, 2, -2)$
5. $(4, 6, -1)$, $(-4, -6, 1)$
6. $(0, 0, -3)$, $(-2, -5, -3)$.

Draw the parallelepiped with edges parallel to the axes and locate the vertices. The ends of a diagonal are:

7. $(0, 0, 0)$, $(2, 3, 1)$
8. $(4, 2, 3)$, $(1, 1, 1)$.

Are the points collinear?

9. $(0, 0, 0)$, $(1, 3, 2)$, $(2, 6, 4)$
10. $(0, 0, 0)$, $(-1, 3, -4)$, $(2, -5, 8)$
11. $(1, 1, 1)$, $(0, 1, 2)$, $(-1, -3, -5)$
12. $(1, -1, -2)$, $(-1, 2, 3)$, $(3, -4, -7)$.

Compute:

13. $(1, 2, -3) + (4, 0, 7)$
14. $(-1, -1, 0) + (3, 5, 2)$
15. $(4, 0, 7) - (1, 2, -3)$
16. $(2, 1, 1) - (3, -1, -2)$
17. $(1, 2, 3) - 6(0, 3, -1)$
18. $4[(1, -2, -7) - (1, 1, 1)]$
19. $3(1, 4, 2) - 2(2, 1, 1)$
20. $4(1, -1, 2) - 3(1, -1, 2)$.

Prove:

21. $\mathbf{u} + \mathbf{v} = \mathbf{v} + \mathbf{u}$
22. $\mathbf{u} + (\mathbf{v} + \mathbf{w}) = (\mathbf{u} + \mathbf{v}) + \mathbf{w}$
23. $(a + b)\mathbf{v} = a\mathbf{v} + b\mathbf{v}$
24. $a(\mathbf{v} + \mathbf{w}) = a\mathbf{v} + a\mathbf{w}$.
25. Show that $\frac{1}{2}(\mathbf{v} + \mathbf{w})$ is the midpoint of the segment from $\mathbf{v}$ to $\mathbf{w}$. [*Hint:* Use the parallelogram law.]
26. (cont.) Use Ex. 25 to show that the segments joining the midpoints of opposite sides of a (skew) quadrilateral bisect each other. [*Hint:* $(\mathbf{u} + \mathbf{v}) + (\mathbf{w} + \mathbf{z}) = (\mathbf{v} + \mathbf{w}) + (\mathbf{u} + \mathbf{z})$.]
27. Show that $\frac{1}{3}(\mathbf{u} + \mathbf{v} + \mathbf{w})$ is the intersection of the medians of the triangle with vertices $\mathbf{u}$, $\mathbf{v}$, and $\mathbf{w}$.

28*. In a tetrahedron, prove that the four lines joining each vertex to the centroid (intersection of the medians) of the opposite face are concurrent.

29*. Space billiards—no gravity. The astronaut cues a ball toward the corner of a rectangular room, with velocity **v**. The ball misses the corner, but rebounds off of each of the three adjacent walls. Find its returning velocity vector.

2. LENGTH AND DOT PRODUCT

The **length** $|\mathbf{v}|$ of a vector $\mathbf{v} = (v_1, v_2, v_3)$ is its distance from the origin. By regarding this distance as the length of the diagonal of a rectangular solid (Fig. 2.1a), we see that

$$|\mathbf{v}|^2 = v_1^2 + v_2^2 + v_3^2.$$

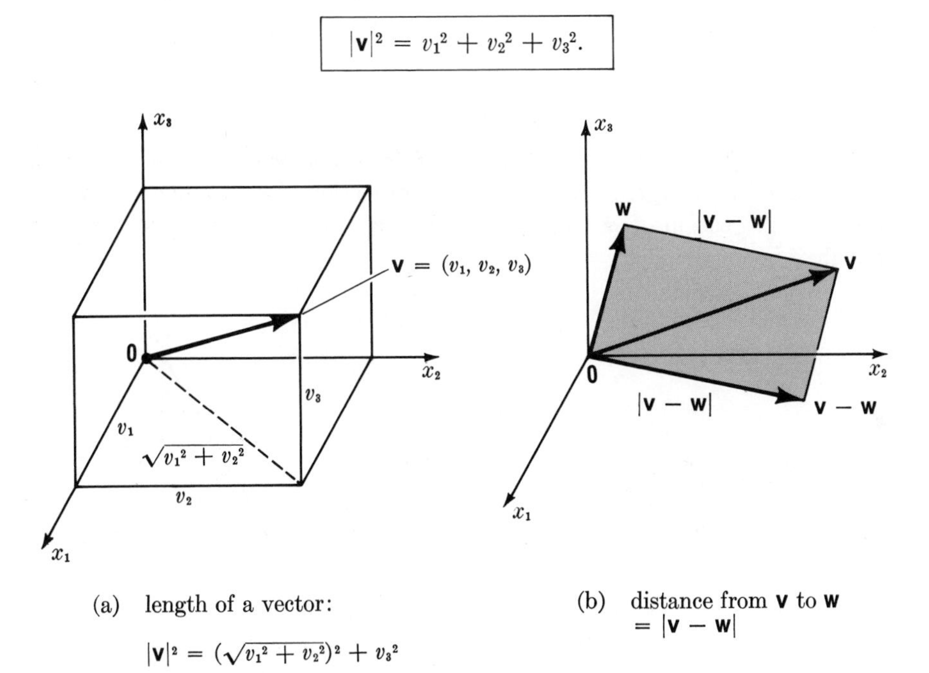

(a) length of a vector:

$$|\mathbf{v}|^2 = (\sqrt{v_1^2 + v_2^2})^2 + v_3^2$$
$$= v_1^2 + v_2^2 + v_3^2$$

(b) distance from **v** to **w** $= |\mathbf{v} - \mathbf{w}|$

FIG. 2.1 length and distance

Vector length has the following properties:

$$|\mathbf{0}| = 0, \qquad |\mathbf{v}| > 0 \quad \text{if} \quad \mathbf{v} \neq \mathbf{0},$$
$$|a\mathbf{v}| = |a| \cdot |\mathbf{v}|,$$
$$|\mathbf{v} + \mathbf{w}| \leq |\mathbf{v}| + |\mathbf{w}| \qquad \text{(triangle inequality)}.$$

The distance between two points **v** and **w** is equal to $|\mathbf{v} - \mathbf{w}|$. See Fig. 2.1b. If $\mathbf{v} = (v_1, v_2, v_3)$ and $\mathbf{w} = (w_1, w_2, w_3)$, then $\mathbf{v} - \mathbf{w} = (v_1 - w_1, v_2 - w_2,$

$v_3 - w_3$). Therefore we have the distance formula:

Distance Formula The distance between two points $\mathbf{v}$ and $\mathbf{w}$ is $|\mathbf{v} - \mathbf{w}|$, where

$$|\mathbf{v} - \mathbf{w}|^2 = (v_1 - w_1)^2 + (v_2 - w_2)^2 + (v_3 - w_3)^2.$$

Dot Product

There is another important vector operation, the **inner product** or **dot product** of two vectors. Let $\mathbf{v}$ and $\mathbf{w}$ be vectors, and let θ be the angle between them (Fig. 2.2a). Define

$$\mathbf{v} \cdot \mathbf{w} = |\mathbf{v}| \cdot |\mathbf{w}| \cos \theta.$$

Since $\cos(-\theta) = \cos \theta$, you can measure θ from $\mathbf{v}$ to $\mathbf{w}$ or from $\mathbf{w}$ to $\mathbf{v}$. Note (Fig. 2.2b, c) that $|\mathbf{w}| \cos \theta$ is the (signed) projection of $\mathbf{w}$ on $\mathbf{v}$, hence $\mathbf{v} \cdot \mathbf{w}$ is $|\mathbf{v}|$ times the projection of $\mathbf{w}$ on $\mathbf{v}$. If $\mathbf{v} = \mathbf{0}$ or $\mathbf{w} = \mathbf{0}$, we define $\mathbf{v} \cdot \mathbf{w} = 0$ even though θ is not defined.

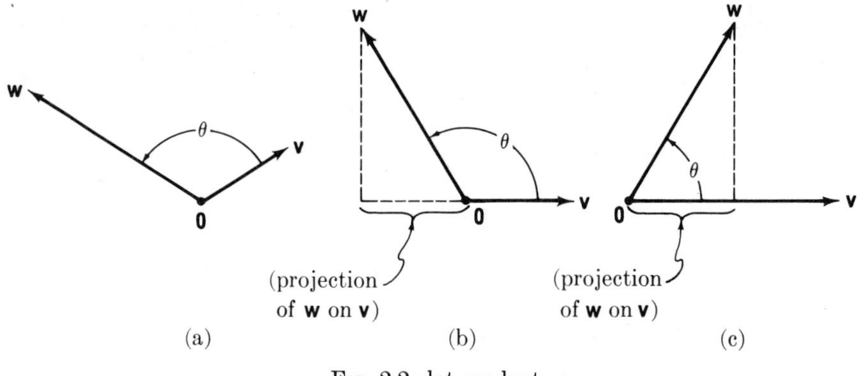

FIG. 2.2 dot product

IMPORTANT: The dot product of two vectors is a *scalar* (number), not a vector.

The numerical rule for computing dot products is

$$\mathbf{v} \cdot \mathbf{w} = v_1 w_1 + v_2 w_2 + v_3 w_3,$$

an important formula. Let us prove it. See Fig. 2.3. By the Law of Cosines,

$$|\mathbf{v} - \mathbf{w}|^2 = |\mathbf{v}|^2 + |\mathbf{w}|^2 - 2\,|\mathbf{v}|\,|\mathbf{w}| \cos \theta = |\mathbf{v}|^2 + |\mathbf{w}|^2 - 2\mathbf{v} \cdot \mathbf{w}.$$

Hence

$$\mathbf{v}\cdot\mathbf{w} = \frac{1}{2}\left[|\mathbf{v}|^2 + |\mathbf{w}|^2 - |\mathbf{v} - \mathbf{w}|^2\right]$$

$$= \frac{1}{2}\left[|(v_1, v_2, v_3)|^2 + |(w_1, w_2, w_3)|^2 - |(v_1 - w_1, v_2 - w_2, v_3 - w_3)|^2\right]$$

$$= \frac{1}{2}\left[(v_1^2 + v_2^2 + v_3^2) + (w_1^2 + w_2^2 + w_3^2) - (v_1 - w_1)^2 - (v_2 - w_2)^2 - (v_3 - w_3)^2\right]$$

$$= v_1w_1 + v_2w_2 + v_3w_3.$$

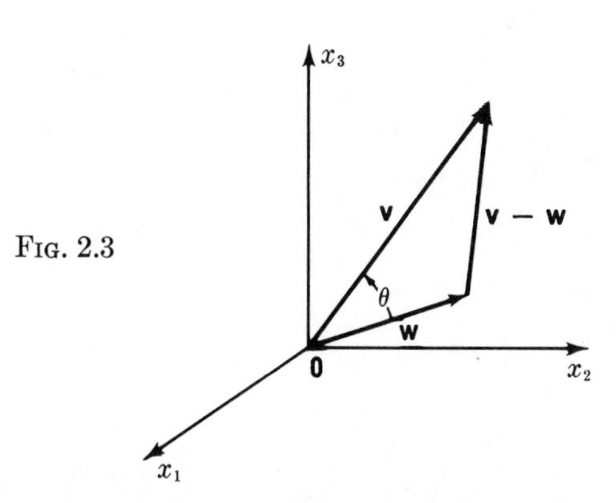

FIG. 2.3

The main algebraic properties of the inner product follow easily from the formula $\mathbf{v}\cdot\mathbf{w} = v_1w_1 + v_2w_2 + v_3w_3$:

$$\mathbf{v}\cdot\mathbf{w} = \mathbf{w}\cdot\mathbf{v} \qquad (a\mathbf{v})\cdot\mathbf{w} = \mathbf{v}\cdot(a\mathbf{w}) = a(\mathbf{v}\cdot\mathbf{w})$$
$$(\mathbf{u} + \mathbf{v})\cdot\mathbf{w} = \mathbf{u}\cdot\mathbf{w} + \mathbf{v}\cdot\mathbf{w}.$$

Two vectors $\mathbf{v}$ and $\mathbf{w}$ are perpendicular if $\theta = \pi/2$, i.e., if $\cos\theta = 0$. This can be expressed very neatly as follows:

The condition for vectors $\mathbf{v}$ and $\mathbf{w}$ to be perpendicular is

$$\mathbf{v}\cdot\mathbf{w} = 0.$$

(The vector $\mathbf{0}$ is considered perpendicular to every vector.)

For example, $(1, 2, 3)$ and $(-1, -1, 1)$ are perpendicular because

$$(1, 2, 3)\cdot(-1, -1, 1) = -1 - 2 + 3 = 0.$$

There is a connection between lengths and dot products. The dot product of a vector $\mathbf{v}$ with itself is $\mathbf{v}\cdot\mathbf{v} = |\mathbf{v}|^2 \cos 0 = |\mathbf{v}|^2$.

For any vector $\mathbf{v}$,

$$\mathbf{v}\cdot\mathbf{v} = |\mathbf{v}|^2 = v_1^2 + v_2^2 + v_3^2.$$

From the dot product can be found the angle θ between any two non-zero vectors $\mathbf{v}$ and $\mathbf{w}$. Indeed,

$$\cos\theta = \frac{\mathbf{v}\cdot\mathbf{w}}{|\mathbf{v}|\,|\mathbf{w}|}.$$

EXAMPLE 2.1

Find the angle between $\mathbf{v} = (1, 2, 1)$ and $\mathbf{w} = (3, -1, 1)$.

Solution:

$$\mathbf{v}\cdot\mathbf{w} = 3 - 2 + 1 = 2,$$

$$|\mathbf{v}|^2 = 1 + 4 + 1 = 6, \qquad |\mathbf{w}|^2 = 9 + 1 + 1 = 11.$$

Hence

$$\cos\theta = \frac{2}{\sqrt{6}\sqrt{11}}.$$

Answer: $\text{arc cos}\,\dfrac{2}{\sqrt{66}}$.

EXAMPLE 2.2

The point $(1, 1, 2)$ is joined to the points $(1, -1, -1)$ and $(3, 0, 4)$ by lines L_1 and L_2. What is the angle θ between these lines?

Solution: The vector

$$\mathbf{v} = (1, -1, -1) - (1, 1, 2) = (0, -2, -3)$$

is parallel to L_1 (but starts at $\mathbf{0}$). Likewise

$$\mathbf{w} = (3, 0, 4) - (1, 1, 2) = (2, -1, 2)$$

is parallel to L_2. Hence

$$\cos\theta = \frac{\mathbf{v}\cdot\mathbf{w}}{|\mathbf{v}|\,|\mathbf{w}|} = \frac{0 + 2 - 6}{\sqrt{0 + 4 + 9}\sqrt{4 + 1 + 4}} = \frac{-4}{\sqrt{13}\sqrt{9}}.$$

Answer: $\theta = \text{arc cos}\left(\dfrac{-4}{3\sqrt{13}}\right)$.

NOTE: When we find $\cos\theta < 0$ for an angle θ between two lines, then

θ is not the smaller angle between the lines, but its supplement. The basic fact here is that $\cos(\pi - \theta) = -\cos\theta$.

Direction Cosines

It is customary to use the notation

$$\mathbf{i} = (1, 0, 0), \qquad \mathbf{j} = (0, 1, 0), \qquad \mathbf{k} = (0, 0, 1)$$

for the three unit-length vectors along the positive coordinate axes (Fig. 2.4). If $\mathbf{v}$ is any vector, then

$$\begin{aligned} \mathbf{v} &= (v_1, v_2, v_3) \\ &= v_1(1, 0, 0) + v_2(0, 1, 0) + v_3(0, 0, 1) \\ &= v_1\mathbf{i} + v_2\mathbf{j} + v_3\mathbf{k}. \end{aligned}$$

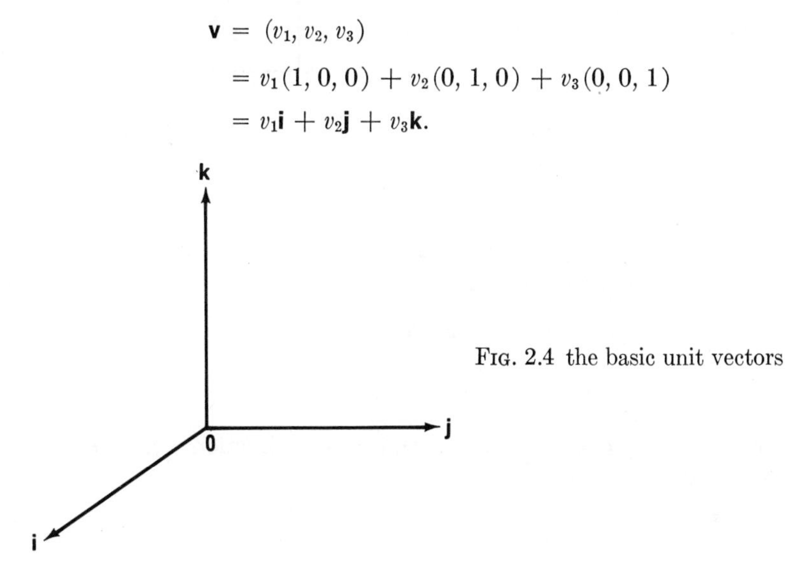

FIG. 2.4 the basic unit vectors

Thus $\mathbf{v}$ is the sum of three vectors $v_1\mathbf{i}$, $v_2\mathbf{j}$, $v_3\mathbf{k}$ which lie along the three coordinate axes. The components v_1, v_2, v_3 can be interpreted as dot products:

$$\mathbf{v}\cdot\mathbf{i} = (v_1, v_2, v_3)\cdot(1, 0, 0) = v_1.$$

Similarly, $v_2 = \mathbf{v}\cdot\mathbf{j}$ and $v_3 = \mathbf{v}\cdot\mathbf{k}$.

Now suppose $\mathbf{u}$ is a **unit vector**, i.e., a vector of length one (Fig. 2.5a). Let α be the angle from $\mathbf{i}$ to $\mathbf{u}$. Define β and γ similarly. Then $\mathbf{u}\cdot\mathbf{i} = \cos\alpha$, $\mathbf{u}\cdot\mathbf{j} = \cos\beta$, and $\mathbf{u}\cdot\mathbf{k} = \cos\gamma$. Hence

$$\mathbf{u} = \cos\alpha\,\mathbf{i} + \cos\beta\,\mathbf{j} + \cos\gamma\,\mathbf{k} = (\cos\alpha, \cos\beta\ \cos\gamma).$$

Since $|\mathbf{u}| = 1$,

$$\cos^2\alpha + \cos^2\beta + \cos^2\gamma = 1.$$

Unit vectors are direction indicators. Any non-zero vector $\mathbf{v}$ is a positive multiple of a unit vector $\mathbf{u}$ in the same direction as $\mathbf{v}$. In fact $\mathbf{v} = |\mathbf{v}|\,\mathbf{u}$, so

$$\mathbf{u} = \frac{1}{|\mathbf{v}|}\mathbf{v} \qquad (\mathbf{v} \neq \mathbf{0}).$$

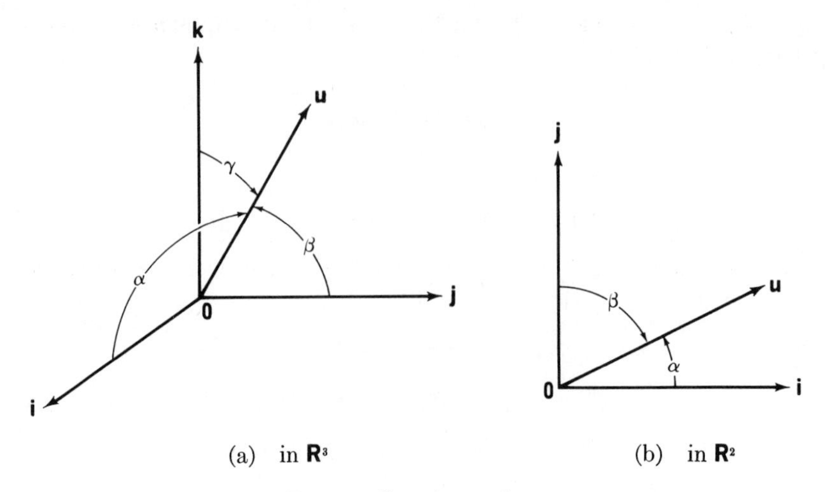

(a) in $\mathbf{R}^3$ (b) in $\mathbf{R}^2$

Fig. 2.5 direction cosines

> Each non-zero vector $\mathbf{v}$ can be expressed as
>
> $$\mathbf{v} = |\mathbf{v}|\,\mathbf{u}, \qquad \mathbf{u} \text{ a unit vector,}$$
>
> or as
>
> $$\mathbf{v} = |\mathbf{v}|\,(\cos\alpha, \cos\beta, \cos\gamma).$$
>
> The numbers $\cos\alpha$, $\cos\beta$, $\cos\gamma$ are called the **direction cosines** of $\mathbf{v}$. They satisfy
>
> $$\cos^2\alpha + \cos^2\beta + \cos^2\gamma = 1.$$

If $\mathbf{u}$ is a unit vector (Fig. 2.5b) in the plane of $\mathbf{i}$ and $\mathbf{j}$, then

$$\mathbf{u} = \cos\alpha\,\mathbf{i} + \cos\beta\,\mathbf{j}.$$

Since $\mathbf{u}$ is a unit vector, $\cos^2\alpha + \cos^2\beta = 1$. But, as is seen in the figure, $\cos\beta = \sin\alpha$. Therefore, the preceding equation simply says

$$\cos^2\alpha + \sin^2\alpha = 1.$$

Summary

Addition of Vectors

$$(v_1, v_2, v_3) + (w_1, w_2, w_3) = (v_1 + w_1, v_2 + w_2, v_3 + w_3).$$

Multiplication by a Scalar:

$$a(v_1, v_2, v_3) = (av_1, av_2, av_3).$$

Length:

$$|\mathbf{v}|^2 = \mathbf{v}\cdot\mathbf{v} = v_1^2 + v_2^2 + v_3^2.$$

Dot Product:

$$\mathbf{v} \cdot \mathbf{w} = |\mathbf{v}|\,|\mathbf{w}| \cos\theta = v_1 w_1 + v_2 w_2 + v_3 w_3.$$

Vectors **i**, **j**, **k**:

These are unit vectors in the direction of the positive x-axis, y-axis, z-axis, respectively. If $\mathbf{v} = (v_1, v_2, v_3)$, then $\mathbf{v} = v_1\mathbf{i} + v_2\mathbf{j} + v_3\mathbf{k}$, where $v_1 = \mathbf{v}\cdot\mathbf{i}$, $v_2 = \mathbf{v}\cdot\mathbf{j}$, $v_3 = \mathbf{v}\cdot\mathbf{k}$.

Direction Cosines:

If $\mathbf{u}$ is a unit vector, then $\mathbf{u} = \cos\alpha\,\mathbf{i} + \cos\beta\,\mathbf{j} + \cos\gamma\,\mathbf{k}$, where α, β, γ are the angles to $\mathbf{u}$ from $\mathbf{i}$, $\mathbf{j}$, $\mathbf{k}$, respectively. Furthermore $\cos^2\alpha + \cos^2\beta + \cos^2\gamma = 1$. Any non-zero vector $\mathbf{v}$ can be written as $\mathbf{v} = |\mathbf{v}|\mathbf{u} = |\mathbf{v}|(\cos\alpha, \cos\beta, \cos\gamma)$. The numbers $\cos\alpha$, $\cos\beta$, $\cos\gamma$ are the direction cosines of $\mathbf{v}$.

EXERCISES

Compute:

1. $(8, 2, 1)\cdot(3, 0, 5)$
2. $(-1, -1, -1)\cdot(1, 2, 3)$
3. $(1, 0, 2)\cdot[(1, 4, 1) + (2, 0, -3)]$
4. $|(2, -4, 7)|$
5. $|3\mathbf{i} - \mathbf{j} + \mathbf{k}|$
6. $|(-1, -1, 0) - (3, 5, 2)|$
7. $|\frac{1}{3}\sqrt{3}(-1, 1, 1)|$
8. $[3\mathbf{j} - (1, 1, 2)]\cdot(4\mathbf{j} - \mathbf{k})$.

Find the angle between the vectors:

9. $(4, 3, 0)$, $(-3, 0, 4)$
10. $(1, 2, 2)$, $(-2, 1, -2)$
11. $(6, 1, 5)$, $(-2, -3, 3)$
12. $(-5, 6, 1)$, $(2, 3, -8)$
13. $(1, 1, -1)$, $(2, 0, 4)$
14. $(2, 2, 2)$, $(-2, 2, -2)$.

Compute the distance between the points:

15. $(0, 1, 2)$, $(5, -3, 1)$
16. $(1, 1, 1)$, $(1, -1, 2)$
17. $(7, 0, 0)$, $(2, 3, 4)$
18. $(8, 5, -1)$, $(7, 9, 3)$.

Find the direction cosines:

19. $(1, 0, 1)$
20. $(-1, -1, -1)$
21. $(2, 1, -3)$
22. $(4, -7, -4)$.

23. Find two non-collinear vectors perpendicular to $(1, -1, 2)$.
24. Find the angle between the line joining $(0, 0, 0)$ to $(1, 1, 1)$ and the line joining $(1, 0, 0)$ to $(0, 1, 0)$.
25. Prove the **Cauchy–Schwarz inequality:**
$$|\mathbf{v}\cdot\mathbf{w}| \le |\mathbf{v}|\cdot|\mathbf{w}|.$$
When does equality hold?
26. (cont.) Now prove the triangle inequality:
$$|\mathbf{v} + \mathbf{w}| \le |\mathbf{v}| + |\mathbf{w}|.$$

[*Hint:*

$$|\mathbf{v}+\mathbf{w}|^2 = |(\mathbf{v}+\mathbf{w})\cdot(\mathbf{v}+\mathbf{w})| = |(\mathbf{v}+\mathbf{w})\cdot\mathbf{v} + (\mathbf{v}+\mathbf{w})\cdot\mathbf{w}|$$
$$\leq |\mathbf{v}+\mathbf{w}|\cdot|\mathbf{v}| + |\mathbf{v}+\mathbf{w}|\cdot|\mathbf{w}|.]$$

27. Prove the identity $|\mathbf{v}+\mathbf{w}|^2 - |\mathbf{v}-\mathbf{w}|^2 = 4\mathbf{v}\cdot\mathbf{w}$.
28. Let $\mathbf{u}$ be a unit vector. Show that the formula $\mathbf{v} = (\mathbf{v}\cdot\mathbf{u})\mathbf{u} - [\mathbf{v} - (\mathbf{v}\cdot\mathbf{u})\mathbf{u}]$ expresses $\mathbf{v}$ as the sum of two vectors, one parallel to $\mathbf{u}$, the other perpendicular to $\mathbf{u}$, and is the only such expression.
29. Let $\mathbf{u} = (\cos\alpha_1, \cos\alpha_2, \cos\alpha_3)$ and $\mathbf{v} = (\cos\beta_1, \cos\beta_2, \cos\beta_3)$ be two unit vectors. Show that the angle θ between them satisfies

$$\cos\theta = \cos\alpha_1\cos\beta_1 + \cos\alpha_2\cos\beta_2 + \cos\alpha_3\cos\beta_3.$$

Interpret the formula when $\alpha_3 = \beta_3 = \frac{1}{2}\pi$.

3. LINES AND PLANES

Take a non-zero vector $\mathbf{v}$. The set of all scalar multiples $\mathbf{x} = t\mathbf{v}$ of $\mathbf{v}$ is a line through the origin (Fig. 3.1a). If $\mathbf{x}_0$ is any point, then the set of all points $\mathbf{x} = \mathbf{x}_0 + t\mathbf{v}$ is a line through $\mathbf{x}_0$ parallel to the first line (Fig. 3.1b). The equation $\mathbf{x} = \mathbf{x}_0 + t\mathbf{v}$ is called a **parametric vector equation** for the line. For example

$$(x_1, x_2, x_3) = (0, 1, -2) + t(2, -1, 3)$$

is a parametric equation for the line through $(0, 1, -2)$ parallel to $(2, -1, 3)$. This vector equation is equivalent to three **parametric scalar equations:**

$$x_1 = 2t, \qquad x_2 = -t + 1, \qquad x_3 = 3t - 2.$$

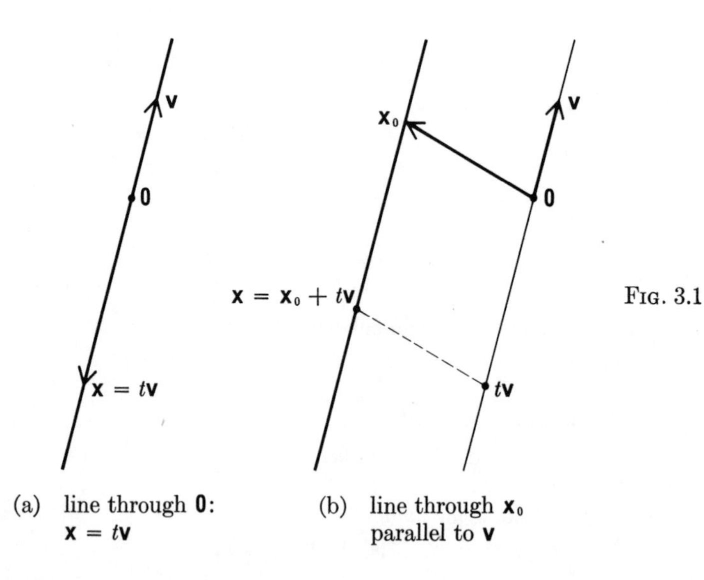

Fig. 3.1

Given a point $\mathbf{x}_0$ of $\mathbf{R}^3$, and a non-zero vector $\mathbf{v}$, the line through $\mathbf{x}_0$ parallel to $\mathbf{v}$ consists of all points

$$\mathbf{x} = \mathbf{x}_0 + t\mathbf{v},$$

where $-\infty < t < \infty$.

EXAMPLE 3.1

Find a parametric vector equation for the line through $(3, -1, 2)$ and $(4, 1, 1)$.

Solution: The line passes through $\mathbf{x}_0 = (3, -1, 2)$ and is parallel to $\mathbf{v} = (4, 1, 1) - (3, -1, 2) = (1, 2, -1)$. Hence $\mathbf{x} = (3, -1, 2) + t(1, 2, -1)$ is a parametric form for the line.

Answer: $\mathbf{x} = (3, -1, 2) + t(1, 2, -1) = (3 + t, -1 + 2t, 2 - t).$

Equation of a Plane

Let P be a plane in $\mathbf{R}^3$. Draw the line L through $\mathbf{0}$ perpendicular to P and take one of the two unit vectors along L; call it $\mathbf{n}$. See Fig. 3.2. The line L meets P in a point $p\mathbf{n}$. If $\mathbf{x}$ is any point of P, then $\mathbf{x} - p\mathbf{n}$ is perpendicular to $\mathbf{n}$, therefore

$$(\mathbf{x} - p\mathbf{n})\cdot\mathbf{n} = 0, \qquad \mathbf{x}\cdot\mathbf{n} = p\mathbf{n}\cdot\mathbf{n} = p.$$

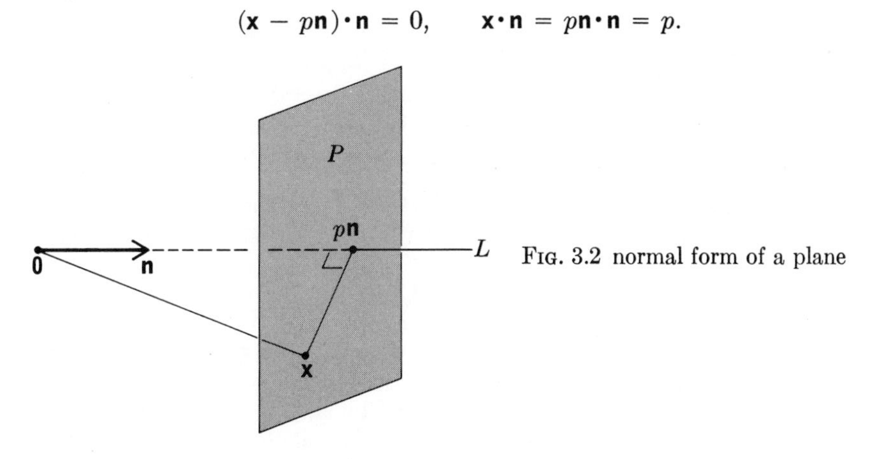

FIG. 3.2 normal form of a plane

Normal Form Each plane in $\mathbf{R}^3$ can be represented by an equation

$$\mathbf{x}\cdot\mathbf{n} = p,$$

where $\mathbf{n}$ is a unit vector perpendicular to the plane and p is a real number.

Conversely, each equation $\mathbf{x}\cdot\mathbf{n} = p$ represents a plane, the one through the point $p\mathbf{n}$ and perpendicular to $\mathbf{n}$.

It follows that each linear equation in 3 variables,

$$a_1x_1 + a_2x_2 + a_3x_3 = b \qquad (a_1^2 + a_2^2 + a_3^3 \neq 0),$$

is the equation of a plane. Just set $\mathbf{a} = (a_1, a_2, a_3)$ and $\mathbf{n} = \mathbf{a}/|\mathbf{a}|$. Then the equation, in vector notation, is

$$\mathbf{a}\cdot\mathbf{x} = b, \qquad \text{or} \qquad \mathbf{n}\cdot\mathbf{x} = \frac{b}{|a|} = p,$$

the normal form of a plane.

REMARK 1: There really are two normal forms of a plane,

$$\mathbf{x}\cdot\mathbf{n} = p, \qquad \mathbf{x}\cdot(-\mathbf{n}) = -p.$$

REMARK 2: In applications it is often more convenient if $\mathbf{n}$ is not a unit vector. Certainly it is simpler to write

$$(x_1, x_2, x_3)\cdot(1, 1, 1) = 5$$

than the equivalent normal form

$$(x_1, x_2, x_3)\cdot\tfrac{1}{3}\sqrt{3}(1, 1, 1) = \tfrac{5}{3}\sqrt{3}.$$

EXERCISES

Express in normal form:

1. $x_1 - 2x_2 + 2x_3 = 1$
2. $2x_1 + 6x_2 - 3x_3 = 14$
3. $-8x_1 + x_2 - 4x_3 = 27$
4. $3x_1 - 2x_2 - 6x_3 = 4$
5. $x_1 + x_2 + x_3 = 3$
6. $x_1 - x_2 + x_3 = -12$.
7. Find a parametric form for the line through two distinct points $\mathbf{x}_0$ and $\mathbf{x}_1$.
8. Let $\mathbf{x} = \mathbf{x}_0 + t\mathbf{u}$ and $\mathbf{x} = \mathbf{x}_1 + t\mathbf{v}$ be two parallel lines. Prove that $\mathbf{v} = c\mathbf{u}$ for some c.
9. Let $\mathbf{x}\cdot\mathbf{n} = p$ be a plane in normal form and $\mathbf{x}_0$ a point. Prove that the distance from $\mathbf{x}_0$ to the plane is $|\mathbf{x}_0\cdot\mathbf{n} - p|$.
10. (cont.) Prove that the point of the plane closest to $\mathbf{x}_0$ is $\mathbf{z} = \mathbf{x}_0 + (p - \mathbf{x}_0\cdot\mathbf{n})\mathbf{n}$.
11. Let $\mathbf{x} = \mathbf{x}_0 + t\mathbf{v}$ be a line in parametric form and let $\mathbf{x}\cdot\mathbf{n} = p$ be a plane in normal form. Prove that the line and plane are parallel if and only if $\mathbf{v}\cdot\mathbf{n} = 0$.
12. (cont.) Prove that the line is on the plane if and only if $\mathbf{v}\cdot\mathbf{n} = 0$ and $\mathbf{x}_0\cdot\mathbf{n} = p$.
13. (cont.) Suppose $\mathbf{v}\cdot\mathbf{n} \neq 0$. Prove that the point of intersection of the line and the plane is
$$\mathbf{x}_0 + [(p - \mathbf{x}_0\cdot\mathbf{n})/(\mathbf{v}\cdot\mathbf{n})]\mathbf{v}.$$
14. Let $\mathbf{x}\cdot\mathbf{m} = p$ and $\mathbf{x}\cdot\mathbf{n} = q$ be two non-parallel planes in normal form. Let θ be one of their (dihedral) angles of intersection. Determine $\cos\theta$.
15. Let $\mathbf{x} = \mathbf{x}_0 + t\mathbf{u}$ be a line in parametric form, where $\mathbf{u}$ is a unit vector, and let $\mathbf{y}_0$ be a point. Find the point on the line closest to $\mathbf{y}_0$.
16. (cont.) Prove that the distance D from $\mathbf{y}_0$ to the line satisfies $D^2 = |\mathbf{x}_0 - \mathbf{y}_0|^2 - [(\mathbf{x}_0 - \mathbf{y}_0)\cdot\mathbf{u}]^2$.

4. LINEAR SYSTEMS AND INTERSECTIONS

Suppose we are given three planes

$$\mathbf{a}_1 \cdot \mathbf{x} = d_1, \qquad \mathbf{a}_2 \cdot \mathbf{x} = d_2, \qquad \mathbf{a}_3 \cdot \mathbf{x} = d_3.$$

In general, they will intersect in a single point $\mathbf{x}$. How can we find this point? In coordinates, the problem is to solve simultaneously three linear equations

$$\begin{cases} a_1x + b_1y + c_1z = d_1 \\ a_2x + b_2y + c_2z = d_2 \\ a_3x + b_3y + c_3z = d_3 \end{cases}$$

for x, y, z, where $a_1, \cdots, d_3$ are given constants.

This is typical of problems in a subject called *Linear Algebra,* which we shall take up in Chapter 7. While it is beyond the scope of this book to go into the general theory, we shall give a practical method of solution.

EXAMPLE 4.1

Find the solution (x, y, z) of the system

$$\begin{cases} 2x - y + z = 4 \\ 3y + 2z = -1 \\ -z = 3. \end{cases}$$

Solution: By the third equation, $z = -3$. Substitute this into the first two equations. The result is a new system of two equations for x and y:

$$\begin{cases} 2x - y = 4 - (-3) = 7 \\ 3y = -1 - 2(-3) = 5. \end{cases}$$

By the second equation, $y = \frac{5}{3}$. Substitute this into the first equation; the result is a single equation for x:

$$2x = 7 + \tfrac{5}{3} = \tfrac{26}{3}.$$

Its solution is $x = \frac{13}{5}$.

Answer: $(\frac{13}{3}, \frac{5}{3}, -3)$.

This example was very easy because we could solve for the unknowns one at a time. To solve a more general system, we reduce it to a system of this type by eliminating the unknowns one by one.

EXAMPLE 4.2

Find the solution of the system

$$\begin{cases} 2x - y + z = 4 \\ 2x + 2y + 3z = 3 \\ 6x - 9y - 2z = 17. \end{cases}$$

Solution: Eliminate x from the second and third equations. Subtract the first equation from the second, and subtract 3 times the first equation from the third; the result is an equivalent system of three equations (the first the same as before):

$$\begin{cases} 2x - y + z = 4 \\ 3y + 2z = -1 \\ -6y - 5z = 5. \end{cases}$$

Now eliminate y from the third equation. Add twice the second equation to the third, but keep the first two equations:

$$\begin{cases} 2x - y + z = 4 \\ 3y + 2z = -1 \\ -z = 3. \end{cases}$$

This is the system in Example 4.1, which we can solve by elimination.

Answer: $x = \frac{13}{3}$, $y = \frac{5}{3}$, $z = -3$.

The method of elimination works for a very simple reason: adding a constant multiple of one equation to another equation does not affect the solution of the system. To verify this, suppose $\mathbf{a}_1 \cdot \mathbf{x} = d_1$ and $\mathbf{a}_2 \cdot \mathbf{x} = d_2$ are two linear equations. If c is any number, then the two systems

$$\begin{cases} \mathbf{a}_1 \cdot \mathbf{x} = d_1 \\ \mathbf{a}_2 \cdot \mathbf{x} = d_2 \end{cases} \quad \text{and} \quad \begin{cases} \mathbf{a}_1 \cdot \mathbf{x} = d_1 \\ (\mathbf{a}_2 + c\mathbf{a}_1) \cdot \mathbf{x} = d_2 + cd_1 \end{cases}$$

have *exactly* the same solutions. For if $\mathbf{x}$ satisfies $\mathbf{a}_1 \cdot \mathbf{x} = d_1$ and $\mathbf{a}_2 \cdot \mathbf{x} = d_2$, then

$$(\mathbf{a}_2 + c\mathbf{a}_1) \cdot \mathbf{x} = \mathbf{a}_2 \cdot \mathbf{x} + c(\mathbf{a}_1 \cdot \mathbf{x}) = d_2 + cd_1.$$

Conversely, if $\mathbf{x}$ satisfies $\mathbf{a}_1 \cdot \mathbf{x} = d_1$ and $(\mathbf{a}_2 + c\mathbf{a}_1) \cdot \mathbf{x} = d_2 + cd_1$, then

$$\mathbf{a}_2 \cdot \mathbf{x} = (\mathbf{a}_2 + c\mathbf{a}_1) \cdot \mathbf{x} - c(\mathbf{a}_1 \cdot \mathbf{x}) = (d_2 + cd_1) - cd_1 = d_2.$$

Thus the two systems are equivalent; they have precisely the same solutions.

PRACTICAL HINT: When you apply the method of elimination, you do not *have* to eliminate first x and then y. Eliminate any two of the unknowns in an order that makes the computation easiest.

EXAMPLE 4.3

Solve the system

$$\begin{cases} 3x + y - 2z = 4 \\ -5x \qquad + 2z = 5 \\ -7x - y + 3z = -2. \end{cases}$$

Solution: Since y is missing from the second equation, add the first to the third; then y is eliminated from two equations:

$$\begin{cases} 3x + y - 2z = 4 \\ -5x \qquad + 2z = 5 \\ -4x \qquad + z = 2. \end{cases}$$

Now add -2 times the third to the second; this eliminates z:

$$\begin{cases} 3x + y - 2z = 4 \\ 3x \qquad\qquad = 1 \\ -4x \qquad + z = 2. \end{cases}$$

By the second equation, $x = \frac{1}{3}$. By the third equation, $z = 2 + 4x = 2 + 4(\frac{1}{3}) = \frac{10}{3}$. Finally,

$$y = 4 - 3x + 2z = 4 - 1 + \tfrac{20}{3} = \tfrac{29}{3}.$$

Answer: $x = \frac{1}{3}$, $y = \frac{29}{3}$, $z = \frac{10}{3}$.

Certain systems of equations do not have precisely one solution. There may be either no solutions at all, or more than one solution. In the latter case it turns out that there are infinitely many. Both cases can be handled by elimination.

Inconsistent Systems

Sometimes the elimination process leads to an equation

$$0x + 0y + 0z = d,$$

where $d \neq 0$. Obviously no choice of x, y, and z will satisfy this equation. Then the system simply has no solution, and it is called an **inconsistent** system.

An example with two unknowns will suffice to illustrate this. Look at the system

$$\begin{cases} x + 3y = -1 \\ 2x + 6y = 3. \end{cases}$$

Add -2 times the first equation to the second. The result is the system

$$\begin{cases} x + 3y = -1. \\ \quad\;\; 0 = 5. \end{cases}$$

There is no solution, for if there were, then $0 = 5$ would be a correct statement.

Underdetermined Systems

Some systems have more than one solution. This happens when one of the equations is a consequence of the other two, so that really there are only two (or fewer) equations. Such systems are called **underdetermined.**

First consider an example in two unknowns:

$$\begin{cases} x + y = 1 \\ 2x + 2y = 2. \end{cases}$$

The second equation is obviously twice the first. If we try to eliminate x by adding -2 times the first to the second, the result is

$$\begin{cases} x + y = 1 \\ \quad\; 0 = 0. \end{cases}$$

The second equation in this equivalent system gives no information whatever. *Any* solution of the first equation is a solution of the system. Thus there are infinitely many: each point (x, y) on the line $x + y = 1$ is a solution. It is convenient to express these solutions in parametric form. A parametric representation of the line is $(x, y) = (t, 1 - t)$, where $-\infty < t < \infty$. Hence, the solution of the system is

$$x = t, \qquad y = 1 - t, \qquad -\infty < t < \infty.$$

Now consider the example

$$\begin{cases} x + y + z = 1 \\ x - 2y + 2z = 4 \\ 2x - y + 3z = 5. \end{cases}$$

Eliminate x from the second and third equations:

$$\begin{cases} x + y + z = 1 \\ -3y + z = 3 \\ -3y + z = 3. \end{cases}$$

The last two equations both say the same thing. Therefore the system is equivalent to the system

$$\begin{cases} x + y + z = 1 \\ -3y + z = 3. \end{cases}$$

and no further elimination is possible. To get a parametric solution, set $y = t$. Then $z = 3 + 3y = 3 + 3t$, and

$$x = 1 - y - z = 1 - t - (3 + 3t) = -2 - 4t.$$

The most general solution is

$$(x, y, z) = (-2 - 4t, t, 3 + 3t),$$

where $-\infty < t < \infty$. The set of solutions is a line, $(x, y, z) = (-2, 0, 3) + t(-4, 1, 3)$.

Finally, consider the system

$$\begin{cases} x + y + z = 1 \\ 2x + 2y + 2z = 2 \\ 3x + 3y + 3z = 3. \end{cases}$$

This system is obviously equivalent to the single equation $x + y + z = 1$. Therefore the set of solutions (x, y, z) is a plane in space. For a parametric solution, set $x = s$ and $y = t$. Then $z = 1 - s - t$. The general solution is

$$(x, y, z) = (s, t, 1 - s - t).$$

Intersections of Planes

Now we can close the books on intersection of planes. If $\mathbf{a}_1 \cdot \mathbf{x} = d_1$, $\mathbf{a}_2 \cdot \mathbf{x} = d_2$, $\mathbf{a}_3 \cdot \mathbf{x} = d_3$ are three planes in space, we find the points common to the three planes by the elimination method. Generally, there is a single common point (Fig. 4.1a). However, if the planes are parallel or if one is parallel to the intersection of the other two, then there is no common point (Fig. 4.1c). In this case, the corresponding system of equations is inconsistent. For example, the system

$$\begin{cases} x + y + z = 1 \\ x + y + z = 2 \\ 3x - 2y + 4z = 7 \end{cases}$$

is obviously inconsistent; the first two equations cannot both be satisfied. Geometrically, the first two planes are parallel.

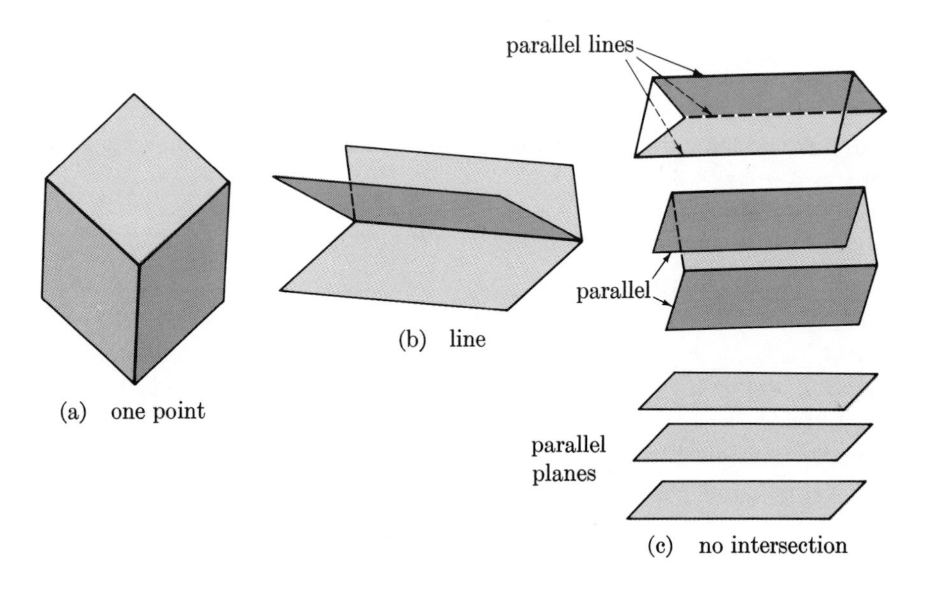

FIG. 4.1 possible intersections of three distinct planes

The planes have more than one common point if they pass through a common line, or if two or all of them coincide. In this case, the corresponding system of equations is underdetermined. For example the system

$$\begin{cases} x - 2y + 3z = 5 \\ 8x + 7y + z = 2 \\ 2x - 4y + 6z = 10 \end{cases}$$

is underdetermined; the third equation is twice the first. Geometrically, the first and third planes coincide. The system represents two distinct planes that have a line in common (Fig. 4.1b).

Review of Determinants

An important computational tool in solving linear systems is the use of determinants. We recall from elementary algebra the definitions of **determinants** of orders two and three:

$$\begin{vmatrix} a_1 & b_1 \\ a_2 & b_2 \end{vmatrix} = a_1b_2 - a_2b_1,$$

$$\begin{vmatrix} a_1 & b_1 & c_1 \\ a_2 & b_2 & c_2 \\ a_3 & b_3 & c_3 \end{vmatrix} = a_1b_2c_3 + a_2b_3c_1 + a_3b_1c_2 - a_1b_3c_2 - a_2b_1c_3 - a_3b_2c_1.$$

From the defining formulas: (1) if two rows (columns) are equal, the determinant is zero; (2) if two rows (columns) are transposed, the determinant changes sign; (3) if a multiple of one row (column) is added to another row (column), the determinant is unchanged; and (4) if all the terms in one row (column) are multiplied by a scalar, the determinant is multiplied by the same scalar.

Also the defining formulas imply various expansions by minors of a row (column), for instance

$$\begin{vmatrix} a_1 & b_1 & c_1 \\ a_2 & b_2 & c_2 \\ a_3 & b_3 & c_3 \end{vmatrix} = a_1 \begin{vmatrix} b_2 & c_2 \\ b_3 & c_3 \end{vmatrix} - b_1 \begin{vmatrix} a_2 & c_2 \\ a_3 & c_3 \end{vmatrix} + c_1 \begin{vmatrix} a_2 & b_2 \\ a_3 & b_3 \end{vmatrix}$$

is the expansion by minors of the first row. Here, for instance,

$$\begin{vmatrix} a_2 & c_2 \\ a_3 & c_3 \end{vmatrix}$$

is the **minor** of b_1. It is the 2×2 determinant remaining after the row and the column containing b_1 are crossed off.

A system of equations

$$\begin{cases} a_1x + b_1y + c_1z = d_1 \\ a_2x + b_2y + c_2z = d_2 \\ a_3x + b_3y + c_3z = d_3 \end{cases}$$

is both consistent (not inconsistent) and determined (not underdetermined) if and only if the **system determinant** $D \neq 0$, where

$$D = \begin{vmatrix} a_1 & b_1 & c_1 \\ a_2 & b_2 & c_2 \\ a_3 & b_3 & c_3 \end{vmatrix}$$

When this is so, the system has a unique solution, given explicitly by **Cramer's Rule:**

$$x = \frac{1}{D}\begin{vmatrix} d_1 & b_1 & c_1 \\ d_2 & b_2 & c_2 \\ d_3 & b_3 & c_3 \end{vmatrix}, \qquad y = \frac{1}{D}\begin{vmatrix} a_1 & d_1 & c_1 \\ a_2 & d_2 & c_2 \\ a_3 & d_3 & c_3 \end{vmatrix}, \qquad z = \frac{1}{D}\begin{vmatrix} a_1 & b_1 & d_1 \\ a_2 & b_2 & d_2 \\ a_3 & b_3 & d_3 \end{vmatrix}$$

Cramer's Rule will be derived in Chapter 7, Section 7.

EXERCISES

Solve by elimination:

1. $\begin{cases} x + 2y = 1 \\ 3y = 2 \end{cases}$

2. $\begin{cases} 2x = 3 \\ -x + y = 0 \end{cases}$

3. $\begin{cases} x + 2y = 1 \\ x + 3y = 2 \end{cases}$

4. $\begin{cases} x + y = a \\ x - y = b \end{cases}$

5. $\begin{cases} 2x - 3y = -1 \\ 3x + 5y = 2 \end{cases}$

6. $\begin{cases} 2x - 3y = -1 \\ -3x + 5y = 2 \end{cases}$

7. $\begin{cases} x + y + z = 0 \\ 2y - 3z = -1 \\ 3y + 5z = 2 \end{cases}$

8. $\begin{cases} 2x - y - z = 1 \\ 2y - 3z = -1 \\ -3y + 5z = 2 \end{cases}$

9. $\begin{cases} x + y - z = 0 \\ x - y + z = 0 \\ -x + y + z = 0 \end{cases}$

10. $\begin{cases} 2x + y + 3z = 1 \\ -x + 4y + 2z = 0 \\ 3x + y + z = -1 \end{cases}$

11. $\begin{cases} 2x - y - 3z = 1 \\ -x - 4y - 2z = 1 \\ 3x - y - z = 1 \end{cases}$

12. $\begin{cases} 4x + 2y - z = 0 \\ x + 3y + 2z = 0 \\ x + y + 3z = 4. \end{cases}$

Show that the system is inconsistent and interpret geometrically:

13. $\begin{cases} x - y = 1 \\ -x + y = 1 \end{cases}$

14. $\begin{cases} x = 2 \\ x = 3 \end{cases}$

15. $\begin{cases} x + y + 2z = 1 \\ 3x + 5y + 7z = 2 \\ -x - y - 2z = 0 \end{cases}$

16. $\begin{cases} x + y + 2z = 1 \\ -x + 2y + z = 3 \\ y + z = 1. \end{cases}$

Find all solutions of each underdetermined system:

17. $\begin{cases} 2x - 3y = 1 \\ -4x + 6y = -2 \end{cases}$

18. $\begin{cases} x + y = 0 \\ x + y = 0 \end{cases}$

19. $\begin{cases} 2x - y + z = 1 \\ 3x + y + z = 0 \\ 7x - y + 3z = 2 \end{cases}$

20. $\begin{cases} 11x + 10y + 9z = 5 \\ x + 2y + 3z = 1 \\ 3x + 2y + z = 1. \end{cases}$

21*. Suppose $a \neq b$, $b \neq c$, $c \neq a$. Show that the system

$$\begin{cases} x + y + z = d_1 \\ ax + by + cz = d_2 \\ a^2x + b^2y + c^2z = d_3 \end{cases}$$

has a unique solution.

22*. Suppose the three planes $\mathbf{a}_1 \cdot \mathbf{x} = 0$, $\mathbf{a}_2 \cdot \mathbf{x} = 0$, $\mathbf{a}_3 \cdot \mathbf{x} = 0$ have only $\mathbf{0}$ in common. Show that the three planes $\mathbf{a}_1 \cdot \mathbf{x} = d_1$, $\mathbf{a}_2 \cdot \mathbf{x} = d_2$, $\mathbf{a}_3 \cdot \mathbf{x} = d_3$ have exactly one point in common. [*Hint:* Apply the *same* elimination process to both systems.]

5. CROSS PRODUCT

Geometric Definition

Given a pair of vectors $\mathbf{v}$ and $\mathbf{w}$, we define a new vector $\mathbf{v} \times \mathbf{w}$.

> The **cross product** of $\mathbf{v}$ and $\mathbf{w}$, written
>
> $$\mathbf{v} \times \mathbf{w},$$
>
> is the vector whose direction is determined by the right-hand rule from the pair $\mathbf{v}$, $\mathbf{w}$, and whose magnitude is the area of the parallelogram based on $\mathbf{v}$ and $\mathbf{w}$. See Fig. 5.1.

Note that $\mathbf{v} \times \mathbf{w}$ is a vector perpendicular both to $\mathbf{v}$ and to $\mathbf{w}$. Note also that if $\mathbf{v}$ and $\mathbf{w}$ are collinear, then the parallelogram collapses, so $\mathbf{v} \times \mathbf{w} = \mathbf{0}$.

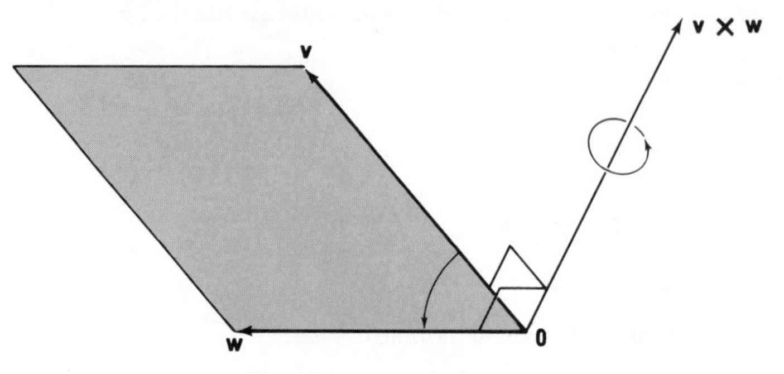

Fig. 5.1 cross product

In particular,

$$\mathbf{v} \times \mathbf{v} = \mathbf{0} \qquad \text{for each vector } \mathbf{v}.$$

If $\mathbf{v}$ and $\mathbf{w}$ are interchanged, the thumb reverses direction, hence

$$\mathbf{w} \times \mathbf{v} = -\mathbf{v} \times \mathbf{w}.$$

For the basic vectors $\mathbf{i}$, $\mathbf{j}$, $\mathbf{k}$, the cross products are simply

$$\mathbf{i} \times \mathbf{j} = \mathbf{k}, \qquad \mathbf{j} \times \mathbf{k} = \mathbf{i}, \qquad \mathbf{k} \times \mathbf{i} = \mathbf{j}.$$

Remark: This definition of cross product is motivated by physics. See the discussion of torque on p. 145.

Analytic Definition

Given a pair of vectors $\mathbf{v} = v_1\mathbf{i} + v_2\mathbf{j} + v_3\mathbf{k}$ and $\mathbf{w} = w_1\mathbf{i} + w_2\mathbf{j} + w_3\mathbf{k}$, it seems reasonable to define their cross product using the cross products of the basic vectors $\mathbf{i}$, $\mathbf{j}$, $\mathbf{k}$ as follows:

$$\begin{aligned}\mathbf{v} \times \mathbf{w} &= (v_1\mathbf{i} + v_2\mathbf{j} + v_3\mathbf{k}) \times (w_1\mathbf{i} + w_2\mathbf{j} + w_3\mathbf{k}) \\ &= v_1w_2\mathbf{i} \times \mathbf{j} + v_2w_1\mathbf{j} \times \mathbf{i} + v_1w_3\mathbf{i} \times \mathbf{k} \\ &\quad + v_3w_1\mathbf{k} \times \mathbf{i} + v_2w_3\mathbf{j} \times \mathbf{k} + v_3w_2\mathbf{k} \times \mathbf{j}.\end{aligned}$$

(We have used $\mathbf{i} \times \mathbf{i} = \mathbf{j} \times \mathbf{j} = \mathbf{k} \times \mathbf{k} = \mathbf{0}$.) Now $\mathbf{i} \times \mathbf{j} = \mathbf{k} = -\mathbf{j} \times \mathbf{i}$, etc. Hence, collecting terms, we have

$$\mathbf{v} \times \mathbf{w} = (v_2w_3 - v_3w_2)\mathbf{i} + (v_3w_1 - v_1w_3)\mathbf{j} + (v_1w_2 - v_2w_1)\mathbf{k}.$$

Note that each coefficient on the right is a 2×2 determinant.

Cross Product Let

$$\mathbf{v} = (v_1, v_2, v_3) \quad \text{and} \quad \mathbf{w} = (w_1, w_2, w_3).$$

Then

$$\mathbf{v} \times \mathbf{w} = \left(\begin{vmatrix} v_2 & v_3 \\ w_2 & w_3 \end{vmatrix}, \begin{vmatrix} v_3 & v_1 \\ w_3 & w_1 \end{vmatrix}, \begin{vmatrix} v_1 & v_2 \\ w_1 & w_2 \end{vmatrix} \right)$$

$$= (v_2w_3 - v_3w_2, v_3w_1 - v_1w_3, v_1w_2 - v_2w_1).$$

Here are two numerical examples:

$$(4, 3, -1) \times (-2, 2, 1) = \left(\begin{vmatrix} 3 & -1 \\ 2 & 1 \end{vmatrix}, \begin{vmatrix} -1 & 4 \\ 1 & -2 \end{vmatrix}, \begin{vmatrix} 4 & 3 \\ -2 & 2 \end{vmatrix} \right)$$

$$= (3 + 2, 2 - 4, 8 + 6) = (5, -2, 14).$$

$$(1, 0, 1) \times (0, 1, 1) = \left(\begin{vmatrix} 0 & 1 \\ 1 & 1 \end{vmatrix}, \begin{vmatrix} 1 & 1 \\ 1 & 0 \end{vmatrix}, \begin{vmatrix} 1 & 0 \\ 0 & 1 \end{vmatrix} \right) = (-1, -1, 1).$$

A device for remembering the cross product is a symbolic determinant, to be expanded by the first row:

$$(v_1, v_2, v_3) \times (w_1, w_2, w_3) = \begin{vmatrix} \mathbf{i} & \mathbf{j} & \mathbf{k} \\ v_1 & v_2 & v_3 \\ w_1 & w_2 & w_3 \end{vmatrix}$$

$$= \begin{vmatrix} v_2 & v_3 \\ w_2 & w_3 \end{vmatrix} \mathbf{i} - \begin{vmatrix} v_1 & v_3 \\ w_1 & w_3 \end{vmatrix} \mathbf{j} + \begin{vmatrix} v_1 & v_2 \\ w_1 & w_2 \end{vmatrix} \mathbf{k}.$$

For the moment we shall take the formula for $\mathbf{v} \times \mathbf{w}$ as *the definition* of cross product. Our problem is to prove that it has the required geometric properties. We begin the proof with a formula interesting in itself:

$$\mathbf{u} \cdot (\mathbf{v} \times \mathbf{w}) = D(\mathbf{u}, \mathbf{v}, \mathbf{w}) = \begin{vmatrix} u_1 & u_2 & u_3 \\ v_1 & v_2 & v_3 \\ w_1 & w_2 & w_3 \end{vmatrix}.$$

To prove it, simply expand the determinant $D(\mathbf{u}, \mathbf{v}, \mathbf{w})$ by the first row:

$$D(\mathbf{u}, \mathbf{v}, \mathbf{w}) = \begin{vmatrix} u_1 & u_2 & u_3 \\ v_1 & v_2 & v_3 \\ w_1 & w_2 & w_3 \end{vmatrix}$$

$$= u_1 \begin{vmatrix} v_2 & v_3 \\ w_2 & w_3 \end{vmatrix} - u_2 \begin{vmatrix} v_1 & v_3 \\ w_1 & w_3 \end{vmatrix} + u_3 \begin{vmatrix} v_1 & v_2 \\ w_1 & w_2 \end{vmatrix}$$

$$= \mathbf{u} \cdot (\mathbf{v} \times \mathbf{w}).$$

As a consequence of the formula, $\mathbf{v} \cdot (\mathbf{v} \times \mathbf{w}) = D(\mathbf{v}, \mathbf{v}, \mathbf{w}) = 0$, because a determinant with two equal rows is zero. Hence $\mathbf{v}$ is perpendicular to $\mathbf{v} \times \mathbf{w}$. Similarly so is $\mathbf{w}$. We have proved

(1) $\mathbf{v} \times \mathbf{w}$ is perpendicular to $\mathbf{v}$ and to $\mathbf{w}$.

This is one of the required geometric conditions. Next, we prove a formula for the length of $\mathbf{v} \times \mathbf{w}$:

$$\boxed{|\mathbf{v} \times \mathbf{w}|^2 = |\mathbf{v}|^2 |\mathbf{w}|^2 - (\mathbf{v} \cdot \mathbf{w})^2.}$$

The left-hand side is

$$(v_2w_3 - v_3w_2)^2 + (v_3w_1 - v_1w_3)^2 + (v_1w_2 - v_2w_1)^2$$

and the right-hand side is

$$(v_1^2 + v_2^2 + v_3^2)(w_1^2 + w_2^2 + w_3^2) - (v_1w_1 + v_2w_2 + v_3w_3)^2.$$

The first product in the right-hand side yields nine terms and the second product, six terms. The three terms like $v_1^2w_1^2$ cancel. There remain six terms like $v_1^2w_2^2$ and three terms like $-2v_1v_2w_1w_2$, exactly what occur on the left-hand side.

From this last formula we deduce

(2) $|\mathbf{v} \times \mathbf{w}|$ is the area of the parallelogram determined by $\mathbf{v}$ and $\mathbf{w}$.

For let θ be the angle between $\mathbf{v}$ and $\mathbf{w}$. Then $\mathbf{v} \cdot \mathbf{w} = |\mathbf{v}| \cdot |\mathbf{w}| \cos\theta$, hence

$$|\mathbf{v} \times \mathbf{w}|^2 = |\mathbf{v}|^2 |\mathbf{w}|^2 - (\mathbf{v} \cdot \mathbf{w})^2 = |\mathbf{v}|^2 |\mathbf{w}|^2 (1 - \cos^2\theta) = |\mathbf{v}|^2 |\mathbf{w}|^2 \sin^2\theta,$$

$$|\mathbf{v} \times \mathbf{w}| = |\mathbf{v}| \cdot |\mathbf{w}| \sin\theta.$$

This last expression is precisely the required area (Fig. 5.2).

Finally we must prove

(3) $\mathbf{v}, \mathbf{w}, \mathbf{v} \times \mathbf{w}$ is a right-handed system (provided $\mathbf{v}$ and $\mathbf{w}$ are not parallel).

In order to do so, we need some analytic way of deciding whether a given triple $\mathbf{u}, \mathbf{v}, \mathbf{w}$ is a right-handed system or not.

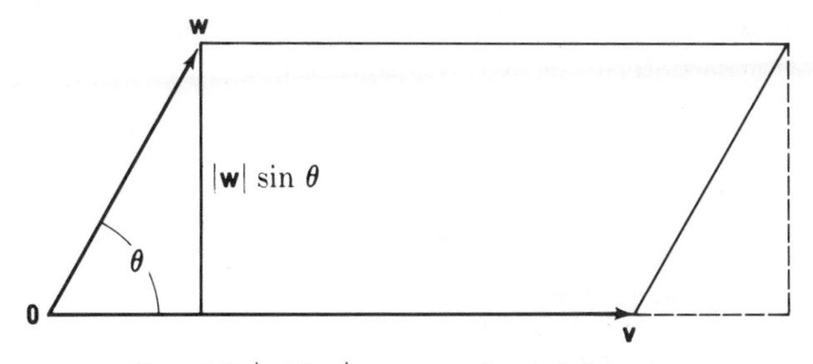

FIG. 5.2 $|\mathbf{v} \times \mathbf{w}| =$ area of parallelogram

Now observe that $\mathbf{u}, \mathbf{v}, \mathbf{w}$ and $\mathbf{v}, \mathbf{u}, \mathbf{w}$ have opposite *orientations*, that is, one is a right-handed system and the other is left-handed. By analogy, the determinants $D(\mathbf{u}, \mathbf{v}, \mathbf{w})$ and $D(\mathbf{v}, \mathbf{u}, \mathbf{w})$ have opposite signs. This suggests that the sign of $D(\mathbf{u}, \mathbf{v}, \mathbf{w})$ corresponds to the orientation of $\mathbf{u}, \mathbf{v}, \mathbf{w}$. Since $\mathbf{i}, \mathbf{j}, \mathbf{k}$ is right-handed and $D(\mathbf{i}, \mathbf{j}, \mathbf{k}) = 1$, we suspect that $\mathbf{u}, \mathbf{v}, \mathbf{w}$ is right-handed if $D(\mathbf{u}, \mathbf{v}, \mathbf{w}) > 0$. This indeed is the case, but instead of proving it, we shall simply take the determinant criterion as the definition of right-handedness.

In view of this definition, we must prove that $D(\mathbf{v}, \mathbf{w}, \mathbf{v} \times \mathbf{w}) > 0$. Now

$$\begin{aligned} D(\mathbf{v}, \mathbf{w}, \mathbf{v} \times \mathbf{w}) &= -D(\mathbf{v}, \mathbf{v} \times \mathbf{w}, \mathbf{w}) = D(\mathbf{v} \times \mathbf{w}, \mathbf{v}, \mathbf{w}) \\ &= (\mathbf{v} \times \mathbf{w}) \cdot (\mathbf{v} \times \mathbf{w}) = |\mathbf{v} \times \mathbf{w}|^2 > 0. \end{aligned}$$

This completes the proof that the analytic and geometric definitions of $\mathbf{v} \times \mathbf{w}$ coincide. We now summarize the main algebraic properties of the cross product. They follow readily from our discussion.

$$\begin{gathered} \mathbf{v} \times \mathbf{v} = \mathbf{0}, \qquad \mathbf{w} \times \mathbf{v} = -\mathbf{v} \times \mathbf{w}, \\ (a\mathbf{u} + b\mathbf{v}) \times \mathbf{w} = a(\mathbf{u} \times \mathbf{w}) + b(\mathbf{v} \times \mathbf{w}), \\ \mathbf{u} \times (a\mathbf{v} + b\mathbf{w}) = a(\mathbf{u} \times \mathbf{v}) + b(\mathbf{u} \times \mathbf{w}), \\ \mathbf{u} \cdot (\mathbf{v} \times \mathbf{w}) = \mathbf{v} \cdot (\mathbf{w} \times \mathbf{u}) = \mathbf{w} \cdot (\mathbf{u} \times \mathbf{v}) = D(\mathbf{u}, \mathbf{v}, \mathbf{w}), \\ \mathbf{v} \times \mathbf{w} = \mathbf{0} \quad \text{if and only if } \mathbf{v} \text{ and } \mathbf{w} \text{ are collinear.} \end{gathered}$$

REMARK: The quantity $\mathbf{u} \cdot (\mathbf{v} \times \mathbf{w}) = D(\mathbf{u}, \mathbf{v}, \mathbf{w})$ is sometimes written $[\mathbf{u}, \mathbf{v}, \mathbf{w}]$ and called the **triple scalar product.**

Torque

The original motivation for the cross product of vectors came from physics. Consider this situation.

A rigid body is free to turn about the origin. A force $\mathbf{F}$ acts at a point $\mathbf{x}$ of the body. As a result the body "wants" to rotate about an axis through $\mathbf{0}$ perpendicular to the plane of $\mathbf{x}$ and $\mathbf{F}$ (unless $\mathbf{x}$ and $\mathbf{F}$ are collinear; then there is no turning). See Fig. 5.3a. As usual, the force vector $\mathbf{F}$ is *drawn* at its point of application **x**. But analytically it starts at **0**. See Fig. 5.3b. The positiv axis of rotation is determined by the right-hand rule as applied to the pair **x**, **F** in that order: **x** first, **F** second.

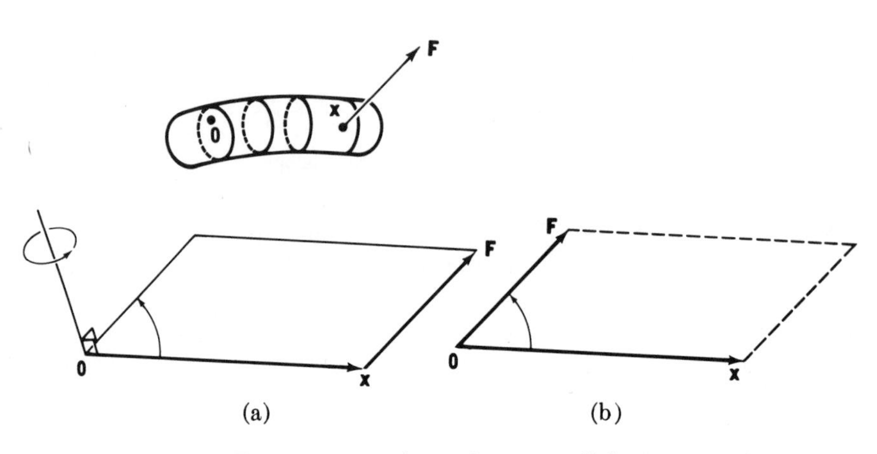

FIG. 5.3 torque due to force **F** applied at **x**

In physics, one speaks of the **torque** (at the origin) resulting from the force **F** applied at **x**. Roughly speaking, torque is a measure of the tendency of a body to rotate under the action of forces. (Torque will be defined precisely in a moment.)

By experiment, if **F** is tripled in magnitude, the torque is tripled; if **x** is moved out twice as far along the same line and the same **F** is applied there, the torque is doubled. Hence the torque is proportional to the length of **x** and to the length of **F**. Therefore (Fig. 5.3b) the torque is proportional to the area of the parallelogram determined by **x** and **F**.

Resolve **F** into components $\mathbf{F}^{\parallel}$ and $\mathbf{F}^{\perp}$, where $\mathbf{F}^{\parallel}$ is parallel to **x** and $\mathbf{F}^{\perp}$ is perpendicular to **x**. See Fig. 5.4a. Only $\mathbf{F}^{\perp}$ produces torque; the amount of torque is the product $|\mathbf{F}^{\perp}|\,|\mathbf{x}|$ of the magnitude of $\mathbf{F}^{\perp}$ by the length $|\mathbf{x}|$ of the lever arm. But this product *is* the area of the parallelogram determined by **x** and **F**. See Fig. 5.4b.

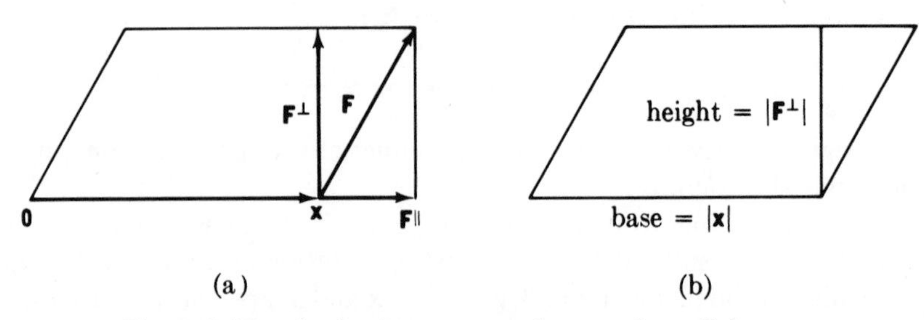

FIG. 5.4 Magnitude of torque equals area of parallelogram.

Therefore the torque about the origin is completely described by the one vector $\mathbf{x} \times \mathbf{F}$. The length of $\mathbf{x} \times \mathbf{F}$ is the magnitude of the torque. The direction of $\mathbf{x} \times \mathbf{F}$ is the positive axis of rotation; with your right thumb along $\mathbf{x} \times \mathbf{F}$, your fingers curl in the direction of turning. In physics, torque about the origin is *defined* to be the vector $\mathbf{x} \times \mathbf{F}$.

EXERCISES

Find the cross product:

1. $(-2, 2, 1) \times (4, 3, -1)$
2. $(1, 0, 1) \times (1, 1, 0)$
3. $(1, 2, 3) \times (3, 2, 1)$
4. $(3, 1, -1) \times (3, -1, -1)$
5. $(-2, -2, -2) \times (1, 1, 0)$
6. $(-1, 2, 2) \times (3, -1, 2)$
7. $(0, 0, 0) \times (1, 1, 2)$
8. $(1, -1, 1) \times (-1, 1, -1)$
9. $(2, 1, 3) \times (2, 2, -1)$
10. $(1, 2, 3) \times (4, 5, 6)$.

A force $\mathbf{F}$ is applied at point $\mathbf{x}$ Find its torque about the origin:

11. $\mathbf{F} = (-1, 1, 1), \quad \mathbf{x} = (10, 0, 0)$
12. $\mathbf{F} = (3, 0, 0), \quad \mathbf{x} = (0, 0, 1)$
13. $\mathbf{F} = (-1, 1, 1), \quad \mathbf{x} = (2, 2, -1)$
14. $\mathbf{F} = (2, -1, 5), \quad \mathbf{x} = (-7, 1, 0)$.

Prove:

15. $\mathbf{u} \cdot (\mathbf{v} \times \mathbf{w}) = \mathbf{v} \cdot (\mathbf{w} \times \mathbf{u})$
16. $(\mathbf{u} + \mathbf{v}) \times \mathbf{w} = \mathbf{u} \times \mathbf{w} + \mathbf{v} \times \mathbf{w}$
17. $(a\mathbf{v}) \times \mathbf{w} = a(\mathbf{v} \times \mathbf{w})$
18. $\mathbf{v} \times (b\mathbf{w}) = b(\mathbf{v} \times \mathbf{w})$.

19*. Prove

$$\begin{vmatrix} \mathbf{p}\cdot\mathbf{u} & \mathbf{p}\cdot\mathbf{v} \\ \mathbf{q}\cdot\mathbf{u} & \mathbf{q}\cdot\mathbf{v} \end{vmatrix} = (\mathbf{p} \times \mathbf{q}) \cdot (\mathbf{u} \times \mathbf{w}).$$

20. (cont.) Use this result for a new proof of
$$|\mathbf{u} \times \mathbf{v}|^2 = |\mathbf{u}|^2\,|\mathbf{v}|^2 - (\mathbf{u}\cdot\mathbf{v})^2.$$
21. Prove the formula $\mathbf{u} \times (\mathbf{v} \times \mathbf{w}) = (\mathbf{u}\cdot\mathbf{w})\mathbf{v} - (\mathbf{u}\cdot\mathbf{v})\mathbf{w}$.
[*Hint:* By the linearity of each side and by symmetry, reduce to the cases where $\mathbf{u}$, $\mathbf{v}$, and $\mathbf{w}$ are chosen from $\mathbf{i}$ and $\mathbf{j}$.]
22. (cont.) Prove the formula
$$\begin{aligned}(\mathbf{a} \times \mathbf{b}) \times (\mathbf{u} \times \mathbf{v}) &= [(\mathbf{a} \times \mathbf{b})\cdot\mathbf{v}]\mathbf{u} - [(\mathbf{a} \times \mathbf{b})\cdot\mathbf{u}]\mathbf{v} \\ &= [(\mathbf{u} \times \mathbf{v})\cdot\mathbf{a}]\mathbf{b} - [(\mathbf{u} \times \mathbf{v})\cdot\mathbf{b}]\mathbf{a}.\end{aligned}$$
Hence show that the left-hand side is a vector along the line of intersection of the plane of $\mathbf{a}$ and $\mathbf{b}$ with the plane of $\mathbf{u}$ and $\mathbf{v}$.
23. (cont.) Show that $(\mathbf{a} \times \mathbf{b}) \times (\mathbf{a} \times \mathbf{c})$ is collinear with $\mathbf{a}$.
24. (cont.) Prove the **Jacobi identity**
$$\mathbf{u} \times (\mathbf{v} \times \mathbf{w}) + \mathbf{v} \times (\mathbf{w} \times \mathbf{u}) + \mathbf{w} \times (\mathbf{u} \times \mathbf{v}) = \mathbf{0}.$$

6. APPLICATIONS

Volume

Two non-collinear vectors determine a parallelogram. Three non-coplanar vectors determine a parallelepiped (Fig. 6.1a) whose volume is given by the formula

$$V = (\text{area of base}) \cdot (\text{height}).$$

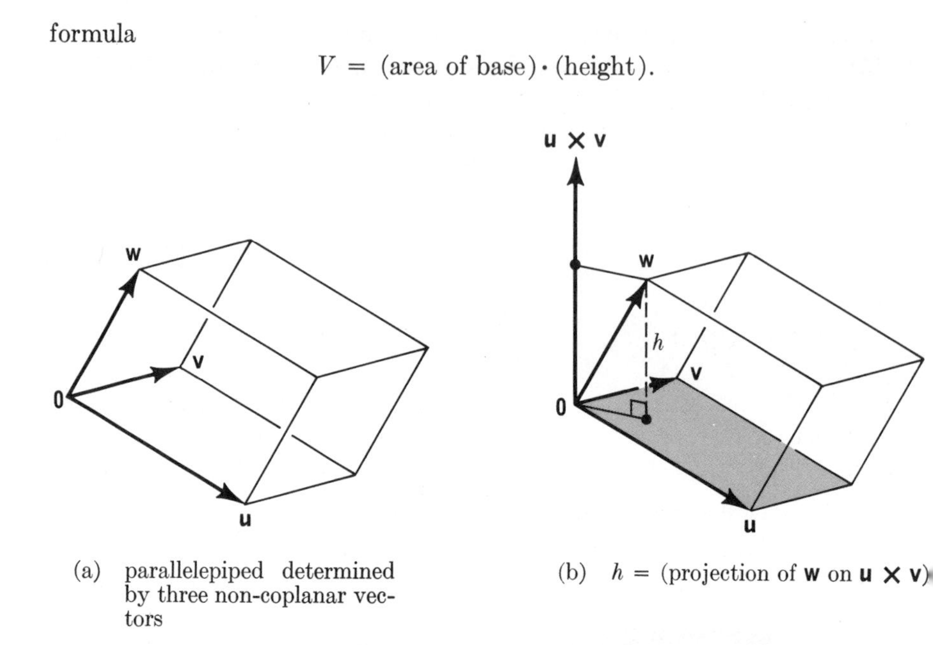

(a) parallelepiped determined by three non-coplanar vectors

(b) h = (projection of **w** on **u** × **v**)

FIG. 6.1 volume of a parallelepiped

Suppose the base is the parallelogram determined by $\mathbf{u}$ and $\mathbf{v}$; its area is $|\mathbf{u} \times \mathbf{v}|$. Assume temporarily that $\mathbf{w}$ lies on the same side of the $\mathbf{u}$, $\mathbf{v}$-plane as $\mathbf{u} \times \mathbf{v}$. Then the height is the projection of $\mathbf{w}$ on $\mathbf{u} \times \mathbf{v}$. See Fig. 6.1b. Therefore

$$V = (\text{projection } \mathbf{w} \text{ on } \mathbf{u} \times \mathbf{v}) \cdot |\mathbf{u} \times \mathbf{v}| = (\mathbf{u} \times \mathbf{v}) \cdot \mathbf{w} = D(\mathbf{u}, \mathbf{v}, \mathbf{w}).$$

If $\mathbf{w}$ is on the other side of the $\mathbf{u}$, $\mathbf{v}$-plane, then $V = -D(\mathbf{u}, \mathbf{v}, \mathbf{w})$. In any case:

> The volume of the parallelepiped determined by three non-coplanar vectors $\mathbf{u}$, $\mathbf{v}$, $\mathbf{w}$ is given by the formula
>
> $$V = |D(\mathbf{u}, \mathbf{v}, \mathbf{w})|.$$

Intersection of Two Planes

Given two planes $\mathbf{x} \cdot \mathbf{m} = p$ and $\mathbf{x} \cdot \mathbf{n} = q$, how can we find their line of intersection? We must assume the planes are not parallel, that is, $\mathbf{m}$ and $\mathbf{n}$ are not collinear. Then $\mathbf{u} = \mathbf{m} \times \mathbf{n}$ is perpendicular to both $\mathbf{m}$ and $\mathbf{n}$, so $\mathbf{u}$ is parallel to the line of intersection (Fig. 6.2).

If we can find a single point $\mathbf{x}_0$ on both planes, then the desired line is

$$\mathbf{x} = \mathbf{x}_0 + t\mathbf{u}$$

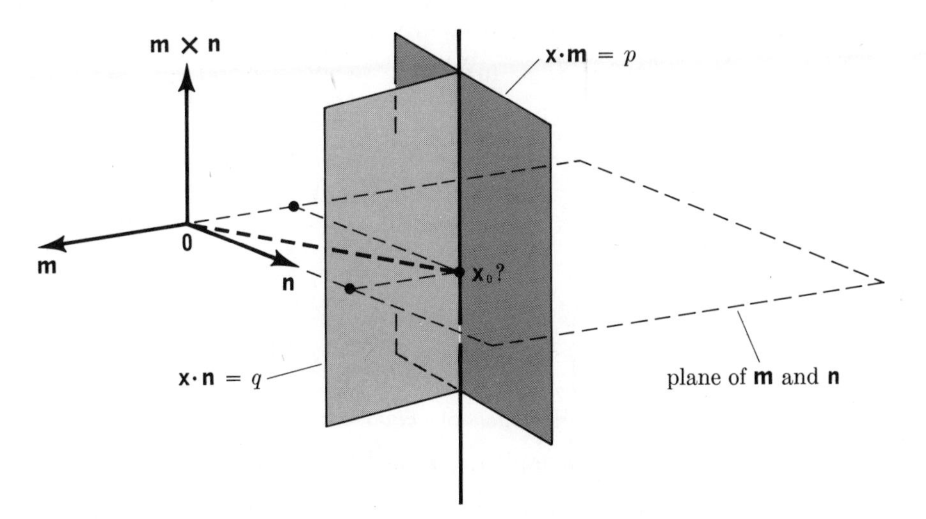

FIG. 6.2 Intersection of two planes: $\mathbf{m} \times \mathbf{n}$ is parallel to their line of intersection. To find $\mathbf{x}_0$ on this line of intersection and in the plane of $\mathbf{m}$ and $\mathbf{n}$.

in parametric form. The figure suggests a point in the plane of $\mathbf{m}$ and $\mathbf{n}$, so let $\mathbf{x}_0 = a\mathbf{m} + b\mathbf{n}$. Then the conditions $\mathbf{x}_0 \cdot \mathbf{m} = p$ and $\mathbf{x}_0 \cdot \mathbf{n} = q$ result in a system of linear equations for a and b:

$$\begin{cases} a\,\mathbf{m}\cdot\mathbf{m} + b\,\mathbf{m}\cdot\mathbf{n} = p \\ a\,\mathbf{m}\cdot\mathbf{n} + b\,\mathbf{n}\cdot\mathbf{n} = q. \end{cases}$$

The determinant of this system is

$$\begin{vmatrix} \mathbf{m}\cdot\mathbf{m} & \mathbf{m}\cdot\mathbf{n} \\ \mathbf{m}\cdot\mathbf{n} & \mathbf{n}\cdot\mathbf{n} \end{vmatrix} = |\mathbf{m}|^2\,|\mathbf{n}|^2 - (\mathbf{m}\cdot\mathbf{n})^2 = |\mathbf{m} \times \mathbf{n}|^2 > 0.$$

Therefore there is a unique solution.

EXAMPLE 6.1

Find the line of intersection of the planes $x + y + z = -1$ and $2x + y - z = 3$.

Solution: The equations of these planes are $\mathbf{x}\cdot\mathbf{m} = p$ and $\mathbf{x}\cdot\mathbf{n} = q$, where

$$\mathbf{m} = (1, 1, 1), \qquad \mathbf{n} = (2, 1, -1), \qquad p = -1, \qquad q = 3.$$

Set

$$\mathbf{u} = \mathbf{m} \times \mathbf{n} = (-2, 3, -1).$$

This vector is perpendicular to $\mathbf{m}$ and $\mathbf{n}$, so it has the direction of the line of intersection.

To find a point on the line of intersection, set $\mathbf{x}_0 = a\mathbf{m} + b\mathbf{n}$ and choose a

and b so that $\mathbf{x}_0 \cdot \mathbf{m} = -1$ and $\mathbf{x}_0 \cdot \mathbf{n} = 3$. Now

$$|\mathbf{m}|^2 = 3, \qquad \mathbf{m} \cdot \mathbf{n} = 2, \qquad |\mathbf{n}|^2 = 6,$$

and the equations $\mathbf{x}_0 \cdot \mathbf{m} = -1$ and $\mathbf{x}_0 \cdot \mathbf{n} = 3$ become

$$\begin{cases} 3a + 2b = -1 \\ 2a + 6b = 3. \end{cases}$$

The solution is $a = -\frac{6}{7}$, $b = \frac{11}{14}$; therefore

$$\mathbf{x}_0 = -\tfrac{6}{7}(1, 1, 1) + \tfrac{11}{14}(2, 1, -1) = (\tfrac{10}{14}, -\tfrac{1}{14}, -\tfrac{23}{14}).$$

Answer: $\mathbf{x} = (\frac{10}{14}, -\frac{1}{14}, -\frac{23}{14}) + t(-2, 3, -1)$.

Homogeneous Equations

Suppose we are given three planes through the origin,

$$\mathbf{x} \cdot \mathbf{u} = 0, \qquad \mathbf{x} \cdot \mathbf{v} = 0, \qquad \mathbf{x} \cdot \mathbf{w} = 0.$$

In general, the planes will only have $\mathbf{0}$ in common. However, it may happen that they have a line in common, or even coincide. This occurs precisely when the three normal vectors $\mathbf{u}$, $\mathbf{v}$, $\mathbf{w}$ lie in the same plane. The situation can be described algebraically:

> A system of three linear homogeneous equations
>
> $$\mathbf{x} \cdot \mathbf{u} = 0, \qquad \mathbf{x} \cdot \mathbf{v} = 0, \qquad \mathbf{x} \cdot \mathbf{w} = 0$$
>
> has a solution $\mathbf{x}_0 \neq \mathbf{0}$ (a non-trivial solution) if and only if
>
> $$\mathbf{u} \cdot (\mathbf{v} \times \mathbf{w}) = D(\mathbf{u}, \mathbf{v}, \mathbf{w}) = 0.$$
>
> If $\mathbf{x}_0$ is any solution, then $t\mathbf{x}_0$ is a solution for each t.

It is useful to restate this result in terms of determinants and linear equations.

> A homogeneous system of linear equations
>
> $$\begin{cases} u_1x_1 + u_2x_2 + u_3x_3 = 0 \\ v_1x_1 + v_2x_2 + v_3x_3 = 0 \\ w_1x_1 + w_2x_2 + w_3x_3 = 0 \end{cases}$$
>
> has a solution $(x_1, x_2, x_3) \neq (0, 0, 0)$ if and only if
>
> $$\begin{vmatrix} u_1 & u_2 & u_3 \\ v_1 & v_2 & v_3 \\ w_1 & w_2 & w_3 \end{vmatrix} = 0.$$
>
> If (x_1, x_2, x_3) is any solution, then (tx_1, tx_2, tx_3) is a solution for each t.

EXAMPLE 6.2

Find a non-trivial solution of

$$\mathbf{x}\cdot\mathbf{u} = \mathbf{x}\cdot\mathbf{v} = \mathbf{x}\cdot\mathbf{w} = 0,$$

where $\mathbf{u} = (1, -2, 2)$, $\mathbf{v} = (3, 1, -2)$, and $\mathbf{w} = (5, -3, 2)$.

Solution: First

$$\mathbf{u}\cdot(\mathbf{v}\times\mathbf{w}) = D(\mathbf{u}, \mathbf{v}, \mathbf{w}) = \begin{vmatrix} 1 & -2 & 2 \\ 3 & 1 & -2 \\ 5 & -3 & 2 \end{vmatrix} = 0.$$

(Note that $\mathbf{w} = 2\mathbf{u} + \mathbf{v}$.) Therefore the vectors $\mathbf{u}$, $\mathbf{v}$, $\mathbf{w}$ are coplanar (the parallelepiped collapses) so the corresponding perpendicular planes have a line L in common. Certainly L contains the point $\mathbf{0}$ which is common to all three planes. Furthermore it contains $\mathbf{u}\times\mathbf{v}$, since this vector starts at $\mathbf{0}$ and is parallel to L. Therefore L is the set of all multiples $t(\mathbf{u}\times\mathbf{v})$, provided $\mathbf{u}\times\mathbf{v} \neq \mathbf{0}$. A similar statement holds for $\mathbf{v}\times\mathbf{w}$ and $\mathbf{w}\times\mathbf{u}$.

Note that $\mathbf{u}\times\mathbf{v} = (2, 8, 7)$, while $\mathbf{v}\times\mathbf{w} = (-4, -16, -14)$ and $\mathbf{w}\times\mathbf{u} = (-2, -8, -7)$, so these three vectors really are collinear.

Answer: $(2, 8, 7)$.

REMARK: Recall that Cramer's Rule (Section 4) guarantees a *unique* solution if $D(\mathbf{u}, \mathbf{v}, \mathbf{w}) \neq 0$. Since $(0, 0, 0)$ is obviously a solution to the homogeneous system, it is the only solution when $D(\mathbf{u}, \mathbf{v}, \mathbf{w}) \neq 0$. This proves again that for a homogeneous system to have a non-trivial solution, its determinant must be zero.

Skew Lines

Let $\mathbf{x} = \mathbf{x}_0 + s\mathbf{u}$ and $\mathbf{x} = \mathbf{y}_0 + t\mathbf{v}$ be two lines in $\mathbf{R}^3$ that do not intersect and are not parallel, i.e., skew lines. We ask how far apart they are (Fig. (6.3a).

The vector $\mathbf{u}\times\mathbf{v}$ is perpendicular to both lines, so $\mathbf{n} = (\mathbf{u}\times\mathbf{v})/|\mathbf{u}\times\mathbf{v}|$ is a unit vector perpendicular to both lines. From Fig. 6.3b we see that the required distance is the length of the projection of $\mathbf{x}_0 - \mathbf{y}_0$ on $\mathbf{n}$, that is, $|(\mathbf{x}_0 - \mathbf{y}_0)\cdot\mathbf{n}|$.

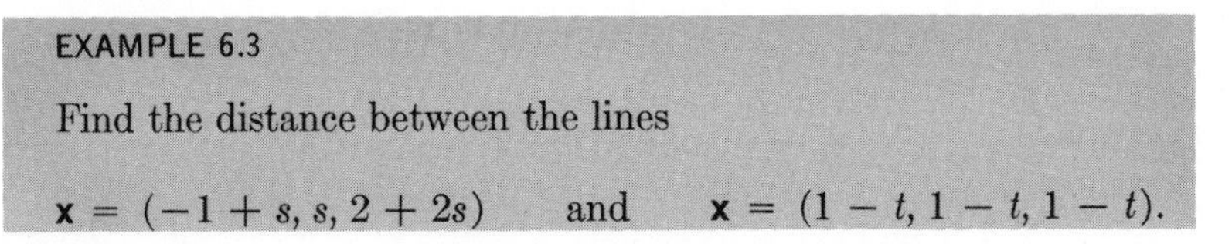

EXAMPLE 6.3

Find the distance between the lines

$$\mathbf{x} = (-1 + s, s, 2 + 2s) \quad \text{and} \quad \mathbf{x} = (1 - t, 1 - t, 1 - t).$$

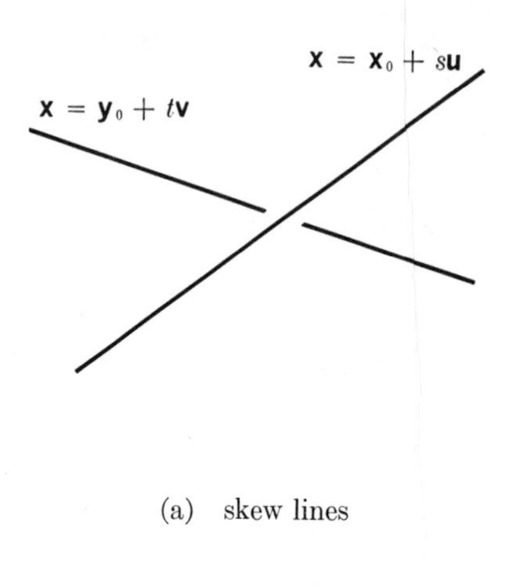

(a) skew lines

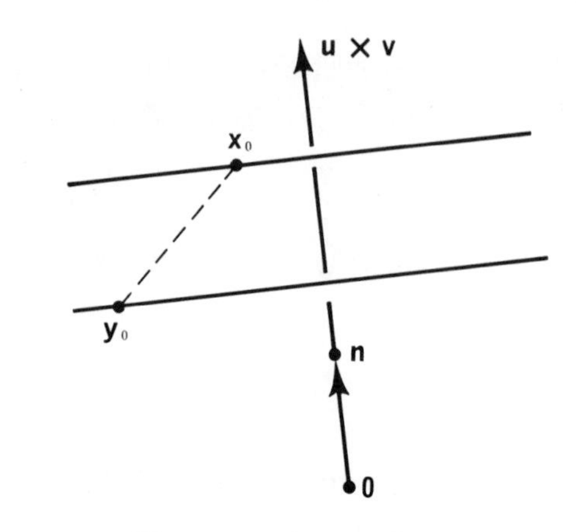

(b) as seen from a direction that makes the lines appear to be parallel

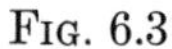
Fig. 6.3

Solution: In this example

$$\mathbf{u} = (1, 1, 2), \qquad \mathbf{v} = (-1, -1, -1), \qquad \mathbf{x}_0 = (-1, 0, 2), \qquad \mathbf{y}_0 = (1, 1, 1).$$

Therefore $\mathbf{u} \times \mathbf{v} = (1, -1, 0)$ and

$$\mathbf{n} = \frac{\mathbf{u} \times \mathbf{v}}{|\mathbf{u} \times \mathbf{v}|} = \frac{1}{2}\sqrt{2}(1, -1, 0).$$

Finally,

$$(\mathbf{x}_0 - \mathbf{y}_0)\cdot\mathbf{n} = (-2, -1, 1)\cdot\tfrac{1}{2}\sqrt{2}(1, -1, 0) = -\tfrac{1}{2}\sqrt{2}.$$

Answer: $\frac{1}{2}\sqrt{2}$.

Parametric Form of a Plane

Two non-collinear vectors $\mathbf{u}$ and $\mathbf{v}$ in $\mathbf{R}^3$ determine a unique plane through $\mathbf{0}$ consisting of all points

$$\mathbf{x} = s\mathbf{u} + t\mathbf{v},$$

where s and t are any real numbers. If $\mathbf{x}_0$ is any point of $\mathbf{R}^3$, then adding $\mathbf{x}_0$ to each point of the plane displaces it to a parallel plane through $\mathbf{x}_0$. See Fig. 6.4.

> Given a point $\mathbf{x}_0$ of $\mathbf{R}^3$ and two non-collinear vectors $\mathbf{u}$ and $\mathbf{v}$, the plane through $\mathbf{x}_0$ parallel to the plane of $\mathbf{u}$ and $\mathbf{v}$ consists of all points
>
> $$\mathbf{x} = \mathbf{x}_0 + s\mathbf{u} + t\mathbf{v},$$
>
> where $-\infty < s < \infty$ and $-\infty < t < \infty$.

The variables s and t are called **parameters,** and a plane presented in this fashion is said to be in **parametric form.**

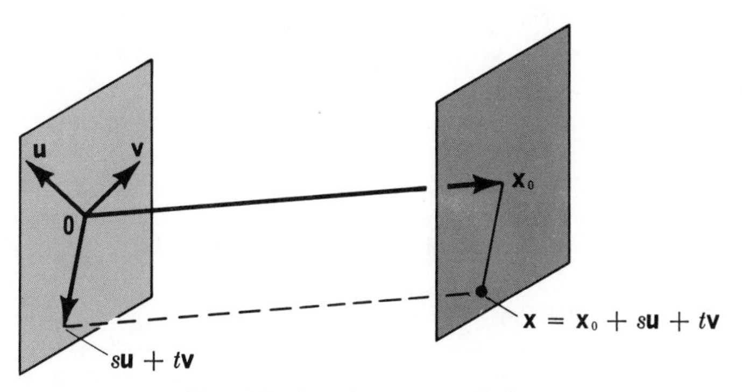

FIG. 6.4 plane in parametric form

Example:

Let $\mathbf{x}_0 = (-1, 1, 2)$, $\mathbf{u} = (1, 0, 1)$, $\mathbf{v} = (1, 1, 0)$. Clearly neither $\mathbf{u}$ nor $\mathbf{v}$ is a multiple of the other, so they are not collinear. Then

$$\begin{aligned} \mathbf{x} &= \mathbf{x}_0 + s\mathbf{u} + t\mathbf{v} = (-1, 1, 2) + s(1, 0, 1) + t(1, 1, 0) \\ &= (-1 + s + t, 1 + t, 2 + s). \end{aligned}$$

In coordinates,

$$x = -1 + s + t, \qquad y = 1 + t, \qquad z = 2 + s.$$

Given a plane in parametric form, how do we put it in normal form? We have $\mathbf{x} = \mathbf{x}_0 + s\mathbf{u} + t\mathbf{v}$, where $\mathbf{u}$, $\mathbf{v}$ are linearly independent. A vector perpendicular to both $\mathbf{u}$ and $\mathbf{v}$ (hence to the plane) is $\mathbf{u} \times \mathbf{v}$. This vector is guaranteed to be non-zero because $\mathbf{u}$ and $\mathbf{v}$ are not collinear. Set

$$\mathbf{n} = (\mathbf{u} \times \mathbf{v})/|\mathbf{u} \times \mathbf{v}|.$$

Then $\mathbf{n}$ is a unit vector perpendicular to the plane. Hence

$$\mathbf{x}\cdot\mathbf{n} = \mathbf{x}_0\cdot\mathbf{n} + s\mathbf{u}\cdot\mathbf{n} + t\mathbf{v}\cdot\mathbf{n} = \mathbf{x}_0\cdot\mathbf{n} = p.$$

This is a normal form.

In the example above,

$$\mathbf{u} \times \mathbf{v} = (1, 0, 1) \times (1, 1, 0) = (-1, 1, 1),$$

$$\mathbf{n} = \tfrac{1}{3}\sqrt{3}(-1, 1, 1), \qquad p = \mathbf{x}_0\cdot\mathbf{n} = \tfrac{4}{3}\sqrt{3},$$

so a normal form is

$$\tfrac{1}{3}\sqrt{3}(-x_1 + x_2 + x_3) = \tfrac{4}{3}\sqrt{3}.$$

Plane through Three Points

Three non-collinear points $\mathbf{x}_0$, $\mathbf{x}_1$, $\mathbf{x}_2$ determine a unique plane. If we want a normal form, we argue that $\mathbf{u} = \mathbf{x}_1 - \mathbf{x}_0$ and $\mathbf{v} = \mathbf{x}_2 - \mathbf{x}_0$ are parallel to the plane, hence $\mathbf{m} = \mathbf{u} \times \mathbf{v}$ is perpendicular to it. Therefore $\mathbf{x} \cdot \mathbf{m} = \mathbf{x}_0 \cdot \mathbf{m}$ is an equation of the plane.

If we want a parametric form, then $\mathbf{x} = \mathbf{x}_0 + s\mathbf{u} + t\mathbf{v}$ does the job. Now

$$\mathbf{x}_0 + s\mathbf{u} + t\mathbf{v} = \mathbf{x}_0 + s(\mathbf{x}_1 - \mathbf{x}_0) + t(\mathbf{x}_2 - \mathbf{x}_0)$$

$$= (1 - s - t)\mathbf{x}_0 + s\mathbf{x}_1 + t\mathbf{x}_2,$$

so we have the alternative symmetric form:

> The plane through three non-collinear points $\mathbf{x}_0$, $\mathbf{x}_1$, $\mathbf{x}_2$ consists of all points
>
> $$\mathbf{x} = s_0\mathbf{x}_0 + s_1\mathbf{x}_1 + s_2\mathbf{x}_2$$
>
> where s_0, s_1, s_2 take on all real values subject to $s_0 + s_1 + s_2 = 1$.

Here is a physical interpretation. Put masses s_0, s_1, s_2 at $\mathbf{x}_0$, $\mathbf{x}_1$, $\mathbf{x}_2$ respectively, where $s_0 + s_1 + s_2 = 1$. Then $\mathbf{x} = s_0\mathbf{x}_0 + s_1\mathbf{x}_1 + s_2\mathbf{x}_2$ is the center of gravity of the masses.

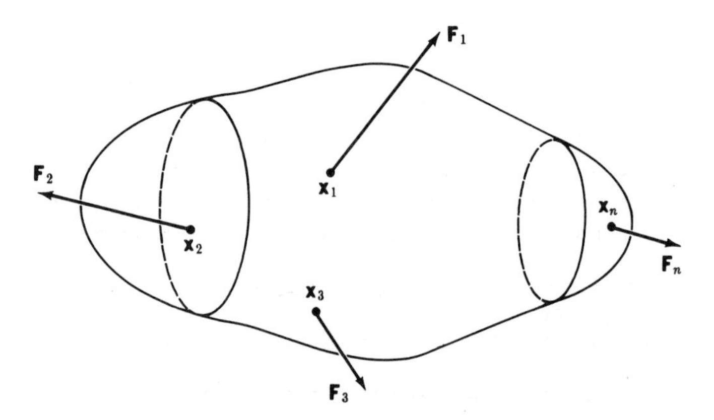

FIG. 6.5 rigid body acted on by forces

Equilibrium

Forces $\mathbf{F}_1, \cdots, \mathbf{F}_n$ are applied at points $\mathbf{x}_1, \cdots, \mathbf{x}_n$ of a rigid body (Fig. 6.5). Now a rigid body is in equilibrium when both the sum of the forces vanishes and the sum of the turning moments (torques) of the forces about $\mathbf{0}$

vanishes. Thus the conditions for equilibrium are the two vector equations:

$$\mathbf{F}_1 + \mathbf{F}_2 + \cdots + \mathbf{F}_n = \mathbf{0},$$

$$\mathbf{x}_1 \times \mathbf{F}_1 + \mathbf{x}_2 \times \mathbf{F}_2 + \cdots + \mathbf{x}_n \times \mathbf{F}_n = \mathbf{0}.$$

EXERCISES

Find the volume of the parallelepiped determined by

1. $(1, 1, 0), (0, 1, 1), (1, 0, 1)$
2. $(4, -1, 0), (3, 0, 2), (1, 1, 1)$.

Find the line of intersection in parametric form of:

3. $x + 2y + 3z = 0, \quad y - z = 1$
4. $x - y + z = 0, \quad x + y + z = 3$
5. $x + 2y + z = 3, \quad 2x - y + z = 4$
6. $x + y = 1, \quad y + z = -1$.

Find a non-trivial solution, if it exists, of

7. $\begin{cases} -2x + 6y = 0 \\ 3x - 9y = 0 \end{cases}$

8. $\begin{cases} -12x + 4y = 0 \\ 3x - y = 0 \end{cases}$

9. $\begin{cases} 5x + 4y + 3z = 0 \\ -x + 2y + z = 0 \\ 3x + y + z = 0 \end{cases}$

10. $\begin{cases} -2x + 2y + 4z = 0 \\ -3x - 8y - 5z = 0 \\ -3x - y + 2z = 0 \end{cases}$

11 $\begin{cases} 3x - 4y + 2z = 0 \\ 5x + 6y = 0 \\ x + 5y - z = 0 \end{cases}$

12. $\begin{cases} 3x + 3y + 2z = 0 \\ 7x + 5y + 12z = 0 \\ x + 2y - 3z = 0 \end{cases}$

13. $\begin{cases} 4x + 3y + 5z = 0 \\ -4x - 3y - 5z = 0 \\ 12x + 9y + 15z = 0 \end{cases}$

14. $\begin{cases} 6x - 9y + 12z = 0 \\ 2x - 3y + 4z = 0 \\ -10x + 15y - 20z = 0. \end{cases}$

Find an equation for the parametric plane:

15. $\mathbf{x} = (1, s, t)$
16. $\mathbf{x} = (s, s + t, -1 + t)$
17. $\mathbf{x} = (2 + s, 1 + s + t, s - t)$
18. $\mathbf{x} = (3s, 2s - t, 1 + 2t)$.

Find an equation for the plane through the three points:

19. $(a, 0, 0), (0, b, 0), (0, 0, c), abc \neq 0$
20. $(1, 1, 0), (1, 0, 1), (0, 1, 1)$.

21. Prove

$$\begin{vmatrix} a_1 & b_1 & c_1 \\ a_2 & b_2 & c_2 \\ a_3 & b_3 & c_3 \end{vmatrix}^2 \leq (a_1^2 + a_2^2 + a_3^2)(b_1^2 + b_2^2 + b_3^2)(c_1^2 + c_2^2 + c_3^2)$$

by interpreting the determinant as a volume.

22. A seesaw with unequal arms of lengths a and b is in horizontal equilibrium. Find the relations between weights A and B at the ends and the upward reaction C at the fulcrum.
23. Unit vertical forces act downward at the points $\mathbf{p}_1, \cdots, \mathbf{p}_n$ of the horizontal x, y-plane. A force $\mathbf{F}$ acts at another point $\mathbf{p}$ of the plane so that the rigid system is in equilibrium. Find $\mathbf{F}$ and $\mathbf{p}$.
24. A force $\mathbf{F}$ is applied at a point $\mathbf{x}$. Its **torque about a point p** is $(\mathbf{x} - \mathbf{p}) \times \mathbf{F}$. Suppose $\mathbf{F}_1, \cdots, \mathbf{F}_n$ are applied at points $\mathbf{x}_1, \cdots, \mathbf{x}_n$ of a rigid body and the body is in equilibrium. Show that the sum of the torques about $\mathbf{p}$ vanishes. (Here $\mathbf{p}$ is any point of space, not just $\mathbf{0}$.)
25. A **couple** consists of a pair of opposite forces $\mathbf{F}$ and $-\mathbf{F}$ applied at two different points $\mathbf{p}$ and $\mathbf{q}$. Show that the total torque is unchanged if $\mathbf{p}$ and $\mathbf{q}$ are displaced the same amount, i.e., replaced by $\mathbf{p} + \mathbf{c}$ and $\mathbf{q} + \mathbf{c}$. Interpret this total torque geometrically.

5. Vector Calculus

1. VECTOR FUNCTIONS

In this chapter we study functions whose values are vectors. For example, the position $\mathbf{x}$ of a moving particle at time t, or the gravitational force $\mathbf{F}$ on an orbiting satellite at time t are vector functions. To indicate that $\mathbf{x}$ is a function of time, we write

$$\mathbf{x} = \mathbf{x}(t);$$

in components,

$$\mathbf{x}(t) = (x(t), y(t), z(t)).$$

Thus a vector function is a single expression for three ordinary (scalar) functions

$$x = x(t), \qquad y = y(t), \qquad z = z(t).$$

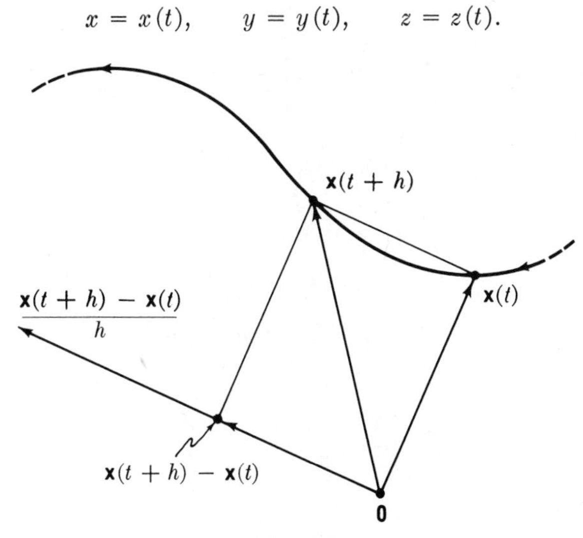

FIG. 1.1

What is the derivative of a vector function? Think of $\mathbf{x} = \mathbf{x}(t)$ as tracing a path in space (Fig. 1.1). For h small, the difference vector

$$\mathbf{x}(t+h) - \mathbf{x}(t)$$

represents the secant from $\mathbf{x}(t)$ to $\mathbf{x}(t+h)$. The difference quotient

$$\frac{\mathbf{x}(t+h) - \mathbf{x}(t)}{h}$$

represents this (short) secant divided by the small number h. The limit as $h \longrightarrow 0$ is called the **derivative** of the vector function:

$$\dot{\mathbf{x}}(t) = \frac{d\mathbf{x}}{dt} = \lim_{h \to 0} \frac{\mathbf{x}(t+h) - \mathbf{x}(t)}{h}.$$

The derivative $\dot{\mathbf{x}}(t)$ is a vector in the direction of the tangent to the curve because the tangent is the limiting position of the secant.

To compute the derivative, express all vectors in components:

$$\frac{\mathbf{x}(t+h) - \mathbf{x}(t)}{h} = \frac{1}{h}[(x(t+h), y(t+h), z(t+h)) - (x(t), y(t), z(t))]$$

$$= \frac{1}{h}(x(t+h) - x(t), y(t+h) - y(t), z(t+h) - z(t))$$

$$= \left(\frac{x(t+h) - x(t)}{h}, \frac{y(t+h) - y(t)}{h}, \frac{z(t+h) - z(t)}{h}\right).$$

It follows that

$$\lim_{h \to 0} \frac{\mathbf{x}(t+h) - \mathbf{x}(t)}{h} = \left(\lim_{h \to 0} \frac{x(t+h) - x(t)}{h}, \cdots, \lim_{h \to 0} \frac{z(t+h) - z(t)}{h}\right)$$

$$= \left(\frac{dx}{dt}, \frac{dy}{dt}, \frac{dz}{dt}\right).$$

The result is

$$\dot{\mathbf{x}} = \frac{d\mathbf{x}}{dt} = \left(\frac{dx}{dt}, \frac{dy}{dt}, \frac{dz}{dt}\right).$$

The derivative of a vector function

$$\mathbf{x}(t) = (x(t), y(t), z(t))$$

is the vector function

$$\frac{d\mathbf{x}}{dt} = \left(\frac{dx}{dt}, \frac{dy}{dt}, \frac{dz}{dt}\right).$$

If a particle moves along a path $\mathbf{x}(t)$, its **velocity** is the vector function

$$\mathbf{v}(t) = \frac{d\mathbf{x}}{dt}.$$

The magnitude $|\mathbf{v}(t)|$ of the velocity is called the **speed.** It is a scalar (numerical) function. The direction of $\mathbf{v}(t)$ is tangential to the path of motion.

EXAMPLE 1.1

The position of a moving particle at time t is (t, t^2, t^3). Find its velocity vector and its speed.

Solution: Let $\mathbf{x}(t) = (t, t^2, t^3)$. Then

$$\mathbf{v}(t) = \dot{\mathbf{x}}(t) = (1, 2t, 3t^2),$$

$$|\mathbf{v}(t)|^2 = 1 + (2t)^2 + (3t^2)^2 = 1 + 4t^2 + 9t^4.$$

Answer: $\mathbf{v}(t) = (1, 2t, 3t^2)$,

speed $= \sqrt{1 + 4t^2 + 9t^4}$.

EXAMPLE 1.2

If $\mathbf{x}(t)$ is a vector function whose derivative is zero, show that $\mathbf{x}(t) = \mathbf{c}$, a constant vector.

Solution:

$$\dot{\mathbf{x}}(t) = (\dot{x}_1(t), \dot{x}_2(t), \dot{x}_3(t)) = (0, 0, 0).$$

Hence $\dot{x}_i(t) = 0$, and so $x_i(t) = c_i$ (constant) for $i = 1, 2, 3$. Therefore

$$\mathbf{x}(t) = (c_1, c_2, c_3) = \mathbf{c}.$$

REMARK: Physically, this example simply says that an object with zero velocity is standing still.

Differentiation Formulas

The following formulas are essential for differentiating vector functions. Each can be verified by differentiating components.

$$\frac{d}{dt}[f(t)\mathbf{x}(t)] = \dot{f}(t)\mathbf{x}(t) + f(t)\dot{\mathbf{x}}(t),$$

$$\frac{d}{dt}[\mathbf{x}(t) + \mathbf{y}(t)] = \dot{\mathbf{x}}(t) + \dot{\mathbf{y}}(t),$$

$$\frac{d}{dt}[\mathbf{x}(t)\cdot\mathbf{y}(t)] = \dot{\mathbf{x}}(t)\cdot\mathbf{y}(t) + \mathbf{x}(t)\cdot\dot{\mathbf{y}}(t).$$

$$\frac{d}{dt}(\mathbf{v} \times \mathbf{w}) = \dot{\mathbf{v}} \times \mathbf{w} + \mathbf{v} \times \dot{\mathbf{w}}.$$

$$\frac{d}{dt}\mathbf{x}[s(t)] = \frac{d\mathbf{x}}{ds}\frac{ds}{dt} \qquad \text{(Chain Rule).}$$

To establish the first formula, for example, write

$$f(t)\mathbf{x}(t) = (f(t)x_1(t), f(t)x_2(t), f(t)x_3(t)).$$

Then

$$\frac{d}{dt}[f(t)\mathbf{x}(t)] = \left(\frac{d}{dt}[f(t)x_1(t)], \frac{d}{dt}[f(t)x_2(t)], \frac{d}{dt}[f(t)x_3(t)]\right)$$

$$= (\dot{f}(t)x_1(t) + f(t)\dot{x}_1(t), \dot{f}(t)x_2(t) + f(t)\dot{x}_2(t), \dot{f}(t)x_3(t) + f(t)\dot{x}_3(t))$$

$$= \dot{f}(t)(x_1(t), x_2(t), x_3(t)) + f(t)(\dot{x}_1(t), \dot{x}_2(t), \dot{x}_3(t))$$

$$= \dot{f}(t)\mathbf{x}(t) + f(t)\dot{\mathbf{x}}(t).$$

EXAMPLE 1.3

Differentiate $t^2\mathbf{x}(t)$, where $\mathbf{x}(t) = (\cos 3t, \sin 3t, t)$.

Solution: Apply the first formula above:

$$\frac{d}{dt}[t^2\mathbf{x}(t)] = 2t\mathbf{x}(t) + t^2\dot{\mathbf{x}}(t)$$

$$= 2t(\cos 3t, \sin 3t, t) + t^2(-3\sin 3t, 3\cos 3t, 1).$$

Answer: $(2t\cos 3t - 3t^2\sin 3t, 2t\sin 3t + 3t^2\cos 3t, 3t^2)$.

EXAMPLE 1.4

Suppose $\mathbf{x}(t)$ is a moving *unit* vector. Show that $\mathbf{x}(t)$ is always perpendicular to its velocity vector $\mathbf{v}(t)$.

Solution: Verify that $\mathbf{x}(t)\cdot\mathbf{v}(t) = 0$:

$$\mathbf{x}\cdot\mathbf{v} = (x_1, x_2, x_3)\cdot\left(\frac{dx_1}{dt}, \frac{dx_2}{dt}, \frac{dx_3}{dt}\right)$$

$$= x_1\frac{dx_1}{dt} + x_2\frac{dx_2}{dt} + x_3\frac{dx_3}{dt} = \frac{1}{2}\frac{d}{dt}[x_1^2 + x_2^2 + x_3^2].$$

But $x_1^2 + x_2^2 + x_3^2 = 1$ for every t, since $\mathbf{x}$ is a unit vector. Hence $\mathbf{x}\cdot\mathbf{v} = 0$.

Alternate Solution:

$$\mathbf{x}(t)\cdot\mathbf{x}(t) = |\mathbf{x}(t)|^2 = 1, \qquad \frac{d}{dt}[\mathbf{x}(t)\cdot\mathbf{x}(t)] = 0.$$

But by the third differentiation formula on the previous page,

$$\frac{d}{dt}[\mathbf{x}(t)\cdot\mathbf{x}(t)] = \dot{\mathbf{x}}(t)\cdot\mathbf{x}(t) + \mathbf{x}(t)\cdot\dot{\mathbf{x}}(t) = 2\mathbf{x}(t)\cdot\dot{\mathbf{x}}(t).$$

Thus $\mathbf{x}(t)\cdot\dot{\mathbf{x}}(t) = 0$, that is, $\mathbf{x}(t)\cdot\mathbf{v}(t) = 0$.

REMARK: This example makes sense geometrically. A moving unit vector represents a particle on the unit sphere $|\mathbf{x}| = 1$. Its velocity vector is tangent to the sphere, i.e., perpendicular to the radius.

EXERCISES

Differentiate:

1. $\mathbf{x}(t) = (e^t, e^{2t}, e^{3t})$
2. $\mathbf{x}(t) = (t^4, t^5, t^6)$
3. $\mathbf{x}(t) = (t+1, 3t-1, 4t)$
4. $\mathbf{x}(t) = (t^2, 0, t^3)$.

Find the velocity and the speed:

5. $\mathbf{x}(t) = (t^2, t^3 + t^4, 1)$
6. $\mathbf{x}(t) = (2t-1, 3t+1, -2t+1)$
7. $\mathbf{x}(t) = (A \cos \omega t, A \sin \omega t, Bt)$
8. $\mathbf{x}(t) = (a_1 t + b_1, a_2 t + b_2, a_3 t + b_3)$.
9. Suppose that $\mathbf{x} = \mathbf{x}(t)$ is a moving point such that $\dot{\mathbf{x}}(t)$ is always perpendicular to $\mathbf{x}(t)$. Show that $\mathbf{x}(t)$ moves on a sphere with center at $\mathbf{0}$.
   [*Hint:* Differentiate $|\mathbf{x}|^2$.]
10. Suppose $\mathbf{x}(t) \neq \mathbf{0}$. Show that $\dfrac{d}{dt}|\mathbf{x}(t)| = \dfrac{1}{|\mathbf{x}|}\mathbf{x}\cdot\dot{\mathbf{x}}$.
11. Prove the formula $\dfrac{d}{dt}[\mathbf{x}(t) + \mathbf{y}(t)] = \dot{\mathbf{x}}(t) + \dot{\mathbf{y}}(t)$.
12. Prove the formula $\dfrac{d}{dt}[\mathbf{x}(t)\cdot\mathbf{y}(t)] = \dot{\mathbf{x}}\cdot\mathbf{y} + \mathbf{x}\cdot\dot{\mathbf{y}}$.
13. Prove the formula $\dfrac{d}{dt}(\mathbf{x}\times\mathbf{y}) = \dot{\mathbf{x}}\times\mathbf{y} + \mathbf{x}\times\dot{\mathbf{y}}$.
14. Suppose $\mathbf{x}(t)$ is a space curve which does not pass through $\mathbf{0}$, and that $\mathbf{x}(t_0)$ is the point of the curve closest to $\mathbf{0}$. Show that $\mathbf{x}(t_0)\cdot\dot{\mathbf{x}}(t_0) = 0$.
15. Suppose that $\mathbf{x}(t)$ and $\mathbf{y}(\tau)$ are two space curves which do not intersect. Suppose the distance $\mathbf{x}(t) - \mathbf{y}(\tau)$ is minimal at $t = t_0$ and $\tau = \tau_0$. Show that the vector $\mathbf{x}(t_0) - \mathbf{y}(\tau_0)$ is perpendicular to the tangents to the two curves at $\mathbf{x}(t_0)$ and $\mathbf{y}(\tau_0)$, respectively.

2. SPACE CURVES

In this section we study the arc lengths and the tangents of curves in the plane and in space. To avoid analytic difficulties, we shall always assume that the vector functions under consideration have as many continuous derivatives as are needed. This applies to the following sections also.

Length of a Curve

Let $\mathbf{x} = \mathbf{x}(t)$ represent a curve in space. How long is the part of the curve between the points $\mathbf{x}(t_0)$ and $\mathbf{x}(t_1)$? To answer this question, we need a reasonable definition of curve length. Intuitively, the velocity vector $\dot{\mathbf{x}}(t)$ is directed tangent to the curve (Fig. 2.1), and its length $|\dot{\mathbf{x}}(t)|$ represents *speed*, the rate at which distance s along the curve increases with respect to time.

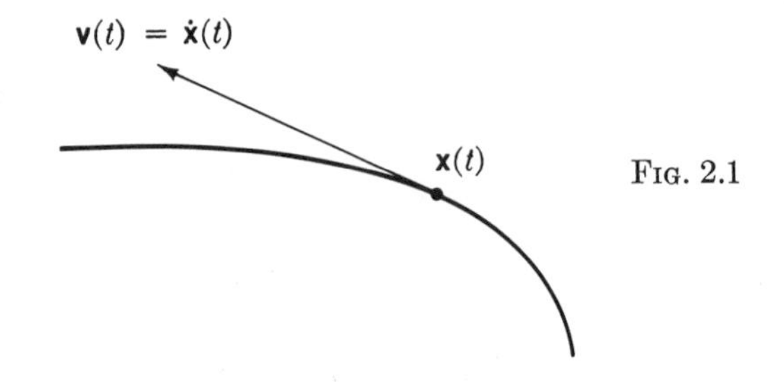

Fig. 2.1

This leads us to *define* arc length s by

$$\frac{ds}{dt} = |\mathbf{v}(t)| = |\dot{\mathbf{x}}(t)|, \qquad s(t_0) = 0.$$

Therefore,

$$\left(\frac{ds}{dt}\right)^2 = |\dot{\mathbf{x}}(t)|^2 = \left|\left(\frac{dx}{dt}, \frac{dy}{dt}, \frac{dz}{dt}\right)\right|^2$$

$$= \left(\frac{dx}{dt}\right)^2 + \left(\frac{dy}{dt}\right)^2 + \left(\frac{dz}{dt}\right)^2.$$

In terms of differentials,

$$ds = \sqrt{\left(\frac{dx}{dt}\right)^2 + \left(\frac{dy}{dt}\right)^2 + \left(\frac{dz}{dt}\right)^2}\, dt.$$

This formula has a simple geometric interpretation. See Fig. 2.2. The tiny bit of arc length ds corresponds to three "displacements" dx, dy, and dz along the coordinate axes. By the Distance Formula,

$$(ds)^2 = (dx)^2 + (dy)^2 + (dz)^2.$$

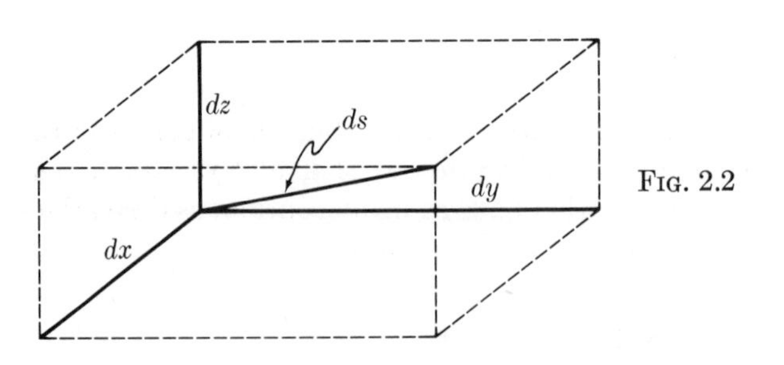

Fig. 2.2

Divide by $(dt)^2$ and take square roots to obtain

$$\frac{ds}{dt} = \sqrt{\left(\frac{dx}{dt}\right)^2 + \left(\frac{dy}{dt}\right)^2 + \left(\frac{dz}{dt}\right)^2}.$$

This is the time derivative of arc length. Integrate it to obtain the arc length itself.

Suppose $\mathbf{x}(t)$ describes a curve in space. Let $s(t)$ denote the length of the curve measured from a fixed initial point. Then

$$\frac{ds}{dt} = \sqrt{\dot{x}^2 + \dot{y}^2 + \dot{z}^2}.$$

The **length** of the curve from $\mathbf{x}(t_0)$ to $\mathbf{x}(t_1)$ is

$$L = \int_{t_0}^{t_1} \sqrt{\dot{x}^2 + \dot{y}^2 + \dot{z}^2}\, dt.$$

For plane curves the formula is slightly simpler because $z = 0$.

Suppose $\mathbf{x}(t) = (x(t), y(t))$ describes a plane curve. The length of the curve from $\mathbf{x}(t_0)$ to $\mathbf{x}(t_1)$ is

$$L = \int_{t_0}^{t_1} \sqrt{\dot{x}^2 + \dot{y}^2}\, dt.$$

If the curve is the graph of a function $y = f(x)$, then its length from $(x_0, f(x_0))$ to $(x_1, f(x_1))$ is

$$L = \int_{x_0}^{x_1} \sqrt{1 + (f')^2}\, dx.$$

The last formula is a special case of the preceding one. Indeed, set $x = t$, $y = f(t)$, where $x_0 \leq t \leq x_1$. Then $\dot{x} = 1$ and $\dot{y} = \dot{f}$, so

$$\frac{ds}{dt} = \sqrt{\dot{x}^2 + \dot{y}^2} = \sqrt{1 + \dot{f}^2} = \sqrt{1 + (f')^2}.$$

The formula for L follows.

EXAMPLE 2.1

Find the length of the parabola $\mathbf{x}(t) = (t, t^2)$, $0 \leq t \leq 1$.

Solution: This plane curve is a parabola because

$$x = t, \qquad y = t^2,$$

hence $y = x^2$. Its length is

$$\int_0^1 \sqrt{\dot{x}^2 + \dot{y}^2}\, dt = \int_0^1 \sqrt{1 + (2t)^2}\, dt.$$

From integral tables,

$$\int_0^1 \sqrt{1+4t^2}\,dt = \frac{\sqrt{5}}{2} + \frac{1}{4}\ln(2+\sqrt{5}).$$

Answer: $\frac{1}{4}[2\sqrt{5} + \ln(2+\sqrt{5})] \approx 1.479.$

EXAMPLE 2.2

Find the length of the curve $y = \sin x$ for $0 \le x \le \pi$.

Solution: The length L is given by

$$L = \int_0^\pi \sqrt{1 + \left(\frac{dy}{dx}\right)^2}\,dx = \int_0^\pi \sqrt{1+\cos^2 x}\,dx.$$

The exact evaluation of this integral is impossible. It can, however, be approximated by Simpson's Rule.

Answer: $L = \int_0^\pi \sqrt{1+\cos^2 x}\,dx \approx 3.820.$

EXAMPLE 2.3

Find the length of the curve $\mathbf{x}(t) = (t\cos t, t\sin t, 2t)$, $0 \le t \le 4\pi$. Sketch the curve.

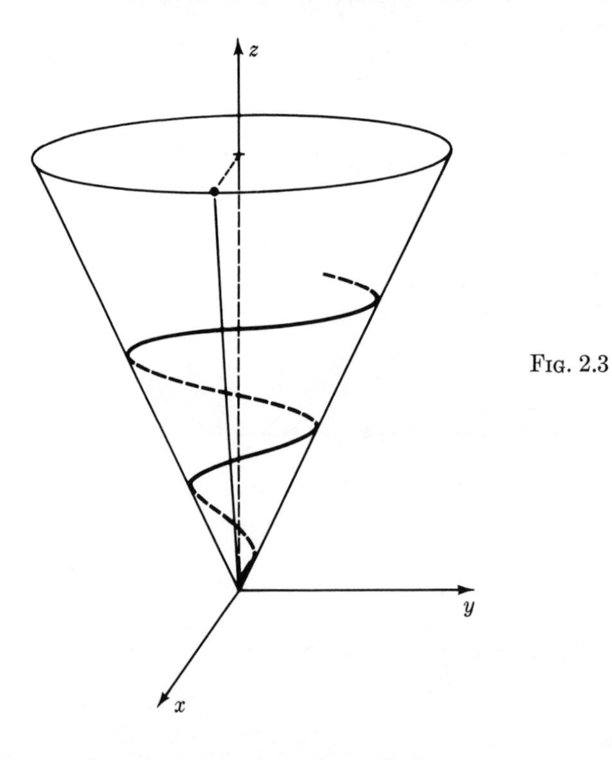

FIG. 2.3

Solution: Since

$$x^2 + y^2 = (t \cos t)^2 + (t \sin t)^2 = t^2 = \frac{1}{4} z^2,$$

the curve lies on the right circular cone $z^2 = 4(x^2 + y^2)$. As t increases, z steadily increases also, while the projection $(t \cos t, t \sin t)$ of $\mathbf{x}(t)$ on the x, y-plane traces a spiral. Hence the space curve $\mathbf{x} = \mathbf{x}(t)$ is a spiral on the surface of the cone (Fig. 2.3).

Compute ds/dt:

$$\begin{aligned}\left(\frac{ds}{dt}\right)^2 &= \left(\frac{dx}{dt}\right)^2 + \left(\frac{dy}{dt}\right)^2 + \left(\frac{dz}{dt}\right)^2 \\ &= (\cos t - t \sin t)^2 + (\sin t + t \cos t)^2 + (2)^2 \\ &= 5 + t^2.\end{aligned}$$

Hence

$$L = \int_0^{4\pi} \sqrt{5 + t^2}\, dt = \frac{1}{2}\left[t\sqrt{5 + t^2} + 5 \ln\left(t + \sqrt{5 + t^2}\right)\right]\Bigg|_0^{4\pi}.$$

Answer: $L = \dfrac{1}{2}\left[4\pi a + 5 \ln(4\pi + a) - \dfrac{5}{2} \ln 5\right],$

where $a = \sqrt{5 + 16\pi^2}$. $L \approx 86.3$.

REMARK: Suppose the same geometric curve has two different parametrizations. How do we know that we get the same length? We might have $\mathbf{x} = \mathbf{x}(t)$ where $t_0 \leq t \leq t_1$, and $\mathbf{x} = \mathbf{x}(u)$ where the corresponding interval on the u-axis is $u_0 \leq u \leq u_1$. We suppose we can obtain either parametrization from the other by a smooth change of variable. Let us take $t = t(u)$ for the change of variable. We assume $t_0 = t(u_0)$, $t_1 = t(u_1)$, and $dt/du > 0$. The t-length and the u-length of the curve are

$$L_t = \int_{t_0}^{t_1} \left|\frac{d\mathbf{x}}{dt}\right| dt \quad \text{and} \quad L_u = \int_{u_0}^{u_1} \left|\frac{d\mathbf{x}}{du}\right| du.$$

By the Chain Rule, $d\mathbf{x}/du = (d\mathbf{x}/dt)(dt/du)$. The formula for change of variable in a definite integral implies

$$L_u = \int_{u_0}^{u_1} \left|\frac{d\mathbf{x}}{dt}\right| \frac{dt}{du}\, du = \int_{t_0}^{t_1} \left|\frac{d\mathbf{x}}{dt}\right| dt = L_t.$$

This proves that the length of a curve is a geometric quantity, independent of the analytic representation of the curve.

Unit Tangent Vector

EXAMPLE 2.4

Plot the locus $\mathbf{x}(t) = (t^2, t^3)$.

Solution: The locus is the plane curve described by

$$x = t^2, \qquad y = t^3.$$

Hence

$$x^3 = y^2, \qquad y = \pm x^{3/2}.$$

The curve is defined only for $x \geq 0$. For each positive value of x, there are two values of y. (See Fig. 2.4.)

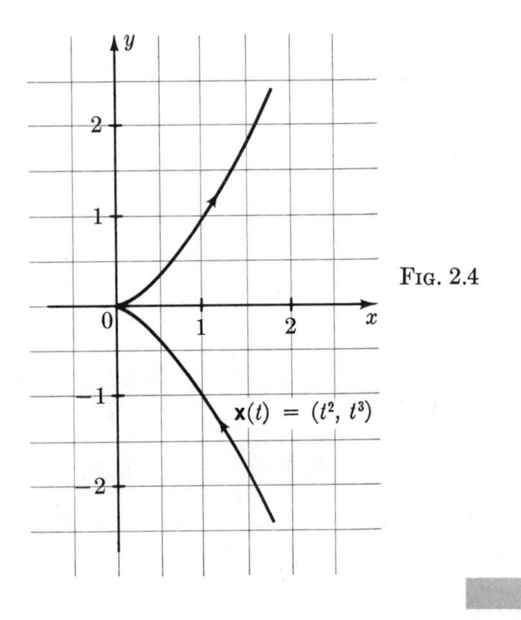

FIG. 2.4

REMARK: The sharp point at the origin is called a **cusp**. At that point, a particle moving along the curve changes direction abruptly. Note that its velocity at the origin is zero since

$$\mathbf{v} = \dot{\mathbf{x}} = (2t, 3t^2), \qquad \mathbf{v}(0) = \mathbf{0}.$$

In fact, an abrupt change in direction can occur only when the velocity vector is zero. Physically, this seems plausible; a moving particle cannot change direction suddenly unless it slows down to an instantaneous stop at the "corner."

To avoid such curves with cusps as shown in Fig. 2.4, we study only curves $\mathbf{x}(t)$ for which the velocity $\dot{\mathbf{x}}(t)$ *never* equals zero.

Suppose a particle moves along a curve $\mathbf{x}(t)$. Its velocity vector $\mathbf{v}(t) = d\mathbf{x}/dt$ has length ds/dt and is directed along the tangent to the curve; hence

$$\mathbf{v} = \frac{d\mathbf{x}}{dt} = \frac{ds}{dt}\mathbf{T},$$

where $\mathbf{T}$ is a unit vector in the tangential direction. But by the Chain Rule,

$$\mathbf{v} = \frac{d\mathbf{x}}{dt} = \frac{d\mathbf{x}}{ds}\frac{ds}{dt}.$$

Compare these two expressions for $\mathbf{v}$; the result is

$$\frac{ds}{dt}\frac{d\mathbf{x}}{ds} = \frac{ds}{dt}\mathbf{T}.$$

Therefore $d\mathbf{x}/ds = \mathbf{T}$ since $ds/dt \neq 0$ is assumed.

If $\mathbf{x}(t)$ represents a space curve, then

$$\frac{d\mathbf{x}}{ds} = \mathbf{T}$$

is the **unit tangent vector** to the curve. In terms of the velocity vector $\mathbf{v}$,

$$\mathbf{T} = \frac{\mathbf{v}}{|\mathbf{v}|}.$$

(It is assumed $\mathbf{v} \neq \mathbf{0}$.)

EXAMPLE 2.5

Find the unit tangent vector to the curve $\mathbf{x}(t) = (t, t^2, t^3)$ at the point $\mathbf{x}(1) = (1, 1, 1)$.

Solution:

$$\mathbf{T} = \frac{\mathbf{v}}{|\mathbf{v}|},$$

where

$$\mathbf{v} = \dot{\mathbf{x}} = (1, 2t, 3t^2), \qquad |\mathbf{v}|^2 = 1 + 4t^2 + 9t^4.$$

Hence

$$\mathbf{T} = \frac{1}{\sqrt{1 + 4t^2 + 9t^4}}(1, 2t, 3t^2).$$

Now substitute $t = 1$.

Answer: $\mathbf{T} = \dfrac{1}{\sqrt{14}}(1, 2, 3).$

EXERCISES

1. Find the arc length of $\mathbf{x}(t) = (a_1t + b_1, a_2t + b_2, a_3t + b_3)$ for $0 \leq t \leq 1$.
2. Find the length of $\mathbf{x}(t) = (t^2, t^3)$ for $0 \leq t \leq a$.

3. Find the length of $\mathbf{x}(t) = (t, \sin t, \cos t)$ for $0 \leq t \leq 2\pi$.
4. Set up the length of $y = x^3$ for $-1 \leq x \leq 1$, but do not evaluate the integral.
5. Set up the length of $y = ax^n$ for $x_0 \leq x \leq x_1$, but do not evaluate the integral.
6. Set up the length of $\mathbf{x}(t) = (t^m, t^n, t^r)$ for $0 \leq t \leq b$, but do not evaluate the integral.
7. Find the length of $y = -x^2 + 2x$ for $-1 \leq x \leq 1$.
8. Carefully plot $\mathbf{x}(t) = (t^2, t^4 + t^5)$ for t near 0.
9. Find the unit tangent $\mathbf{T}$ to the curve $\mathbf{x}(t) = (t, \cos t, \sin t)$ at $t = 0$.
10. Find the unit tangent $\mathbf{T}$ to the curve $\mathbf{x}(t) = (3t - 1, 4t, -2t + 1)$ at any point.
11. Find the unit tangent $\mathbf{T}$ to the curve $\mathbf{x}(t) = (a_1t + b_1, a_2t + b_2, a_3t + b_3)$ at any point.
12. Find the unit tangent $\mathbf{T}$ to the curve $\mathbf{x}(t) = (t \cos t, t \sin t, 2t)$ at any point.

3. CURVATURE

The curvature of a curve is a quantity which tells how fast the direction of the curve is changing relative to arc length.

The magnitude of the rate of change of the unit tangent $\mathbf{T}$ with respect to arc length is called the **curvature** of a curve, and is denoted by k:

$$k = \left| \frac{d\mathbf{T}}{ds} \right| .$$

Since the length of $\mathbf{T}$ is constant, $\mathbf{T}$ changes in direction only. Thus the curvature k measures its rate of change of direction. The curvature is a geometric quantity; it does not depend on how the curve is parametrized.

EXAMPLE 3.1

A curve has curvature zero. What is the curve?

Solution: A natural guess is a straight line. Let is prove this is so. We are given $k = 0$. Therefore,

$$\left| \frac{d\mathbf{T}}{ds} \right| = 0,$$

hence

$$\frac{d\mathbf{T}}{ds} = \mathbf{0}.$$

Consequently $\mathbf{T}$ is constant,

$$\mathbf{T} = \mathbf{T}_0 = (t_1, t_2, t_3).$$

But $d\mathbf{x}/ds = \mathbf{T}_0$, which means

$$\frac{dx}{ds} = t_1, \qquad \frac{dy}{ds} = t_2, \qquad \frac{dz}{ds} = t_3.$$

Integrating, we have

$$x = a + t_1 s, \qquad y = b + t_2 s, \qquad z = c + t_3 s.$$

In vector notation,

$$\mathbf{x} = \mathbf{x}_0 + s\mathbf{T}_0.$$

But this is the vector equation of the line through $\mathbf{x}_0$ parallel to $\mathbf{T}_0$.

Answer: A straight line.

Computation of Curvature

The following three formulas are needed to compute curvature. The first two apply to curves given in parametric form, the third to the graph of a function.

If $\mathbf{x} = \mathbf{x}(t)$ is a space curve, then

$$k = \frac{[|\dot{\mathbf{x}}|^2 |\ddot{\mathbf{x}}|^2 - (\dot{\mathbf{x}} \cdot \ddot{\mathbf{x}})^2]^{1/2}}{|\dot{\mathbf{x}}|^3}.$$

If $\mathbf{x} = (x(t), y(t))$ is a plane curve, then

$$k = \frac{\dot{x}\ddot{y} - \dot{y}\ddot{x}}{(\dot{x}^2 + \dot{y}^2)^{3/2}}.$$

If a plane curve is the graph of a function $y = f(x)$, then

$$k = \frac{|f''(x)|}{[1 + f'(x)^2]^{3/2}}.$$

Proof: By the Chain Rule,

$$\dot{\mathbf{x}} = \frac{ds}{dt}\frac{d\mathbf{x}}{ds} = \frac{ds}{dt}\mathbf{T}, \qquad \ddot{\mathbf{x}} = \frac{d^2s}{dt^2}\mathbf{T} + \frac{ds}{dt}\frac{d\mathbf{T}}{dt} = \frac{d^2s}{dt^2}\mathbf{T} + \left(\frac{ds}{dt}\right)^2 \frac{d\mathbf{T}}{ds}.$$

Hence

$$|\dot{\mathbf{x}}|^2 = \dot{\mathbf{x}} \cdot \dot{\mathbf{x}} = \left(\frac{ds}{dt}\right)^2, \qquad \dot{\mathbf{x}} \cdot \ddot{\mathbf{x}} = \frac{ds}{dt}\frac{d^2s}{dt^2},$$

$$|\ddot{\mathbf{x}}|^2 = \ddot{\mathbf{x}} \cdot \ddot{\mathbf{x}} = \left(\frac{d^2s}{dt^2}\right)^2 + \left(\frac{ds}{dt}\right)^4 \left|\frac{d\mathbf{T}}{ds}\right|^2 = \left(\frac{d^2s}{dt^2}\right)^2 + \left(\frac{ds}{dt}\right)^4 k^2.$$

Consequently

$$|\dot{\mathbf{x}}|^2\,|\ddot{\mathbf{x}}|^2 - (\dot{\mathbf{x}}\cdot\ddot{\mathbf{x}})^2 = \left(\frac{ds}{dt}\right)^6 k^2 = |\dot{\mathbf{x}}|^6\,k^2.$$

The first formula for k follows.

If $\mathbf{x} = (x(t), y(t))$ is a plane curve, then

$$\dot{\mathbf{x}} = (\dot{x}, \dot{y}) \quad \text{and} \quad \ddot{\mathbf{x}} = (\ddot{x}, \ddot{y}),$$

hence

$$\begin{aligned}|\dot{\mathbf{x}}|^2\,|\ddot{\mathbf{x}}|^2 - (\dot{\mathbf{x}}\cdot\ddot{\mathbf{x}})^2 &= (\dot{x}^2 + \dot{y}^2)(\ddot{x}^2 + \ddot{y}^2) - (\dot{x}\ddot{x} + \dot{y}\ddot{y})^2\\ &= (\dot{x}\ddot{y} - \dot{y}\ddot{x})^2,\end{aligned}$$

so the second formula follows.

Finally, if the plane curve is the graph of $y = f(x)$, apply the second formula with $t = x$ and $\mathbf{x} = (t, f(t)) = (x, f(x))$. Then $\dot{x} = 1$, $\ddot{x} = 0$, $\dot{y} = f'(x)$, and $\ddot{y} = f''(x)$, so the third formula follows by direct substitution.

EXAMPLE 3.2

Find the curvature of a circle of radius a.

Solution: Let the equation of the circle be $x^2 + y^2 = a^2$. Thus

$$y = \pm\sqrt{a^2 - x^2}.$$

(This equation describes either the upper or lower half of the circle depending on whether the positive or negative square root is chosen.) Differentiate:

$$y' = \frac{-x}{\pm\sqrt{a^2 - x^2}} = -\frac{x}{y}.$$

Differentiate again:

$$y'' = -\frac{y - xy'}{y^2} = -\frac{y - x(-x/y)}{y^2} = -\frac{x^2 + y^2}{y^3} = -\frac{a^2}{y^3}.$$

Now

$$1 + y'^2 = 1 + \left(-\frac{x}{y}\right)^2 = \frac{y^2 + x^2}{y^2} = \frac{a^2}{y^2}.$$

Hence by the formula for curvature,

$$k = \frac{|-a^2/y^3|}{(a^2/y^2)^{3/2}} = \frac{a^2}{a^3} = \frac{1}{a}.$$

Alternate Solution: Write

$$\mathbf{x}(t) = (a\cos t, a\sin t).$$

(This describes the circle by its central angle t.) Then

$$\dot{\mathbf{x}} = (-a \sin t,\ a \cos t), \qquad |\dot{\mathbf{x}}|^2 = \left(\frac{ds}{dt}\right)^2 = (-a \sin t)^2 + (a \cos t)^2 = a^2,$$

$$|\dot{\mathbf{x}}| = \frac{ds}{dt} = a.$$

Hence

$$\mathbf{T} = \frac{\dot{\mathbf{x}}}{|\dot{\mathbf{x}}|} = (-\sin t,\ \cos t).$$

Differentiate:

$$\frac{ds}{dt}\frac{d\mathbf{T}}{ds} = \frac{d\mathbf{T}}{dt} = (-\cos t,\ -\sin t).$$

Take lengths, substituting $a = ds/dt$:

$$a\left|\frac{d\mathbf{T}}{ds}\right| = [(-\cos t)^2 + (-\sin t)^2]^{1/2} = 1, \qquad k = \left|\frac{d\mathbf{T}}{ds}\right| = \frac{1}{a}.$$

Answer: $k = \dfrac{1}{a}$.

REMARK: The curvature of a circle is the reciprocal of its radius. This is reasonable on two counts. First, the curvature is the same at all points of a circle. Second, it is small for large circles, since the larger the circle the more slowly its direction changes per unit of arc length.

The Unit Normal

The vector $d\mathbf{T}/ds$ has length k, the curvature. Therefore

$$\frac{d\mathbf{T}}{ds} = k\mathbf{N},$$

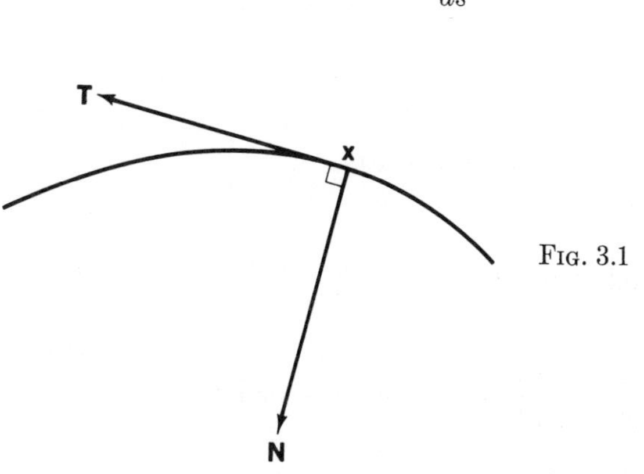

FIG. 3.1

where $\mathbf{N}$ is a unit vector in the direction of $d\mathbf{T}/ds$. (We assume $k \neq 0$.) Since $\mathbf{T}$ is a unit vector, $\mathbf{T}$ is perpendicular to $d\mathbf{T}/ds$; this was shown in Example 1.4. The vector $\mathbf{N}$ is called the **unit normal vector** to the curve (Fig. 3.1).

We summarize:

Let $\mathbf{x}(t)$ represent a curve in space.

$$\frac{d\mathbf{x}}{ds} = \mathbf{T}.$$

$$\frac{d\mathbf{T}}{ds} = k\mathbf{N}, \qquad k = k(s).$$

$$|\mathbf{T}| = |\mathbf{N}| = 1, \qquad \mathbf{T}\cdot\mathbf{N} = 0.$$

The further study of space curves, not pursued here, begins with an analysis of $d\mathbf{N}/ds$. That leads to another quantity, the torsion, which measures how fast the plane of $\mathbf{T}$ and $\mathbf{N}$ is turning around the tangent line.

EXAMPLE 3.3

Compute $\mathbf{T}$, $\mathbf{N}$ and k for the circular spiral (helix) $\mathbf{x}(t) = (a\cos t, a\sin t, bt)$. Assume $a > 0$ and $b > 0$.

Solution: The projection of $\mathbf{x}(t)$ on the x, y-plane is $(a\cos t, a\sin t, 0)$. As a particle describes the curve $\mathbf{x}(t)$, its projection describes a circle of radius a. The third component of $\mathbf{x}(t)$ is bt; the particle moves upward at a steady rate. Thus, the curve is a spiral; it is circular but steadily rising (Fig. 3.2). Differentiate $\mathbf{x} = (a\cos t, a\sin t, bt)$:

$$\dot{\mathbf{x}} = (-a\sin t, a\cos t, b).$$

Introduce $c > 0$ by $c^2 = a^2 + b^2$. Then

$$|\dot{\mathbf{x}}|^2 = a^2 + b^2 = c^2, \qquad \frac{ds}{dt} = |\dot{\mathbf{x}}| = c,$$

and

$$\mathbf{T} = \frac{1}{c}\dot{\mathbf{x}} = \frac{1}{c}(-a\sin t, a\cos t, b).$$

To find k and $\mathbf{N}$, use the relation $k\mathbf{N} = \dfrac{d\mathbf{T}}{ds}$:

$$k\mathbf{N} = \frac{d\mathbf{T}}{ds} = \frac{d\mathbf{T}/dt}{ds/dt} = \frac{1}{c}\frac{d\mathbf{T}}{dt} = \frac{a}{c^2}(-\cos t, -\sin t, 0).$$

Since $k \geq 0$ and $\mathbf{N}$ is a unit vector,

$$k = \frac{a}{c^2}, \qquad \mathbf{N} = (-\cos t, -\sin t, 0).$$

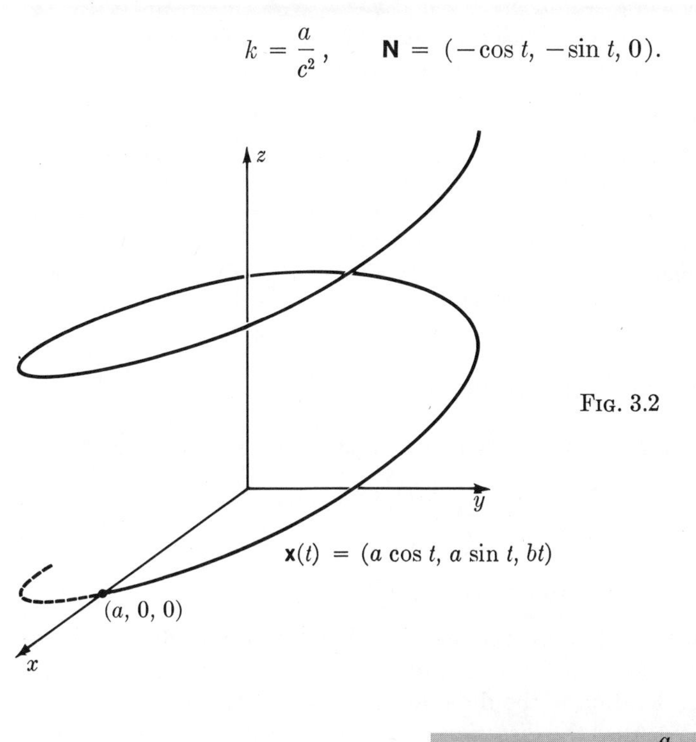

Fig. 3.2

> *Answer:* $k = \dfrac{a}{a^2 + b^2}$,
>
> $$\mathbf{T} = \frac{1}{\sqrt{a^2 + b^2}}(-a \sin t, a \cos t, b),$$
>
> $\mathbf{N} = (-\cos t, -\sin t, 0)$.

Remark: If $b = 0$, the spiral degenerates into a circle of radius a and the curvature k reduces to $1/a$, which agrees with Example 3.2.

EXERCISES

Find the curvature:

1. $y = x^2$; at $x = 1$
2. $\mathbf{x}(t) = (t^3, t^2)$; at $t = 1$
3. $\mathbf{x}(t) = (t, t^2, t^3)$; at $t = -1$
4. $\mathbf{x}(t) = (a_1t + a_2t^3, b_1t + b_2t^3, c_1t + c_2t^3)$; at $t = 0$.
5. Let $\mathbf{x} = \mathbf{x}(s)$ be a plane curve. Show that $d\mathbf{N}/ds = -k\mathbf{T}$. [*Hint:* Differentiate $\mathbf{T}\cdot\mathbf{N} = 0$ and $\mathbf{N}\cdot\mathbf{N} = 1$.]
6. Find the point of the plane curve $y = x^2$ where k is maximum.

7. Find the point of $y = \sin x$ where k is maximum, $0 < x < \pi$.
8. Find the curvature of $y = x^3$ at $x = 0$ and at $x = 1$.
9. Show that the curvature of a plane curve at an inflection point is zero.
10. Let the tangent line of a plane curve intersect the x-axis with angle α. Show that $k = |d\alpha/ds|$.
   [*Hint:* Write $\mathbf{T} = (\cos\alpha, \sin\alpha)$.]
11. A point moves along the curve $y = e^x$ at the rate of 3 in./sec. How fast is the tangent turning when the point is at $(2, e^2)$?
12. Compute the maximum and minimum curvature of an ellipse with semimajor axis a and semiminor axis b. Check the case $a = b$.
13. From a graph, predict the behavior of the curvature of $y = 1/x$ as $x \longrightarrow 0$ and as $x \longrightarrow \infty$. Verify your prediction.

4. VELOCITY AND ACCELERATION

If $\mathbf{x} = \mathbf{x}(t)$ represents the position of a moving particle, its **velocity** is

$$\mathbf{v}(t) = \dot{\mathbf{x}}(t)$$

and its **acceleration** is

$$\mathbf{a}(t) = \dot{\mathbf{v}}(t) = \ddot{\mathbf{x}}(t).$$

Velocity and acceleration are vectors, each having magnitude and direction. The direction of the velocity is the direction the particle is moving. The direction of the acceleration is the direction the particle is turning. The following example shows that the direction of the acceleration is not necessarily that of the velocity; it may even be perpendicular to the velocity.

EXAMPLE 4.1

The path of a particle moving around the circle $x^2 + y^2 = r^2$ is given by $\mathbf{x}(t) = (r\cos\omega t, r\sin\omega t)$, where ω is a constant. Find its velocity and acceleration vectors.

Solution: Differentiate to find $\mathbf{v}$ and $\mathbf{a}$:

$$\mathbf{v}(t) = \dot{\mathbf{x}}(t) = r\omega(-\sin\omega t, \cos\omega t),$$

$$\mathbf{a}(t) = \dot{\mathbf{v}}(t) = r\omega^2(-\cos\omega t, -\sin\omega t) = -\omega^2\mathbf{x}(t).$$

Answer: $\mathbf{v}(t) = r\omega(-\sin\omega t, \cos\omega t)$,

$\mathbf{a}(t) = r\omega^2(-\cos\omega t, -\sin\omega t)$.

REMARK: The speed, $|\mathbf{v}| = r\omega$, is constant; the motion is uniform circular motion. The velocity $\mathbf{v}(t)$ is perpendicular to the position vector $\mathbf{x}(t)$ since $\mathbf{x}(t)\cdot\mathbf{v}(t) = 0$. This is expected since each tangent to a circle is perpendicular to the corresponding radius. But $\mathbf{a}(t) = -\omega^2\mathbf{x}(t)$, so the acceleration vector $\mathbf{a}(t)$ is directed opposite to the position vector $\mathbf{x}(t)$. See Fig. 4.1. What is the physical meaning of this phenomenon?

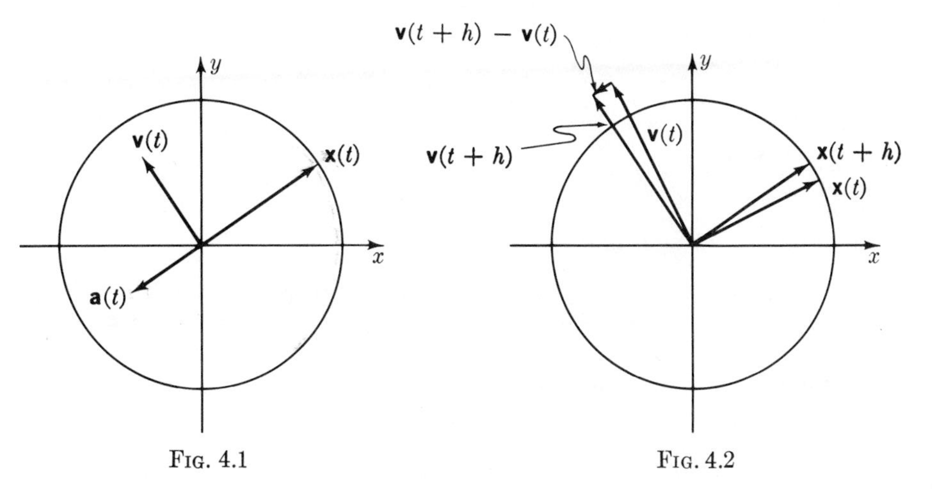

FIG. 4.1

FIG. 4.2

Remember that $\mathbf{a}(t)$ is the rate of change of the velocity vector. Observe the velocity vectors at t and an instant later at $t + h$. See Fig. 4.2. The difference $\mathbf{v}(t + h) - \mathbf{v}(t)$ is nearly parallel, but oppositely directed, to $\mathbf{x}(t)$. Thus the velocity is changing in a direction opposite to $\mathbf{x}(t)$. It seems reasonable, therefore, that $\mathbf{a}(t) = -c\mathbf{x}(t)$, where $c > 0$.

Newton's Law of Motion

This famous principle states that

$$\text{force} = \text{mass} \times \text{acceleration}.$$

But force and acceleration are vectors, both have magnitude and direction. Thus Newton's Law is a vector equation:

$$\mathbf{F} = m\ddot{\mathbf{x}}.$$

It is equivalent to three scalar equations for the components:

$$F_1 = m\ddot{x}_1, \qquad F_2 = m\ddot{x}_2, \qquad F_3 = m\ddot{x}_3.$$

EXAMPLE 4.2

A particle of mass m is subject to zero force. What is its trajectory?

Solution: By Newton's Law,

$$m\ddot{\mathbf{x}} = \mathbf{0}, \qquad \ddot{\mathbf{x}} = \mathbf{0}.$$

Since $\ddot{\mathbf{x}} = \dot{\mathbf{v}}$,

$$\frac{d}{dt}(\mathbf{v}) = \mathbf{0}.$$

Integrate once; $\mathbf{v}$ is constant:

$$\mathbf{v} = \mathbf{v}_0, \qquad \frac{d\mathbf{x}}{dt} = \mathbf{v}_0.$$

Integrate again:

$$\mathbf{x} = t\mathbf{v}_0 + \mathbf{x}_0.$$

The result is a straight line.

Answer: The trajectory is a straight line, traversed at constant speed.

REMARK: Let us check the second integration in components. The equation

$$\frac{d\mathbf{x}}{dt} = \mathbf{v}_0$$

means

$$\dot{x}_1 = v_{01}, \qquad \dot{x}_2 = v_{02}, \qquad \dot{x}_3 = v_{03},$$

where the v_{0j} are constants. Integrating,

$$x_1 = tv_{01} + x_{01}, \qquad x_2 = tv_{02} + x_{02}, \qquad x_3 = tv_{03} + x_{03}.$$

Written as a vector equation, this is simply $\mathbf{x} = t\mathbf{v}_0 + \mathbf{x}_0$.

EXAMPLE 4.3

A shell is fired at an angle α with the ground. What is its path? Neglect air resistance.

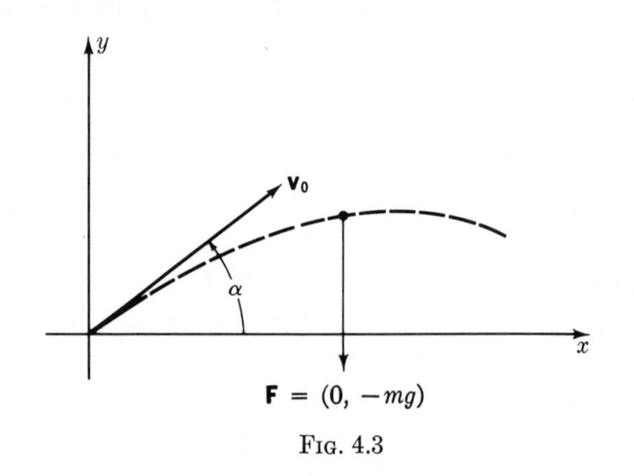

FIG. 4.3

Solution: Draw a figure, taking the axes as indicated (Fig. 4.3). Let $\mathbf{v}_0$ be the initial velocity vector, so $\mathbf{v}_0 = v_0(\cos\alpha, \sin\alpha)$, where v_0 is the initial speed. Let m denote the mass of the shell. The force of gravity at each point is constant,

$$\mathbf{F} = (0, -mg).$$

The equation of motion is

$$m\mathbf{a} = \mathbf{F}, \qquad \text{that is,} \qquad \frac{d^2\mathbf{x}}{dt^2} = (0, -g).$$

Integrate:

$$\frac{d\mathbf{x}}{dt} = (0, -gt) + \mathbf{v}_0.$$

Integrate again, noting that $\mathbf{x}_0 = \mathbf{0}$ by the choice of axes:

$$\mathbf{x} = \left(0, -\frac{1}{2}gt^2\right) + t\mathbf{v}_0.$$

Hence

$$(x(t), y(t)) = \left(0, -\frac{1}{2}gt^2\right) + tv_0(\cos\alpha, \sin\alpha)$$

$$= \left(v_0t\cos\alpha, v_0t\sin\alpha - \frac{1}{2}gt^2\right).$$

To describe the path, eliminate t:

$$x = v_0t\cos\alpha, \qquad t = \frac{x}{v_0\cos\alpha},$$

$$y = v_0t\sin\alpha - \frac{1}{2}gt^2 = x\tan\alpha - \frac{g}{2v_0^2\cos^2\alpha}x^2.$$

The graph of this quadratic is a parabola.

Answer: $x = (v_0\cos\alpha)t, \quad y = (v_0\sin\alpha)t - \frac{1}{2}gt^2,$

where v_0 is the initial speed. The path is a parabola:

$$y = x\tan\alpha - \frac{g}{2v_0^2\cos^2\alpha}x^2.$$

EXAMPLE 4.4

In Example 4.3, what is the maximum range?

Solution: The shell hits ground when $y = 0$:

$$\left(v_0\sin\alpha - \frac{1}{2}gt\right)t = 0.$$

This equation has two roots. The root $t = 0$ indicates the initial point. We want the other root,

$$t = \frac{2v_0 \sin \alpha}{g}.$$

The range is the value of x at this time:

$$x = (v_0 \cos \alpha)\left(\frac{2v_0 \sin \alpha}{g}\right) = \frac{v_0^2}{g} \sin 2\alpha.$$

Clearly x is maximum when $\sin 2\alpha = 1$, or $\alpha = \pi/4$. The maximum range is v_0^2/g.

Answer: The maximum range is v_0^2/g. It is obtained by firing at 45°.

REMARK: If the initial speed is doubled, the maximum range is quadrupled. Is this reasonable? (By what factor must the gunpowder be increased to double the initial speed?)

Components of Acceleration

The arc length s, the unit tangent $\mathbf{T}$, the unit normal $\mathbf{N}$, and the curvature k are geometric properties of a curve. If a particle moves on the curve, it is useful to express its velocity and acceleration in terms of these quantities. We already know

$$\mathbf{v} = \frac{ds}{dt}\mathbf{T},$$

which says that the motion is directed along the tangent with speed ds/dt.

For further information, differentiate $\mathbf{v}$ with respect to time, using the Chain Rule carefully:

$$\mathbf{a} = \dot{\mathbf{v}} = \frac{d^2s}{dt^2}\mathbf{T} + \frac{ds}{dt}\dot{\mathbf{T}}.$$

But

$$\dot{\mathbf{T}} = \frac{d\mathbf{T}}{dt} = \frac{ds}{dt}\frac{d\mathbf{T}}{ds} = \frac{ds}{dt}k\mathbf{N},$$

where k is the curvature. Therefore

$$\mathbf{a} = \frac{d^2s}{dt^2}\mathbf{T} + k\left(\frac{ds}{dt}\right)^2\mathbf{N}.$$

This is an important equation in mechanics. It says that the acceleration is composed of two perpendicular components. The first is a tangential component with magnitude $\ddot{s}$, the rate of change of the speed. The second is a normal component, directed along $\mathbf{N}$ with magnitude $k\dot{s}^2$. It is called the **centripetal acceleration.**

EXAMPLE 4.5

A particle moves along a circle. Find its velocity and acceleration.

Solution: Let r denote the radius and let $\theta = \theta(t)$ denote the central angle at time t. Place the circle in the x, y-plane with center at $\mathbf{0}$. Then the path is given by

$$\mathbf{x}(t) = r(\cos\theta, \sin\theta).$$

Differentiate:

$$\mathbf{v} = \dot{\mathbf{x}} = r\dot{\theta}(-\sin\theta, \cos\theta).$$

It follows that

$$\mathbf{T} = (-\sin\theta, \cos\theta), \qquad \frac{ds}{dt} = r\dot{\theta} = r\omega(t).$$

Here $\omega(t) = \dot{\theta}(t)$ represents the instantaneous angular speed. Thus

$$\mathbf{v} = r\omega(-\sin\theta, \cos\theta) = r\omega\mathbf{T}.$$

Differentiate:

$$\begin{aligned}\mathbf{a} = \dot{\mathbf{v}} &= r\dot{\omega}\mathbf{T} + r\omega\dot{\mathbf{T}} = r\dot{\omega}(-\sin\theta, \cos\theta) + r\omega^2(-\cos\theta, -\sin\theta)\\ &= r\dot{\omega}\mathbf{T} + r\omega^2\mathbf{N}.\end{aligned}$$

Answer: $\mathbf{v} = r\omega\mathbf{T}$, $\mathbf{a} = r\dot{\omega}\mathbf{T} + r\omega^2\mathbf{N}$, where $\omega = \dot{\theta}$ is the angular speed.

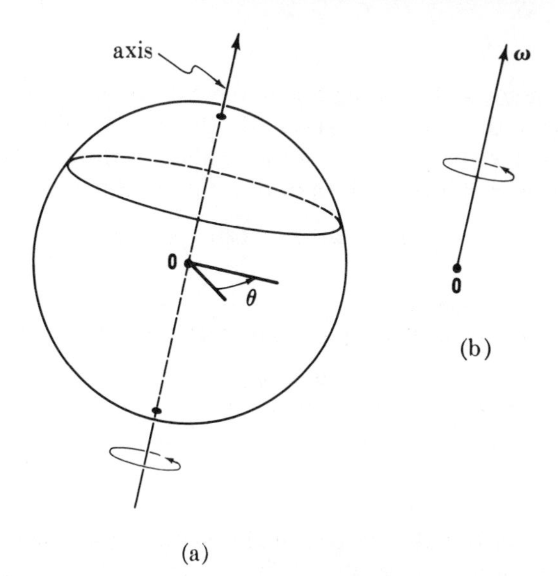

Fig. 4.4

REMARK: When the motion is uniform (ω constant), then $\mathbf{a} = r\omega^2\mathbf{N}$, so the acceleration is all centripetal, perpendicular to the direction of motion. This agrees with the answer to Example 4.1.

Angular Velocity

A rigid body rotates about an axis a through $\mathbf{0}$. See Fig. 4.4a. The central angle is $\theta = \theta(t)$, so $\omega = \dot{\theta}$ is the **angular speed,** the rate of rotation in radians per second.

The **angular velocity** is defined to be the vector $\boldsymbol{\omega}$ having magnitude $\dot{\theta}$ and pointing along the (positive) axis of rotation according to the right-hand rule (Fig. 4.4b).

Suppose the actual velocity $\mathbf{v}$ of a point $\mathbf{x}$ in the rigid body is required. How can it be expressed in terms of $\mathbf{x}$ and the angular velocity $\boldsymbol{\omega}$? See Fig. 4.5.

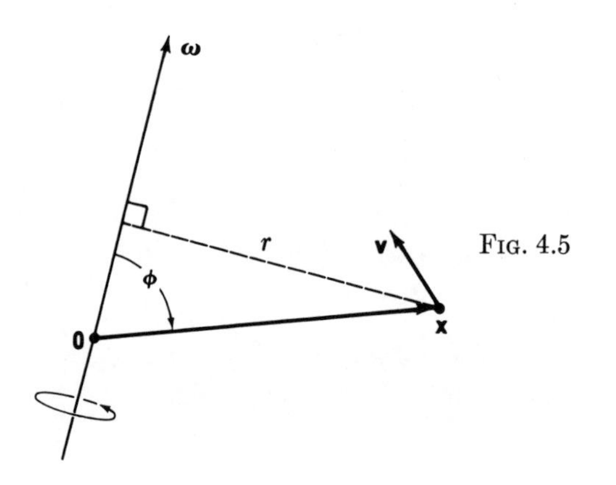

FIG. 4.5

Since the point $\mathbf{x}$ is rotating about the axis of $\boldsymbol{\omega}$, its velocity vector $\mathbf{v}$ is perpendicular to the plane of $\boldsymbol{\omega}$ and $\mathbf{x}$. By the right-hand rule, $\mathbf{v}$ points in the direction of $\boldsymbol{\omega} \times \mathbf{x}$. The speed $|\mathbf{v}|$ is the product of the angular speed $\omega = |\boldsymbol{\omega}|$ and the distance r of $\mathbf{x}$ from the axis of rotation. But $r = |\mathbf{x}| \sin\phi$, hence

$$|\mathbf{v}| = |\boldsymbol{\omega}|\cdot|\mathbf{x}| \sin\phi = |\boldsymbol{\omega} \times \mathbf{x}|.$$

Therefore:

> The velocity of a point $\mathbf{x}$ in a rigid body rotating with angular velocity $\boldsymbol{\omega}$ is
>
> $$\mathbf{v} = \boldsymbol{\omega} \times \mathbf{x}.$$

EXERCISES

1. A hill makes angle β with the ground (Fig. 4.6). A shell is fired from the base of the hill at angle α with the ground. Show that the x-component of the position where the shell strikes the hill is $x = (2v_0^2/g)(\sin\alpha\cos\alpha - \tan\beta\cos^2\alpha)$.

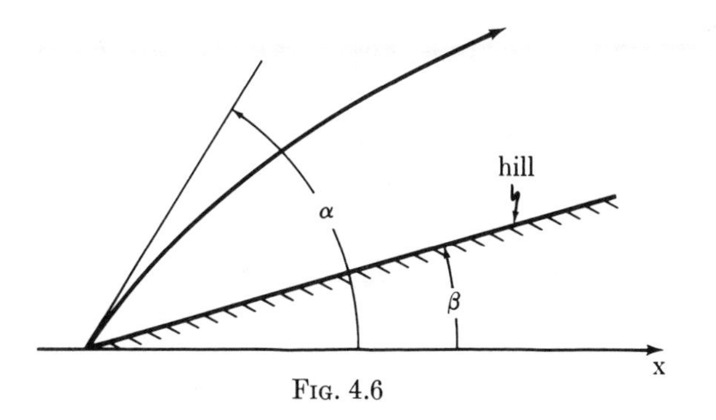

FIG. 4.6

2. (cont.) Find the maximum of x as a function of α. Show that it occurs for

$$\frac{\pi}{2} - \alpha = \alpha - \beta, \quad \text{that is,} \quad \alpha = \frac{1}{2}\left(\frac{\pi}{2} + \beta\right).$$

3. Let $\mathbf{x}(t) = (t, t^2)$. Find $\mathbf{v}(t)$ and $\mathbf{a}(t)$.
4. (cont.) Find the tangential and normal components of $\mathbf{a}$ at $t = 0$ and $t = -1$.
5. A particle moves along the curve $y = \sin x$ with constant speed 1. Find the tangential and normal components of $\mathbf{a}$ at $x = 0$ and at $x = \pi/2$.
6. Find the tangential and normal components of acceleration for $\mathbf{x}(t) = (\cos t^2, \sin t^2)$.
7. Find the tangential and normal components of acceleration for $\mathbf{x}(t) = (a \cos \omega t, a \sin \omega t, bt)$, where ω is constant.
8. A particle moves with constant speed 1 on the surface of the unit sphere $|\mathbf{x}| = 1$. Show that the normal component of the acceleration has magnitude at least 1.
9. A particle moves on the surface $z = x^2 + y^2$ with constant speed 1. At a certain instant t_0 it passes through $\mathbf{0}$. Show that the tangential component of $\mathbf{a}$ is $\mathbf{0}$ and the normal component is $(\ddot{x}(t_0), \ddot{y}(t_0), 2)$. Show also with $\dot{x}\ddot{x} + \dot{y}\ddot{y} = 0$ at t_0.
10. Let $\mathbf{x} = \mathbf{x}(t)$ be a space curve. Show that its curvature is $k = |\mathbf{v} \times \mathbf{a}|/|\mathbf{v}|^3$.
11. The earth turns on its axis with angular velocity 360° per day. Find the actual speed (mph) of a point on the surface (a) at the equator, (b) at the 40-th parallel, (c) at the south pole. Approximate the earth by a sphere of radius 4000 miles.

5. INTEGRALS

Suppose a particle moves along a path $\mathbf{x} = \mathbf{x}(t)$ from $\mathbf{x}(t_0)$ to $\mathbf{x}(t_1)$, acted on by a force $\mathbf{F} = \mathbf{F}(t)$. How much work is done by the force?

From physics we learn that only the component of the force in the direction of motion does work, and that the amount of work done in a small displacement of length ds is

$$dW = F_a \, ds,$$

where F_a is the average component of force in the direction of motion.

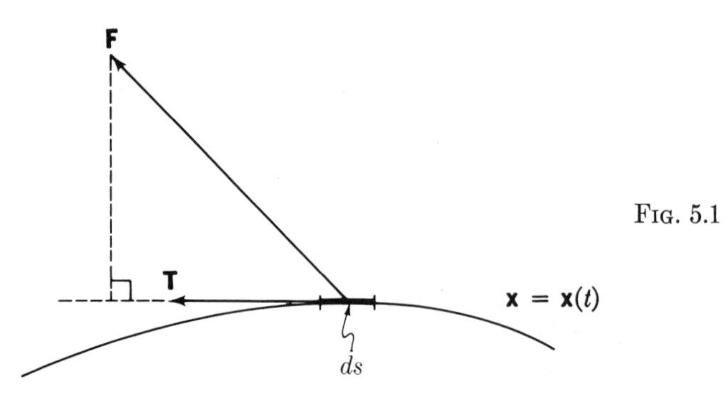

FIG. 5.1

Draw the unit tangent $\mathbf{T}$, the force $\mathbf{F}$, and a small portion of the path of length ds. See Fig. 5.1. Since $\mathbf{T}$ is a unit vector in the direction of motion, the component of the force $\mathbf{F}$ in the direction of motion is the dot product

$$\mathbf{F}\cdot\mathbf{T}.$$

Hence

$$dW = (\mathbf{F}\cdot\mathbf{T})\, ds.$$

Replace ds by $(ds/dt)\, dt$. This makes good physical sense: The length ds traveled in a short time period dt is given by speed $\times$ time $= (ds/dt)\, dt$. The result is

$$dW = \mathbf{F}\cdot\mathbf{T}\,\frac{ds}{dt}\, dt.$$

But

$$\mathbf{T}\,\frac{ds}{dt} = \mathbf{v} = \frac{d\mathbf{x}}{dt},$$

hence

$$dW = (\mathbf{F}\cdot\mathbf{v})\, dt = \left(\mathbf{F}\cdot\frac{d\mathbf{x}}{dt}\right) dt.$$

We define the total work done by an integral that "adds up" these small bits of work from $\mathbf{x}(t_0)$ to $\mathbf{x}(t_1)$:

$$W = \int_{t_0}^{t_1} (\mathbf{F}\cdot\mathbf{v})\, dt = \int_{t_0}^{t_1} \left(\mathbf{F}\cdot\frac{d\mathbf{x}}{dt}\right) dt.$$

Since $(d\mathbf{x}/dt)\, dt = d\mathbf{x}$, the integral can be written

$$W = \int_{\mathbf{x}(t_0)}^{\mathbf{x}(t_1)} \mathbf{F}\cdot d\mathbf{x}.$$

This type of integral is called a **line integral.** It arises naturally in connection with work, but has many other practical applications in physics. The evaluation of a line integral involves nothing more than the evaluation of an ordinary integral.

Suppose a particle moves on a curve $\mathbf{x}(t)$ from $\mathbf{x}(t_0)$ to $\mathbf{x}(t_1)$ and is subject to a force $\mathbf{F}(t)$. Then the work done by the force is given by the line integral

$$\int_{t_0}^{t_1} \left(\mathbf{F}\cdot\frac{d\mathbf{x}}{dt}\right) dt = \int_{\mathbf{x}(t_0)}^{\mathbf{x}(t_1)} \mathbf{F}\cdot d\mathbf{x}.$$

In the integral on the right,

$$d\mathbf{x} = \left(\frac{dx}{dt}, \frac{dy}{dt}, \frac{dz}{dt}\right) dt.$$

Let $\mathbf{F}(t) = (F_1(t), F_2(t), F_3(t))$. Then the line integral is evaluated as an ordinary integral:

$$\int_{\mathbf{x}(t_0)}^{\mathbf{x}(t_1)} \mathbf{F}\cdot d\mathbf{x} = \int_{t_0}^{t_1} \left[F_1(t)\frac{dx}{dt} + F_2(t)\frac{dy}{dt} + F_3(t)\frac{dz}{dt}\right] dt.$$

EXAMPLE 5.1

Evaluate the line integral $\int_{\mathbf{x}(0)}^{\mathbf{x}(3)} \mathbf{F}\cdot d\mathbf{x}$, where $\mathbf{F} = (3, -1, 2)$ and $\mathbf{x}(t) = (t, t^2, t^3)$.

Solution:

$$d\mathbf{x} = (1, 2t, 3t^2)\, dt,$$

$$\mathbf{F}\cdot d\mathbf{x} = (3, -1, 2)\cdot(1, 2t, 3t^2)\, dt = (3 - 2t + 6t^2)\, dt.$$

Therefore

$$\int_{\mathbf{x}(0)}^{\mathbf{x}(3)} \mathbf{F}\cdot d\mathbf{x} = \int_0^3 (3 - 2t + 6t^2)\, dt = (3t - t^2 + 2t^3)\Big|_0^3 = (9 - 9 + 54).$$

Answer: 54.

EXAMPLE 5.2

Under the action of a force $\mathbf{F}(t)$, a particle moves on a path $\mathbf{x}(t)$ from $\mathbf{x}(t_0)$ to $\mathbf{x}(t_1)$. Let W denote the work done by the force. From Newton's Law, show that $W = \frac{1}{2}m\,|\mathbf{v}(t_1)|^2 - \frac{1}{2}m\,|\mathbf{v}(t_0)|^2$.

Solution:

$$W = \int_{t_0}^{t_1} \mathbf{F}\cdot\dot{\mathbf{x}}\, dt.$$

According to Newton's Law, $\mathbf{F} = m\ddot{\mathbf{x}}$, so

$$\mathbf{F}\cdot\dot{\mathbf{x}} = m\ddot{\mathbf{x}}\cdot\dot{\mathbf{x}}.$$

But observe that

$$2\ddot{\mathbf{x}} \cdot \dot{\mathbf{x}} = \frac{d}{dt}(\dot{\mathbf{x}} \cdot \dot{\mathbf{x}}).$$

Therefore

$$\mathbf{F} \cdot \dot{\mathbf{x}} = \frac{1}{2} m \frac{d}{dt}(\dot{\mathbf{x}} \cdot \dot{\mathbf{x}}) = \frac{1}{2} m \frac{d}{dt} |\mathbf{v}|^2,$$

so

$$W = \int_{t_0}^{t_1} \mathbf{F} \cdot \dot{\mathbf{x}}\, dt = \int_{t_0}^{t_1} \frac{1}{2} m \frac{d}{dt} |\mathbf{v}|^2\, dt = \frac{1}{2} m\, |\mathbf{v}(t_1)|^2 - \frac{1}{2} m\, |\mathbf{v}(t_0)|^2.$$

REMARK: The quantity $\frac{1}{2}m|\mathbf{v}|^2$ is the **kinetic energy** of the particle. The result of this example is the Law of Conservation of Energy: work done equals change in kinetic energy.

Integrals of Vector Functions

Next we define the integral of a vector function as a componentwise operation.

Suppose $\mathbf{u}(t) = (u_1(t), u_2(t), u_3(t))$ is defined for $a \le t \le b$. Then

$$\int_a^b \mathbf{u}(t)\, dt = \left(\int_a^b u_1(t)\, dt,\ \int_a^b u_2(t)\, dt,\ \int_a^b u_3(t)\, dt \right).$$

Notice that the integral of a vector function is a *vector*, whereas a line integral is a *scalar*.

EXAMPLE 5.3

Let $\mathbf{u}(t) = (1, t - 1, t^2)$. Find $\displaystyle\int_{-2}^{3} \mathbf{u}(t)\, dt$.

Solution:

$$\int_{-2}^{3} \mathbf{u}(t)\, dt = \int_{-2}^{3} (1, t - 1, t^2)\, dt$$

$$= \left(\int_{-2}^{3} dt,\ \int_{-2}^{3} (t - 1)\, dt,\ \int_{-2}^{3} t^2\, dt \right) = \left(5, -\frac{5}{2}, \frac{35}{3}\right).$$

Answer: $\left(5, -\dfrac{5}{2}, \dfrac{35}{3}\right)$.

The integral of a vector function $\mathbf{u}(t)$ is particularly easy to evaluate if an antiderivative of $\mathbf{u}(t)$ is known.

> If
>
> $$\mathbf{u}(t) = \frac{d}{dt}\mathbf{w}(t),$$
>
> then
>
> $$\int_a^b \mathbf{u}(t)\,dt = \int_a^b \frac{d\mathbf{w}}{dt}\,dt = \mathbf{w}(b) - \mathbf{w}(a).$$

To prove this, simply check the three components; each is the integral of a derivative. Here is an example:

$$\int_0^2 (2t, 3t^2, 4t^3)\,dt = \int_0^2 \frac{d}{dt}(t^2, t^3, t^4)\,dt = (t^2, t^3, t^4)\Big|_0^2 = (4, 8, 16).$$

Momentum

The following applications show the importance in physics of vector-valued integrals.

A particle of mass m has position $\mathbf{x} = \mathbf{x}(t)$ at time t and moves under the action of a (variable) force $\mathbf{F}$, so that

$$\mathbf{F} = m\ddot{\mathbf{x}} = m\frac{d\dot{\mathbf{x}}}{dt}.$$

Integrate with respect to t on the interval $t_0 \leq t \leq t_1$:

$$\int_{t_0}^{t_1} \mathbf{F}\,dt = \int_{t_0}^{t_1} m\frac{d\dot{\mathbf{x}}}{dt}\,dt,$$

hence,

$$\int_{t_0}^{t_1} \mathbf{F}\,dt = m\dot{\mathbf{x}}(t_1) - m\dot{\mathbf{x}}(t_0).$$

The quantity $m\dot{\mathbf{x}}$ is the **momentum** of the particle, so the right-hand side is its change in momentum. The left-hand side is called the **impulse** of the force during the time from t_0 to t_1. This equation is a form of the Law of Conservation of Momentum: impulse equals change in momentum.

The **angular momentum** of the particle with respect to the origin $\mathbf{0}$ is defined as

$$m\mathbf{x} \times \dot{\mathbf{x}}.$$

Now

$$\frac{d}{dt}[m\mathbf{x} \times \dot{\mathbf{x}}] = m\dot{\mathbf{x}} \times \dot{\mathbf{x}} + m\mathbf{x} \times \ddot{\mathbf{x}} = m\mathbf{x} \times \ddot{\mathbf{x}}$$

since $\dot{\mathbf{x}} \times \dot{\mathbf{x}} = \mathbf{0}$. But

$$m\ddot{\mathbf{x}} = \mathbf{F}$$

by Newton's Law of Motion; hence

$$m\mathbf{x} \times \ddot{\mathbf{x}} = \mathbf{x} \times m\ddot{\mathbf{x}} = \mathbf{x} \times \mathbf{F},$$

which is the torque of $\mathbf{F}$ at $\mathbf{x}$. The result is that

$$\frac{d}{dt}[m\mathbf{x} \times \dot{\mathbf{x}}] = \mathbf{x} \times \mathbf{F}.$$

Integrating,

$$m\mathbf{x} \times \dot{\mathbf{x}}\Big|_{t_0}^{t_1} = \int_{t_0}^{t_1} \mathbf{x} \times \mathbf{F}\, dt.$$

This result, called the law of Conservation of Angular Momentum, asserts that the change in angular momentum during a motion is the time integral of the torque.

EXERCISES

Evaluate:

1. $\displaystyle\int_{\mathbf{x}(0)}^{\mathbf{x}(1)} (t, 2t, 3t)\cdot d\mathbf{x}; \quad \mathbf{x}(t) = (1, t, t^2)$

2. $\displaystyle\int_{\mathbf{x}(0)}^{\mathbf{x}(\pi)} (\cos t, \sin t)\cdot d\mathbf{x}; \quad \mathbf{x}(t) = (\sin t, -\cos t)$

3. $\displaystyle\int_{\mathbf{x}(0)}^{\mathbf{x}(2\pi)} (0, 0, 3t)\cdot d\mathbf{x}; \quad \mathbf{x}(t) = (\cos t, \sin t, t + 1)$

4. $\displaystyle\int_{\mathbf{x}(-1)}^{\mathbf{x}(1)} (e^t, e^t, e^t)\cdot d\mathbf{x}; \quad \mathbf{x}(t) = (t + 1, t - 1, t)$

5. $\displaystyle\int_{(0,0,0)}^{(1,1,1)} (t, -t^2, t)\cdot d\mathbf{x};$ straight path

6. $\displaystyle\int_{\mathbf{x}(0)}^{\mathbf{x}(2\pi)} \left(\frac{-y}{x^2 + y^2}, \frac{x}{x^2 + y^2}\right)\cdot d\mathbf{x}; \quad \mathbf{x}(t) = (a \cos t, a \sin t)$

7. $\displaystyle\int_{\mathbf{x}(0)}^{\mathbf{x}(1)} x\, dx + y\, dy; \quad \mathbf{x}(t) = (t^2, t^3)$

8. $\displaystyle\int_{(1,1,2)}^{(-1,0\ 4)} x\, dy + y\, dz + z\, dx;$ straight path.

9. Find the work done by the uniform gravitational field $\mathbf{F} = (0, 0, -g)$ in moving a particle from $(0, 0, 1)$ to $(1, 1, 0)$ along a straight path.

10. Find the work done by the central force field $\mathbf{F} = -\dfrac{1}{|\mathbf{x}|^3}\mathbf{x}$ in moving a particle from (2, 2, 2) to (1, 1, 1) along a straight path.

Evaluate:

11. $\displaystyle\int_0^1 (1+t,\ 1+2t,\ 1+3t)\,dt$

12. $\displaystyle\int_0^{2\pi} (\cos t,\ \sin t,\ 1)\,dt$

13. $\displaystyle\int_{-1}^1 (t^3,\ t^4,\ t^5)\,dt$

14. $\displaystyle\int_1^4 \left(\frac{1}{t},\ \frac{1}{t^2},\ \frac{1}{t^3}\right)dt.$

15*. Let $\mathbf{x} = \mathbf{x}(t)$, where $a \le t \le b$, be a plane curve which does not pass through (0, 0).

Show that $\displaystyle\frac{1}{2}\int_{\mathbf{x}(a)}^{\mathbf{x}(b)} (x\,dy - y\,dx)$ is the area in Fig. 5.2.

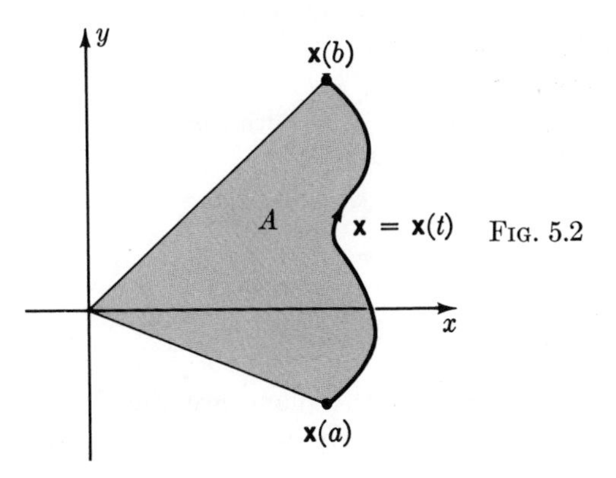

Fig. 5.2

16. The hyperbola $\dfrac{x^2}{a^2} - \dfrac{y^2}{b^2} = 1$ is parameterized by $\mathbf{x}(t) = (a\cosh t,\ b\sinh t)$.

Show that t is related to the area A by $t = 2A/ab$. See Fig. 5.3.
[*Hint:* Use Ex. 15.]

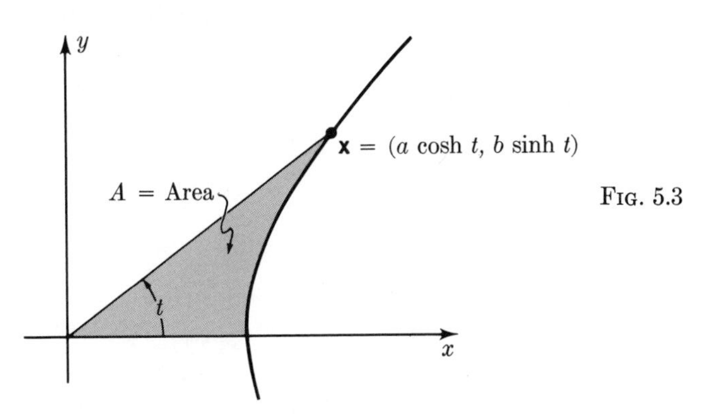

Fig. 5.3

17. A circle of radius a rolls along the x-axis. Initially its center is $(0, a)$. The point on the circle initially at $(0, 0)$ traces a curve called the **cycloid.** Show that $\mathbf{x}(\theta) = a(\theta - \sin\theta,\ 1 - \cos\theta)$ is a parametrization of the cycloid, where θ is the angle at the center of the rolling circle, measured clockwise from the downward vertical to the moving point. Graph the curve.
18. (cont.) Show that the area under one arch of the cycloid is 3 times the area of the circle.
[*Hint:* Express $y\,dx$ in terms of θ and integrate.]
19. (cont.) Find the arc length of one arch of the cycloid.
20. Show that the "witch of Agnesi" $y = 1/(1 + x^2)$ is parameterized by $\mathbf{x} = \mathbf{x}(\theta) = 2(\cot\theta,\ \sin^2\theta)$. Sketch the curve.
21. Show that the "folium of Descartes" $x^3 + y^3 = 3axy$ is parameterized by $\mathbf{x} = \mathbf{x}(t) = (1 + t^3)^{-1}(3at,\ 3at^2)$. Sketch the curve.
22. The force $\mathbf{F}(t) = (1 - t,\ 1 - t^2,\ 1 - t^3)$ acts from $t = 0$ to $t = 2$. Find its impulse.
23. The force $\mathbf{F}(t) = (e^t,\ e^{2t},\ e^{3t})$ acts from $t = -1$ to $t = 0$. Find its impulse.
24. An electron in a uniform magnetic field follows the spiral path
$$\mathbf{x}(t) = (a\cos t,\ a\sin t,\ bt).$$
Find its angular momentum with respect to $\mathbf{0}$.
25. A particle of unit mass moves on the unit sphere $|\mathbf{x}| = 1$ with unit speed. Show that its angular momentum with respect to $\mathbf{0}$ is a unit vector.

6. POLAR COORDINATES

Review

The **polar coordinates** of a point $\mathbf{x} \neq \mathbf{0}$ in the plane are the distance $r = |\mathbf{x}|$ of $\mathbf{x}$ from $\mathbf{0}$, and the angle θ from the positive x-axis to the vector $\mathbf{x}$, measured counterclockwise. The angle θ is determined up to a multiple of 2π.

We shall write polar coordinates $\{r, \theta\}$ with curly braces to distinguish them from rectangular coordinates (x, y).

Any value of r is allowed, even $r = 0$ (the origin) and negative r; the point $\{-r, \theta\}$ is the reflection of $\{r, \theta\}$ through the origin and is identical with $\{r, \theta + \pi\}$. Note that θ is undefined at the origin.

The rectangular coordinates (x, y) of the point with polar coordinates $\{r, \theta\}$ are given by

$$\begin{cases} x = r\cos\theta \\ y = r\sin\theta. \end{cases}$$

Conversely, given (x, y) we find $\{r, \theta\}$ from

$$r^2 = x^2 + y^2, \qquad \begin{cases} \cos\theta = \dfrac{x}{r} \\ \sin\theta = \dfrac{y}{r}. \end{cases}$$

(The single formula for the angle, $\tan\theta = y/x$, is not adequate to distinguish quadrants. For example, $\tan 0 = \tan\pi = 0$.)

The number r is called the **radius** of the point $\mathbf{x} = \{r, \theta\}$ and θ is called the **polar angle**.

A curve may be presented by a relation between r and θ, often in the form $r = f(\theta)$. For example, $r = a$ is the equation of the circle of radius a with center $\mathbf{0}$. The graph of $r = \cos\theta$ is also a circle, but with center $(\frac{1}{2}, 0)$ and radius $\frac{1}{2}$, because of the computation

$$r = \cos\theta, \qquad r^2 = r\cos\theta, \qquad x^2 + y^2 = x,$$

$$(x^2 - x) + y^2 = 0, \qquad \left(x - \frac{1}{2}\right)^2 + y^2 = \frac{1}{4}.$$

Length

Suppose a curve is given in the form

$$r = r(t), \qquad \theta = \theta(t), \qquad t_0 \leq t \leq t_1.$$

What is the length? Write

$$\mathbf{x} = (r\cos\theta, r\sin\theta) = r(\cos\theta, \sin\theta).$$

Differentiate:

$$\dot{\mathbf{x}} = \dot{r}(\cos\theta, \sin\theta) + r\dot{\theta}(-\sin\theta, \cos\theta) = \dot{r}\mathbf{u} + r\dot{\theta}\mathbf{w},$$

where

$$\mathbf{u} = (\cos\theta, \sin\theta) \qquad \text{and} \qquad \mathbf{w} = (-\sin\theta, \cos\theta).$$

Now

$$\begin{aligned}\left(\frac{ds}{dt}\right)^2 = |\dot{\mathbf{x}}|^2 &= \dot{\mathbf{x}}\cdot\dot{\mathbf{x}} \\ &= (\dot{r}\mathbf{u} + r\dot{\theta}\mathbf{w})\cdot(\dot{r}\mathbf{u} + r\dot{\theta}\mathbf{w}) \\ &= \dot{r}^2\mathbf{u}\cdot\mathbf{u} + 2r\dot{r}\dot{\theta}\mathbf{u}\cdot\mathbf{w} + r^2\dot{\theta}^2\mathbf{w}\cdot\mathbf{w}.\end{aligned}$$

But $\mathbf{u}\cdot\mathbf{u} = 1$, $\mathbf{w}\cdot\mathbf{w} = 1$, and $\mathbf{u}\cdot\mathbf{w} = 0$. Hence,

$$\left(\frac{ds}{dt}\right)^2 = \dot{r}^2 + r^2\dot{\theta}^2.$$

It follows that

$$\boxed{\text{Length} = \int_{t_0}^{t_1} \sqrt{\dot{r}^2 + r^2\dot{\theta}^2}\, dt.}$$

Figure 6.1 provides an aid to memory. The "right triangle" has sides dr, $r\,d\theta$, and ds, so the Pythagorean Theorem suggests

$$(ds)^2 = (dr)^2 + r^2(d\theta)^2.$$

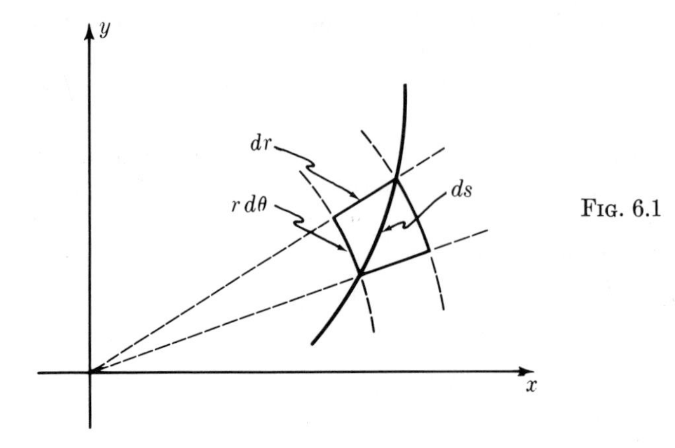

FIG. 6.1

Suppose a curve is given by $r = r(\theta)$. This is a special case of the previous situation with $\theta = t$ and $r = r(t)$. The length formula specializes to

$$L = \int_{\theta_0}^{\theta_1} \sqrt{r^2 + \left(\frac{dr}{d\theta}\right)^2}\, d\theta.$$

EXAMPLE 6.1

Find the length of the spiral $r = \theta^2$, $0 \le \theta \le 2\pi$.

Solution:

$$L = \int_0^{2\pi} \sqrt{r^2 + \left(\frac{dr}{d\theta}\right)^2}\, d\theta = \int_0^{2\pi} \sqrt{\theta^4 + (2\theta)^2}\, d\theta$$

$$= \int_0^{2\pi} \theta\sqrt{\theta^2 + 4}\, d\theta = \frac{1}{3}(\theta^2 + 4)^{3/2}\Big|_0^{2\pi}.$$

Answer: $L = \dfrac{8}{3}(\pi^2 + 1)^{3/2} - \dfrac{8}{3}.$

Area

There are problems that require the area swept out by the segment joining **0** to a moving point on a curve (Fig. 6.2).

Suppose the curve is given by

$$r = r(t), \qquad \theta = \theta(t), \qquad t_0 \le t \le t_1.$$

In a small time interval dt, a thin triangle of base $r\, d\theta$ and height r (ignoring

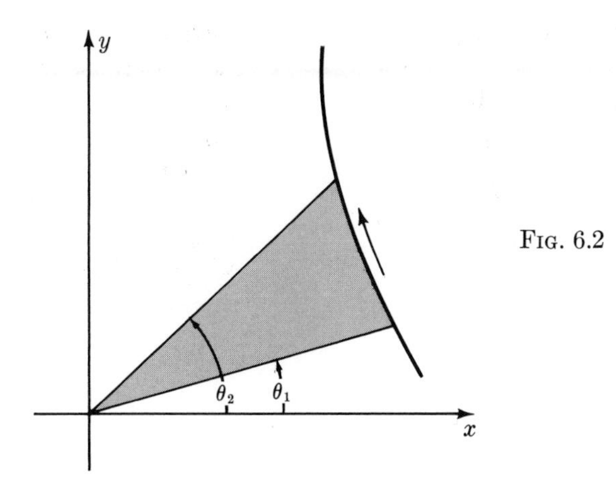

Fig. 6.2

negligible errors) is swept out (Fig. 6.3). Hence

$$dA = \frac{1}{2} r^2 \, d\theta = \frac{1}{2} r^2 \frac{d\theta}{dt} \, dt,$$

$$A = \int_{t_0}^{t_1} \frac{1}{2} r^2 \frac{d\theta}{dt} \, dt.$$

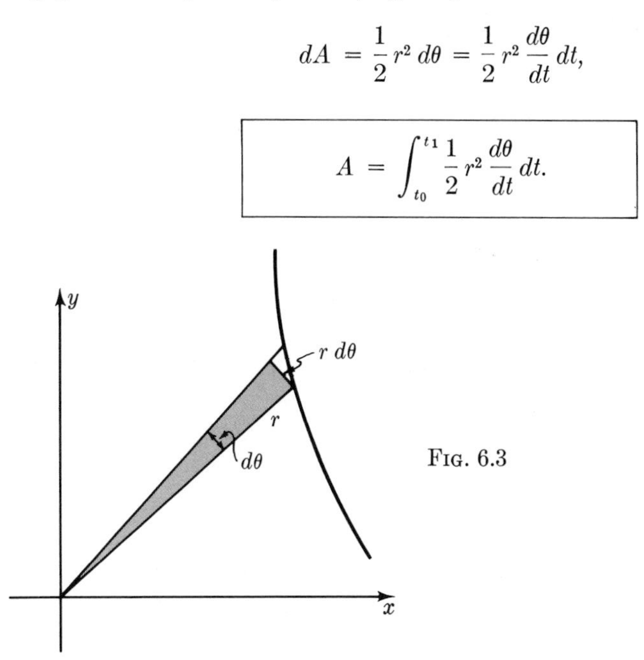

Fig. 6.3

In the special case that the curve is given by $r = r(\theta)$ for $\theta_0 \leq \theta \leq \theta_1$, choose $t = \theta$. The formula specializes to

$$A = \int_{\theta_0}^{\theta_1} \frac{1}{2} r(\theta)^2 \, d\theta.$$

EXAMPLE 6.2

Find the area of the "four-petal rose" $r = a \cos 2\theta$.

Solution: Graph the curve carefully (Fig. 6.4). The portion on which $0 \leq \theta \leq \pi/2$ is emphasized. Note that $r < 0$ for $\pi/4 < \theta < \pi/2$. Because of symmetry it suffices to find the area of half of one petal. Thus

$$A = 8 \int_0^{\pi/4} \frac{1}{2} (a \cos 2\theta)^2 \, d\theta = 4a^2 \int_0^{\pi/4} \cos^2 2\theta \, d\theta = 2a^2 \int_0^{\pi/2} \cos^2 t \, dt.$$

Answer: $A = \dfrac{\pi a^2}{2}$.

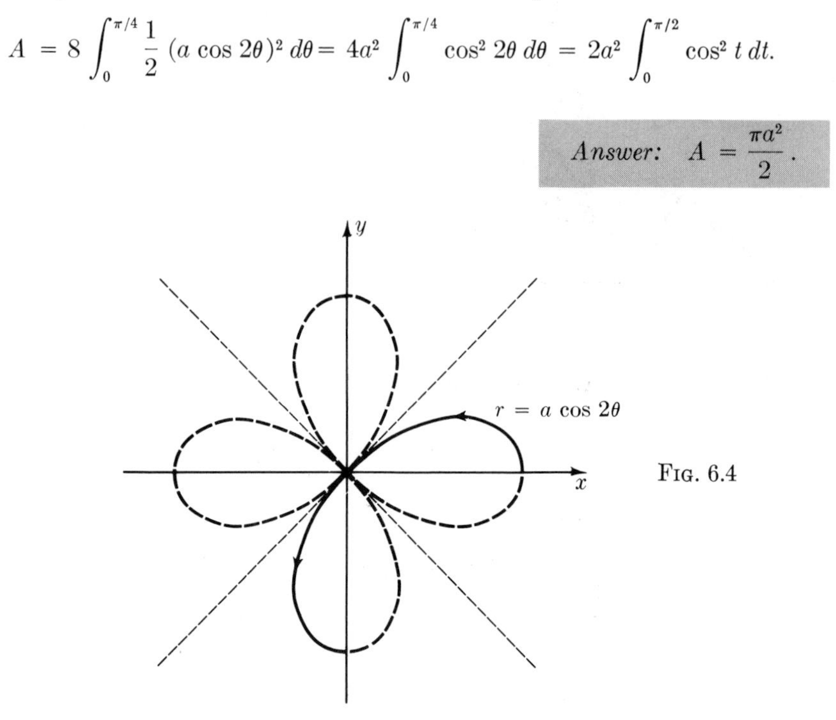

FIG. 6.4

Summary

Suppose a curve is given in polar coordinates by

$$r = r(t), \qquad \theta = \theta(t), \qquad t_0 \leq t \leq t_1 \qquad r = r(\theta), \qquad \theta_0 \leq \theta \leq \theta_1.$$

ARC LENGTH:

$$L = \int_{t_0}^{t_1} \sqrt{\left(\frac{dr}{dt}\right)^2 + r^2 \left(\frac{d\theta}{dt}\right)^2} \, dt \qquad L = \int_{\theta_0}^{\theta_1} \sqrt{r^2 + \left(\frac{dr}{d\theta}\right)^2} \, d\theta.$$

AREA:

$$A = \int_{t_0}^{t_1} \frac{1}{2} r^2 \frac{d\theta}{dt} \, dt \qquad A = \int_{\theta_0}^{\theta_1} \frac{1}{2} r^2(\theta) \, d\theta.$$

EXERCISES

Express in rectangular coordinates:

1. $\{1, \pi/2\}$
2. $\{-1, 3\pi/2\}$

3. $\{2, \pi/4\}$
4. $\{-4, 3\pi/4\}$
5. $\{1, \pi/6\}$
6. $\{2, 5\pi/6\}$.

Express in polar coordinates:

7. $(1, -1)$
8. $(-1, -1)$
9. $(-\sqrt{3}, 1)$
10. $(-\frac{1}{2}, \sqrt{3}/2)$.

11. Find the length of the "spiral of Archimedes" $r = a\theta$ from $\theta = 0$ to $\theta = 1$.
12. Set up an integral for the length of the "four-petal rose" $r = a \cos 2\theta$.
13. Set up an integral for the length of the "three-petal rose" $r = a \cos 3\theta$.
14. Find the length of the "one-petal rose" $r = a \sin \theta$. Precisely what is this curve?
15. Find the area enclosed by the "three-petal rose" $r = a \cos 3\theta$.
16. Find the area enclosed by the "$(2n + 1)$-petal rose" $r = a \cos (2n + 1)\theta$.
17. Find the area enclosed by the "$4n$-petal rose" $r = a \cos 2n\theta$.
18. Find the area enclosed by the curve $r = a \cos^2 2n\theta$.
19. Find the area enclosed by the "cardioid" $r = a(1 - \cos \theta)$.
20. Show that the "cissoid of Diocles" $y^2 = x^3/(a - x)$ can be expressed in polar coordinates by $r = a(\sec \theta - \cos \theta)$; sketch the curve.
21. Find the area enclosed by the figure eight, the "lemniscate of Bernoulli" $r^2 = a^2 \cos 2\theta$. Sketch the curve.
22. Sketch the "strophoid" $r = a \cos 2\theta \sec \theta$. Find the area of the closed loop.
23. Sketch the "limaçon of Pascal" $r = b + a \cos \theta$ in the three cases $0 < a < b$, $0 < a = b$, and $0 < b < a$. In the third case compute the area between the two loops.

7. POLAR VELOCITY AND ACCELERATION [optional]

Let us think of a curve given by

$$r = r(t), \qquad \theta = \theta(t)$$

as the path of a particle. What are its velocity and acceleration vectors?

In the above discussion of length, the perpendicular unit vectors

$$\mathbf{u} = (\cos \theta, \sin \theta) \qquad \text{and} \qquad \mathbf{w} = (-\sin \theta, \cos \theta)$$

were introduced. In using polar coordinates, it is natural to express the velocity and acceleration in terms of these vectors (Fig. 7.1). Note that $\mathbf{x} = r\mathbf{u}$ and

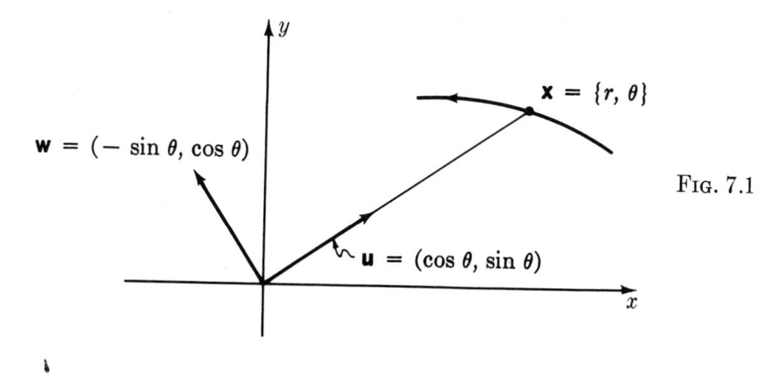

FIG. 7.1

that

$$\dot{\mathbf{u}} = \dot{\theta}\mathbf{w}, \qquad \dot{\mathbf{w}} = -\dot{\theta}\mathbf{u}.$$

Differentiate $\mathbf{x} = r\mathbf{u}$:

$$\mathbf{v} = \dot{\mathbf{x}} = \dot{r}\mathbf{u} + r\dot{\theta}\mathbf{w}.$$

Differentiate again:

$$\mathbf{a} = \dot{\mathbf{v}} = (\ddot{r}\mathbf{u} + \dot{r}\dot{\theta}\mathbf{w}) + (\dot{r}\dot{\theta}\mathbf{w} + r\ddot{\theta}\mathbf{w} - r\dot{\theta}^2\mathbf{u}).$$

Hence

$$\mathbf{a} = (\ddot{r} - r\dot{\theta}^2)\mathbf{u} + (r\ddot{\theta} + 2\dot{r}\dot{\theta})\mathbf{w}.$$

Let us apply this formula to motion involving a **central force**

$$\mathbf{F} = f(t)\mathbf{u}.$$

At each instant, the force is directed toward or away from the origin. Since $m\mathbf{a} = \mathbf{F}$, the component of $\mathbf{a}$ in the direction of $\mathbf{w}$ is zero:

$$r\ddot{\theta} + 2\dot{r}\dot{\theta} = 0, \qquad \text{that is,} \qquad \frac{1}{2}r^2\ddot{\theta} + r\dot{r}\dot{\theta} = 0.$$

This is the same as

$$\frac{d}{dt}\left(\frac{1}{2}r^2\dot{\theta}\right) = 0, \qquad \text{that is,} \qquad \frac{1}{2}r^2\dot{\theta} = \text{constant}.$$

But

$$\frac{1}{2}r^2\dot{\theta} = \frac{dA}{dt},$$

the rate at which central area is swept out by the curve. It follows that the same area is swept out in equal time anywhere along the path. This is Kepler's Second Planetary Law; it is a case of the Law of Conservation of Angular Momentum.

Summary

Suppose a curve is given in polar coordinates by

$$r = r(t), \qquad \theta = \theta(t).$$

VELOCITY:

$$\mathbf{v} = \frac{dr}{dt}\mathbf{u} + r\frac{d\theta}{dt}\mathbf{w};$$

ACCELERATION:

$$\mathbf{a} = \left[\frac{d^2r}{dt^2} - r\left(\frac{d\theta}{dt}\right)^2\right]\mathbf{u} + \left[r\frac{d^2\theta}{dt^2} + 2\frac{dr}{dt}\frac{d\theta}{dt}\right]\mathbf{w},$$

where $\mathbf{u} = (\cos\theta, \sin\theta)$ and $\mathbf{w} = (-\sin\theta, \cos\theta)$.

EXERCISES

This set of exercises develops Kepler's First and Third Laws of Planetary Motion. Assume a particle of unit mass is moving in a central force field given by the inverse square law:

$$\mathbf{F} = -\frac{1}{r^2}\mathbf{u}.$$

1. Show that the equations of motion are $r^2\dot{\theta} = J$, $\ddot{r} - \dfrac{J^2}{r^3} = -\dfrac{1}{r^2}$, where J is a constant.
2. Show that $\dot{r}^2 + \dfrac{J^2}{r^2} = \dfrac{2}{r} + C$, where C is a constant. This equation is essentially the Law of Conservation of Energy.
   [*Hint:* Multiply the second equation in Ex. 1 by $\dot{r}$ and integrate.]
3. Set $p = \dfrac{1}{r}$. Show that $\dfrac{\dot{p}^2}{p^4} + J^2p^2 = 2p + C$.
4. Imagine $\theta = \theta(t)$ solved for t as a function of θ and this substituted into $p = p(t)$. Thus p may be considered as a function of θ. Show that

   $$J^2\left(\frac{dp}{d\theta}\right)^2 = \frac{\dot{p}^2}{p^4}, \text{ and conclude that } J^2\left[\left(\frac{dp}{d\theta}\right)^2 + p^2\right] = 2p + C.$$

5. Show that $\dfrac{d^2p}{d\theta^2} + p = \dfrac{1}{J^2}$. [*Hint:* Differentiate the previous relation.]
6. Show that $p = A\cos\theta + B\sin\theta + 1/J^2$, where A and B are constants, is a solution of the preceding differential equation. (In Chapter 14 it is shown that every solution is of this type.)
7. Show that by a suitable choice of the x-axis, the solution may be written

   $$\frac{1}{r} = \frac{1}{J^2}(1 - e\cos\theta), \text{ where } e \text{ is a constant, the } \textbf{eccentricity} \text{ of the orbit, } e \geq 0.$$

8. By passing to rectangular coordinates, show that the orbit is a conic section.
9. Suppose $e = 0$. Show that the orbit is a circle with center at $\mathbf{0}$, and that the speed is constant.
10. Suppose $e = 1$. Show the orbit is a parabola with focus at the origin and opening in the positive x-direction.

11. Suppose $e > 1$. Show that the orbit is a branch of a hyperbola with one focus at the origin.

12. Suppose $0 < e < 1$. Show that the orbit is the ellipse $\dfrac{(x-c)^2}{a^2} + \dfrac{y^2}{b^2} = 1$, where

where $a = \dfrac{J^2}{1-e^2}$, $b = \dfrac{J^2}{\sqrt{1-e^2}}$, and $c = ae$.

[By Ex. 8–12, each closed orbit is an ellipse (or circle), Kepler's First Law.]

13. (cont.) Show that $a^2 = b^2 + c^2$. Conclude that the foci of the ellipse are $(0, 0)$ and $(2c, 0)$.

14. (cont.) Let T denote the **period** of the orbit, the time necessary for a complete revolution. Show that $\dfrac{J}{2}T = \pi ab$. [*Hint:* Use Kepler's Second Law.]

15. Conclude that $T^2 = 4\pi^2 a^3$. This is Kepler's Third Law: The square of the period of a planetary orbit is proportional to the cube of its semimajor axis.

6. Functions of Several Variables

1. INTRODUCTION

Elementary calculus is concerned with functions such as $y = f(x)$, where one quantity y depends on another quantity x. In all sorts of situations, however, a quantity may depend on several variables. Here are two examples:

(1) The speed v of sound in an ideal gas is

$$v = \sqrt{\gamma \frac{p}{D}},$$

where D is the density of the gas, p is the pressure, and γ is a constant characteristic of the gas. Then v depends on (is a function of) the two variables p and D. We may write

$$v = f(p, D), \qquad \text{or,} \qquad v = v(p, D).$$

(2) The area of a triangle with sides x, y, z is

$$A = \sqrt{s(s-x)(s-y)(s-z)},$$

where s is the semiperimeter $\frac{1}{2}(x + y + z)$. Then A depends on the three variables x, y, and z. We may write

$$A = f(x, y, z), \qquad \text{or} \qquad A = A(x, y, z).$$

Note that x, y, z are not three arbitrary numbers but must satisfy the inequalities $x > 0$, $y > 0$, $z > 0$ and $z < x + y$, $x < y + z$, $y < z + x$.

In Example (1), the quantity v is a function of p and D defined for a certain set of pairs (p, D), which we can think of as a subset of the p, D-plane. In Example (2), the area A is a function of x, y, z defined for a certain set of points (x, y, z) in space.

In general, let $\mathbf{S}$ be any subset of the plane $\mathbf{R}^2$ or space $\mathbf{R}^3$, and let f be a real-valued function defined on $\mathbf{S}$. Thus f assigns a real number to each point in $\mathbf{S}$. We write

$$f\colon \mathbf{S} \longrightarrow \mathbf{R}$$

and we say that $\mathbf{S}$ is the **domain** of f. Alternatively, we say that f is a function with domain $\mathbf{S}$. Here $\mathbf{R}$ denotes as usual the set of all real numbers.

We want to extend the concepts of one-variable calculus to functions of several variables, concepts such as continuity, derivative, and integral. Now in the one-variable situation, in order for these concepts to be meaningful, the domain of a function has to be a reasonably nice set, generally an interval or the union of several intervals. For example, the integral is usually defined for a function on a closed interval.

For a function of several variables, the nature of the domain is equally important. Therefore, we devote the next section to the study of properties of useful domains.

Set Notation

Let us review some standard notation that is useful in dealing with domains.

The notation $\mathbf{S} = \{x \mid P(x)\}$ is read "$\mathbf{S}$ is the set of all x with property $P(x)$". For example, an **open interval** is defined by

$$(a, b) = \{x \mid a < x < b\}.$$

We even allow infinite open intervals such as

$$(a, \infty) = \{x \mid a < x\}.$$

A **closed interval** is defined by

$$[a, b] = \{x \mid a \leq x \leq b\}.$$

The square brackets indicate that end points are *included*, the round brackets that they are *excluded*. Sometimes we used mixed types such as

$$[a, b) = \{x \mid a \leq x < b\}.$$

Each of these sets is a subset of the reals $\mathbf{R}$, and we write, for instance, $(a, b) \subseteq \mathbf{R}$. In general, $\mathbf{S} \subseteq \mathbf{T}$ is read "$\mathbf{S}$ is a subset of $\mathbf{T}$", and it means that each point of $\mathbf{S}$ is a point of $\mathbf{T}$.

If x is a point of $\mathbf{S}$, we write $x \in \mathbf{S}$ and read "x belongs to $\mathbf{S}$".

Given two sets $\mathbf{S}$ and $\mathbf{T}$, we can form two other sets from them. First, their **intersection** is

$$\mathbf{S} \cap \mathbf{T} = \{x \mid x \in \mathbf{S} \quad \text{and} \quad x \in \mathbf{T}\}.$$

It consists of all points *both* in $\mathbf{S}$ and $\mathbf{T}$. For example, let $a < b < c < d$. Then

$$(a, c) \cap (b, d) = (b, c).$$

This means that both $a < x < c$ *and* $b < x < d$ if and only if $b < x < c$.

The **union** (also called **join**) of $\mathbf{S}$ and $\mathbf{T}$ is

$$\mathbf{S} \cup \mathbf{T} = \{x \mid x \in \mathbf{S} \quad \text{or} \quad x \in \mathbf{T} \quad \text{or both}\}.$$

It consists of all the points of $\mathbf{S}$ and all the points of $\mathbf{T}$ thrown together. For example, the domain of $F(x) = \sqrt{x^2 - 1}$ is

$$\{x \mid x \leq -1 \quad \text{or} \quad x \geq 1\} = (-\infty, -1] \cup [1, \infty).$$

Another example is

$$(-\infty, 1) \cup (-1, \infty) = (-\infty, \infty) = \mathbf{R}.$$

In this case the two sets overlap, indeed, $(-\infty, 1) \cap (-1, \infty) = (-1, 1)$. Still, each point of $\mathbf{R}$ is in either $(-\infty, 1)$ or $(-1, \infty)$, some are in both.

In set theory there is something called the empty set, the set with no points at all. We have no real use for this here, so it will be understood, without further notice, that whenever we say "set" we mean a "non-empty set", a set with *at least one point*.

2. DOMAINS

In this section, we discuss closed sets and open sets in the plane and in space. These are sets that share certain basic properties with closed and open intervals on the line.

A closed interval $\mathbf{D} = [a, b]$ has this property: if $x_n \in \mathbf{D}$ and $x_n \longrightarrow x$, then $x \in \mathbf{D}$. In other words if a sequence of points in $\mathbf{D}$ converges, it converges to a point in $\mathbf{D}$. Not all intervals have this property. For example, take $\mathbf{D} = (0, 1)$ and $x_n = 1/n$. Then $x_n \in \mathbf{D}$, the sequence $\{x_n\}$ converges, but $\lim x_n \notin \mathbf{D}$.

Convergence

We need the notion of convergence of a sequence of points in space. Henceforth, to avoid repetition, we shall use the word "space" to mean either two-space (plane) $\mathbf{R}^2$ or three-space $\mathbf{R}^3$.

The definition of convergence in space looks just like the definition on the line with the "nearness" of points $\mathbf{x}$ and $\mathbf{y}$ measured by $|\mathbf{x} - \mathbf{y}|$, the distance between the points.

Definition Let $\{\mathbf{x}_n\}$ be a sequence of points in space and $\mathbf{a}$ another point. We say $\{\mathbf{x}_n\}$ **converges** to $\mathbf{a}$, and write

$$\mathbf{x}_n \longrightarrow \mathbf{a} \qquad \text{or} \qquad \mathbf{a} = \lim_{n\to\infty} \mathbf{x}_n$$

provided $|\mathbf{x}_n - \mathbf{a}| \longrightarrow 0$ as $n \longrightarrow \infty$.

From the given sequence $\{\mathbf{x}_n\}$ and point $\mathbf{a}$, we construct a new sequence $\{|\mathbf{x}_n - \mathbf{a}|\}$ of real numbers and we ask whether this sequence converges to 0. If yes, we say $\mathbf{x}_n \longrightarrow \mathbf{a}$. Just how close this definition is to the old definition of convergence for sequences of real numbers is seen in the following result, which interprets convergence in terms of coordinates.

Theorem Let $\mathbf{x}_n = (x_n, y_n, z_n)$ and $\mathbf{a} = (a, b, c)$. Then $\mathbf{x}_n \longrightarrow \mathbf{a}$ if and only if $x_n \longrightarrow a$, $y_n \longrightarrow b$, and $z_n \longrightarrow c$.

Proof: Suppose $x_n \longrightarrow a$, $y_n \longrightarrow b$, and $z_n \longrightarrow c$. Then $|x_n - a| \longrightarrow 0$ so $(x_n - a)^2 \longrightarrow 0$. Similarly, $(y_n - b)^2 \longrightarrow 0$ and $(z_n - c)^2 \longrightarrow 0$. Therefore

$$|\mathbf{x}_n - \mathbf{a}|^2 = (x_n - a)^2 + (y_n - b)^2 + (z_n - c)^2 \longrightarrow 0,$$

and it follows that $|\mathbf{x}_n - \mathbf{a}| \longrightarrow 0$ so $\mathbf{x}_n \longrightarrow \mathbf{a}$.

Conversely, let $\mathbf{x}_n \longrightarrow \mathbf{a}$. Then $|\mathbf{x}_n - \mathbf{a}| \longrightarrow 0$. But

$$|\mathbf{x}_n - \mathbf{a}|^2 = (x_n - a)^2 + (y_n - b)^2 + (z_n - c)^2 \geq (x_n - a)^2,$$

so $|\mathbf{x}_n - \mathbf{a}| \geq |x_n - a| \geq 0$ and it follows that $|x_n - a| \longrightarrow 0$, that is, $x_n \longrightarrow a$. Similarly, $y_n \longrightarrow b$ and $z_n \longrightarrow c$.

Closed Sets

Now that we know what it means for a sequence of points to converge, we can define closed sets.

Definition A set $\mathbf{S}$ in space is **closed** provided each limit of a sequence of points taken from $\mathbf{S}$ is itself in $\mathbf{S}$. That is, if $\mathbf{x}_n \in \mathbf{S}$ for $n = 1, 2, 3, \cdots$ and if $\mathbf{x}_n \longrightarrow \mathbf{x}$, then $\mathbf{x} \in \mathbf{S}$.

The intuitive idea of a closed set is a clump $\mathbf{S}$ of points in space which include all of its boundary points. Any point of space that you can "sneak up on" by points of $\mathbf{S}$ must be a point of $\mathbf{S}$. The following examples should help.

EXAMPLE 2.1 (Closed Half-Plane)

Let a, b, c be given with $a^2 + b^2 \neq 0$, and let

$$\mathbf{H} = \{(x, y) \mid ax + by \geq c\}.$$

Prove that $\mathbf{H}$ is a closed set.

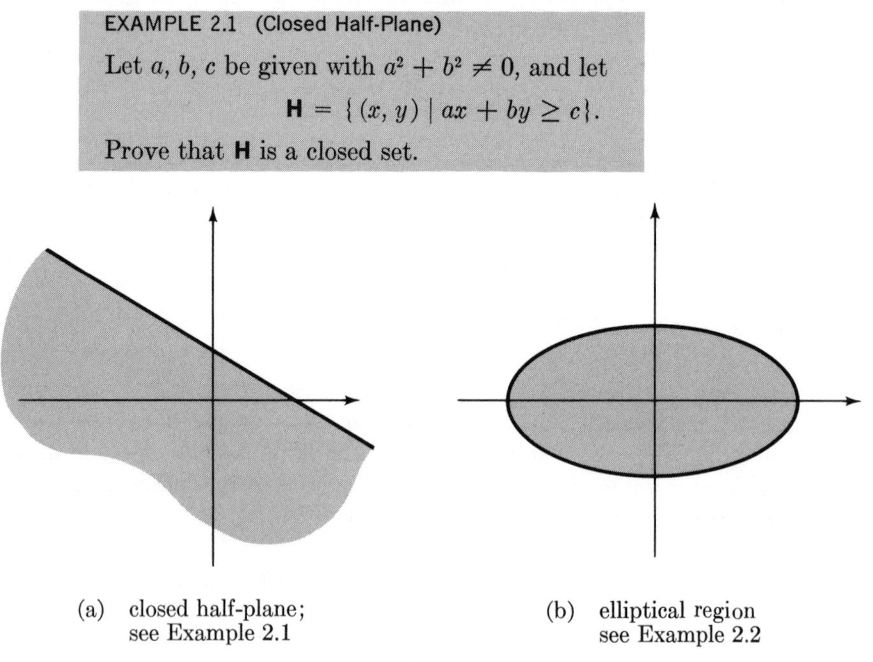

(a) closed half-plane; see Example 2.1

(b) elliptical region see Example 2.2

FIG. 2.1

Solution: Suppose $\mathbf{x}_n \in \mathbf{H}$ and $\mathbf{x}_n \longrightarrow \mathbf{x}$. Prove that $\mathbf{x} \in \mathbf{H}$. Now $\mathbf{x}_n = (x_n, y_n) \longrightarrow \mathbf{x} = (x, y)$ and $ax_n + by_n \geq c$. Since $\mathbf{x}_n \longrightarrow \mathbf{x}$, we have $x_n \longrightarrow x$ and $y_n \longrightarrow y$. Therefore $ax_n + by_n \longrightarrow ax + by$. But $ax_n + by_n \geq c$, hence $ax + by \geq c$, that is, $(x, y) \in \mathbf{H}$. See Fig. 2.1a.

EXAMPLE 2.2

Let $a > 0$ and $b > 0$ and set

$$\mathbf{E} = \left\{(x, y) \,\middle|\, \frac{x^2}{a^2} + \frac{y^2}{b^2} \leq 1\right\}.$$

Prove that $\mathbf{E}$ is a closed set.

Solution: Let $(x_n, y_n) \in \mathbf{E}$ and $(x_n, y_n) \longrightarrow (x, y)$. To prove: $(x, y) \in \mathbf{E}$. Now $x_n \longrightarrow x$ and $y_n \longrightarrow y$, hence

$$\frac{x_n^2}{a^2} + \frac{y_n^2}{b^2} \longrightarrow \frac{x^2}{a^2} + \frac{y^2}{b^2}.$$

But $x_n^2/a^2 + y_n^2/b^2 \leq 1$. Therefore $x^2/a^2 + y^2/b^2 \leq 1$ so $(x, y) \in \mathbf{E}$. See Fig. 2.1b.

EXAMPLE 2.3

Let $\mathbf{S}$ and $\mathbf{T}$ be closed sets that have common points. Prove that $\mathbf{S} \cap \mathbf{T}$ is closed.

Solution: Let $\mathbf{x}_n \in \mathbf{S} \cap \mathbf{T}$ and $\mathbf{x}_n \longrightarrow \mathbf{x}$. To prove: $\mathbf{x} \in \mathbf{S} \cap \mathbf{T}$. Now $\mathbf{x}_n \in \mathbf{S}$ and $\mathbf{S}$ is closed, so $\mathbf{x} \in \mathbf{S}$. Likewise $\mathbf{x} \in \mathbf{T}$, so $\mathbf{x} \in \mathbf{S} \cap \mathbf{T}$.

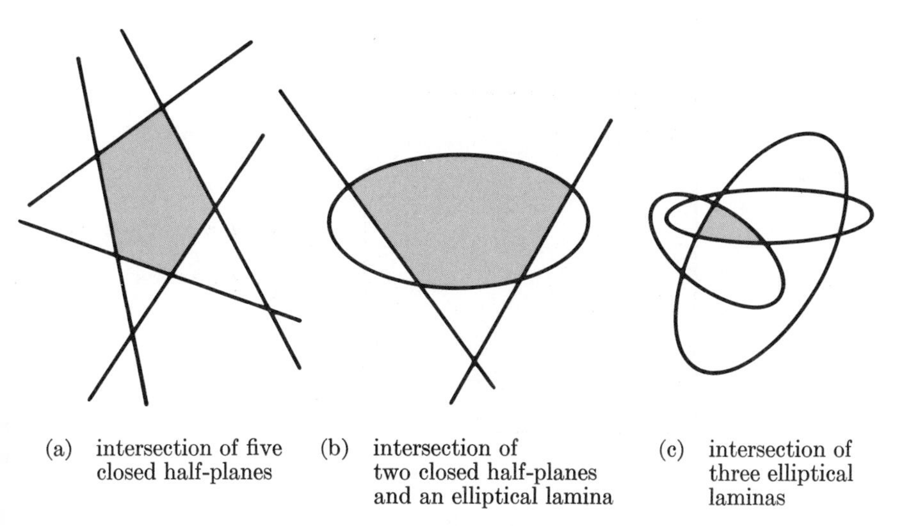

(a) intersection of five closed half-planes (b) intersection of two closed half-planes and an elliptical lamina (c) intersection of three elliptical laminas

FIG. 2.2 An intersection of closed sets—again a closed set.

REMARK: It follows easily that the intersection of any number of closed sets, if non-empty, is another closed set. We can use this result to construct many closed sets. See Fig. 2.2 for some examples.

EXAMPLE 2.4

Let $\mathbf{S}$ and $\mathbf{T}$ be closed sets. Prove that $\mathbf{S} \cup \mathbf{T}$ is a closed set.

Solution: Let $\mathbf{x}_n \in \mathbf{S} \cup \mathbf{T}$ and $\mathbf{x}_n \longrightarrow \mathbf{x}$. To prove: $\mathbf{x} \in \mathbf{S} \cup \mathbf{T}$. Now either infinitely many of the points $\mathbf{x}_n$ belong to $\mathbf{S}$ or infinitely many belong to $\mathbf{T}$ (or both). For otherwise there would be only a finite number of points $\mathbf{x}_n$. Suppose infinitely many belong to $\mathbf{S}$. That means there is a subsequence $\{\mathbf{x}_{n_j}\}$ of $\{\mathbf{x}_n\}$ with $\mathbf{x}_{n_j} \in \mathbf{S}$. Since $\mathbf{x}_n \longrightarrow \mathbf{x}$, then also $\mathbf{x}_{n_j} \longrightarrow \mathbf{x}$. But $\mathbf{S}$ is closed, so $\mathbf{x} \in \mathbf{S}$. Hence $\mathbf{x} \in \mathbf{S} \cup \mathbf{T}$.

Open Sets

Definition Let $\mathbf{S}$ be a set in space and $\mathbf{x}_0$ a point of $\mathbf{S}$. Then $\mathbf{x}_0$ is called an **interior point** of $\mathbf{S}$ provided for some $\delta > 0$,

$$\{\mathbf{x} \mid |\mathbf{x} - \mathbf{x}_0| < \delta\} \subseteq \mathbf{S}.$$

Interior points are important because we have a fighting chance to define derivatives at interior points of the domain of a function. See Fig. 2.3 for some plane examples that illustrate the concept. Note that if $\mathbf{x}_0$ is fixed, then $\{\mathbf{x} \mid |\mathbf{x} - \mathbf{x}_0| < \delta\}$ is the circular disk of radius δ and center $\mathbf{x}_0$, without the boundary circle.

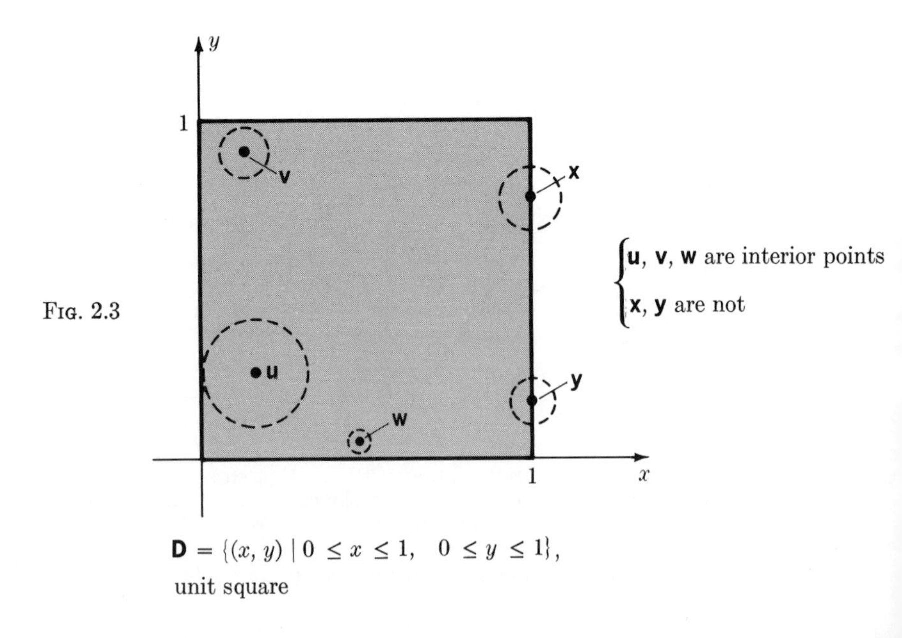

FIG. 2.3

$\mathbf{D} = \{(x, y) \mid 0 \leq x \leq 1, \quad 0 \leq y \leq 1\}$, unit square

In the figure, $\mathbf{u}$ is an interior point of $\mathbf{D}$ because there is a circular disk centered at $\mathbf{u}$ and entirely inside $\mathbf{D}$. Likewise $\mathbf{v}$ and $\mathbf{w}$ are interior points of $\mathbf{D}$. But $\mathbf{x}$ is not an interior point of $\mathbf{D}$ because *any* circular disk centered at $\mathbf{x}$ has points *outside* of $\mathbf{D}$. Likewise $\mathbf{y}$ is not an interior point. The interior points are precisely those points (x, y) where $0 < x < 1$ *and* $0 < y < 1$. The remaining points of $\mathbf{D}$ are *not* interior points. They are the points of $\mathbf{D}$ where $x = 0$ or $x = 1$ or $y = 0$ or $y = 1$, that is, the points on the boundary of the square.

Now consider Fig. 2.4. The set $\mathbf{D}$ is the unit disk *without* its boundary. Every point of $\mathbf{D}$ is an interior point! For if $|\mathbf{x}_0| < 1$, then $\delta = 1 - |\mathbf{x}_0| > 0$, and if $|\mathbf{x} - \mathbf{x}_0| < \delta$, then

$$|\mathbf{x}| = |(\mathbf{x} - \mathbf{x}_0) + \mathbf{x}_0| \leq |\mathbf{x} - \mathbf{x}_0| + |\mathbf{x}_0| < \delta + |\mathbf{x}_0| = 1,$$

so $\mathbf{x} \in \mathbf{D}$.

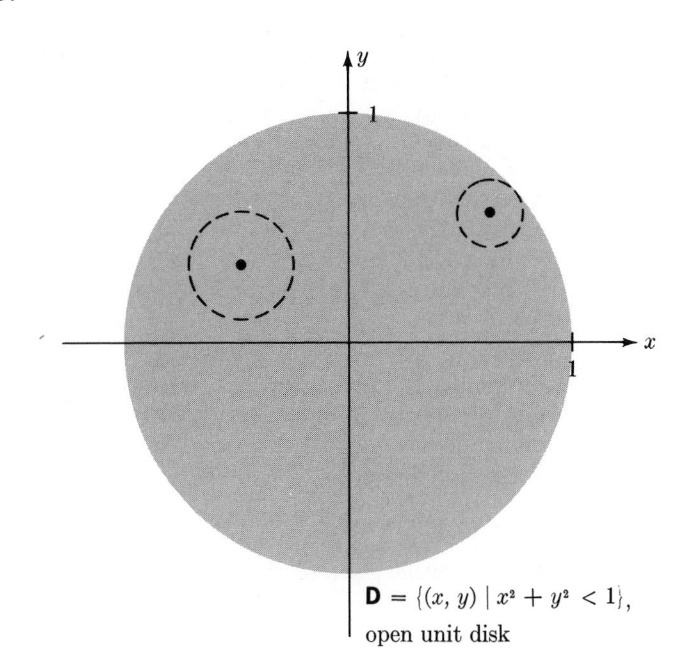

FIG. 2.4 Every point of $\mathbf{D}$ is an interior point.

In $\mathbf{R}^3$, the set $\{\mathbf{x} \mid |\mathbf{x} - \mathbf{x}_0| < \delta\}$ is a ball of radius δ and center $\mathbf{x}_0$, consisting of all points inside the spherical surface $|\mathbf{x} - \mathbf{x}_0| = \delta$.

Definition A set $\mathbf{D}$ is **open** if *every* point of $\mathbf{D}$ is an interior point.

The set in Fig. 2.4 is open; that in Fig. 2.3 is not open because the points on the boundary of the square are *not* interior points, but they are points of the set.

EXAMPLE 2.5

Let $a > 0$ and $b > 0$, and set

$$\mathbf{D} = \left\{ (x, y) \,\middle|\, \frac{x^2}{a^2} + \frac{y^2}{b^2} < 1 \right\}.$$

Prove that **D** is open.

Solution: Suppose $(x_0, y_0) \in \mathbf{D}$. To prove: (x_0, y_0) is an interior point of **D**, that is, that there is a positive δ so small that $(x, y) \in \mathbf{D}$ whenever $|(x, y) - (x_0, y_0)| < \delta$.

We have $x_0^2/a^2 + y_0^2/b^2 < 1$. The idea is to choose δ so small that $x^2/a^2 + y^2/b^2$ is very close to $x_0^2/a^2 + y_0^2/b^2$ whenever $|(x, y) - (x_0, y_0)| < \delta$. How close? Closer than $\epsilon = 1 - (x_0^2/a^2 + y_0^2/b^2)$. Then $x^2/a^2 + y^2/b^2$ must be less than 1.

Now $f(x) = x^2/a^2$ and $g(y) = y^2/b^2$ are continuous, so we can make $|f(x) - f(x_0)| < \frac{1}{2}\epsilon$ and $|g(y) - g(y_0)| < \frac{1}{2}\epsilon$ by demanding that x be sufficiently near to x_0 and y to y_0, say $|x - x_0| < \delta$ and $|y - y_0| < \delta$. Then

$$\begin{aligned} \left| \left(\frac{x^2}{a^2} + \frac{y^2}{b^2} \right) - \left(\frac{x_0^2}{a^2} + \frac{y_0^2}{b^2} \right) \right| &= |f(x) - f(x_0) + g(y) - g(y_0)| \\ &\leq |f(x) - f(x_0)| + |g(y) - g(y_0)| \\ &< \tfrac{1}{2}\epsilon + \tfrac{1}{2}\epsilon = \epsilon. \end{aligned}$$

WARNING: Despite what the words seem to suggest, not every set is either open or closed. For example, take a disk in the plane together with some, but not all, of its boundary points.

For a more unusual example let

$$\mathbf{S} = \{ (x, y) \mid x \text{ and } y \text{ are rational numbers} \}.$$

Then **S** is very far from being either open or closed. Can you see why?

EXERCISES

1. Let $\mathbf{x}_n \longrightarrow \mathbf{a}$ and $\mathbf{y}_n \longrightarrow \mathbf{b}$. Prove that $\mathbf{x}_n + \mathbf{y}_n \longrightarrow \mathbf{a} + \mathbf{b}$.
2. Let $c_n \longrightarrow c$ and $\mathbf{x}_n \longrightarrow \mathbf{a}$. Prove that $c_n\mathbf{x}_n \longrightarrow c\mathbf{a}$.
3. Let $\mathbf{x}_n \longrightarrow \mathbf{x}$ and $\mathbf{y}_n \longrightarrow \mathbf{y}$. Prove $\mathbf{x}_n \cdot \mathbf{y}_n \longrightarrow \mathbf{x} \cdot \mathbf{y}$.
4. Let $\mathbf{x}_n \longrightarrow \mathbf{x}$. Prove $|\mathbf{x}_n| \longrightarrow |\mathbf{x}|$.
5. Let $\mathbf{x}_n \longrightarrow \mathbf{x}$ and $\mathbf{y}_n \longrightarrow \mathbf{y}$. Prove $\mathbf{x}_n \times \mathbf{y}_n \longrightarrow \mathbf{x} \times \mathbf{y}$.

Prove each of the following sets is closed and sketch:

6. $\{ (x, y) \mid y \leq x^2 \}$
7. $\{ (x, y) \mid y \geq x^2 \}$
8. $\{ (x, y) \mid x^2/a^2 + y^2/b^2 \geq 1 \}$

9. $\{(x, y) \mid x^2/a^2 - y^2/b^2 \le 1\}$
10. $\{(x, y, z) \mid x^2 + y^2 + z^2 \le 1\}$.
11. Prove that $\mathbf{R}^3$ is a closed subset of $\mathbf{R}^3$.
12. Show that $\{\mathbf{x} \mid 0 < |\mathbf{x}| \le 1\}$ is not a closed set.
13. Prove that $\{(x, y) \mid 0 < x < 1\}$ is an open set. Sketch it.
14. Prove that $\mathbf{S} \cup \mathbf{T}$ is open if $\mathbf{S}$ and $\mathbf{T}$ are.
15. Suppose $\mathbf{S}$ and $\mathbf{T}$ are open and have common points. Prove that $\mathbf{S} \cap \mathbf{T}$ is open.
16*. Prove that $\mathbf{S}$ is closed if and only if its complement (the rest of space) is open or empty.

Is the set open or closed or neither?

17. $\{(x, y) \mid 2x + 3y > 1\}$
18. the domain of the function $f(x, y) = \dfrac{1}{x - y}$
19. $\{(x, y) \mid 0 < x < 4,\ 2 \le y \le 3\}$
20. $\{(x, y) \mid 0 \le y \le e^x\}$
21. $\{(x, y) \mid x \text{ is an integer}\}$
22. $\{(x, y) \mid \text{neither } x \text{ nor } y \text{ is an integer}\}$.

23*. Prove the **Cauchy criterion:** A sequence $\{\mathbf{x}_n\}$ of points in $\mathbf{R}^3$ converges if and only if for each $\epsilon > 0$, there exists a positive integer N such that $|\mathbf{x}_m - \mathbf{x}_n| < \epsilon$ whenever $m, n \ge N$. [*Hint:* Use the one variable case, p. 5.]
24*. Prove the **Bolzano–Weierstrass Theorem:** If $\{\mathbf{x}_n\}$ is a bounded sequence of points in $\mathbf{R}^3$, then there is a convergent subsequence. You may presuppose the result for $\mathbf{R}$.

3. CONTINUITY

For a function of several variables to be useful, it must have some reasonable properties. The most basic of such properties is continuity. Here is the formal definition, a direct generalization of the definition of continuity for a function of one variable.

> **Definition** Let $f\colon \mathbf{D} \longrightarrow \mathbf{R}$, where $\mathbf{D}$, the domain of f, is a subset of $\mathbf{R}^2$ or $\mathbf{R}^3$. Let $\mathbf{a}$ be a point of $\mathbf{D}$. We say f is **continuous** at $\mathbf{a}$ if $f(\mathbf{x}) \longrightarrow f(\mathbf{a})$ as $\mathbf{x} \longrightarrow \mathbf{a}$. Precisely, for each $\epsilon > 0$ there exists $\delta > 0$ such that $|f(\mathbf{x}) - f(\mathbf{a})| < \epsilon$ whenever $\mathbf{x} \in \mathbf{D}$ and $|\mathbf{x} - \mathbf{a}| < \delta$.
> We say f is **continuous** on $\mathbf{D}$ if f is continuous at each point of $\mathbf{D}$.

As for functions of one variable, this definition requires that a continuous function be "predictable"; you should be able to predict the value of the function at $\mathbf{a}$ from its values near $\mathbf{a}$.

The elementary properties of continuous functions of one variable carry over to this case easily. In particular, *sums, products, and quotients* (with nonzero denominator) *of continuous functions are continuous.*

Obviously the functions defined by $f(x, y) = x$ and $g(x, y) = y$ are continuous. By forming products and sums we conclude that *each polynomial is a continuous function on* $\mathbf{R}^2$. (Of course the same holds on $\mathbf{R}^3$.) From this we deduce that *each rational function is continuous wherever the denominator is not zero.* (Recall that a rational function is a quotient of polynomials.)

It is also not hard to see that if $f(x)$ and $g(y)$ are continuous functions of one variable, then $h(x, y) = f(x)g(y)$ is a continuous function of two variables. (See Ex. 2.) For example, the continuity of $h(x, y) = x \ln y$ follows from that of $f(x) = x$ and $g(y) = \ln y$.

Composite Functions

Suppose we wanted to prove that $f(x, y) = y^x$ is continuous on the domain $x > 0$, $y > 0$. We could write

$$f(x, y) = e^{x \ln y} = e^{g(x,y)}.$$

Thus $f(x, y)$ the composite of the continuous functions $h(t) = e^t$ and $t = x \ln y$. It seems reasonable that $f(x, y)$ is continuous also.

Here is a more complicated type of example. Suppose we somehow manage to prove that

$$K(x, y, z) = \int_0^x (y^3 + t^4) \sin(zt^2)\, dt$$

is continuous in x, y, z-space. We want to conclude that $K(u^v, v^u, uv)$ is continuous on the domain $u > 0$, $v > 0$. What we need is the following theorem.

Theorem Let $K(x, y, z)$ be continuous on a domain $\mathbf{D}$ in x, y, z-space. Let f, g, h be continuous on a domain $\mathbf{E}$ of the u, v-plane, and suppose that $(f(u, v), g(u, v), h(u, v)) \in \mathbf{D}$ whenever $(u, v) \in \mathbf{E}$. Then the composite function

$$k(u, v) = K[f(u, v), g(u, v), h(u, v)]$$

is continuous on $\mathbf{E}$.

Proof: Let $(u, v) \longrightarrow (u_0, v_0)$. Then $f(u, v) \longrightarrow f(u_0, v_0)$, $g(u, v) \longrightarrow g(u_0, v_0)$, and $h(u, v) \longrightarrow h(u_0, v_0)$. Hence

$$(f(u, v), g(u, v), h(u, v)) \longrightarrow (f(u_0, v_0), g(u_0, v_0), h(u_0, v_0)).$$

But K is continuous, so

$$k(u, v) = K[f(u, v), g(u, v), h(u, v)]$$
$$\longrightarrow K[f(u_0, v_0), g(u_0, v_0), h(u_0, v_0)] = k(u_0, v_0),$$

therefore k is continuous.

NOTE: The theorem above is stated for a function of three variables, where each variable is replaced by a function of two variables. Clearly, there is nothing special about three and two, and the result may be modified as needed.

Maxima and Minima

One of the main concerns of calculus is maximum and minimum values of functions. Recall one of the basic facts about continuous functions of one variable:

> If f is continuous on a closed interval $[a, b]$, then there exist points x_0 and x_1 in the interval such that
>
> $$f(x_0) \leq f(x) \leq f(x_1)$$
>
> for all $x \in [a, b]$.

The result says that

$$f(x_0) = \min\{ f(x) \mid a \leq x \leq b\},$$

$$f(x_1) = \max\{ f(x) \mid a \leq x \leq b\}.$$

If the interval is not closed, then f need not have a maximum or a minimum. For example, $f(x) = x$ has neither a maximum nor a minimum on the *open* interval (a, b). The same holds for any continuous increasing or decreasing function.

Furthermore, the result is not true on a domain which is unbounded, that is, contains points arbitrarily far from the origin. For example, on the domain $[0, \infty)$, the function $f(x) = e^{-x}$ has a maximum but no minimum; on $(0, \infty)$ it has neither.

The correct generalization of the preceding theorem requires a domain that is both closed and bounded.

> **Definition** A subset $\mathbf{S}$ of space is **bounded** if there is a number B such that $|\mathbf{x}| \leq B$ for all $\mathbf{x} \in \mathbf{S}$.

Thus a set is bounded provided it is contained in some sphere of finite radius centered at the origin.

> **Theorem** Let $\mathbf{S}$ be a bounded and closed subset of space and let f have domain $\mathbf{S}$. Then there exist points $\mathbf{x}_0$ and $\mathbf{x}_1$ of $\mathbf{S}$ such that
>
> $$f(\mathbf{x}_0) \leq f(\mathbf{x}) \leq f(\mathbf{x}_1)$$
>
> for all $\mathbf{x} \in \mathbf{S}$.

A complete proof of this fundamental result is best postponed to a more advanced course; here is a brief sketch of a proof. Let $M = \sup\{ f(\mathbf{x}) \mid \mathbf{x} \in \mathbf{S}\}$. (Possibly $M = \infty$.) Choose $\mathbf{x}_n = (x_n, y_n, z_n) \in \mathbf{S}$ so that $f(\mathbf{x}_n) \longrightarrow M$.

The sequence $\{x_n\}$ is bounded (because each $\mathbf{x}_n$ is in $\mathbf{S}$ and $\mathbf{S}$ is bounded). Therefore $\{x_n\}$ has a convergent subsequence. We may restrict attention to this subsequence only. Hence we may assume $x_n \longrightarrow a$.

Next we look at $\{y_n\}$. Again, passing to a subsequence, we may assume $y_n \longrightarrow b$. Similarly we may assume $z_n \longrightarrow c$ and hence $\mathbf{x}_n \longrightarrow \mathbf{a} = (a, b, c)$. Then $\mathbf{a} \in \mathbf{S}$ because $\mathbf{S}$ is closed, and $f(\mathbf{x}_n) \longrightarrow f(\mathbf{a})$ because f is continuous. But $f(\mathbf{x}_n) \longrightarrow M$, therefore $M = f(\mathbf{a})$. A similar argument applies to the minimum.

Uniform Continuity

The final property of continuous functions we shall consider is the property of uniform continuity. It is important in the theory of integration.

Definition Let f be defined on a set $\mathbf{S}$. Then f is called **uniformly continuous** on $\mathbf{S}$ if for each $\epsilon > 0$ there is a $\delta > 0$ such that

$$|f(\mathbf{x}) - f(\mathbf{z})| < \epsilon$$

whenever $\mathbf{x}$ and $\mathbf{z}$ are in $\mathbf{S}$ and $|\mathbf{x} - \mathbf{z}| < \delta$.

At first reading, this definition may appear to be the same as the definition of continuity. The point, however, is that given ϵ, the *same* δ works throughout the domain of f. In other words, the δ given by the definition of continuity is independent of the point $\mathbf{x}$.

It is rather obvious that each uniformly continuous function is continuous. The converse is not true in general (see Ex. 5). The converse *is* true, however, provided the domain is closed and bounded.

Theorem Let $\mathbf{S}$ be a bounded and closed subset of space and let f be continuous on $\mathbf{S}$. Then f is uniformly continuous on $\mathbf{S}$.

The proof of this theorem, like the last one, is best postponed to a later course. However, for the brave we offer the following: Suppose for some $\epsilon > 0$ there is no $\delta > 0$ that fills the bill. Then we can choose $\mathbf{x}_n \in \mathbf{S}$ and $\mathbf{y}_n \in \mathbf{S}$ such that $|\mathbf{x}_n - \mathbf{y}_n| \longrightarrow 0$ and $|f(\mathbf{x}_n) - f(\mathbf{y}_n)| \geq \epsilon$. As in the proof sketched for the previous theorem, by passing to a subsequence we may assume $\mathbf{x}_n \longrightarrow \mathbf{a} \in \mathbf{S}$. Then $\mathbf{y}_n \longrightarrow \mathbf{a}$ also, since $\mathbf{y}_n = \mathbf{x}_n + (\mathbf{y}_n - \mathbf{x}_n)$ and $|\mathbf{y}_n - \mathbf{x}_n| \longrightarrow 0$. Therefore $f(\mathbf{x}_n) \longrightarrow f(\mathbf{a})$ and $f(\mathbf{y}_n) \longrightarrow f(\mathbf{a})$ since f is continuous at $\mathbf{a}$. This contradicts $|f(\mathbf{x}_n) - f(\mathbf{y}_n)| \geq \epsilon$ for all n.

EXERCISES

1. Prove that the sum of two continuous functions is continuous.
2. If $f(x)$ and $g(y)$ are continuous, prove that $h(x, y) = f(x)g(y)$ is continuous.
3. Let $f(\mathbf{x}) = |\mathbf{x}|$ for $\mathbf{x} \in \mathbf{R}^3$. Prove that $f(\mathbf{x})$ is continuous using properties of length.

4. (cont.) Do Ex. 3 using composite functions.
5. Prove that $f(x, y) = 1/(x + y)$ is continuous on the open first quadrant $x > 0$, $y > 0$, but is not uniformly continuous.
6*. (cont.) Prove that $f(x, y)$ is uniformly continuous on the domain $x > 1$, $y > 1$.

Suppose $f(x, y)$ is continuous on $\mathbf{R}^2$ and c is constant. Prove:

7. The set $\{(x, y) \mid f(x, y) > c\}$ is open (if non-empty).
8. The set $\{(x, y) \mid f(x, y) = c\}$ is closed (if non-empty).
9. By what general principles do you know that $\sin \frac{1}{6}(x + y)$ has a minimum value on the disk $x^2 + y^2 \leq 1$?
10. Let $\mathbf{S}$ be a closed bounded set in $\mathbf{R}^3$ and let $\mathbf{x}_0$ be a point not in $\mathbf{S}$. Prove there is a point in $\mathbf{S}$ closest to $\mathbf{x}_0$. [*Hint:* Ex. 3.]
11. (cont.) Show by example that there may be more than one point of $\mathbf{S}$ "closest" to $\mathbf{x}_0$.
12*. Let $f(\mathbf{x})$ be continuous for all $\mathbf{x} \in \mathbf{R}^3$. Suppose that $f(\mathbf{x}) = 0$ for some $\mathbf{x}$ but $f(\mathbf{0}) \neq 0$. Prove there is a point nearest to $\mathbf{0}$ where $f(\mathbf{x}) = 0$.
13. Prove that $f(x, y) = x^2 - 6xy + 10y^2$ has a positive minimum value p on the circle $x^2 + y^2 = 1$.
14. (cont.) If $f(x, y) = ax^2 + bxy + cy^2$ has a positive minimum p on $x^2 + y^2 = 1$, prove that $f(x, y) > 0$ for all (x, y) except $(0, 0)$. [*Hint:* Find the minimum of $f(x, y)$ on $x^2 + y^2 = r^2$.]

Let $f(x, y) = e^{-(x^2+y^2)}$:

15. Find the maximum of $f(x, y)$ on $\mathbf{R}^2$
16. Find the minimum of $f(x, y)$ on $\mathbf{R}^2$
17. Is $f(x, y)$ continuous on $\mathbf{R}^2$?
18*. Is $f(x, y)$ uniformly continuous on $\mathbf{R}^2$?

19. Let $f(x, y)$ be continuous on $\mathbf{R}^2$ and let b be a real number. Prove that $g(x) = f(x, b)$ is continuous on $\mathbf{R}$.
20. Let $f(x, y)$ be continuous on $\mathbf{R}^2$. Prove that $g(x) = f(x, x)$ is continuous on $\mathbf{R}$.
21*. Let $g(t)$ have a continuous derivative for all t. Define $f(x, y)$ by

$$\begin{cases} f(x, y) = \dfrac{g(x) - g(y)}{x - y} & x \neq y \\ f(x, x) = g'(x). & \end{cases}$$

Prove that $f(x, y)$ is continuous on $\mathbf{R}^2$. [*Hint:* Use the Mean Value Theorem. It is stated on p. 285.]

4. GRAPHS

Suppose a function $z = f(x, y)$ is defined for all points (x, y) in some domain in the plane. Its **graph** is the surface in space consisting of all points

$$(x, y, f(x, y)),$$

where (x, y) is in the domain of f. Figure 4.1 illustrates the graph of a function defined for all points (x, y) in a circular disk.

To get an idea of the shape of the surface, draw several sections by planes perpendicular to the x-axis or the y-axis.

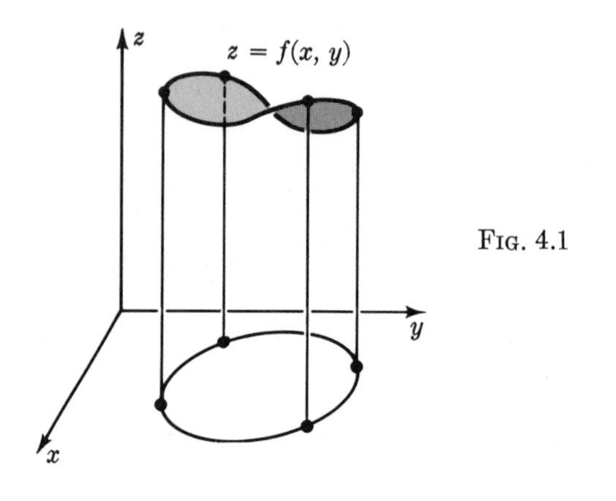

Fig. 4.1

> EXAMPLE 4.1
>
> Graph the function $z = f(x, y) = 1 - x^2$.

Solution: The function $f(x, y)$ is independent of y. Its graph is a cylinder with generators parallel to the y-axis. To see this, first graph the parabola $z = 1 - x^2$ in the x, z-plane (Fig. 4.2a). If (a, c) is any point on this parabola and b is any value of y whatsoever, then (a, b, c) is on the graph of $z = f(x, y)$.

> *Answer:* The graph is a parabolic cylinder with generators parallel to the y-axis. See Fig. 4.2b.

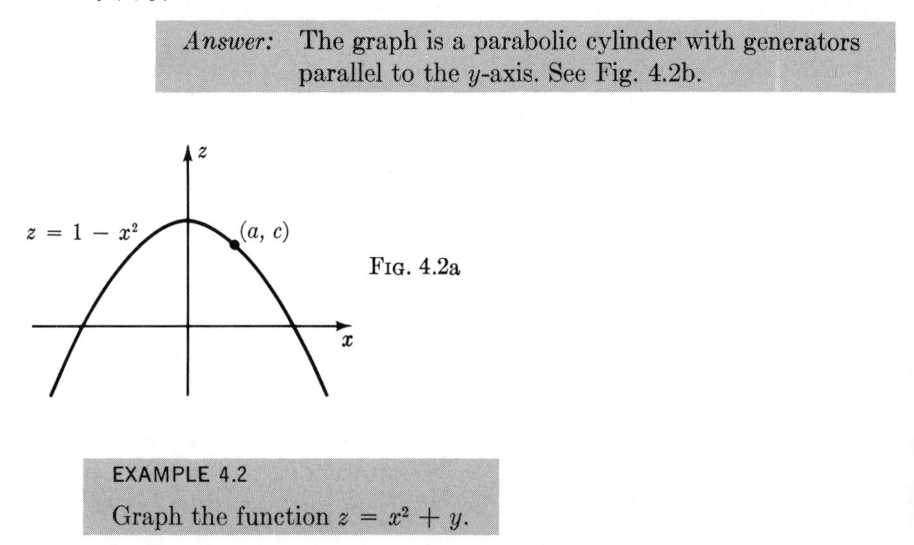

Fig. 4.2a

> EXAMPLE 4.2
>
> Graph the function $z = x^2 + y$.

Solution: Each cross-section by a plane $x = x_0$ is a straight line $z = y + x_0^2$ of slope 1. The surface meets the x, z-plane in the parabola $z = x^2$. See Fig. 4.3a. The figure does not yet convey the shape of the surface, so look at

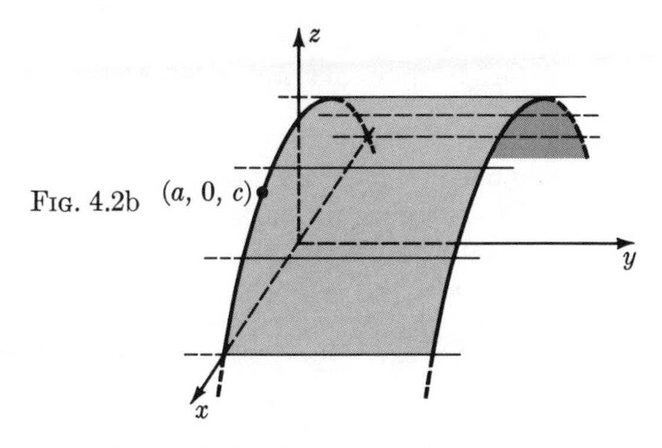

Fig. 4.2b

cross-sections by planes $y = y_0$. See Fig. 4.3b. Each is a parabola $z = x^2 + y_0$. Now the picture is clearer.

> *Answer:* The surface is a cylinder oblique to the x, z-plane; it intersects the x, z-plane in the parabola $z = x^2$.

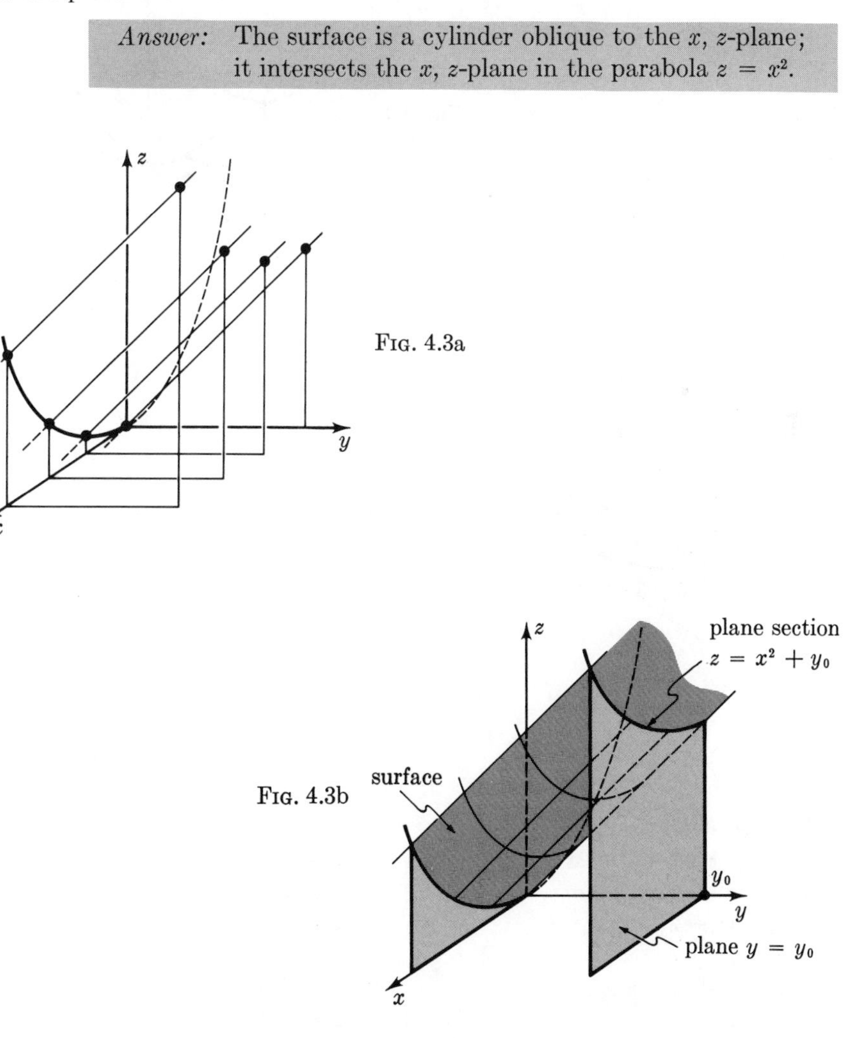

Fig. 4.3a

Fig. 4.3b

EXAMPLE 4.3

Graph the function $z = x^2 + y^2$.

Solution: Each cross-section by a plane perpendicular to the x-axis is a parabola $z = x_0^2 + y^2$. This is shown in Fig. 4.4a. But from this figure, it is hard to visualize the surface. You learn more studying the cross-sections of the graph by planes $z = z_0$ perpendicular to the z-axis. Each cross-section is a circle $x^2 + y^2 = z_0$.

Answer: The surface is a paraboloid of revolution (Fig. 4.4b).

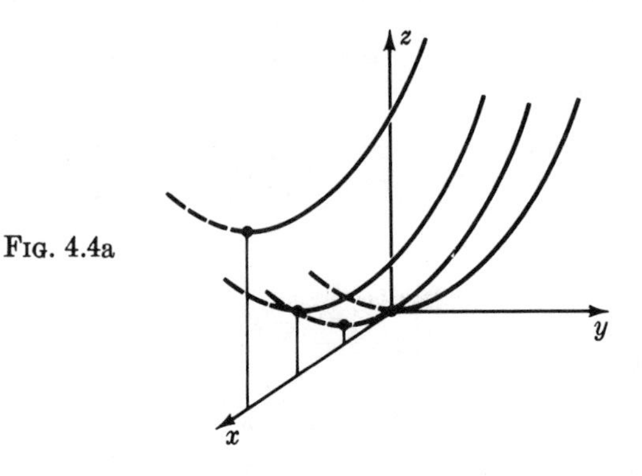

FIG. 4.4a

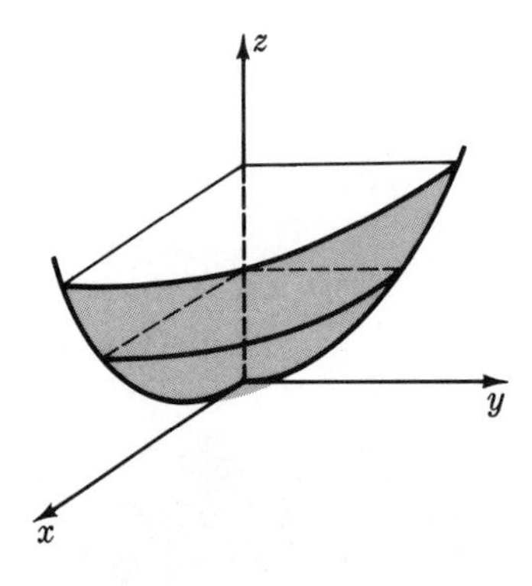

FIG. 4.4b

EXERCISES

Sketch the graphs:

1. $z = f(x, y) = 1 - 2x$
2. $z = f(x, y) = 3 + y$
3. $z = x + y$
4. $z = 3x + 2y$
5. $z = (x - 2)^2$
6. $z = (y + 1)^3$
7. $z = x^3 + y$
8. $z = x^2 - y^2$
9. $z = x^2 + x - y$
10. $z = x^3 + y^3$
11. $z = xy$
12. $z = x^2 + xy + y^2$.

13. Graph the function $z = \sqrt{9 - x^2 - y^2}$ for all points (x, y) in the circle of radius 3 and center $(0, 0)$.
14. Graph the function $z = x^4 + y^4$ for all points (x, y) in the domain $x^2 + y^2 \leq 1$, $x \geq 0$, $y \geq 0$.
15. Find the curves formed when planes parallel to the x, z-plane intersect the graph of $z = 2x^2 + 3y^2$.
16. Find all points common to both the plane $x = 3$ and the graph of $z = x^2y$.
17. Let $f(x, y)$ be a continuous function defined on a closed domain $\mathbf{D}$. Prove that the graph of f is a closed subset of $\mathbf{R}^3$.
18. Prove that the graph of $f(x, y)$ is never an open subset of $\mathbf{R}^3$.

5. PARTIAL DERIVATIVES

Let $z = f(x, y)$ be a function of two variables and $\mathbf{a} = (a, b)$ an interior point of its domain. Suppose we set $y = b$ and allow only x to vary. Then $f(x, b)$ is a function of the single variable x, defined at least in some open interval including a. We define

$$\frac{\partial z}{\partial x}(a, b) = \frac{d}{dx} f(x, b)\bigg|_{x=a}.$$

This is called the **partial derivative** (or simply **partial**) of z with respect to x. (The "∂" is a curly "d".) It measures the rate of change of z with respect to x while y is held constant.

Similarly, we define the partial derivative of z with respect to y:

$$\frac{\partial z}{\partial y}(a, b) = \frac{d}{dy} f(a, y)\bigg|_{y=b}.$$

In like manner, given a function $w = f(x, y, z)$ of three variables, we may define the three partial derivatives $\partial w/\partial x$, $\partial w/\partial y$, and $\partial w/\partial z$. For instance,

$$\frac{\partial w}{\partial y}(a, b, c) = \frac{d}{dy} f(a, y, c)\bigg|_{y=b}.$$

Each of the partials is the derivative of w with respect to the variable in question, taken while all other variables are held fixed.

NOTE: Our definition of partial derivatives applies only at interior points of the domain.

EXAMPLE 5.1

Let $z = f(x, y) = xy^2$. Find

$$\frac{\partial z}{\partial x} \text{ for } y = 3, \qquad \frac{\partial z}{\partial y} \text{ for } x = -4,$$

$$\frac{\partial z}{\partial x} \text{ and } \frac{\partial z}{\partial y} \text{ in general.}$$

Solution: If $y = 3$, then $z = 9x$, and

$$\left.\frac{\partial z}{\partial x}\right|_{y=3} = \frac{d}{dx}(9x) = 9.$$

Likewise, if $x = -4$, then $z = -4y^2$, and

$$\left.\frac{\partial z}{\partial y}\right|_{x=-4} = \frac{d}{dy}(-4y^2) = -8y.$$

In general, to compute $\partial z/\partial x$ just differentiate as usual, pretending y is a constant:

$$\frac{\partial z}{\partial x} = \frac{\partial(xy^2)}{\partial x} = y^2\frac{d}{dx}(x) = y^2.$$

To compute $\partial z/\partial y$, differentiate, pretending x is a constant:

$$\frac{\partial z}{\partial y} = \frac{\partial(xy^2)}{\partial y} = x\frac{d}{dy}(y^2) = 2xy.$$

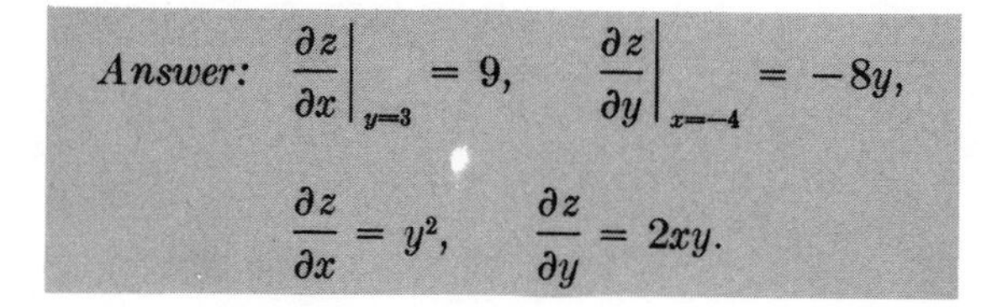

Answer: $\left.\frac{\partial z}{\partial x}\right|_{y=3} = 9, \qquad \left.\frac{\partial z}{\partial y}\right|_{x=-4} = -8y,$

$$\frac{\partial z}{\partial x} = y^2, \qquad \frac{\partial z}{\partial y} = 2xy.$$

We consider two further examples of partial derivatives.

(1) The gas law for a fixed mass of n moles of an ideal gas is

$$P = nR\frac{T}{V},$$

where R is the universal gas constant. Thus P is a function of the two variables T and V:

$$\frac{\partial P}{\partial T} = nR\frac{1}{V}, \qquad \frac{\partial P}{\partial V} = -nR\frac{T}{V^2}.$$

(2) The area A of a parallelogram of base b, slant height s, and angle α is $A = sb \sin \alpha$. The partial derivatives are

$$\frac{\partial A}{\partial s} = b \sin \alpha, \qquad \frac{\partial A}{\partial b} = s \sin \alpha, \qquad \frac{\partial A}{\partial \alpha} = sb \cos \alpha.$$

Geometric Interpretation

The graph of $z = f(x, y)$ is a surface in three dimensions. A plane $x = x_0$ cuts the graph in a plane curve $x = x_0$, $z = f(x_0, y)$. See Fig. 5.1a. If this curve is projected straight back onto the y, z-plane, the graph of the function $z = f(x_0, y)$ is obtained (Fig. 5.1b). The partial derivative

$$\frac{\partial f}{\partial y}(x_0, y)$$

is the slope of this graph.

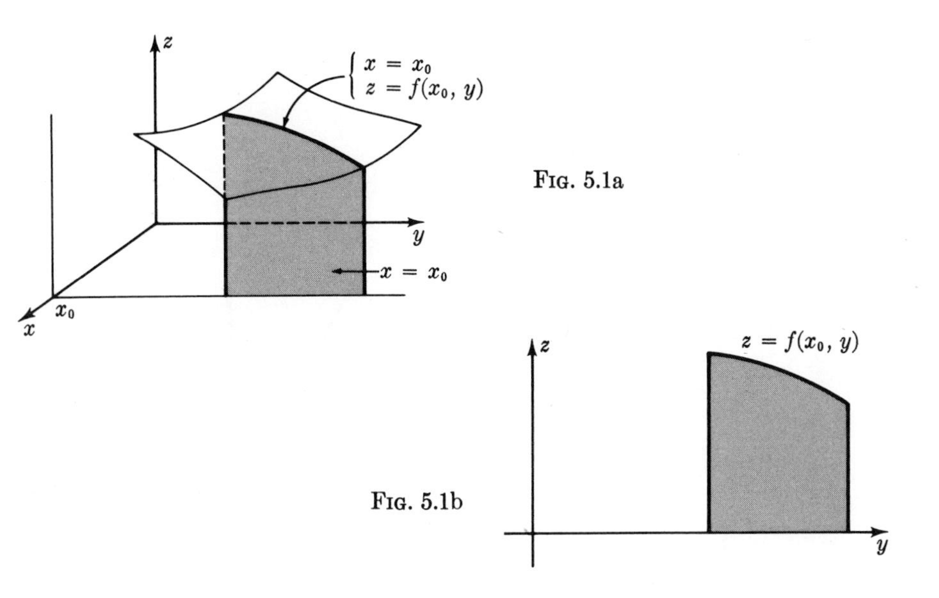

FIG. 5.1a

FIG. 5.1b

For example, suppose the graph of the function $z = x + y^2$ is sliced by the plane $x = x_0$. See Fig. 5.2a. The resulting curve is the parabola $x = x_0$, $z = y^2 + x_0$. If this is projected onto the y, z-plane, a parabola is obtained (Fig. 5.2b). Its slope is

$$\frac{\partial z}{\partial y} = 2y.$$

Notation

Unfortunately there are several different notations for partial derivatives in common use. Become familiar with them; they come up again and again in applications.

Suppose

$$w = f(x, y, z).$$

Common notations for $\partial w/\partial z$ are:

$$f_x, \quad f_x(x, y, z), \quad w_x, \quad w_x(x, y, z), \quad D_x f.$$

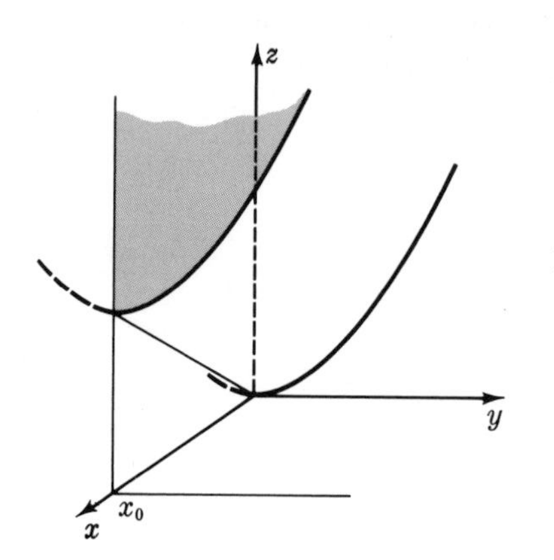

Fig. 5.2a

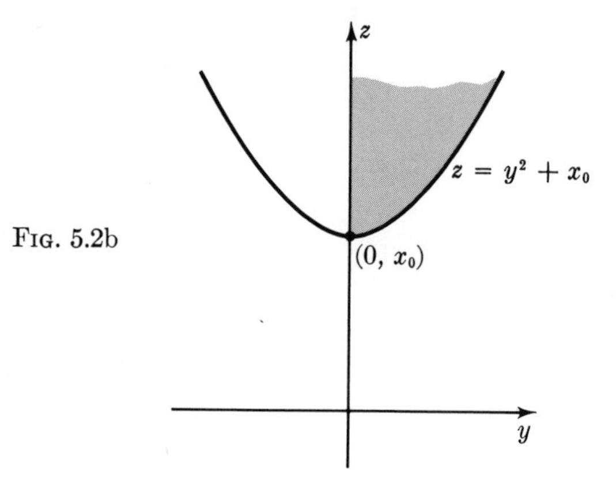

Fig. 5.2b

For example, if

$$w = f(x, y, z) = x^3y^2 \sin z,$$

then

$$f_x = 3x^2y^2 \sin z, \qquad w_y = 2x^3y \sin z, \qquad D_zf = x^3y^2 \cos z.$$

EXERCISES

Find $\dfrac{\partial z}{\partial x}$ and $\dfrac{\partial z}{\partial y}$:

1. $z = f(x, y) = x + 2y$
2. $z = f(x, y) = 3x + 4y$
3. $z = 3xy$
4. $z = x^2y$
5. $z = \dfrac{2x^2}{y + 1}$
6. $z = x^2y + xy^2$
7. $z = x \sin y$
8. $z = y^2 \cos x$
9. $z = \sin 2x + \cos 3y$
10. $z = \sin x - \cos y$
11. $z = \sin 2xy$
12. $z = \cos (2x + y)$

13. $z = \dfrac{x}{y} + \dfrac{y}{x}$

14. $z = \dfrac{x^3}{y}$

15. $z = xe^y$

16. $z = \dfrac{1}{x + 2y + 5}$

17. $z = e^{xy}$

18. $z = -3e^{x+y}$

19. $z = e^{2x} \sin y$

20. $z = e^{-y} \cos x$.

21. Let $z = x^2y$. Find $\partial z/\partial x$ for $y = 2$, and $\partial z/\partial y$ for $x = -1$.
22. Let $z = y^2/x$. Find z_x for $y = 3$.
23. Let $w = xy^2z^3$. Find w_x for $y = 2$ and $z = 2$, find w_y for $x = 1$ and $z = 0$, and find w_z for $x = y$.
24. Let $w = xy - xz - yz$. Find $\dfrac{\partial w}{\partial x} + \dfrac{\partial w}{\partial y} + \dfrac{\partial w}{\partial z}$.
25. Show that $z = (3x - y)^2$ satisfies $\partial z/\partial x + 3\, \partial z/\partial y = 0$.
26. Show that $z = f(x) + y^2$ satisfies $\partial z/\partial y = 2y$.
27. Show that $z = x^2 - y^2$ satisfies $(\partial z/\partial x)^2 - (\partial z/\partial y)^2 = 4z$.
28. Show that $z = x^6 - x^5y + 7x^3y^3$ satisfies $x(\partial z/\partial x) + y(\partial z/\partial y) = 6z$.

6. MAXIMA AND MINIMA

Suppose

$$z = z(x, y)$$

is a function of two variables defined on the domain

$$x_0 < x < x_1, \qquad y_0 < y < y_1.$$

Suppose z takes on its minimum value at $(x, y) = (a, b)$. By holding y fixed at $y = b$, the function z becomes a function of x alone with minimum at $x = a$. Hence

$$\frac{\partial z}{\partial x}(a, b) = 0.$$

Similarly

$$\frac{\partial z}{\partial y}(a, b) = 0.$$

These two relations are often enough to locate the points where a function takes on its minimum (or maximum) value. See Fig. 6.1.

Geometrically the idea is simple: If the graph of $z(x, y)$ has a high (low) point at (a, b, c), then so do its cross-sectional curves by planes through (a, b, c) parallel to the x, z-plane and to the y, z-plane. The slopes of these curves are $\partial z/\partial x$ and $\partial z/\partial y$ respectively.

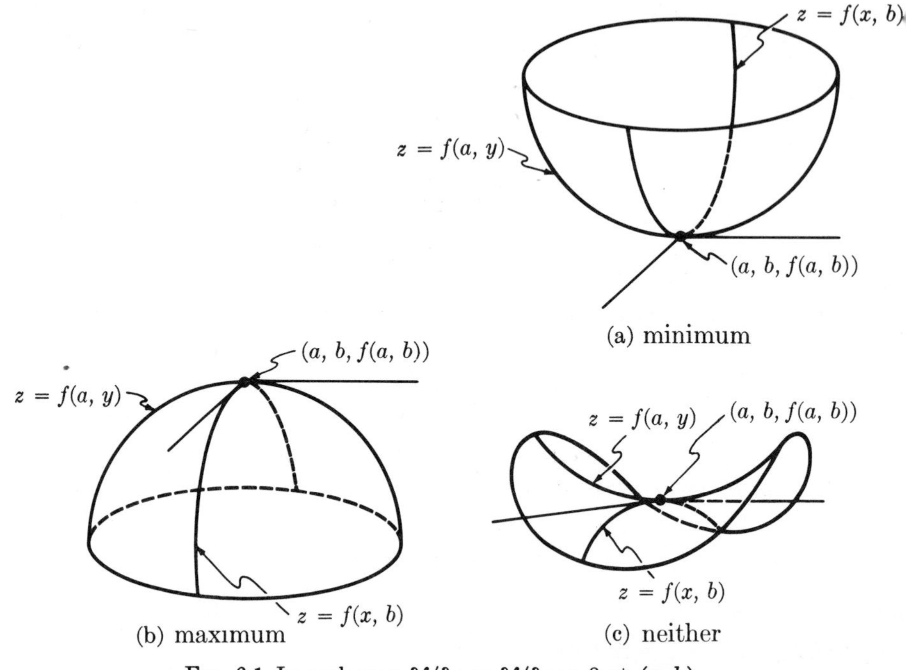

FIG. 6.1 In each case $\partial f/\partial x = \partial f/\partial y = 0$ at (a, b).

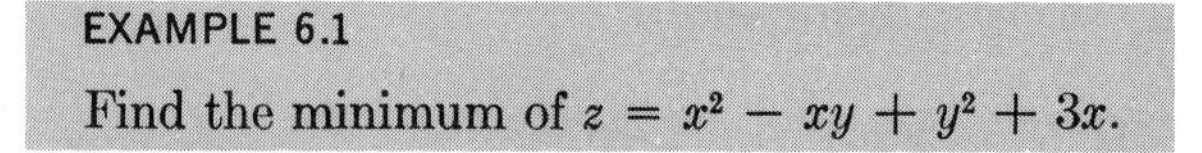

EXAMPLE 6.1

Find the minimum of $z = x^2 - xy + y^2 + 3x$.

Discussion: Is the question reasonable? For a fixed value of y, say $y = b$, the resulting function of x,

$$z(x, b) = x^2 + (3 - b)x + b^2,$$

is a quadratic polynomial whose graph is a parabola turned upward; it has a minimum. Similarly for each fixed $x = a$, the function of y,

$$z(a, y) = y^2 - ay + (a^2 + 3a),$$

has a minimum.

This is fairly good experimental evidence that the function $z(x, y)$ ought to have at least one minimum, probably no maximum.

Solution: The function is defined for all values of x and y. Begin by finding all points (x, y) at which both

$$\frac{\partial z}{\partial x} = 0 \quad \text{and} \quad \frac{\partial z}{\partial y} = 0.$$

Now

$$\frac{\partial z}{\partial x} = \frac{\partial}{\partial x}(x^2 - xy + y^2 + 3x) = 2x - y + 3,$$

$$\frac{\partial z}{\partial y} = \frac{\partial}{\partial y}(x^2 - xy + y^2 + 3x) = -x + 2y,$$

so the conditions are

$$\begin{cases} 2x - y + 3 = 0 \\ \quad -x + 2y = 0. \end{cases}$$

Solve:

$$x = -2, \qquad y = -1.$$

The corresponding value of z is

$$\begin{aligned} z(-2, -1) &= (-2)^2 - (-2)(-1) + (-1)^2 + 3(-2) \\ &= 4 - 2 + 1 - 6 = -3. \end{aligned}$$

Is this value a maximum, a minimum, or neither? (Recall that the vanishing of the derivative of a function $f(x)$ does not guarantee a maximum or minimum, e.g., $f(x) = x^3$ at $x = 0$.)

In this case you can prove that the value $z(-2, -1) = -3$ is the minimum value of z by these algebraic steps. First set up a u, v-coordinate system with its origin at $(x, y) = (-2, -1)$. Take

$$x = u - 2, \qquad y = v - 1.$$

Then

$$\begin{aligned} z &= (u - 2)^2 - (u - 2)(v - 1) + (v - 1)^2 + 3(u - 2) \\ &= u^2 - uv + v^2 - 3. \end{aligned}$$

Next, complete the square:

$$z = \left(u - \frac{v}{2}\right)^2 - \frac{v^2}{4} + v^2 - 3 = \left(u - \frac{v}{2}\right)^2 + \frac{3}{4}v^2 - 3.$$

Since squares are nonnegative, $z(x, y) \geq -3$.

Answer: -3.

In general, to find possible maximum and minimum values of a function, locate points where all its partial derivatives are zero. Whether a particular one of these actually yields a maximum or a minimum may not be easy to determine. (Later we shall study a second derivative test which sometimes helps.)

EXAMPLE 6.2

Find the rectangular solid of maximum volume whose total edge length is a given constant.

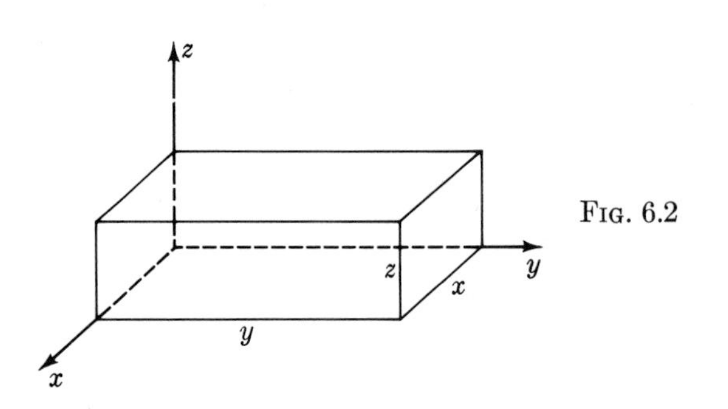

FIG. 6.2

Solution: As drawn in Fig. 6.2, the total length of the 12 edges is $4x + 4y + 4z$. Thus $4x + 4y + 4z = 4k$, where k is a constant, so

$$x + y + z = k.$$

The volume is

$$\begin{aligned} V = xyz &= xy(k - x - y) \\ &= kxy - x^2y - xy^2. \end{aligned}$$

The conditions for a maximum,

$$\frac{\partial V}{\partial x} = 0, \qquad \frac{\partial V}{\partial y} = 0,$$

are

$$\begin{cases} ky - 2xy - y^2 = 0 \\ kx - x^2 - 2xy = 0. \end{cases}$$

The nature of the geometric problem requires $x > 0$ and $y > 0$. Thus we may cancel y from the first equation and x from the second:

$$\begin{cases} 2x + y = k \\ x + 2y = k. \end{cases}$$

This pair of simultaneous linear equations has the unique solution

$$x = \frac{k}{3}, \qquad y = \frac{k}{3}.$$

Hence also $z = k/3$; the solid is a cube.

Answer: A cube.

EXAMPLE 6.3

What is the largest possible volume, and what are the dimensions of an open rectangular aquarium constructed from 12 ft^2 of Plexiglas? Ignore the thickness of the plastic.

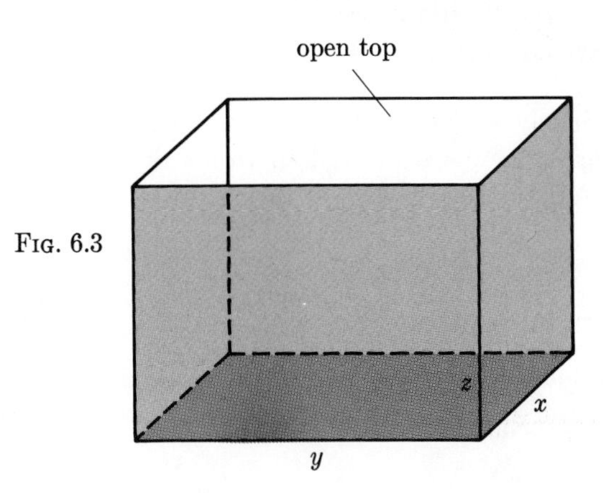

FIG. 6.3

Solution: See Fig. 6.3. The volume is $V = xyz$. The total surface area of the bottom and four sides is

$$xy + 2yz + 2zx = 12.$$

Solve for z, then substitute into the formula for V:

$$z = \frac{12 - xy}{2(x + y)}, \qquad V = \frac{(12 - xy)xy}{2(x + y)}.$$

Compute the partials carefully (only one computation is needed because of the symmetry in x and y):

$$\frac{\partial V}{\partial x} = \frac{y^2(-x^2 - 2xy + 12)}{2(x + y)^2}, \qquad \frac{\partial V}{\partial y} = \frac{x^2(-y^2 - 2xy + 12)}{2(x + y)^2}.$$

Now find all points (x, y) where both partials are zero. Such points must satisfy

$$\begin{cases} y^2(-x^2 - 2xy + 12) = 0 \\ x^2(-y^2 - 2xy + 12) = 0. \end{cases}$$

By the nature of the problem, both x and y are positive quantities. Therefore, the factors y^2 and x^2 may safely be canceled:

$$\begin{cases} -x^2 - 2xy + 12 = 0 \\ -y^2 - 2xy + 12 = 0. \end{cases}$$

It is easy to solve these equations for x and y. Subtract: $x^2 - y^2 = 0$, hence $y = \pm x$. Since x and y are both positive, only $y = x$ applies. Now substitute $y = x$ into either of the two equations:

$$-3x^2 + 12 = 0, \qquad x = \pm 2.$$

The only feasible choice is $x = y = 2$. It follows that $z = 1$. The optimal dimensions are $2 \times 2 \times 1$; the maximal volume 4 ft^3.

Alternate Solution: Previously we chose the side lengths as variables because volume and surface area can be expressed in terms of these side lengths. Is this the only way? Perhaps we can use the face areas as variables. They are

$$u = yz, \qquad v = zx, \qquad w = xy.$$

The total surface area, in terms of u, v, w, is

$$w + 2u + 2v = 12.$$

Can we express the volume in terms of u, v, and w? It is unnecessary to solve the system

$$yz = u, \qquad zx = v, \qquad xy = w$$

for x, y, z in terms of u, v, w, and then substitute the resulting expressions in $V = xyz$. We can express V directly in u, v, w:

$$V^2 = (xyz)^2 = x^2y^2z^2 = (yz)(zx)(xy) = uvw.$$

The maximum value of V corresponds to the maximum value of V^2. Now we have formulated a reasonable problem: Maximize uvw subject to the constraint $2u + 2v + w = 12$.

Set

$$g(u, v) = uv(12 - 2u - 2v) = 12uv - 2u^2v - 2uv^2.$$

Take partials:

$$g_u = 12v - 4uv - 2v^2,$$

$$g_v = 12u - 2u^2 - 4uv.$$

These partials are 0 where the equations

$$\begin{cases} 12v - 4uv - 2v^2 = 0 \\ 12u - 2u^2 - 4uv = 0 \end{cases}$$

are satisfied. Cancel the factor v in the first and u in the second ($v = 0$ or $u = 0$ doesn't hold water). The equations reduce to

$$\begin{cases} 2u + v = 6 \\ u + 2v = 6. \end{cases}$$

Hence $u = v = 2$. It follows that $w = 4$ and $V^2 = uvw = 16$. The answer, however, should be in terms of x, y, and z. Recall

$$yz = u = 2, \qquad zx = v = 2, \qquad xy = w = 4.$$

Hence $x = y$ and $xy = x^2 = w = 4$. Therefore $x = y = 2$ and $z = 1$.

Answer: The maximum volume is 4 ft^3, achieved by a tank of base 2 ft $\times$ 2 ft and height 1 ft.

REMARK: The theory of maxima and minima will be treated with more care in Chapter 9, Section 4.

EXERCISES

Find the possible maximum and minimum values:

1. $z = 4 - 2x^2 - y^2$
2. $z = x^2 + y^2 - 1$
3. $z = (x - 2)^2 + (y + 3)^2$
4. $z = (x - 1)^2 + y^2 + 3$
5. $z = x^2 - 2xy + 2y^2 + 4$
6. $z = xy - x^2 - 2y^2 + x + 2y$
7. $z = x - y^2 - x^3$
8. $z = 3x + 12y - x^3 - y^3$.
9. Find the rectangular solid of greatest volume whose total surface area is 24 ft^2.
10. Find the dimensions of an open-top rectangular box of given volume V, if the surface area is to be a minimum.
11. Find the dimensions of the cheapest open-top rectangular box of given volume V, if the base material costs 3 times as much per ft^2 as the side material.

7. Linear Functions and Matrices

1. INTRODUCTION

In this chapter we shall discuss some parts of linear algebra that are useful in our further study of the calculus of functions of several variables.

Linear algebra can be looked at as the study of one basic functional equation,

$$F(a\mathbf{u} + b\mathbf{v}) = aF(\mathbf{u}) + bF(\mathbf{v}), \tag{1}$$

where a and b are scalars and $\mathbf{u}$ and $\mathbf{v}$ are vectors.

This equation is fundamental in many branches of mathematics. We can think of F as an operation that transforms an "input" $\mathbf{u}$ into an "output" $F(\mathbf{u})$. Then equation (1) says that for input

$$a(\text{input}_1) + b(\text{input}_2),$$

the output is

$$a(\text{output}_1) + b(\text{output}_2).$$

Not every F has this property. Examples: $F(x) = x^2$ for each real number x and $F(\mathbf{u}) = |\mathbf{u}|$ for each vector $\mathbf{u}$ in $\mathbf{R}^3$.

REMARK: If F satisfies (1), then it also satisfies

$$F(a\mathbf{u} + b\mathbf{v} + c\mathbf{w}) = aF(\mathbf{u}) + bF(\mathbf{v}) + cF(\mathbf{w}) \tag{2}$$

(and similar equations for four or more summands). This equation follows in two steps from (1):

$$\begin{aligned} F(a\mathbf{u} + b\mathbf{v} + c\mathbf{w}) &= F(a\mathbf{u} + b\mathbf{v}) + cF(\mathbf{w}) \\ &= [aF(\mathbf{u}) + bF(\mathbf{v})] + cF(\mathbf{w}). \end{aligned}$$

An operation F satisfying (1) is called **linear**. In the most general setting, the inputs (domain) of F may come from any mathematical system in which expressions $a\mathbf{u} + b\mathbf{v}$ make sense. The outputs may be in any other systems with the same property. Such systems are called abstract vector spaces; their study belongs to a course in linear algebra. We shall consider only linear operations with domain one of the spaces $\mathbf{R}$, $\mathbf{R}^2$, $\mathbf{R}^3$ and with values also in one of these spaces, not necessarily the same one. We shall call such operations **linear functions**.

Real-Valued Linear Functions

Let us consider a linear function F defined on three-space $\mathbf{R}^3$, with real values. Thus

$$F\colon \mathbf{R}^3 \longrightarrow \mathbf{R},$$

so F is a function of three variables of the type studied in Chapter 6. However F is subject to an important restriction: it must satisfy (1).

Now what does F look like? From previous experience, we would guess that $F(x, y, z) = px + qy + rz$ is a reasonable possibility. Let us verify that this function is indeed linear.

We set $\mathbf{x}_1 = (x_1, y_1, z_1)$ and $\mathbf{x}_2 = (x_2, y_2, z_2)$. Then

$$\begin{aligned} F(a\mathbf{x}_1 + b\mathbf{x}_2) &= F(ax_1 + bx_2, ay_1 + by_2, az_1 + bz_2) \\ &= p(ax_1 + bx_2) + q(ay_1 + by_2) + r(az_1 + bz_2) \\ &= a(px_1 + qy_1 + rz_1) + b(px_2 + qy_2 + rz_2) \\ &= aF(\mathbf{x}_1) + bF(\mathbf{x}_2), \end{aligned}$$

so F does satisfy (1).

It is important that the converse is also true.

> **Theorem** The most general linear function
>
> $$F\colon \mathbf{R}^3 \longrightarrow \mathbf{R}$$
>
> is given by
>
> $$F(x, y, z) = px + qy + rz,$$
>
> where p, q, r are constants.

Proof: We use the decomposition of an arbitrary vector into a combination of the basic unit vectors:

$$(x, y, z) = x\mathbf{i} + y\mathbf{j} + z\mathbf{k}.$$

From (2) it follows that

$$F(x, y, z) = F(x\mathbf{i} + y\mathbf{j} + z\mathbf{k}) = xF(\mathbf{i}) + yF(\mathbf{j}) + zF(\mathbf{k}).$$

Set $p = F(\mathbf{i})$, $q = F(\mathbf{j})$, $r = F(\mathbf{k})$, three scalars which do not depend on (x, y, z). We have proved that

$$F(x, y, z) = px + qy + rz.$$

REMARK: The proof illustrates an important principle: a linear function $F\colon \mathbf{R}^3 \longrightarrow \mathbf{R}$ is completely determined by its effect on the three vectors $\mathbf{i}$, $\mathbf{j}$, $\mathbf{k}$.

As a corollary of the preceding theorem, we obtain a useful alternative representation for linear functions.

Theorem Let $F\colon \mathbf{R}^3 \longrightarrow \mathbf{R}$ be linear. Then there is a vector $\mathbf{p}$ such that

$$F(\mathbf{x}) = \mathbf{p}\cdot\mathbf{x}.$$

Proof: Just observe that

$$F(\mathbf{x}) = px + qy + rz = (p, q, r)\cdot(x, y, z) = \mathbf{p}\cdot\mathbf{x},$$

where $\mathbf{p} = (p, q, r)$.

An interesting consequence is the following estimate on the order of magnitude of a linear function. It will be useful in the next chapter when we study composite functions.

Theorem Let $F(\mathbf{x}) = \mathbf{p}\cdot\mathbf{x}$. Then

$$|F(\mathbf{x})| \leq |\mathbf{p}|\,|\mathbf{x}|.$$

Proof: Given $\mathbf{x}$, let θ be the angle between $\mathbf{p}$ and $\mathbf{x}$. Then $\mathbf{p}\cdot\mathbf{x} = |\mathbf{p}|\,|\mathbf{x}|\cos\theta$, hence

$$|F(\mathbf{x})| = |\mathbf{p}\cdot\mathbf{x}| = |\mathbf{p}|\,|\mathbf{x}|\,|\cos\theta| \leq |\mathbf{p}|\,|\mathbf{x}|.$$

REMARK: In some discussions it is common to call a function of the form

$$f(x, y, z) = Ax + By + Cz + D$$

a linear function (A, B, C, D are constants) or a linear polynomial. If $D = 0$, then $f(x, y, z)$ is called a *homogeneous* linear function. In this chapter, we shall always mean homogeneous when we refer to linear functions.

EXERCISES

Let $F\colon \mathbf{R}^3 \longrightarrow \mathbf{R}$ be linear:

1. Find $F(x, y, z)$ if $F(1, 0, 0) = 3$, $F(0, 1, 0) = 2$, $F(0, 0, 1) = -5$.
2. Find $F(x, y, z)$ if $F(1, 1, 0) = 0$, $F(0, 1, 1) = 1$, $F(1, 1, 1) = 3$.
3. Show that the set of $\mathbf{x}$ for which $F(\mathbf{x}) = 0$ is a plane through $\mathbf{0}$, provided $F \neq 0$.
4. Prove that $F(\mathbf{x})$ is determined by its values at $\mathbf{i}$, $\mathbf{i} + \mathbf{j}$, and $\mathbf{i} + \mathbf{j} + \mathbf{k}$.
5. Prove that $F(\mathbf{x})$ is not determined by its values at $\mathbf{i}$ and $\mathbf{j}$.
6. Prove that $F(\mathbf{x})$ is continuous.
7. Prove that the sum of two linear functions is linear.
8. Prove that cF is linear if F is linear and c is a scalar.
9. Let g vary over continuous functions on $[0, 1]$ and set

$$L(g) = \int_0^1 g(t)\,dt.$$

Show that L is linear.

10. Let g vary over polynomials and set

$$D(g) = g',$$

the derivative. Show that D is linear.

11. Prove the identity

$$(a_1x_1 + a_2x_2 + a_3x_3)^2 + (a_1x_2 - a_2x_1)^2 + (a_2x_3 - a_3x_2)^2 + (a_3x_1 - a_1x_3)^2$$
$$= (a_1^2 + a_2^2 + a_3^2)(x_1^2 + x_2^2 + x_3^2).$$

12. (cont.) Hence give another proof, free of cosines, that $|\mathbf{a} \cdot \mathbf{x}| \leq |\mathbf{a}|\,|\mathbf{x}|$.

Let $F\colon \mathbf{R}^3 \longrightarrow \mathbf{R}$. Call F an **affine function** if

$$F(a\mathbf{u} + b\mathbf{v}) = aF(\mathbf{u}) + bF(\mathbf{v})$$

whenever $a + b = 1$.

13. Prove that each non-homogeneous linear function is an affine function.

14*. Let F be an affine function. Prove that

$$F(a\mathbf{u} + b\mathbf{v} + c\mathbf{w}) = aF(\mathbf{u}) + bF(\mathbf{v}) + cF(\mathbf{w})$$

if $a + b + c = 1$.

15*. (cont.) Prove that F is a non-homogeneous linear function.

2. LINEAR TRANSFORMATIONS

In this section we study linear functions $F\colon \mathbf{R}^3 \longrightarrow \mathbf{R}^3$. (Linear functions taking values in a space of dimension greater than one are often called **linear transformations.**) Let us begin by describing the most general function of $\mathbf{R}^3$ into $\mathbf{R}^3$, linear or not.

> If $F\colon \mathbf{R}^3 \longrightarrow \mathbf{R}^3$, then for each $\mathbf{x} \in \mathbf{R}^3$,
>
> $$F(\mathbf{x}) = (F_1(\mathbf{x}), F_2(\mathbf{x}), F_3(\mathbf{x})).$$

Thus F is equivalent to a triple F_1, F_2, F_3 of real-valued functions called **coordinate functions** associated with F.

Now, if $F\colon \mathbf{R}^3 \longrightarrow \mathbf{R}^3$ is linear, we can show that each coordinate function $F_j\colon \mathbf{R}^3 \longrightarrow \mathbf{R}$ is linear. We write the basic equation $F(a\mathbf{u} + b\mathbf{v}) = aF(\mathbf{u}) + bF(\mathbf{v})$ in coordinates:

$$(F_1(a\mathbf{u} + b\mathbf{v}), F_2(a\mathbf{u} + b\mathbf{v}), F_3(a\mathbf{u} + b\mathbf{v}))$$
$$= a(F_1(\mathbf{u}), F_2(\mathbf{u}), F_3(\mathbf{u})) + b(F_1(\mathbf{v}), F_2(\mathbf{v}), F_3(\mathbf{v})).$$

Now equate the j-th coordinates:

$$F_j(a\mathbf{u} + b\mathbf{v}) = aF_j(\mathbf{u}) + bF_j(\mathbf{v}).$$

Therefore $F_j\colon \mathbf{R}^3 \longrightarrow \mathbf{R}$ and F_j is linear. By the second theorem of Section 1, there is a vector $\mathbf{p}_j$ such that $F_j(\mathbf{x}) = \mathbf{p}_j \cdot \mathbf{x}$.

Theorem Let $\mathbf{R}^3 \longrightarrow \mathbf{R}^3$ be linear. Then there are vectors $\mathbf{p}_1$, $\mathbf{p}_2$, $\mathbf{p}_3$ such that

$$F(\mathbf{x}) = (\mathbf{p}_1 \cdot \mathbf{x},\ \mathbf{p}_2 \cdot \mathbf{x},\ \mathbf{p}_3 \cdot \mathbf{x}).$$

A Bound for F

By analogy with our procedure in Section 1, we may use the preceding theorem to obtain a bound for $|F(\mathbf{x})|$.

Theorem Let $F: \mathbf{R}^3 \longrightarrow \mathbf{R}^3$ be a linear function. Then there is a constant c such that

$$|F(\mathbf{x})| \leq c\,|\mathbf{x}|.$$

Proof: We have

$$F(\mathbf{x}) = (\mathbf{p}_1 \cdot \mathbf{x},\ \mathbf{p}_2 \cdot \mathbf{x},\ \mathbf{p}_3 \cdot \mathbf{x})$$

and we know that $|\mathbf{p}_j \cdot \mathbf{x}| \leq |\mathbf{p}_j|\,|\mathbf{x}|$. Therefore

$$\begin{aligned} |F(\mathbf{x})|^2 &= |\mathbf{p}_1 \cdot \mathbf{x}|^2 + |\mathbf{p}_2 \cdot \mathbf{x}|^2 + |\mathbf{p}_3 \cdot \mathbf{x}|^2 \\ &\leq |\mathbf{p}_1|^2\,|\mathbf{x}|^2 + |\mathbf{p}_2|^2\,|\mathbf{x}|^2 + |\mathbf{p}_3|^2\,|\mathbf{x}|^2 = c^2\,|\mathbf{x}|^2, \end{aligned}$$

where $c^2 = |\mathbf{p}_1|^2 + |\mathbf{p}_2|^2 + |\mathbf{p}_3|^2$. Hence $|F(\mathbf{x})| \leq c\,|\mathbf{x}|$.

The Matrix of F

Again consider a linear function $F: \mathbf{R}^3 \longrightarrow \mathbf{R}^3$. Let

$$F(\mathbf{x}) = (\mathbf{p}_1 \cdot \mathbf{x},\ \mathbf{p}_2 \cdot \mathbf{x},\ \mathbf{p}_3 \cdot \mathbf{x}).$$

Each vector $\mathbf{p}_j$ has three components:

$$\begin{cases} \mathbf{p}_1 = (p_1, q_1, r_1) \\ \mathbf{p}_2 = (p_2, q_2, r_2) \\ \mathbf{p}_3 = (p_3, q_3, r_3). \end{cases}$$

The function F is specified by the three vectors $\mathbf{p}_1$, $\mathbf{p}_2$, $\mathbf{p}_3$, and each of these is specified by its three components. Therefore F is determined by the square array of 9 numbers

$$A = \begin{bmatrix} p_1 & q_1 & r_1 \\ p_2 & q_2 & r_2 \\ p_3 & q_3 & r_3 \end{bmatrix}.$$

Such an array is called a 3×3 **matrix.** In this way each linear function $F: \mathbf{R}^3 \longrightarrow \mathbf{R}^3$ gives rise to a 3×3 matrix A.

We have seen that the *rows* of the matrix A are the vectors $\mathbf{p}_1, \mathbf{p}_2, \mathbf{p}_3$ written in components. There is also an interesting interpretation of the *columns* of A.

Recall that $F(\mathbf{x}) = (\mathbf{p}_1 \cdot \mathbf{x}, \mathbf{p}_2 \cdot \mathbf{x}, \mathbf{p}_3 \cdot \mathbf{x})$. In particular $F(\mathbf{i}) = (\mathbf{p}_1 \cdot \mathbf{i}, \mathbf{p}_2 \cdot \mathbf{i}, \mathbf{p}_3 \cdot \mathbf{i})$. Now

$$\mathbf{p}_1 = p_1\mathbf{i} + p_2\mathbf{j} + p_3\mathbf{k}, \qquad \text{so} \quad \mathbf{p}_1 \cdot \mathbf{i} = p_1.$$

Similarly, $\mathbf{p}_2 \cdot \mathbf{i} = p_2$ and $\mathbf{p}_3 \cdot \mathbf{i} = p_3$. Hence, $F(\mathbf{i}) = (p_1, p_2, p_3)$. If this vector were written as a column

$$\begin{bmatrix} p_1 \\ p_2 \\ p_3 \end{bmatrix}$$

instead of a row, it would match the first column of A.

Similarly we find that $F(\mathbf{j})$ and $F(\mathbf{k})$ correspond to

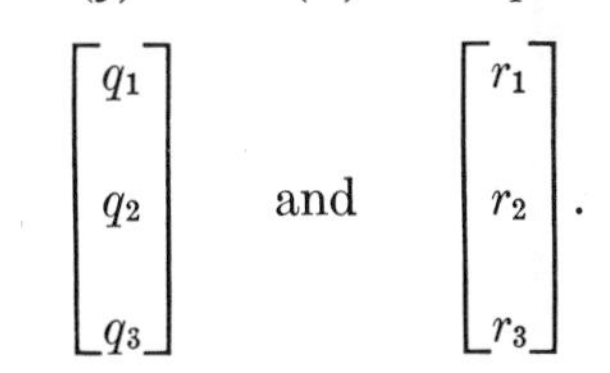

$$\begin{bmatrix} q_1 \\ q_2 \\ q_3 \end{bmatrix} \quad \text{and} \quad \begin{bmatrix} r_1 \\ r_2 \\ r_3 \end{bmatrix}.$$

> Let $F: \mathbf{R}^3 \longrightarrow \mathbf{R}^3$ be a linear function. Then F determines a 3×3 matrix A whose columns represent the vectors $F(\mathbf{i}), F(\mathbf{j}), F(\mathbf{k})$ respectively.
>
> Conversely, each 3×3 matrix A determines a linear function $F: \mathbf{R}^3 \longrightarrow \mathbf{R}^3$ as follows:
>
> If $\mathbf{c}_1, \mathbf{c}_2, \mathbf{c}_3$ are the vectors represented by the columns of A, then F is determined by $F(\mathbf{i}) = \mathbf{c}_1$, $F(\mathbf{j}) = \mathbf{c}_2$, $F(\mathbf{k}) = \mathbf{c}_3$.

Geometric Interpretation

Let $F: \mathbf{R}^3 \longrightarrow \mathbf{R}^3$ be linear, so F assigns to each point $\mathbf{x}$ in 3-space another point $F(\mathbf{x})$ in 3-space. Thus F can be looked at geometrically as a mapping of space into itself.

EXAMPLE 2.1

Describe geometrically the transformations

(a) $F(x, y, z) = (x, y, 0)$

(b) $F(x, y, z) = (x, 2y, z)$.

In each case, find the associated matrix.

Solution: (a) The transformation assigns to $x\mathbf{i} + y\mathbf{j} + z\mathbf{k}$ the vector $x\mathbf{i} + y\mathbf{j}$. In other words F projects the vector $\mathbf{x}$ perpendicularly onto the x, y-plane. See Fig. 2.1a. Clearly $F(\mathbf{i}) = (1, 0, 0)$, $F(\mathbf{j}) = (0, 1, 0)$ and $F(\mathbf{k}) = (0, 0, 0)$. The corresponding column vectors are the columns of the associated matrix:

$$\begin{bmatrix} 1 & 0 & 0 \\ 0 & 1 & 0 \\ 0 & 0 & 0 \end{bmatrix}.$$

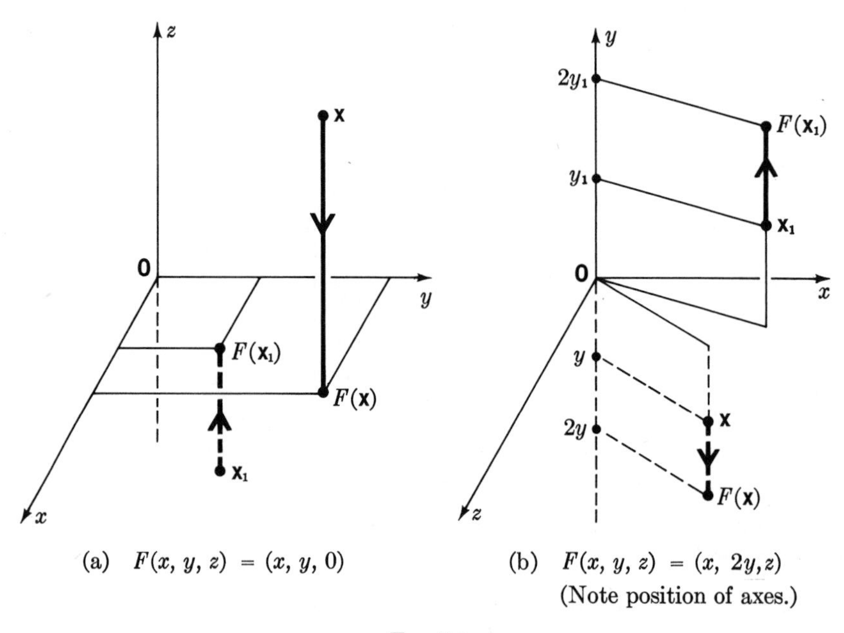

(a) $F(x, y, z) = (x, y, 0)$

(b) $F(x, y, z) = (x, 2y, z)$
(Note position of axes.)

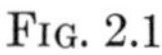

FIG. 2.1

(b) The transformation doubles the y-component of $\mathbf{x}$. Geometrically this amounts to a stretching of space by a factor of 2 in the y-direction. See Fig. 2.1b. We note that $F(\mathbf{i}) = (1, 0, 0)$, $F(\mathbf{j}) = (0, 2, 0)$, $F(\mathbf{k}) = (0, 0, 1)$, so the associated matrix is

$$\begin{bmatrix} 1 & 0 & 0 \\ 0 & 2 & 0 \\ 0 & 0 & 1 \end{bmatrix}.$$

Answer: (a) projection onto the x, y-plane; (b) stretching parallel to the y-axis by a factor of 2;

$$\begin{bmatrix} 1 & 0 & 0 \\ 0 & 1 & 0 \\ 0 & 0 & 0 \end{bmatrix}, \qquad \begin{bmatrix} 1 & 0 & 0 \\ 0 & 2 & 0 \\ 0 & 0 & 1 \end{bmatrix}.$$

EXERCISES

1. Let F be linear with matrix $A = [a_{ij}]$. Prove that $|F(\mathbf{x})| \leq c\,|\mathbf{x}|$ with $c^2 = \sum a_{ij}^2$.
2. Let F and G be linear with respective matrices A and B. Prove that $F + G$ is linear and find its matrix.
3. Let $\mathbf{a}$ be a vector. Prove that $F(\mathbf{x}) = \mathbf{a} \times \mathbf{x}$ is linear.
4. (cont.) Find the corresponding matrix.

Show that F is not linear:

5. $F(x, y, z) = (x, y, 1)$
6. $F(x, y, z) = (xy, 0, 0)$.

Find the matrix corresponding to the linear function:

7. $F(x, y, z) = (x, y, x + y + z)$
8. $F(x, y, z) = (2x - y - 3z, 4x + 5y + 6z, y - 2z)$
9. $F(\mathbf{i}) = \mathbf{0}$, $F(\mathbf{j}) = \mathbf{i} + \mathbf{j} + 2\mathbf{k}$, $F(\mathbf{k}) = 4\mathbf{k}$
10. $F(\mathbf{i}) = \mathbf{j}$, $F(\mathbf{j}) = \mathbf{k}$, $F(\mathbf{k}) = \mathbf{i}$.

For each matrix describe geometrically the corresponding linear function:

11. $\begin{bmatrix} 1 & 0 & 0 \\ 0 & 1 & 0 \\ 0 & 0 & 1 \end{bmatrix}$
12. $\begin{bmatrix} 0 & 0 & 0 \\ 0 & 0 & 0 \\ 0 & 0 & 0 \end{bmatrix}$
13. $\begin{bmatrix} 1 & 0 & 0 \\ 0 & 0 & 0 \\ 0 & 0 & 1 \end{bmatrix}$
14. $\begin{bmatrix} 0 & 0 & 0 \\ 0 & 0 & 0 \\ 0 & 0 & 1 \end{bmatrix}$
15. $\begin{bmatrix} -1 & 0 & 0 \\ 0 & -1 & 0 \\ 0 & 0 & -1 \end{bmatrix}$
16. $\begin{bmatrix} -1 & 0 & 0 \\ 0 & -1 & 0 \\ 0 & 0 & 1 \end{bmatrix}$
17. $\begin{bmatrix} 1 & 0 & 0 \\ 0 & 1 & 0 \\ 0 & 0 & -1 \end{bmatrix}$
18. $\begin{bmatrix} \cos\alpha & -\sin\alpha & 0 \\ \sin\alpha & \cos\alpha & 0 \\ 0 & 0 & 1 \end{bmatrix}$

19. $\begin{bmatrix} 2 & 0 & 0 \\ 0 & 3 & 0 \\ 0 & 0 & -1 \end{bmatrix}$ 20. $\begin{bmatrix} 0 & 1 & 0 \\ 1 & 0 & 0 \\ 0 & 0 & 1 \end{bmatrix}$.

Let $F\colon \mathbf{R}^3 \longrightarrow \mathbf{R}^3$ be linear:

21. If $F(\mathbf{i})$, $F(\mathbf{j})$, $F(\mathbf{k})$ all lie in a plane $\mathbf{P}$ passing through $\mathbf{0}$, show that $F(\mathbf{x})$ lies in $\mathbf{P}$ for every $\mathbf{x} \in \mathbf{R}^3$.
22. If $F(\mathbf{i})$, $F(\mathbf{j})$, $F(\mathbf{k})$ all lie on a line $\mathbf{L}$ passing through $\mathbf{0}$, show that $F(\mathbf{x})$ lies on $\mathbf{L}$ for every $\mathbf{x} \in \mathbf{R}^3$.
23. Let $F\colon \mathbf{R}^3 \longrightarrow \mathbf{R}^3$ be linear and let $\mathbf{a} \in \mathbf{R}^3$. Prove that $\mathbf{x} \longrightarrow \mathbf{a} \cdot F(\mathbf{x})$ is linear on $\mathbf{R}^3$ to $\mathbf{R}$.
24. (cont.) Suppose $F\colon \mathbf{R}^3 \longrightarrow \mathbf{R}^3$ and $\mathbf{x} \longrightarrow \mathbf{a} \cdot F(\mathbf{x})$ is linear for each $\mathbf{a} \in \mathbf{R}^3$. Prove that F is linear.

3. MATRIX CALCULATIONS

In Section 2, we saw the advantage of expressing vectors not only as row vectors,

$$\mathbf{a} = (a_1, a_2, a_3),$$

but as column vectors

$$\mathbf{a}' = \begin{bmatrix} a_1 \\ a_2 \\ a_3 \end{bmatrix}.$$

We shall denote vectors in column form by primed letters.

In this section we introduce various kinds of operations involving rows, columns, and matrices. These facilitate the study of linear functions in Section 4.

Row-by-Column Multiplication

The product of a row vector by a column vector is the scalar defined by

$$\mathbf{a}\mathbf{b}' = (a_1, a_2, a_3) \begin{bmatrix} b_1 \\ b_2 \\ b_3 \end{bmatrix} = a_1b_1 + a_2b_2 + a_3b_3.$$

It is an alternative form of the dot product of two vectors. Note that this product is taken with the row first, then the column.

Now let

$$B = \begin{bmatrix} b_{11} & b_{12} & b_{13} \\ b_{21} & b_{22} & b_{23} \\ b_{31} & b_{32} & b_{33} \end{bmatrix}$$

be a 3×3 matrix and $\mathbf{a} = (a_1, a_2, a_3)$ a row vector. We can multiply B on the left by $\mathbf{a}$, taking the product of $\mathbf{a}$ with each column of B:

$$\begin{aligned} \mathbf{a}B &= (a_1, a_2, a_3) \begin{bmatrix} b_{11} & b_{12} & b_{13} \\ b_{21} & b_{22} & b_{23} \\ b_{31} & b_{32} & b_{33} \end{bmatrix} \\ &= (a_1b_{11} + a_2b_{21} + a_3b_{31}, a_1b_{12} + a_2b_{22} + a_3b_{32}, a_1b_{13} + a_2b_{23} + a_3b_{33}). \end{aligned}$$

Similarly we can multiply B on the *right* by a *column* vector $\mathbf{c}'$, taking the product of each row of B with $\mathbf{c}'$:

$$B\mathbf{c}' = \begin{bmatrix} b_{11} & b_{12} & b_{13} \\ b_{21} & b_{22} & b_{23} \\ b_{31} & b_{32} & b_{33} \end{bmatrix} \begin{bmatrix} c_1 \\ c_2 \\ c_3 \end{bmatrix} = \begin{bmatrix} b_{11}c_1 + b_{12}c_2 + b_{13}c_3 \\ b_{21}c_1 + b_{22}c_2 + b_{23}c_3 \\ b_{31}c_1 + b_{32}c_2 + b_{33}c_3 \end{bmatrix}.$$

The answer is a row vector in the first case, a column vector in the second.

Examples:

$$(3, -1, 1) \begin{bmatrix} 1 & 2 & 0 \\ -1 & -1 & 2 \\ 4 & -2 & -1 \end{bmatrix} = (8, 5, -3),$$

$$\begin{bmatrix} 1 & 2 & 0 \\ -1 & -1 & 2 \\ 4 & -2 & -1 \end{bmatrix} \begin{bmatrix} 3 \\ -1 \\ 1 \end{bmatrix} = \begin{bmatrix} 1 \\ 0 \\ 13 \end{bmatrix}.$$

Sometimes it is convenient to abbreviate the notation. We can write B as a *row* of three *column vectors*:

$$B = (\mathbf{b}_1', \mathbf{b}_2', \mathbf{b}_3'), \qquad \text{where } \mathbf{b}_1' = \begin{bmatrix} b_{11} \\ b_{21} \\ b_{31} \end{bmatrix}, \quad \mathbf{b}_2' = \begin{bmatrix} b_{12} \\ b_{22} \\ b_{32} \end{bmatrix}, \quad \mathbf{b}_3' = \begin{bmatrix} b_{13} \\ b_{23} \\ b_{33} \end{bmatrix}.$$

Then $\mathbf{a}B$ is a row vector:

$$\mathbf{a}B = \mathbf{a}(\mathbf{b}_1', \mathbf{b}_2', \mathbf{b}_3') = (\mathbf{a}\mathbf{b}_1', \mathbf{a}\mathbf{b}_2', \mathbf{a}\mathbf{b}_3').$$

Each $\mathbf{a}\mathbf{b}_j'$ is a scalar, a row-by-column product. Similarly we can write B as a *column* of three *row vectors*:

$$B = \begin{bmatrix} \mathbf{b}_1 \\ \mathbf{b}_2 \\ \mathbf{b}_3 \end{bmatrix}, \qquad \text{where} \quad \begin{cases} \mathbf{b}_1 = (b_{11}, b_{12}, b_{13}) \\ \mathbf{b}_2 = (b_{21}, b_{22}, b_{23}). \\ \mathbf{b}_3 = (b_{31}, b_{32}, b_{33}) \end{cases}$$

Then $B\mathbf{c}'$ is a column vector:

$$B\mathbf{c} = \begin{bmatrix} \mathbf{b}_1 \\ \mathbf{b}_2 \\ \mathbf{b}_3 \end{bmatrix} \mathbf{c}' = \begin{bmatrix} \mathbf{b}_1\mathbf{c}' \\ \mathbf{b}_2\mathbf{c}' \\ \mathbf{b}_3\mathbf{c}' \end{bmatrix}$$

Row × Matrix × Column

Let $\mathbf{a}$ be a row vector, B a matrix, and $\mathbf{c}'$ a column vector. Then $B\mathbf{c}'$ is a column vector, so we can form the scalar $\mathbf{a}(B\mathbf{c}')$. We can also form the scalar $(\mathbf{a}B)\mathbf{c}'$. It is an important fact (an associative law) that these two are equal:

$$\mathbf{a}(B\mathbf{c}') = (\mathbf{a}B)\mathbf{c}'.$$

To prove it, we simply compute the product in either order. With the notation above,

$$\mathbf{a}(B\mathbf{c}') = (a_1, a_2, a_3) \begin{bmatrix} \sum b_{1j}c_j \\ \sum b_{2j}c_j \\ \sum b_{3j}c_j \end{bmatrix} = a_1 \sum b_{1j}c_j + a_2 \sum b_{2j}c_j + a_3 \sum b_{3j}c_j$$

$$= \sum_{i=1}^{3} a_i \sum_{j=1}^{3} b_{ij}c_j = \sum_{i,j} a_i b_{ij} c_j.$$

The double sum contains 9 summands, and it does not depend on whether we sum first on j or on i. If we sum first on i, then on j, we obtain

$$\mathbf{a}(B\mathbf{c}') = \sum_{i,j} a_i b_{ij} c_j = \sum_j \left(\sum_i a_i b_{ij} \right) c_j$$

$$= \left(\sum a_i b_{i1}, \sum a_i b_{i2}, \sum a_i b_{i3} \right) \begin{bmatrix} c_1 \\ c_2 \\ c_3 \end{bmatrix} = (\mathbf{a}B)\mathbf{c}'.$$

If $\mathbf{a}$ is a row vector, $\mathbf{c}'$ is a column vector, and B is a matrix, then

$$\mathbf{a}(B\mathbf{c}') = (\mathbf{a}B)\mathbf{c}' = \sum_{i,j} a_i b_{ij} c_j.$$

Product of Matrices

It is time to define the product of two 3×3 matrices A and B. As you might suspect by now, we shall define AB in terms of row-by-column products. Let $\mathbf{r}_1$, $\mathbf{r}_2$, $\mathbf{r}_3$ be the rows of A and let $\mathbf{c}_1'$, $\mathbf{c}_2'$, $\mathbf{c}_3'$ be the columns of B, so

$$A = \begin{bmatrix} \mathbf{r}_1 \\ \mathbf{r}_2 \\ \mathbf{r}_3 \end{bmatrix} \quad \text{and} \quad B = (\mathbf{c}_1', \mathbf{c}_2', \mathbf{c}_3').$$

We define

$$AB = \begin{bmatrix} \mathbf{r}_1 \\ \mathbf{r}_2 \\ \mathbf{r}_3 \end{bmatrix} (\mathbf{c}_1', \mathbf{c}_2', \mathbf{c}_3') = \begin{bmatrix} \mathbf{r}_1\mathbf{c}_1' & \mathbf{r}_1\mathbf{c}_2' & \mathbf{r}_1\mathbf{c}_3' \\ \mathbf{r}_2\mathbf{c}_1' & \mathbf{r}_2\mathbf{c}_2' & \mathbf{r}_2\mathbf{c}_3' \\ \mathbf{r}_3\mathbf{c}_1' & \mathbf{r}_3\mathbf{c}_2' & \mathbf{r}_3\mathbf{c}_3' \end{bmatrix}.$$

Thus the product AB is another 3×3 matrix. The number in the i, j-th position of AB is $\mathbf{r}_i\mathbf{c}_j'$, the product of the i-th row of A by the j-th column of B.

Example: $C =$

$$\begin{bmatrix} 2 & 0 & -1 \\ -1 & 3 & -2 \\ 2 & 4 & 4 \end{bmatrix} \begin{bmatrix} 3 & -3 & -3 \\ 1 & 2 & 1 \\ 5 & 1 & -4 \end{bmatrix} = \begin{bmatrix} (\ 2, & 0, & -1) \\ (-1, & 3, & -2) \\ (\ 2, & 4, & 4) \end{bmatrix} \left[\begin{bmatrix} 3 \\ 1 \\ 5 \end{bmatrix} \begin{bmatrix} -3 \\ 2 \\ 1 \end{bmatrix} \begin{bmatrix} -3 \\ 1 \\ -4 \end{bmatrix} \right].$$

Then

$$c_{11} = (2, 0, -1)\begin{bmatrix} 3 \\ 1 \\ 5 \end{bmatrix} = 1, \qquad c_{32} = (2, 4, 4)\begin{bmatrix} -3 \\ 2 \\ 1 \end{bmatrix} = 6, \qquad \text{etc.}$$

and we find

$$C = \begin{bmatrix} 1 & -7 & -2 \\ -10 & 7 & 14 \\ 30 & 6 & -18 \end{bmatrix}.$$

WARNING: Multiplication of matrices is *not commutative*! In general, $AB \neq BA$. For example, multiply the matrices of the example above in the opposite order:

$$\begin{bmatrix} 3 & -3 & 3 \\ 1 & 2 & 1 \\ 5 & 1 & -4 \end{bmatrix}\begin{bmatrix} 2 & 0 & -1 \\ -1 & 3 & -2 \\ 2 & 4 & 4 \end{bmatrix} = \begin{bmatrix} 15 & 3 & 15 \\ 2 & 10 & -1 \\ 1 & -13 & -23 \end{bmatrix}.$$

The product is totally different.

Despite the failure of the commutative law for multiplication of matrices, the associative law does hold.

> **Associative Law** If A, B, C are 3×3 matrices, then
>
> $$A(BC) = (AB)C.$$

The proof is given in the next section.

Transpose

The **transpose** A' of a matrix A is obtained by flipping A across its main diagonal (mirror reflection in the main diagonal; interchange of rows and columns):

$$\begin{bmatrix} a & b \\ c & d \end{bmatrix}' = \begin{bmatrix} a & c \\ b & d \end{bmatrix}, \qquad \begin{bmatrix} a_{11} & a_{12} & a_{13} \\ a_{21} & a_{22} & a_{23} \\ a_{31} & a_{32} & a_{33} \end{bmatrix}' = \begin{bmatrix} a_{11} & a_{21} & a_{31} \\ a_{12} & a_{22} & a_{32} \\ a_{13} & a_{23} & a_{33} \end{bmatrix}.$$

We also define transpose for row and column vectors:

$$(a, b, c)' = \begin{bmatrix} a \\ b \\ c \end{bmatrix}, \qquad \begin{bmatrix} a \\ b \\ c \end{bmatrix}' = (a, b, c).$$

In any case, if the transpose operation is performed twice in succession, we return to where we started:

$$(\mathbf{x}')' = \mathbf{x}, \qquad (A')' = A, \qquad [(\mathbf{y}')']' = \mathbf{y}'.$$

There is an important connection between matrix multiplication and transposing.

Let $\mathbf{x}$ and $\mathbf{y}$ be row vectors and A and B matrices. Then

$$\mathbf{x}\mathbf{y}' = \mathbf{y}\mathbf{x}', \qquad (\mathbf{x}A)' = A'\mathbf{x}', \qquad (A\mathbf{y}')' = \mathbf{y}A',$$

$$\mathbf{x}A\mathbf{y}' = \mathbf{y}A'\mathbf{x}', \qquad (AB)' = B'A'.$$

The proofs of these relations are straightforward and are left as exercises.

Further Operations

Sometimes it is necessary to add two 3×3 matrices or multiply a matrix by a scalar. The definitions of these operations are quite natural:

$$\begin{bmatrix} a_{11} & a_{12} & a_{13} \\ a_{21} & a_{22} & a_{23} \\ a_{31} & a_{32} & a_{33} \end{bmatrix} + \begin{bmatrix} b_{11} & b_{12} & b_{13} \\ b_{21} & b_{22} & b_{23} \\ b_{31} & b_{32} & b_{33} \end{bmatrix} = \begin{bmatrix} a_{11}+b_{11} & a_{12}+b_{12} & a_{13}+b_{13} \\ a_{21}+b_{21} & a_{22}+b_{22} & a_{23}+b_{23} \\ a_{31}+b_{31} & a_{32}+b_{32} & a_{33}+b_{33} \end{bmatrix},$$

$$c\begin{bmatrix} a_{11} & a_{12} & a_{13} \\ a_{21} & a_{22} & a_{23} \\ a_{31} & a_{32} & a_{33} \end{bmatrix} = \begin{bmatrix} ca_{11} & ca_{12} & ca_{13} \\ ca_{21} & ca_{22} & ca_{23} \\ ca_{31} & ca_{32} & ca_{33} \end{bmatrix}.$$

WARNING: In the second definition, each one of the 9 entries of the matrix is multiplied by c. This is different from the corresponding rule for determinants, where multiplying by c is equivalent to multiplying just one row or one column by c.

REMARK: It should be clear that everything in this section applies not only to 3×3 matrices, but also to 2×2 or $n \times n$ matrices as well.

EXERCISES

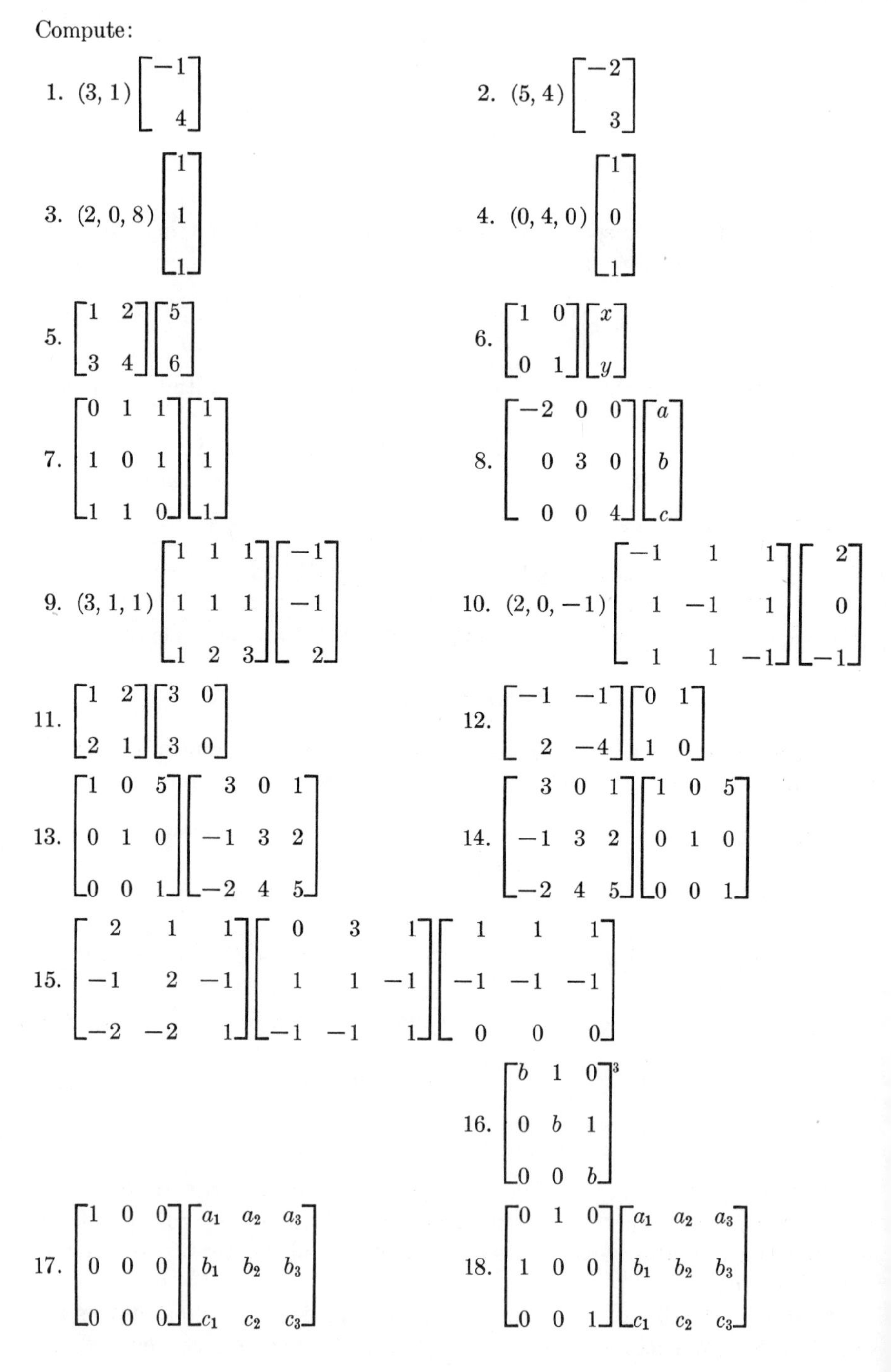

Compute:

1. $(3, 1)\begin{bmatrix} -1 \\ 4 \end{bmatrix}$

2. $(5, 4)\begin{bmatrix} -2 \\ 3 \end{bmatrix}$

3. $(2, 0, 8)\begin{bmatrix} 1 \\ 1 \\ 1 \end{bmatrix}$

4. $(0, 4, 0)\begin{bmatrix} 1 \\ 0 \\ 1 \end{bmatrix}$

5. $\begin{bmatrix} 1 & 2 \\ 3 & 4 \end{bmatrix}\begin{bmatrix} 5 \\ 6 \end{bmatrix}$

6. $\begin{bmatrix} 1 & 0 \\ 0 & 1 \end{bmatrix}\begin{bmatrix} x \\ y \end{bmatrix}$

7. $\begin{bmatrix} 0 & 1 & 1 \\ 1 & 0 & 1 \\ 1 & 1 & 0 \end{bmatrix}\begin{bmatrix} 1 \\ 1 \\ 1 \end{bmatrix}$

8. $\begin{bmatrix} -2 & 0 & 0 \\ 0 & 3 & 0 \\ 0 & 0 & 4 \end{bmatrix}\begin{bmatrix} a \\ b \\ c \end{bmatrix}$

9. $(3, 1, 1)\begin{bmatrix} 1 & 1 & 1 \\ 1 & 1 & 1 \\ 1 & 2 & 3 \end{bmatrix}\begin{bmatrix} -1 \\ -1 \\ 2 \end{bmatrix}$

10. $(2, 0, -1)\begin{bmatrix} -1 & 1 & 1 \\ 1 & -1 & 1 \\ 1 & 1 & -1 \end{bmatrix}\begin{bmatrix} 2 \\ 0 \\ -1 \end{bmatrix}$

11. $\begin{bmatrix} 1 & 2 \\ 2 & 1 \end{bmatrix}\begin{bmatrix} 3 & 0 \\ 3 & 0 \end{bmatrix}$

12. $\begin{bmatrix} -1 & -1 \\ 2 & -4 \end{bmatrix}\begin{bmatrix} 0 & 1 \\ 1 & 0 \end{bmatrix}$

13. $\begin{bmatrix} 1 & 0 & 5 \\ 0 & 1 & 0 \\ 0 & 0 & 1 \end{bmatrix}\begin{bmatrix} 3 & 0 & 1 \\ -1 & 3 & 2 \\ -2 & 4 & 5 \end{bmatrix}$

14. $\begin{bmatrix} 3 & 0 & 1 \\ -1 & 3 & 2 \\ -2 & 4 & 5 \end{bmatrix}\begin{bmatrix} 1 & 0 & 5 \\ 0 & 1 & 0 \\ 0 & 0 & 1 \end{bmatrix}$

15. $\begin{bmatrix} 2 & 1 & 1 \\ -1 & 2 & -1 \\ -2 & -2 & 1 \end{bmatrix}\begin{bmatrix} 0 & 3 & 1 \\ 1 & 1 & -1 \\ -1 & -1 & 1 \end{bmatrix}\begin{bmatrix} 1 & 1 & 1 \\ -1 & -1 & -1 \\ 0 & 0 & 0 \end{bmatrix}$

16. $\begin{bmatrix} b & 1 & 0 \\ 0 & b & 1 \\ 0 & 0 & b \end{bmatrix}^3$

17. $\begin{bmatrix} 1 & 0 & 0 \\ 0 & 0 & 0 \\ 0 & 0 & 0 \end{bmatrix}\begin{bmatrix} a_1 & a_2 & a_3 \\ b_1 & b_2 & b_3 \\ c_1 & c_2 & c_3 \end{bmatrix}$

18. $\begin{bmatrix} 0 & 1 & 0 \\ 1 & 0 & 0 \\ 0 & 0 & 1 \end{bmatrix}\begin{bmatrix} a_1 & a_2 & a_3 \\ b_1 & b_2 & b_3 \\ c_1 & c_2 & c_3 \end{bmatrix}$

19. $\begin{bmatrix} 0 & a & b \\ 0 & 0 & c \\ 0 & 0 & 0 \end{bmatrix}^3$

20. $\begin{bmatrix} a & 0 & 0 \\ 0 & b & 0 \\ 0 & 0 & c \end{bmatrix}^5$

21. $(x, y, z)\begin{bmatrix} 1 & 0 & 0 \\ 0 & 2 & 0 \\ 0 & 0 & 3 \end{bmatrix}\begin{bmatrix} x \\ y \\ z \end{bmatrix}$

22. $(x, y, z)\begin{bmatrix} 1 & 2 & 0 \\ 2 & 3 & 0 \\ 0 & 0 & 4 \end{bmatrix}\begin{bmatrix} x \\ y \\ z \end{bmatrix}.$

Compute $A^2 + 3A$:

23 $A = \begin{bmatrix} -3 & 0 \\ 0 & -3 \end{bmatrix}$

24. $A = \begin{bmatrix} 1 & 2 & 0 \\ 0 & -3 & 1 \\ 1 & 1 & -1 \end{bmatrix}.$

25. Let 0 be the 3×3 matrix all of whose entries are zero. If A and B are 3×3 matrices and $AB = 0$, does it follow that either $A = 0$ or $B = 0$?

Prove

26. $\mathbf{x}\mathbf{y}' = \mathbf{y}\mathbf{x}'$

27. $(\mathbf{x}A)' = A'\mathbf{x}'$

28 $(A\mathbf{y}')' = \mathbf{y}A'$

29. $\mathbf{x}A\mathbf{y}' = \mathbf{y}A'\mathbf{x}'$

30. $(AB)' = B'A'$

31. $(A + B)' = A' + B'.$

4. APPLICATIONS

We return to linear functions of $\mathbf{R}^3$ into $\mathbf{R}^3$ and apply vector and matrix techniques. It will be convenient to write all vectors in columns. Thus we shall think of $\mathbf{R}^3$ as the set of all column vectors with three elements.

Suppose $F: \mathbf{R}^3 \longrightarrow \mathbf{R}^3$ is linear. According to Section 2, there exist vectors $\mathbf{p}_1, \mathbf{p}_2, \mathbf{p}_3$ such that $F(\mathbf{x}) = (\mathbf{p}_1 \cdot \mathbf{x}, \mathbf{p}_2 \cdot \mathbf{x}, \mathbf{p}_3 \cdot \mathbf{x})$. But we were going to use column vectors instead of row vectors, so we re-write this formula as follows:

$$F(\mathbf{x}') = \begin{bmatrix} \mathbf{p}_1\mathbf{x}' \\ \mathbf{p}_2\mathbf{x}' \\ \mathbf{p}_3\mathbf{x}' \end{bmatrix},$$

where $\mathbf{p}_1, \mathbf{p}_2, \mathbf{p}_3$ are *row* vectors. Recall that in Section 2 we associated with F the 3×3 matrix A whose rows are $\mathbf{p}_1, \mathbf{p}_2, \mathbf{p}_3$. Therefore the last formula is simply $F(\mathbf{x}') = A\mathbf{x}'$.

> If $F\colon \mathbf{R}^3 \longrightarrow \mathbf{R}^3$ is linear and if A is the matrix of F, then
> $$F(\mathbf{x}') = A\mathbf{x}'$$
> for each column vector $\mathbf{x}'$.

REMARK: The bound $|F(\mathbf{x}')| \le c\,|\mathbf{x}'|$ that we derived before can now be written for the product $A\mathbf{x}'$ of a matrix by a vector:

$$|A\mathbf{x}'| \le c\,|\mathbf{x}'|,$$

where c is a constant independent of $\mathbf{x}'$.

The formula $F(\mathbf{x}') = A\mathbf{x}'$ provides a practical algorithm for computing the values of a linear function.

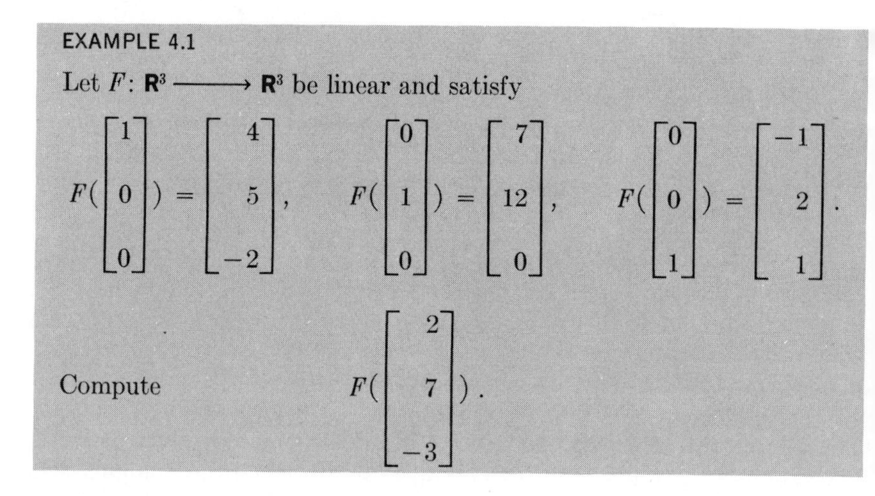

EXAMPLE 4.1

Let $F\colon \mathbf{R}^3 \longrightarrow \mathbf{R}^3$ be linear and satisfy

$$F\left(\begin{bmatrix}1\\0\\0\end{bmatrix}\right) = \begin{bmatrix}4\\5\\-2\end{bmatrix}, \qquad F\left(\begin{bmatrix}0\\1\\0\end{bmatrix}\right) = \begin{bmatrix}7\\12\\0\end{bmatrix}, \qquad F\left(\begin{bmatrix}0\\0\\1\end{bmatrix}\right) = \begin{bmatrix}-1\\2\\1\end{bmatrix}.$$

Compute
$$F\left(\begin{bmatrix}2\\7\\-3\end{bmatrix}\right).$$

Solution: Let A be the matrix of F. In Section 2, we showed that the columns of A are

$$F\left(\begin{bmatrix}1\\0\\0\end{bmatrix}\right), \qquad F\left(\begin{bmatrix}0\\1\\0\end{bmatrix}\right), \qquad F\left(\begin{bmatrix}0\\0\\1\end{bmatrix}\right)$$

(only there we used the notation $F(\mathbf{i})$, $F(\mathbf{j})$, $F(\mathbf{k})$).

Consequently

$$A = \begin{bmatrix}4 & 7 & -1\\5 & 12 & 2\\-2 & 0 & 1\end{bmatrix}, \text{ so } F\left(\begin{bmatrix}2\\7\\-3\end{bmatrix}\right) = \begin{bmatrix}4 & 7 & -1\\5 & 12 & 2\\-2 & 0 & 1\end{bmatrix}\begin{bmatrix}2\\7\\-3\end{bmatrix} = \begin{bmatrix}60\\88\\-7\end{bmatrix}.$$

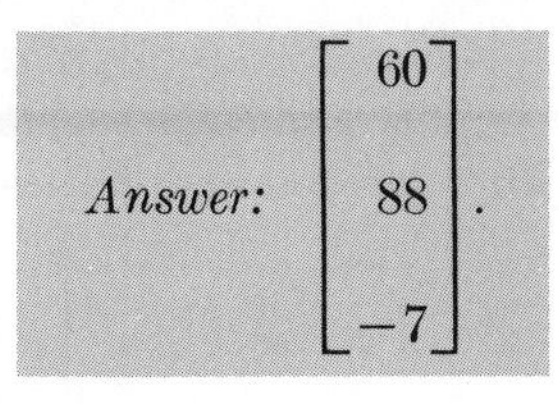

Answer: $\begin{bmatrix} 60 \\ 88 \\ -7 \end{bmatrix}$.

Composition of Linear Functions

Let F and G be linear functions with corresponding matrices A and B:

$$F(\mathbf{x}') = A\mathbf{x}', \qquad G(\mathbf{x}') = B\mathbf{x}'.$$

As usual, the **composite** function $F \circ G$ is defined by

$$(F \circ G)(\mathbf{x}') = F[G(\mathbf{x}')].$$

> **Theorem** The composite of linear functions is again linear. If $F(\mathbf{x}) = A\mathbf{x}'$ and $G(\mathbf{x}) = B\mathbf{x}'$, then
>
> $$(F \circ G)(\mathbf{x}') = (AB)\mathbf{x}'.$$

Proof: To prove that $F \circ G$ is linear, we go right back to the definition:

$$F[G(a\mathbf{u}' + b\mathbf{v}')] = F[aG(\mathbf{u}') + bG(\mathbf{v}')] = aF[G(\mathbf{u}')] + bF[G(\mathbf{v}')],$$

the first equality because G is linear, the second because F is linear. It follows that

$$(F \circ G)(a\mathbf{u}' + b\mathbf{v}') = a(F \circ G)(\mathbf{u}') + b(F \circ G)(\mathbf{v}'),$$

that is, $F \circ G$ is linear.

We must compute the matrix corresponding to $F \circ G$. We have

$$(F \circ G)(\mathbf{x}') = F[G(\mathbf{x}')] = F[B\mathbf{x}'] = A(Bx').$$

To finish the proof, we must establish the associativity relation

$$A(B\mathbf{x}') = (AB)\mathbf{x}'.$$

If $\mathbf{a}_i$ denotes the i-th row of A, then the i-th element of the *column* vector $A(B\mathbf{x}')$ is $\mathbf{a}_i(B\mathbf{x}')$. But we know that $\mathbf{a}_i(B\mathbf{x}') = (\mathbf{a}_iB)\mathbf{x}'$. However, $\mathbf{a}_iB$ is the i-th row of the *matrix* AB. Consequently $A(B\mathbf{x}')$ and $(AB)\mathbf{x}'$ have the same elements, i.e., they are equal, $A(B\mathbf{x}') = (AB)\mathbf{x}'$. This completes the proof that $(F \circ G)(\mathbf{x}') = (AB)\mathbf{x}'$.

Here is a concrete application of composition of linear functions, showing how matrix multiplication can be a labor saving device. Consider this problem in algebraic manipulation: Given

$$(1)\quad \begin{cases} r = 3x + 4y + 5z \\ s = 2x - y - z \\ t = 5x - 6y + 3z \end{cases} \qquad (2)\quad \begin{cases} x = u + v + 4w \\ y = 3u - 2v - w, \\ z = 6u + 5v + 6w \end{cases}$$

express r, s, t in terms of u, v, w.

The obvious approach is to substitute equations (2) into equations (1), simplify and collect terms. But this is tedious, while matrix multiplication will do it for you automatically. Write

$$\begin{bmatrix} r \\ s \\ t \end{bmatrix} = A \begin{bmatrix} x \\ y \\ z \end{bmatrix}, \qquad \begin{bmatrix} x \\ y \\ z \end{bmatrix} = B \begin{bmatrix} u \\ v \\ w \end{bmatrix},$$

where

$$A = \begin{bmatrix} 3 & 4 & 5 \\ 2 & -1 & -1 \\ 5 & -6 & 3 \end{bmatrix}, \qquad B = \begin{bmatrix} 1 & 1 & 4 \\ 3 & -2 & -1 \\ 6 & 5 & 6 \end{bmatrix}.$$

Then

$$\begin{bmatrix} r \\ s \\ t \end{bmatrix} = (AB) \begin{bmatrix} u \\ v \\ w \end{bmatrix} = \begin{bmatrix} 45 & 20 & 38 \\ -7 & -1 & 3 \\ 5 & 32 & 44 \end{bmatrix} \begin{bmatrix} u \\ v \\ w \end{bmatrix}.$$

Hence

$$\begin{cases} r = 45u + 20v + 38w \\ s = -7u - v + 3w \\ t = 5u + 32v + 44w. \end{cases}$$

Associative Law

We can now prove, with a minimum of computation, the associative law for multiplication of matrices given in Section 3.

Let A, B, C be 3×3 matrices and let F, G, H be the corresponding linear functions:

$$F(\mathbf{x}') = A\mathbf{x}', \qquad G(\mathbf{x}') = B\mathbf{x}', \qquad H(\mathbf{x}') = C\mathbf{x}'.$$

We compute $F\{G[H(\mathbf{x}')]\}$ in two ways. First, $G[H(\mathbf{x}')] = (G \circ H)(\mathbf{x}')$, so

$$F\{G[H(\mathbf{x}')]\} = F[(G \circ H)(\mathbf{x}')] = [F \circ (G \circ H)](\mathbf{x}').$$

Second, $F[G(\mathbf{y}')] = (F \circ G)(\mathbf{y}')$. Replace $\mathbf{y}'$ by $H(\mathbf{x}')$:

$$F\{G[H(\mathbf{x}')]\} = (F \circ G)[H(\mathbf{x}')] = [(F \circ G) \circ H](\mathbf{x}').$$

We conclude that

$$F \circ (G \circ H) = (F \circ G) \circ H.$$

Now we translate this into matrices. The matrix corresponding to $G \circ H$ is

BC, hence the matrix corresponding to $F \circ (G \circ H)$ is $A(BC)$. Similarly the matrix corresponding to the linear function $(F \circ G) \circ H$ is $(AB)C$. Therefore $A(BC) = (AB)C$.

EXERCISES

1. Let F: $\mathbf{R}^3 \longrightarrow \mathbf{R}^3$ be linear. If

$$F\left(\begin{bmatrix} 1 \\ 0 \\ 0 \end{bmatrix}\right) = \begin{bmatrix} 6 \\ 2 \\ 7 \end{bmatrix}, \qquad F\left(\begin{bmatrix} 0 \\ 1 \\ 0 \end{bmatrix}\right) = \begin{bmatrix} -4 \\ 3 \\ 0 \end{bmatrix}, \qquad F\left(\begin{bmatrix} 0 \\ 0 \\ 1 \end{bmatrix}\right) = \begin{bmatrix} 8 \\ -1 \\ -3 \end{bmatrix},$$

compute

$$F\left(\begin{bmatrix} 5 \\ -1 \\ 8 \end{bmatrix}\right).$$

2. Given a 3×3 matrix A, verify that $\mathbf{x}' \longrightarrow A\mathbf{x}'$ defines a linear function F: $\mathbf{R}^3 \longrightarrow \mathbf{R}^3$.

Let F and G be linear functions $\mathbf{R}^3 \longrightarrow \mathbf{R}^3$ with associated matrices A and B. Describe $F \circ G$:

3. $A = \begin{bmatrix} 0 & 0 & 1 \\ 0 & 1 & 1 \\ 0 & 0 & 0 \end{bmatrix}, \qquad B = \begin{bmatrix} 2 & 0 & 0 \\ 3 & 0 & 0 \\ 5 & 1 & 0 \end{bmatrix}$

4. $A = \begin{bmatrix} \cos\theta & \sin\theta \\ -\sin\theta & \cos\theta \end{bmatrix}, \qquad B = \begin{bmatrix} \cos\theta & -\sin\theta \\ \sin\theta & \cos\theta \end{bmatrix}$; interpret geometrically.

5. Use matrix and vector notation to write the linear system

$$\begin{cases} a_1x + b_1y + c_1z = d_1 \\ a_2x + b_2y + c_2z = d_2 \\ a_3x + b_3y + c_3z = d_3. \end{cases}$$

6. Let F: $\mathbf{R}^3 \longrightarrow \mathbf{R}^3$ be linear with matrix A. Given $\mathbf{y}'$, prove that there is a unique vector $\mathbf{x}'$ such that $F(\mathbf{x}') = \mathbf{y}'$ if and only if the determinant of A is not zero.

7. Given

$$\begin{cases} x = 3u + 4v + 5w \\ y = u - v + 2w \\ z = 2u + 2v - w \end{cases} \qquad \begin{cases} u = r - s - t \\ v = 2r + 6s - 7t \\ w = 3r - 2s + t, \end{cases}$$

express x, y, z in terms of r, s, t.

8. Given

$$\begin{cases} x = 2u - 3v \\ y = 6u + v \end{cases} \qquad \begin{cases} u = 4r - 5s \\ v = r + 2s \end{cases} \qquad \begin{cases} r = 5p + 2q \\ s = -p + 3q, \end{cases}$$

express x and y in terms of p and q.

9. Do Ex. 4 of Section 2 in column notation. That is, prove

$$\mathbf{a}' \times \mathbf{x}' = A\mathbf{x}', \qquad \text{where} \quad A = \begin{bmatrix} 0 & -a_3 & a_2 \\ a_3 & 0 & -a_1 \\ -a_2 & a_1 & 0 \end{bmatrix}.$$

10. (cont.) Use this result to prove

$$\mathbf{a}' \cdot (\mathbf{b}' \times \mathbf{c}') = -\mathbf{c}' \cdot (\mathbf{b}' \times \mathbf{a}').$$

11. Use Ex. 9 to show that the matrix of

$$F(\mathbf{x}') = \mathbf{a}' \times (\mathbf{b}' \times \mathbf{x}') \qquad \text{is} \qquad -(\mathbf{a}' \cdot \mathbf{b}')I + \mathbf{b}'\mathbf{a}.$$

12. (cont.) Now prove the identity

$$\mathbf{a}' \times (\mathbf{b}' \times \mathbf{c}') = -(\mathbf{a}' \cdot \mathbf{b}')\mathbf{c}' + (\mathbf{a}' \cdot \mathbf{c}')\mathbf{b}'.$$

(Compare Ex. 21, p. 147.)

13. Let A be a 3×3 matrix. We proved there is a constant c such that $|A\mathbf{x}'| \leq c\,|\mathbf{x}'|$ for all vectors $\mathbf{x}'$. Prove that $|A^n\mathbf{x}'| \leq c^n\,|\mathbf{x}'|$ for $n = 1, 2, 3, \cdots$.

14*. (cont.) Prove that

$$\sum_{n=0}^{\infty} \frac{1}{n!} A^n\mathbf{x}'$$

converges for each $\mathbf{x}'$. [*Hint:* Use Ex. 23, p. 205.]

15*. (cont.) Prove that

$$\mathbf{x}' \longrightarrow \sum_{n=0}^{\infty} \frac{1}{n!} A^n\mathbf{x}'$$

is a linear function. (It defines a matrix called e^A or $\exp A$.)

5. QUADRATIC FORMS

This section paves the way for the study of maxima and minima of functions of several variables in Chapter 9. In one variable calculus, for $f(x)$ to have a minimum at a point c where $f'(c) = 0$, a sufficient condition is that $f''(c) > 0$. In several variables, the analogous second derivative test is more complicated, and requires some knowledge of quadratic forms.

A **quadratic form** is simply a homogeneous quadratic polynomial. The general quadratic form in two variables is

$$f(x, y) = ax^2 + 2bxy + cy^2.$$

We note immediately a useful expression for f in terms of matrices:

$$f(x, y) = (x, y)\begin{bmatrix} a & b \\ b & c \end{bmatrix}\begin{bmatrix} x \\ y \end{bmatrix}.$$

This expression justifies using $2b$ instead of b for the middle coefficient.

The general quadratic form in three variables is

$$f(x, y, z) = (x, y, z)\begin{bmatrix} a & b & d \\ b & c & e \\ d & e & f \end{bmatrix}\begin{bmatrix} x \\ y \\ z \end{bmatrix}$$

$$= ax^2 + 2bxy + cy^2 + 2dzx + 2eyz + fz^2.$$

The matrices

$$\begin{bmatrix} a & b \\ b & c \end{bmatrix} \quad \text{and} \quad \begin{bmatrix} a & b & d \\ b & c & e \\ d & e & f \end{bmatrix}$$

have mirror symmetry in the main diagonal (top left to bottom right) and so are called **symmetric** matrices. This property is concisely stated in terms of transposes:

A matrix A is symmetric if and only if $A' = A$.

To each symmetric matrix corresponds a unique quadratic form. Conversely, to each quadratic form corresponds a unique symmetric matrix. This converse statement may seem obvious, but it does require a proof, which we give for the two variable case.

Suppose

$$A_1 = \begin{bmatrix} a_1 & b_1 \\ b_1 & c_1 \end{bmatrix} \quad \text{and} \quad A_2 = \begin{bmatrix} a_2 & b_2 \\ b_2 & c_2 \end{bmatrix}$$

determine the *same* quadratic form. Then

$$a_1x^2 + 2b_1xy + c_1y^2 = a_2x^2 + 2b_2xy + c_2y^2$$

for all x and y. The choices $(x, y) = (1, 0)$, $(0, 1)$, and $(1, 1)$ yield

$$a_1 = a_2, \qquad c_1 = c_2, \qquad b_1 = b_2.$$

Hence $A_1 = A_2$. A similar easy proof applies for three variables.

> The general quadratic form may be written as $\mathbf{x}A\mathbf{x}'$, where A is a unique symmetric matrix.

Examples of quadratic forms:

$$x^2 + y^2 = (x, y)\begin{bmatrix}1 & 0\\ 0 & 1\end{bmatrix}\begin{bmatrix}x\\ y\end{bmatrix}, \qquad 2xy = (x, y)\begin{bmatrix}0 & 1\\ 1 & 0\end{bmatrix}\begin{bmatrix}x\\ y\end{bmatrix},$$

$$ax^2 + by^2 + cz^2 = (x, y, z)\begin{bmatrix}a & 0 & 0\\ 0 & b & 0\\ 0 & 0 & c\end{bmatrix}\begin{bmatrix}x\\ y\\ z\end{bmatrix},$$

$$2cxy + 2ayz + 2bzx = (x, y, z)\begin{bmatrix}0 & c & b\\ c & 0 & a\\ b & a & 0\end{bmatrix}\begin{bmatrix}x\\ y\\ z\end{bmatrix}.$$

Positive Definite Matrices

Take a quadratic form in two variables,

$$f(x, y) = \mathbf{x}A\mathbf{x}' = ax^2 + 2bxy + cy^2,$$

where

$$\mathbf{x} = (x, y) \qquad \text{and} \qquad A = \begin{bmatrix}a & b\\ b & c\end{bmatrix}.$$

We want to find conditions on the coefficients a, b, c so that

$$f(x, y) > 0 \qquad \text{for all} \qquad (x, y) \neq (0, 0).$$

When this is the case both f and A are called **positive definite.**

The simplest case of a positive definite quadratic form is $f(x, y) = x^2 + y^2$. The associated positive definite matrix is

$$A = \begin{bmatrix}1 & 0\\ 0 & 1\end{bmatrix}.$$

Other examples:

$$f(x, y) = x^2 + 2xy + 5y^2 = (x + y)^2 + 4y^2, \qquad A = \begin{bmatrix}1 & 1\\ 1 & 5\end{bmatrix};$$

$$f(x, y) = x^2 - 6xy + 11y^2 = (x - 3y)^2 + 2y^2, \qquad A = \begin{bmatrix}1 & -3\\ -3 & 11\end{bmatrix}.$$

To check that $(x - 3y)^2 + 2y^2$ is positive definite, suppose $(x, y) \neq (0, 0)$. If $y \neq 0$, then

$$(x - 3y)^2 + 2y^2 \geq 2y^2 > 0.$$

If $y = 0$, then $x \neq 0$ and

$$(x - 3y)^2 + 2y^2 = x^2 > 0.$$

We are ready to state the basic facts about positive definite matrices. We begin with the 2×2 case:

Theorem 5.1 The symmetric matrix

$$\begin{bmatrix} a & b \\ b & c \end{bmatrix}$$

is positive definite if and only if

$$a > 0 \qquad \text{and} \qquad \begin{vmatrix} a & b \\ b & c \end{vmatrix} > 0.$$

Proof: Suppose the matrix is positive definite. Then

$$f(x, y) = ax^2 + 2bxy + cy^2 > 0$$

whenever $(x, y) \neq (0, 0)$. In particular $a = f(1, 0) > 0$. Now complete the square:

$$f(x, y) = a\left(x + \frac{b}{a}y\right)^2 + \left(\frac{ac - b^2}{a}\right)y^2.$$

Then $f(-b/a, 1) > 0$, hence

$$\frac{ac - b^2}{a} > 0.$$

Therefore $a > 0$ and $ac - b^2 > 0$.

Conversely, suppose $a > 0$ and $ac - b^2 > 0$. Then certainly

$$f(x, y) = a\left(x + \frac{b}{a}y\right)^2 + \left(\frac{ac - b^2}{a}\right)y^2 \geq 0$$

for any (x, y). Furthermore, $f(x, y) = 0$ only if each of the squared quantities is zero:

$$x + \frac{b}{a}y = 0, \qquad y = 0,$$

hence only for $(x, y) = (0, 0)$. This completes the proof.

There is a corresponding theorem for 3×3 symmetric matrices. A proof based on "completing the square" is possible, but it is long and tedious, so we shall omit it. In linear algebra courses other types of proofs are often developed. (Compare the remark after Theorem 8.8.)

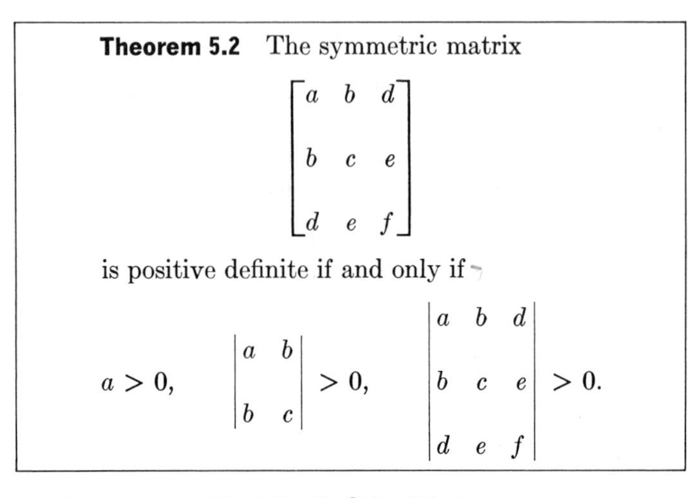

Theorem 5.2 The symmetric matrix

$$\begin{bmatrix} a & b & d \\ b & c & e \\ d & e & f \end{bmatrix}$$

is positive definite if and only if

$$a > 0, \qquad \begin{vmatrix} a & b \\ b & c \end{vmatrix} > 0, \qquad \begin{vmatrix} a & b & d \\ b & c & e \\ d & e & f \end{vmatrix} > 0.$$

Negative Definite Matrices

Let A be a symmetric matrix and $f(\mathbf{x}) = \mathbf{x}A\mathbf{x}'$ the corresponding quadratic form. We say that A and f are **negative definite** if $f(\mathbf{x}) < 0$ except for $\mathbf{x} = \mathbf{0}$. It is pretty clear that $f(\mathbf{x}) < 0$ is the same thing as $-f(\mathbf{x}) > 0$, hence A is negative definite if and only if $-A$ is positive definite, where $-A$ denotes the matrix whose elements are the negatives of the elements of A (it is the matrix of $-f$). A corresponding modification of Theorem 5.2 gives the following test for negative definiteness:

Theorem 5.3 The symmetric matrix

$$\begin{bmatrix} a & b & d \\ b & c & e \\ d & e & f \end{bmatrix}$$

is negative definite if and only if

$$a < 0, \qquad \begin{vmatrix} a & b \\ b & c \end{vmatrix} > 0, \qquad \begin{vmatrix} a & b & d \\ b & c & e \\ d & e & f \end{vmatrix} < 0.$$

The corresponding conditions for a quadratic form $ax^2 + 2bxy + cy^2$ in two variables are simply $a < 0$, $ac - b^2 > 0$.

A Lower Bound

The following result is essential in our discussion later of maxima and minima. We prove it for three variables, but an immediate modification covers the two variable case.

Theorem 5.4 Let $f(\mathbf{x})$ be a positive definite quadratic form. Then there exists a constant $k > 0$ such that

$$f(\mathbf{x}) \geq k\,|\mathbf{x}|^2$$

for all $\mathbf{x}$.

Proof: Let $\mathbf{S} = \{\mathbf{x} \mid |\mathbf{x}| = 1\}$ denote the unit sphere, a closed and bounded set. The function f is continuous on $\mathbf{S}$, so it has a minimum on $\mathbf{S}$. That is, there is a point $\mathbf{x}_0 \in \mathbf{S}$ such that $f(\mathbf{x}) \geq f(\mathbf{x}_0)$ for all $\mathbf{x} \in \mathbf{S}$. We set $k = f(\mathbf{x}_0)$. Since f is positive definite, we have $k > 0$.

Let $\mathbf{x} \neq \mathbf{0}$. Then $\mathbf{x}/|\mathbf{x}| \in \mathbf{S}$, hence $f(\mathbf{x}/|\mathbf{x}|) \geq k$. But $f(\mathbf{x}/|\mathbf{x}|) = f(\mathbf{x})/|\mathbf{x}|^2$ since f is a quadratic form. Therefore $f(\mathbf{x})/|\mathbf{x}|^2 \geq k$, and finally $f(\mathbf{x}) \geq k\,|\mathbf{x}|^2$.

Semi-definite Forms

Positive definiteness is defined by $f(\mathbf{x}) > 0$ for all $\mathbf{x} \neq \mathbf{0}$. A quadratic form $f(\mathbf{x})$ is called **positive semi-definite** if

$$f(\mathbf{x}) \geq 0 \qquad \text{for all} \quad \mathbf{x}.$$

The property **negative semi-definite** is defined similarly. We use the same adjectives for the symmetric matrix corresponding to the form $f(\mathbf{x})$.

Theorem 5.5 The quadratic form $f(\mathbf{x}) = ax^2 + 2bxy + cy^2$ is positive semi-definite if and only if

$$a \geq 0, \qquad c \geq 0, \qquad ac - b^2 \geq 0.$$

Proof: Suppose $f(\mathbf{x})$ is positive semi-definite. Let $h > 0$. Then

$$f(\mathbf{x}) + h(x^2 + y^2) = (a + h)x^2 + 2bxy + (c + h)y^2$$

is positive definite. By Theorem 5.1,

$$a + h > 0, \qquad (a + h)(c + h) - b^2 > 0.$$

Also $c + h > 0$ as we see, for instance, by interchanging x and y. Now let $h \longrightarrow 0$. Then we conclude that

$$a \geq 0, \qquad c \geq 0, \qquad ac - b^2 \geq 0.$$

Conversely, suppose these three inequalities hold. Let $h > 0$. Then $a + h > 0$, $c + h > 0$, and

$$(a + h)(c + h) - b^2 = (ac - b^2) + (a + c)h + h^2 \geq h^2 > 0.$$

Hence $(a + h)x^2 + 2bxy + (c + h)y^2$ is positive definite. Let $(x, y) \neq 0$. Then

$$(a + h)x^2 + 2bxy + (c + h)y^2 > 0.$$

Let $h \longrightarrow 0$:

$$f(\mathbf{x}) = ax^2 + 2bxy + cy^2 \geq 0.$$

Therefore $f(\mathbf{x})$ is positive semi-definite.

REMARK 1: An alternate proof can be constructed by modifying the proof of Theorem 5.1.

REMARK 2: The corresponding conditions for $f(\mathbf{x})$ to be negative semi-definite are

$$a \leq 0, \qquad c \leq 0, \qquad ac - b^2 \geq 0.$$

We shall state without proof the corresponding results for three dimensions.

Theorem 5.6 The symmetric matrix

$$A = \begin{bmatrix} a & b & d \\ b & c & e \\ d & e & f \end{bmatrix}$$

is positive semi-definite if and only if all of its **principal minors** are nonnegative, that is,

$$a \geq 0, \qquad c \geq 0, \qquad f \geq 0,$$

$$ac - b^2 \geq 0, \qquad cf - e^2 \geq 0, \qquad af - d^2 \geq 0,$$

$$|A| = \begin{vmatrix} a & b & d \\ b & c & e \\ d & e & f \end{vmatrix} \geq 0.$$

The conditions for A to be negative semi-definite are

$$a \leq 0, \qquad c \leq 0, \qquad f \leq 0,$$

$$ac - b^2 \geq 0, \qquad cf - e^2 \geq 0, \qquad af - d^2 \geq 0$$

$$|A| \leq 0.$$

For instance, consider

$$A = \begin{bmatrix} 1 & 1 & 1 \\ 1 & 1 & 1 \\ 1 & 1 & 1 \end{bmatrix}.$$

Then $a = c = f = 1 \geq 0$, $ac - b^2 = cf - e^2 = af - d^2 = 0 \geq 0$, and $|A| = 0 \geq 0$, so A is positive semi-definite. If $\mathbf{x} = (x, y, z)$, note that $f(\mathbf{x}) = \mathbf{x}A\mathbf{x}' = (x + y + z)^2 \geq 0$, confirming the diagnosis. (But A is not positive *definite*; for instance $f(1, 1, -2) = 0$.)

Indefinite Forms

A quadratic form $f(\mathbf{x}) = \mathbf{x}A\mathbf{x}'$ is called **indefinite** if it is neither positive nor negative semi-definite, that is, if there are points $\mathbf{x}_1$ and $\mathbf{x}_2$ such that $f(\mathbf{x}_1) < 0$ and $f(\mathbf{x}_2) > 0$.

Theorem 5.7 A form $f(\mathbf{x}) = ax^2 + 2bxy + cy^2$ in two variables is indefinite if and only if

$$\begin{vmatrix} a & b \\ b & c \end{vmatrix} < 0.$$

Proof: First suppose $ac - b^2 < 0$. By Theorem 5.5, $f(\mathbf{x})$ is not positive semi-definite. By the corresponding result mentioned in Remark 2 above, $f(\mathbf{x})$ is not negative semi-definite. Therefore $f(\mathbf{x})$ is indefinite.

Now suppose $f(\mathbf{x})$ is indefinite. We examine three cases: (1) $a \geq 0$, $c \geq 0$. Then $ac - b^2 < 0$, otherwise $f(\mathbf{x})$ is positive semi-definite by Theorem 5.5. (2) $a \leq 0$, $c \leq 0$. Then $ac - b^2 < 0$, otherwise $f(\mathbf{x})$ is negative semi-definite. (3) $ac < 0$. Then $ac - b^2 \leq ac < 0$. This exhausts all possibilities; the proof is complete.

There is not such a simple criterion for indefiniteness of a 3×3 matrix

$$A = \begin{bmatrix} a & b & d \\ b & c & e \\ d & e & f \end{bmatrix}.$$

The best we can do is try the test in Theorem 5.6. If A fails to be positive semi-definite, and if it fails to be negative semi-definite, then it *must* be indefinite. Thus we first inspect a, c, f. If two have different signs, A is indefinite; if not we inspect the 2×2 principal minors

$$ac - b^2, \qquad cf - e^2, \qquad af - d^2.$$

If one of these is negative, A is indefinite; if not we inspect the determinant $|A|$. If its sign differs from that of a, c, and e, then A is indefinite; if not, A is positive or negative semi-definite.

For instance, consider

$$A = \begin{bmatrix} -1 & 1 & 1 \\ 1 & -1 & 1 \\ 1 & 1 & -1 \end{bmatrix}.$$

Then $a = c = f = -1 < 0$, so we pass to 2×2 principal minors: $ac - b^2 = cf - e^2 = af - d^2 = 0$, so we still do not know if A is or isn't indefinite. But $|A| = 4 > 0$, so A is indefinite. If this is not a convincing argument, then here is an overwhelming one: set $\mathbf{x} = (1, 0, 0)$ and $\mathbf{y} = (1, 1, 1)$. Then $\mathbf{x}A\mathbf{x}' = -1 < 0$ and $\mathbf{y}A\mathbf{y}' = 3 > 0$.

EXERCISES

Write out the quadratic form corresponding to the symmetric matrix:

1. $\begin{bmatrix} 4 & -1 \\ -1 & 3 \end{bmatrix}$
2. $\begin{bmatrix} -1 & 5 \\ 5 & -2 \end{bmatrix}$
3. $\begin{bmatrix} 1 & 0 & 0 \\ 0 & 2 & 4 \\ 0 & 4 & 3 \end{bmatrix}$
4. $\begin{bmatrix} 1 & 2 & 3 \\ 2 & 4 & 6 \\ 3 & 6 & 5 \end{bmatrix}$.

Find the symmetric matrix of the quadratic form:

5. $x^2 - y^2$
6. xy
7. $(x + y)^2$
8. $(x - 2y)^2$
9. $(ax + by)^2$
10. $(ax + by)(cx + dy)$
11. $(x + y)(y + z)$
12. $(x + 2y + 3z)^2$
13. $(ax + by + cz)^2$
14. $(ax + by + cz)(\alpha x + \beta y + \gamma z)$.

Determine whether the quadratic form is positive definite:

15. $x^2 + 4xy + 2y^2$
16 $9x^2 - 12xy + y^2$
17. $x^2 + 6xy + 10y^2$
18. $3x^2 + 2xy + y^2 - 2zx + 3zy + z^2$.
19. Let A be any 3×3 matrix, not necessarily symmetric. Then $f(\mathbf{x}) = \mathbf{x}A\mathbf{x}'$ is a quadratic form. Find its matrix.
20. (cont.) Do the special case

$$A = \begin{bmatrix} 0 & 0 & 0 \\ 2 & 0 & 0 \\ 4 & 6 & 0 \end{bmatrix}.$$

21. Prove that the sum of two positive definite quadratic forms is again positive definite. Find the corresponding matrix.

22. Let A be symmetric, $\mathbf{x} = (x_1, x_2, x_3)$, and $f(\mathbf{x}) = \mathbf{x}A\mathbf{x}'$. Prove that
$$\frac{\partial f}{\partial x_1}y_1 + \frac{\partial f}{\partial x_2}y_2 + \frac{\partial f}{\partial x_3}y_3 = 2\mathbf{x}A\mathbf{y}'.$$
23. Let A be any 3×3 matrix. Prove that $\frac{1}{2}(A + A')$ is symmetric.
24. Let A be any 3×3 matrix. Prove that AA' is symmetric.
25. Suppose A and B are symmetric 3×3 matrices that commute, that is, $AB = BA$. Prove that AB is symmetric.
26*. Suppose A and B are positive definite 2×2 matrices. Is the quadratic form $f(\mathbf{x}) = \mathbf{x}AB\mathbf{x}'$ positive definite? (Note that AB need not be symmetric.)
27. The form $x^2 + 2xy + 2y^2$ is positive definite so there is a $k > 0$ such that $x^2 + 2xy + 2y^2 \geq k(x^2 + y^2)$ for all (x, y). Find the largest such k.
28*. (cont.) Solve this problem in general for a positive definite $f(x) = ax^2 + 2bxy + cy^2$.
29. Prove the statement in Remark 2, p. 250.
30. Carry out the proof suggested in Remark 1, p. 250.
31*. Let $f(x, y) = ax^2 + 2bxy + cy^2$ be an indefinite form. Prove that $f(x, y)$ is the product of two non-proportional linear functions. [*Hint:* Rotate coordinates.]
32. (cont.) Sketch in the plane the regions $f > 0$, $f < 0$, $f = 0$.

6. QUADRIC SURFACES

A **quadric surface** is the graph of an equation $f(x, y, z) = 0$, where $f(x, y, z)$ is a quadratic polynomial. The most general quadratic polynomial f is the sum of a quadratic form and a linear function,
$$f(x, y, z) = Q(x) + px + qy + rz + k.$$
The quadratic form Q involves the pure terms x^2, y^2, z^2 and the mixed terms xy, yz, zx. According to a fairly deep result in linear algebra, the mixed terms can be eliminated by rotating the coordinate system. We shall accept this result without proof. It means that we may confine our study of quadric surfaces to the cases where there are no mixed quadratic terms.

Thus we consider $f(x, y, z) = 0$, where
$$f(x, y, z) = Ax^2 + By^2 + Cz^2 + px + qy + rz + k.$$
Of course, if there are no quadratic terms, then $f = 0$ represents a plane, which doesn't interest us here.

If $A \neq 0$, then a translation in the x-direction eliminates the term px, and similarly if $B \neq 0$ or $C \neq 0$. This reduces our study to the following types of f:

(i) $Ax^2 + By^2 + Cz^2 + k$

(ii) $Ax^2 + By^2 + rz + k$

(iii) $Ax^2 + qy + rz + k$.

These include all possibilities, provided we are willing to permute the variables.

For instance, the polynomial $Cz^2 + px + qy + k$ becomes type (iii) when x and z are interchanged.

In type (ii), if $r \neq 0$, then a translation in the z-direction eliminates k. In type (iii), a rotation in the y, z-plane, taken so that $qy + rz = 0$ is the new z-axis, changes the function to the form $ax^2 + qy + k$. Again, k can be eliminated by translation if $q \neq 0$.

This reduces our study to the following types of f:

(1) $Ax^2 + By^2 + Cz^2 + k$

(2) $Ax^2 + By^2 + rz$ (2′) $Ax^2 + By^2 + k$

(3) $Ax^2 + qy$ (3′) $Ax^2 + k$.

We begin with (1) in case $ABC \neq 0$ and $k \neq 0$. Then $f = 0$ can be written in the form

$$\pm\frac{x^2}{a^2} \pm \frac{y^2}{b^2} \pm \frac{z^2}{c^2} = 1.$$

If all three signs are minus, the graph is empty. Otherwise the graph is symmetric in each coordinate plane because if (x, y, z) is on the graph, so are all eight points $(\pm x, \pm y, \pm z)$. Therefore it suffices to draw the first octant portion of the graph and determine the rest by symmetry.

Ellipsoids

Consider the graph of

$$\frac{x^2}{a^2} + \frac{y^2}{b^2} + \frac{z^2}{c^2} = 1, \qquad a, b, c > 0.$$

Since squares are non-negative, each point of the graph satisfies

$$\frac{x^2}{a^2} \leq 1, \qquad \frac{y^2}{b^2} \leq 1, \qquad \frac{z^2}{c^2} \leq 1.$$

This means the graph is confined to the box

$$-a \leq x \leq a, \qquad -b \leq y \leq b, \qquad -c \leq z \leq c.$$

Suppose $-c < z_0 < c$. The intersection of the graph and the horizontal plane $z = z_0$ consists of all points (x, y, z_0) that satisfy

$$\frac{x^2}{a^2} + \frac{y^2}{b^2} = 1 - \frac{z_0^2}{c^2}.$$

This curve is an ellipse. It is as large as possible when $z_0 = 0$, and it becomes smaller and smaller as $z_0 \longrightarrow c$ or $z_0 \longrightarrow -c$. Thus each such cross-section by a horizontal plane is an ellipse, except at the extremes $z_0 = \pm c$, where it is a single point.

The same argument applies to plane sections parallel to the other coordinate planes. This gives us enough information for a sketch. The surface is called an **ellipsoid** (Fig. 6.1). In the special case $a = b = c$, it is a sphere.

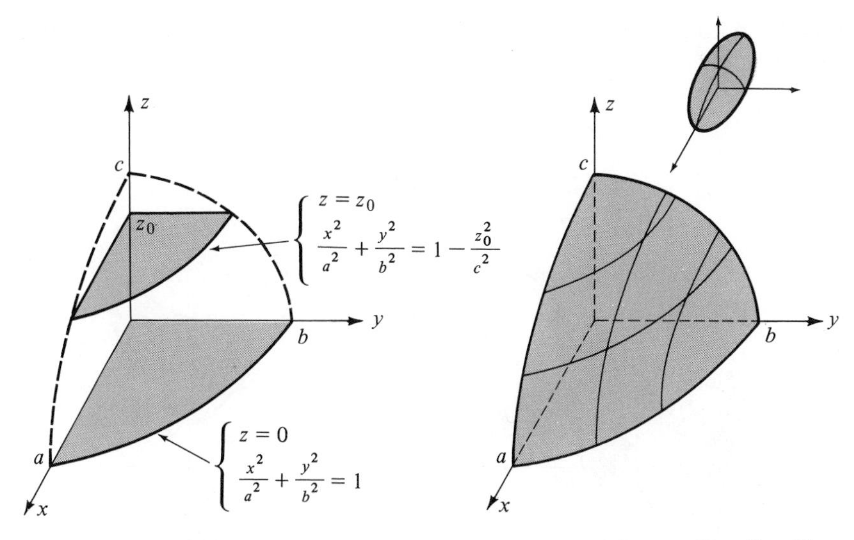

(a) Horizontal cross-sections are ellipses.

(b) the complete ellipsoid

FIG. 6.1 graph of $\frac{x^2}{a^2} + \frac{y^2}{b^2} + \frac{z^2}{c^2} = 1$

Hyperboloids of One Sheet

Consider the graph of

$$\frac{x^2}{a^2} + \frac{y^2}{b^2} - \frac{z^2}{c^2} = 1, \qquad a, b, c > 0.$$

Each horizontal cross-section is an ellipse

$$\begin{cases} z = z_0 \\ \dfrac{x^2}{a^2} + \dfrac{y^2}{b^2} = 1 + \dfrac{{z_0}^2}{c^2}, \end{cases}$$

no matter what value z_0 is. The smallest ellipse occurs for $z_0 = 0$; as $z_0 \longrightarrow \infty$ or $z_0 \longrightarrow -\infty$, the ellipses get larger and larger.

The surface meets the y, z-plane in the hyperbola

$$\frac{y^2}{b^2} - \frac{z^2}{c^2} = 1,$$

and it meets the z, x-plane in the hyperbola

$$\frac{x^2}{a^2} - \frac{z^2}{c^2} = 1.$$

This information is enough to sketch the surface, called a **hyperboloid of one sheet** (Fig. 6.2a).

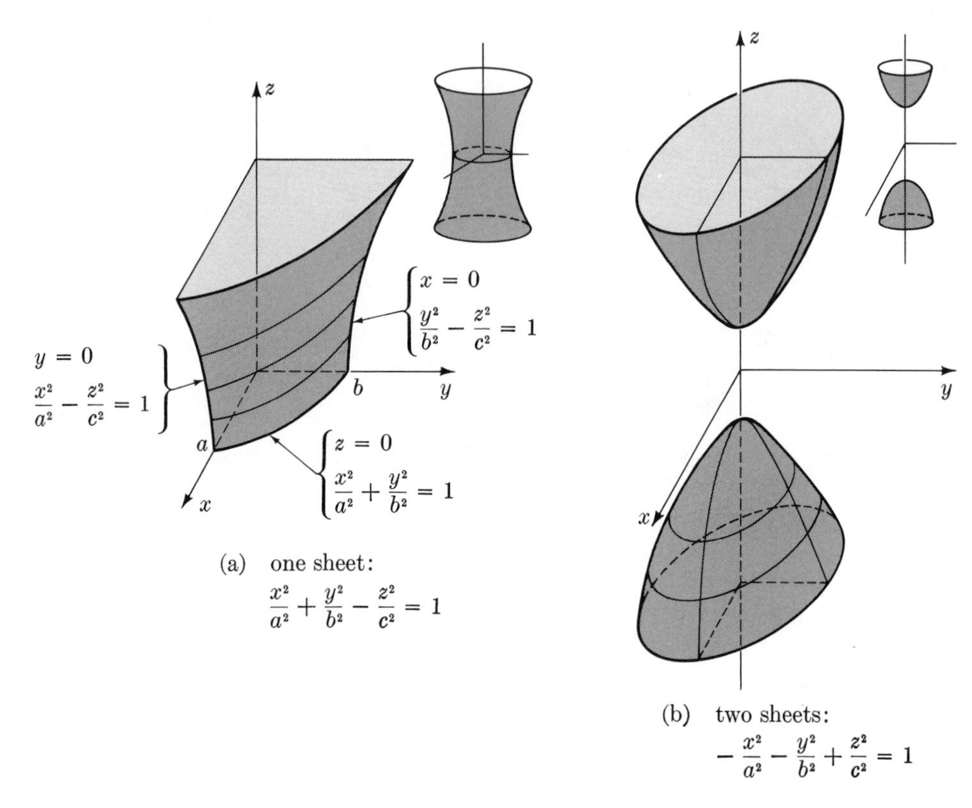

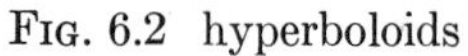
FIG. 6.2 hyperboloids

Hyperboloids of Two Sheets

Consider the equation

$$-\frac{x^2}{a^2} - \frac{y^2}{b^2} + \frac{z^2}{c^2} = 1, \qquad a, b, c > 0.$$

If (x, y, z) is a point on the surface, then

$$\frac{z^2}{c^2} = 1 + \frac{x^2}{a^2} + \frac{y^2}{b^2} \geq 1,$$

hence $z^2 \geq c^2$. This means either $z \geq c$ or $z \leq -c$, that is, there are no points of the surface between the horizontal planes $z = c$ and $z = -c$.

If $z_0^2 > c^2$, the horizontal plane $z = z_0$ meets the surface in the curve

$$\begin{cases} z = z_0 \\ \dfrac{x^2}{a^2} + \dfrac{y^2}{b^2} = \dfrac{z_0^2}{c^2} - 1 > 0, \end{cases}$$

an ellipse. Also the surface meets the y, z-plane and the z, x-plane in the hyperbolas

$$-\frac{y^2}{b^2} + \frac{z^2}{c^2} = 1 \qquad \text{and} \qquad -\frac{x^2}{a^2} + \frac{z^2}{c^2} = 1$$

respectively. The surface breaks into two parts, and it is called a **hyperboloid of two sheets** (Fig. 6.2b).

Cones

Now we complete the study of type (1) on p. 254 for the case $ABC \neq 0$, but $k = 0$. Then $f = 0$ can be written

$$\pm\frac{x^2}{a^2} \pm \frac{y^2}{b^2} \pm \frac{z^2}{c^2} = 0, \qquad a, b, c > 0.$$

If the signs are all the same, then $(0, 0, 0)$ is the only point on the graph; not interesting. Otherwise two signs are equal, the other opposite. Changing signs if necessary, we are reduced to

$$z^2 = \frac{x^2}{a^2} + \frac{y^2}{b^2}, \qquad a, b > 0.$$

It is a surface with the following property. For each point $\mathbf{x}_0$ on the surface, the entire line $\mathbf{x} = t\mathbf{x}_0$ lies on the surface. Such a surface is called a **cone**, and the lines $\mathbf{x} = t\mathbf{x}_0$ are called **generators** of the cone.

To check that the graph of

$$z^2 = \frac{x^2}{a^2} + \frac{y^2}{b^2}$$

is a cone, we take any point $\mathbf{x}_0$ on the graph and check that $t\mathbf{x}_0$ is also on the graph. If $\mathbf{x}_0 = (x_0, y_0, z_0)$, then $t\mathbf{x}_0 = (tx_0, ty_0, tz_0)$ and

$$(tz_0)^2 = t^2 z_0^2 = t^2\left(\frac{x_0^2}{a^2} + \frac{y_0^2}{b^2}\right) = \frac{(tx_0)^2}{a^2} + \frac{(ty_0)^2}{b^2}.$$

Thus $t\mathbf{x}_0$ is on the graph.

To sketch the cone, we note that it meets the horizontal plane $z = 1$ in the ellipse

$$\frac{x^2}{a^2} + \frac{y^2}{b^2} = 1.$$

For each point on this ellipse, we draw the line through the point and **0**. See Fig. 6.3.

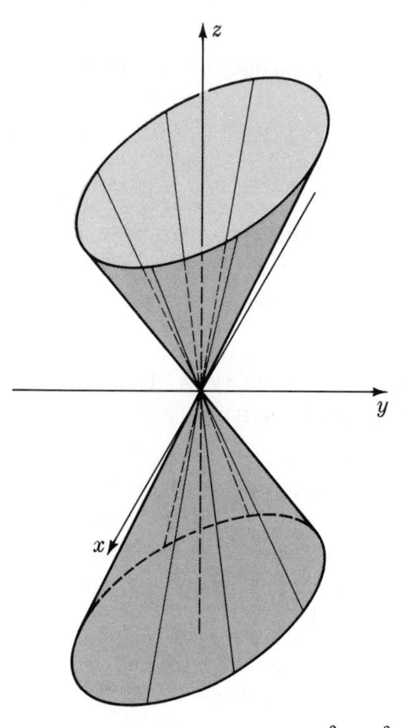

FIG. 6.3 quadratic cone: $z^2 = \frac{x^2}{a^2} + \frac{y^2}{b^2}$

Paraboloids

Next we take up case (2) on p. 254 and study the graph of $f = 0$, where

$$f(x, y, z) = Ax^2 + By^2 + rz, \qquad AB \neq 0, \qquad r \neq 0.$$

We may write the equation $f = 0$ in the form

$$z = \pm\frac{x^2}{a^2} \pm \frac{y^2}{b^2}, \qquad a, b > 0.$$

If both signs are minus, replacing z by $-z$ changes both signs to plus; therefore we are reduced to two cases.

The first case is the surface

$$z = \frac{x^2}{a^2} + \frac{y^2}{b^2}, \qquad a, b > 0,$$

called an **elliptic paraboloid**. It lies above the x, y-plane, and it is symmetric in the y, z- and z, x-planes. Each horizontal cross-section

$$\frac{x^2}{a^2} + \frac{y^2}{b^2} = z_0 > 0$$

is an ellipse, and these ellipses grow larger as z_0 increases. The graph meets the y, z- and z, x-planes in parabolas $z = y^2/b^2$ and $z = x^2/a^2$ respectively (Fig. 6.4a).

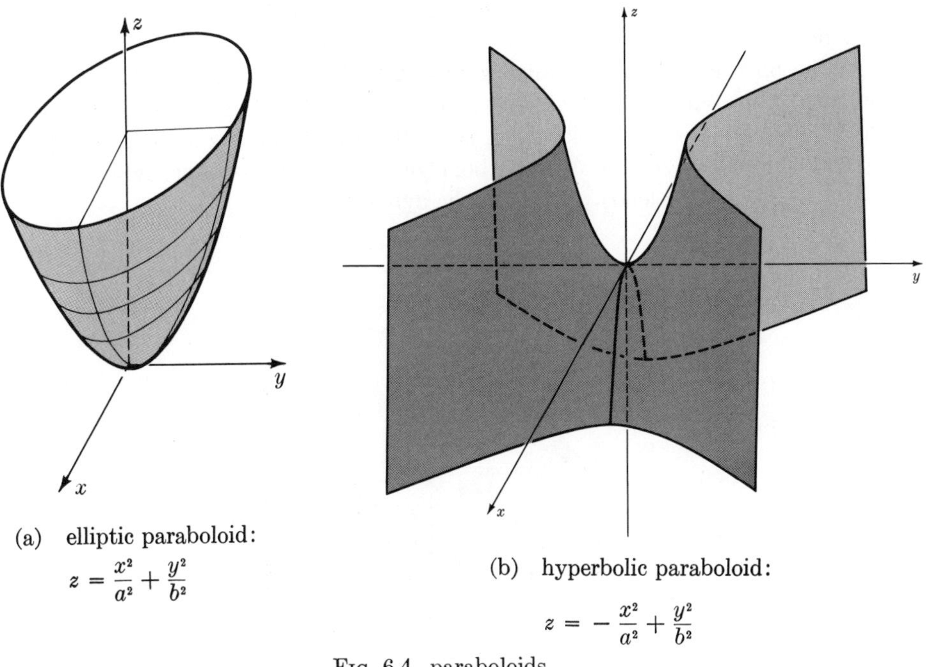

(a) elliptic paraboloid: $z = \dfrac{x^2}{a^2} + \dfrac{y^2}{b^2}$

(b) hyperbolic paraboloid: $z = -\dfrac{x^2}{a^2} + \dfrac{y^2}{b^2}$

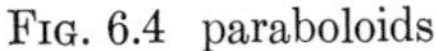

FIG. 6.4 paraboloids

The second case is the **hyperbolic paraboloid**, the locus of

$$z = -\frac{x^2}{a^2} + \frac{y^2}{b^2}, \qquad a, b > 0.$$

It is symmetric in the y, z- and z, x-planes. The horizontal planes $z = z_0 > 0$ meet it in hyperbolas whose branches open out in the y-direction. The horizontal planes $z = z_0 < 0$ meet it in hyperbolas that open out in the x-direction. The y, z-plane meets the locus in the parabola $z = y^2/b^2$, which opens upwards;

and the z, x-plane meets it in the parabola, $z = -x^2/a^2$, which opens downwards. The best description is "saddle-shaped". See Fig. 6.4b.

Cylinders

In cases (2′) and (3) on p. 254, the variable z is missing. In general, when one variable is missing the locus is a cylinder. Take for example $Ax^2 + By^2 + k = 0$, where $A > 0$, $B > 0$, and $k < 0$. This can be written in the form

$$\frac{x^2}{a^2} + \frac{y^2}{b^2} = 1, \qquad a, b > 0.$$

The surface meets each horizontal plane $z = z_0$ in the same ellipse. If (x_0, y_0, z_0) is any point of the surface, the whole vertical line (x_0, y_0, z) for $-\infty < z < \infty$ lies on the surface. The surface is an **elliptic cylinder** and these vertical lines that lie on the surface are called **generators** of the cylinder (Fig. 6.5). Any curve $f(x, y) = 0$ in the x, y-plane generates a cylinder in space consisting of all points (x_0, y_0, z) for which $f(x_0, y_0) = 0$. In particular, a circle generates a (right) circular cylinder, a hyperbola generates a hyperbolic cylinder, etc. Case (3) on p. 254 leads to a parabolic cylinder.

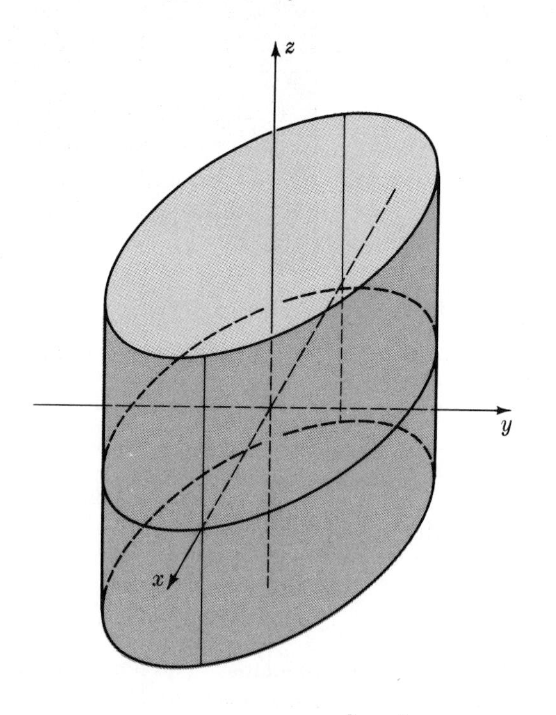

FIG. 6.5 elliptic cylinder: $\dfrac{x^2}{a^2} + \dfrac{y^2}{b^2} = 1$

In the final case (3′) both y and z are missing. Depending on the signs of A and k, the locus of $Ax^2 + k = 0$ is empty or consists of one plane or two planes parallel to the y, z-plane. In general, the locus in $\mathbf{R}^3$ of $f(x) = 0$ is a set of planes parallel to the y, z-plane. For each zero x_0 of $f(x)$, the plane $x = x_0$ is included in the locus.

EXERCISES

Sketch the first octant portion:

1. $\frac{1}{4}x^2 + y^2 + \frac{1}{4}z^2 = 1$
2. $x^2 + \frac{1}{9}y^2 + \frac{1}{4}z^2 = 1$
3. $x^2 + y^2 - z^2 = 1$
4. $-x^2 - y^2 + z^2 = 1$
5. $x^2 - y^2 + z^2 = 1$
6. $-x^2 + y^2 - z^2 = 1$
7. $z = x^2 + y^2$
8. $z = \frac{1}{4}x^2 + y^2$
9. $z = -x^2 + y^2$
10. $z = x^2 - y^2$.
11. Identify the quadric surface $z = x^2 + 2x + y^2$. [*Hint:* Complete the square.]
12. Identify the quadric surface $z = xy$. [*Hint:* Rotate 45° around the z-axis.]

Sketch the paraboloids:

13. $x = y^2 + z^2$
14. $y = x^2 - z^2$.

Sketch the surface in x, y, z-space:

15. $x - z = 1$
16. $y = x^2$
17. $xy = 1$
18. $-x^2 + y^2 = 1$
19. $x = z^2$
20. $y^2 + 4z^2 = 1$
21. $z^2 = x^2 + y^2$
22. $x^2 + 4z^2 = 1$
23. $y^2 = z^2 + 4x^2$
24. $z^2 = xy$.
25. Suppose $f(x, y) = 0$, $z = 1$ is a curve on the plane $z = 1$. Find an equation for the cone obtained by taking all points on all lines through $\mathbf{0}$ and points of the curve.
26. (cont.) Test your result on $x^2 + y^2 = 1$, $z = 1$.
27. Let $f(y, z) = 0$ be a curve in the y, z-plane. Find an equation for the surface of revolution obtained by revolving the curve around the y-axis.
28. (cont.) Test your result on the curve $y^2 + 9z^2 = 1$.

7. INVERSES

The square matrix I with ones on the main diagonal and zeros everywhere else is called the **identity matrix**. The 2×2 and 3×3 identity matrices are

$$\begin{bmatrix} 1 & 0 \\ 0 & 1 \end{bmatrix}, \quad \begin{bmatrix} 1 & 0 & 0 \\ 0 & 1 & 0 \\ 0 & 0 & 1 \end{bmatrix}.$$

The matrix I behaves very much like the number 1. If $\mathbf{x}'$ is a column vector, then $I\mathbf{x}' = \mathbf{x}'$; if A is a matrix (the same size as I) then $IA = AI = A$.

Here is an important practical question. Given a 2×2 or 3×3 matrix A, is there a matrix B (of the same size) such that $AB = I$ and $BA = I$? Not necessarily; for example, if

$$A = \begin{bmatrix} 1 & 0 \\ 0 & 0 \end{bmatrix} \quad \text{and} \quad B = \begin{bmatrix} b_{11} & b_{12} \\ b_{21} & b_{22} \end{bmatrix},$$

then

$$AB = \begin{bmatrix} b_{11} & b_{12} \\ 0 & 0 \end{bmatrix}, \qquad BA = \begin{bmatrix} b_{11} & 0 \\ b_{21} & 0 \end{bmatrix}.$$

Therefore, both equations $AB = I$ and $BA = I$ are impossible for any choice of B.

> Let A be a square matrix for which there exists a matrix B satisfying $AB = I$ and $BA = I$. Then we write $B = A^{-1}$ and call A^{-1} the **inverse** of A. Thus
>
> $$AA^{-1} = A^{-1}A = I.$$

Remark: It is reasonable to say "the" inverse. For if $AB = BA = I$, and also $AC = CA = I$, then

$$B = BI = B(AC) = (BA)C = IC = C.$$

Some matrices have inverses and some do not. How can we tell whether A^{-1} exists? We shall derive a simple test using determinants.

Recall that to each square matrix A is associated a real number $|A|$, the determinant of A. For instance $|I| = 1$, as is easily checked. One of the most important properties of determinants concerns the determinant of a product of matrices.

> **Theorem** If A and B are square matrices of the same size, then
>
> $$|AB| = |A|\,|B|.$$

The proof is not easy; we shall omit it.

Suppose A has an inverse, B. Then $AB = I$, so by the theorem above

$$|AB| = |I|, \qquad |A|\,|B| = 1.$$

Conclusion: $|A| \neq 0$. Thus for A to have an inverse, it is necessary that $|A| \neq 0$. We are going to show that this condition is sufficient as well. It will then follow that the matrices possessing inverses are precisely those whose determinants are non-zero.

Remark: The situation is similar to that for reciprocals of real numbers. The real number a has an inverse a^{-1} such that $aa^{-1} = a^{-1}a = 1$ if and only if $a \neq 0$.

Cofactor Matrix

Suppose $|A| \neq 0$. We shall prove that A^{-1} exists by actually computing it. The **cofactor matrix** cof A of a 2×2 matrix A is defined by

$$A = \begin{bmatrix} a & b \\ c & d \end{bmatrix}, \qquad \text{cof } A = \begin{bmatrix} d & -b \\ -c & a \end{bmatrix}.$$

For a 3×3 matrix, cof A is defined by

$$\text{cof } A = \begin{bmatrix} m_{11} & -m_{21} & m_{31} \\ -m_{12} & m_{22} & -m_{32} \\ m_{13} & -m_{23} & m_{33} \end{bmatrix}.$$

Here m_{ij} is the minor of a_{ij} in A. Note that the signs alternate and that the i, j-element of cof A is $\pm m_{ji}$, not $\pm m_{ij}$. Precisely, cof $A = [c_{ij}]$ where $c_{ij} = (-1)^{i+j} m_{ji}$.

The cofactor matrix cof A has an important relation to the original matrix A. We shall prove the basic formula:

$$\boxed{A(\text{cof } A) = (\text{cof } A)A = |A|\, I.}$$

This is easy for a 2×2 matrix:

$$A(\text{cof } A) = \begin{bmatrix} a & b \\ c & d \end{bmatrix} \begin{bmatrix} d & -b \\ -c & a \end{bmatrix} = \begin{bmatrix} ad - bc & 0 \\ 0 & ad - bc \end{bmatrix} = |A|\, I,$$

and similarly $(\text{cof } A)A = |A|\, I$.

For a 3×3 matrix we have

$$(\text{cof } A)A = \begin{bmatrix} m_{11} & -m_{21} & m_{31} \\ -m_{12} & m_{22} & -m_{32} \\ m_{13} & -m_{23} & m_{33} \end{bmatrix} \begin{bmatrix} a_{11} & a_{12} & a_{13} \\ a_{21} & a_{22} & a_{23} \\ a_{31} & a_{32} & a_{33} \end{bmatrix}.$$

Call the product $[b_{ij}]$. A typical element on the principal diagonal is

$$b_{11} = m_{11}a_{11} - m_{21}a_{21} + m_{31}a_{31}.$$

This is nothing else but the expansion of $|A|$ by minors of the first column. Therefore $b_{11} = |A|$.

A typical element off the diagonal is

$$b_{12} = m_{11}a_{12} - m_{21}a_{22} + m_{31}a_{32}.$$

It is the expansion by minors (of the first column) of

$$\begin{vmatrix} a_{12} & a_{12} & a_{13} \\ a_{22} & a_{22} & a_{23} \\ a_{32} & a_{32} & a_{33} \end{vmatrix}.$$

But this is the determinant of a matrix with two equal columns. Therefore $b_{12} = 0$.

We see that all elements on the principal diagonal of $(\operatorname{cof} A)A$ are $|A|$, and all elements off are 0. Therefore $(\operatorname{cof} A)A = |A|\, I$. Using expansion by minors of rows, we can prove $A(\operatorname{cof} A) = |A|\, I$ similarly.

Example:

$$A = \begin{bmatrix} 3 & 0 & -1 \\ 2 & 1 & -2 \\ 4 & -3 & 1 \end{bmatrix},$$

$$\operatorname{cof} A = \begin{bmatrix} +\begin{vmatrix} 1 & -2 \\ -3 & 1 \end{vmatrix} & -\begin{vmatrix} 0 & -1 \\ -3 & 1 \end{vmatrix} & +\begin{vmatrix} 0 & -1 \\ 1 & -2 \end{vmatrix} \\ -\begin{vmatrix} 2 & -2 \\ 4 & 1 \end{vmatrix} & +\begin{vmatrix} 3 & -1 \\ 4 & 1 \end{vmatrix} & -\begin{vmatrix} 3 & -1 \\ 2 & -2 \end{vmatrix} \\ +\begin{vmatrix} 2 & 1 \\ 4 & -3 \end{vmatrix} & -\begin{vmatrix} 3 & 0 \\ 4 & -3 \end{vmatrix} & +\begin{vmatrix} 3 & 0 \\ 2 & 1 \end{vmatrix} \end{bmatrix} = \begin{bmatrix} -5 & 3 & 1 \\ -10 & 7 & 4 \\ -10 & 9 & 3 \end{bmatrix},$$

$$A(\operatorname{cof} A) = -5I = (\operatorname{cof} A)A, \qquad |A| = -5.$$

Let us return to the question of inverses. Assume $|A| \neq 0$. From the rule

$$A(\operatorname{cof} A) = (\operatorname{cof} A)A = |A|\, I,$$

follows

$$A\left(\frac{1}{|A|}\operatorname{cof} A\right) = \left(\frac{1}{|A|}\operatorname{cof} A\right)A = I.$$

> A matrix A has an inverse if and only if $|A| \neq 0$. If $|A| \neq 0$, then
>
> $$A^{-1} = \frac{1}{|A|}\operatorname{cof} A.$$

Example 1:

$$A = \begin{bmatrix} 5 & 3 \\ -3 & 1 \end{bmatrix}, \qquad |A| = -4 \neq 0.$$

$$\text{cof } A = \begin{bmatrix} 1 & -3 \\ -3 & 5 \end{bmatrix}, \qquad A^{-1} = -\frac{1}{4}\begin{bmatrix} 1 & -3 \\ -3 & 5 \end{bmatrix}.$$

Check:

$$AA^{-1} = -\frac{1}{4}\begin{bmatrix} 5 & 3 \\ 3 & 1 \end{bmatrix}\begin{bmatrix} 1 & -3 \\ -3 & 5 \end{bmatrix} = -\frac{1}{4}\begin{bmatrix} -4 & 0 \\ 0 & -4 \end{bmatrix} = I.$$

Example 2:

$$A = \begin{bmatrix} 1 & 0 & 1 \\ 2 & -1 & 1 \\ -2 & -2 & 1 \end{bmatrix}, \qquad |A| = -5 \neq 0.$$

$$\text{cof } A = \begin{bmatrix} 1 & -2 & 1 \\ -4 & 3 & 1 \\ -6 & 2 & -1 \end{bmatrix}, \qquad A^{-1} = -\frac{1}{5}\begin{bmatrix} 1 & -2 & 1 \\ -4 & 3 & 1 \\ -6 & 2 & 1 \end{bmatrix}.$$

Check: $A^{-1}A =$

$$-\frac{1}{5}\begin{bmatrix} 1 & -2 & 1 \\ -4 & 3 & 1 \\ -6 & 2 & -1 \end{bmatrix}\begin{bmatrix} 1 & 0 & 1 \\ 2 & -1 & 1 \\ -2 & -2 & 1 \end{bmatrix} = -\frac{1}{5}\begin{bmatrix} -5 & 0 & 0 \\ 0 & -5 & 0 \\ 0 & 0 & -5 \end{bmatrix} = I.$$

TERMINOLOGY: A square matrix that *has* an inverse is called **non-singular**. A square matrix that does *not* have an inverse is called **singular**. Thus a square matrix A is non-singular if and only if $|A| \neq 0$, and A is singular if and only if $|A| = 0$.

REMARK: Given A, if there is a B with $AB = I$, then $|A| \neq 0$ so A has an inverse. This inverse must be B because,

$$A^{-1} = A^{-1}(I) = A^{-1}(AB) = (A^{-1}A)B = IB = B.$$

Similarly if $CA = I$, then A has an inverse and $A^{-1} = C$.

Linear Systems

A 3×3 linear system (three equations, three unknowns)

$$\begin{cases} a_{11}x_1 + a_{12}x_2 + a_{13}x_3 = b_1 \\ a_{21}x_1 + a_{22}x_2 + a_{23}x_3 = b_2 \\ a_{31}x_1 + a_{32}x_2 + a_{33}x_3 = b_3 \end{cases}$$

can be written neatly in matrix form as

$$A\mathbf{x}' = \mathbf{b}',$$

where

$$A = \begin{bmatrix} a_{11} & a_{12} & a_{13} \\ a_{21} & a_{22} & a_{23} \\ a_{31} & a_{32} & a_{33} \end{bmatrix}, \qquad \mathbf{x}' = \begin{bmatrix} x_1 \\ x_2 \\ x_3 \end{bmatrix}, \qquad \mathbf{b}' = \begin{bmatrix} b_1 \\ b_2 \\ b_3 \end{bmatrix}.$$

Suppose the matrix A is non-singular. Then we can multiply both sides of the equation by A^{-1}:

$$A^{-1}(A\mathbf{x}') = A^{-1}\mathbf{b}', \qquad I\mathbf{x}' = A^{-1}\mathbf{b}', \qquad \mathbf{x}' = A^{-1}\mathbf{b}'.$$

Therefore, if there is a solution, it can only be the vector $\mathbf{x}' = A^{-1}\mathbf{b}'$. But this vector is in fact a solution. Check:

$$A\mathbf{x}' = A(A^{-1}\mathbf{b}') = (AA^{-1})\mathbf{b}' = I\mathbf{b}' = \mathbf{b}'.$$

A linear system $A\mathbf{x}' = \mathbf{b}'$ with A non-singular has the unique solution

$$\mathbf{x}' = A^{-1}\mathbf{b}'.$$

REMARK: Note the formal similarity between the linear system $A\mathbf{x}' = \mathbf{b}'$ and the elementary equation $ax = b$. The first is solved by multiplying both sides by A^{-1}, the second by multiplying both sides by $1/a = a^{-1}$.

EXAMPLE 7.1

Solve

$$\begin{bmatrix} 3 & 0 & -1 \\ 2 & 1 & -2 \\ 4 & -3 & 1 \end{bmatrix} \begin{bmatrix} x_1 \\ x_2 \\ x_3 \end{bmatrix} = \begin{bmatrix} 15 \\ 0 \\ 5 \end{bmatrix}.$$

Solution: Evaluate $|A|$, say by minors of the first row:

$$|A| = 3(-5) + 0 + (-1)(-10) = -5.$$

Since $|A| \neq 0$, the matrix A is non-singular, and

$$A^{-1} = \frac{1}{|A|}\operatorname{cof} A = -\frac{1}{5}\begin{bmatrix} -5 & 3 & 1 \\ -10 & 7 & 4 \\ -10 & 9 & 3 \end{bmatrix}.$$

Therefore the solution is

$$\mathbf{x}' = A^{-1}\mathbf{b}' = -\frac{1}{5}\begin{bmatrix} -5 & 3 & 1 \\ -10 & 7 & 4 \\ -10 & 9 & 3 \end{bmatrix}\begin{bmatrix} 15 \\ 0 \\ 5 \end{bmatrix} = \begin{bmatrix} 14 \\ 26 \\ 27 \end{bmatrix}.$$

Answer: $x_1 = 14,\ x_2 = 26,\ x_3 = 27.$

The solution of $A\mathbf{x}' = \mathbf{b}'$ is

$$\mathbf{x}' = A^{-1}\mathbf{b}' = \frac{1}{|A|}(\operatorname{cof} A)\mathbf{b}'.$$

From this formula follows Cramer's Rule for the solution of a linear system. If m_{ij} denotes the minor of a_{ij} in A, then

$$(\operatorname{cof} A)\mathbf{b}' = \begin{bmatrix} m_{11} & -m_{21} & m_{31} \\ -m_{12} & m_{22} & -m_{32} \\ m_{13} & -m_{23} & m_{33} \end{bmatrix}\begin{bmatrix} b_1 \\ b_2 \\ b_3 \end{bmatrix} = \begin{bmatrix} m_{11}b_1 - m_{21}b_2 + m_{31}b_3 \\ -m_{12}b_1 + m_{22}b_2 - m_{32}b_3 \\ m_{13}b_1 - m_{23}b_2 + m_{33}b_3 \end{bmatrix}.$$

Look carefully at the first element, $m_{11}b_1 - m_{21}b_2 + m_{31}b_3$. It is the value of the determinant

$$\begin{vmatrix} b_1 & a_{12} & a_{13} \\ b_2 & a_{22} & a_{23} \\ b_2 & a_{32} & a_{33} \end{vmatrix}$$

obtained as the expansion by minors of the first column. A similar statement applies to the other elements. The result is a set of explicit formulas:

> Let $A = (\mathbf{c}_1', \mathbf{c}_2', \mathbf{c}_3')$ be a 3×3 non-singular matrix. Then the solution of $A\mathbf{x}' = \mathbf{b}'$ is given by $[x_1, x_2, x_3]'$, where
>
> $$x_1 = \frac{D(\mathbf{b}', \mathbf{c}_2', \mathbf{c}_3')}{|A|}, \qquad x_2 = \frac{D(\mathbf{c}_1', \mathbf{b}', \mathbf{c}_3')}{|A|}, \qquad x_3 = \frac{D(\mathbf{c}_1', \mathbf{c}_2', \mathbf{b}')}{|A|}.$$
>
> In other words, to obtain x_i, replace the i-th column of A by $\mathbf{b}'$, take the determinant, then divide by $|A|$.

EXERCISES

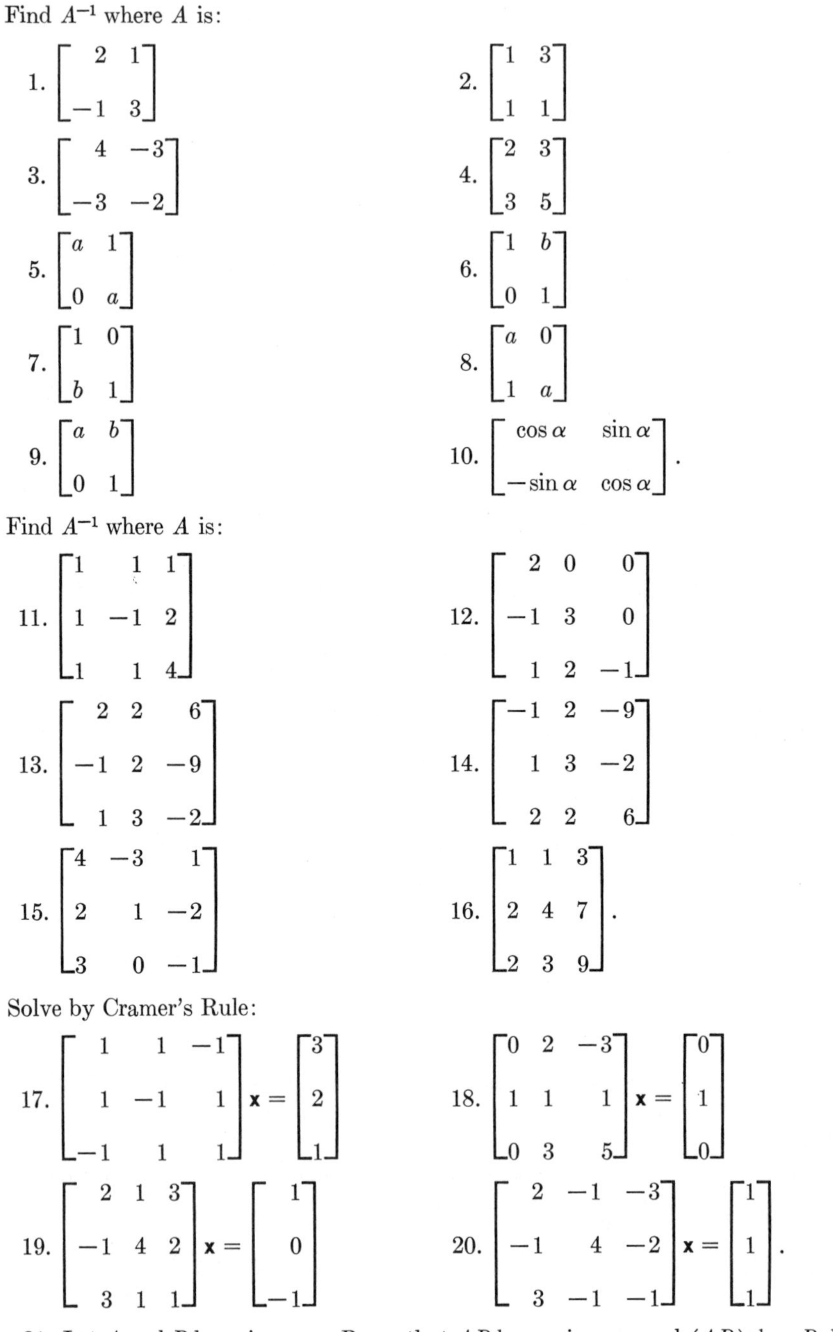

Find A^{-1} where A is:

1. $\begin{bmatrix} 2 & 1 \\ -1 & 3 \end{bmatrix}$

2. $\begin{bmatrix} 1 & 3 \\ 1 & 1 \end{bmatrix}$

3. $\begin{bmatrix} 4 & -3 \\ -3 & -2 \end{bmatrix}$

4. $\begin{bmatrix} 2 & 3 \\ 3 & 5 \end{bmatrix}$

5. $\begin{bmatrix} a & 1 \\ 0 & a \end{bmatrix}$

6. $\begin{bmatrix} 1 & b \\ 0 & 1 \end{bmatrix}$

7. $\begin{bmatrix} 1 & 0 \\ b & 1 \end{bmatrix}$

8. $\begin{bmatrix} a & 0 \\ 1 & a \end{bmatrix}$

9. $\begin{bmatrix} a & b \\ 0 & 1 \end{bmatrix}$

10. $\begin{bmatrix} \cos\alpha & \sin\alpha \\ -\sin\alpha & \cos\alpha \end{bmatrix}$.

Find A^{-1} where A is:

11. $\begin{bmatrix} 1 & 1 & 1 \\ 1 & -1 & 2 \\ 1 & 1 & 4 \end{bmatrix}$

12. $\begin{bmatrix} 2 & 0 & 0 \\ -1 & 3 & 0 \\ 1 & 2 & -1 \end{bmatrix}$

13. $\begin{bmatrix} 2 & 2 & 6 \\ -1 & 2 & -9 \\ 1 & 3 & -2 \end{bmatrix}$

14. $\begin{bmatrix} -1 & 2 & -9 \\ 1 & 3 & -2 \\ 2 & 2 & 6 \end{bmatrix}$

15. $\begin{bmatrix} 4 & -3 & 1 \\ 2 & 1 & -2 \\ 3 & 0 & -1 \end{bmatrix}$

16. $\begin{bmatrix} 1 & 1 & 3 \\ 2 & 4 & 7 \\ 2 & 3 & 9 \end{bmatrix}$.

Solve by Cramer's Rule:

17. $\begin{bmatrix} 1 & 1 & -1 \\ 1 & -1 & 1 \\ -1 & 1 & 1 \end{bmatrix} \mathbf{x} = \begin{bmatrix} 3 \\ 2 \\ 1 \end{bmatrix}$

18. $\begin{bmatrix} 0 & 2 & -3 \\ 1 & 1 & 1 \\ 0 & 3 & 5 \end{bmatrix} \mathbf{x} = \begin{bmatrix} 0 \\ 1 \\ 0 \end{bmatrix}$

19. $\begin{bmatrix} 2 & 1 & 3 \\ -1 & 4 & 2 \\ 3 & 1 & 1 \end{bmatrix} \mathbf{x} = \begin{bmatrix} 1 \\ 0 \\ -1 \end{bmatrix}$

20. $\begin{bmatrix} 2 & -1 & -3 \\ -1 & 4 & -2 \\ 3 & -1 & -1 \end{bmatrix} \mathbf{x} = \begin{bmatrix} 1 \\ 1 \\ 1 \end{bmatrix}$.

21. Let A and B have inverses. Prove that AB has an inverse and $(AB)^{-1} = B^{-1}A^{-1}$.

22. (cont.) Prove that $\operatorname{cof}(AB) = (\operatorname{cof} B)(\operatorname{cof} A)$ if A and B are non-singular.

23. Let A have an inverse. Prove that its transpose A' has an inverse and $(A')^{-1} = (A^{-1})'$.
24. Prove that a triangular matrix

$$A = \begin{bmatrix} a_{11} & a_{12} & a_{13} \\ 0 & a_{22} & a_{23} \\ 0 & 0 & a_{33} \end{bmatrix}$$

has an inverse if and only if $a_{11}a_{22}a_{33} \neq 0$. If this condition is satisfied, prove that A^{-1} is also triangular.

The following exercises outline another method for inverting matrices. Set

$$P_{12} = \begin{bmatrix} 0 & 1 & 0 \\ 1 & 0 & 0 \\ 0 & 0 & 1 \end{bmatrix}, \quad D_3(c) = \begin{bmatrix} 1 & 0 & 0 \\ 0 & 1 & 0 \\ 0 & 0 & c \end{bmatrix}, \quad R_{31}(c) = \begin{bmatrix} 1 & 0 & c \\ 0 & 1 & 0 \\ 0 & 0 & 1 \end{bmatrix}.$$

25. Show that $P_{12}A$ is the result of interchanging row 1 and row 2 of A.
26. Show that $D_3(c)A$ is the result of multiplying row 3 of A by c.
27. Show that $R_{31}(c)A$ is the result of adding c times row 3 of A to row 1.
28. Write out P_{23}, $D_1(c)$, and $R_{12}(c)$.

29*. Let A be non-singular. Show there are matrices $Q_1, Q_2, \cdots, Q_n$ such that
 (1) $Q_nQ_{n-1}\cdots Q_2Q_1A = I$,
 (2) each Q is a P_{ij} or a $D_i(c)$ with $c \neq 0$ or an $R_{ij}(c)$.

30. (cont.) Interpret this statement as follows:
Apply a sequence of elementary row operations to A (interchange rows, multiply a row by a non-zero constant, add a multiple of one row to another) so as to change A to I. Apply the same sequence of operations to I. The result is A^{-1}.

Try the method of Ex. 30 on the A of:

31. Ex. 1
32. Ex. 2
33. Ex. 11
34. Ex. 12
35. Ex. 13
36. Ex. 16.

Let $A = A(t) = [a_{ij}(t)]$ be a matrix whose elements are differentiable functions. Define its **derivative** by $A^{\cdot} = [\dot{a}_{ij}(t)]$. Prove:

37. $(A + B)^{\cdot} = A^{\cdot} + B^{\cdot}$
38. $(fA)^{\cdot} = f^{\cdot}A + fA^{\cdot}$
39. $(AB)^{\cdot} = A^{\cdot}B + AB^{\cdot}$
40. $(A^{-1})^{\cdot} = -A^{-1}A^{\cdot}A^{-1}$.

8. CHARACTERISTIC ROOTS [optional]

Let A be a 2×2 matrix. A **characteristic vector** of A is a non-zero column vector $\mathbf{x}'$ such that $A\mathbf{x}'$ is a multiple of $\mathbf{x}'$:

$$A\mathbf{x}' = \lambda\mathbf{x}',$$

where λ is a scalar.

Given a vector $\mathbf{x}'$, the vector $A\mathbf{x}'$ generally does not resemble $\mathbf{x}'$ at all, whereas for a characteristic vector, $A\mathbf{x}'$ is simply a multiple of $\mathbf{x}'$. Therefore a characteristic vector is a rare animal, in fact, it is not obvious whether A has any characteristic vectors at all. One thing is clear though: if $\mathbf{x}'$ *is* a characteristic vector, then so is any non-zero multiple $\mathbf{y}' = c\mathbf{x}'$. For

$$A\mathbf{y}' = A(c\mathbf{x}') = c(A\mathbf{x}') = c(\lambda\mathbf{x}') = \lambda(c\mathbf{x}') = \lambda\mathbf{y}'.$$

Let us investigate the existence of characteristic vectors. We write the condition $A\mathbf{x}' = \lambda\mathbf{x}'$ in the form

$$(\lambda I - A)\mathbf{x}' = \mathbf{0},$$

and ask whether there is a non-zero $\mathbf{x}'$ that satisfies this homogeneous system for some scalar λ. Set

$$A = \begin{bmatrix} a & b \\ c & d \end{bmatrix} \quad \text{and} \quad \mathbf{x}' = \begin{bmatrix} x \\ y \end{bmatrix}.$$

The question we ask is whether there is a scalar λ for which the homogeneous linear system

$$\begin{bmatrix} \lambda - a & b \\ c & \lambda - d \end{bmatrix} \begin{bmatrix} x \\ y \end{bmatrix} = \begin{bmatrix} 0 \\ 0 \end{bmatrix}$$

has a non-trivial solution. The answer is yes if and only if there is a λ for which the determinant of the system is 0. Thus λ must satisfy $f(\lambda) = 0$, where

$$f(t) = |tI - A| = \begin{vmatrix} t - a & -b \\ -c & t - d \end{vmatrix}$$

$$= (t - a)(t - d) - bc = t^2 - (a + d)t + (ad - bc).$$

The quadratic polynomial $f(t)$ is called the **characteristic polynomial** of A, and its zeros are called the **characteristic roots** of A.

Conclusion: For each characteristic root λ of A, there is a characteristic vector $\mathbf{x}'$ with $A\mathbf{x}' = \lambda\mathbf{x}'$; conversely, each characteristic vector corresponds in this way to a characteristic root.

Terminology: Several adjectives are often used where we have used "characteristic". The most common ones are "proper", "eigen" (from German), and "latent". Characteristic roots are also called characteristic "values".

EXAMPLE 8.1

Find all characteristic roots and vectors of

$$A = \begin{bmatrix} 2 & -1 \\ 4 & -3 \end{bmatrix}.$$

Solution:

$$f(t) = |tI - A| = \begin{vmatrix} t-2 & 1 \\ -4 & t+3 \end{vmatrix} = (t-2)(t+3) + 4 = t^2 + t - 2$$

$$= (t-1)(t+2).$$

The characteristic roots are $\lambda = 1$ and $\mu = -2$. The corresponding homogeneous systems $(\lambda I - A)\mathbf{x}' = \mathbf{0}$ and $(\mu I - A)\mathbf{y}' = \mathbf{0}$ are

$$\begin{bmatrix} -1 & 1 \\ -4 & 4 \end{bmatrix} \mathbf{x}' = \mathbf{0} \quad \text{and} \quad \begin{bmatrix} -4 & 1 \\ -4 & 1 \end{bmatrix} \mathbf{y}' = \mathbf{0}.$$

Obvious non-trivial solutions are

$$\mathbf{x}' = \begin{bmatrix} 1 \\ 1 \end{bmatrix} \quad \text{and} \quad \mathbf{y}' = \begin{bmatrix} 1 \\ 4 \end{bmatrix}.$$

Any non-zero multiple is also a characteristic vector.

Answer: $\lambda = 1, \quad \mathbf{x}' = a \begin{bmatrix} 1 \\ 1 \end{bmatrix}; \quad \mu = -2, \quad \mathbf{y}' = b \begin{bmatrix} 1 \\ 4 \end{bmatrix}.$

The characteristic polynomial $f(t)$ of a 2×2 matrix A is quadratic. It has two complex roots which may or may not be real. Hence A will always have a characteristic vector if we admit complex scalars and vectors with complex entries. Otherwise, the discussion that follows applies only to matrices with real characteristic roots. (See Chapter 15 for complex numbers.)

Now $f(t) = t^2 - (a + d)t + (ad - bc)$, so the two characteristic roots are real provided

$$\Delta = (a + d)^2 - 4(ad - bc) \geq 0.$$

The roots are distinct if $\Delta > 0$, equal if $\Delta = 0$.

Basic Structure Theorems

Suppose a certain problem involves a 2×2 matrix about which we know nothing. Now the general 2×2 matrix has four arbitrary elements, so is awkward to handle. The following results show that in many situations a general 2×2 matrix can be replaced by one of two simple types,

$$\begin{bmatrix} \lambda & 0 \\ 0 & \mu \end{bmatrix} \quad \text{or} \quad \begin{bmatrix} \lambda & 1 \\ 0 & \lambda \end{bmatrix}.$$

There are many practical applications.

Theorem 8.1 Let A be a 2×2 matrix with *distinct* characteristic roots λ and μ. Then there is a non-singular matrix P such that

$$P^{-1}AP = \begin{bmatrix} \lambda & 0 \\ 0 & \mu \end{bmatrix}.$$

Proof: Let $\mathbf{x}'$ and $\mathbf{y}'$ be corresponding characteristic vectors. Then

$$A\mathbf{x}' = \lambda\mathbf{x}', \qquad A\mathbf{y}' = \mu\mathbf{y}',$$

$\mathbf{x}' \neq \mathbf{0}$, and $\mathbf{y}' \neq \mathbf{0}$. Set $P = (\mathbf{x}', \mathbf{y}')$, considered as a 2×2 matrix. Then

$$AP = A(\mathbf{x}', \mathbf{y}') = (A\mathbf{x}', A\mathbf{y}') = (\lambda\mathbf{x}', \mu\mathbf{y}') = (\mathbf{x}', \mathbf{y}')\begin{bmatrix} \lambda & 0 \\ 0 & \mu \end{bmatrix} = P\begin{bmatrix} \lambda & 0 \\ 0 & \mu \end{bmatrix}.$$

If P is non-singular, then we multiply the last equation by P^{-1}:

$$P^{-1}AP = P^{-1}(AP) = P^{-1}\left(P\begin{bmatrix} \lambda & 0 \\ 0 & \mu \end{bmatrix}\right)$$

$$= (P^{-1}P)\begin{bmatrix} \lambda & 0 \\ 0 & \mu \end{bmatrix} = I\begin{bmatrix} \lambda & 0 \\ 0 & \mu \end{bmatrix} = \begin{bmatrix} \lambda & 0 \\ 0 & \mu \end{bmatrix}.$$

It remains to prove that P is non-singular. If not, then the homogeneous system

$$P\begin{bmatrix} a \\ b \end{bmatrix} = 0$$

has a non-trivial solution. That is, there exist a and b, not both zero, such that $a\mathbf{x}' + b\mathbf{y}' = \mathbf{0}$. Multiply by A:

$$aA\mathbf{x}' + bA\mathbf{y}' = \mathbf{0}, \qquad a\lambda\mathbf{x}' + b\mu\mathbf{y}' = \mathbf{0}.$$

Now eliminate $\mathbf{x}'$ from $a\mathbf{x}' + b\mathbf{y}' = \mathbf{0}$ and $a\lambda\mathbf{x}' + b\mu\mathbf{y}' = \mathbf{0}$:

$$b(\lambda - \mu)\mathbf{y}' = \mathbf{0}.$$

Since $\lambda \neq \mu$ and $\mathbf{y}' \neq \mathbf{0}$, we have $b = 0$. Similarly $a = 0$, a contradiction.

EXAMPLE 8.2

Apply the theorem to the matrix of Example 8.1.

Solution: The matrix is

$$A = \begin{bmatrix} 2 & -1 \\ 4 & -3 \end{bmatrix},$$

and we found the characteristic vectors

$$\mathbf{x}' = \begin{bmatrix} 1 \\ 1 \end{bmatrix} \quad \text{and} \quad \mathbf{y}' = \begin{bmatrix} 1 \\ 4 \end{bmatrix}$$

corresponding to $\lambda = 1$ and $\mu = -2$. Set

$$P = (\mathbf{x}', \mathbf{y}') = \begin{bmatrix} 1 & 1 \\ 1 & 4 \end{bmatrix}.$$

Then $|P| = 3 \neq 0$ and

$$AP = \begin{bmatrix} 2 & -1 \\ 4 & -3 \end{bmatrix} \begin{bmatrix} 1 & 1 \\ 1 & 4 \end{bmatrix} = \begin{bmatrix} 1 & -2 \\ 1 & -8 \end{bmatrix},$$

$$P \begin{bmatrix} 1 & 0 \\ 0 & -2 \end{bmatrix} = \begin{bmatrix} 1 & 1 \\ 1 & 4 \end{bmatrix} \begin{bmatrix} 1 & 0 \\ 0 & -2 \end{bmatrix} = \begin{bmatrix} 1 & -2 \\ 1 & -8 \end{bmatrix},$$

$$AP = P \begin{bmatrix} 1 & 0 \\ 0 & -2 \end{bmatrix}, \qquad P^{-1}AP = \begin{bmatrix} 1 & 0 \\ 0 & -2 \end{bmatrix}.$$

Theorem 8.2 Let A be a 2×2 matrix with *equal* characteristic roots λ, λ. Then there is a non-singular matrix P such that either

$$P^{-1}AP = \begin{bmatrix} \lambda & 0 \\ 0 & \lambda \end{bmatrix} \quad \text{or} \quad P^{-1}AP = \begin{bmatrix} \lambda & 1 \\ 0 & \lambda \end{bmatrix}.$$

Proof: Let $\mathbf{x}'$ be a characteristic vector, so $\mathbf{x}' \neq \mathbf{0}$ and $A\mathbf{x}' = \lambda\mathbf{x}'$. Choose *any* vector $\mathbf{y}'$ so that $P = (\mathbf{x}', \mathbf{y}')$ is non-singular. Then $\mathbf{x}'$ and $\mathbf{y}'$ are non-collinear with $\mathbf{0}$, so the vector $A\mathbf{y}'$ can be expressed as $A\mathbf{y}' = a\mathbf{x}' + b\mathbf{y}'$. In fact, if we set

$$P^{-1}A\mathbf{y}' = \begin{bmatrix} a \\ b \end{bmatrix},$$

then

$$A\mathbf{y}' = (PP^{-1})(A\mathbf{y}') = P(P^{-1}A\mathbf{y}') = P \begin{bmatrix} a \\ b \end{bmatrix} = a\mathbf{x}' + b\mathbf{y}'.$$

Therefore

$$AP = A(\mathbf{x}', \mathbf{y}') = (A\mathbf{x}', A\mathbf{y}') = (\lambda\mathbf{x}', a\mathbf{x}' + b\mathbf{y}')$$

$$= (\mathbf{x}', \mathbf{y}')\begin{bmatrix} \lambda & a \\ 0 & b \end{bmatrix} = P\begin{bmatrix} \lambda & a \\ 0 & b \end{bmatrix},$$

hence $P^{-1}AP = \begin{bmatrix} \lambda & a \\ 0 & b \end{bmatrix}.$

Now an important observation: $P^{-1}AP$ and A have the same characteristic polynomial. For

$$|tI - P^{-1}AP| = |tP^{-1}P - P^{-1}AP|$$

$$= |P^{-1}(tI - A)P| = |P^{-1}|\,|tI - A|\,|P|$$

$$= |tI - A|\,|P^{-1}|\,|P| = |tI - A|\,|P^{-1}P| = |tI - A|.$$

Let $f(t)$ denote this common characteristic polynomial. On the one hand, the zeros of $f(t)$ are λ, λ by hypothesis. On the other hand,

$$f(t) = |tI - P^{-1}AP| = \begin{vmatrix} t-\lambda & -a \\ 0 & t-b \end{vmatrix} = (t-\lambda)(t-b),$$

so the zeros are λ, b. Hence $\lambda = b$. Therefore

$$\begin{cases} A\mathbf{x}' = \lambda\mathbf{x}' \\ A\mathbf{y}' = a\mathbf{x}' + \lambda\mathbf{y}' \end{cases} \qquad \text{and} \qquad P^{-1}AP = \begin{bmatrix} \lambda & a \\ 0 & \lambda \end{bmatrix}.$$

If $a = 0$, we are done. If $a \neq 0$, we simply repeat the argument with $\mathbf{x}'$ replaced by $a\mathbf{x}'$. Then

$$A\mathbf{x}' = \lambda\mathbf{x}', \qquad A\mathbf{y}' = \mathbf{x}' + \lambda\mathbf{y}'$$

so, with P modified accordingly,

$$P^{-1}AP = \begin{bmatrix} \lambda & 1 \\ 0 & \lambda \end{bmatrix}.$$

This completes the proof.

EXAMPLE 8.3

Apply the theorem to

$$A = \begin{bmatrix} -1 & 0 \\ 3 & -1 \end{bmatrix}.$$

Solution: The characteristic polynomial is $f(t) = (t+1)^2$ so the characteristic roots are $-1, -1$. Let $\mathbf{x}'$ be a characteristic vector. Then

$$\lambda I - A = -I - A = \begin{bmatrix} 0 & 0 \\ -3 & 0 \end{bmatrix},$$

so we must have

$$\begin{bmatrix} 0 & 0 \\ -3 & 0 \end{bmatrix} \mathbf{x}' = \mathbf{0}.$$

An obvious solution is

$$\mathbf{x}' = \begin{bmatrix} 0 \\ 1 \end{bmatrix}.$$

The easiest choice of $\mathbf{y}'$ is

$$\mathbf{y}' = \begin{bmatrix} 1 \\ 0 \end{bmatrix}, \quad \text{so} \quad P = (\mathbf{x}', \mathbf{y}') = \begin{bmatrix} 0 & 1 \\ 1 & 0 \end{bmatrix}.$$

Then

$$A\mathbf{x}' = -\mathbf{x}' \quad \text{and} \quad A\mathbf{y}' = \begin{bmatrix} -1 & 0 \\ 3 & -1 \end{bmatrix}\begin{bmatrix} 1 \\ 0 \end{bmatrix} = \begin{bmatrix} -1 \\ 3 \end{bmatrix} = 3\mathbf{x}' - \mathbf{y}'.$$

Therefore

$$AP = A(\mathbf{x}', \mathbf{y}') = (A\mathbf{x}', A\mathbf{y}') = (-\mathbf{x}', 3\mathbf{x}' - \mathbf{y}')$$

$$= (\mathbf{x}', \mathbf{y}')\begin{bmatrix} -1 & 3 \\ 0 & -1 \end{bmatrix} = P\begin{bmatrix} -1 & 3 \\ 0 & -1 \end{bmatrix}.$$

We replace $\mathbf{x}'$ by $3\mathbf{x}'$, that is, we now take

$$\mathbf{x}' = \begin{bmatrix} 0 \\ 3 \end{bmatrix}, \quad \mathbf{y}' = \begin{bmatrix} 1 \\ 0 \end{bmatrix}, \quad P = \begin{bmatrix} 0 & 1 \\ 3 & 0 \end{bmatrix}.$$

After these changes,

$$A\mathbf{x}' = -\mathbf{x}' \quad \text{and} \quad A\mathbf{y}' = \begin{bmatrix} -1 \\ 3 \end{bmatrix} = \mathbf{x}' - \mathbf{y}',$$

so

$$AP = P\begin{bmatrix} -1 & 1 \\ 0 & -1 \end{bmatrix}, \quad P^{-1}AP = \begin{bmatrix} -1 & 1 \\ 0 & -1 \end{bmatrix}.$$

We shall apply these results in Chapter 15, Section 6 to systems of differential equations.

3×3 Matrices

For a 3×3 matrix A, characteristic vectors and roots are defined just as for 2×2 matrices. A **characteristic vector** of A is a column vector $\mathbf{x}'$ such that $A\mathbf{x}' = \lambda\mathbf{x}'$ for some scalar λ. Searching for characteristic vectors leads to the characteristic polynomial $f(t) = |tI - A|$, whose zeros are the characteristic roots of A.

Now

$$f(t) = \begin{vmatrix} t - a_{11} & -a_{12} & -a_{13} \\ -a_{21} & t - a_{22} & -a_{23} \\ -a_{31} & -a_{32} & t - a_{33} \end{vmatrix} = t^3 + bt^2 + ct + d,$$

a cubic polynomial with real coefficients. Since f has odd degree, there must be at least one *real* characteristic root. In general, the cubic has either one real and two complex zeros or three real zeros.

Theorem 8.3 Each 3×3 matrix has a real characteristic root.

Corresponding to the structure theorems above is a more complicated result which we state without proof.

Theorem 8.4 Let A be a 3×3 matrix, and let λ, μ, ν be the characteristic roots of A, not necessarily all real or distinct. Then there is a non-singular matrix P such that $P^{-1}AP$ has one of the following forms:

$$\begin{bmatrix} \lambda & 0 & 0 \\ 0 & \mu & 0 \\ 0 & 0 & \nu \end{bmatrix}, \qquad \begin{bmatrix} \lambda & 1 & 0 \\ 0 & \lambda & 0 \\ 0 & 0 & \mu \end{bmatrix}, \qquad \begin{bmatrix} \lambda & 1 & 0 \\ 0 & \lambda & 1 \\ 0 & 0 & \lambda \end{bmatrix}.$$

REMARK: There is only one possible basic form if all three characteristic roots are distinct, but three basic forms if the three characteristic roots are equal, and two basic forms if two roots are equal and the third distinct.

Symmetric Matrices

All characteristic roots of a symmetric matrix are real. That is, if A is an $n \times n$ symmetric matrix (with real coefficients), then the characteristic polynomial $f(t)$ factors completely,

$$f(t) = (t - \lambda_1)(t - \lambda_2) \cdots (t - \lambda_n),$$

where $\lambda_1, \cdots, \lambda_n$ are *real*.

The proof of this basic fact, while not hard, belongs to a course in linear algebra. The 2×2 case is easy. For if

$$A = \begin{bmatrix} a & b \\ b & c \end{bmatrix},$$

then

$$f(t) = \begin{vmatrix} t - a & -b \\ -b & t - c \end{vmatrix} = (t - a)(t - c) - b^2$$

$$= t^2 - (a + c)t + (ac - b^2),$$

and the discriminant is

$$\Delta = (a + c)^2 - 4(ac - b^2) = (a - c)^2 + 4b^2 \geq 0.$$

Therefore the roots are real.

We shall prove the 3×3 case in Chapter 9, Section 8, but the analytic proof there does not generalize to 4×4 or higher cases. We state the result formally.

Theorem 8.5 All characteristic roots of a 2×2 or 3×3 symmetric matrix are real.

The basic structure theorems take a particularly simple form for *symmetric* matrices. The possibilities

$$\begin{bmatrix} \lambda & 1 \\ 0 & \lambda \end{bmatrix}, \quad \begin{bmatrix} \lambda & 1 & 0 \\ 0 & \lambda & 0 \\ 0 & 0 & \mu \end{bmatrix}, \quad \begin{bmatrix} \lambda & 1 & 0 \\ 0 & \lambda & 1 \\ 0 & 0 & \lambda \end{bmatrix}$$

simply never occur! The following theorem summarizes the situation.

Theorem 8.6 If A is a 2×2 symmetric matrix, then there is a non-singular matrix P such that

$$P^{-1}AP = \begin{bmatrix} \lambda & 0 \\ 0 & \mu \end{bmatrix}.$$

If A is a 3×3 symmetric matrix, then there is a non-singular matrix P such that

$$P^{-1}AP = \begin{bmatrix} \lambda & 0 & 0 \\ 0 & \mu & 0 \\ 0 & 0 & \nu \end{bmatrix}.$$

The proof for 3×3 matrices is beyond the scope of this course. See Ex. 35 for the 2×2 case.

Now we come to an important relation between characteristic roots and definiteness.

Theorem 8.7 Let A be a symmetric matrix with characteristic roots $\lambda_1, \lambda_2, \cdots$. Then

(1) A is positive (negative) definite if and only if all $\lambda_i > 0$ (all $\lambda_i < 0$);

(2) A is positive (negative) semi-definite if and only if all $\lambda_i \geq 0$ (all $\lambda_i \leq 0$);

(3) A is indefinite if and only if some $\lambda_i > 0$ and some $\lambda_j < 0$.

We shall prove only a small part of this result because the complete proof, especially for 3×3 and higher, is hard. We shall show that if A is positive definite, then all $\lambda_i > 0$. The negative definite case is similar.

Thus let A be positive definite and let λ be any characteristic root. There is an $\mathbf{x}' \neq \mathbf{0}$ such that $A\mathbf{x}' = \lambda\mathbf{x}'$. Therefore

$$\mathbf{x}A\mathbf{x}' = \lambda\mathbf{x}\mathbf{x}'.$$

But $\mathbf{x}A\mathbf{x}' > 0$ because A is positive definite and $\mathbf{x}' \neq \mathbf{0}$; also $\mathbf{x}\mathbf{x}' > 0$ because $\mathbf{x}' \neq \mathbf{0}$. Therefore $\lambda > 0$.

EXERCISES

Find all characteristic roots and vectors:

1. $\begin{bmatrix} 2 & 0 \\ 0 & -3 \end{bmatrix}$

2. $\begin{bmatrix} 3 & 0 \\ 0 & 1 \end{bmatrix}$

3. $\begin{bmatrix} 3 & 1 \\ 0 & 2 \end{bmatrix}$

4. $\begin{bmatrix} 1 & 1 \\ 0 & 1 \end{bmatrix}$

5. $\begin{bmatrix} -1 & 0 \\ 1 & -1 \end{bmatrix}$

6. $\begin{bmatrix} 1 & 0 \\ -1 & -1 \end{bmatrix}$

7. $\begin{bmatrix} 1 & 2 \\ 1 & 2 \end{bmatrix}$

8. $\begin{bmatrix} 2 & 2 \\ -1 & -1 \end{bmatrix}$

9. $\begin{bmatrix} 3 & 1 \\ 4 & 3 \end{bmatrix}$

10. $\begin{bmatrix} 5 & -1 \\ -4 & 2 \end{bmatrix}$

11. $\begin{bmatrix} -3 & 1 \\ -1 & -1 \end{bmatrix}$

12. $\begin{bmatrix} 4 & -9 \\ 1 & -2 \end{bmatrix}$

13. $\begin{bmatrix} 6 & 2 \\ -3 & 1 \end{bmatrix}$

14. $\begin{bmatrix} 6 & -3 \\ 2 & 1 \end{bmatrix}$

15. $\begin{bmatrix} 1 & 2 \\ 1 & -1 \end{bmatrix}$

16. $\begin{bmatrix} 1 & 3 \\ 2 & 1 \end{bmatrix}$.

Set $A = \begin{bmatrix} a & b \\ c & d \end{bmatrix}$:

17. Suppose $bc \geq 0$. Prove that A has real characteristic roots.
18. (cont.) Under the same hypothesis, find when the roots coincide.
19. Prove $\lambda + \mu = a + d$ and $\lambda\mu = ad - bc$.
20. Prove A', the transpose, has the same roots as A.

Find P so that $P^{-1}AP = \begin{bmatrix} \lambda & 0 \\ 0 & \mu \end{bmatrix}$:

21. $A = \begin{bmatrix} 3 & 1 \\ 4 & 3 \end{bmatrix}$

22. $A = \begin{bmatrix} 1 & 0 \\ -1 & -1 \end{bmatrix}$

23. $A = \begin{bmatrix} 1 & 2 \\ 1 & 2 \end{bmatrix}$

24. $A = \begin{bmatrix} 6 & 2 \\ -3 & 1 \end{bmatrix}$.

Find P so that $P^{-1}AP = \begin{bmatrix} \lambda & 1 \\ 0 & \lambda \end{bmatrix}$:

25. $\begin{bmatrix} -1 & 0 \\ 1 & -1 \end{bmatrix}$

26. $\begin{bmatrix} 4 & -9 \\ 1 & -2 \end{bmatrix}$.

Find all characteristic roots and vectors:

27. $\begin{bmatrix} 3 & 0 & 0 \\ 0 & 1 & 0 \\ 0 & 0 & 2 \end{bmatrix}$

28. $\begin{bmatrix} 2 & 1 & 0 \\ 0 & 2 & 0 \\ 0 & 0 & -1 \end{bmatrix}$

29. $\begin{bmatrix} -2 & 0 & 0 \\ 1 & -2 & 0 \\ 0 & 1 & -2 \end{bmatrix}$

30. $\begin{bmatrix} 3 & 1 & 0 \\ 0 & 3 & 1 \\ 0 & 0 & 3 \end{bmatrix}$

31. $\begin{bmatrix} 4 & 0 & -9 \\ 0 & 0 & 0 \\ 1 & 0 & -2 \end{bmatrix}$

32. $\begin{bmatrix} 3 & 0 & 0 \\ 0 & 1 & 2 \\ 0 & 1 & 2 \end{bmatrix}$.

33*. Let $A = \begin{bmatrix} a & b \\ b & c \end{bmatrix}$ have characteristic roots $\lambda \leq \mu$. Prove that $\lambda \leq a \leq \mu$ and $\lambda \leq c \leq \mu$.

34*. Let A be a symmetric matrix and $\mathbf{x}'$ a characteristic vector. Let $\mathbf{y}' \cdot \mathbf{x}' = 0$. Prove that $(A\mathbf{y}') \cdot \mathbf{x}' = 0$.

35*. (cont.) Prove Theorem 8.6 for 2×2 symmetric matrices.

36. Let A be a 3×3 matrix with characteristic roots λ, μ, ν. Prove that $|A| = \lambda\mu\nu$. [*Hint:* Set $t = 0$ in the identity $|tI - A| = (t - \lambda)(t - \mu)(t - \nu)$.]

37*. (cont.) Prove $\lambda + \mu + \nu = a_{11} + a_{22} + a_{33}$.

38*. Let A be a 3×3 symmetric matrix. Prove that $A = 0$ if and only if 0 is the only characteristic root of A.

39. (cont.) Show this is false for some non-symmetric A.

40. Let λ be a characteristic root of A. Prove that λ^2 is a characteristic root of A^2.

8. Several Variable Differential Calculus

1. DIFFERENTIABLE FUNCTIONS

It may be said that differential calculus is the study of functions which have a linear approximation at each point of their domains. We shall extend this point of view to functions of several variables, but first let us review the one variable situation.

Suppose $f(x)$ has derivative $f'(c)$ at $x = c$. Then for x near c, the approximation $f(x) \approx f(c) + f'(c)(x - c)$ is quite accurate. The term "accurate" is not well defined; let us express what we mean geometrically. We know that $y = f(c) + f'(c)(x - c)$ is the equation of the tangent line to the graph of $y = f(x)$ at $(c, f(c))$. See Fig. 1.1. Geometrically, the tangent "hugs" the curve near the point of tangency. Therefore if $e(x)$ is the vertical distance between the curve and the tangent line, $e(x)$ ought to be small compared to $|x - c|$. In fact, the closer x is to c, the smaller the ratio $e(x)/|x - c|$ should be.

Let us now state these ideas more precisely.

> Suppose f is differentiable at a point c of an open interval **D**. Set
> $$f(x) = f(c) + f'(c)(x - c) + e(x).$$
> Then
> $$\frac{e(x)}{|x - c|} \longrightarrow 0 \quad \text{as} \quad x \longrightarrow c.$$

This statement characterizes the number $f'(c)$. It suggests that the derivative could have been *defined* by an approximation property:

> Suppose there exists a number k such that
> $$f(x) = f(c) + k(x - c) + e(x),$$
> where
> $$\frac{e(x)}{|x - c|} \longrightarrow 0 \quad \text{as} \quad x \longrightarrow c.$$
> Then f is differentiable at c and $k = f'(c)$.

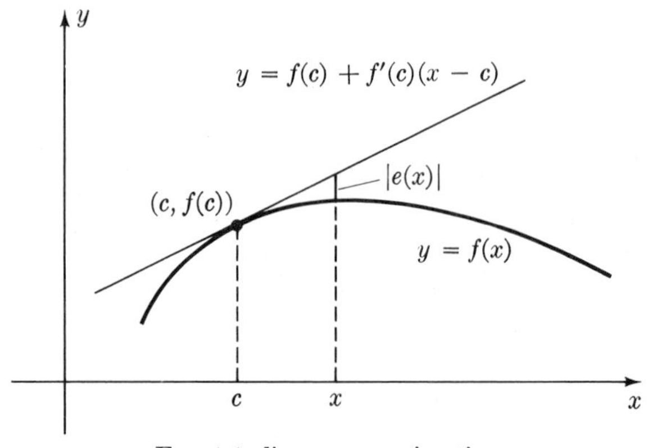

FIG. 1.1 linear approximation

This definition of derivative implies the usual one. For if the condition is satisfied, then

$$\frac{f(x) - f(c)}{x - c} = k + \frac{e(x)}{x - c} \longrightarrow k + 0 \qquad \text{as} \quad x \longrightarrow c.$$

Hence $f'(c) = k$.

Let us change our notation slightly. We shall always assume that c is an interior point of a domain **D**. Given such a c, there is a small interval $(c - \delta, c + \delta)$ contained in **D**. Hence we can denote points in **D** near c by $c + x$ where $|x| < \delta$. Using this notation and writing $e(x)$ instead of $e(x + c)$, we may restate the basic facts about differentiability as follows:

> Suppose f is differentiable at a point c of an open interval **D**. Set
>
> $$f(c + x) = f(c) + f'(c)x + e(x).$$
>
> Then $e(x)/|x| \longrightarrow 0$ as $x \longrightarrow 0$.
>
> Conversely, suppose there exists a number k and $\delta > 0$ such that
>
> $$f(c + x) = f(c) + kx + e(x) \qquad \text{for} \quad |x| < \delta,$$
>
> where $e(x)/|x| \longrightarrow 0$. Then f is differentiable at c and $k = f'(c)$.

For a function of *one* variable, the relation is clear: if the derivative exists, the linear approximation exists; if the linear approximation exists, the derivative exists.

For a function of *several* variables, the situation is not at all like this. It can happen that $\partial f/\partial x$ and $\partial f/\partial y$ both exist at a point (a, b), but no linear approximation $c + hx + ky$ exists!

EXAMPLE 1.1

Define f by

$$\begin{cases} f(0,0) = 0 \\ f(x,y) = \dfrac{2xy}{x^2+y^2}, & (x,y) \neq (0,0). \end{cases}$$

Prove that $\partial f/\partial x$ and $\partial f/\partial y$ both exist at $(0,0)$, but f does not have a linear approximation at $(0,0)$; indeed, f is not even continuous at $(0,0)$.

Solution: Clearly $f(x,0) = 0$ for all x. Hence

$$\frac{\partial f}{\partial x}(0,0) = \left.\frac{df(x,0)}{dx}\right|_{(0,0)} = 0.$$

Similarly $\partial f/\partial y = 0$ at $(0,0)$, so both partials exist and are zero. Now $f(x,y) = 0$ at each point of the x- and y-axes, so the only conceivable linear approximation is $0 + 0x + 0y = 0$. But for $x = y \neq 0$,

$$f(x,x) = \frac{2x^2}{x^2+x^2} = 1.$$

Hence $f(x,y) = 1$ at all points of the line $y = x$ except $(0,0)$ so f is not continuous at $(0,0)$ and certainly 0 does not begin to approximate f.

REMARK: It is also possible to construct an example of a *continuous* function whose first partials exist, but which fails to have a linear approximation. See Exs. 1–3.

Differentiable Functions

As we have just seen, the mere existence of both partials of $f(x,y)$ may not guarantee that $f(x,y)$ is a reasonably behaved function. We shall therefore limit our attention to functions that have good linear approximations at each point.

Recall that each linear function $L: \mathbf{R}^2 \longrightarrow \mathbf{R}$ is given by $L(\mathbf{x}) = \mathbf{k}\cdot\mathbf{x}$ for some fixed vector $\mathbf{k}$.

Let us now define differentiability for functions of two variables, that is, functions $f: \mathbf{D} \longrightarrow \mathbf{R}$, where $\mathbf{D}$ is a subset of $\mathbf{R}^2$. The discussion will extend easily to functions of three or more variables.

Differentiable Function Let f be defined on a domain $\mathbf{D}$ of the plane $\mathbf{R}^2$ and let $\mathbf{c} \in \mathbf{D}$. Then f is **differentiable** at $\mathbf{c}$ if there exist a linear function $\mathbf{k} \cdot \mathbf{x}$ and a function $e(\mathbf{x})$ such that

(i) $f(\mathbf{c} + \mathbf{x}) = f(\mathbf{c}) + \mathbf{k} \cdot \mathbf{x} + e(\mathbf{x})$ whenever $\mathbf{c} + \mathbf{x} \in \mathbf{D}$,

(ii) $\dfrac{e(\mathbf{x})}{|\mathbf{x}|} \longrightarrow 0$ as $\mathbf{x} \longrightarrow \mathbf{0}$.

The definition is completely analogous to the one for functions of a single variable. Just compare the relations

$$f(c + x) = f(c) + kx + e(x), \qquad e(x)/|x| \longrightarrow 0,$$

and

$$f(\mathbf{c} + \mathbf{x}) = f(\mathbf{c}) + \mathbf{k} \cdot \mathbf{x} + e(\mathbf{x}), \qquad e(\mathbf{x})/|\mathbf{x}| \longrightarrow 0.$$

It is instructive to write out the definition in coordinate form:

A function f is **differentiable** at (a, b) if there exists a linear function $hx + ky$ such that

$$f(a + x, b + y) = f(a, b) + hx + ky + e(x, y),$$

where

$$\frac{e(x, y)}{\sqrt{x^2 + y^2}} \longrightarrow 0 \quad \text{as} \quad (x, y) \longrightarrow (0, 0).$$

An immediate consequence of differentiability is continuity.

Theorem 1.1 If f is differentiable at (a, b), then f is continuous at (a, b).

Proof: Let $f(a + x, b + y) = f(a, b) + hx + ky + e(x, y)$ as in the definition. If $(x, y) \longrightarrow (0, 0)$, then $x \longrightarrow 0$, $y \longrightarrow 0$, and $e(x, y) \longrightarrow 0$. Consequently $f(a + x, b + y) \longrightarrow f(a, b)$. This is continuity.

The mere existence of partial derivatives at (a, b) does not guarantee differentiability, as we noted in Example 1.1. However differentiability does guarantee the existence of partial derivatives.

Theorem 1.2 Let f be differentiable at an *interior* point (a, b) of its domain, and let

$$f(a + x, b + y) = f(a, b) + hx + ky + e(x, y),$$

where $e(x, y)/|(x, y)| \longrightarrow 0$ as $(x, y) \longrightarrow (0, 0)$. Then the first partials of f exist at (a, b) and

$$\frac{\partial f}{\partial x}(a, b) = h, \qquad \frac{\partial f}{\partial y}(a, b) = k.$$

Proof:

$$\frac{\partial f}{\partial x}(a, b) = \lim_{x \to 0} \frac{f(a + x, b) - f(a, b)}{x}$$

$$= \lim_{x \to 0} \frac{hx + e(x, 0)}{x} = h + \lim_{x \to 0} \frac{e(x, 0)}{x} = h.$$

Similarly $f_y(a, b) = k$.

Again there is a strong analogy between the preceding formula and the corresponding one for functions of one variable. Just compare the relations

$$f(c + x) = f(c) + \frac{df(c)}{dx} x + e(x)$$

and

$$f(a + x, b + y) = f(a, b) + \frac{\partial f(a, b)}{\partial x} x + \frac{\partial f(a, b)}{\partial y} y + e(x, y).$$

A Test for Differentiability

The next proof uses the Mean Value Theorem for functions of one variable. Let us review its content:

Mean Value Theorem Suppose f is continuous on a closed interval $a \leq x \leq b$ and differentiable on the open interior $a < x < b$. Then there exists c such that $a < c < x$ and

$$f(b) - f(a) = f'(c)(b - a).$$

We are prepared for a practical test for the differentiability of a function.

Theorem 1.3 Suppose f has domain **D** and (a, b) is an interior point of **D**. Suppose (1) $\partial f/\partial x$ and $\partial f/\partial y$ exist at each interior point of **D** and (2) these partials are continuous at (a, b). Then f is differentiable at (a, b).

Proof: Write

$$f(a + x, b + y) - f(a, b)$$
$$= [f(a + x, b + y) - f(a, b + y)] + [f(a, b + y) - f(a, b)].$$

By the Mean Value Theorem, applied twice,

$$\begin{cases} f(a + x, b + y) - f(a, b + y) = xf_x(a + \theta x, b + y) \\ f(a, b + y) - f(a, b) = yf_y(a, b + \lambda y), \end{cases}$$

where $0 < \theta < 1$ and $0 < \lambda < 1$. (But θ and λ depend on x and y.) Set

$$(1) \qquad f(a + x, b + y) = f(a, b) + x f_x(a, b) + y f_y(a, b) + e(x, y),$$

so that

$$(2) \qquad e(x, y) = x g(x, y) + y h(x, y),$$

where

$$(3) \qquad \begin{cases} g(x, y) = f_x(a + \theta x, b + y) - f_x(a, b) \\ h(x, y) = f_y(a, b + \lambda y) - f_y(a, b). \end{cases}$$

Relation (1) implies that f is differentiable at (a, b), *provided* we prove that

$$\frac{e(x, y)}{\sqrt{x^2 + y^2}} \longrightarrow 0 \quad \text{as} \quad (x, y) \longrightarrow (0, 0).$$

We obviously have

$$(a + \theta x, b + y) \longrightarrow (a, b) \quad \text{and} \quad (a, b + \lambda y) \longrightarrow (a, b)$$

as $(x, y) \longrightarrow (0, 0)$. This implies

$$g(x, y) \longrightarrow 0 \quad \text{and} \quad h(x, y) \longrightarrow 0$$

as $(x, y) \longrightarrow (0, 0)$ by (3) and the assumption that f_x and f_y are continuous at (a, b).

Now divide both sides of (2) by $\sqrt{x^2 + y^2}$, take absolute values, and apply the triangle inequality:

$$\begin{aligned} \left| \frac{e(x, y)}{\sqrt{x^2 + y^2}} \right| &\le \frac{|x|}{\sqrt{x^2 + y^2}} |g(x, y)| + \frac{|y|}{\sqrt{x^2 + y^2}} |h(x, y)| \\ &\le |g(x, y)| + |h(x, y)|. \end{aligned}$$

Clearly $|g(x, y)| + |h(x, y)| \longrightarrow 0$ as $(x, y) \longrightarrow (0, 0)$. This completes the proof.

Three Variables

For three or more variables, the definitions and theorems are easy extensions of those for two variables. We shall merely state the definition of differentiability for three variables and leave the extensions of Theorem 1.1–1.3 as exercises.

Differentiable Function Let f be defined on a domain $\mathbf{D}$ of three-space $\mathbf{R}^3$, and let $\mathbf{c} \in \mathbf{D}$. Then f is called **differentiable** at $\mathbf{c}$ if there exist a linear function $\mathbf{k} \cdot \mathbf{x}$ and a function $e(\mathbf{x})$ such that

(i) $f(\mathbf{c} + \mathbf{x}) = f(\mathbf{c}) + \mathbf{k} \cdot \mathbf{x} + e(\mathbf{x}) \quad$ whenever $\quad \mathbf{c} + \mathbf{x} \in \mathbf{D}$,

(ii) $\dfrac{e(\mathbf{x})}{|\mathbf{x}|} \longrightarrow 0 \quad$ as $\quad \mathbf{x} \longrightarrow \mathbf{0}$.

The function f is called **differentiable** on $\mathbf{D}$ provided it is differentiable at each point of $\mathbf{D}$.

Exactly as for two variables, it follows from this definition that

$$\mathbf{k} = \left(\frac{\partial f}{\partial x}, \frac{\partial f}{\partial y}, \frac{\partial f}{\partial z}\right)\Big|_{\mathbf{x}=\mathbf{c}}$$

The proof is left as an exercise.

EXERCISES

Show that $f(x, y)$ is differentiable at (a, b) by finding a linear function $hx + ky$ that satisfies the definition of differentiability:

1. $f(x, y) = xy^2$ at $(0, 0)$

2*. $f(x, y) = 1/xy$ at $(1, 2)$.

3. Define f on $\mathbf{R}^2$ by $f(0, 0) = 0$ and

$$f(x, y) = \frac{xy(x + y)}{x^2 + y^2} \qquad \text{for} \quad (x, y) \neq (0, 0).$$

Prove f is continuous on $\mathbf{R}^2$.

4*. (cont.) Prove that $\partial f/\partial x$ and $\partial f/\partial y$ exist on $\mathbf{R}^2$ and are continuous except at $(0, 0)$.

5. (cont.) Prove that f is not differentiable at $(0, 0)$.

6. Prove that the sum of two differentiable functions is differentiable.

7. Prove that the product of differentiable functions is differentiable.

8. Let g be differentiable at (a, b) and $g(a, b) \neq 0$. Prove that $1/g$ is differentiable at (a, b).

9. (cont.) Prove that the quotient of differentiable functions is differentiable whenever the denominator is not zero.

State and prove for three variables:

10. Theorem 1.1

11. Theorem 1.2.

2. CHAIN RULE

The Chain Rule for functions of one variable gives the derivative of a composite function: if $y = f(x)$ where $x = x(t)$, then

$$\frac{dy}{dt} = \frac{dy}{dx}\frac{dx}{dt}.$$

The Chain Rule for functions of several variables really has two parts, a theoretical part, which says that a composite function built out of differentiable functions is itself differentiable, and a practical part, which is a formula for computing partial derivatives of such a composite function. We shall postpone a precise statement and proof until Section 9. First we shall work intuitively in order to gain a feeling for how the Chain Rule is used in practice.

Suppose $z = f(x, y)$ where $x = x(t)$ and $y = y(t)$. Thus z is indirectly a function of t. The Chain Rule asserts that

$$\frac{dz}{dt} = \frac{\partial z}{\partial x}\frac{dx}{dt} + \frac{\partial z}{\partial y}\frac{dy}{dt}.$$

This Chain Rule can be better understood in terms of vectors. The composite function $z(t) = f[x(t), y(t)]$ can be thought of as $z(t) = f[\mathbf{x}(t)]$. If t is time, then $\mathbf{x}(t)$ represents the path of a moving particle in the plane, and the composite function $f[\mathbf{x}(t)]$ assigns a number z to each value of t. The Chain Rule is a formula for the rate of change of $f[\mathbf{x}(t)]$ with respect to t. For instance $f(\mathbf{x})$ might be the temperature at position $\mathbf{x}$. Then the Chain Rule tells how fast the temperature is changing as the particle moves along the curve $\mathbf{x}(t)$.

Chain Rule Let $z = f(x, y)$, where $x = x(t)$ and $y = y(t)$. Then

$$\frac{dz}{dt} = \frac{\partial z}{\partial x}\frac{dx}{dt} + \frac{\partial z}{\partial y}\frac{dy}{dt},$$

where $\dfrac{\partial z}{\partial x}$ and $\dfrac{\partial z}{\partial y}$ are evaluated at $(x(t), y(t))$. In briefer notation,

$$\dot{z} = f_x\dot{x} + f_y\dot{y}.$$

In terms of vectors, $z = f[\mathbf{x}(t)]$,

$$\frac{dz}{dt} = f_x[\mathbf{x}(t)]\dot{x}(t) + f_y[\mathbf{x}(t)]\dot{y}(t).$$

Similar rules hold for functions of more than two variables. For instance, if $w = f(x, y, z)$, where $x = x(t)$, $y = y(t)$, $z = z(t)$, then

$$\frac{dw}{dt} = \frac{\partial w}{\partial x}\frac{dx}{dt} + \frac{\partial w}{\partial y}\frac{dy}{dt} + \frac{\partial w}{\partial z}\frac{dz}{dt}.$$

EXAMPLE 2.1

Let $w = f(x, y, z) = xy^2z^3$, where $x = t\cos t$, $y = e^t$, and $z = \ln(t^2 + 2)$. Compute $\dfrac{dw}{dt}$ at $t = 0$.

Solution: There is a direct but tedious way to do the problem. Write

$$w = (t\cos t)e^{2t}[\ln(t^2 + 2)]^3.$$

Differentiate, then set $t = 0$. That's quite a job!

Use of the Chain Rule is much simpler:

$$\dot{w}(0) = \frac{\partial w}{\partial x}\dot{x}(0) + \frac{\partial w}{\partial y}\dot{y}(0) + \frac{\partial w}{\partial z}\dot{z}(0),$$

where the partial derivatives are evaluated at

$$\mathbf{x}_0 = (x(0), y(0), z(0)) = (0, 1, \ln 2).$$

Since $w = xy^2z^3$,

$$\left.\frac{\partial w}{\partial x}\right|_{\mathbf{x}_0} = y^2z^3\Big|_{\mathbf{x}_0} = (\ln 2)^3, \qquad \left.\frac{\partial w}{\partial y}\right|_{\mathbf{x}_0} = 2xyz^3\Big|_{\mathbf{x}_0} = 0,$$

$$\left.\frac{\partial w}{\partial z}\right|_{\mathbf{x}_0} = 3xy^2z^2\Big|_{\mathbf{x}_0} = 0,$$

hence

$$\dot{w}(0) = (\ln 2)^3\dot{x}(0) + 0 + 0.$$

But

$$\dot{x}(0) = (\cos t - t\sin t)|_0 = 1,$$

therefore $\dot{w}(0) = (\ln 2)^3$.

Answer: $(\ln 2)^3$.

Another Version of the Chain Rule

Suppose $z = f(x, y)$, where this time x and y are functions of *two* variables, $x = x(s, t)$ and $y = y(s, t)$. Then indirectly, z is a function of the variables s and t. There is a chain rule for computing $\partial z/\partial s$ and $\partial z/\partial t$:

Chain Rule If $z = f(x, y)$ is a function of two variables x and y, where $x = x(s, t)$ and $y = y(s, t)$, then

$$\begin{cases} \dfrac{\partial z}{\partial s} = \dfrac{\partial z}{\partial x}\dfrac{\partial x}{\partial s} + \dfrac{\partial z}{\partial y}\dfrac{\partial y}{\partial s} \\[2ex] \dfrac{\partial z}{\partial t} = \dfrac{\partial z}{\partial x}\dfrac{\partial x}{\partial t} + \dfrac{\partial z}{\partial y}\dfrac{\partial y}{\partial t}, \end{cases}$$

where $\dfrac{\partial z}{\partial x}$ and $\dfrac{\partial z}{\partial y}$ are evaluated at $(x(s, t), y(s, t))$.

This Chain Rule is a consequence of the previous one. For instance, to compute $\partial z/\partial s$, hold t fixed, making $x(s, t)$ and $y(s, t)$ effectively functions of the one variable s. Then apply the previous Chain Rule.

EXAMPLE 2.2

Let $w = x^2y$, where $x = s^2 + t^2$ and $y = \cos st$. Compute $\dfrac{\partial w}{\partial s}$.

Solution:

$$\frac{\partial w}{\partial s} = \frac{\partial w}{\partial x}\frac{\partial x}{\partial s} + \frac{\partial w}{\partial y}\frac{\partial y}{\partial s}$$

$$= (2xy)2s + x^2(-t \sin st)$$

$$= 2(s^2 + t^2)(\cos st)2s + (s^2 + t^2)^2(-t \sin st).$$

Answer: $(s^2 + t^2)[4s \cos st - t(s^2 + t^2) \sin st]$.

The next example is important in physical applications.

EXAMPLE 2.3

If $w = f(x, y)$, where $x = r \cos \theta$ and $y = r \sin \theta$, show that

$$\left(\frac{\partial w}{\partial x}\right)^2 + \left(\frac{\partial w}{\partial y}\right)^2 = \left(\frac{\partial w}{\partial r}\right)^2 + \frac{1}{r^2}\left(\frac{\partial w}{\partial \theta}\right)^2.$$

Solution: Use the Chain Rule to compute $\partial w/\partial r$ and $\partial w/\partial \theta$:

$$\frac{\partial w}{\partial r} = \frac{\partial w}{\partial x}\frac{\partial x}{\partial r} + \frac{\partial w}{\partial y}\frac{\partial y}{\partial r} = \frac{\partial w}{\partial x}\cos\theta + \frac{\partial w}{\partial y}\sin\theta;$$

$$\frac{\partial w}{\partial \theta} = \frac{\partial w}{\partial x}\frac{\partial x}{\partial \theta} + \frac{\partial w}{\partial y}\frac{\partial y}{\partial \theta} = \frac{\partial w}{\partial x}(-r\sin\theta) + \frac{\partial w}{\partial y}r\cos\theta$$

$$= r\left(-\frac{\partial w}{\partial x}\sin\theta + \frac{\partial w}{\partial y}\cos\theta\right).$$

From these formulas

$$\left(\frac{\partial w}{\partial r}\right)^2 = \left(\frac{\partial w}{\partial x}\right)^2\cos^2\theta + 2\frac{\partial w}{\partial x}\frac{\partial w}{\partial y}\sin\theta\cos\theta + \left(\frac{\partial w}{\partial y}\right)^2\sin^2\theta,$$

$$\frac{1}{r^2}\left(\frac{\partial w}{\partial \theta}\right)^2 = \left(\frac{\partial w}{\partial x}\right)^2\sin^2\theta - 2\frac{\partial w}{\partial x}\frac{\partial w}{\partial y}\sin\theta\cos\theta + \left(\frac{\partial w}{\partial y}\right)^2\cos^2\theta.$$

Add:

$$\left(\frac{\partial w}{\partial r}\right)^2 + \frac{1}{r^2}\left(\frac{\partial w}{\partial \theta}\right)^2 = \left[\left(\frac{\partial w}{\partial x}\right)^2 + \left(\frac{\partial w}{\partial y}\right)^2\right](\cos^2\theta + \sin^2\theta)$$

$$= \left(\frac{\partial w}{\partial x}\right)^2 + \left(\frac{\partial w}{\partial y}\right)^2.$$

EXERCISES

Find dz/dt by the Chain Rule:

1. $z = e^{xy}; \quad x = 3t + 1, y = t^2$
2. $z = x/y; \quad x = t + 1, y = t - 1$
3. $z = x^2 \cos y - x; \quad x = t^2, y = 1/t$
4. $z = x/y; \quad x = \cos t, y = 1 + t^2$.

Find dw/dt by the Chain Rule:

5. $w = xyz; \quad x = t^2, y = t^3, z = t^4$
6. $w = e^x \cos(y + z); \quad x = 1/t, y = t^2, z = -t$
7. $w = e^{-x}y^2 \sin z; \quad x = t, y = 2t, z = 4t$
8. $w = (e^{-x} \sec z)/y^2; \quad x = t^2, y = 1 + t, z = t^3$.

Find $\partial z/\partial s$ and $\partial z/\partial t$ by the Chain Rule:

9. $z = x^3/y^2; \quad x = s^2 - t, y = 2st$
10. $z = (x + y^2)^4; \quad x = se^t, y = se^{-t}$
11. $z = \sqrt{1 + x^2 + y^2}; \quad x = st^2, y = 1 + st$
12. $z = e^{x^2 y}; \quad x = \dfrac{s}{\sqrt{1 + t^2}}, y = st$.
13. The radius r and height h of a conical tank increase at rates $\dot{r} = 0.3$ in./hr and $\dot{h} = 0.5$ in./hr. Find the rate of increase $\dot{V}$ of the volume when $r = 6$ ft and $h = 30$ ft.
14. Prove that
$$\frac{d}{dx}\int_{g(x)}^{h(x)} F(t)\,dt = F[h(x)]h'(x) - F[g(x)]g'(x).$$
15. Given $F(x, y)$, show that
$$\frac{\partial}{\partial u} F(u + v, u - v) + \frac{\partial}{\partial v} F(u + v, u - v) = 2F_x(u + v, u - v).$$

A function $w = f(x, y, z)$ is **homogeneous of degree** n if $f(tx, ty, tz) = t^n f(x, y, z)$ for all $t > 0$. The condition of homogeneity can be written vectorially:

$$f(t\mathbf{x}) = t^n f(\mathbf{x}).$$

Show that the function is homogeneous: What degree?

16. $x^2 + yz$
17. $x - y + 2z$
18. $x^3 + y^3 + z^3 - 3xyz$
19. $x^2 e^{-y/z}$
20. $\dfrac{xyz}{x^4 + y^4 + z^4}$
21. $\dfrac{1}{x + y}$.
22. Suppose f and g are homogeneous of degree m and n respectively. Show that fg is homogeneous of degree mn.
23. Let $f(x, y, z)$ be homogeneous of degree n. Show that f_x is homogeneous of degree $n - 1$. (Exception: $n = 0$ and f constant.)
24. Let $f(x, y, z)$ be homogeneous of degree n. Prove **Euler's Relation:**
$xf_x + yf_y + zf_z = nf$.
[*Hint:* Differentiate $f(tx, ty, tz) = t^n f(x, y, z)$ with respect to t, using the Chain Rule; then set $t = 1$.]

25. Verify Euler's Relation for the functions in Exs. 18 and 19.

26*. (Converse of Euler's Relation) Let $f(\mathbf{x})$ be differentiable for $\mathbf{x} \neq \mathbf{0}$, and suppose $\mathbf{x} \cdot \operatorname{grad} f = nf$. Prove f is homogeneous of degree n.
[*Hint:* Show that $\partial[t^{-n}f(t\mathbf{x})]/\partial t = 0$.]

3. TANGENT PLANE

The graph of

$$z = f(x, y)$$

is a surface. Given a point $\mathbf{x}_0 = (x_0, y_0, z_0)$ on this surface, we are going to describe the plane tangent to the surface at $\mathbf{x}_0$.

> Let $z = f(x, y)$ represent a surface and let $\mathbf{x}_0 = (x_0, y_0, z_0)$ be a fixed point on it. Consider all curves lying on the surface and passing through $\mathbf{x}_0$. Their velocity vectors fill out a plane through the origin. The parallel plane through $\mathbf{x}_0$ is called the **tangent plane** at $\mathbf{x}_0$.

Let us see why this is so. Any curve on the surface (Fig. 3.1) is given by

$$\mathbf{x}(t) = (x(t), y(t), z(t)) = (x(t), y(t), f[x(t), y(t)]).$$

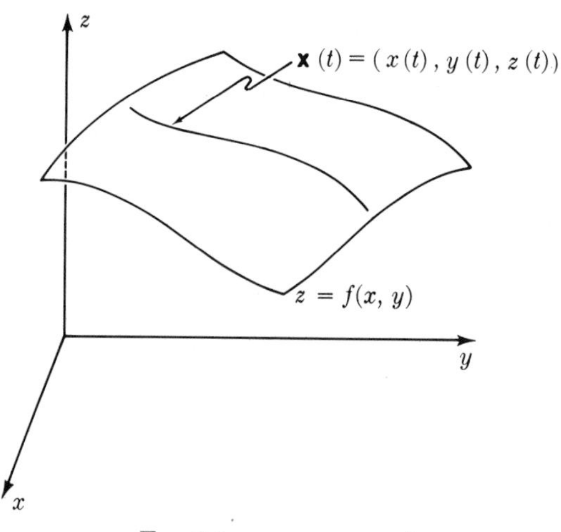

FIG. 3.1 curve on a surface

Its velocity vector is found by the Chain Rule:

$$\mathbf{v}(t) = \dot{\mathbf{x}}(t) = (\dot{x}(t), \dot{y}(t), f_x\dot{x} + f_y\dot{y}).$$

Let us assume time is measured so that $\mathbf{x}(0) = \mathbf{x}_0$, that is,

$$x(0) = x_0, \qquad y(0) = y_0, \qquad z(0) = z_0.$$

Then

$$\mathbf{v}(0) = (\dot{x}(0), \dot{y}(0), f_x(x_0, y_0)\dot{x}(0) + f_y(x_0, y_0)\dot{y}(0))$$
$$= \dot{x}(0)(1, 0, f_x(x_0, y_0)) + \dot{y}(0)(0, 1, f_y(x_0, y_0)).$$

The vectors

$$\mathbf{w}_1 = (1, 0, f_x(x_0, y_0)) \qquad \text{and} \qquad \mathbf{w}_2 = (0, 1, f_y(x_0, y_0))$$

depend only on the function f and the point $\mathbf{x}_0$, not on the particular curve (Fig. 3.2).

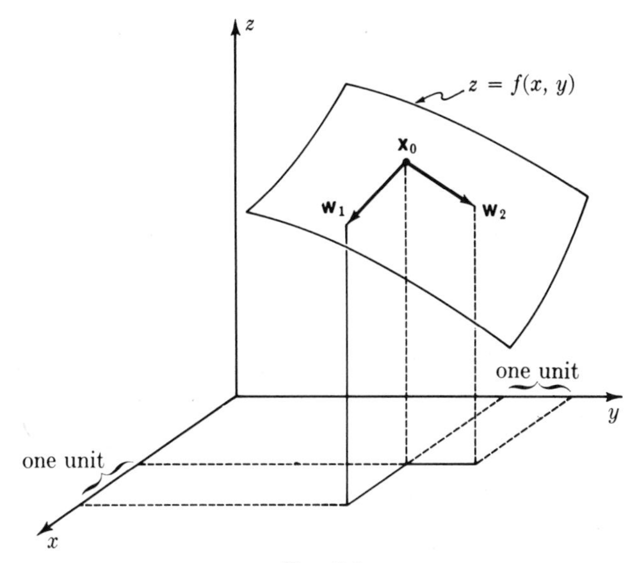

FIG. 3.2

Suppose a curve lies on the surface $z = f(x, y)$ and passes through

$$\mathbf{x}_0 = (x_0, y_0, z_0).$$

Then its velocity vector has the form

$$\mathbf{v} = a\mathbf{w}_1 + b\mathbf{w}_2,$$

where

$$\mathbf{w}_1 = (1, 0, f_x(x_0, y_0)), \qquad \mathbf{w}_2 = (0, 1, f_y(x_0, y_0)),$$

and a and b are constants, $a = \dot{x}(0)$ and $b = \dot{y}(0)$.

Given any pair of numbers a and b, there are curves $\mathbf{x}(t)$ which lie on the surface, pass through $\mathbf{x}_0$ when $t = 0$, and have $\dot{x}(0) = a$ and $\dot{y}(0) = b$. One such is

$$\mathbf{x}(t) = (x_0 + at, y_0 + bt, f(x_0 + at, y_0 + bt)).$$

Its velocity vector is $a\mathbf{w}_1 + b\mathbf{w}_2$. Therefore, all vectors $a\mathbf{w}_1 + b\mathbf{w}_2$ actually occur as velocity vectors. It follows that the velocity vectors fill out a plane through the origin. The parallel plane through $\mathbf{x}_0$ is the tangent plane to the surface at $\mathbf{x}_0$. It consists of all points

$$\mathbf{x} = \mathbf{x}_0 + a\mathbf{w}_1 + b\mathbf{w}_2, \qquad a \text{ and } b \text{ arbitrary.}$$

EXAMPLE 3.1

Find the tangent plane to $z = x^2 + y$ at $(1, 1, 2)$.

Solution:

$$\mathbf{w}_1 = \left(1, 0, \frac{\partial z}{\partial x}(1, 1)\right) = (1, 0, 2), \qquad \mathbf{w}_2 = \left(0, 1, \frac{\partial z}{\partial y}(1, 1)\right) = (0, 1, 1).$$

Hence the typical velocity vector is

$$\mathbf{v} = a\mathbf{w}_1 + b\mathbf{w}_2 = a(1, 0, 2) + b(0, 1, 1) = (a, b, 2a + b).$$

The tangent plane consists of all points

$$(1, 1, 2) + (a, b, 2a + b) = (a + 1, b + 1, 2a + b + 2),$$

where a and b are arbitrary.

Answer: All points $(a + 1, b + 1, 2a + b + 2)$.

REMARK: The typical point on this tangent plane is

$$\mathbf{x} = (x, y, z) = (a + 1, b + 1, 2a + b + 2).$$

Thus

$$a = x - 1, \qquad b = y - 1,$$

and

$$z = 2a + b + 2 = 2(x - 1) + (y - 1) + 2 = 2x + y - 1.$$

Consequently an *equation* for the tangent plane is

$$z = 2x + y - 1.$$

The Normal

The vector

$$\mathbf{n} = (-f_x, -f_y, 1)$$

is perpendicular to both tangent vectors $\mathbf{w}_1$ and $\mathbf{w}_2$:

$$\mathbf{n}\cdot\mathbf{w}_1 = (-f_x, -f_y, 1)\cdot(1, 0, f_x) = -f_x + 0 + f_x = 0,$$

$$\mathbf{n}\cdot\mathbf{w}_2 = (-f_x, -f_y, 1)\cdot(0, 1, f_y) = 0 - f_y + f_y = 0.$$

Consequently $\mathbf{n}$ is perpendicular to each vector $a\mathbf{w}_1 + b\mathbf{w}_2$:

$$\mathbf{n}\cdot(a\mathbf{w}_1 + b\mathbf{w}_2) = a(\mathbf{n}\cdot\mathbf{w}_1) + b(\mathbf{n}\cdot\mathbf{w}_2) = 0.$$

In other words, **n** is perpendicular to the tangent plane at $\mathbf{x}_0$, that is, **n** is a normal, so we may write a normal form of the plane.

> The tangent plane to the surface $z = f(x, y)$ at the point $\mathbf{x}_0 = (x_0, y_0, z_0)$ is given by
> $$-f_x \cdot (x - x_0) - f_y \cdot (y - y_0) + (z - z_0) = 0,$$
> where f_x and f_y are evaluated at (x_0, y_0). The vector form of this equation is
> $$(\mathbf{x} - \mathbf{x}_0) \cdot \mathbf{n} = 0, \qquad \mathbf{n} = (-f_x, -f_y, 1).$$

EXAMPLE 3.2

Find an equation for the tangent plane to $z = x^2 + y$ at $(1, 1, 2)$.

Solution: Write $f(x, y) = x^2 + y$. Then

$$f_x(x, y) = 2x, \qquad f_y(x, y) = 1.$$

At $(1, 1, 2)$,

$$f_x = 2, \qquad f_y = 1.$$

Hence the equation is

$$-2(x - 1) - (y - 1) + (z - 2) = 0.$$

Answer: $2x + y - z = 1.$

The vector $(-f_x, -f_y, 1)$ is perpendicular to the tangent plane and has a positive z-component (it is the upward normal rather than the downward normal). We introduce the unit vector in the same direction.

> The **unit normal** at $\mathbf{x}_0$ to the surface $z = f(x, y)$ is the vector
> $$\mathbf{N} = \frac{1}{\sqrt{f_x^2 + f_y^2 + 1}} (-f_x, -f_y, 1),$$
> where f_x and f_y are evaluated at (x_0, y_0).

EXAMPLE 3.3

Find the unit normal at $(1, 1, 2)$ to the surface $z = x^2 + y$.

Solution: We have $(-f_x, -f_y, 1) = (-2, -1, 1)$, hence

$$\mathbf{N} = \frac{1}{\sqrt{2^2 + 1^2 + 1^2}} (-2, -2, 1).$$

Answer: $\mathbf{N} = \dfrac{1}{\sqrt{6}} (-2, -1, 1).$

EXAMPLE 3.4

How far is **0** from the tangent plane to $z = x^2 + y$ at $(1, 1, 2)$?

Solution: The distance is the length of the projection of $\mathbf{x}_0 = (1, 1, 2)$ on $\mathbf{N}$. See Fig. 3.3.

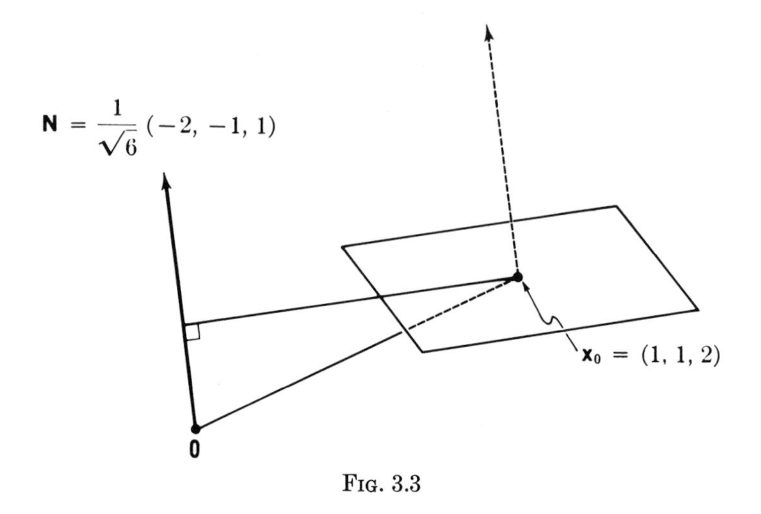

FIG. 3.3

$$\begin{aligned}\text{Distance} &= |\mathbf{x}_0 \cdot \mathbf{N}| \\ &= \left| (1, 1, 2) \cdot \frac{1}{\sqrt{6}} (-2, -1, 1) \right| \\ &= \frac{1}{\sqrt{6}} |-2 - 1 + 2| = \frac{1}{\sqrt{6}}.\end{aligned}$$

Answer: $\dfrac{1}{\sqrt{6}}$.

EXERCISES

Give the equation $z = ax + by + c$ of the tangent plane to the surface at the indicated point:

1. $z = x^2 - y^2$; $x = 0, y = 0$
2. $z = x^2 - y^2$; $x = 1, y = -1$
3. $z = x^2 + 4y^2$; $x = 2, y = 1$
4. $z = x^2e^y$; $x = -1, y = 2$
5. $z = x^2y + y^3$; $x = -1, y = 2$
6. $z = x \cos y + y \cos x$; $x = 0, y = 0$.

Find the unit normal to the surface at the indicated point:

7. $z = x + x^2y^3 + y$; $x = 0, y = 0$
8. $z = x^3 + y^3$; $x = 1, y = -1$
9. $z = x^2 + xy + y^2$; $x = -1, y = 2$
10. $z = \sqrt{1 - x^2 - y^2}$; $x = \frac{1}{2}, y = \frac{1}{2}$.

11. Show that the tangent plane to the hyperbolic paraboloid $z = x^2 - y^2$ at $(0, 0, 0)$ intersects the surface in a pair of straight lines.
12. (cont.) Show that the conclusion is valid for the tangent plane at *any* point of the surface.
13*. (cont.) Show that the property of the tangent planes in Ex. 12 is also valid for the hyperboloid of one sheet $z^2 = x^2 + y^2 - 1$.

4. GRADIENT

The gradient field of a function f on a region is the assignment of a certain vector, called grad f, to each point of that region.

Suppose $z = z(x, y)$ is defined on a region **D** of the x, y-plane. The **gradient** of f is the vector

$$\text{grad } f = (f_x, f_y).$$

Likewise, if $w = f(x, y, z)$ is defined on a region **D** of space, the **gradient** of f is the vector

$$\text{grad } f = (f_x, f_y, f_z).$$

The **gradient field** of a function f is the assignment of the vector grad f to each point of the region **D**.

For example, if

$$f(x, y) = x^2 + y, \qquad \text{grad } f = (2x, 1);$$

if

$$f(x, y, z) = |\mathbf{x}|^2 = x^2 + y^2 + z^2, \qquad \text{grad } f = (2x, 2y, 2z) = 2\mathbf{x}.$$

In this section, we discuss several uses of the gradient. The first of these concerns notation. Certain formulas are simplified if expressed in terms of gradients. An example is the Chain Rule, which asserts

$$\dot{z} = f_x\dot{x} + f_y\dot{y}$$

if $z = f(x, y)$ and $x = x(t)$, $y = y(t)$. But grad $f = (f_x, f_y)$ and $\dot{\mathbf{x}} = (\dot{x}, \dot{y})$. Therefore, in vector notation

$$\dot{z} = (\text{grad } f)\cdot\dot{\mathbf{x}}.$$

Level Curves

Imagine a surface $z = f(x, y)$ above a portion of the x, y-plane. In the plane we can draw a contour map of the surface by indicating curves of constant altitude. These are called **contour lines** or **level curves**. Figure 4.1

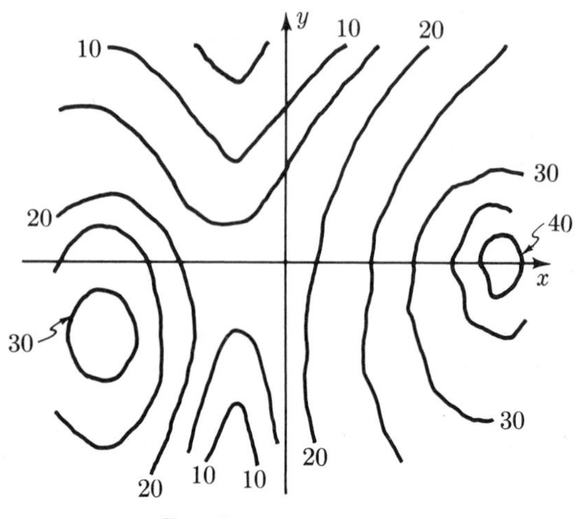

FIG. 4.1 contour map

shows a contour map with two hills and a pass between them. Level curves are obtained by slicing the surface $z = f(x, y)$ with planes $z = c$ for various constants c. Each plane $z = c$ intersects the surface in a plane curve (Fig. 4.2). The projection of this curve onto the x, y-plane is the **level curve at level** c. It is the graph of $f(x, y) = c$. Where level curves are close together the surface is steep; where they are far apart it is relatively flat.

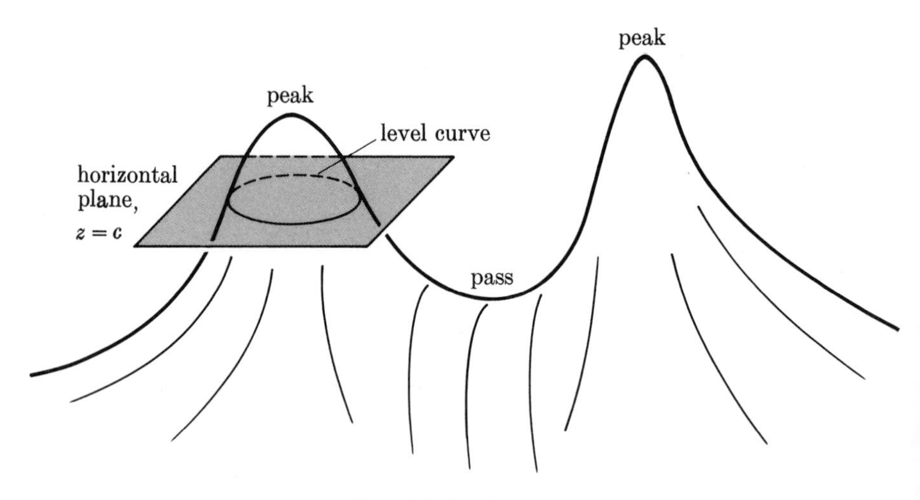

FIG. 4.2 level curves

An important relation exists between the gradient field and the level curves of a function.

The gradient field of f is **orthogonal** (perpendicular) to the level curves of f.

For suppose a particle moves along the level curve $f(x, y) = c$. Let its position at time t be $\mathbf{x}(t) = (x(t), y(t))$. Then

$$f[x(t), y(t)] = c.$$

Therefore the time derivative is zero:

$$f_x\dot{x} + f_y\dot{y} = 0, \qquad (\text{grad } f)\cdot\dot{\mathbf{x}} = 0.$$

Hence grad f is orthogonal to the tangent to the level curve.

Suppose we want an explicit equation for the tangent to a level curve $f(x, y) = c$ at a point $\mathbf{x}_0 = (x_0, y_0)$. We know a point, $\mathbf{x}_0$, on the tangent and we know a normal, grad $f(\mathbf{x}_0)$, so the answer is

$$(\mathbf{x} - \mathbf{x}_0)\cdot\text{grad } f(\mathbf{x}_0) = 0.$$

EXAMPLE 4.1

Find the tangent to the cubic $x^2 = y^3$ at $(-1, 1)$.

Solution: Set $f(x, y) = x^2 - y^3$. Then the cubic is the level curve $f = 0$. We have

$$\text{grad } f(\mathbf{x}_0) = (2x, -3y^2)|_{(-1,1)} = (-2, -3).$$

Hence the tangent is

$$[(x, y) - (-1, 1)]\cdot(-2, -3) = 0,$$

$$-2(x + 1) - 3(y - 1) = 0.$$

Answer: $2x + 3y = 1$.

Level Surfaces

We cannot graph a function of three variables

$$w = f(x, y, z)$$

(the graph would be four-dimensional). We can, however, learn a good deal about the function by plotting in three-space the **level surfaces**

$$f(x, y, z) = \text{constant}.$$

For example, the level surfaces of

$$f(x, y, z) = x^2 + y^2 + z^2$$

are the spheres

$$x^2 + y^2 + z^2 = c^2$$

centered at the origin.

The gradient field of f is orthogonal to the level surfaces of f.

The proof of this statement is practically identical to the proof of the analogous statement in two variables.

Just as for curves, the gradient provides a convenient tool for finding explicitly tangents to level surfaces. Suppose we want the tangent plane at a point $\mathbf{x}_0$ of the level surface $f(\mathbf{x}) = c$. The answer is

$$(\mathbf{x} - \mathbf{x}_0)\cdot \operatorname{grad} f(\mathbf{x}_0) = 0.$$

Note that this method fails at points where $\operatorname{grad} f = 0$. At such points there usually is not a clearly defined tangent plane anyhow.

EXAMPLE 4.2

Let A be a non-zero 3×3 symmetric matrix and let $\mathbf{x}_0$ be a point on the quadric surface

$$\mathbf{x}A\mathbf{x}' = 1.$$

Find the tangent plane to the surface at $\mathbf{x}_0$.

Solution: Let $f(\mathbf{x}) = \mathbf{x}A\mathbf{x}'$. Then the quadric is the level surface $f = 1$, and since $\mathbf{x}_0$ is on the surface, $\mathbf{x}_0 A\mathbf{x}_0' = 1$.

Our problem is to compute $\operatorname{grad} f$. To be definite, set

$$A = \begin{bmatrix} a & b & d \\ b & c & e \\ d & e & f \end{bmatrix}.$$

The terms involving x in $\mathbf{x}A\mathbf{x}'$ are $ax^2 + 2bxy + 2dzx$, so $\partial f/\partial x = 2(ax + by + dz)$. Similarly $\partial f/\partial y = 2(bx + cy + ez)$ and $\partial f/\partial z = 2(dx + ey + fz)$. It follows that

$$\operatorname{grad} f = 2\mathbf{x}A.$$

Therefore, the tangent plane at $\mathbf{x}_0$ is given by

$$(\mathbf{x} - \mathbf{x}_0)\cdot(2\mathbf{x}_0 A) = 0.$$

(Note that $\mathbf{x}_0 A \neq 0$ since $\mathbf{x}_0 A\mathbf{x}_0' = 1$.) By the symmetry of A, the equation can be written

$$(\mathbf{x} - \mathbf{x}_0)(\mathbf{x}_0 A)' = 0, \qquad (\mathbf{x} - \mathbf{x}_0)A\mathbf{x}_0' = 0, \qquad \mathbf{x}A\mathbf{x}_0' = \mathbf{x}_0 A\mathbf{x}_0' = 1.$$

Answer: $\mathbf{x}A\mathbf{x}_0' = 1.$

REMARK: This answer is particularly simple. You just replace one of the $\mathbf{x}$'s in $\mathbf{x}A\mathbf{x}' = 1$ by $\mathbf{x}_0$. For instance, the tangent plane to the hyperboloid $2x^2 + 3y^2 - 4z^2 = 1$ at a point (x_0, y_0, z_0) on the surface is $2xx_0 + 3yy_0 - 4zz_0 = 1$, by inspection.

EXERCISES

Plot the level curves and the gradient field in the region $|x| \leq 3$, $|y| \leq 3$:

1. $z = x - 2y$
2. $z = x^2 - y^2$
3. $z = x^2 + y$
4. $z = x^2 + 4y^2$.

Describe the level surfaces:

5. $w = x + y + z$
6. $w = x^2 + 4y^2 + 9z^2$
7. $w = xyz$
8. $w = x^2 + y^2 - z^2$.

9. For each function in Exs. 5–8, find the gradient field.
10. Suppose $z = f(r, \theta)$ is given in terms of polar coordinates. Show that

$$\operatorname{grad} z = f_r \mathbf{u} + \frac{1}{r} f_\theta \mathbf{w},$$

where $\mathbf{u} = (\cos\theta, \sin\theta)$ and $\mathbf{w} = (-\sin\theta, \cos\theta)$.
11. Find $\operatorname{grad}(r^{-2}\cos 2\theta)$.

Find the tangent line to

12. $x^2 - 3xy + y^2 = -1$ at $(1, 2)$
13. $x + x^3y^4 - y = 0$ at $(0, 0)$
14. $y + \sin xy = 1$ at $(0, 1)$.

Find the tangent plane to

15. $z = x^2 - y^2$ at $(1, 0, 1)$
16. $xyz = 1$ at $(1, 1, 1)$.

17. Let $\mathbf{a}$ be a constant vector. Find the gradient of $f(\mathbf{x}) = \mathbf{a} \cdot \mathbf{x}$.
18*. Let $\mathbf{x}_0$ be a point of the quadric $\mathbf{x}A\mathbf{x}' + 2\mathbf{a}\mathbf{x}' = 1$, where A is a 3×3 symmetric matrix and $\mathbf{a}$ a vector. Assuming $\mathbf{x}_0 A + \mathbf{a} \neq \mathbf{0}$, prove that the tangent plane at $\mathbf{x}_0$ is

$$\mathbf{x}A\mathbf{x}_0' + \mathbf{a}(\mathbf{x}' + \mathbf{x}_0') = 1.$$

Apply this to $z = xy$.

Let $\mathbf{u} = (u, v, w)$ be a vector field (p. 306) on an open domain of $\mathbf{R}^3$ with differentiable coordinate functions. Define the **divergence** and **curl** of $\mathbf{u}$ by

$$\operatorname{div} \mathbf{u} = \frac{\partial u}{\partial x} + \frac{\partial v}{\partial y} + \frac{\partial w}{\partial z}, \qquad \operatorname{curl} \mathbf{u} = \left(\frac{\partial w}{\partial y} - \frac{\partial v}{\partial x}, \frac{\partial u}{\partial z} - \frac{\partial w}{\partial x}, \frac{\partial v}{\partial x} - \frac{\partial u}{\partial y}\right).$$

Prove:

19. $\operatorname{div}(f\mathbf{u}) = (\operatorname{grad} f) \cdot \mathbf{u} + f(\operatorname{div} \mathbf{u})$
20. $\operatorname{div}(f(\rho)\mathbf{x}) = [\rho^3 f(\rho)]'/\rho^2$, where $\rho^2 = x^2 + y^2 + z^2$
21. $\operatorname{curl}(f\mathbf{u}) = (\operatorname{grad} f) \times \mathbf{u} + f(\operatorname{curl} \mathbf{u})$
22. $\operatorname{curl}(\operatorname{grad} f) = \mathbf{0}$
23. $\operatorname{curl}[f(\rho)\mathbf{x}] = \mathbf{0}$, where $\rho^2 = x^2 + y^2 + z^2$
24. $\operatorname{div}(\operatorname{curl} \mathbf{u}) = 0$
25. $\operatorname{curl}(\mathbf{a} \times \mathbf{x}) = 2\mathbf{a}$
26*. $\operatorname{div}[\operatorname{grad} f(\rho)] = [\rho f(\rho)]''/\rho$, where $\rho^2 = x^2 + y^2 + z^2$.

5. DIRECTIONAL DERIVATIVE

Given a function $w = f(x, y, z)$ and a point $\mathbf{x}$ in space, we may ask how fast the function is changing at $\mathbf{x}$ in various directions. (A direction is indicated by a unit vector $\mathbf{u}$.)

The **directional derivative** of $f(x, y, z)$ at a point $\mathbf{x}$ in the direction $\mathbf{u}$ is

$$D_{\mathbf{u}} f(\mathbf{x}) = \frac{d}{dt} f(\mathbf{x} + t\mathbf{u}) \bigg|_{t=0}.$$

Think of a directional derivative this way. Imagine a particle moving along a straight line with constant velocity $\mathbf{u}$, passing through the point $\mathbf{x}$ when $t = 0$. See Fig. 5.1. To each point $\mathbf{x} + t\mathbf{u}$ of its path is assigned the number

$$w(t) = f(\mathbf{x} + t\mathbf{u}).$$

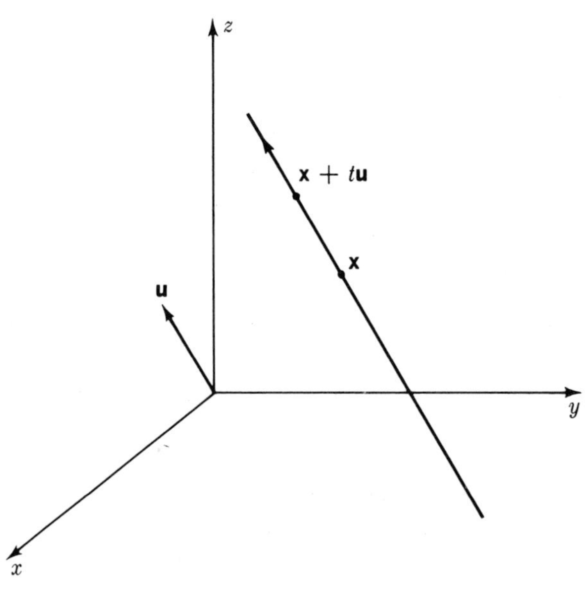

FIG. 5.1

Then

$$D_{\mathbf{u}} f(\mathbf{x}) = w'(0),$$

the rate of change of $w(t)$ as the particle moves through the point $\mathbf{x}$. For example, suppose $f(x, y, z)$ is the steady temperature at each point (x, y, z) of a fluid. Suppose a particle moves with unit speed through a point $\mathbf{x}$ in the direction $\mathbf{u}$. Then $D_{\mathbf{u}} f(\mathbf{x})$ measures the time rate of change of the particle's temperature.

There is a handy formula for directional derivatives. Let

$$\mathbf{p}(t) = \mathbf{x} + t\mathbf{u}.$$

Then $w(t) = f[\mathbf{p}(t)]$, a composite function. By the Chain Rule,

$$D_{\mathbf{u}} f(\mathbf{x}) = \dot{w}(0) = \frac{d}{dt} f[\mathbf{p}(t)]\bigg|_{t=0} = (\text{grad } f)\cdot\dot{\mathbf{p}}(0) = (\text{grad } f)\cdot\mathbf{u}.$$

Since $\mathbf{u}$ is a unit vector, $(\text{grad } f)\cdot\mathbf{u}$ is simply the projection of grad f on $\mathbf{u}$.

> The derivative of f in the direction $\mathbf{u}$ is the projection of grad f on $\mathbf{u}$:
>
> $$D_{\mathbf{u}} f(\mathbf{x}) = (\text{grad } f)\cdot\mathbf{u}.$$

In particular if $\mathbf{u} = \mathbf{i} = (1, 0, 0)$, then

$$D_{\mathbf{i}} f(\mathbf{x}) = (\text{grad } f)\cdot(1, 0, 0) = \left(\frac{\partial f}{\partial x}, \frac{\partial f}{\partial y}, \frac{\partial f}{\partial z}\right)\cdot(1, 0, 0) = \frac{\partial f}{\partial x}.$$

A similar situation holds for $\mathbf{j} = (0, 1, 0)$ and $\mathbf{k} = (0, 0, 1)$.

> The directional derivatives of $f(x, y, z)$ in the directions $\mathbf{i}$, $\mathbf{j}$, and $\mathbf{k}$ are the partial derivatives:
>
> $$D_{\mathbf{i}} f = \frac{\partial f}{\partial x}, \qquad D_{\mathbf{j}} f = \frac{\partial f}{\partial y}, \qquad D_{\mathbf{k}} f = \frac{\partial f}{\partial z}.$$

EXAMPLE 5.1

Compute the directional derivatives of $f(x, y, z) = xy^2z^3$ at $(3, 2, 1)$, in the direction of the vectors

(a) $(-2, -1, 0)$, (b) $(5, 4, 1)$.

Solution:

$$D_{\mathbf{u}} f(\mathbf{x}) = (\text{grad } f)\cdot\mathbf{u},$$

where $\mathbf{u}$ is a unit vector in the desired direction, and grad f is evaluated at $(3, 2, 1)$. Since

$$\frac{\partial f}{\partial x} = y^2z^3, \qquad \frac{\partial f}{\partial y} = 2xyz^3, \qquad \text{and} \qquad \frac{\partial f}{\partial z} = 3xy^2z^2,$$

$$\text{grad } f\bigg|_{(3,2,1)} = (4, 12, 36).$$

Thus

$$D_{\mathbf{u}} f(\mathbf{x}) = (4, 12, 36) \cdot \mathbf{u}.$$

(a) $\mathbf{u} = \dfrac{1}{\sqrt{5}}(-2, -1, 0),$

$$D_{\mathbf{u}} f(\mathbf{x}) = (4, 12, 36) \cdot \frac{1}{\sqrt{5}}(-2, -1, 0) = \frac{-20}{\sqrt{5}}.$$

(b) $\mathbf{u} = \dfrac{1}{\sqrt{42}}(5, 4, 1),$

$$D_{\mathbf{u}} f(\mathbf{x}) = (4, 12, 36) \cdot \frac{1}{\sqrt{42}}(5, 4, 1) = \frac{104}{\sqrt{42}}.$$

Answer: (a) $\dfrac{-20}{\sqrt{5}}$; (b) $\dfrac{104}{\sqrt{42}}$.

Question: In what direction is a given function f increasing fastest? In other words, at a fixed point $\mathbf{x}$ in space, for which unit vector $\mathbf{u}$ is

$$D_{\mathbf{u}} f(\mathbf{x})$$

largest? Now

$$D_{\mathbf{u}} f(\mathbf{x}) = (\operatorname{grad} f) \cdot \mathbf{u} = |\operatorname{grad} f| \cos\theta,$$

where θ is the angle between $\operatorname{grad} f$ and $\mathbf{u}$. Therefore, the largest value of $D_{\mathbf{u}} f(\mathbf{x})$ is $|\operatorname{grad} f|$, taken where $\cos\theta = 1$, that is, $\theta = 0$.

The direction of most rapid increase of $f(x, y, z)$ at a point $\mathbf{x}$ is the direction of the gradient. The derivative in that direction is $|\operatorname{grad} f|$.

The direction of most rapid decrease is opposite to the direction of the gradient.

EXAMPLE 5.2

Find the direction of most rapid increase of the function

$$f(x, y, z) = x^2 + yz$$

at $(1, 1, 1)$ and give the rate of increase in this direction.

Solution:

$$\operatorname{grad} f \Big|_{(1,1,1)} = (2x, z, y) \Big|_{(1,1,1)} = (2, 1, 1).$$

The most rapid increase is

$$D_{\mathbf{u}} f = |\operatorname{grad} f| = \sqrt{2^2 + 1^2 + 1^2} = \sqrt{6},$$

where $\mathbf{u}$ is the direction of grad f:

$$\mathbf{u} = \frac{\operatorname{grad} f}{|\operatorname{grad} f|} = \frac{1}{\sqrt{6}}(2, 1, 1).$$

Answer: $D_{\mathbf{u}} f \Big|_{(1,1,1)} = \sqrt{6}$ for $\mathbf{u} = \dfrac{1}{\sqrt{6}}(2, 1, 1).$

We conclude this section with two examples involving directions of most rapid change of a function.

Consider water running down a hill from a spring at a point P. See Fig. 5.2. The water descends as quickly as possible. What is the path of the stream?

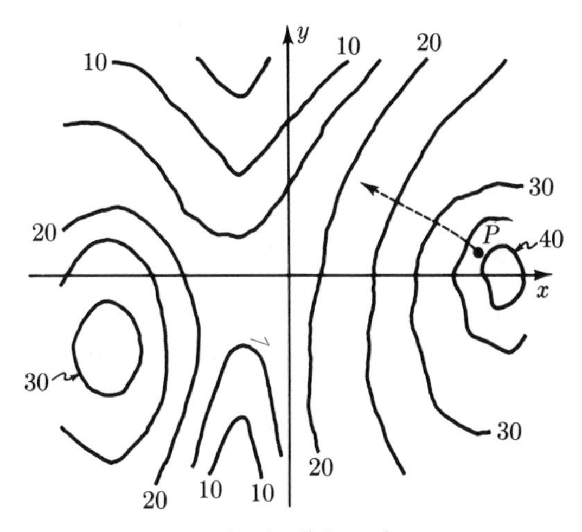

FIG. 5.2 path of quickest descent

From physics, change in kinetic energy $[\frac{1}{2}m(\text{speed})^2]$ equals change in potential energy [height]. Hence the speed of a water particle depends only on how far it has descended (its altitude). Since the speed at a given time does not depend on direction, the particle "chooses" the direction of steepest descent (most rapid change of altitude). Let the hill be represented by the surface $z = f(x, y)$. Then water will flow in the direction of $-\operatorname{grad} f$, that is, perpendicular to the level curves.

Next, consider a function $z = f(x, y)$ and all curves $\mathbf{x}(t) = (x(t), y(t))$ in the x, y-plane which pass through a fixed point $\mathbf{x}_0$ with speed 1. Along which of these is $f(x, y)$ increasing the fastest at $\mathbf{x}_0$?

To find the direction of most rapid increase, write

$$z(t) = f[\mathbf{x}(t)].$$

By the Chain Rule,

$$\dot{z} = (\operatorname{grad} f) \cdot \dot{\mathbf{x}} = (\operatorname{grad} f) \cdot \mathbf{v},$$

where $\mathbf{v}$ is the velocity vector. But $\mathbf{v}$ is a unit vector; hence $\dot{z}$ is the directional derivative in the direction of $\mathbf{v}$ (tangential to the curve $\mathbf{x}(t)$). Thus $\dot{z}$ is greatest for those curves whose tangents at $\mathbf{x}_0$ point in the direction of grad f; such curves are orthogonal to the level curves.

EXERCISES

Find the directional derivative of $f(x, y, z)$ at $\mathbf{x}_0$ in the directions of $\mathbf{v}_1$, $\mathbf{v}_2$, and $\mathbf{v}_3$:

1. $f = x + y + z$; $\mathbf{x}_0 = \mathbf{0}$, $\mathbf{v}_1 = (1, 0, 0)$, $\mathbf{v}_2 = (0, 1, 0)$, $\mathbf{v}_3 = (0, 0, 1)$
2. $f = xy + yz + zx$; $\mathbf{x}_0 = (1, 1, 1)$, $\mathbf{v}_1 = (1, 1, 1)$, $\mathbf{v}_2 = (1, -1, 1)$, $\mathbf{v}_3 = (-1, -1, -1)$
3. $f = xyz$; $\mathbf{x}_0 = (1, -1, 2)$, $\mathbf{v}_1 = (1, 1, 0)$, $\mathbf{v}_2 = (1, 0, 1)$, $\mathbf{v}_3 = (0, 1, 1)$
4. $f = x^2y^2z^2$; $\mathbf{x}_0 = (-1, -1, -1)$, $\mathbf{v}_1 = (1, 2, 3)$, $\mathbf{v}_2 = (1, 1, 0)$, $\mathbf{v}_3 = (3, 2, 1)$.

Find the largest directional derivative of $f(x, y, z)$ at $\mathbf{x}_0$:

5. $f = x^3 + y^2 + z$; $\mathbf{x}_0 = \mathbf{0}$
6. $f = xyz$; $\mathbf{x}_0 = (-1, -1, -1)$
7. $f = x^2 + y^2 + z^2$; $\mathbf{x}_0 = (1, 2, 2)$
8. $f = x^2 - y^2 + 4z^2$; $\mathbf{x}_0 = (-1, -1, 1)$.

6. APPLICATIONS

A **vector field** is the assignment of a vector $\mathbf{F}(\mathbf{x})$ to each point $\mathbf{x}$ of a region $\mathbf{D}$ in space. (The gradient of a function is one example.)

Let $\mathbf{F}(\mathbf{x})$ be a vector field on $\mathbf{D}$ and suppose $\mathbf{x}(t)$ is a path in $\mathbf{D}$ from $\mathbf{x}_0 = \mathbf{x}(t_0)$ to $\mathbf{x}_1 = \mathbf{x}(t_1)$. The line integral

$$\int_{\mathbf{x}_0}^{\mathbf{x}_1} \mathbf{F} \cdot d\mathbf{x}$$

is defined over this path and is computed by the ordinary integral

$$\int_{t_0}^{t_1} \mathbf{F}[\mathbf{x}(t)] \cdot \dot{\mathbf{x}}(t)\, dt.$$

Its value generally depends on the path $\mathbf{x}(t)$ connecting $\mathbf{x}_0$ and $\mathbf{x}_1$.

EXAMPLE 6.1

Let $\mathbf{F}(\mathbf{x})$ be the vector field $\mathbf{F}(\mathbf{x}) = (x, y, x + y + z)$. Let $\mathbf{x}_0 = (0, 0, 0)$ and $\mathbf{x}_1 = (1, 1, 1)$. Compute the line integral $\int_{\mathbf{x}_0}^{\mathbf{x}_1} \mathbf{F} \cdot d\mathbf{x}$ over the paths

(a) $\mathbf{x}(t) = (t, t, t)$, (b) $\mathbf{x}(t) = (t, t^2, t^3)$.

Solution: Notice that for both paths, $\mathbf{x}_0 = \mathbf{x}(0)$ and $\mathbf{x}_1 = \mathbf{x}(1)$.

(a) On this path $\mathbf{x} = (t, t, t)$ and $\dot{\mathbf{x}}(t) = (1, 1, 1)$.

$$\int_{\mathbf{x}_0}^{\mathbf{x}_1} \mathbf{F} \cdot d\mathbf{x} = \int_0^1 \mathbf{F}[\mathbf{x}(t)] \cdot \dot{\mathbf{x}}(t)\, dt = \int_0^1 (t, t, 3t) \cdot (1, 1, 1)\, dt = \int_0^1 5t\, dt = \frac{5}{2}.$$

(b) This time $\mathbf{x} = (t, t^2, t^3)$ and $\dot{\mathbf{x}}(t) = (1, 2t, 3t^2)$.

$$\int_{\mathbf{x}_0}^{\mathbf{x}_1} \mathbf{F} \cdot d\mathbf{x} = \int_0^1 (t, t^2, t + t^2 + t^3) \cdot (1, 2t, 3t^2)\, dt$$

$$= \int_0^1 [t + 2t^3 + 3(t^3 + t^4 + t^5)]\, dt$$

$$= \int_0^1 (t + 5t^3 + 3t^4 + 3t^5)\, dt = \frac{1}{2} + \frac{5}{4} + \frac{3}{5} + \frac{1}{2} = \frac{57}{20}$$

Answer: (a) $\frac{5}{2}$; (b) $\frac{57}{20}$.

In an important special case, the line integral does not depend on the path $\mathbf{x}(t)$, but only on the initial and terminal points $\mathbf{x}_0$ and $\mathbf{x}_1$:

> If the vector field $\mathbf{F}(\mathbf{x})$ is the gradient of some function f,
>
> $$\mathbf{F} = \operatorname{grad} f,$$
>
> then
>
> $$\int_{\mathbf{x}_0}^{\mathbf{x}_1} \mathbf{F} \cdot d\mathbf{x} = f(\mathbf{x}_1) - f(\mathbf{x}_0).$$
>
> Thus the value of the line integral is independent of the path connecting $\mathbf{x}_0$ and $\mathbf{x}_1$.

This assertion is easily verified by means of the Chain Rule,

$$\frac{d}{dt} f[\mathbf{x}(t)] = (\operatorname{grad} f) \cdot \dot{\mathbf{x}}.$$

If $\mathbf{F} = \operatorname{grad} f$, then

$$\int_{\mathbf{x}_0}^{\mathbf{x}_1} \mathbf{F} \cdot d\mathbf{x} = \int_{t_0}^{t_1} (\operatorname{grad} f) \cdot \dot{\mathbf{x}}(t)\, dt$$

$$= \int_{t_0}^{t_1} \frac{d}{dt} f[\mathbf{x}(t)]\, dt = f[\mathbf{x}(t)] \Big|_{t_0}^{t_1} = f(\mathbf{x}_1) - f(\mathbf{x}_0).$$

REMARK: Not every vector field is the gradient of some function. For example, the field $\mathbf{F}(\mathbf{x}) = (x, y, x + y + z)$ is not a gradient;

$$\int_{\mathbf{x}_0}^{\mathbf{x}_1} \mathbf{F} \cdot d\mathbf{x}$$

is *not* independent of the path (see Example 6.1).

Application to Physics

If $\mathbf{F} = \operatorname{grad} f$ is a force, the net work done by this force in moving a particle from $\mathbf{x}_0$ to $\mathbf{x}_1$ is $f(\mathbf{x}_1) - f(\mathbf{x}_0)$, independent of the path. The function f, which is unique up to an additive constant, is called the **potential** of the force.

An important example is that of a central force subject to the inverse square law, for instance, the electric force $\mathbf{E}$ on a unit charge at $\mathbf{x}$ due to a unit charge of the same sign at the origin. The magnitude of the vector $\mathbf{E}$ is inversely proportional to $|\mathbf{x}|^2$. Its direction is the same as that of $\mathbf{x}$. See Fig. 6.1.

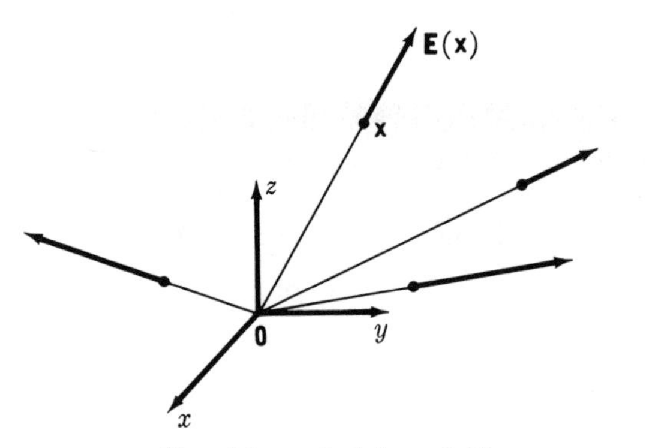

FIG. 6.1 central force field

The unit vector in the direction of $\mathbf{x}$ is

$$\frac{\mathbf{x}}{|\mathbf{x}|} = \frac{\mathbf{x}}{\rho}, \qquad \rho = |\mathbf{x}|.$$

Therefore, expressed in suitable units,

$$\mathbf{E} = \frac{1}{\rho^2}\frac{\mathbf{x}}{\rho} = \frac{\mathbf{x}}{\rho^3}.$$

The force field $\mathbf{E}$ is defined at all points of space except the origin. We shall prove that $\mathbf{E}$ is the gradient of a function, in fact that

$$\mathbf{E} = \operatorname{grad} f, \qquad \text{where } f(x, y, z) = -\frac{1}{\rho} = \frac{-1}{\sqrt{x^2 + y^2 + z^2}}.$$

Let us compute the gradient of $f = -1/\rho$:

$$\frac{\partial f}{\partial x} = \frac{x}{(x^2 + y^2 + z^2)^{3/2}} = \frac{x}{\rho^3},$$

and similarly

$$\frac{\partial f}{\partial y} = \frac{y}{\rho^3}, \qquad \frac{\partial f}{\partial z} = \frac{z}{\rho^3}.$$

Therefore

$$\operatorname{grad} f = \left(\frac{x}{\rho^3}, \frac{y}{\rho^3}, \frac{z}{\rho^3}\right) = \frac{1}{\rho^3}(x, y, z)$$

$$= \frac{\mathbf{x}}{\rho^3} = \mathbf{E}.$$

Since $\mathbf{E} = \operatorname{grad} f$, it follows from our discussion of line integrals that

$$\int_{\mathbf{x}_0}^{\mathbf{x}_1} \mathbf{E} \cdot d\mathbf{x} = f(\mathbf{x}_1) - f(\mathbf{x}_0) = \frac{1}{|\mathbf{x}_0|} - \frac{1}{|\mathbf{x}_1|}.$$

The right-hand side is the **potential difference** or **voltage**. It represents the work done by the electric force when a unit charge moves from $\mathbf{x}_0$ to $\mathbf{x}_1$ *along any path.*

If $\mathbf{x}_1$ is far out, then $1/|\mathbf{x}_1|$ is small, so

$$\int_{\mathbf{x}_0}^{\mathbf{x}_1} \mathbf{E} \cdot d\mathbf{x} \approx \frac{1}{|\mathbf{x}_0|}.$$

As $\mathbf{x}_1$ moves farther out, the approximation improves. In mathematical shorthand,

$$\int_{\mathbf{x}_0}^{\mathbf{x}_1} \mathbf{E} \cdot d\mathbf{x} \longrightarrow \frac{1}{|\mathbf{x}_0|} \qquad \text{as} \quad |\mathbf{x}_1| \longrightarrow \infty,$$

or

$$\int_{\mathbf{x}_0}^{\infty} \mathbf{E} \cdot d\mathbf{x} = \frac{1}{|\mathbf{x}_0|}.$$

Physical conservation laws are usually derived by identifying an appropriate vector field with the gradient of a function and then evaluating a line integral.

EXERCISES

1. Compute $\displaystyle\int_{(0,1)}^{(0,-1)} (xy, 1 + 2y) \cdot d\mathbf{x}$
   (a) along the straight path,
   (b) along the semicircular path passing through $(-1, 0)$.

2. Let $\mathbf{F} = (3x^2y^2z,\ 2x^3yz,\ x^3y^2)$. Show that $\int_{(0,0,0)}^{(1,1,1)} \mathbf{F}\cdot d\mathbf{x}$ is independent of the path, and evaluate it.

3. Let $\mathbf{F} = (x^2 + yz,\ y^2 + zx,\ z^2 + xy)$. Show that $\int_{(0,0,0)}^{(a,b,c)} \mathbf{F}\cdot d\mathbf{x}$ is independent of the path, and evaluate it.

4. Let θ denote the polar angle in the plane. Show that $\operatorname{grad}\theta = \left(\dfrac{-y}{x^2+y^2},\ \dfrac{x}{x^2+y^2}\right)$.

5. Find $\displaystyle\int \frac{-y\,dx + x\,dy}{x^2+y^2}$ over the circle $|\mathbf{x}| = a$.

6. Let $\mathbf{F} = \mathbf{x}/|\mathbf{x}|^5$ and suppose $\mathbf{a} \neq \mathbf{0}$. Show that $\int_{\mathbf{a}}^{\infty} \mathbf{F}\cdot d\mathbf{x}$, taken along any path from $\mathbf{a}$ which does not pass through $\mathbf{0}$ and goes out indefinitely, depends only on $a = |\mathbf{a}|$. Evaluate the integral.
7. (cont.) Do the same for $\mathbf{F} = \mathbf{x}/|\mathbf{x}|^n$, for any $n > 2$,
8. (cont.) Show that $\mathbf{F} = \mathbf{x}/|\mathbf{x}|^2$ is a gradient.

7. IMPLICIT FUNCTIONS

Often a function $y = g(x)$ is defined only as the root of an equation

$$f(x, y) = 0,$$

which may be hard or impossible to solve explicitly. In such a case, the equation is said to define an **implicit function** $y = g(x)$. For example, Fig. 7.1 shows part of the graph of

$$y^6 + y + xy - x = 0.$$

Near the origin, this equation defines y as an implicit function of x. (It is hopeless to express y as an *explicit* function of x.)

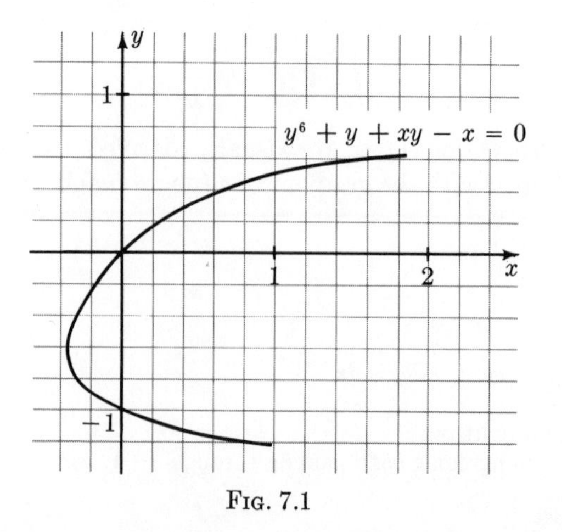

FIG. 7.1

What is the derivative of an implicit function $y = g(x)$ defined by $f(x, y) = 0$? Substitute $y = g(x)$:

$$f[x, g(x)] = 0.$$

Differentiate with respect to x using the Chain Rule:

$$f_x + f_y \cdot g' = 0, \qquad g' = -\frac{f_x}{f_y}.$$

> If y is an implicit function of x defined by
>
> $$f(x, y) = 0,$$
>
> then
>
> $$\frac{dy}{dx} = -\frac{f_x(x, y)}{f_y(x, y)}$$
>
> at each point (x, y) where $f(x, y) = 0$ and $f_y(x, y) \neq 0$.

Differentiation of implicit functions is called **implicit differentiation.**

REMARK: The minus sign in this formula may appear puzzling. It seems to contradict the natural procedure of "canceling" differentials. For instance, if $g = g[y(x)]$, the Chain Rule says $dg/dx = (dg/dy)(dy/dx)$. Hence

$$\frac{dy}{dx} = \frac{dg/dx}{dg/dy},$$

so dg appears to cancel out.

The reason for the minus sign is this. When we write $f(x, y) = 0$, we have "taken y to the other side". The equation $y = g(x)$ is equivalent to $f(x, y) = y - g(x) = 0$. Now "canceling" differentials fails because

$$\frac{f_x}{f_y} = \frac{-g'(x)}{1} = -g'(x).$$

There is the minus sign!

EXAMPLE 7.1

Find $\left.\frac{dy}{dx}\right|_{(0,0)}$ where $y = g(x)$ is defined by $y^6 + y + xy - x = 0$.

Solution:

$$f(x, y) = y^6 + y + xy - x,$$

$$f_x = y - 1, \qquad f_y = 6y^5 + 1 + x.$$

Hence

$$\frac{dy}{dx} = -\frac{f_x}{f_y} = -\frac{y - 1}{6y^5 + 1 + x}.$$

At $(0, 0)$,

$$\frac{dy}{dx} = -\frac{-1}{1} = 1.$$

Alternate Solution: Differentiate the equation

$$y^6 + y + xy - x = 0,$$

treating y as a function of x:

$$6y^5 \frac{dy}{dx} + \frac{dy}{dx} + x\frac{dy}{dx} + y - 1 = 0,$$

$$\frac{dy}{dx} = -\frac{y - 1}{6y^5 + 1 + x}.$$

Answer: $\left.\frac{dy}{dx}\right|_{(0,0)} = 1.$

REMARK: The technique in the alternate solution is equivalent to use of the rule

$$\frac{dy}{dx} = -\frac{f_x}{f_y}$$

because the rule was derived by that very technique.

EXAMPLE 7.2

Let $y = \sqrt{1 - x^2}$. Compute y' and y'' by differentiating implicitly $x^2 + y^2 - 1 = 0$.

Solution: Differentiate:

$$2x + 2yy' = 0, \qquad y' = -\frac{x}{y}.$$

Differentiate again:

$$y'' = -\frac{y - xy'}{y^2} = -\frac{y - x\left(-\dfrac{x}{y}\right)}{y^2} = -\frac{y^2 + x^2}{y^3} = -\frac{1}{y^3},$$

since $x^2 + y^2 = 1$.

Answer:

$$y' = -\frac{x}{y} = \frac{-x}{\sqrt{1 - x^2}},$$

$$y'' = -\frac{1}{y^3} = \frac{-1}{(1 - x^2)^{3/2}}.$$

Applications

Implicit differentiation is useful when a function must be maximized or minimized subject to certain restrictions.

EXAMPLE 7.3

A cylindrical container (right circular) is required to have a given volume V. The material on the top and bottom is k times as expensive as the material on the sides. What are the proportions of the most economical container?

Solution: The cost C of the container is proportional to

$$\text{(area of side)} + k\,\text{(area of top + area of bottom)}.$$

Let r and h denote the radius and height of the container. In the proper units,

$$C = 2\pi rh + k(2\pi r^2),$$

where

$$\pi r^2 h = V, \qquad \text{a constant.}$$

We must minimize C subject to this restriction.

One approach is obvious: solve the last equation for h and substitute into the equation for C. Then C is an explicit function of r which can be minimized.

It is simpler, however, not to make the substitution, but to consider C as a function of r anyway (as if the substitution had been made). Differentiate implicitly:

$$\frac{dC}{dr} = 2\pi\left(r\frac{dh}{dr} + h + 2kr\right).$$

Now differentiate the equation for V with respect to r:

$$2\pi rh + \pi r^2\frac{dh}{dr} = 0, \qquad \frac{dh}{dr} = -\frac{2h}{r}.$$

Substitute this value of dh/dr into the preceding equation:

$$\frac{dC}{dr} = 2\pi\left[r\left(\frac{-2h}{r}\right) + h + 2kr\right] = 2\pi(2kr - h).$$

Hence

$$\frac{dC}{dr} = 0 \qquad \text{when} \quad h = 2kr.$$

It is easily verified that C is minimal for $h = 2kr$. Since $\pi r^2 h$ is constant, h is large if r is small and decreases as r increases. Therefore $(2kr - h)$ increases from negative to positive as r increases. Thus dC/dr satisfies the conditions for C to have a minimum at $h = 2kr$.

Answer: height $= 2k \times$ radius.

REMARK: The special case $k = 1$ is interesting. All parts of the cylinder are equally expensive; the cheapest cylinder is the one with least surface area. Conclusion: Of all cylinders with fixed volume, the one with least surface area is the one whose height is twice its radius.

EXAMPLE 7.4

Find the greatest distance between the origin and a point of the curve $x^4 + y^4 = 1$.

Solution: Draw a graph (Fig. 7.2). Because of symmetry we need consider only $x \geq 0$ and $y \geq 0$. Since the curve lies outside of the circle $x^2 + y^2 = 1$, the maximum distance is greater than 1 and occurs at some point (x, y) where $x > 0$ and $y > 0$.

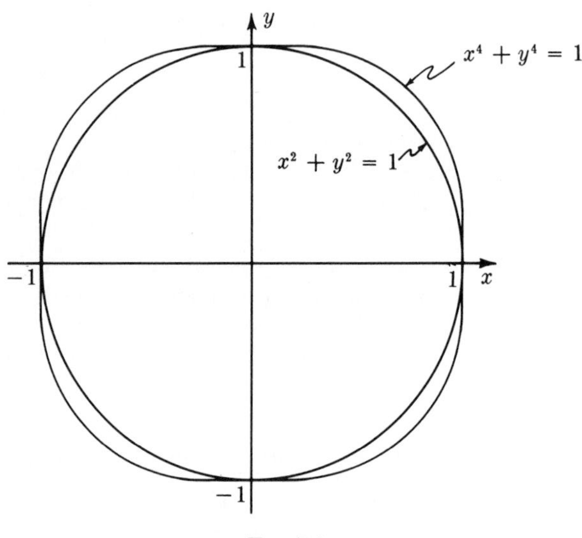

FIG. 7.2

The square of the distance from any point (x, y) to the origin is $x^2 + y^2$. Hence, we must maximize

$$L^2 = x^2 + y^2$$

subject to

$$x^4 + y^4 - 1 = 0.$$

Differentiate both relations with respect to x:

$$\frac{d}{dx}(L^2) = 2x + 2yy', \qquad 4x^3 + 4y^3y' = 0.$$

It follows that

$$\frac{d}{dx}(L^2) = 2x + 2y\left(-\frac{x^3}{y^3}\right) = 2x\left(\frac{y^2 - x^2}{y^2}\right).$$

This derivative vanishes in the first quadrant only for $x = y$. Hence the maximum distance occurs at the point (x, y) of the curve in the first quadrant for which $x = y$. Thus

$$x^4 + y^4 = 1, \qquad x = y,$$

from which

$$x = y = \frac{1}{\sqrt[4]{2}},$$

and

$$L^2 = x^2 + y^2 = \frac{1}{\sqrt{2}} + \frac{1}{\sqrt{2}} = \sqrt{2}$$

Answer: $\sqrt[4]{2}$.

EXERCISES

Find dy/dx:

1. $x + y = x \sin y$
2. $x^2 + y^3 = xy$
3. $e^{xy} = 3xy^2$
4. $x^4 - y^4 = 3x^2y^3$
5. $e^x \sin y = e^y \cos x$
6. $x^3y^3 = x^2 - y^2 + 1$
7. $x^4 + 3y^6 = 1$
8. $x^5 + y^5 = xy + 1$.

9*. Find the maximum and minimum values of $f(x, y, z) = x^4 + y^4 + z^4$ on the surface of the unit sphere $x^2 + y^2 + z^2 = 1$.

10*. (cont.) Deduce that $\frac{1}{3}(x^2 + y^2 + z^2)^2 \leq x^4 + y^4 + z^4 \leq (x^2 + y^2 + z^2)^2$ for *any* (x, y, z).

11*. (cont.) Find the corresponding inequalities for $n \geq 3$ relating $x^n + y^n + z^n$ and $x^2 + y^2 + z^2$. (Assume $x \geq 0$, $y \geq 0$, $z \geq 0$ if n is odd.)

12*. (cont.) Find the largest A and the smallest B so that $A(x^4 + y^4 + z^4)^3 \leq (x^6 + y^6 + z^6)^2 \leq B(x^4 + y^4 + z^4)^3$ for all (x, y, z).

13. Suppose $z(x, y)$ is defined implicitly by $F(x, y, z) = 0$. Assuming $F_z(x, y, z) \neq 0$, prove

$$\frac{\partial z}{\partial x} = -\frac{F_x[x, y, z(x, y)]}{F_z[x, y, z(x, y)]}, \qquad \frac{\partial z}{\partial y} = -\frac{F_y[x, y, z(x, y)]}{F_z[x, y, z(x, y)]}.$$

[*Hint:* Use the Chain Rule.]

14*. (cont.) Suppose $z(x, y)$ and $w(x, y)$ are defined implicitly by

$$\begin{cases} F(x, y, z, w) = 0 \\ G(x, y, z, w) = 0. \end{cases}$$

Find formulas for $\partial z/\partial x$, $\partial z/\partial y$, $\partial w/\partial x$, $\partial w/\partial y$ under suitable hypotheses.

8. DIFFERENTIALS

We shall re-examine our basic approximation formula

$$f(\mathbf{c}+\mathbf{x}) = f(\mathbf{c}) + \mathbf{k}\cdot\mathbf{x} + e(\mathbf{x})$$

with error $e(\mathbf{x})$. We assume that the domain $\mathbf{D}$ of f is an open set and that f is differentiable at every point $\mathbf{c}$ of $\mathbf{D}$. Hence

$$\mathbf{k} = \left(\frac{\partial f}{\partial x}, \frac{\partial f}{\partial y}, \frac{\partial f}{\partial z}\right)\bigg|_{\mathbf{c}}.$$

Suppose we allow *both* $\mathbf{c}$ and $\mathbf{x}$ to vary. It is customary to replace $\mathbf{c}$ by $\mathbf{x}$ and $\mathbf{x}$ by $d\mathbf{x} = (dx, dy, dz)$, a new quantity. In this notation,

$$f(\mathbf{x} + d\mathbf{x}) = f(\mathbf{x}) + \boxed{\mathbf{k}\cdot d\mathbf{x}} + e(d\mathbf{x}).$$

The boxed quantity is particularly significant. For each $\mathbf{x}$ it is a linear function of $d\mathbf{x}$. We therefore give it a name, df, the **differential** of f, and write

$$\boxed{df = \frac{\partial f}{\partial x}\,dx + \frac{\partial f}{\partial y}\,dy + \frac{\partial f}{\partial z}\,dz.}$$

The differential is really a function of the six variables x, y, z, dx, dy, dz, where $\mathbf{x}$ is confined to $\mathbf{D}$ and $d\mathbf{x}$ is arbitrary. We could even write $df = df(\mathbf{x}, d\mathbf{x})$ but this is a cumbersome notation.

The differential has elementary algebraic properties that correspond to analogous results for derivatives:

$$\boxed{\begin{array}{ll} d(f+g) = df + dg & d(af) = a\,df \\[2ex] d(fg) = (df)g + f\,dg & d(f/g) = \dfrac{(df)g - f\,dg}{g^2}, \quad g \neq 0. \end{array}}$$

Substitution

The differential has a useful formal property of remaining unchanged when the variables are replaced by functions of other variables. This is another form of the Chain Rule.

For instance, suppose $f = f(x, y, z)$ so

$$df = f_x\,dx + f_y\,dy + f_z\,dz.$$

Now suppose $x = x(u, v)$, $y = y(u, v)$, $z = z(u, v)$. We can look at the composite function $f[x(u, v)]$ and compute its differential, another "df":

$$df = f_u\,du + f_v\,dv.$$

Is this different? No, because by the Chain Rule

$$\begin{aligned} f_u\,du + f_v\,dv &= (f_x x_u + f_y y_u + f_z z_u)\,du + (f_x x_v + f_y y_v + f_z z_v)\,dv \\ &= f_x(x_u\,du + x_v\,dv) + f_y(y_u\,du + y_v\,dv) + f_z(z_u\,du + z_v\,dv) \\ &= f_x\,dx + f_y\,dy + f_z\,dz. \end{aligned}$$

This may appear as simply a consequence of sloppy notation for composite functions. Still there is something more to it.

Suppose we have a function f of independent variables u, v, but there are some intermediate variables in the way. After some computation we arrive at an expression

$$df = M\,du + N\,dv. \tag{1}$$

Then we know automatically that $M = f_u$ and $N = f_v$ no matter how we obtained (1).

For example, suppose $z = z(x, y)$, and we know that

$$x^2 + y^2 + z^2 = 1.$$

Then, since the differential of a constant function is 0,

$$x\,dx + y\,dy + z\,dz = 0,$$

hence

$$dz = -\frac{x}{z}\,dx - \frac{y}{z}\,dy.$$

We conclude that $\partial z/\partial x = -x/z$ and $\partial z/\partial y = -y/z$.

More generally, if $z = z(x, y)$ is constrained by a relation

$$F(x, y, z) = 0,$$

then

$$F_x\,dx + F_y\,dy + F_z\,dz = 0,$$

which implies

$$dz = -\frac{F_x}{F_z}\,dx - \frac{F_y}{F_z}\,dy,$$

whenever $F_z \neq 0$ at $(x, y, z(x, y))$. Therefore

$$\frac{\partial z}{\partial x} = -\frac{F_x}{F_z}, \qquad \frac{\partial z}{\partial y} = -\frac{F_y}{F_z}.$$

Here is a situation where an intermediate variable is in the way. Suppose $z = z(x, y)$ is subject to two constraints,

$$\begin{cases} F(x, y, z, u) = 0 \\ G(x, y, u) = 0. \end{cases}$$

If we could solve the second relation for u and substitute the solution in the first, the result would be a relation $H(x, y, z) = 0$. But that might be impractical to carry out, so we proceed indirectly by forming differentials:

$$\begin{cases} F_x\,dx + F_y\,dy + F_z\,dz + F_u\,du = 0 \\ G_x\,dx + G_y\,dy \qquad\qquad + G_u\,du = 0. \end{cases}$$

Now we eliminate du:

$$(G_uF_x - F_uG_x)\,dx + (G_uF_y - F_uG_y)\,dy + G_uF_z\,dz = 0.$$

At points where $G_u \neq 0$ and $F_z \neq 0$, we have

$$\frac{\partial z}{\partial x} = -\frac{G_uF_x - F_uG_x}{G_uF_z}, \qquad \frac{\partial z}{\partial y} = -\frac{G_uF_y - F_uG_y}{G_uF_z}.$$

Numerical Approximations

The approximation

$$f(\mathbf{x} + d\mathbf{x}) \approx f(\mathbf{x}) + df$$

can supply quick numerical estimates.

EXAMPLE 8.1

Estimate $(2.01)^{0.98}$.

Solution: Set $f(x, y) = x^y$. Then

$$\ln f = y \ln x$$

$$\frac{df}{f} = \frac{y}{x}\,dx + (\ln x)\,dy.$$

Set $x = 2$, $y = 1$, $dx = 0.01$, $dy = -0.02$. Also use the estimate $\ln 2 \approx 0.69$. Then $f(2, 1) = 2$ and

$$\frac{df}{2} \approx \frac{1}{2}(0.01) - (0.69)(0.02) = -0.0088,$$

$$df \approx -0.0176, \qquad f \approx 2 + df \approx 1.9824.$$

Answer: 1.9824.

REMARK: By 4-place logs, $(2.01)^{0.98} \approx 1.982$.

EXERCISES

Compute dz:

1. $xyz = 1$

2. $\dfrac{x^2}{a^2} + \dfrac{y^2}{b^2} + \dfrac{z^2}{c^2} = 1$

3. $x = z^2 - y^2$

4. $ze^{-xy} + xe^{-yz} = 0.$

5. For $x = r\cos\theta$, $y = r\sin\theta$, prove $r\,dr = x\,dx + y\,dy$ and $x\,dy - y\,dx = r^2\,d\theta$.
6. Let $z = z(x, y)$, $p = \partial z/\partial x$, $q = \partial z/\partial y$, $r = \partial^2 z/\partial x^2$, $s = \partial^2 z/\partial x\,\partial y$, $t = \partial^2 z/\partial y^2$. Express dz, dp, dq in terms of dx, dy.
7. Let f be homogeneous of degree n, that is, $f(tx, ty, tz) = t^n f(x, y, z)$ for $t > 0$. Compute "d" of both sides of this relation and equate coefficients of dx, dy, dz, dt. What results? (Compare Ex. 24, p. 291.)
8. Let $z = z(x, y)$ be defined by

$$\begin{cases} x^2 + y^2 + z^2 = t^2 \\ y = tx. \end{cases}$$

Find z_x and z_y.

If $\mathbf{v} = \mathbf{v}(s, t) = (f(s, t), g(s, t), h(s, t))$ is a vector function of several variables, then we define $d\mathbf{v}$ by $d\mathbf{v} = (df, dg, dh) = \mathbf{v}_s\,ds + \mathbf{v}_t\,dt$.

9. Consider the vectors $\mathbf{x} = (x, y)$, $\mathbf{u} = (\cos\theta, \sin\theta)$, $\mathbf{w} = (-\sin\theta, \cos\theta)$ of Chapter 5, Section 7, p. 193. Prove that

$$\begin{cases} d\mathbf{u} = \mathbf{w}\,d\theta, \qquad d\mathbf{w} = -\mathbf{u}\,d\theta, \\ d\mathbf{x} = \mathbf{u}\,dr + r\mathbf{w}\,d\theta. \end{cases}$$

Estimate using differentials:

10. $5.1 \times 7.1 \times 9.9$

11. $\sqrt{(5.99)^2 - (3.02)^3}$.

9. PROOF OF THE CHAIN RULE [optional]

Vector-valued Functions

In Chapter 7 we discussed linear functions $L\colon \mathbf{R}^3 \longrightarrow \mathbf{R}^3$. In column vector notation, each linear L has the form

$$L(\mathbf{x}') = A\mathbf{x}',$$

where A is a 3×3 matrix. It will be convenient throughout this section to think of $\mathbf{R}^3$ as the space of column vectors, even when we deal with more general functions, not necessarily linear.

We shall consider functions $F\colon \mathbf{D} \longrightarrow \mathbf{R}^3$, where $\mathbf{D} \subseteq \mathbf{R}^3$. Such a function assigns to each $\mathbf{x}'$ in the domain $\mathbf{D}$ a column vector $F(\mathbf{x}')$ in $\mathbf{R}^3$. We may write

$$F(\mathbf{x}') = \begin{bmatrix} f_1(\mathbf{x}') \\ f_2(\mathbf{x}') \\ f_3(\mathbf{x}') \end{bmatrix}.$$

Thus the vector-valued function F is equivalent to a triple f_1, f_2, f_3 of real-valued functions, the components of F.

Our first chore is to say when a vector-valued function is differentiable.

Differentiable Vector-valued Function Let $F: \mathbf{D} \longrightarrow \mathbf{R}^3$, where $\mathbf{D} \subseteq \mathbf{R}^3$. Then F is **differentiable** at a point $\mathbf{c}'$ of $\mathbf{D}$ if there are a matrix A and a vector-valued function E such that

(i) $F(\mathbf{c}' + \mathbf{x}') = F(\mathbf{c}') + A\mathbf{x}' + E(\mathbf{x}')$ whenever $\mathbf{c}' + \mathbf{x}' \in \mathbf{D}$,

(ii) $\dfrac{1}{|\mathbf{x}'|} E(\mathbf{x}') \longrightarrow \mathbf{0}$ as $\mathbf{x}' \longrightarrow \mathbf{0}$.

It is understood in (ii) that $\mathbf{c}' + \mathbf{x}' \in \mathbf{D}$ when $\mathbf{x}' \longrightarrow \mathbf{0}$.

We want to identify the matrix A in (i). If we compare this situation with that of a differentiable $f: \mathbf{R}^3 \longrightarrow \mathbf{R}$, as discussed in Section 1, we suspect that A must involve partial derivatives. That this is so will fall out of the following characterization of differentiability.

Theorem Let $F: \mathbf{D} \longrightarrow \mathbf{R}^3$ and let f_1, f_2, f_3 be the components of F, so $f_i: \mathbf{D} \longrightarrow \mathbf{R}$. Then F is differentiable at a point $\mathbf{c}'$ if and only if f_1, f_2, and f_3 are differentiable at $\mathbf{c}'$.

Proof: Suppose F is differentiable. Write

$$(1) \qquad F(\mathbf{x}') = \begin{bmatrix} f_1(\mathbf{x}') \\ f_2(\mathbf{x}') \\ f_3(\mathbf{x}') \end{bmatrix}, \qquad A = \begin{bmatrix} \mathbf{a}_1 \\ \mathbf{a}_2 \\ \mathbf{a}_3 \end{bmatrix}, \qquad E(\mathbf{x}') = \begin{bmatrix} e_1(\mathbf{x}') \\ e_2(\mathbf{x}') \\ e_3(\mathbf{x}') \end{bmatrix},$$

where $\mathbf{a}_1, \mathbf{a}_2, \mathbf{a}_3$ are the rows of A. From

$$(2) \qquad F(\mathbf{c}' + \mathbf{x}') = F(\mathbf{c}') + A\mathbf{x}' + E(\mathbf{x}')$$

we have

$$(3) \qquad f_i(\mathbf{c}' + \mathbf{x}') = f_i(\mathbf{c}') + \mathbf{a}_i\mathbf{x}' + e_i(\mathbf{x}').$$

The vector function $|\mathbf{x}'|^{-1} E(\mathbf{x}') \longrightarrow \mathbf{0}$ as $\mathbf{x}' \longrightarrow \mathbf{0}$, hence each of its components approaches 0, that is, $|\mathbf{x}'|^{-1} e_i(\mathbf{x}') \longrightarrow 0$. Hence by (3), each f_i is differentiable.

Conversely, if each f_i is differentiable at $\mathbf{c}'$, then (3) holds and $|\mathbf{x}'|^{-1} e_i(\mathbf{x}') \longrightarrow 0$ for $i = 1, 2, 3$. Consequently (2) holds with A and $E(\mathbf{x}')$ as in (1); furthermore $|\mathbf{x}'|^{-1} E(\mathbf{x}') \longrightarrow \mathbf{0}$ as $\mathbf{x}' \longrightarrow \mathbf{0}$ because each component of $|\mathbf{x}'|^{-1} E(\mathbf{x}')$ approaches 0. Therefore F is differentiable.

Theorem Let F be differentiable at an interior point $\mathbf{c}'$ of its domain and let

$$F(\mathbf{c}' + \mathbf{x}') = F(\mathbf{c}') + A\mathbf{x}' + E(\mathbf{x}')$$

where $|\mathbf{x}'|^{-1} E(\mathbf{x}') \longrightarrow \mathbf{0}$ as $\mathbf{x}' \longrightarrow \mathbf{0}$. Then

$$A = \begin{bmatrix} \dfrac{\partial f_1}{\partial x_1} & \dfrac{\partial f_1}{\partial x_2} & \dfrac{\partial f_1}{\partial x_3} \\[2ex] \dfrac{\partial f_2}{\partial x_1} & \dfrac{\partial f_2}{\partial x_2} & \dfrac{\partial f_2}{\partial x_3} \\[2ex] \dfrac{\partial f_3}{\partial x_1} & \dfrac{\partial f_3}{\partial x_2} & \dfrac{\partial f_3}{\partial x_3} \end{bmatrix},$$

where f_1, f_2, f_3 are the components of F and all partials are evaluated at $\mathbf{c}$.

This is an immediate consequence of the corresponding result for the functions $f_i\colon \mathbf{R}^3 \longrightarrow \mathbf{R}$ and the previous theorem.

The matrix A written above is called the **Jacobian matrix** of F. It is important in the study of transformations of regions of space and in change of variables in multiple integrals.

Chain Rule

We first consider a composite function $h = f \circ G$, where G is a differentiable function on a domain $\mathbf{S}$ of $\mathbf{u}$-space into $\mathbf{x}$-space and f is a differentiable function on a domain $\mathbf{D}$ of $\mathbf{x}$-space into $\mathbf{R}$. We assume G sends $\mathbf{S}$ into $\mathbf{D}$. We want to prove that h is differentiable and to compute its partials. See Fig. 9.1.

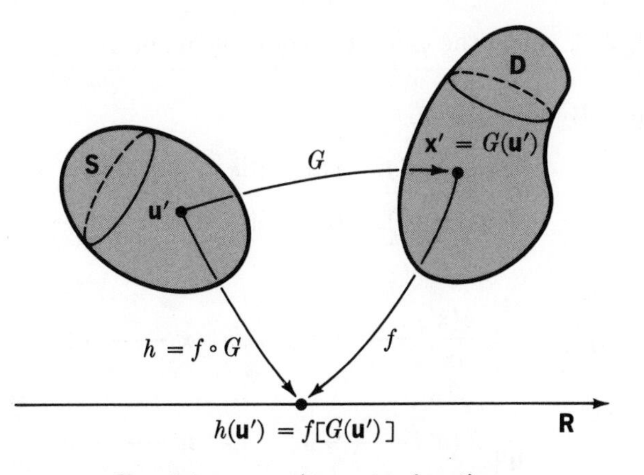

FIG. 9.1 composite vector function

Let us experiment with the special case in which both f and G are non-homogeneous linear:

$$f(\mathbf{x}') = \mathbf{d} + \mathbf{k}\mathbf{x}', \qquad G(\mathbf{u}') = \mathbf{b}' + A\mathbf{u}'.$$

Then

$$h(\mathbf{u}') = f[G(\mathbf{u}')] = \mathbf{d} + \mathbf{k}\mathbf{b}' + \mathbf{k}A\mathbf{u}'.$$

For any $\mathbf{c}'$,

$$h(\mathbf{c}' + \mathbf{u}') = \mathbf{d} + \mathbf{k}\mathbf{b}' + \mathbf{k}A(\mathbf{c}' + \mathbf{u}') = h(\mathbf{c}') + \mathbf{k}A\mathbf{u}'.$$

Therefore h is also non-homogeneous linear and

$$\left(\frac{\partial h}{\partial u_1}, \frac{\partial h}{\partial u_2}, \frac{\partial h}{\partial u_3}\right) = \mathbf{k}A = \left(\frac{\partial f}{\partial x_1}, \frac{\partial f}{\partial x_2}, \frac{\partial f}{\partial x_3}\right)\left[\frac{\partial g_i}{\partial u_j}\right].$$

With this easy case as a guide, we are ready to formulate the general result.

Chain Rule Let $f\colon \mathbf{D} \longrightarrow \mathbf{R}$ be defined on a domain $\mathbf{D}$ in $\mathbf{x}$-space, differentiable at a point $\mathbf{a}'$. Let $G\colon \mathbf{S} \longrightarrow \mathbf{R}^3$ be defined on a domain $\mathbf{S}$ of $\mathbf{u}$-space into $\mathbf{x}$-space, differentiable at a point $\mathbf{c}'$ such that $G(\mathbf{c}') = \mathbf{a}'$. Assume $G(\mathbf{u}') \in \mathbf{D}$ whenever $\mathbf{u}' \in \mathbf{S}$.

Then the composite function $h = f \circ G$ is differentiable at $\mathbf{c}'$.

If

$$f(\mathbf{a}' + \mathbf{x}') \approx f(\mathbf{a}') + \mathbf{k}\mathbf{x}' \quad \text{and} \quad G(\mathbf{c}' + \mathbf{u}') \approx G(\mathbf{c}') + A\mathbf{u}',$$

then

$$h(\mathbf{c}' + \mathbf{u}') \approx h(\mathbf{c}') + \mathbf{k}A\mathbf{u}'.$$

Finally, if $\mathbf{a}'$ and $\mathbf{c}'$ are interior points of $\mathbf{D}$ and $\mathbf{S}$ respectively, then

$$\frac{\partial h(\mathbf{c}')}{\partial u_j} = \sum_{i=1}^{3} \frac{\partial f(\mathbf{a}')}{\partial x_i} \frac{\partial g_i(\mathbf{c}')}{\partial u_j}$$

for $j = 1, 2, 3$, where the g_i are the components of G.

Proof: By hypothesis,

$$f(\mathbf{a}' + \mathbf{x}') = f(\mathbf{a}') + \mathbf{k}\mathbf{x}' + e(\mathbf{x}') \tag{1}$$

$$G(\mathbf{c}' + \mathbf{u}') = G(\mathbf{c}') + A\mathbf{u}' + E(\mathbf{u}'), \tag{2}$$

where

$$\frac{e(\mathbf{x}')}{|\mathbf{x}'|} \longrightarrow 0 \quad \text{as} \quad \mathbf{x}' \longrightarrow \mathbf{0} \tag{3}$$

$$\frac{1}{|\mathbf{u}'|} E(\mathbf{u}') \longrightarrow \mathbf{0} \quad \text{as} \quad \mathbf{u}' \longrightarrow \mathbf{0}. \tag{4}$$

Substitute:

$$\begin{aligned} h(\mathbf{c}' + \mathbf{u}') &= f[G(\mathbf{c}' + \mathbf{u}')] \\ &= f[G(\mathbf{c}') + A\mathbf{u}' + E(\mathbf{u}')] \\ &= f[G(\mathbf{c}')] + \mathbf{k}[A\mathbf{u}' + E(\mathbf{u}')] + e[A\mathbf{u}' + E(\mathbf{u}')], \end{aligned}$$

hence

$$(5) \qquad h(\mathbf{c}' + \mathbf{u}') = h(\mathbf{c}') + \mathbf{k}A\mathbf{u}' + \{\mathbf{k}E(\mathbf{u}') + e[A\mathbf{u}' + E(\mathbf{u}')]\}.$$

We must prove that the term in braces goes to zero faster than $\mathbf{u}'$, that is,

$$\frac{\mathbf{k}E(\mathbf{u}') + e[A\mathbf{u}' + E(\mathbf{u}')]}{|\mathbf{u}'|} \longrightarrow 0 \quad \text{as} \quad \mathbf{u}' \longrightarrow \mathbf{0}.$$

For then (5) says that h is differentiable at $\mathbf{c}'$ and that $h(\mathbf{c}' + \mathbf{u}') \approx h(\mathbf{c}') + \mathbf{k}A\mathbf{u}'$.

Clearly

$$\left| \frac{\mathbf{k}E(\mathbf{u}')}{|\mathbf{u}'|} \right| \le |k| \left| \frac{1}{|\mathbf{u}'|} E(\mathbf{u}') \right| \longrightarrow 0$$

by (4). It remains to prove that

$$\frac{e[A\mathbf{u}' + E(\mathbf{u}')]}{|\mathbf{u}'|} \longrightarrow 0 \quad \text{as} \quad \mathbf{u}' \longrightarrow \mathbf{0}.$$

This is the crux of the proof, and it requires some care and patience.

First, by the triangle inequality

$$|A\mathbf{u}' + E(\mathbf{u}')| \le |A\mathbf{u}'| + |E(\mathbf{u}')|.$$

We proved in Chapter 7, Section 4, that $|A\mathbf{u}'| \le b_1 |\mathbf{u}'|$ where b_1 is a constant. We also have $|E(\mathbf{u}')| \le |\mathbf{u}'|$ for $|\mathbf{u}'|$ sufficiently small since $|\mathbf{u}'|^{-1}E(\mathbf{u}') \longrightarrow \mathbf{0}$ by (2). Therefore

$$|A\mathbf{u}' + E(\mathbf{u}')| \le b\,|\mathbf{u}'|,$$

where $b = b_1 + 1$ is a constant.

Now write $e(\mathbf{x}') = |\mathbf{x}'|\, e_1(\mathbf{x}')$, where by (3) we have $e_1(\mathbf{x}') \longrightarrow 0$ as $\mathbf{x}' \longrightarrow \mathbf{0}$. Then

$$\begin{aligned} |e[A\mathbf{u}' + E(\mathbf{u}')]| &= |A\mathbf{u}' + E(\mathbf{u}')|\, |e_1[A\mathbf{u}' + E(\mathbf{u}')]| \\ &\le b\,|\mathbf{u}'|\, |e_1[A\mathbf{u}' + E(\mathbf{u}')]|. \end{aligned}$$

But $A\mathbf{u}' + E(\mathbf{u}') \longrightarrow \mathbf{0}$ as $\mathbf{u}' \longrightarrow \mathbf{0}$, hence

$$\left| \frac{e[A\mathbf{u}' + E(\mathbf{u}')]}{|\mathbf{u}'|} \right| \le b\, |e_1[A\mathbf{u}' + E(\mathbf{u}')]| \longrightarrow 0$$

as $\mathbf{u}' \longrightarrow \mathbf{0}$. This completes the proof of the first part of the Chain Rule.

If $\mathbf{c}'$ and $\mathbf{a}'$ are interior points, then

$$k = \left(\frac{\partial f}{\partial x_1}, \frac{\partial f}{\partial x_2}, \frac{\partial f}{\partial x_3}\right)\bigg|_{\mathbf{a}'} \quad \text{and} \quad A = \begin{bmatrix} \dfrac{\partial g_1}{\partial u_1} & \dfrac{\partial g_1}{\partial u_2} & \dfrac{\partial g_1}{\partial u_3} \\ \dfrac{\partial g_2}{\partial u_1} & \dfrac{\partial g_2}{\partial u_2} & \dfrac{\partial g_2}{\partial u_3} \\ \dfrac{\partial g_3}{\partial u_1} & \dfrac{\partial g_3}{\partial u_2} & \dfrac{\partial g_3}{\partial u_3} \end{bmatrix}_{\mathbf{c}'}.$$

We have proved that

$$h(\mathbf{c}' + \mathbf{u}') = h(\mathbf{c}') + \mathbf{k}A\mathbf{u}' + E(\mathbf{u}')$$

where $|\mathbf{u}'|^{-1} E(\mathbf{u}') \longrightarrow \mathbf{0}$ as $\mathbf{u}' \longrightarrow \mathbf{0}$. Hence the elements of the row vector $\mathbf{k}A$ are the partials of h at $\mathbf{c}'$. But the j-th element of kA is precisely

$$\sum_{i=1}^{3} \frac{\partial f}{\partial x_i} \frac{\partial g_i}{\partial u_j}$$

as required.

Composite of Vector-valued Functions

We next consider a more general form of the Chain Rule. It turns out to be an easy consequence of the previous one.

Theorem Let $F\colon \mathbf{D} \longrightarrow \mathbf{R}^3$ and $G\colon \mathbf{S} \longrightarrow \mathbf{R}^3$, where $G(\mathbf{u}') \in \mathbf{D}$ for each $\mathbf{u}' \in \mathbf{S}$. Suppose that $G(\mathbf{c}') = \mathbf{a}'$, that G is differentiable at $\mathbf{c}'$, and that F is differentiable at $\mathbf{a}'$. Then the composite function $H = F \circ G$ is differentiable at $\mathbf{c}'$. If

$$F(\mathbf{x}') \approx F(\mathbf{a}') + A\mathbf{x}' \quad \text{and} \quad G(\mathbf{u}') \approx G(\mathbf{c}') + B\mathbf{u}',$$

then

$$H(\mathbf{u}') \approx H(\mathbf{c}') + (AB)\mathbf{u}'.$$

If $\mathbf{a}'$ and $\mathbf{c}'$ are interior points, then

$$\frac{h_i(\mathbf{c}')}{\partial u_k} = \sum_{j=1}^{3} \frac{\partial f_i(\mathbf{a}')}{\partial x_j} \frac{\partial g_j(\mathbf{c}')}{\partial u_k}.$$

Proof: Write

$$H(\mathbf{u}') = \begin{bmatrix} h_1(\mathbf{u}') \\ h_2(\mathbf{u}') \\ h_3(\mathbf{u}') \end{bmatrix} \quad \text{and} \quad F(\mathbf{x}') = \begin{bmatrix} f_1(\mathbf{x}') \\ f_2(\mathbf{x}') \\ f_3(\mathbf{x}') \end{bmatrix}.$$

Then $h_i = f_i \circ G$ because $H = F \circ G$. Since F is differentiable, each f_i is differentiable. Hence, by the previous theorem, each h_i is differentiable and

$$\frac{\partial h_i}{\partial u_k} = \sum_{j=1}^{3} \frac{\partial f_i}{\partial x_j} \frac{\partial g_j}{\partial u_k}$$

in the case of interior points. But this implies H is differentiable since a vector-valued function is differentiable if and only if its component functions are differentiable.

REMARK: The theorem asserts that the Jacobian matrix of the composite function $F \circ G$ built from F and G is the matrix product AB of their respective Jacobian matrices A and B.

Other Chain Rules

The definitions, theorems, and proofs in this section can be modified with little trouble to cover all useful versions of the Chain Rule. The most general case might be indicated schematically as follows

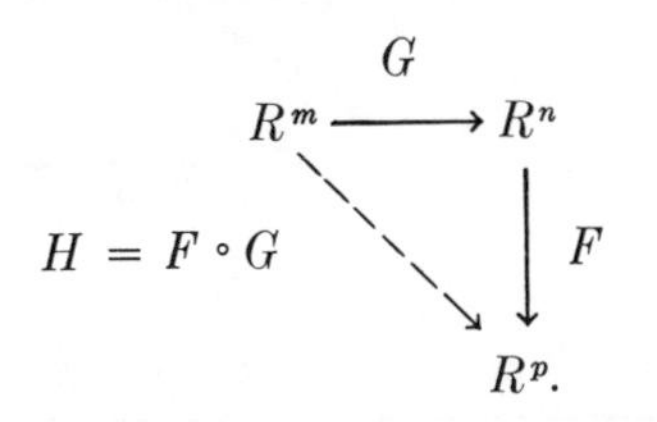

Here F and G are differentiable on suitable domains. Then H is differentiable and

$$\frac{\partial h_i}{\partial u_k} = \sum_{j=1}^{n} \frac{\partial h_i}{\partial x_j} \frac{\partial g_j}{\partial u_k}$$

at interior points. The long proof starting on page 322 covers the case $p = 1$ almost verbatim, and the last proof above extends to arbitrary p.

9. Higher Partial Derivatives

1. MIXED PARTIALS

Differentiate a function $f(x, y)$ of two variables. There are two first derivatives,

$$f_x(x, y) \qquad \text{and} \qquad f_y(x, y),$$

each itself a function of two variables. Each in turn has two first partial derivatives; these four new functions are the second derivatives of $f(x, y)$. Figure 1.1 shows their evolution:

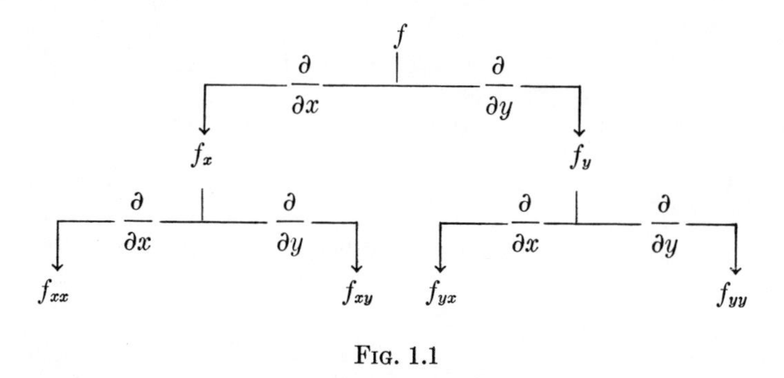

Fig. 1.1

The **pure** second partials

$$f_{xx} \qquad \text{and} \qquad f_{yy}$$

represent nothing really new. Each is found by holding one variable constant and differentiating twice with respect to the other variable.

Alternate notation:

$$f_{xx} = \frac{\partial^2 f}{\partial x^2}, \qquad f_{yy} = \frac{\partial^2 f}{\partial y^2}.$$

For example, if $f(x, y) = x^3y^4 + \cos 5y$,

$$f_x = 3x^2y^4, \qquad f_y = 4x^3y^3 - 5 \sin 5y,$$

$$f_{xx} = 6xy^4, \qquad f_{yy} = 12x^3y^2 - 25 \cos 5y.$$

The **mixed** second partials

$$f_{xy} = \frac{\partial}{\partial y}\left(\frac{\partial f}{\partial x}\right) = \frac{\partial^2 f}{\partial y\, \partial x} \qquad \text{and} \qquad f_{yx} = \frac{\partial}{\partial x}\left(\frac{\partial f}{\partial y}\right) = \frac{\partial^2 f}{\partial x\, \partial y}$$

are new. The mixed partial f_{xy} measures the rate of change in the y-direction of the rate of change of f in the x-direction. The other mixed partial f_{yx} measures the rate of change in the x-direction of the rate of change of f in the y-direction. It is not easy to see how, if at all, the two mixed partials are related to each other.

Let us compute the mixed partials of the function $f(x, y) = x^3y^4 + \cos 5y$:

$$f_x = 3x^2y^4, \qquad f_{xy} = 3\cdot 4x^2y^3.$$

$$f_y = 4x^3y^3 - 5\sin 5y, \qquad f_{yx} = 4\cdot 3x^2y^3 + 0.$$

The mixed partials are equal! This is not an accident but a special case of a general phenomenon, true for functions normally encountered in applications.

Theorem Let $f(x, y)$ be defined on an open set $\mathbf{D}$. If f_{xy} and f_{yx} exist at each point of $\mathbf{D}$ and are continuous at (a, b), then

$$\frac{\partial^2 f}{\partial x\, \partial y} = \frac{\partial^2 f}{\partial y\, \partial x} \qquad \text{at} \quad (a, b).$$

Proof: We consider the mixed second difference

$$(1) \quad \Delta = [f(a + h, b + k) - f(a + h, b)] - [f(a, b + k) - f(a, b)]$$

and its alternate form

$$(2) \quad \Delta = [f(a + h, b + k) - f(a, b + k)] - [f(a + h, b) - f(a, b)].$$

We shall apply the Mean Value Theorem (twice) to (1). To do this, we set

$$g(x) = f(x, b + k) - f(x, b).$$

Then

$$\Delta = g(a + h) - g(a) = hg'(x_1),$$

where x_1 is between a and $a + h$. Next,

$$g'(x_1) = f_x(x_1, b + k) - f_x(x_1, b) = kf_{xy}(x_1, y_1),$$

where y_1 is between b and $b + k$. Hence

$$(3) \qquad \Delta = hkf_{xy}(x_1, y_1).$$

We apply similar reasoning to the second expression for Δ to obtain

$$(4) \qquad \Delta = hkf_{yx}(x_2, y_2),$$

where x_2 is between a and $a + h$ and y_2 is between b and $b + k$.

Now we take $h = k \neq 0$. By (3) and (4),

$$f_{xy}(x_1, y_1) = f_{yx}(x_2, y_2). \tag{5}$$

Let $h \longrightarrow 0$. Then $(x_1, y_1) \longrightarrow (a, b)$ and $(x_2, y_2) \longrightarrow (a, b)$. Sinc
f_{xy} and f_{yx} are continuous at (a, b), we deduce from (5) that

$$f_{xy}(a, b) = f_{yx}(a, b).$$

EXERCISE

Compute $\dfrac{\partial^2 f}{\partial x\,\partial y}$ and $\dfrac{\partial^2 f}{\partial y\,\partial x}$; verify that they are equal:

1. x^4y^5
2. xy^6
3. x/y^2
4. $x + x^3y + y^4$
5. $\sin(x + y)$
6. $\cos(xy)$
7. $e^{x/y}$
8. $\arctan(x + 2y)$
9. x^my^n
10. $g(x) + h(y)$
11. x^y
12. y^x
13. $\dfrac{x + y}{x - y}$
14. $\dfrac{2xy}{x^2 + y^2}$
15. $\dfrac{x - y}{1 + xy}$
16. $(x - y)(x - 2y)(x - 3y)$
17. $\dfrac{1}{(x - y)(x - 2y)}$
18. $\dfrac{x^2 + y}{y^2 + x}$
19. $ax^2 + 2bxy + cy^2$
20. $\sinh(x + y^2)$.
21. Show that each function of the form $f(x, y) = g(x + y) + h(x - y)$ satisfies th partial differential equation $\dfrac{\partial^2 f}{\partial x^2} - \dfrac{\partial^2 f}{\partial y^2} = 0$.
22. Show that each function of the form $f(x, y) = g(x) + h(y)$ satisfies the partia differential equation $\dfrac{\partial^2 f}{\partial x\,\partial y} = 0$.
23. Prove the converse of Ex. 22: each solution of $f_{xy} = 0$ has the form $f(x, y) = g(x) + h(y)$.

2. HIGHER PARTIALS

A function $z = f(x, y)$ has two distinct first partials:

$$\frac{\partial z}{\partial x}, \qquad \frac{\partial z}{\partial y},$$

and three distinct second partials:

$$\frac{\partial^2 z}{\partial x^2}, \qquad \frac{\partial^2 z}{\partial x\, \partial y}, \qquad \frac{\partial^2 z}{\partial y^2}.$$

Because the mixed second partials are equal, so are certain mixed third partials. For example,

$$\frac{\partial^2}{\partial x^2}\left(\frac{\partial z}{\partial y}\right) = \frac{\partial}{\partial y}\left(\frac{\partial^2 z}{\partial x^2}\right),$$

since

$$\frac{\partial^2}{\partial x^2}\left(\frac{\partial z}{\partial y}\right) = \frac{\partial}{\partial x}\left[\frac{\partial}{\partial x}\left(\frac{\partial z}{\partial y}\right)\right] = \frac{\partial}{\partial x}\left[\frac{\partial}{\partial y}\left(\frac{\partial z}{\partial x}\right)\right] = \frac{\partial}{\partial y}\left[\frac{\partial}{\partial x}\left(\frac{\partial z}{\partial x}\right)\right] = \frac{\partial}{\partial y}\left(\frac{\partial^2 z}{\partial x^2}\right).$$

Thus there are precisely four distinct third partials:

$$\frac{\partial^2 z}{\partial x^3}, \qquad \frac{\partial^3 z}{\partial x^2\, \partial y}, \qquad \frac{\partial^3 z}{\partial x\, \partial y^2}, \qquad \frac{\partial^3 z}{\partial y^3}.$$

In general there are $n + 1$ distinct partials of order n:

$$\frac{\partial^n z}{\partial x^k\, \partial y^{n-k}}, \qquad k = 0, 1, 2, \cdots, n.$$

EXAMPLE 2.1

Find all functions $z = f(x, y)$ which satisfy the system of partial differential equations $\dfrac{\partial^2 z}{\partial x^2} = 0, \quad \dfrac{\partial^2 z}{\partial x\, \partial y} = 0, \quad \dfrac{\partial^2 z}{\partial y^2} = 0.$

Solution: If both first partials of a function are 0 everywhere, then the function is constant. (Why?) This applies to the function $\partial z/\partial x$ since

$$\frac{\partial}{\partial x}\left(\frac{\partial z}{\partial x}\right) = \frac{\partial^2 z}{\partial x^2} = 0 \qquad \text{and} \qquad \frac{\partial}{\partial y}\left(\frac{\partial z}{\partial x}\right) = \frac{\partial^2 z}{\partial y\, \partial x} = \frac{\partial^2 z}{\partial x\, \partial y} = 0.$$

Hence

$$\frac{\partial z}{\partial x} = A.$$

Integrate with respect to x, holding y constant:

$$z = Ax + g(y),$$

where the "constant of integration" is an arbitrary function of y. Since $\partial^2 z/\partial y^2 = 0$,

$$0 = \frac{\partial^2 z}{\partial y^2} = \frac{d^2 g}{dy^2}.$$

Consequently $g(y)$ is a linear function,

$$g(y) = By + C.$$

Answer: $z = Ax + By + C.$

EXAMPLE 2.2

Find all functions $z = f(x, y)$ whose third partials are all 0.

Solution: The second partials of $\partial z/\partial x$ are all 0. By the last example,

$$\frac{\partial z}{\partial x} = Ax + By + C.$$

Integrate:

$$z = \frac{1}{2}Ax^2 + Bxy + Cx + g(y).$$

But

$$0 = \frac{\partial^3 z}{\partial y^3} = \frac{d^3 g}{dy^3}.$$

Consequently $g(y)$ is a quadratic polynomial in y.

Answer: Any quadratic polynomial in x and y.

EXAMPLE 2.3

Find all functions $z = f(x, y)$ that satisfy $\dfrac{\partial^2 z}{\partial x^2} = 0.$

Solution: Write the equation

$$\frac{\partial}{\partial x}\left(\frac{\partial z}{\partial x}\right) = 0,$$

then integrate:

$$\frac{\partial z}{\partial x} = g(y).$$

Integrate again:

$$z = g(y)x + h(y).$$

Answer: $z(x, y) = g(y)x + h(y)$, where $g(y)$ and $h(y)$ are arbitrary functions of y.

EXAMPLE 2.4

Find all functions $z = f(x, y)$ that satisfy $\dfrac{\partial^2 z}{\partial x\, \partial y} = 0.$

Solution: Write the equation

$$\frac{\partial}{\partial y}\left(\frac{\partial z}{\partial x}\right) = 0,$$

then integrate:

$$\frac{\partial z}{\partial x} = p(x).$$

Integrate again:

$$z = g(x) + h(y),$$

where $g(x)$ is an antiderivative of $p(x)$. Note that $g(x)$ is an arbitrary function of x since $p(x)$ is.

> *Answer:* $z(x, y) = g(x) + h(y)$, where $g(x)$ and $h(y)$ are arbitrary differentiable functions of one variable.

CHECK:

$$\frac{\partial^2}{\partial x\,\partial y}[g(x) + h(y)] = \frac{\partial}{\partial x}\left(\frac{\partial}{\partial y}[g(x) + h(y)]\right) = \frac{\partial}{\partial x}[h'(y)] = 0.$$

EXAMPLE 2.5

Find all functions $z = f(x, y)$ that satisfy the system of partial differential equations $\dfrac{\partial z}{\partial x} = y, \quad \dfrac{\partial z}{\partial y} = 1.$

Solution: Integrate the first equation:

$$z = xy + g(y).$$

Substitute this into the second equation:

$$\frac{\partial}{\partial y}[xy + g(y)] = 1, \qquad x + g'(y) = 1, \qquad g'(y) = 1 - x.$$

This is impossible since the left-hand side is a function of y alone.

> *Answer:* No solution.

This example illustrates an important point. A system of partial differential equations may have no solution at all! Could we have forseen this catastrophe for the system above? Yes; for suppose there *were* a function $f(x, y)$ satisfying

$$\frac{\partial f}{\partial x} = y \qquad \text{and} \qquad \frac{\partial f}{\partial y} = 1.$$

Then

$$\frac{\partial^2 f}{\partial y\,\partial x} = \frac{\partial}{\partial y}(y) = 1, \qquad \frac{\partial^2 f}{\partial x\,\partial y} = \frac{\partial}{\partial x}(1) = 0,$$

so the mixed partials would be unequal, a contradiction.

> If the system of equations $\dfrac{\partial z}{\partial x} = p(x, y), \qquad \dfrac{\partial z}{\partial y} = q(x, y)$
>
> has a solution, then $\dfrac{\partial p}{\partial y} = \dfrac{\partial q}{\partial x}$.

Indeed,

$$\frac{\partial p}{\partial y} = \frac{\partial}{\partial y}\left(\frac{\partial z}{\partial x}\right) = \frac{\partial}{\partial x}\left(\frac{\partial z}{\partial y}\right) = \frac{\partial q}{\partial x}.$$

More Variables

All that has been said applies to functions of three or more variables. For example, suppose $w = f(x, y, z)$. Then w has three first partials:

$$\frac{\partial w}{\partial x}, \qquad \frac{\partial w}{\partial y}, \qquad \frac{\partial w}{\partial z}.$$

The nine possible second partials may be written in matrix form:

$$\begin{bmatrix} \dfrac{\partial^2 w}{\partial x^2} & \dfrac{\partial^2 w}{\partial x\,\partial y} & \dfrac{\partial^2 w}{\partial x\,\partial z} \\[2ex] \dfrac{\partial^2 w}{\partial y\,\partial x} & \dfrac{\partial^2 w}{\partial y^2} & \dfrac{\partial^2 w}{\partial y\,\partial z} \\[2ex] \dfrac{\partial^2 w}{\partial z\,\partial x} & \dfrac{\partial^2 w}{\partial z\,\partial y} & \dfrac{\partial^2 w}{\partial z^2} \end{bmatrix}.$$

This matrix is symmetric since the mixed second partials are equal in pairs:

$$\frac{\partial^2 w}{\partial y\,\partial x} = \frac{\partial^2 w}{\partial x\,\partial y}, \qquad \frac{\partial^2 w}{\partial x\,\partial z} = \frac{\partial^2 w}{\partial z\,\partial x}, \qquad \frac{\partial^2 w}{\partial z\,\partial y} = \frac{\partial^2 w}{\partial y\,\partial z}.$$

EXERCISES

Compute $\dfrac{\partial^3 f}{\partial x^2\,\partial y}$ and $\dfrac{\partial^3 f}{\partial x\,\partial y^2}$:

1. x^3y^3
2. x^4y^5
3. x^2y^4
4. x^my^n
5. $\cos(xy)$
6. $\sin(x^2y)$
7. $e^{xy}\sin x$
8. x^y
9. $x^{1/y}$
10. $\dfrac{x-y}{x+y}$
11. $\dfrac{1}{x^2+y^2}$
12. $\dfrac{xy}{x^2+y^2}$.
13. Find all functions $f(x, y)$ such that $\dfrac{\partial^3 f}{\partial x^2\,\partial y} = 0$.
14. Find all functions $f(x, y)$ such that $\dfrac{\partial^3 f}{\partial x^2\,\partial y} = 0$ and $\dfrac{\partial^3 f}{\partial x\,\partial y^2} = 0$.
15. Find all functions $f(x, y)$ whose 4-th partial derivatives all equal 0.
16. Find all functions $f(x, y)$ such that $\dfrac{\partial^4 f}{\partial x^2\,\partial y^2} = 0$.

Find all functions $f(x, y)$ that satisfy the system of partial differential equations:

17. $\dfrac{\partial f}{\partial x} = a,\quad \dfrac{\partial f}{\partial y} = b$
18. $\dfrac{\partial f}{\partial x} = y,\quad \dfrac{\partial f}{\partial y} = x$
19. $\dfrac{\partial f}{\partial x} = y^2,\quad \dfrac{\partial f}{\partial y} = x^2$
20. $\dfrac{\partial^2 f}{\partial x^2} = 2y^3,\quad \dfrac{\partial f}{\partial y} = 3x^2y^2$.

Write the matrix of 9 second partials:

21. $x^my^nz^p$
22. $xy + yz + zx$
23. $\sin(x + 2y + 3z)$
24. $x^2 + yz$.
25. How many distinct third partials does $f(x, y, z)$ have?
26. (cont.) Find an explicit function for which they really are distinct.
27. Find all functions $f(x, y, z)$ satisfying $\dfrac{\partial^3 f}{\partial x\,\partial y\,\partial z} = 0$.
28. How many distinct second partials does a function $f(x, y, z, w)$ of 4 variables have? How many distinct third partials?

3. TAYLOR POLYNOMIALS

Let us recall some facts about Taylor polynomials. (See Chapter 2, Section 3.) If $y = f(x)$, then

$$f(x) = f(a) + f'(a)(x - a) + r_1(x),$$

and

$$f(x) = f(a) + f'(a)(x - a) + \frac{1}{2}f''(a)(x - a)^2 + r_2(x),$$

where

$$|r_1(x)| \leq \frac{M_2}{2!}(x - a)^2, \qquad |r_2(x)| \leq \frac{M_3}{3!}|x - a|^3,$$

and where M_2 and M_3 are bounds for $|f''(x)|$ and $|f'''(x)|$ respectively.

The Taylor polynomial

$$p_1(x) = f(a) + f'(a)(x - a)$$

is constructed so that $p_1(a) = f(a)$ and $p_1'(a) = f'(a)$. The Taylor polynomial

$$p_2(x) = f(a) + f'(a)(x - a) + \frac{1}{2}f''(a)(x - a)^2$$

is constructed so that $p_2(a) = f(a)$, $p_2'(a) = f'(a)$, $p_2''(a) = f''(a)$.

In a similar way, one can construct linear and quadratic polynomials in two variables approximating a given function of two variables.

Taylor polynomials Let $f(x, y)$ have continuous first and second partials on an open domain **D**. The **first degree** and **second degree Taylor polynomials** of f at (a, b) are

$$p_1(x, y) = f(a, b) + f_x\cdot(x - a) + f_y\cdot(y - b),$$

$$p_2(x, y) = p_1(x, y) + \tfrac{1}{2}[f_{xx}\cdot(x - a)^2 + 2f_{xy}\cdot(x - a)(y - b) + f_{yy}\cdot(y - b)^2],$$

where all the partials are evaluated at (a, b).

It is easy to check that $p_1(a, b) = f(a, b)$ and that the first partials of p_1 agree with those of f at (a, b). Similarly, $p_2(a, b) = f(a, b)$ and all first and second partials of p_2 agree with the corresponding partials of f at (a, b).

Now we ask how closely these Taylor polynomials approximate $f(x, y)$ for (x, y) near (a, b). In other words, we want estimates for the errors in the approximations $f(x, y) \approx p_1(x, y)$ and $f(x, y) \approx p_2(x, y)$. We shall obtain error estimates subject to the mild condition that f is defined on a convex domain. A set **S** in the plane or space is called **convex** if it contains the whole segment joining any two of its points.

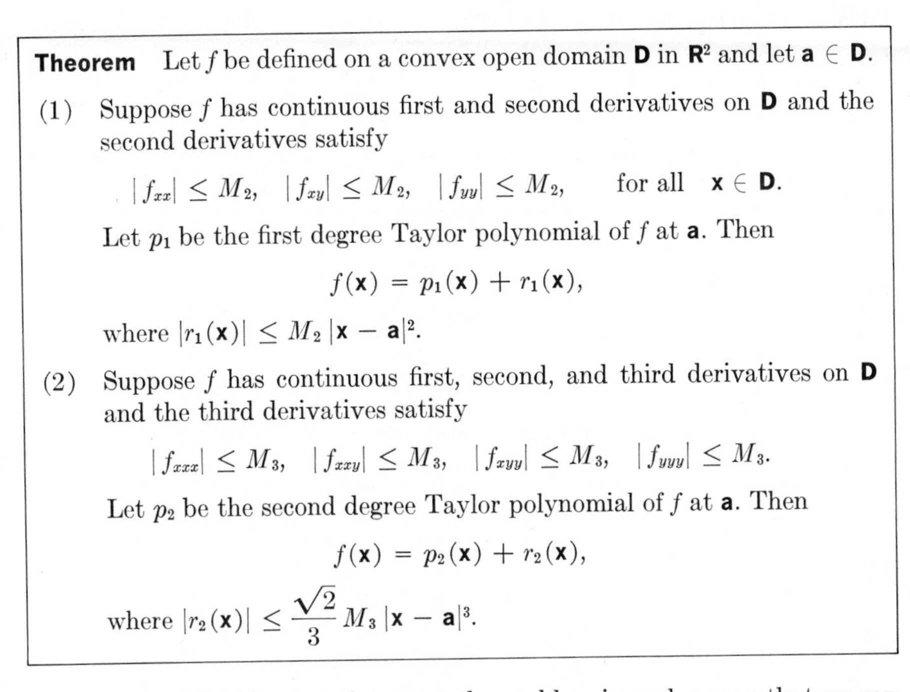

Theorem Let f be defined on a convex open domain $\mathbf{D}$ in $\mathbf{R}^2$ and let $\mathbf{a} \in \mathbf{D}$.

(1) Suppose f has continuous first and second derivatives on $\mathbf{D}$ and the second derivatives satisfy

$$|f_{xx}| \leq M_2, \quad |f_{xy}| \leq M_2, \quad |f_{yy}| \leq M_2, \qquad \text{for all} \quad \mathbf{x} \in \mathbf{D}.$$

Let p_1 be the first degree Taylor polynomial of f at $\mathbf{a}$. Then

$$f(\mathbf{x}) = p_1(\mathbf{x}) + r_1(\mathbf{x}),$$

where $|r_1(\mathbf{x})| \leq M_2 |\mathbf{x} - \mathbf{a}|^2$.

(2) Suppose f has continuous first, second, and third derivatives on $\mathbf{D}$ and the third derivatives satisfy

$$|f_{xxx}| \leq M_3, \quad |f_{xxy}| \leq M_3, \quad |f_{xyy}| \leq M_3, \quad |f_{yyy}| \leq M_3.$$

Let p_2 be the second degree Taylor polynomial of f at $\mathbf{a}$. Then

$$f(\mathbf{x}) = p_2(\mathbf{x}) + r_2(\mathbf{x}),$$

where $|r_2(\mathbf{x})| \leq \dfrac{\sqrt{2}}{3} M_3 |\mathbf{x} - \mathbf{a}|^3$.

Proof: The idea is to interpret the problem in such a way that we can use the error estimates on the previous page for a function of one variable.

Assume at first that $\mathbf{a} = \mathbf{0}$; this will simplify the notation considerably. Now fix a point $\mathbf{x} \in \mathbf{D}$. By convexity, $\mathbf{D}$ contains the entire line segment connecting $\mathbf{0}$ and $\mathbf{x}$, that is, $\mathbf{D}$ contains all points $t\mathbf{x}$ for $0 \leq t \leq 1$. (Actually because it is open, $\mathbf{D}$ contains a slightly larger open segment.)

Set $g(t) = f(t\mathbf{x})$. Then $g(t)$ is a function of one variable defined for $0 \leq t \leq 1$ and

$$g(0) = f(\mathbf{0}), \qquad g(1) = f(\mathbf{x}).$$

Let us compute the first and second degree Taylor polynomials of g at $t = 0$. By the Chain Rule,

$$g'(t) = f_x(t\mathbf{x})x + f_y(t\mathbf{x})y,$$
$$g'(0) = f_x(\mathbf{0})x + f_y(\mathbf{0})y,$$
$$g''(t) = f_{xx}(t\mathbf{x})x^2 + 2f_{xy}(t\mathbf{x})xy + f_{yy}(t\mathbf{x})y^2,$$
$$g''(0) = f_{xx}(\mathbf{0})x^2 + 2f_{xy}(\mathbf{0})xy + f_{yy}(\mathbf{0})y^2,$$
$$g'''(t) = f_{xxx}(t\mathbf{x})x^3 + 3f_{xxy}(t\mathbf{x})x^2y + 3f_{xyy}(t\mathbf{x})xy^2 + f_{yyy}(t\mathbf{x})y^3.$$

From these calculations we see that

$$p_1(\mathbf{x}) = g(0) + g'(0), \qquad p_2(\mathbf{x}) = g(0) + g'(0) + \frac{1}{2}g''(0).$$

Thus $p_1(\mathbf{x})$ and $p_2(\mathbf{x})$ are the first and second degree Taylor polynomials of $g(t)$ evaluated at $t = 1$. Therefore

$$|r_1(\mathbf{x})| = |f(\mathbf{x}) - p_1(\mathbf{x})| \le \frac{\max |g''(t)|}{2!},$$

$$|r_2(x)| = |f(x) - p_2(x)| \le \frac{\max |g'''(t)|}{3!}.$$

It remains to estimate $|g''(t)|$ and $|g'''(t)|$. We have

$$|g''(t)| \le M_2 |x|^2 + 2M_2 |x| |y| + M_2 |y|^2 = M_2(|x| + |y|)^2,$$

$$|g'''(t)| \le M_3 |x|^3 + 3M_3 |x|^2 |y| + 3M_3 |x| |y|^2 + M_3 |y|^3 = M_3(|x| + |y|)^3.$$

Now we modify these estimates slightly as follows. From $(|x| - |y|)^2 \ge 0$ we have $2 |x| |y| \le |x|^2 + |y|^2$, hence

$$(|x| + |y|)^2 = |x|^2 + 2 |x| |y| + |y|^2 \le 2(|x|^2 + |y|^2) = 2|\mathbf{x}|^2.$$

Take the $\frac{3}{2}$ power:

$$(|x| + |y|)^3 \le 2^{3/2}|\mathbf{x}|^3.$$

Therefore,

$$|g''(t)| \le 2M_2 |\mathbf{x}|^2, \qquad |g'''(t)| \le 2\sqrt{2}M_3 |\mathbf{x}|^3.$$

Combining results we obtain

$$|r_1(\mathbf{x})| \le \frac{2M_2 |\mathbf{x}|^2}{2!}, \qquad |r_2(\mathbf{x})| \le \frac{2\sqrt{2}M_3 |\mathbf{x}|^3}{3!}.$$

This completes the proof assuming $\mathbf{a} = \mathbf{0}$. In the general case, we define $g(t) = f[\mathbf{a} + t(\mathbf{x} - \mathbf{a})]$. The proof proceeds as before, except that (x, y) is replaced by $(x - a, y - b)$ and the partials of f are all evaluated at $\mathbf{a}$.

REMARK: There are Taylor polynomials of higher degree and corresponding error estimates. The notation for these polynomials is complicated, and since we shall not need them, we leave their study to an advanced calculus course.

EXAMPLE 3.1

Compute the Taylor polynomials $p_1(x, y)$ and $p_2(x, y)$ of the function $f(x, y) = \sqrt{x^2 + y^2}$ at $(3, 4)$.

Solution:

$$\frac{\partial f}{\partial x} = \frac{x}{\sqrt{x^2 + y^2}}, \qquad \frac{\partial f}{\partial y} = \frac{y}{\sqrt{x^2 + y^2}},$$

$$\frac{\partial^2 f}{\partial x^2} = \frac{y^2}{(x^2 + y^2)^{3/2}}, \qquad \frac{\partial^2 f}{\partial x\, \partial y} = \frac{-xy}{(x^2 + y^2)^{3/2}}, \qquad \frac{\partial^2 f}{\partial y^2} = \frac{x^2}{(x^2 + y^2)^{3/2}}.$$

At (3, 4),

$$\frac{\partial f}{\partial x} = \frac{3}{5}, \qquad \frac{\partial f}{\partial y} = \frac{4}{5}, \qquad \frac{\partial^2 f}{\partial x^2} = \frac{16}{125}, \qquad \frac{\partial^2 f}{\partial x\, \partial y} = -\frac{12}{125}, \qquad \frac{\partial^2 f}{\partial y^2} = \frac{9}{125}.$$

Therefore

$$p_2(x, y) = 5 + \frac{3}{5}(x-3) + \frac{4}{5}(y-4)$$

$$+ \frac{1}{2}\left[\frac{16}{125}(x-3)^2 - \frac{24}{125}(x-3)(y-4) + \frac{9}{125}(y-4)^2\right].$$

The polynomial $p_1(x, y)$ is just the linear (first degree) part of $p_2(x, y)$.

Answer:

$$p_1(x, y) = \frac{1}{5}[25 + 3(x-3) + 4(y-4)];$$

$$p_2(x, y) = p_1(x, y) + \frac{1}{250}[16(x-3)^2$$

$$-24(x-3)(y-4) + 9(y-4)^2].$$

EXAMPLE 3.2

Estimate $\sqrt{(3.1)^2 + (4.02)^2}$ by

(a) $p_1(x, y)$, (b) $p_2(x, y)$,

the Taylor polynomials in the last example.

Solution: Let $f(x, y) = \sqrt{x^2 + y^2}$.

(a) Near (3, 4),

$$f(x, y) \approx \frac{1}{5}[25 + 3(x-3) + 4(y-4)],$$

$$f(3.1, 4.02) \approx \frac{1}{5}[25 + 3(0.1) + 4(0.02)] = 5.076.$$

(b)

$$f(x, y) \approx p_1(x, y) + \frac{1}{250}[16(x-3)^2 - 24(x-3)(y-4) + 9(y-4)^2],$$

$$f(3.1, 4.02) \approx p_1(3.1, 4.02) + \frac{1}{250}[16(0.1)^2 - 24(0.1)(0.02) + 9(0.02)^2]$$

$$= 5.076 + \frac{0.1156}{250} = 5.0764624.$$

Answer: (a) Approximately 5.076; (b) Approximately 5.0764624. (Actual value to 7 places: 5.0764555.)

EXAMPLE 3.3

If $|x| < 0.1$ and $|y| < 0.1$, prove that

$$|e^x \sin(x+y) - (x+y)| < 0.05.$$

Solution: Let $f(x, y) = e^x \sin(x+y)$. Then as is easily checked, $x+y$ is simply $p_1(x, y)$, the first degree Taylor polynomial of $f(x, y)$ at $(0, 0)$. Thus we are asked to verify that $|r_1(\mathbf{x})| < 0.05$ for points $\mathbf{x} = (x, y)$ with $|x| < 0.1$ and $|y| < 0.1$. Such points satisfy $|\mathbf{x}|^2 = (0.1)^2 + (0.1)^2$, so we may restrict the domain of f to the open disk $|\mathbf{x}|^2 < 0.02$.

Our error estimate yields

$$|r_1(\mathbf{x})| \leq M_2 |\mathbf{x}|^2 = (0.02)M_2,$$

where M_2 is a bound for $|f_{xx}|$, $|f_{xy}|$, $|f_{yy}|$. To find a suitable value for M_2, compute the second partials:

$$f_{xx} = 2e^x \cos(x+y), \qquad f_{xy} = e^x[\cos(x+y) - \sin(x+y)],$$

$$f_{yy} = -e^x \sin(x+y).$$

The elementary estimates $|\sin(x+y)| \leq 1$ and $|\cos(x+y)| \leq 1$ show that

$$|f_{xx}| \leq 2e^x, \qquad |f_{xy}| \leq 2e^x, \qquad |f_{yy}| \leq e^x.$$

Now $|x| < 0.1$, so

$$e^x < e^{0.1} < 1.11,$$

as is seen from a table. Therefore we take $M_2 = 2(1.11) = 2.22$, and it follows that

$$|r_1(\mathbf{x})| < (0.02)(2.22) = 0.0444 < 0.05.$$

EXERCISES

Compute the Taylor polynomials $p_1(x, y)$ and $p_2(x, y)$:

1. x^2y^2; at $(1, 1)$
2. x^4y^3; at $(2, -1)$
3. $\sin(xy)$; at $(0, 0)$
4. e^{xy}; at $(0, 0)$
5. x^y; at $(1, 0)$
6. x^y; at $(1, 1)$
7. $\cos(x+y)$; at $(0, \pi/2)$
8. $1 + xy$; at $(1, 1)$
9. $\ln(x+2y)$; at $(\frac{1}{2}, \frac{1}{4})$
10. x^2e^y; at $(1, 0)$.

Estimate using the second degree Taylor polynomial. Carry your work to 5 significant figures:

11. $(1.1)^{1.2}$
12. $[(1.2)^2 + 7.2]^{1/3}$

13. $f(1.01, 2.01)$, where $f(x, y) = x^3y^2 - 2xy^4 + y^5$
14. $\sqrt{(1.99)^2 + (3.01)^2 + (6.01)^2}$.
15. Let $p_1(x, y)$ be the first degree Taylor polynomial of $f(x, y)$ at (a, b). Identify the graph of $z = p_1(x, y)$.
16. Let $p_1(x, y, z)$ be the first degree Taylor polynomial of $f(x, y, z)$ at (a, b, c). Assume grad $f \neq 0$ at (a, b, c). Identify the graph of the equation $p_1(x, y, z) = f(a, b, c)$.

Prove the inequality, given $|x| < 0.1$ and $|y| < 0.1$:

17. $|\sqrt{1 + x + 2y} - (1 + \frac{1}{2}x + y)| < 0.04$
18. $|e^x \sin(x + y) - (1 + x)(1 + y)| < 0.01$ (even <0.005 with more careful estimates).

4. MAXIMA AND MINIMA

In this section we shall develop second derivative tests for maxima and minima.

We shall assume here and in the rest of this chapter that all functions have continuous first, second, and third partial derivatives. With this assumption, the Taylor approximations of the previous section apply.

Let us begin with a brief review of the one-variable case. We consider a function $g(t)$ and a point c where $g'(c) = 0$. Suppose $g''(c) > 0$. We want to conclude that $g(c)$ is a relative minimum of g. For this purpose, an excellent tool is the second degree Taylor approximation of g at c:

$$\begin{cases} g(t) = g(c) + \dfrac{1}{2}g''(c)(t - c) + r_2(t), \\ |r_2(t)| \leq k\,|t - c|^3, \qquad k > 0. \end{cases}$$

It follows that

$$\begin{aligned} g(t) - g(c) &= \frac{1}{2}g''(c)(t - c)^2 + r_2(t) \geq \frac{1}{2}g''(c)(t - c)^2 - k\,|t - c|^3 \\ &= (t - c)^2\left[\frac{1}{2}g''(c) - k\,|t - c|\right]. \end{aligned}$$

Since $g''(c) > 0$, the quantity on the right is positive if $0 < |t - c| < \frac{1}{2}g''(c)/k$. Thus there is a positive number $\delta = \frac{1}{2}g''(c)/k$ such that $g(t) - g(c) > 0$ when $0 < |t - c| < \delta$. In other words, $g(c)$ is smaller than any other value of g in an interval of radius δ and center c. Hence $g(c)$ is a relative minimum of g.

Hessian Matrix

Let us try to generalize these ideas to the two variable case. Given $f(x, y)$, suppose (a, b) is an interior point of the domain of f where $f_x(a, b) =$

$f_y(a, b) = 0$. We would like a condition on the second derivatives of f, analogous to $g''(c) > 0$, guaranteeing that $f(a, b)$ is a relative minimum of f. Now the condition $g''(c) > 0$ can be interpreted as meaning that

$$g''(c)(t - c)^2$$

is a positive definite quadratic form in the one variable $t - c$. It is natural, therefore, to examine the quadratic part of the second degree Taylor polynomial of $f(x, y)$ at (a, b). If it is positive definite, we suspect that $f(a, b)$ is a minimum of f.

The quadratic part in question is

$$\frac{1}{2}[f_{xx} \cdot (x - a)^2 + 2f_{xy} \cdot (x - a)(x - b) + f_{yy} \cdot (y - b)^2].$$

In matrix–vector notation, this can be expressed as

$$\frac{1}{2}(\mathbf{x} - \mathbf{a})H_f(\mathbf{x} - \mathbf{a})',$$

where

$$H_f = \begin{bmatrix} f_{xx} & f_{xy} \\ f_{yx} & f_{yy} \end{bmatrix}.$$

The matrix H_f is called the **Hessian** matrix of f. Our assumption on the partial derivatives of f ensure that $f_{xy} = f_{yx}$. Hence H_f is a symmetric matrix.

NOTE: It is more accurate to write $H_f(x, y)$ since the Hessian matrix generally depends on the point (x, y). When the meaning is clear from the context, however, we shall use the simpler notation H_f.

Second Derivative Test

Theorem 4.1 Let $\mathbf{a} = (a, b)$ be an interior point of the domain of f. Suppose that

$$f_x(a, b) = 0, \qquad f_y(a, b) = 0,$$

and that the Hessian matrix $H_f(a, b)$ is positive definite.

Then $f(a, b)$ is a relative minimum of f. That is, there is $\delta > 0$ such that

$$f(x, y) > f(a, b)$$

whenever

$$0 < |\mathbf{x} - \mathbf{a}| < \delta.$$

Proof: To make the notation simple, let us take $(a, b) = (0, 0)$. Then the second degree Taylor approximation of f at $(0, 0)$ is

$$\begin{cases} f(x, y) = f(0, 0) + \frac{1}{2}[Ax^2 + 2Bxy + Cy^2] + r(x, y), \\ |r(x, y)| < h\,|\mathbf{x}|^3, \end{cases}$$

where

$$H_f(0,0) = \begin{bmatrix} A & B \\ B & C \end{bmatrix} = \begin{bmatrix} f_{xx} & f_{xy} \\ f_{yx} & f_{yy} \end{bmatrix}_{(0,0)}$$

and h is a positive constant. Since H_f is positive definite, we can apply Theorem 5.4, p. 249; there is another positive constant k such that

$$Ax^2 + 2Bxy + Cy^2 \geq k\,|\mathbf{x}|^2.$$

Consequently

$$f(x,y) - f(0,0) \geq \tfrac{1}{2}k\,|\mathbf{x}|^2 - h\,|\mathbf{x}|^3 = |\mathbf{x}|^2\,(\tfrac{1}{2}k - h\,|\mathbf{x}|).$$

This implies $f(x,y) - f(0,0) > 0$, that is $f(x,y) > f(0,0)$, provided $0 < |\mathbf{x}| < \frac{1}{2}k/h$. Finally, for δ we may choose any number such that first $0 < \delta \leq \frac{1}{2}k/h$ and second δ is so small that the disk $|\mathbf{x}| < \delta$ is contained in the domain $\mathbf{D}$.

REMARK: The essential point in the proof is that for $|\mathbf{x}|$ small, the positive definite quadratic form $Ax^2 + 2Bxy + Cy^2$ is much larger than the remainder $r(x,y)$ because r is of third order: $|r(x,y)| \leq h\,|\mathbf{x}|^3$.

By Theorem 5.1, p. 247, the matrix $H_f(a,b)$ is positive definite if and only if

$$f_{xx}(a,b) > 0, \qquad \begin{vmatrix} f_{xx} & f_{xy} \\ f_{yx} & f_{yy} \end{vmatrix}_{(a,b)} > 0,$$

that is,

$$f_{xx}(a,b) > 0, \qquad (f_{xx}f_{yy} - f_{xy}^2)\Big|_{(a,b)} > 0.$$

Suppose we want a maximum instead of a minimum. We reason that finding a maximum of f is the same as finding a minimum of $-f$. Therefore the following conditions are sufficient for $f(a,b)$ to be a relative maximum of f at a point of its (open) domain:

$$f_x(a,b) = 0, \qquad f_y(a,b) = 0,$$

$$f_{xx}(a,b) < 0, \qquad (f_{xx}f_{yy} - f_{xy}^2)\Big|_{(a,b)} > 0.$$

The two inequalities make up a necessary and sufficient condition for the Hessian matrix of $-f$,

$$H_{-f} = \begin{bmatrix} -f_{xx} & -f_{xy} \\ -f_{yx} & -f_{yy} \end{bmatrix},$$

to be positive definite, namely, $-f_{xx} > 0$ and $\det H_{-f} > 0$. Note particularly that $\det H_{-f} = \det H_f$, *not* the negative.

Let us review our procedures. We start with f on an open domain **D** We find a point (a, b) where $f_x = 0$ and $f_y = 0$. We compute the Hessian matrix

$$H_f = \begin{bmatrix} f_{xx} & f_{xy} \\ f_{yx} & f_{yy} \end{bmatrix}$$

at (a, b). There are three possibilities for $\det H_f = f_{xx} f_{yy} - f_{xy}^2$:

$$(1)\quad \det H_f > 0, \qquad (2)\quad \det H_f = 0, \qquad (3)\quad \det H_f < 0.$$

In case (1), we have

$$f_{xx} f_{yy} > f_{xy}^2 \geq 0,$$

so $f_{xx} \neq 0$. Either $f_{xx} > 0$ and $f(a, b)$ is a relative minimum, or $f_{xx} < 0$ and $f(a, b)$ is a relative maximum.

In case (2), no conclusion can be drawn. For example take the functions $f(x, y) = x^3y^3$ and $g(x, y) = x^4y^4$. Then $f_x = f_y = 0$ and $H_f = 0$ at $(0, 0)$, and similarly for g. Yet g has a minimum at $(0, 0)$ whereas f has neither a minimum nor a maximum.

In case (3) there is useful information and we state it as a formal theorem.

Theorem 4.2 Suppose $f_x(a, b) = 0$ and $f_y(a, b) = 0$ at an interior point of the domain of f. Suppose

$$\begin{vmatrix} f_{xx} & f_{xy} \\ f_{yx} & f_{yy} \end{vmatrix}_{(a,b)} < 0.$$

Then $f(a, b)$ is neither a relative maximum nor a relative minimum of f. That is, there are points (x, y) arbitrarily close to (a, b) where $f(x, y) < f(a, b)$ and there are points (x, y) arbitrarily close to (a, b) where $f(x, y) > f(a, b)$.

Proof: We take $(a, b) = (0, 0)$ for simplicity and consider the Taylor approximation

$$f(x, y) = f(0, 0) + \frac{1}{2}(Ax^2 + 2Bxy + Cy^2) + r(x, y), \qquad |r(x, y)| < h\,|\mathbf{x}|^3.$$

Since

$$\begin{vmatrix} A & B \\ B & C \end{vmatrix} = \begin{vmatrix} f_{xx} & f_{xy} \\ f_{yx} & f_{yy} \end{vmatrix}_{(0,0)} < 0,$$

the quadratic form $Q(x, y) = Ax^2 + 2Bxy + Cy^2$ is indefinite by Theorem 5.7, p. 251. Therefore there are points (x_1, y_1) and (x_2, y_2) such that $Q(x_1, y_1) < 0$ and $Q(x_2, y_2) > 0$.

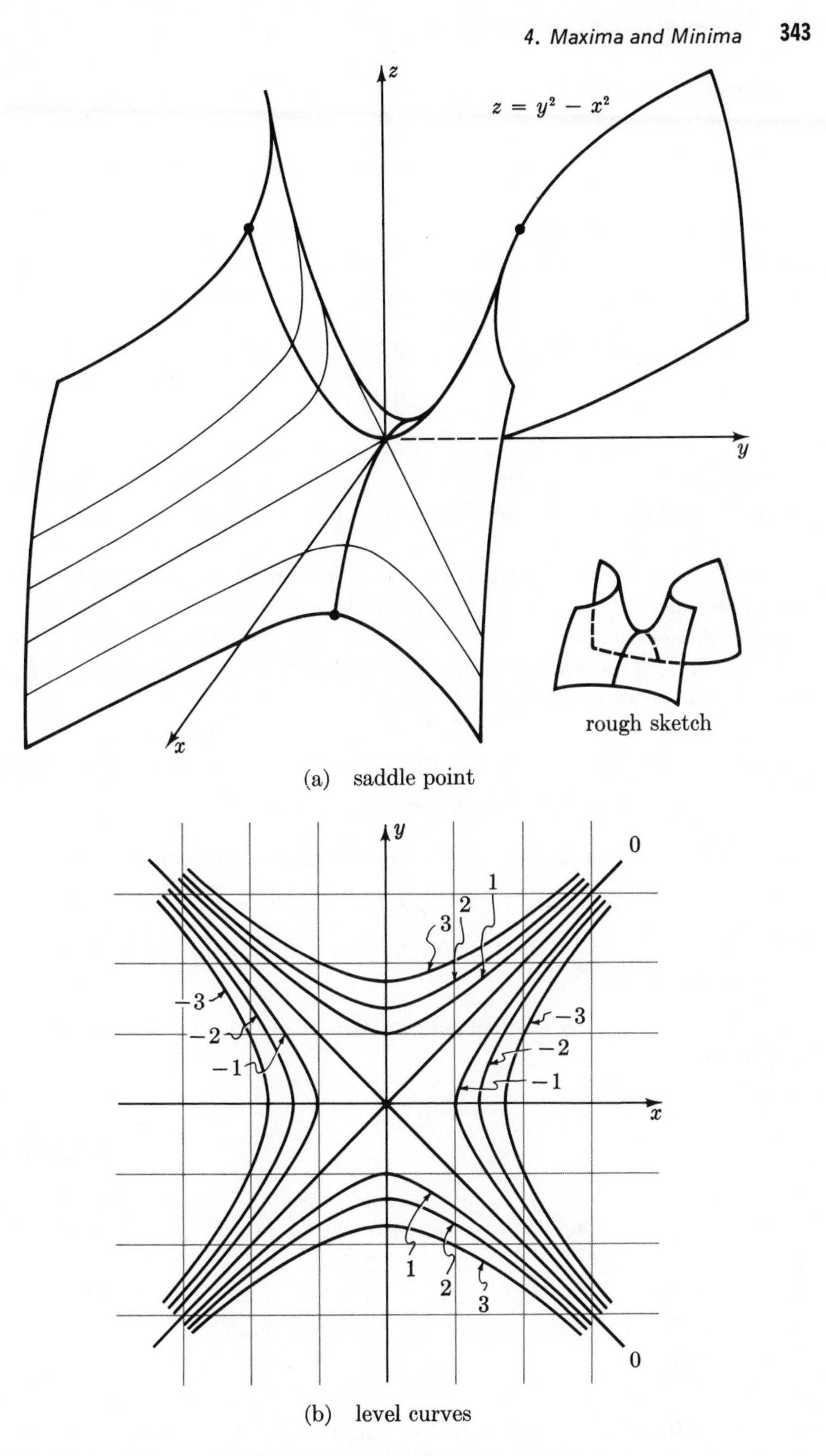

(a) saddle point

(b) level curves

FIG. 4.1 graph of $z = y^2 - x^2$

Consider $(x, y) = (tx_1, ty_1)$, where $t > 0$ and small. Then

$$\frac{1}{2}Q(tx_1, ty_1) = \frac{1}{2}Q(x_1, y_1)t^2 = kt^2, \qquad k < 0,$$

$$|r(tx_1, ty_1)| < h(x_1^3 + y_1^3)^{1/2}t^3 = h_1t^3,$$

hence

$$f(tx_1, ty_1) - f(0, 0) \leq kt^2 + h_1t^3 = t^2(k + h_1t).$$

Now for t sufficiently small, $t^2(k + h_1t) < 0$, hence

$$f(tx_1, ty_1) < f(0, 0)$$

for points (tx_1, ty_1) as close to $(0, 0)$ as we please.

Similarly $f(tx_2, ty_2) > f(0, 0)$ for points (tx_2, ty_2) as close to $(0, 0)$ as we please. Therefore $f(0, 0)$ is neither a max nor a min of f.

In the case we have just discussed, f is said to have a **saddle point** at (a, b); the surface $z = f(x, y)$ is shaped like a saddle near $(a, b, f(a, b))$. The tangent plane is horizontal; the surface rises in some directions, falls in others, so that it crosses its tangent plane. This is like the summit of a mountain pass. See Fig. 4.1 for an example.

Summary Let (a, b) be an interior point of the domain of f and suppose

$$f_x(a, b) = f_y(a, b) = 0.$$

(1) If $f_{xx} > 0$ and $f_{xx}f_{yy} - f_{xy}^2 > 0$, then $f(a, b)$ is a relative min.

(2) If $f_{xx} < 0$ and $f_{xx}f_{yy} - f_{xy}^2 > 0$, then $f(a, b)$ is a relative max.

(3) If $f_{xx}f_{yy} - f_{xy}^2 < 0$, then $f(a, b)$ is a saddle point, neither a relative max nor min.

(4) If $f_{xx}f_{yy} - f_{xy}^2 = 0$, no conclusion can be drawn from this information alone.

EXERCISES

Does the function $z = f(x, y)$ have a possible maximum or minimum at the origin? Try the first and second derivative tests. Sketch a few level curves near the origin:

1. $z = x^2$
2. $z = xy$
3. $z = x^2 - 4y^2$
4. $z = x^2 + 2xy + y^2$
5. $z = -x^2 + 2xy - y^2$
6. $z = x^4 + y^2$
7. $z = x^4 + y^4$
8. $z = -x^2y^2$
9. $z = x^3 + y^3$
10. $z = x^2 + y^5$.
11. Suppose $f(x, y)$ is a function of two variables and $x = x(t)$ and $y = y(t)$ are functions of time. Form the composite function $g(t) = f[x(t), y(t)]$. By the Chain

Rule, $\dot{g}(t) = f_x[x(t), y(t)]\dot{x}(t) + f_y[x(t), y(t)]\dot{y}(t)$. Show that

$$\ddot{g} = f_{xx}\dot{x}^2 + 2f_{xy}\dot{x}\dot{y} + f_{yy}\dot{y}^2 + f_x\ddot{x} + f_y\ddot{y}.$$

12. (cont.) Suppose also that $f_x(0, 0) = f_y(0, 0) = 0$, and that only curves $\mathbf{x}(t) = (x(t), y(t))$ are allowed which pass through $(0, 0)$ with speed 1 at $t = 0$. Suppose for each such curve $\ddot{g}(0) > 0$. Show that $\dot{g}(0) = 0$ and $f_{xx}(0, 0) > 0$ and $f_{yy}(0, 0) > 0$.
13. (cont.) Show also that $f_{xx}(0, 0)f_{yy}(0, 0) - f_{xy}(0, 0)^2 > 0$.
14. (cont.) Find a formula for d^3g/dt^3.
15. (cont.) Suppose $g(x, y) = Ax^2 + 2Bxy + Cy^2$ and $AC - B^2 = 0$. Conclude that $g(x, y) = \pm(ax + by)^2$.

5. APPLICATIONS

EXAMPLE 5.1

Show that $f(x, y) = 4x^2 - xy + y^2$ has a minimum at $(0, 0)$.

Solution:

$$f_x = 8x - y, \qquad f_y = -x + 2y.$$

Both partials are 0 at $(0, 0)$. Furthermore,

$$f_{xx} = 8, \qquad f_{xy} = -1, \qquad f_{yy} = 2.$$

Hence,

$$f_{xx} > 0, \qquad f_{xx}f_{yy} - f_{xy}^2 = 15 > 0.$$

These conditions ensure a minimum at the origin.

EXAMPLE 5.2

Of all triangles of fixed perimeter, find the one with maximum area.

Solution: If the sides are a, b, c, then the area A is found by Heron's formula:

$$A^2 = s(s - a)(s - b)(s - c),$$

where $s = \frac{1}{2}(a + b + c)$ is the semiperimeter. Thus

$$2s = a + b + c, \qquad s - c = a + b - s,$$

$$A^2 = s(s - a)(s - b)(a + b - s).$$

Here s is constant and the variables are a and b. In order to maximize A, it suffices to maximize

$$f(a, b) = (s - a)(s - b)(a + b - s).$$

Now

$$f_a = (s - b)(2s - 2a - b), \qquad f_b = (s - a)(2s - a - 2b),$$

$$f_{aa} = -2(s - b), \qquad f_{ab} = -3s + 2a + 2b, \qquad f_{bb} = -2(s - a).$$

The equations $f_a = 0$ and $f_b = 0$ imply

$$2a + b = 2s, \qquad a + 2b = 2s,$$

since $s = a$ or $s = b$ is impossible in a triangle. (Why?) It follows that

$$a = b = \frac{2}{3}s.$$

For these values of a and b,

$$f_{aa} = f_{bb} = -\frac{2s}{3} < 0, \qquad f_{ab} = -\frac{s}{3},$$

$$f_{aa}f_{bb} - f_{ab}{}^2 = \frac{4s^2}{9} - \frac{s^2}{9} = \frac{s^2}{3} > 0.$$

These conditions ensure a maximum. Now compute c:

$$c = 2s - a - b = 2s - \frac{2}{3}s - \frac{2}{3}s = \frac{2}{3}s.$$

Hence $a = b = c$.

Answer: The equilateral triangle.

EXAMPLE 5.3

Find the point on the paraboloid $z = \dfrac{x^2}{4} + \dfrac{y^2}{9}$ closest to $\mathbf{i} = (1, 0, 0)$.

Solution: Let $\mathbf{x} = (x, y, z)$ be a point on the surface. Then

$$|\mathbf{x} - \mathbf{i}|^2 = (x-1)^2 + y^2 + z^2$$

$$= (x-1)^2 + y^2 + \left(\frac{x^2}{4} + \frac{y^2}{9}\right)^2 = f(x, y).$$

The function $f(x, y)$ must be minimized. Now

$$f_x = 2(x-1) + x\left(\frac{x^2}{4} + \frac{y^2}{9}\right),$$

$$f_y = 2y + \frac{4y}{9}\left(\frac{x^2}{4} + \frac{y^2}{9}\right) = 2y\left[1 + \frac{2}{9}\left(\frac{x^2}{4} + \frac{y^2}{9}\right)\right].$$

The condition $f_y = 0$ is satisfied if either

$$y = 0 \qquad \text{or} \qquad 1 + \frac{2}{9}\left(\frac{x^2}{4} + \frac{y^2}{9}\right) = 0.$$

The latter is impossible. Therefore $f_y = 0$ implies $y = 0$. But if $y = 0$, the condition $f_x = 0$ means

$$2(x-1) + \frac{x^3}{4} = 0, \qquad \text{that is,} \qquad x^3 + 8x - 8 = 0.$$

A rough sketch shows this cubic has only one real root, and the root is near $x = 1$. By Newton's Method iterated twice, $x \approx 0.9068$.

Thus $f(x, y)$ has a possible minimum only at the point $(a, 0)$, where $a^3 + 8a - 8 = 0$. Test the second derivatives at $(a, 0)$:

$$f_{xx} = 2 + \frac{3}{4}a^2 > 0, \qquad f_{xy} = 0, \qquad f_{yy} = 2 + \frac{1}{9}a^2 > 0,$$

$$f_{xx}f_{yy} - f_{xy}{}^2 > 0.$$

Therefore the minimum does occur at $(a, 0)$.

Answer: $\left(a, 0, \frac{1}{4}a^2\right)$, where $a^3 + 8a - 8 = 0$, so $a \approx 0.9068$.

The following example shows that you must be careful when the variables are restricted.

EXAMPLE 5.4

Find the points on the ellipsoid $x^2 + \frac{y^2}{9} + \frac{z^2}{4} = 1$ nearest to and farthest from the origin.

Solution: The square of the distance from (x, y, z) to $(0, 0, 0)$ is

$$f(x, y) = x^2 + y^2 + z^2 = x^2 + y^2 + 4\left(1 - x^2 - \frac{y^2}{9}\right) = 4 - 3x^2 + \frac{5}{9}y^2.$$

Since

$$\frac{z}{2} = \pm\sqrt{1 - \left(x^2 + \frac{y^2}{9}\right)},$$

there is a natural restriction on x and y:

$$x^2 + \frac{y^2}{9} \leq 1.$$

Thus the domain of $f(x, y)$ is the closed set bounded by the ellipse $x^2 + \frac{1}{9}y^2 = 1$.

The first partials are

$$f_x = -6x, \qquad f_y = \frac{10}{9}y.$$

These vanish at $(0, 0)$ only. However, at $(0, 0)$,

$$f_{xx}f_{yy} - f_{xy}{}^2 = -6\cdot\frac{10}{9} - 0 < 0.$$

Therefore $f(x, y)$ has neither a maximum nor a minimum at $(0, 0)$. There seems to be no possible maximum or minimum.

We have forgotten the boundary! The continuous function f has both a maximum and a minimum in its *bounded closed* domain, and since they do not occur inside the domain, they must occur on the boundary curve (Fig. 5.1).

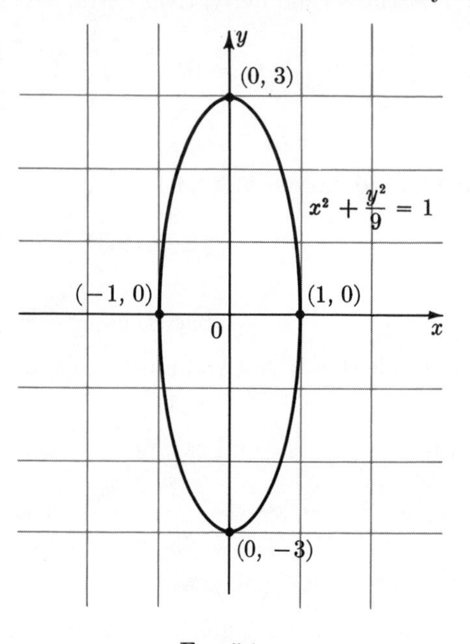

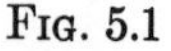
FIG. 5.1

Notice that if (x, y) is on the ellipse, then

$$x^2 + \frac{y^2}{9} = 1 \qquad \text{and} \qquad z = 0.$$

Thus the nearest and farthest points from the origin are actually points on the ellipse. By inspection (Fig. 5.1), they are the points $(\pm1, 0, 0)$ and $(0, \pm3, 0)$. [The points $(0, 0, \pm2)$ are saddle points for $f(x, y)$.]

Answer: Nearest: $(\pm1, 0, 0)$; farthest: $(0, \pm3, 0)$.

EXERCISES

Find the point nearest the origin in the first octant and on the surface:

1. $xyz = 1$
2. $xy^2z = 1$.

3. Find the maximum and minimum of $\sin(xy)$ for $0 \leq x, y \leq \pi$.
4. Find the points on the hyperboloid $x^2 + y^2/4 - z^2/9 = 1$ nearest the origin.
5. Given n points $P_1, \cdots, P_n$ of the plane, find the point P such that $\sum_{j=1}^{n} \overline{PP_j}^2$ is least.
6. For each point (x, y, z), let $f(x, y, z)$ denote the sum of the squares of the distances of (x, y, z) to the three coordinate axes. Find the maximum of $f(x, y, z)$ on the unit sphere $x^2 + y^2 + z^2 = 1$.
7. Find the least distance between the line $\mathbf{x}(t) = (t + 1, t, -t - 3)$ and the line $\mathbf{y}(\tau) = (-2\tau + 3, \tau + 1, -\tau - 2)$.
8. Give an alternate solution to Example 5.3. By inspection, find the point closest to $\mathbf{i}$ on each cross-section $z =$ constant.
9. Approximate to 3 places the minimum first octant distance from the origin to the surface $xe^y z = 1$.

6. THREE VARIABLES

The extension to three variables of the second derivative test for maxima and minima is straightforward.

Theorem Let $\mathbf{a}$ be an interior point of the domain of f and suppose

$$\operatorname{grad} f \Big|_{\mathbf{a}} = \mathbf{0}.$$

Let

$$H_f(\mathbf{a}) = \begin{bmatrix} f_{xx} & f_{xy} & f_{xz} \\ f_{yx} & f_{yy} & f_{yz} \\ f_{zx} & f_{zy} & f_{zz} \end{bmatrix}_{\mathbf{a}}$$

denote the Hessian matrix of f at $\mathbf{a}$.

(1) If $H_f(\mathbf{a})$ is positive definite, then $f(\mathbf{a})$ is a relative minimum of f, that is, there is $\delta > 0$ such that $f(\mathbf{x}) > f(\mathbf{a})$ whenever $0 < |\mathbf{x} - \mathbf{a}| < \delta$.

(2) If $H_f(\mathbf{a})$ is negative definite, then $f(\mathbf{a})$ is a relative maximum of f, that is, there is $\delta > 0$ such that $f(\mathbf{x}) < f(\mathbf{a})$ whenever $0 < |\mathbf{x} - \mathbf{a}| < \delta$.

(3) If $H_f(\mathbf{a})$ is indefinite, then $f(\mathbf{a})$ is neither a relative max nor min.

(4) Except in these cases, no conclusion can be deduced from $H_f(\mathbf{a})$ alone.

The proof of (1) is essentially the same as the proof of Theorem 4.1. As we know, (2) follows by applying (1) to $-f$. The proof of (3) is essentially the proof of Theorem 4.2.

We can use the practical tests in Chapter 7. For instance Theorem 5.2, p. 248, tells us that H_f is positive definite if and only if

$$f_{xx} > 0, \qquad \begin{vmatrix} f_{xx} & f_{xy} \\ f_{yx} & f_{yy} \end{vmatrix} > 0, \qquad \text{and} \qquad |H_f| > 0.$$

Let us take a very simple example. Suppose

$$f(x, y, z) = ax^2 + by^2 + cz^2,$$

where a, b, c are nonzero. Then

$$\operatorname{grad} f = (2ax, 2by, 2cz),$$

so $\operatorname{grad} f(\mathbf{x}) = \mathbf{0}$ only at $\mathbf{x} = \mathbf{0}$. The matrix A in this case is

$$A = \begin{bmatrix} f_{xx} & f_{xy} & f_{xz} \\ f_{yx} & f_{yy} & f_{yz} \\ f_{zx} & f_{zy} & f_{zz} \end{bmatrix} = \begin{bmatrix} 2a & 0 & 0 \\ 0 & 2b & 0 \\ 0 & 0 & 2c \end{bmatrix}.$$

According to the test, $f(\mathbf{0})$ is a minimum if the three determinants are positive:

$$2a > 0, \qquad \begin{vmatrix} 2a & 0 \\ 0 & 2b \end{vmatrix} > 0, \qquad \begin{vmatrix} 2a & 0 & 0 \\ 0 & 2b & 0 \\ 0 & 0 & 2c \end{vmatrix} > 0,$$

that is, if

$$2a > 0, \qquad 4ab > 0, \qquad 8abc > 0.$$

This is so precisely if $a > 0$, $b > 0$, and $c > 0$.

Similarly $f(\mathbf{0})$ is a maximum if

$$2a < 0, \qquad 4ab > 0, \qquad 8abc < 0,$$

which is so precisely if $a < 0$, $b < 0$, and $c < 0$.

These results agree with common sense:

$$f(\mathbf{x}) = ax^2 + by^2 + cz^2$$

so obviously $f(\mathbf{0}) = 0$. If a, b, c are all positive and $\mathbf{x} \neq \mathbf{0}$, then $f(\mathbf{x}) > 0$. If a, b, c are all negative and $\mathbf{x} \neq \mathbf{0}$, then $f(\mathbf{x}) < 0$. (If a, b, c are not all of the same sign, then $f(\mathbf{0})$ is neither a maximum nor a minimum. Why?)

EXAMPLE 6.1

Find all maxima and minima of the function

$$f(\mathbf{x}) = x^2 + 3y^2 + 4z^2 - 2xy - 2yz + 2zx.$$

Solution:

$$\operatorname{grad} f = (2x - 2y + 2z, -2x + 6y - 2z, 2x - 2y + 8z).$$

The vector equation $\operatorname{grad} f = \mathbf{0}$ amounts to the system of scalar equations (divide by 2)

$$\begin{cases} x - y + z = 0 \\ -x + 3y - z = 0 \\ x - y + 4z = 0, \end{cases}$$

whose only solution is $\mathbf{x} = \mathbf{0}$.

The matrix A is

$$\begin{bmatrix} f_{xx} & f_{xy} & f_{xz} \\ f_{yx} & f_{yy} & f_{yz} \\ f_{zx} & f_{zy} & f_{zz} \end{bmatrix} = \begin{bmatrix} 2 & -2 & 2 \\ -2 & 6 & -2 \\ 2 & -2 & 8 \end{bmatrix},$$

so the three relevant determinants are

$$2, \qquad \begin{vmatrix} 2 & -2 \\ -2 & 6 \end{vmatrix} = 8, \qquad \text{and} \qquad \begin{vmatrix} 2 & -2 & 2 \\ -2 & 6 & -2 \\ 2 & -2 & 8 \end{vmatrix} = 48.$$

All are positive. Hence by the test above, $f(\mathbf{0})$ is a minimum of f, the only one; there are no maxima.

Answer: The only extreme value of f is the minimum $f(\mathbf{0}) = 0$.

EXERCISES

The second derivative test is inconclusive at (0, 0, 0) for the given function. Determine nonetheless if the function has a maximum, a minimum, or neither at the origin:

1. $x^2 + y^2 + z^4$
2. $x^2 + y^2z^2$
3. $x^2 + y^2$
4. $x^4 + y^2 - z^6$
5. $x^2 + y^4 + z^6$
6. $x^3y^3z^3$
7. $x^4 + y^3z^3$
8. $x^4y^4 - z^5$.
9. $x^4 + y^2z^2$
10. $x^3 + y^3 + z^3$
11. $x^4y^6z^3$
12. $x^2y^2z^2$.

Find the extreme values:

13. $-2x^2 - y^2 - 3z^2 + 2xy - 2xz$

14. $x^2 + 2y^2 + z^2 + 2xy - 4yz$
15. $2x^2 + y^2 + 2z^2 + 2xy + 2yz + 2zx + x - 3z$
16. $x^2 + 3xy + y^2 - z^2 - x - 2y + z + 3$.

Determine if the function has a maximum, a minimum, or neither at the origin:

17. $x^2 + y^2 + z^2 + xy + yz + zx$
18. $x^2 + 4y^2 + 9z^2 - xy - 2yz$
19. $-x^2 - 2y^2 - z^2 + yz$
20. $x^2 + y^2 + 2z^2 - 10yz$
21. $x^2 - y^2 + 3z^2 + 12xy$
22. $3x^2 + y^2 + 4z^2 - xy - yz - zx$.
23. A surface $\mathbf{x} = \mathbf{x}(u, v)$ and a curve $\mathbf{p} = \mathbf{p}(t)$ are given. Suppose the distance $|\mathbf{x}(u, v) - \mathbf{p}(t)|$ of a point on the surface to a point of the curve is minimized (or maximized) for $\mathbf{x}_0$ on the surface and $\mathbf{p}_0$ on the curve. Show that the segment from $\mathbf{x}_0$ to $\mathbf{p}_0$ is normal to both the surface and the curve. Find the one exception to this assertion.
24. (cont.) Formulate the corresponding statement for two surfaces. (This is a four-variable problem!)

7. MAXIMA WITH CONSTRAINTS [optional]

Here are several problems that have a common feature.

(a) Of all rectangles with perimeter one, which has the shortest diagonal? That is, minimize $(x^2 + y^2)^{1/2}$ subject to $2x + 2y = 1$.

(b) Of all right triangles with perimeter one, which has largest area? That is, maximize $xy/2$ subject to $x + y + (x^2 + y^2)^{1/2} = 1$.

(c) Find the largest value of $x + 2y + 3z$ for points (x, y, z) on the unit sphere $x^2 + y^2 + z^2 = 1$.

(d) Of all rectangular boxes with fixed surface area, which has greatest volume? That is, maximize xyz subject to $xy + yz + zx = c$.

Each of these problems asks for the maximum (or minimum) of a function of several variables, where the variables must satisfy an equation (constraint).

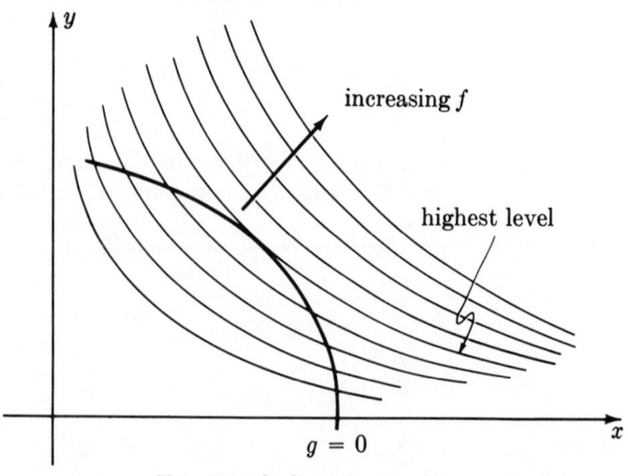

FIG. 7.1 finding the maximum

For example, in (a) you are asked to minimize

$$f(x, y) = x^2 + y^2,$$

where x and y must satisfy

$$g(x, y) = 2x + 2y - 1 = 0.$$

Such problems may be analyzed geometrically. Suppose you are asked to maximize a function $f(x, y)$, subject to a constraint $g(x, y) = 0$. On the same graph plot $g(x, y) = 0$ and several level curves of $f(x, y)$, noting the direction of increase of the level (Fig. 7.1). To find the largest value of $f(x, y)$ on the curve $g(x, y) = 0$, find the highest level curve that intersects $g = 0$. If there is a highest one and the intersection does not take place at an end point, this level curve and the graph $g = 0$ are tangent.

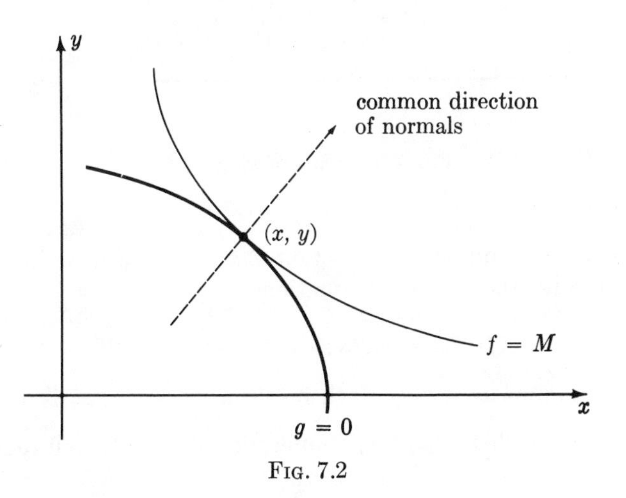

Fig. 7.2

Suppose $f(x, y) = M$ is a level curve tangent to $g(x, y) = 0$ at a point (x, y). See Fig. 7.2. Since the two graphs are tangent at (x, y), their normals at (x, y) are parallel. But the vectors

$$\text{grad} f(x, y), \qquad \text{grad}\, g(x, y)$$

point in the respective normal directions (see p. 298), hence one is a multiple of the other:

$$\text{grad} f(x, y) = \lambda \,\text{grad}\, g(x, y)$$

for some number λ. (The argument presupposes that $\text{grad}\, g \neq \mathbf{0}$ at the point in question.)

This geometric argument yields a practical rule for locating points on $g(x, y) = 0$ where $f(x, y)$ may have a maximum or minimum. Note that where the condition of tangency is satisfied, there may be a maximum, a minimum, or neither (Fig. 7.3).

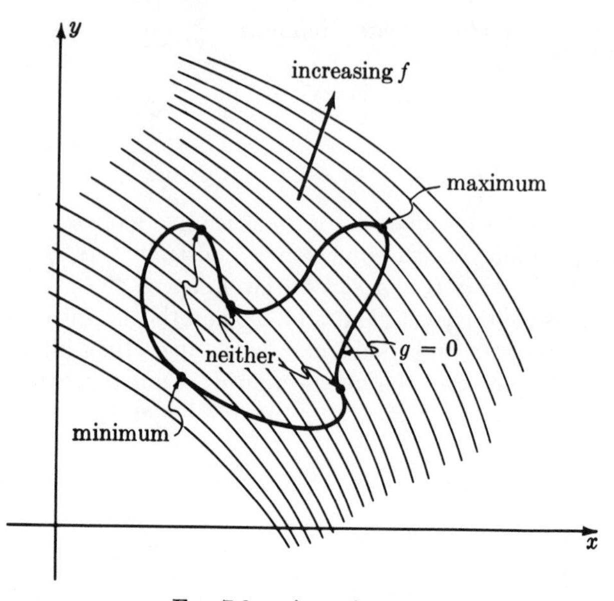

FIG. 7.3 points of tangency

> To maximize or minimize a function $f(x, y)$ subject to a constraint $g(x, y) = 0$, solve the system of equations
>
> $$(f_x, f_y) = \lambda(g_x, g_y), \qquad g(x, y) = 0$$
>
> in the three unknowns x, y, λ. Each resulting point (x, y) is a candidate. The number λ is called a **Lagrange multiplier**, or simply **multiplier**.

To apply this rule, three simultaneous equations

$$\begin{cases} f_x(x, y) = \lambda g_x(x, y) \\ f_y(x, y) = \lambda g_y(x, y) \\ g(x, y) = 0, \end{cases}$$

must be solved for three unknowns x, y, λ.

EXAMPLE 7.1

Find the largest and smallest values of $f(x, y) = x + 2y$ on the circle $x^2 + y^2 = 1$.

Solution: Draw a figure (Fig. 7.4). As seen from the figure, f takes its maximum at a point in the first quadrant, and its minimum at a point in the

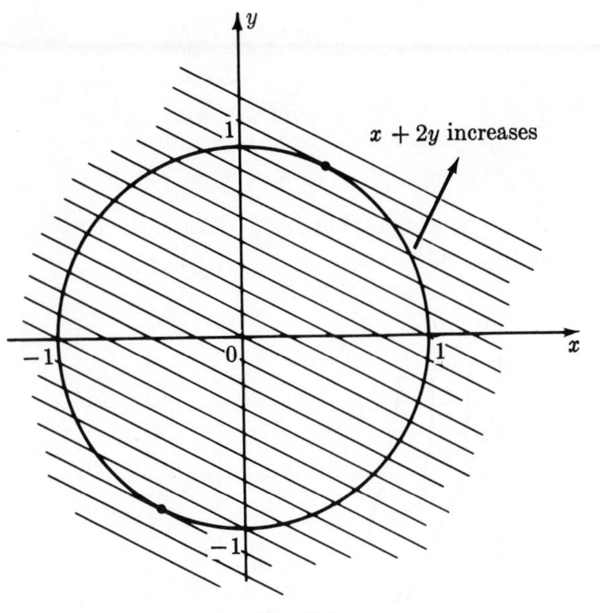

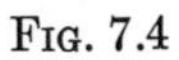

FIG. 7.4

third quadrant. Here

$$f(x, y) = x + 2y, \qquad g(x, y) = x^2 + y^2 - 1;$$
$$\text{grad} f = (1, 2), \qquad \text{grad} g = (2x, 2y).$$

The conditions

$$(f_x, f_y) = \lambda(g_x, g_y), \qquad g(x, y) = 0$$

become

$$(1, 2) = \lambda(2x, 2y), \qquad x^2 + y^2 = 1.$$

Thus

$$x = \frac{1}{2\lambda}, \qquad y = \frac{1}{\lambda}, \qquad \left(\frac{1}{2\lambda}\right)^2 + \left(\frac{1}{\lambda}\right)^2 = 1.$$

By the third equation,

$$\lambda^2 = \frac{5}{4}, \qquad \lambda = \pm\frac{1}{2}\sqrt{5}.$$

The value $\lambda = \frac{1}{2}\sqrt{5}$ yields

$$x = \frac{1}{\sqrt{5}}, \qquad y = \frac{2}{\sqrt{5}}, \qquad f(x, y) = \frac{5}{\sqrt{5}} = \sqrt{5};$$

the value $\lambda = -\frac{1}{2}\sqrt{5}$ yields

$$x = -\frac{1}{\sqrt{5}}, \qquad y = -\frac{2}{\sqrt{5}}, \qquad f(x, y) = -\frac{5}{\sqrt{5}} = -\sqrt{5}.$$

Answer: Largest $\sqrt{5}$; smallest $-\sqrt{5}$.

EXAMPLE 7.2

Find the largest and smallest values of xy on the segment $2x + y = 2, \quad x \geq 0, y \geq 0$.

Solution: Draw a graph (Fig. 7.5). Evidently the smallest value of xy is 0, taken at either end point.

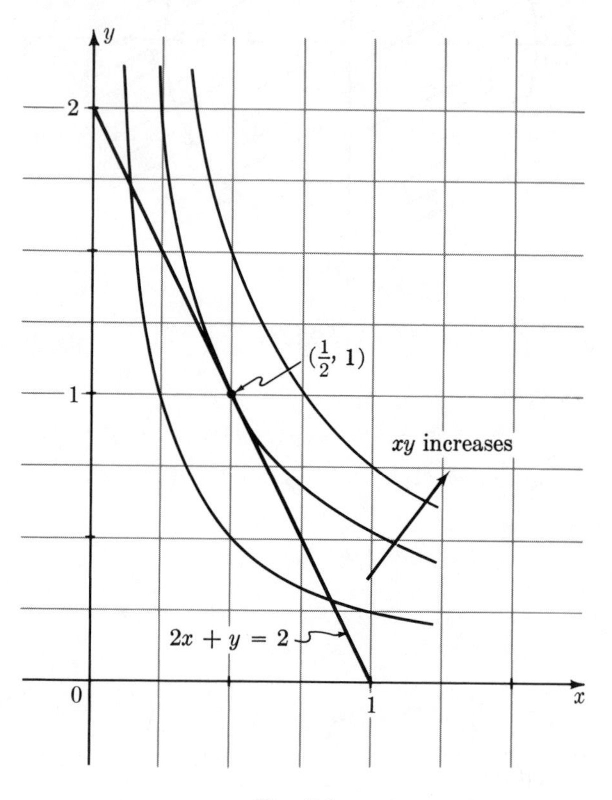

FIG. 7.5

To find the largest value, use the multiplier technique with

$$f(x, y) = xy, \qquad g(x, y) = 2x + y - 2.$$

The relevant system of equations is

$$\begin{cases} (y, x) = \lambda(2, 1) \\ 2x + y - 2 = 0. \end{cases}$$

Thus $x = \lambda$, $y = 2\lambda$, and

$$2\lambda + 2\lambda - 2 = 0, \qquad \lambda = \frac{1}{2}.$$

Therefore

$$(x, y) = \left(\frac{1}{2}, 1\right), \qquad f\left(\frac{1}{2}, 1\right) = \frac{1}{2}.$$

Answer: Largest $\frac{1}{2}$; smallest 0.

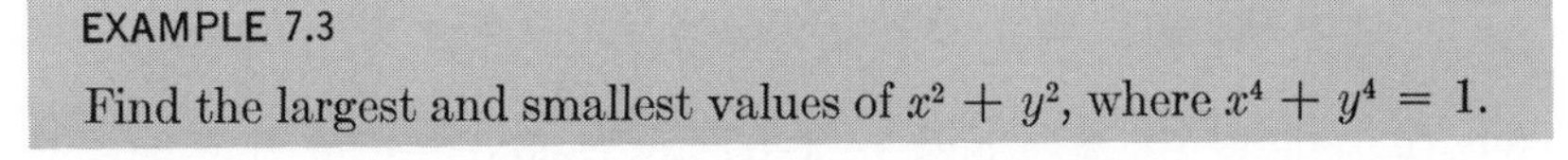

EXAMPLE 7.3

Find the largest and smallest values of $x^2 + y^2$, where $x^4 + y^4 = 1$.

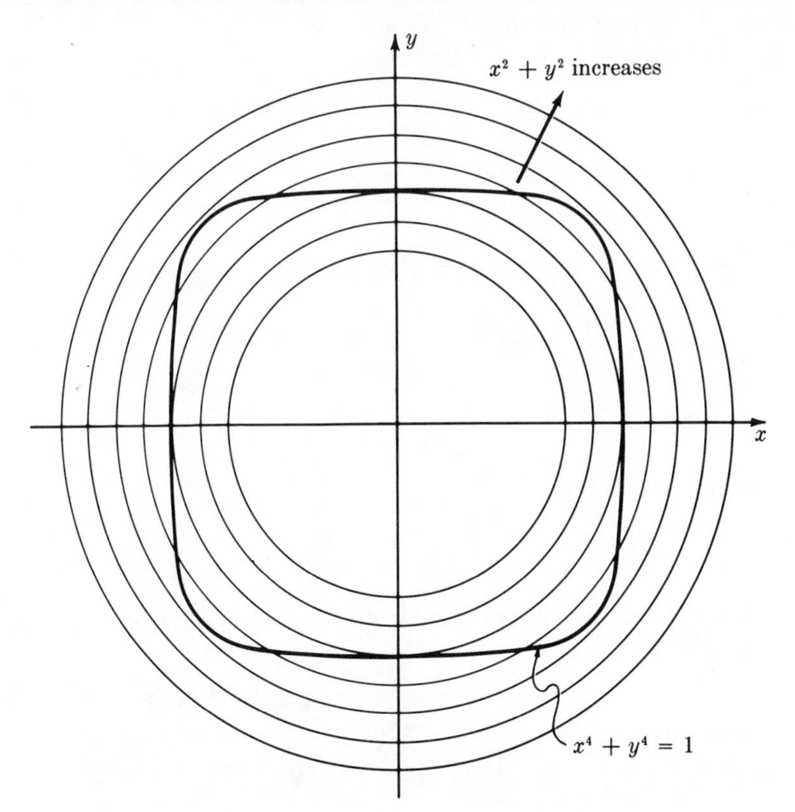

FIG. 7.6 extrema of $x^2 + y^2$ on $x^4 + y^4 = 1$

Solution: Graph the curve $x^4 + y^4 = 1$ and the level curves of $f(x, y) = x^2 + y^2$. See Fig. 7.6. By drawing $x^4 + y^4 = 1$ *accurately*, you see that the graph is quite flat where it crosses the axes and most sharply curved where it crosses the 45° lines $y = \pm x$. It is closest to the origin ($x^2 + y^2$ is least) at $(\pm 1, 0)$ and $(0, \pm 1)$, and farthest where $y = \pm x$.

The analysis confirms this. Use the multiplier technique with

$$f(x, y) = x^2 + y^2, \qquad g(x, y) = x^4 + y^4 - 1.$$

The relevant equations are

$$\begin{cases} (2x, 2y) = \lambda(4x^3, 4y^3) \\ x^4 + y^4 = 1. \end{cases}$$

Obvious solutions are

$$x = 0, \quad y = \pm 1, \quad \lambda = \frac{1}{2}; \qquad y = 0, \quad x = \pm 1, \quad \lambda = \frac{1}{2}.$$

Thus the points $(0, \pm 1)$ and $(\pm 1, 0)$ are candidates for the maximum or minimum. At each of these points $f(x, y) = 1$.

Suppose both $x \neq 0$ and $y \neq 0$. From

$$2x = 4\lambda x^3, \qquad 2y = 4\lambda y^3$$

follows

$$x^2 = y^2 = \frac{1}{2\lambda}.$$

Hence $\lambda = 1/(2x^2) > 0$. From $x^4 + y^4 = 1$ follows

$$\left(\frac{1}{2\lambda}\right)^2 + \left(\frac{1}{2\lambda}\right)^2 = 1, \qquad \lambda^2 = \frac{1}{2}, \qquad \lambda = \frac{1}{\sqrt{2}},$$

$$x^2 = y^2 = \frac{1}{2\lambda} = \frac{\sqrt{2}}{2} = \frac{1}{\sqrt{2}}.$$

Consequently, the four points

$$\left(\pm\frac{1}{\sqrt[4]{2}}, \ \pm\frac{1}{\sqrt[4]{2}}\right)$$

are candidates for the maximum or minimum. At each of these points $f(x, y) = x^2 + y^2 = 2/\sqrt{2} = \sqrt{2}$.

Answer: Largest $\sqrt{2}$; smallest 1.

EXAMPLE 7.4

Maximize $2y - x$ on the curve $y = \sin x$ for $0 \leq x \leq 2\pi$.

Solution: Here $f(x, y) = 2y - x$ and $g(x, y) = y - \sin x$. The equations are

$$\begin{cases} (-1, 2) = \lambda(-\cos x, 1) \\ y = \sin x, \end{cases}$$

from which follow $\lambda = 2$, $\cos x = \frac{1}{2}$, $x = \pi/3$ or $5\pi/3$. The maximum must occur at

$$\left(\frac{\pi}{3}, \frac{\sqrt{3}}{2}\right), \quad \left(\frac{5\pi}{3}, -\frac{\sqrt{3}}{2}\right),$$

or at one of the end points $(0, 0)$ and $(2\pi, 0)$. But

$$f\left(\frac{\pi}{3}, \frac{\sqrt{3}}{2}\right) = \sqrt{3} - \frac{\pi}{3}, \qquad f\left(\frac{5\pi}{3}, -\frac{\sqrt{3}}{2}\right) = -\sqrt{3} - \frac{5\pi}{3},$$

$$f(0, 0) = 0, \qquad f(2\pi, 0) = -2\pi.$$

Of these numbers, $\sqrt{3} - \pi/3$ is the largest, being the only positive one.

Answer: $\sqrt{3} - \dfrac{\pi}{3}$.

REMARK: The preceding example is illustrated in Fig. 7.7. From the figure, can you tell where the maximum occurs if x is restricted to the interval $\pi/2 \leq x \leq 2\pi$?

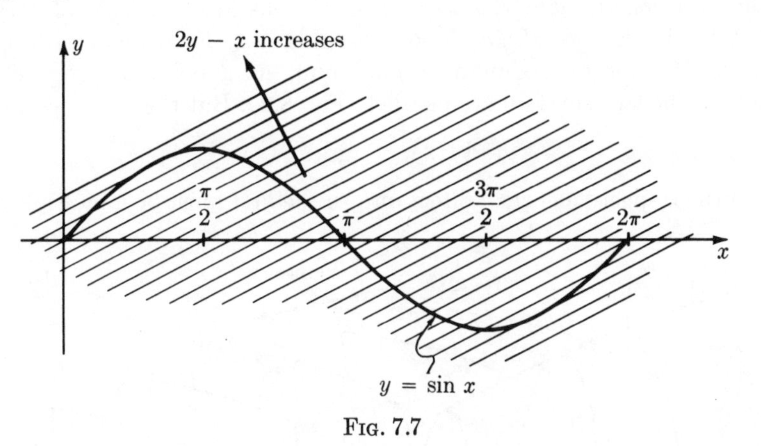

FIG. 7.7

EXERCISES

1. Find the maximum and minimum of $x + y$ on the ellipse $(x^2/4) + (y^2/9) = 1$.
2. Find the extreme values of $x - y$ on the branch $x > 0$ of the hyperbola $(x^2/9) - (y^2/4) = 1$.
3. Find the extreme values of $x - y$ on the branch $x > 0$ of the hyperbola $(x^2/4) - (y^2/9) = 1$. Explain.
4. Find the extreme values of $x - y$ on the branch $x > 0$ of the hyperbola $x^2 - y^2 = 1$. Explain.
5. Find the maximum and minimum of xy on the circle $x^2 + y^2 = 1$.
6. Find the maximum and minimum of xy on the ellipse $(x^2/4) + (y^2/9) = 1$.
7. Find the rectangle of perimeter 1 with shortest diagonal.

8. Find the right triangle of perimeter 1 with greatest area.
   [*Hint:* Eliminate λ.]
9. Find the right circular cone of fixed lateral area with maximum volume.
10. Find the right circular cone of fixed total surface area with maximum volume.
11. Find the right circular cylinder of fixed lateral area with maximum volume.
12. Find the right circular cylinder of fixed total surface area with maximum volume.
13. Let $0 < p < q$. Find the maximum and minimum of $x^p + y^p$ on $x^q + y^q = 1$, where $x \geq 0$ and $y \geq 0$.
14. (cont.) Let $0 < p < q$ and $x \geq 0$, $y \geq 0$. Show that

$$\left(\frac{x^p + y^p}{2}\right)^{1/p} \leq \left(\frac{x^q + y^q}{2}\right)^{1/q}.$$

15. Find the maximum and minimum of x^2y on the short arc of circle $x^2 + y^2 = 1$ between $(\frac{1}{2}\sqrt{3}, \frac{1}{2})$ and $(\frac{1}{2}\sqrt{2}, \frac{1}{2}\sqrt{2})$.

8. FURTHER CONSTRAINT PROBLEMS [optional]

In space, the problem is to maximize (minimize) a function $f(x, y, z)$ subject to a constraint $g(x, y, z) = 0$. One seeks level surfaces of $f(x, y, z)$ tangent to the surface $g(x, y, z) = 0$. See Fig. 8.1. Each point of tangency is a candidate for a maximum or minimum. At a point of tangency, the normals to the two surfaces are parallel (Fig. 8.2). But the two vectors

$$\operatorname{grad} f(\mathbf{x}), \qquad \operatorname{grad} g(\mathbf{x})$$

point in the respective normal directions, so the first must be a multiple of the

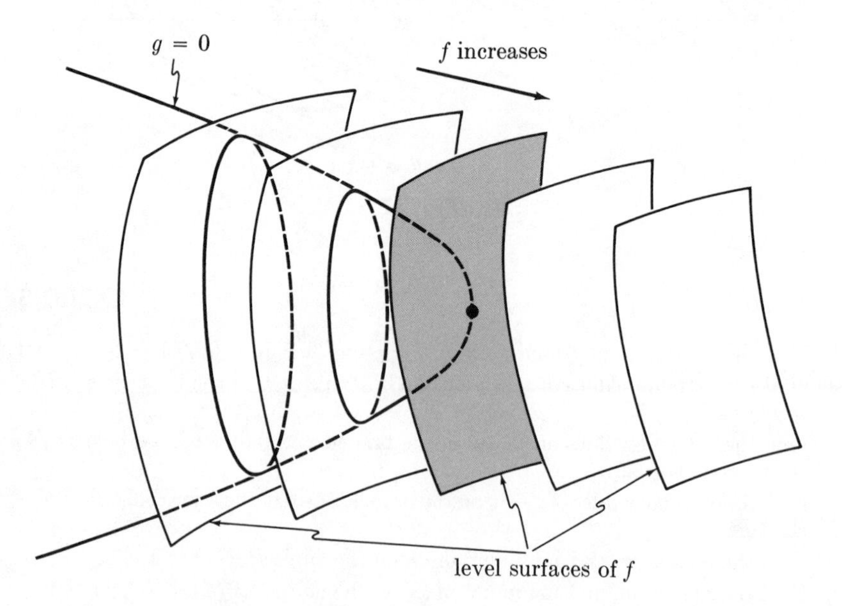

FIG. 8.1 level surface tangent to $g = 0$

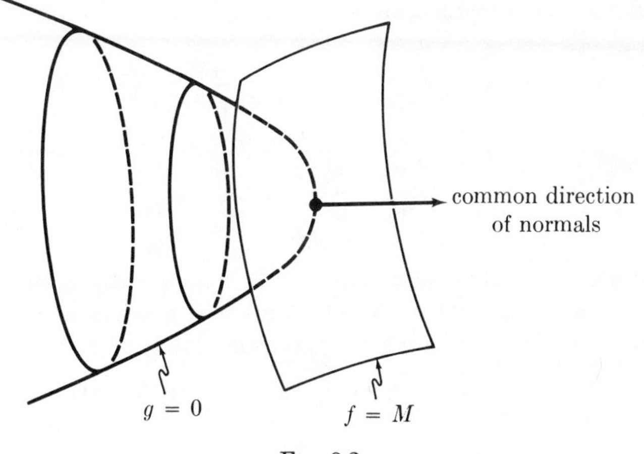

FIG. 8.2

second, provided grad $g(\mathbf{x}) \neq \mathbf{0}$. This observation leads to a practical method for locating possible maxima and minima (proof omitted).

> To maximize (minimize) a function $f(x, y, z)$ subject to a constraint $g(x, y, z) = 0$, solve the system of equations
>
> $$(f_x, f_y, f_z) = \lambda(g_x, g_y, g_z), \qquad g(\mathbf{x}) = 0$$
>
> in the four unknowns x, y, z, λ. Each resulting point (x, y, z) is a candidate.

In applications, the usual precautions concerning the boundary must be observed.

EXAMPLE 8.1

Find the longest and shortest distance from the origin to the ellipsoid $x^2 + \dfrac{y^2}{9} + \dfrac{z^2}{4} = 1$.

Solution: Set $f(x, y, z)$ equal to the square of the distance,

$$f(x, y, z) = x^2 + y^2 + z^2,$$

and set

$$g(x, y, z) = x^2 + \frac{1}{9}y^2 + \frac{1}{4}z^2 - 1.$$

Then

$$\operatorname{grad} f = (2x, 2y, 2z), \qquad \operatorname{grad} g = \left(2x, \frac{2}{9}y, \frac{1}{2}z\right).$$

The required system of equations is

$$\begin{cases} (2x,\ 2y,\ 2z) = \lambda\left(2x,\ \dfrac{2}{9}\,y,\ \dfrac{1}{2}\,z\right) \\[2ex] x^2 + \dfrac{y^2}{9} + \dfrac{z^2}{4} = 1. \end{cases}$$

If $x \neq 0$, then by the first equation $\lambda = 1$; consequently $y = z = 0$, so by the second equation $x^2 = 1$, $x = \pm 1$. Thus two candidates are the points $(\pm 1, 0, 0)$. Similarly, there are four other candidates, namely

$$\lambda = 9:\quad (0, \pm 3, 0); \qquad \lambda = 4:\quad (0, 0, \pm 2).$$

Since

$$f(\pm 1, 0, 0) = 1, \qquad f(0, \pm 3, 0) = 9, \qquad f(0, 0, \pm 2) = 4,$$

the maximum distance is $\sqrt{9}$ and the minimum distance is 1.

Answer: Longest 3; shortest 1.

REMARK: Compare this procedure with the previous solution of the same problem in Section 5, p. 347. The advantage of the present method will be crystal clear.

EXAMPLE 8.2

Find the volume of the largest rectangular solid with sides parallel to the coordinate axes that can be inscribed in the ellipsoid

$$x^2 + \frac{y^2}{9} + \frac{z^2}{4} = 1.$$

Solution: One-eighth of the volume is

$$f(x, y, z) = xyz,$$

where $x > 0$, $y > 0$, $z > 0$. See Fig. 8.3. The constraint is $g(x, y, z) = 0$, where

$$g(x, y, z) = x^2 + \frac{y^2}{9} + \frac{z^2}{4} - 1.$$

Set $\operatorname{grad} f = \lambda \operatorname{grad} g$ and $g = 0$:

$$\begin{cases} (yz,\ zx,\ xy) = \lambda\left(2x,\ \dfrac{2}{9}\,y,\ \dfrac{1}{2}\,z\right) \\[2ex] x^2 + \dfrac{y^2}{9} + \dfrac{z^2}{4} = 1, \end{cases}$$

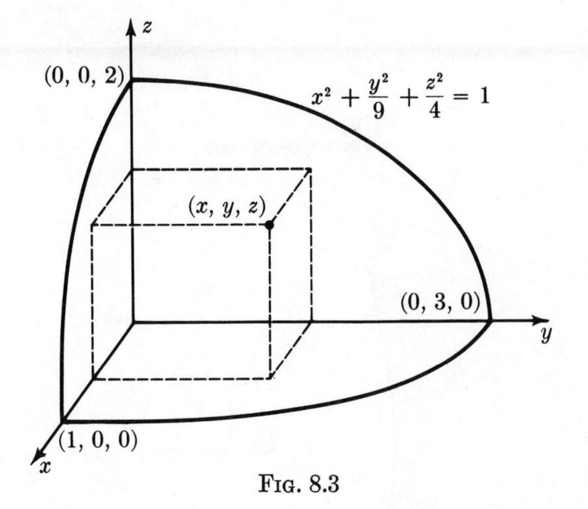

FIG. 8.3

that is,

$$\begin{cases} yz = 2\lambda x \\ zx = \dfrac{2}{9}\lambda y \qquad x^2 + \dfrac{y^2}{9} + \dfrac{z^2}{4} = 1. \\ xy = \dfrac{1}{2}\lambda z \end{cases}$$

Multiply the first two equations and cancel xy:

$$z^2 = \frac{4}{9}\lambda^2.$$

Likewise

$$x^2 = \frac{1}{9}\lambda^2, \qquad y^2 = \lambda^2.$$

Substitute in the fourth equation:

$$\frac{1}{9}\lambda^2 + \frac{1}{9}\lambda^2 + \frac{1}{9}\lambda^2 = 1, \qquad \lambda^2 = 3,$$

$$x^2 = \frac{1}{3}, \qquad y^2 = 3, \qquad z^2 = \frac{4}{3}.$$

Hence

$$f(x, y, z)^2 = x^2y^2z^2 = \frac{4}{3}, \qquad f_{\max} = \frac{2\sqrt{3}}{3}.$$

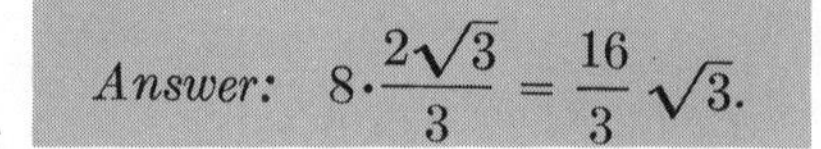
Answer: $8 \cdot \frac{2\sqrt{3}}{3} = \frac{16}{3}\sqrt{3}.$

Two Constraints

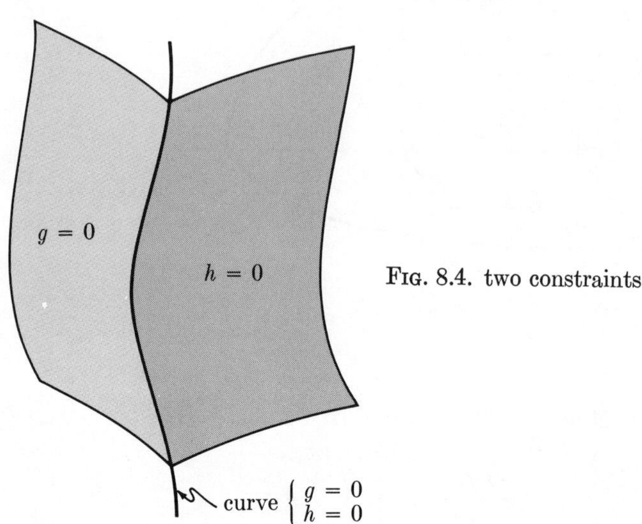

FIG. 8.4. two constraints

Suppose the problem is to maximize (minimize) $f(x, y, z)$, where (x, y, z) is subject to *two* constraints, $g(x, y, z) = 0$ and $h(x, y, z) = 0$. Each constraint defines a surface, and these two surfaces in general have a curve of intersection (Fig. 8.4). A candidate for a maximum or minimum of $f(\mathbf{x})$ is a point $\mathbf{x}$ where a level surface of f is tangent to this curve of intersection (Fig. 8.5). The vector grad $f(\mathbf{x})$ is normal to the level surface at $\mathbf{x}$, hence normal to

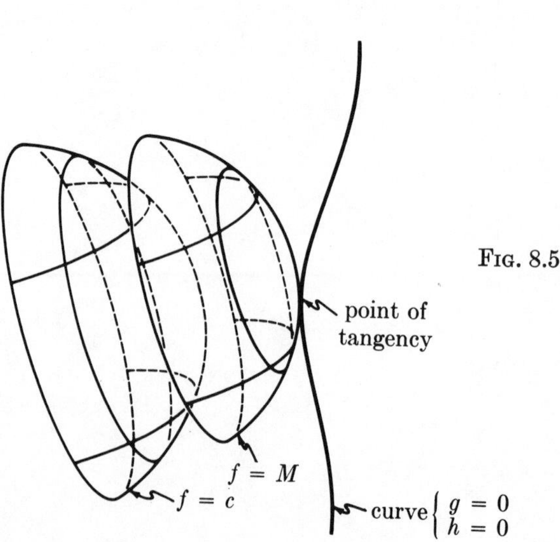

FIG. 8.5

the curve. But the vectors $\operatorname{grad} g(\mathbf{x})$ and $\operatorname{grad} h(\mathbf{x})$ *determine* the normal plane to the curve at $\mathbf{x}$. Hence for some constants λ and μ,

$$\operatorname{grad} f(\mathbf{x}) = \lambda \operatorname{grad} g(\mathbf{x}) + \mu \operatorname{grad} h(\mathbf{x}).$$

REMARK: The existence of such **multipliers** λ and μ presupposes that $\operatorname{grad} g \neq \mathbf{0}$, $\operatorname{grad} h \neq \mathbf{0}$, and that neither is a multiple of the other.

To maximize (minimize) a function $f(x, y, z)$ subject to two constraints $g(x, y, z) = 0$ and $h(x, y, z) = 0$, solve the system of five equations

$$\begin{cases} (f_x, f_y, f_z) = \lambda(g_x, g_y, g_z) + \mu(h_x, h_y, h_z) \\ g(\mathbf{x}) = 0, \qquad h(\mathbf{x}) = 0 \end{cases}$$

in five unknowns x, y, z, λ, μ. Each resulting point (x, y, z) is a candidate.

EXAMPLE 8.3

Find the maximum and minimum of $f(x, y, z) = x + 2y + z$ on the ellipse $x^2 + y^2 = 1, \quad y + z = 1$.

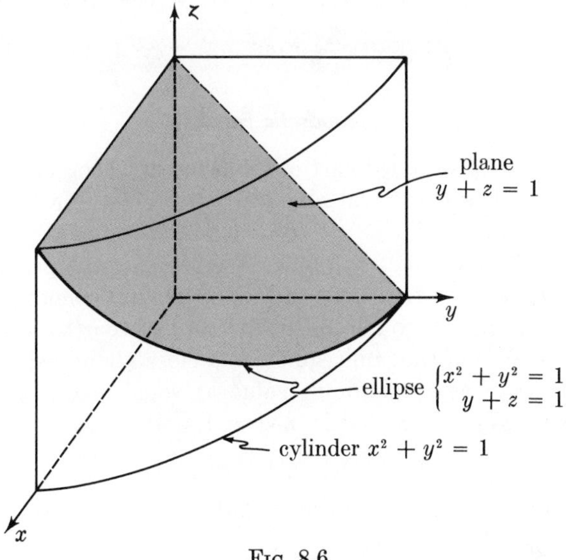

FIG. 8.6

Solution: Here

$$g(x, y, z) = x^2 + y^2 - 1, \qquad h(x, y, z) = y + z - 1;$$

Fig. 8.6 shows the curve of intersection of the surfaces $g(\mathbf{x}) = 0$ and $h(\mathbf{x}) = 0$, a plane section of a right circular cylinder. The equations to be solved are

$$\begin{cases} (1, 2, 1) = \lambda(2x, 2y, 0) + \mu(0, 1, 1) \\ x^2 + y^2 = 1, \qquad y + z = 1. \end{cases}$$

Hence

$$1 = 2\lambda x, \qquad 2 = 2\lambda y + \mu, \qquad 1 = \mu; \qquad x = \frac{1}{2\lambda}, \qquad y = \frac{1}{2\lambda};$$

$$\left(\frac{1}{2\lambda}\right)^2 + \left(\frac{1}{2\lambda}\right)^2 = 1, \qquad \lambda^2 = \frac{1}{2}, \qquad \lambda = \pm\frac{\sqrt{2}}{2}.$$

The solution $\lambda = \sqrt{2}/2$ leads to the point

$$\mathbf{x}_1 = \left(\frac{\sqrt{2}}{2}, \frac{\sqrt{2}}{2}, 1 - \frac{\sqrt{2}}{2}\right),$$

and the solution $\lambda = -\sqrt{2}/2$ leads to the point

$$\mathbf{x}_2 = \left(-\frac{\sqrt{2}}{2}, -\frac{\sqrt{2}}{2}, 1 + \frac{\sqrt{2}}{2}\right).$$

These are the only candidates for maxima or minima. But

$$f(\mathbf{x}_1) = 1 + \sqrt{2}, \qquad f(\mathbf{x}_2) = 1 - \sqrt{2}.$$

Answer: Maximum $1 + \sqrt{2}$; minimum $1 - \sqrt{2}$.

Quadratic Forms

In Chapter 7, we postponed part of the proof of Theorem 8.5 that a 3×3 symmetric matrix has three real characteristic roots. We fill this gap now using Lagrange multipliers.

Let A be a 3×3 symmetric matrix. We consider the problem of finding the minimum of $f(\mathbf{x}) = \mathbf{x}A\mathbf{x}'$ on the surface of the unit sphere. We set $g(\mathbf{x}) = \mathbf{x}\mathbf{x}' - 1$, so the problem is to minimize $f(\mathbf{x})$ subject to $g(\mathbf{x}) = 0$.

Since the (surface of the) unit sphere is a closed bounded set, the continuous function $f(\mathbf{x})$ has a minimum value at some point $\mathbf{x}_0'$. By Lagrange multiplier theory, $g(\mathbf{x}_0) = 0$, that is, $\mathbf{x}_0\mathbf{x}_0' = 1$, and

$$\operatorname{grad} f(\mathbf{x}_0) = \lambda \operatorname{grad} g(\mathbf{x}_0),$$

for some λ. A direct computation shows that

$$\operatorname{grad} f(\mathbf{x}) = 2A\mathbf{x}' \qquad \text{and} \qquad \operatorname{grad} g(\mathbf{x}) = 2\mathbf{x}',$$

hence the condition is

$$A\mathbf{x}_0' = \lambda\mathbf{x}_0'.$$

Thus $\mathbf{x}_0'$ is a characteristic vector of A and λ is the corresponding characteristic root. Furthermore,

$$f(\mathbf{x}_0) = \mathbf{x}_0A\mathbf{x}_0' = \lambda\mathbf{x}_0\mathbf{x}_0' = \lambda,$$

so λ actually is the minimum value. It follows that λ is the smallest real characteristic root of A.

Similarly, $f(\mathbf{x})$ has a maximum value ν on the unit sphere; this value is the largest real characteristic root of A.

It may be that $\lambda = \nu$. If so, then $f(\mathbf{x})$ has the constant value λ on the sphere. Hence $f(\mathbf{x}) = \lambda \mathbf{x}\mathbf{x}'$ and it follows that $A = \lambda I$. Thus the characteristic roots of A are $\lambda, \lambda, \lambda$, all real.

If $\lambda < \nu$, then A has two distinct real characteristic roots. It follows that the third characteristic root must also be real. For the characteristic polynomial of A is a cubic, and if a cubic has two real zeros, the third zero is also real, since complex zeros of polynomials with real coefficients come in conjugate pairs.

A Second Derivative Test

Suppose we wish to minimize $f(x, y, z)$ subject to the constraint $g(x, y, z) = 0$. Suppose we have found a solution (x, y, z, λ) to the system of equations

$$\begin{cases} \operatorname{grad} f = \lambda \operatorname{grad} g \\ \qquad g = 0. \end{cases}$$

It would be nice to have a test for a minimum analogous to the second derivative test in Section 6. There is such a test, but its proof is beyond the scope of this course.

Theorem Let $\mathbf{x}_0, \lambda$ be a solution of

$$\operatorname{grad} f \Big|_{\mathbf{x}_0} = \operatorname{grad} g \Big|_{\mathbf{x}_0}$$

where $\mathbf{x}_0$ is interior to the common domain of f and g, and $(\operatorname{grad} g)(\mathbf{x}_0) \neq \mathbf{0}$. If the matrix

$$(H_f - \lambda H_g)\Big|_{\mathbf{x}_0}$$

is positive definite, then there is a $\delta > 0$ such that $f(\mathbf{x}) > f(\mathbf{x}_0)$ for all $\mathbf{x}$ satisfying $g(\mathbf{x}) = 0$ and $0 < |\mathbf{x} - \mathbf{x}_0| < \delta$.

EXERCISES

1. Assume $a, b, c > 0$. Find the volume of the largest rectangular solid (with sides parallel to the coordinate planes) inscribed in the ellipsoid

$$\frac{x^2}{a^2} + \frac{y^2}{b^2} + \frac{z^2}{c^2} = 1.$$

2. Find the triangle of largest area with fixed perimeter.
[*Hint:* Use Heron's formula $A^2 = s(s-x)(s-y)(s-z)$, where x, y, z are the sides and s is the semiperimeter.]
3. Find the maximum and minimum of the function $x + 2y + 3z$ on the sphere $x^2 + y^2 + z^2 = 1$.
4. Find the rectangular solid of fixed surface area with maximum volume.
5. Find the rectangular solid of fixed total edge length with maximum surface area.
6. Find the rectangular solid of fixed total edge length with maximum volume.
7. Maximize xyz on $x + y + z = 1$.
8. (cont.) Conclude that $\sqrt[3]{xyz} \leq \dfrac{x+y+z}{3}$ for $x \geq 0$, $y \geq 0$, and $z \geq 0$.
9. Let $0 < p < q$. Find the maximum and minimum of $x^p + y^p + z^p$ on the surface $x^q + y^q + z^q = 1$, where $x \geq 0$, $y \geq 0$, and $z \geq 0$.
10. (cont.) Let $0 < p < q$ and $x \geq 0$, $y \geq 0$, $z \geq 0$. Show that
$$\left(\frac{x^p + y^p + z^p}{3}\right)^{1/p} \leq \left(\frac{x^q + y^q + z^q}{3}\right)^{1/q}.$$
11. Show that $xy + yz + zx \leq x^2 + y^2 + z^2$ for $x \geq 0$, $y \geq 0$, and $z \geq 0$.
12. Find the minimum of $x^2 + y^2 + z^2$ on the line $x + y + z = 1$, $x + 2y + 3z = 1$.
13. Find the maximum and minimum volumes of a rectangular solid whose total edge length is 24 ft and whose surface area is 22 ft^2.
14. Find the maximum and minimum of $x + y + z$ on the first octant portion of the curve $xyz = 1$, $x^2 + y^2 + z^2 = \frac{17}{4}$.
[*Hint:* Show that x, y, z are roots of the same quadratic equation; conclude that two of them are equal.]

15* Let A and B be 3×3 matrices with B positive definite. Show that each Lagrange multiplier λ in the problem of maximizing $\mathbf{x}A\mathbf{x}'$ on the ellipsoid $\mathbf{x}B\mathbf{x}' = 1$ is a characteristic root of AB^{-1}. Assume A and B are symmetric.

16. Let $f(x, y, z) = x^2 + y^2 + z - z^2$ and $g(x, y, z) = z$. Prove that the minimum of f on $g = 0$ is $0 = f(0, 0, 0)$.
17. (cont.) Show however that the second derivative test at the end of the section is inconclusive for this example.

10. Double Integrals

1. INTRODUCTION

A geometric motivation for the definite integral (also called **simple integral**)

$$\int_a^b f(x)\,dx$$

is the problem of finding the area under a curve $y = f(x)$. Suppose we consider instead the problem of finding the volume under a surface $z = f(x, y)$, where $f(x, y)$ is a continuous (positive) function defined on a rectangular domain

$$a \leq x \leq b, \qquad c \leq x \leq d.$$

This leads us to a new kind of integral called a double integral. Now the theory behind the double integral (and the triple integral of the next chapter) is rather technical and lengthy, so we shall postpone it until Chapter 12. Meanwhile we shall proceed intuitively, taking a lot of things for granted. Our present aim is to develop a working knowledge of how to set up, evaluate, and apply double and triple integrals (called collectively **multiple integrals**).

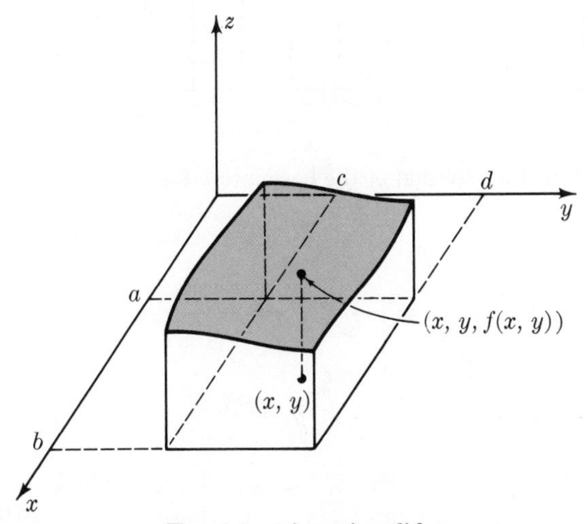

FIG. 1.1 prismatic solid

Volume

Consider the prismatic solid (Fig. 1.1) bounded above by the surface $z = f(x, y)$ and below by the rectangular domain $a \leq x \leq b$, $c \leq y \leq d$.

Intuitively it seems obvious that the solid has a well-defined volume. For we can carve the solid out of a homogeneous material like clay, plaster, or steel and then weigh it or submerge it in a tank of water to find its volume. This is fine, but how do we compute the volume? We shall use a procedure analogous to rectangular approximations for plane areas.

We shall approximate the given solid by a large number of long, thin rectangular solids, and add up their volumes to approximate the desired volume.

Partition the base of the solid into mn equal small rectangles by drawing segments parallel to the x- and y-axes that divide $[a, b]$ into m equal parts and $[c, d]$ into n equal parts (see Fig. 1.2).

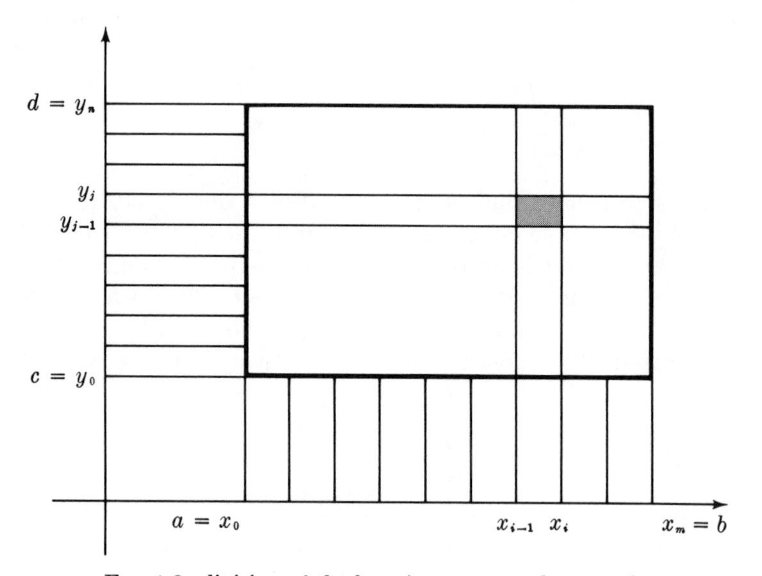

FIG. 1.2 division of the base into mn equal rectangles

Thus

$$\begin{cases} a = x_0 < x_1 < x_2 < \cdots < x_m = b, & x_i - x_{i-1} = \dfrac{b - a}{m}, \\ c = y_0 < y_1 < y_2 < \cdots < y_n = d, & y_j - y_{j-1} = \dfrac{d - c}{n}. \end{cases}$$

The part of the solid above the i, j-th small rectangle is approximately a thin rectangular column of height $f(\bar{x}_i, \bar{y}_j)$, where $(\bar{x}_i, \bar{y}_j)$ is the midpoint of the base (Fig. 1.3).

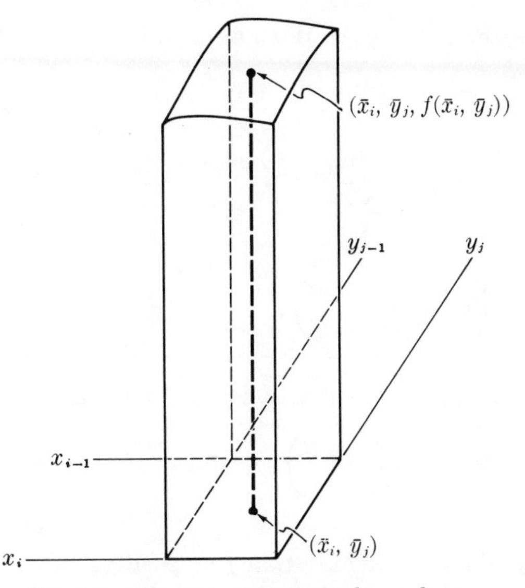

FIG. 1.3 thin, almost rectangular, column

The volume of the column is

$$f(\bar{x}_i, \bar{y}_j)\left(\frac{b-a}{m}\right)\left(\frac{d-c}{n}\right).$$

Add to approximate the total volume:

$$V \approx \sum_{\substack{i=1,2,\cdots,m \\ j=1,2,\cdots,n}} f(\bar{x}_i, \bar{y}_j)\left(\frac{b-a}{m}\right)\left(\frac{d-c}{n}\right)$$

$$= \sum_{i=1}^{m}\sum_{j=1}^{n} f(\bar{x}_i, \bar{y}_j)\left(\frac{b-a}{m}\right)\left(\frac{d-c}{n}\right).$$

Now partition the base into more and more rectangles by letting m and n *both* increase to infinity, so $(b-a)/m$ and $(d-c)/n$ both decrease to zero. Then under reasonable conditions the double sum converges to a limit:

$$V = \lim \sum_{i=1}^{m}\sum_{j=1}^{n} f(\bar{x}_i, \bar{y}_j)\left(\frac{b-a}{m}\right)\left(\frac{d-c}{n}\right), \qquad m \longrightarrow \infty, \quad n \longrightarrow \infty.$$

One reasonable condition is that f be continuous, as will be shown in Chapter 12. The double sum expression suggests the double integral notation

$$V = \iint\limits_{\substack{a \le x \le b \\ c \le y \le d}} f(x, y)\, dx\, dy = \iint\limits_{\mathbf{D}} f(x, y)\, dx\, dy,$$

where **D** denotes the rectangular domain of f. Thus the full definition of the **double integral** is

$$\iint_{\mathbf{D}} f(x, y)\, dx\, dy = \lim_{\substack{m\to\infty \\ n\to\infty}} \sum_{\substack{i=1,\cdots,m \\ j=1,\cdots,n}} f(\bar{x}_i, \bar{y}_j)\left(\frac{b-a}{m}\right)\left(\frac{d-c}{n}\right),$$

where

$$\bar{x}_i = a + \left(i - \frac{1}{2}\right)\left(\frac{b-a}{m}\right), \qquad \bar{y}_j = c + \left(j - \frac{1}{2}\right)\left(\frac{d-c}{n}\right).$$

The formula for $\bar{x}_i$ simply says that $\bar{x}_i$ is the midpoint (average) of

$$x_{i-1} = a + (i-1)\left(\frac{b-a}{m}\right) \qquad \text{and} \qquad x_i = a + i\left(\frac{b-a}{m}\right).$$

Likewise $\bar{y}_j = \frac{1}{2}(y_{j-1} + y_j)$.

The definition does not require that f be positive. If f takes both positive and negative values, the double integral represents an *algebraic volume* rather than a geometric volume. The volume between the surface $z = f(x, y)$ and the x, y-plane counts positively where $f > 0$ and negatively where $f < 0$.

Assuming the double integral exists, the sum formulas

$$\sum\sum kf(\bar{x}_i, \bar{y}_j) = k\sum\sum f(\bar{x}_i, \bar{y}_j),$$

$$\sum\sum [f(\bar{x}_i, \bar{y}_j) + g(\bar{x}_i, \bar{y}_j)] = \sum\sum f(\bar{x}_i, \bar{y}_j) + \sum\sum g(\bar{x}_i, \bar{y}_j)$$

imply the relations

$$\iint_{\mathbf{D}} kf(x, y)\, dx\, dy = k\iint_{\mathbf{D}} f(x, y)\, dx\, dy,$$

$$\iint_{\mathbf{D}} [f(x, y) + g(x, y)]\, dx\, dy = \iint_{\mathbf{D}} f(x, y)\, dx\, dy + \iint_{\mathbf{D}} g(x, y)\, dx\, dy.$$

Once the double integral has been defined, the immediate problem is how to evaluate it. Sections 2 and 3 provide a solution to this problem.

2. SPECIAL CASES

In certain cases where $f(x, y)$ has a particularly simple form, it is easy to evaluate the double integral

$$\iint_{\mathbf{D}} f(x, y)\, dx\, dy.$$

As before, **D** denotes the rectangle $a \le x \le b$, $c \le y \le d$.

Suppose first that $f(x, y) = h$, a constant. Then the double integral represents the volume of a rectangular solid of height h. This volume is h times the area of the base:

$$\iint_{\mathbf{D}} h\, dx\, dy = h \times (\text{area } \mathbf{D}) = h(b - a)(d - c).$$

Suppose next that $f(x, y)$ is a function of x alone,

$$f(x, y) = g(x).$$

Then each cross-section of the solid by a plane parallel to the z, x-plane is an identical plane region (Fig. 2.1a, b). The volume is the area of this region times the y-length of the rectangle:

$$\iint_{\mathbf{D}} g(x)\, dx\, dy = \left(\int_a^b g(x)\, dx \right) (d - c).$$

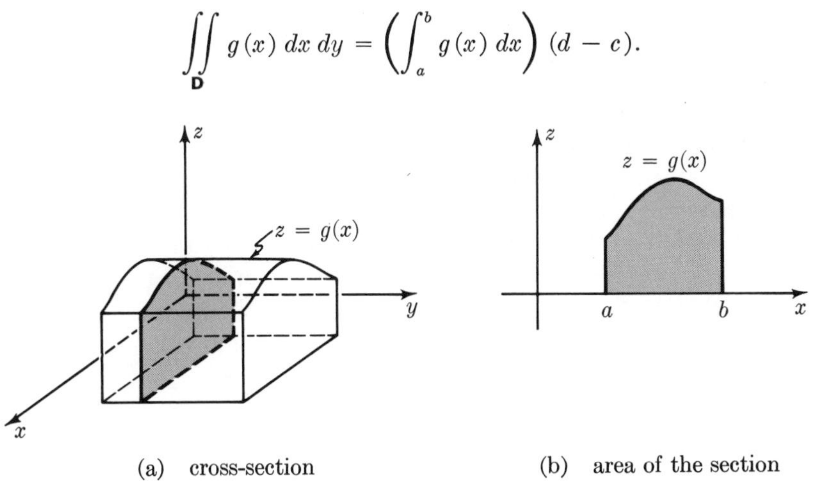

(a) cross-section (b) area of the section

FIG. 2.1

Similarly if $f(x, y) = h(y)$,

$$\iint_{\mathbf{D}} h(y)\, dx\, dy = \left(\int_c^d h(y)\, dy \right) (b - a).$$

These formulas are special cases of the following one:

$$\boxed{\iint_{\mathbf{D}} g(x)\, h(y)\, dx\, dy = \left(\int_a^b g(x)\, dx \right) \left(\int_c^d h(y)\, dy \right).}$$

This formula gives a double integral as the product of two ordinary (single) integrals. It applies to such functions as

$$x^5 y^7, \qquad e^{x-y}, \qquad e^x \cos y, \qquad \text{etc.}$$

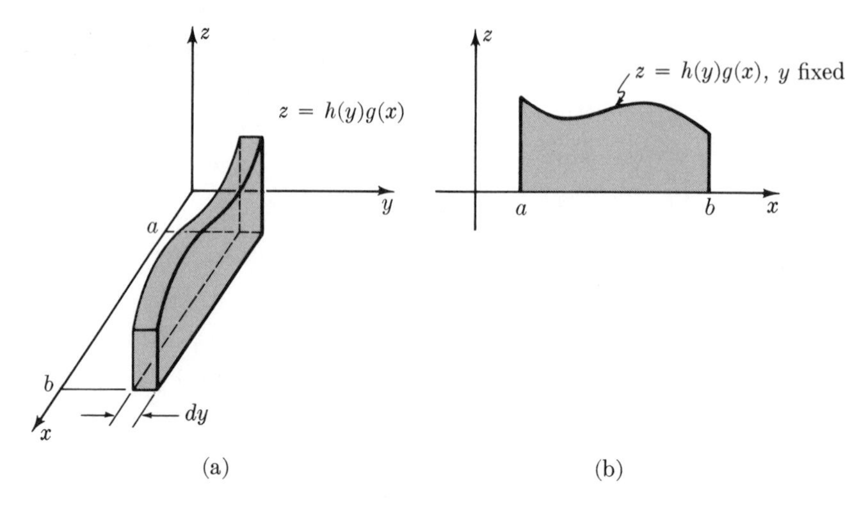

FIG. 2.2 cross-sectional slab

Let us try to justify the formula intuitively. The volume of a solid (the left-hand side) is evaluated by slicing the solid into slabs, then adding up (Fig. 2.2a). Fix y, then slice the solid by the planes parallel to the z, x-plane at y and at $y + dy$. The result is a slab of thickness dy. The projection (Fig. 2.2b) of this slab on the z, x-plane is the region in that plane under the curve $z = z(x) = h(y)g(x)$. (Remember y is fixed!) Hence its area is

$$\int_a^b z(x)\,dx = \int_a^b h(y)\,g(x)\,dx = h(y)\int_a^b g(x)\,dx.$$

The volume of the slab is its area times its thickness:

$$dV = \left(h(y)\int_a^b g(x)\,dx\right)dy = \left(\int_a^b g(x)\,dx\right)h(y)\,dy.$$

On the right-hand side, the first factor is a constant, therefore adding the slabs yields

$$V = \int_c^d\left(\int_a^b g(x)\,dx\right)h(y)\,dy = \left(\int_a^b g(x)\,dx\right)\left(\int_c^d h(y)\,dy\right).$$

This is the desired formula.

EXAMPLE 2.1

Find $\displaystyle\iint_{0\le x,\, y\le 1} x^4y^6\,dx\,dy.$

Solution:

$$\iint_{0\le x,\, y\le 1} x^4y^6\,dx\,dy = \left(\int_0^1 x^4\,dx\right)\left(\int_0^1 y^6\,dy\right) = \frac{1}{5}\cdot\frac{1}{7}.$$

Answer: $\dfrac{1}{35}$.

EXAMPLE 2.2

Find $\displaystyle\iint\limits_{\substack{0\le x\le 1\\ \pi/2\le y\le\pi}} e^x \cos y\, dx\, dy.$

Solution: Let **D** denote the rectangle, $0 \le x \le 1$ and $\pi/2 \le y \le \pi$. Then

$$\iint\limits_{\mathbf{D}} = \left(\int_0^1 e^x\, dx\right)\left(\int_{\pi/2}^{\pi} \cos y\, dy\right) = \left(e^x\Big|_0^1\right)\left(\sin y\Big|_{\pi/2}^{\pi}\right) = (e-1)(-1).$$

Answer: $1 - e$.

QUESTION: The answer is negative. Why?

EXAMPLE 2.3

Find $\displaystyle\iint\limits_{\substack{0\le x\le 1\\ -2\le y\le -1}} e^{x-y}\, dx\, dy.$

Solution: Since $e^{x-y} = e^x e^{-y}$,

$$\iint\limits_{\mathbf{D}} e^{x-y}\, dx\, dy = \left(\int_0^1 e^x\, dx\right)\left(\int_{-2}^{-1} e^{-y}\, dy\right)$$

$$= \left(e^x\Big|_0^1\right)\left(-e^{-y}\Big|_{-2}^{-1}\right) = (e-1)(e^2-e).$$

Answer: $e(e-1)^2$.

EXAMPLE 2.4

Find $\displaystyle\iint\limits_{\substack{1\le x\le 2\\ -1\le y\le 1}} (x^2y - 3xy^2)\, dx\, dy.$

Solution: Use the linear property of the double integral:

$$\iint\limits_{\mathbf{D}} (x^2y - 3xy^2)\, dx\, dy = \iint\limits_{\mathbf{D}} x^2y\, dx\, dy - 3\iint\limits_{\mathbf{D}} xy^2\, dx\, dy.$$

Evaluate these two integrals separately:

$$\iint\limits_{\mathbf{D}} x^2y\, dx\, dy = \left(\int_1^2 x^2\, dx\right)\left(\int_{-1}^{1} y\, dy\right) = 0;$$

$$\iint\limits_{\mathbf{D}} xy^2\, dx\, dy = \left(\int_1^2 x\, dx\right)\left(\int_{-1}^{1} y^2\, dy\right) = \frac{3}{2}\cdot\frac{2}{3} = 1.$$

Answer: $-3.$

EXERCISES

Evaluate:

1. $\iint (3x - 1)\,dx\,dy;\qquad -1 \le x \le 2,\quad 0 \le y \le 5$

2. $\iint e^y\,dx\,dy;\qquad -1 \le x \le 1,\quad 0 \le y \le \ln 2$

3. $\iint x^2y^2\,dx\,dy;\qquad -1 \le x \le 0,\quad 0 \le y \le 1$

4. $\iint x^2y^2\,dx\,dy;\qquad -1 \le x, y \le 1$

5. $\iint x^3y^3\,dx\,dy;\qquad 0 \le x, y \le 1$

6. $\iint (x - y)\,dx\,dy;\qquad 0 \le x, y \le 1$

7. $\iint (x^2 - y^2)\,dx\,dy;\qquad 0 \le x, y \le 1$

8. $\iint (x^{17} - y^{17})\,dx\,dy;\qquad 0 \le x, y \le 1$

9. $\iint (e^{x^2} - e^{y^2})\,dx\,dy;\qquad 0 \le x, y \le 1$

10. $\iint \cos x \cos y\,dx\,dy;\qquad 0 \le x \le \frac{\pi}{4},\quad 0 \le y \le \frac{\pi}{2}$

11. $\iint (x^2 + y^2)\,dx\,dy;\qquad -1 \le x, y \le 1$

12. $\iint (x - y)^2\,dx\,dy;\qquad -1 \le x, y \le 1$

13. $\iint (x - y)^3\,dx\,dy;\qquad -1 \le x, y \le 1$

14. $\iint (x - y)^{75}\,dx\,dy;\qquad -1 \le x, y \le 1$

15. $\iint (x + y)^{93}\,dx\,dy;\qquad -1 \le x, y \le 1$

16. $\displaystyle\iint \frac{x^2}{y^3}\,dx\,dy;$ $\qquad 1 \le x \le 2, \qquad 1 \le y \le 4$

17. $\displaystyle\iint \frac{x}{1+y^2}\,dx\,dy;$ $\qquad 0 \le x \le 2, \qquad 0 \le y \le 1$

18. $\displaystyle\iint xy \ln x\,dx\,dy;$ $\qquad 1 \le x \le 4, \qquad -1 \le y \le 2$

19. $\displaystyle\iint x \ln(xy)\,dx\,dy;$ $\qquad 2 \le x \le 3, \qquad 1 \le y \le 2$

20. $\displaystyle\iint e^{x+y} \cos 2x\,dx\,dy;$ $\qquad 0 \le x \le \pi, \qquad 1 \le y \le 2.$

3. ITERATED INTEGRALS

The formula in the last section for integrating products $g(x)h(y)$ is a useful one, but it is inadequate for many functions, such as $f(x, y) = 1/(x + y)$ and $f(x, y) = y \cos(xy)$. In this section we derive the most general method for evaluating double integrals, the method of iterated integration. This method includes the previous rule for products as a special case.

The problem is to compute

$$V = \iint\limits_{\substack{a \le x \le b \\ c \le y \le d}} f(x, y)\,dx\,dy.$$

Consider the integral as a volume to be found by slicing.

Fix a value of y and slice the solid by the corresponding plane parallel to the x, z-plane (Fig. 3.1).

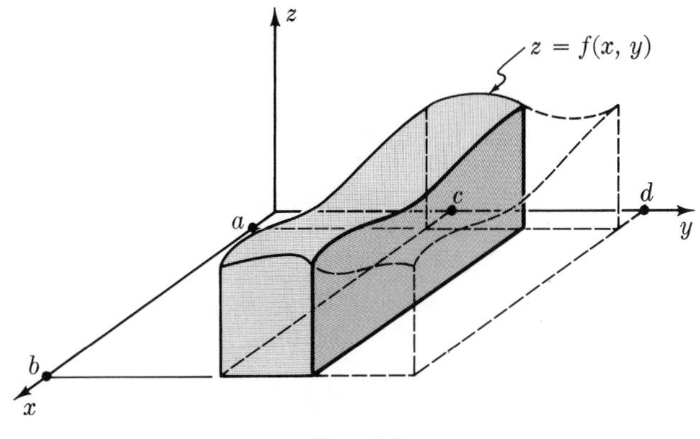

FIG. 3.1

The resulting cross-section has area $A(y)$. Denote the volume to the left of this slice by $V(y)$. (Thus $V(c) = 0$, $V(d) =$ the desired volume.) The fundamental fact needed here is that

$$\frac{dV}{dy} = A(y).$$

Intuitively this is fairly clear. The derivative dV/dy is the limit as $h \longrightarrow 0$ of

$$\frac{V(y+h) - V(y)}{h}.$$

For h very small, the numerator is the volume of a slab of width h and cross-sectional area approximately $A(y)$. Hence the quotient is approximately $A(y)$.

Now integrate $dV/dy = A(y)$ to find V:

$$V = \int_c^d A(y)\, dy.$$

But $A(y)$, the area of the cross-section (Fig. 3.2), can be expressed as a simple integral. Indeed, $A(y)$ is the area under the curve $z = f(x, y)$ from $x = a$ to $x = b$. (Remember y is *fixed* in this process.) Thus

$$A(y) = \int_a^b f(x, y)\, dx,$$

where y is treated as a constant in computing the integral.

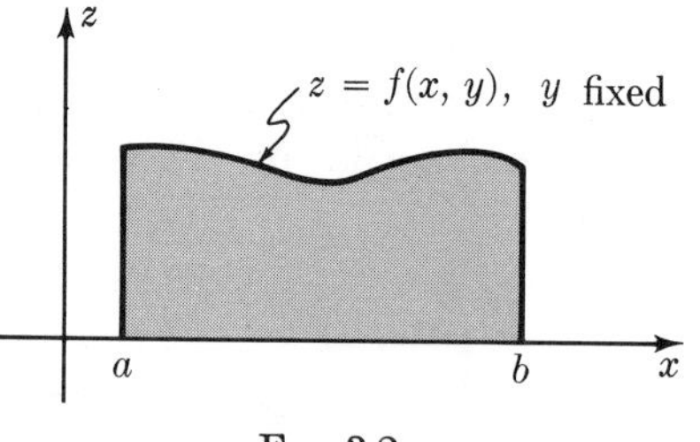

FIG. 3.2

The final result is the following pair of formulas.

Iteration Formulas

$$\iint\limits_{\substack{a \le x \le b \\ c \le y \le d}} f(x, y)\, dx\, dy = \int_c^d \left(\int_a^b f(x, y)\, dx \right) dy$$

$$= \int_a^b \left(\int_c^d f(x, y)\, dy \right) dx.$$

(The second form is obtained by reversing the roles of x and y in the argument.)

Our intuitive justification of the formulas will be replaced by an accurate proof in Chapter 12 for continuous, and even somewhat more general, integrands.

REMARK: The expressions on the right are **iterated integrals**, i.e., repetitions of simple integrals. They are often written this way:

$$\int_c^d dy \int_a^b f(x, y)\, dx, \qquad \int_a^b dx \int_c^d f(x, y)\, dy.$$

EXAMPLE 3.1

Find $\displaystyle\iint\limits_{\substack{0 \leq x \leq 1 \\ 1 \leq y \leq 2}} \frac{dx\, dy}{x + y}$.

Solution:

$$\iint = \int_0^1 \left(\int_1^2 \frac{dy}{x + y} \right) dx.$$

Now for fixed x,

$$\int_1^2 \frac{dy}{x + y} = \ln(x + y) \Bigg|_{y=1}^{y=2} = \ln(2 + x) - \ln(1 + x).$$

Hence

$$\iint = \int_0^1 [\ln(2 + x) - \ln(1 + x)]\, dx.$$

But

$$\int \ln u\, du = u \ln u - u + C,$$

therefore,

$$\int_0^1 [\ln(2 + x) - \ln(1 + x)]\, dx$$
$$= [(2 + x)\ln(2 + x) - (2 + x) - (1 + x)\ln(1 + x) + (1 + x)] \Bigg|_0^1$$
$$= 3 \ln 3 - 2 \ln 2 - 2 \ln 2 = 3 \ln 3 - 4 \ln 2.$$

Answer: $\ln \dfrac{27}{16}$.

An important feature of the Iteration Formulas above is that the iteration may be done in either order. Sometimes the computation is difficult in one order but relatively easy in the opposite order.

EXAMPLE 3.2

Find $\displaystyle\iint\limits_{\substack{0\le x\le 1\\0\le y\le\pi}} y\cos(xy)\,dx\,dy.$

Solution: Here is one set-up:

$$\iint = \int_0^1\left(\int_0^\pi y\cos(xy)\,dy\right)dx.$$

The inner integral,

$$\int_0^\pi y\cos(xy)\,dy,$$

while not impossible to integrate (by parts), is tricky. The alternate procedure is iteration in the other order:

$$\iint = \int_0^\pi\left(\int_0^1 y\cos(xy)\,dx\right)dy.$$

Since y is constant in the inner integration, this can be rewritten as

$$\iint = \int_0^\pi y\left(\int_0^1 \cos(xy)\,dx\right)dy.$$

Now

$$\int_0^1 \cos(xy)\,dx = \frac{1}{y}\sin(xy)\Bigg|_{x=0}^{x=1} = \frac{\sin y}{y}.$$

Hence

$$\iint = \int_0^\pi y\,\frac{\sin y}{y}\,dy = \int_0^\pi \sin y\,dy = 2.$$

Alternate writing of the solution:

$$\iint\limits_{\mathbf{D}} y\cos(xy)\,dx\,dy = \int_0^\pi y\,dy\int_0^1\cos(xy)\,dx$$

$$= \int_0^\pi y\left(\frac{\sin(xy)}{y}\Bigg|_0^1\right)dy = \int_0^\pi \sin y\,dy = 2.$$

Answer: 2.

EXERCISES

Evaluate:

1. $\displaystyle\iint \frac{dx\,dy}{(x+y)^2}$; $\qquad 0\le x\le 1, \quad 1\le y\le 2$

2. $\displaystyle\iint \frac{x}{y}\,dx\,dy;$ $\qquad 0 \le x \le 1, \quad 1 \le y \le 5$

3. $\displaystyle\iint \frac{x^2}{y^2}\,dx\,dy;$ $\qquad -1 \le x \le 1, \quad 1 \le y \le 3$

4. $\displaystyle\iint e^{x+y}\,dx\,dy;$ $\qquad -1 \le x, y \le 0$

5. $\displaystyle\iint y^2 \sin(xy)\,dx\,dy;$ $\qquad 0 \le x \le 2\pi, \quad 0 \le y \le 1$

6. $\displaystyle\iint (1 - 2x)\sin(y^2)\,dx\,dy;$ $\qquad 0 \le x, y \le 1$

7. $\displaystyle\iint e^y \sin(x/y)\,dx\,dy;$ $\qquad -\pi/2 \le x \le \pi/2, \quad 1 \le y \le 2$

8. $\displaystyle\iint (1 - x + 2y)^2\,dx\,dy;$ $\qquad 3 \le x \le 4, \quad 1 \le y \le 2$

9. $\displaystyle\iint (1 + x + y)(3 + x - y)\,dx\,dy;$ $\qquad 2 \le x, y \le 3$

10. $\displaystyle\iint \sin(x + y)\,dx\,dy;$ $\qquad 0 \le x, y \le \pi/2$

11. $\displaystyle\iint (x + y)^n\,dx\,dy;$ $\qquad 0 \le x \le 1, \quad 1 \le y \le 2$

12. $\displaystyle\iint (x - y)^n\,dx\,dy;$ $\qquad 0 \le x, y \le 1.$

13. Suppose $f(-x, -y) = -f(x, y)$. Prove $\displaystyle\iint f(x, y)\,dx\,dy = 0; \quad -1 \le x, y \le 1.$

14. Suppose $f(x, -y) = -f(x, y)$. Prove $\displaystyle\iint f(x, y)\,dx\,dy = 0; \quad -1 \le x, y \le 1.$

15. Find the constant A that best approximates $f(x, y)$ on the square $0 \le x, y \le 1$ in the least squares sense. In other words, minimize

$$\iint [f(x, y) - A]^2\,dx\,dy; \qquad 0 \le x, y \le 1.$$

16. (cont.) Find the least squares linear approximation

$$A + Bx + Cy$$

to $f(x, y) = xy$ on the square $0 \le x, y \le 1$.

17. (cont.) Show that the coefficients of the least squares linear approximation

$A + Bx + Cy$ to $f(x, y)$ on the square $0 \le x, y \le 1$ satisfy

$$\begin{cases} A + \frac{1}{2}B + \frac{1}{2}C = \iint f\,dx\,dy \\ \frac{1}{2}A + \frac{1}{3}B + \frac{1}{4}C = \iint xf\,dx\,dy \\ \frac{1}{2}A + \frac{1}{4}B + \frac{1}{3}C = \iint yf\,dx\,dy. \end{cases}$$

4. APPLICATIONS

Mass and Density

Suppose a sheet of non-homogeneous material covers the rectangle $a \le x \le b$ and $c \le y \le d$. See Fig. 4.1. At each point (x, y), let $\rho(x, y)$ denote the **density** of the material, i.e., the mass per unit area. (Dimensionally, planar density is mass divided by length squared. Common units are gm/cm² and lb/ft².)

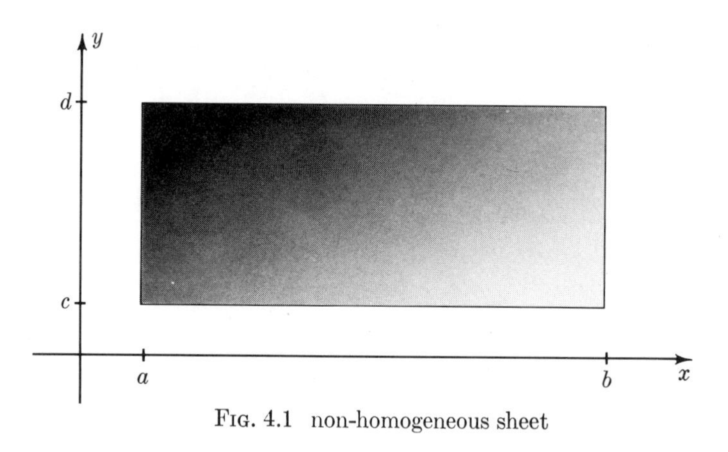

FIG. 4.1 non-homogeneous sheet

The mass of a small rectangular portion of the sheet (Fig. 4.2) is

$$dM \approx \rho(x, y)\,dx\,dy.$$

Therefore the total mass of the sheet is

$$M = \iint\limits_{\substack{a \le x \le b \\ c \le y \le d}} \rho(x, y)\,dx\,dy.$$

EXAMPLE 4.1

The density (lb/ft²) at each point of a one-foot square of plastic is the product of the four distances of the point from the sides of the square. Find the total mass.

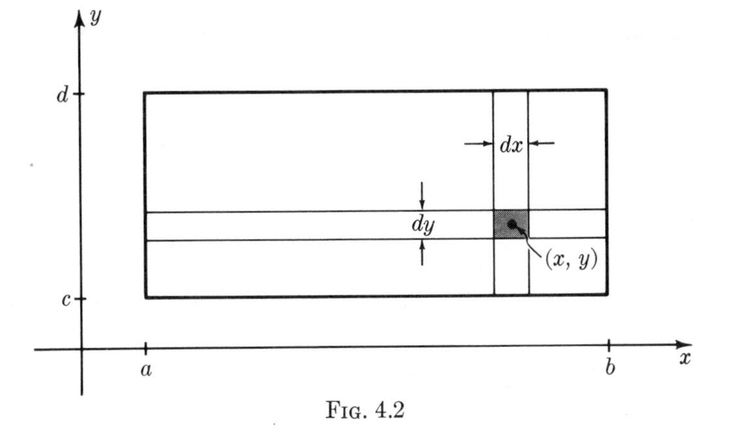

FIG. 4.2

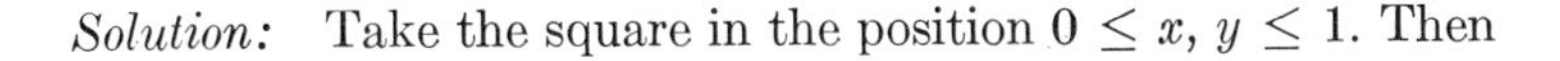

Solution: Take the square in the position $0 \le x,\ y \le 1$. Then

$$\rho(x, y) = x(1 - x)y(1 - y),$$

$$M = \iint \rho(x, y)\,dx\,dy$$

$$= \left(\int_0^1 x(1 - x)\,dx\right)\left(\int_0^1 y(1 - y)\,dy\right) = \frac{1}{6}\cdot\frac{1}{6} = \frac{1}{36}.$$

Answer: $\dfrac{1}{36}$ lb.

Moment and Center of Gravity

Suppose gravity (perpendicular to the plane of the figure) acts on the rectangular sheet of Fig. 4.1. The sheet is to be suspended by a single point so that it balances parallel to the floor. This point of balance is the **center of gravity** of the sheet and is denoted $\bar{\mathbf{x}} = (\bar{x}, \bar{y})$.

The center of gravity $\bar{\mathbf{x}}$ is found in three steps:

(1) Find the mass M.

(2) Find the **moment** of the sheet with respect to the origin. This is the vector

$$\mathbf{m} = \iint_{\mathbf{D}} \rho\mathbf{x}\,dx\,dy = \left(\iint_{\mathbf{D}} \rho x\,dx\,dy,\ \iint_{\mathbf{D}} \rho y\,dx\,dy\right).$$

Here $\mathbf{D}$ denotes the region the sheet occupies, $a \le x \le b$ and $c \le y \le d$, and $\rho = \rho(x, y)$ is the density.

(3) Divide the moment by the mass to obtain the center of gravity:

$$\bar{\mathbf{x}} = \frac{1}{M}\,\mathbf{m}.$$

This formula will be derived after two examples.

EXAMPLE 4.2

Find the center of gravity of a homogeneous rectangular sheet.

Solution: "Homogeneous" means the density ρ is constant. Take the sheet in the position $0 \leq x \leq a$ and $0 \leq y \leq b$. The mass is $M = \rho ab$, and the moment is

$$\mathbf{m} = \iint \rho\mathbf{x}\,dx\,dy = \rho\iint \mathbf{x}\,dx\,dy = \rho\left(\iint x\,dx\,dy,\ \iint y\,dx\,dy\right)$$

$$= \rho\left(\int_0^a x\,dx\int_0^b dy,\ \int_0^a dx\int_0^b y\,dy\right)$$

$$= \rho\left(\frac{1}{2}\,a^2b,\ \frac{1}{2}\,ab^2\right).$$

The center of gravity is

$$\bar{\mathbf{x}} = \frac{1}{M}\,\mathbf{m} = \frac{1}{\rho ab}\,\rho\left(\frac{1}{2}\,a^2b,\ \frac{1}{2}\,ab^2\right) = \frac{1}{2}\,(a, b).$$

This is the midpoint (intersection of the diagonals) of the rectangle. (Of course the rectangle balances on its midpoint; no one needs calculus for this, but it is reassuring that the analytic method gives the right answer.)

Answer: The midpoint of the rectangle.

EXAMPLE 4.3

A rectangular sheet over the region $1 \leq x \leq 2$ and $1 \leq y \leq 3$ has density $\rho(x, y) = xy$. Find its center of gravity.

Solution: The mass is

$$M = \iint xy\,dx\,dy = \int_1^2 x\,dx\int_1^3 y\,dy = \frac{3}{2}\cdot 4 = 6.$$

The moment is

$$\mathbf{m} = \iint xy\,\mathbf{x}\,dx\,dy = \left(\iint x^2y\,dx\,dy,\ \iint xy^2\,dx\,dy\right)$$

$$= \left(\int_1^2 x^2\,dx \int_1^3 y\,dy,\ \int_1^2 x\,dx \int_1^3 y^2\,dy\right)$$

$$= \left(\frac{7}{3}\cdot 4,\ \frac{3}{2}\cdot\frac{26}{3}\right) = \frac{1}{3}\,(28, 39).$$

Therefore

$$\bar{\mathbf{x}} = \frac{1}{M}\,\mathbf{m} = \frac{1}{6}\cdot\frac{1}{3}\,(28, 39) = \left(\frac{14}{9}, \frac{13}{6}\right).$$

Note that $\bar{\mathbf{x}}$ is inside the rectangle, a little northeast of center. Could you have predicted this?

Answer: $\bar{\mathbf{x}} = \left(\dfrac{14}{9}, \dfrac{13}{6}\right).$

The formula for center of gravity is derived by balancing the rectangular sheet on various knife edges.

Suppose a knife edge passes through $\bar{\mathbf{x}}$ and the sheet balances (Fig. 4.3). Divide the sheet into many small rectangles. The turning moments of these pieces about the knife edge must add up to zero.

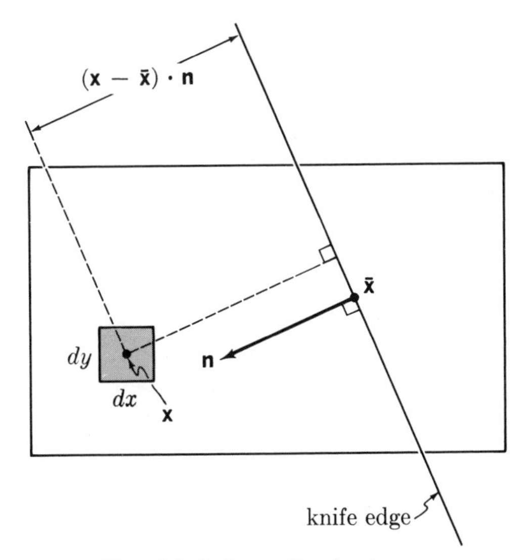

FIG. 4.3 balance the sheet

Let **n** be a unit vector in the plane of the rectangle perpendicular to the knife edge. A small rectangle with sides dx and dy located at **x** has (signed) distance $(\mathbf{x} - \bar{\mathbf{x}})\cdot\mathbf{n}$ from the knife edge and has mass $\rho\, dx\, dy$. Hence its turning moment is

$$\rho(\mathbf{x} - \bar{\mathbf{x}})\cdot\mathbf{n}\, dx\, dy.$$

The sum of all such turning moments must be zero:

$$\iint \rho(\mathbf{x} - \bar{\mathbf{x}})\cdot\mathbf{n}\, dx\, dy = 0.$$

Since $\bar{\mathbf{x}}$ and **n** are constant, this relation may be written

$$\mathbf{n}\cdot\iint \rho\mathbf{x}\, dx\, dy = (\mathbf{n}\cdot\bar{\mathbf{x}})\iint \rho\, dx\, dy,$$

or

$$\mathbf{n}\cdot\mathbf{m} = M\mathbf{n}\cdot\bar{\mathbf{x}}.$$

This equation of balance is true for each choice of the knife edge (each choice of the unit vector **n**). Hence,

$$\mathbf{n}\cdot(\mathbf{m} - M\bar{\mathbf{x}}) = 0$$

for each unit vector **n**. This means the component of $\mathbf{m} - M\bar{\mathbf{x}}$ in each direction is zero. Therefore,

$$\mathbf{m} - M\bar{\mathbf{x}} = \mathbf{0},$$

$$\mathbf{m} = M\bar{\mathbf{x}}.$$

EXERCISES

Find the volume under the surface and over the portion of the x, y-plane indicated. Draw a figure in each case:

1. $z = 2 - (x^2 + y^2)$; $\quad -1 \le x, y \le 1$
2. $z = 1 - xy$; $\quad 0 \le x, y \le 1$
3. $z = x^2 + 4y^2$; $\quad 0 \le x \le 2,\ 0 \le y \le 1$
4. $z = \sin x \sin y$; $\quad 0 \le x, y \le \pi$
5. $z = x^2y + y^2x$; $\quad 1 \le x \le 2,\ 2 \le y \le 3$
6. $z = (1 + x^3)y^2$; $\quad -1 \le x, y \le 1$.

A sheet of non-homogeneous material of density ρ gm/cm^2 covers the indicated rectangle. Find the mass of the sheet, assuming lengths are measured in centimeters:

7. $\rho = 3(1 + x)(1 + y)$; $\quad 0 \le x, y \le 1$
8. $\rho = 1 - 0.2xy$; $\quad 0 \le x \le 1,\ 1 \le y \le 1.5$
9. $\rho = 3 + 0.1x$; $\quad 2 \le x \le 3,\ -1 \le y \le 1$
10. $\rho = 4e^{x+y} - 2$; $\quad 0 \le x \le 1,\ 0 \le y \le 0.5$.

Find the center of gravity of each rectangular sheet, density as given:

11. $\rho = (1 - x)(1 - y) + 1$; $\quad 0 \le x, y \le 1$
12. $\rho = \sin x$; $\quad 0 \le x \le \pi,\ 0 \le y \le 1$

13. $\rho = \sin x\,(1 - \sin y)$; $\quad \pi/2 \le x, y \le \pi$
14. $\rho = 10 - e^{x+y}$; $\quad 0 \le x, y \le 1$
15. $\rho = 1 + x^2 + y^2$; $\quad -1 \le x \le 1, \quad 1 \le y \le 4$
16. $\rho = 2 + x^2y^2$; $\quad -1 \le x \le 1, \quad 0 \le y \le 1.$

5. GENERAL DOMAINS

Suppose we want the volume of a solid (Fig. 5.1) bounded on top by a surface $z = f(x, y)$ defined over a non-rectangular domain **D**. Here the bottom of the solid is **D** itself, and the side of the solid is the cylindrical wall with generator parallel to the z-axis and base the boundary of **D**. The solid is the set

$$\{ (x, y, z) \mid (x, y) \in \mathbf{D} \quad \text{and} \quad 0 \le z \le f(x, y) \}.$$

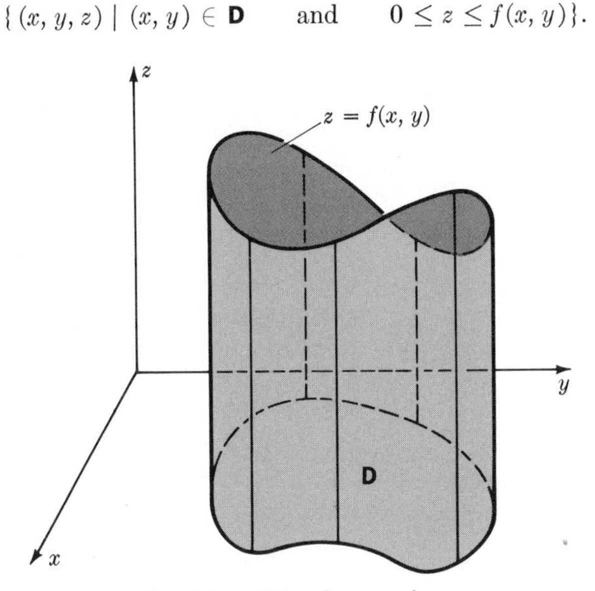

FIG. 5.1 solid under a surface

We shall suppose the volume is

$$V = \iint_{\mathbf{D}} f(x, y)\, dx\, dy,$$

where this double integral over **D** has properties consistent with our intuitive ideas about volume. Let us consider some examples to see how we can compute such volumes and what properties the double integral ought to have.

EXAMPLE 5.1

Find the volume under the surface $z = e^{-(x+y)}$, and over the domain of the x, y-plane bounded by the x-axis, the line $y = x$, and the lines $x = \frac{1}{2}$ and $x = 1$.

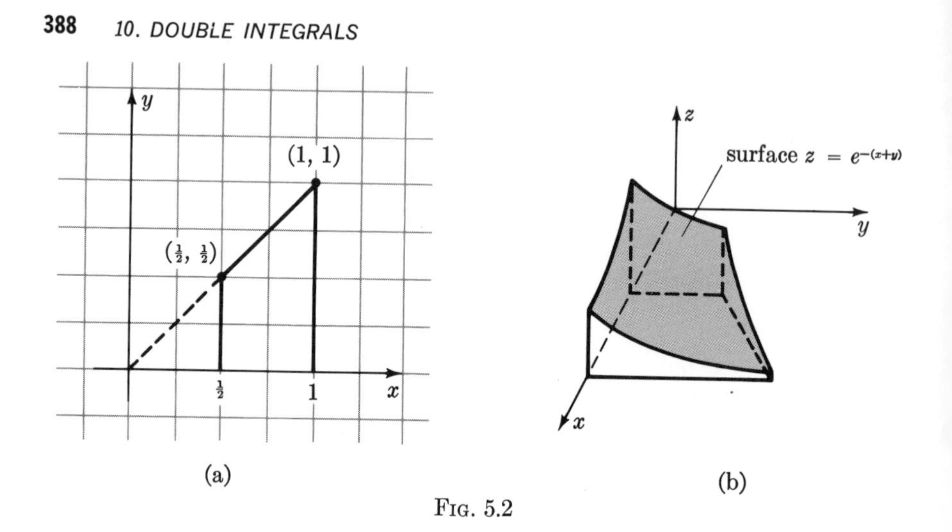

FIG. 5.2

Solution: Draw figures of the domain (Fig. 5.2a) and the solid (Fig. 5.2b). The problem is solved by the slicing method. Slice the solid into slabs by planes perpendicular to the x-axis. Let the area of the slab at x be denoted by $A(x)$. See Fig. 5.3a. The volume of the slab is

$$dV = A(x)\,dx,$$

so the total volume is

$$V = \int_{1/2}^{1} A(x)\,dx.$$

Now carefully draw the cross-section (Fig. 5.3b). Its base has length x, and it is bounded above by the curve $z = e^{-x}e^{-y}$. Note that x is *constant*; in this plane section z is a function of y alone. This is the absolute crux of the matter!

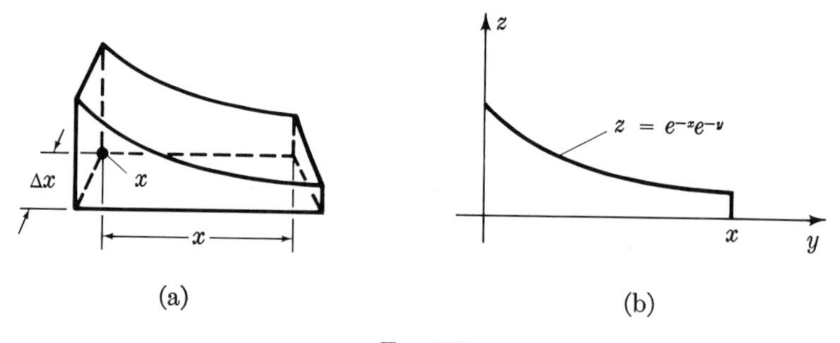

FIG. 5.3

The area of the cross-section at x is

$$A(x) = \int_0^x e^{-x}e^{-y}\,dy = e^{-x}\int_0^x e^{-y}\,dy = -e^{-x}(e^{-y})\Big|_{y=0}^{y=x} = e^{-x} - e^{-2x}.$$

Thus

$$V = \int_{1/2}^{1} (e^{-x} - e^{-2x})\,dx = \left(\frac{1}{2}e^{-2x} - e^{-x}\right)\Bigg|_{1/2}^{1} = \frac{1}{2}e^{-2} - \frac{3}{2}e^{-1} + e^{-1/2}.$$

NOTE: The solution can be set up as follows:

$$V = \int_{1/2}^{1}\left(\int_{0}^{x} e^{-(x+y)}\,dy\right)dx.$$

In the inner integral the variable of integration is y, while x is treated like a constant, both in the integrand and in the upper limit. But once the inner integral is completely evaluated, x becomes the variable of integration for the outer integral.

Answer: $\frac{1}{2}e^{-2}(1 - 3e + 2e^{3/2})$.

EXAMPLE 5.2

Find the volume under the surface $z = 1 - x^2 - y^2$, lying over the square with vertices $(\pm 1, 0)$ and $(0, \pm 1)$.

Solution: First draw the square (Fig. 5.4a). Observe that by symmetry, it suffices to find the volume over the triangular portion in the first quadrant, and then to quadruple it.

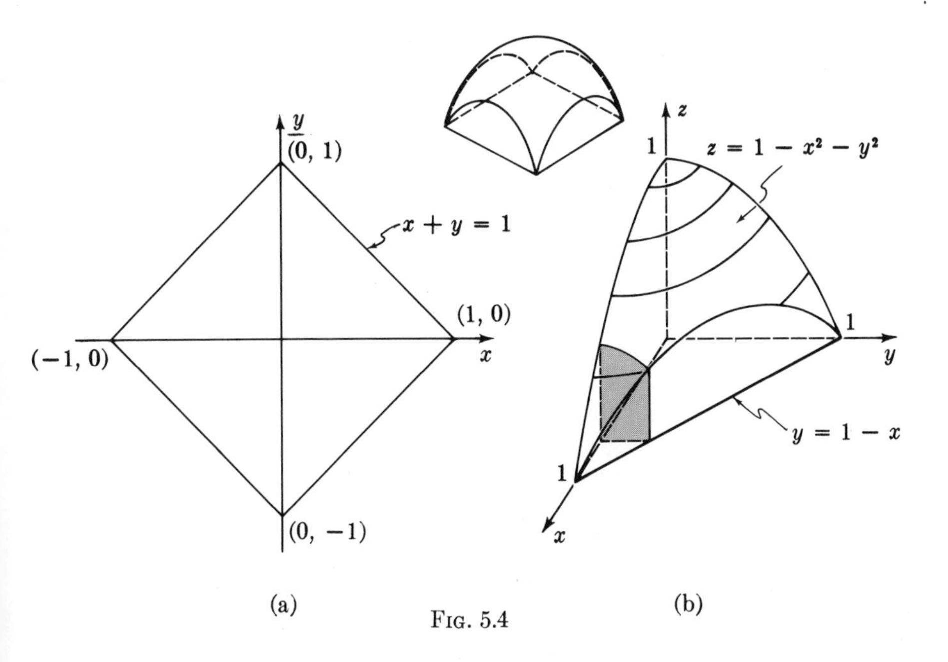

FIG. 5.4

Draw the corresponding portion of the solid (Fig. 5.4b). Slice by planes perpendicular to the x-axis. For each x, the plane cuts the solid in a cross-section (Fig. 5.5) whose area is

$$A(x) = \int_0^{1-x} (1 - x^2 - y^2)\, dy.$$

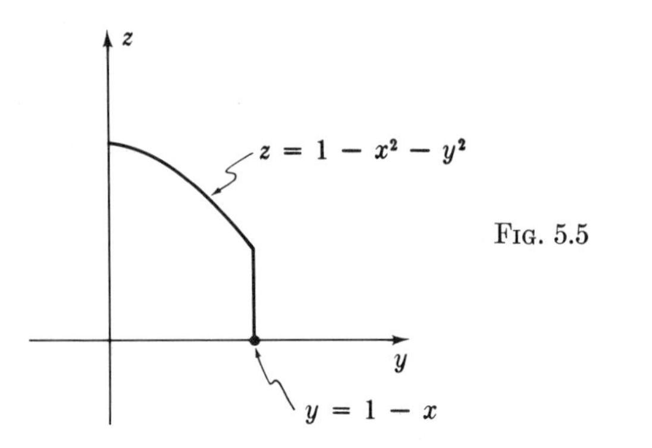

FIG. 5.5

The volume of the solid is

$$V = 4\int_0^1 A(x)\, dx = 4\int_0^1 \left(\int_0^{1-x} (1 - x^2 - y^2)\, dy\right) dx$$

$$= 4\int_0^1 \left[\left(y - x^2 y - \frac{1}{3}y^3\right)\Bigg|_{y=0}^{y=1-x}\right] dx$$

$$= 4\int_0^1 \left[(1 - x) - x^2(1 - x) - \frac{1}{3}(1 - x)^3\right] dx$$

$$= 4\int_0^1 \left[1 - x - x^2 + x^3 - \frac{1}{3}(1 - x)^3\right] dx$$

$$= 4\left[1 - \frac{1}{2} - \frac{1}{3} + \frac{1}{4} - \frac{1}{12}\right] = 4\cdot\frac{4}{12} = \frac{4}{3}.$$

Answer: $\frac{4}{3}$.

EXAMPLE 5.3

Find the volume under the plane $z = 1 + x + y$, and over the domain bounded by the curves $x = \frac{1}{2}$, $x = 1$, $y = x^2$, $y = 2x^2$.

Solution: The domain and the solid are drawn in Fig. 5.6. The cross-section by a plane perpendicular to the x-axis at x is a trapezoid (Fig. 5.7).

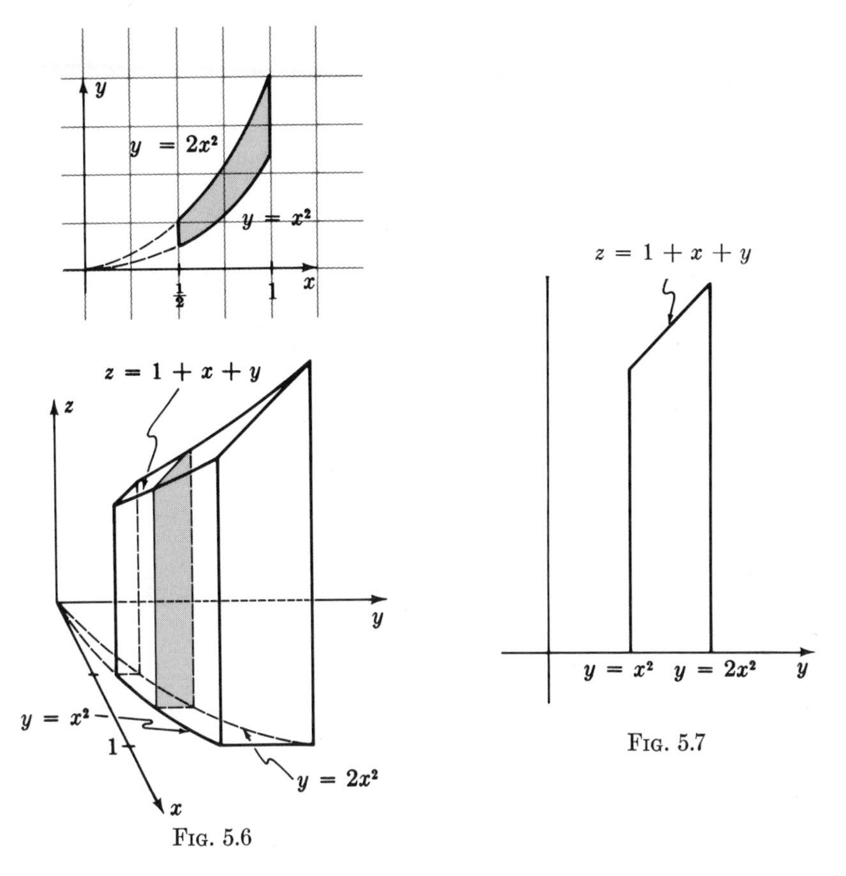

Fig. 5.6

Fig. 5.7

Note the range of y, namely $x^2 \le y \le 2x^2$. Thus

$$V = \int_{1/2}^{1} A(x)\, dx,$$

where

$$A(x) = \int_{x^2}^{2x^2} (1 + x + y)\, dy = \left(y + xy + \frac{1}{2}y^2 \right)\Big|_{y=x^2}^{y=2x^2}$$

$$= (2x^2 + 2x^3 + 2x^4) - \left(x^2 + x^3 + \frac{1}{2}x^4 \right) = x^2 + x^3 + \frac{3}{2}x^4.$$

Finally,

$$V = \int_{1/2}^{1} \left(x^2 + x^3 + \frac{3}{2}x^4 \right) dx = \left(\frac{1}{3}x^3 + \frac{1}{4}x^4 + \frac{3}{10}x^5 \right)\Big|_{1/2}^{1}$$

$$= \left(\frac{1}{3} + \frac{1}{4} + \frac{3}{10} \right) - \frac{1}{8}\left(\frac{1}{3} + \frac{1}{8} + \frac{3}{40} \right) = \frac{49}{60}.$$

Answer: $\dfrac{49}{60}$.

Iteration

With these examples behind us, we are prepared for the statement and solution of the double integration problem. Suppose a domain **D** in the x, y-plane is bounded by lines $x = a$ and $x = b$, and by two curves $y = g(x)$ and $y = f(x)$. Assume $a < b$ and $g(x) < f(x)$ for each x. See Fig. 5.8. Suppose

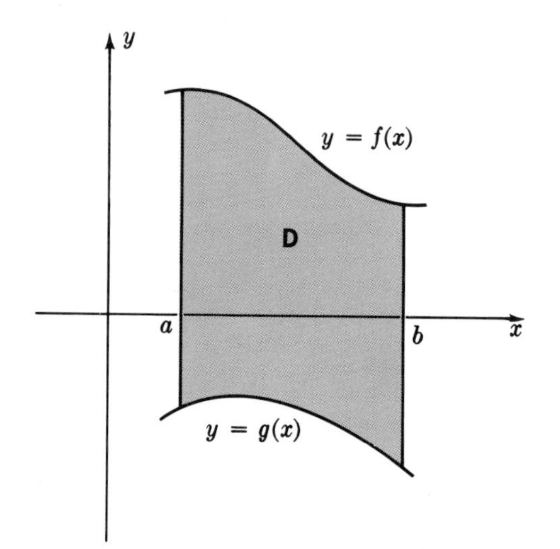

FIG. 5.8. domain between two curves

a surface $z = H(x, y)$ is given, defined over **D**. See Fig. 5.9. The volume of the solid column over **D** and under the surface is then

$$V = \iint_{\mathbf{D}} H(x, y)\, dx\, dy.$$

(As is usual with integrals, the portion where $H < 0$ is counted negative.)

To evaluate the double integral, consider a slab parallel to the y, z-plane at x. Its face area is

$$A(x) = \int_{g(x)}^{f(x)} H(x, y)\, dy.$$

See Fig. 5.10. In this integration, x is constant. Notice that y, the variable of integration, disappears when the definite integral $A(x)$ is evaluated.

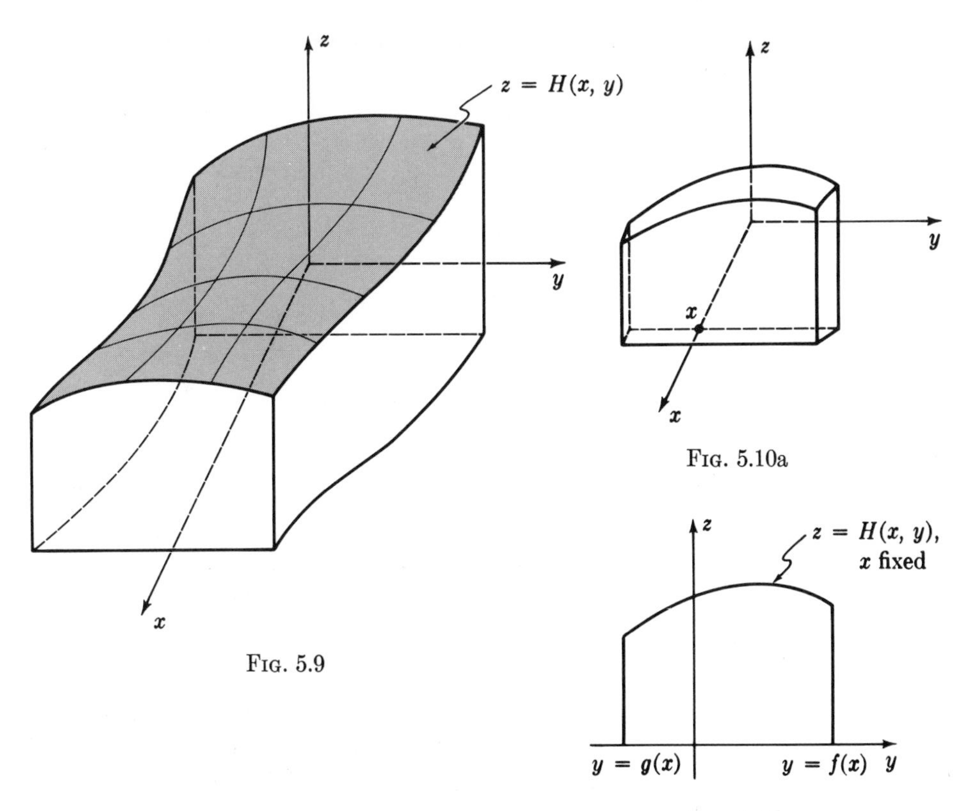

Fig. 5.9

Fig. 5.10a

Fig. 5.10b

Conclusion: the volume integral is

$$V = \iint_{\mathbf{D}} H(x, y)\, dx\, dy = \int_a^b A(x)\, dx = \int_a^b \left(\int_{g(x)}^{f(x)} H(x, y)\, dy \right) dx.$$

This is called the **iteration formula**. In Chapter 12, we shall prove that it holds for any continuous function, not necessarily positive.

> **EXAMPLE 5.4**
>
> Find the volume under the surface $z = xy$, and over the domain **D** bounded by $y = x$ and $y = x^2$.

Solution: The line and parabola intersect at $(0, 0)$ and $(1, 1)$. See Fig. 5.11. For each value of x, the range of y is

$$x^2 \le y \le x.$$

Hence

$$V = \iint_{\mathbf{D}} xy\,dx\,dy = \int_0^1 \left(\int_{x^2}^{x} xy\,dy\right) dx = \int_0^1 \left(\frac{1}{2}xy^2 \Big|_{y=x^2}^{y=x}\right) dx$$

$$= \int_0^1 \frac{1}{2}(x^3 - x^5)\,dx = \frac{1}{2}\left(\frac{1}{4} - \frac{1}{6}\right) = \frac{1}{24}.$$

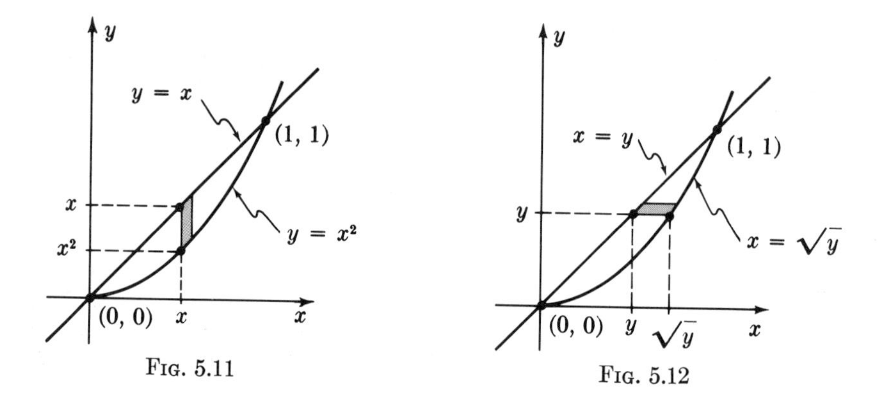

Fig. 5.11

Fig. 5.12

Alternate Solution: The domain $\mathbf{D}$ may be thought of as bounded by the curves $x = y$ (below) and $x = \sqrt{y}$ (above), where $0 \le y \le 1$. See Fig. 5.12. For each y, the range of x is $y \le x \le \sqrt{y}$. Therefore the set-up of the iteration is

$$V = \int_0^1 \left(\int_y^{\sqrt{y}} xy\,dx\right) dy = \int_0^1 \left(\frac{1}{2}x^2y \Big|_{x=y}^{x=\sqrt{y}}\right) dy$$

$$= \int_0^1 \frac{1}{2}(y^2 - y^3)\,dy = \frac{1}{2}\left(\frac{1}{3} - \frac{1}{4}\right) = \frac{1}{24}.$$

Answer: $\dfrac{1}{24}$.

The iteration method does not apply directly to every example. The boundary of $\mathbf{D}$ may be too complicated, in which case you must break the domain into several smaller regions, and deal with each as a separate problem. In Fig. 5.13 two examples are shown. The set-up for the domain in Fig. 5.13a is

$$\iint_{\mathbf{D}} H(x, y)\,dx\,dy = \iint_{\mathbf{D}_1} H(x, y)\,dx\,dy + \iint_{\mathbf{D}_2} H(x, y)\,dx\,dy$$

$$= \int_a^c \left(\int_{g(x)}^{f_1(x)} H(x, y)\,dy\right) dx + \int_c^b \left(\int_{g(x)}^{f_2(x)} H(x, y)\,dy\right) dx.$$

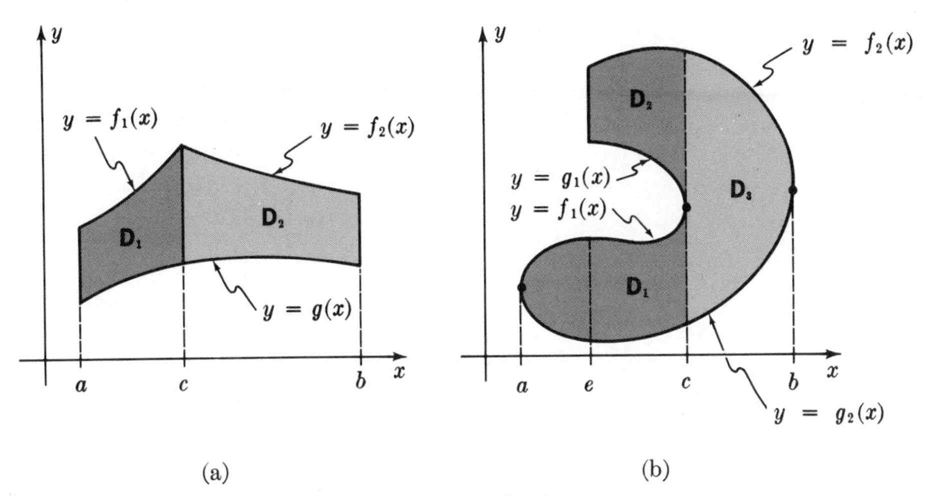

FIG. 5.13 cut up the domain

The set-up for the domain in Fig. 5.13b is

$$\iint_{\mathbf{D}} H(x, y)\, dx\, dy$$

$$= \iint_{\mathbf{D}_1} H(x, y)\, dx\, dy + \iint_{\mathbf{D}_2} H(x, y)\, dx\, dy + \iint_{\mathbf{D}_3} H(x, y)\, dx\, dy$$

$$= \int_a^c \left(\int_{g_2(x)}^{f_1(x)} H(x, y)\, dy \right) dx + \int_e^c \left(\int_{g_1(x)}^{f_2(x)} H(x, y)\, dy \right) dx$$

$$+ \int_c^b \left(\int_{g_2(x)}^{f_2(x)} H(x, y)\, dy \right) dx.$$

EXAMPLE 5.5

Without evaluating, set up the calculation for

$$\iint_{\mathbf{D}} \frac{dx\, dy}{x^2 + y^2}$$

over the region indicated in Fig. 5.14.

Solution: Draw a vertical line from $(1, 0)$ to $(1, 2)$. Call the left-hand part $\mathbf{D}_1$ and the right-hand part $\mathbf{D}_2$. See Fig. 5.15. Then

$$\iint_{\mathbf{D}} \frac{dx\, dy}{x^2 + y^2} = \iint_{\mathbf{D}_1} \frac{dx\, dy}{x^2 + y^2} + \iint_{\mathbf{D}_2} \frac{dx\, dy}{x^2 + y^2}$$

$$= \int_0^1 \left(\int_{\sqrt{1-x^2}}^2 \frac{dy}{x^2 + y^2} \right) dx + \int_1^2 \left(\int_0^2 \frac{dy}{x^2 + y^2} \right) dx.$$

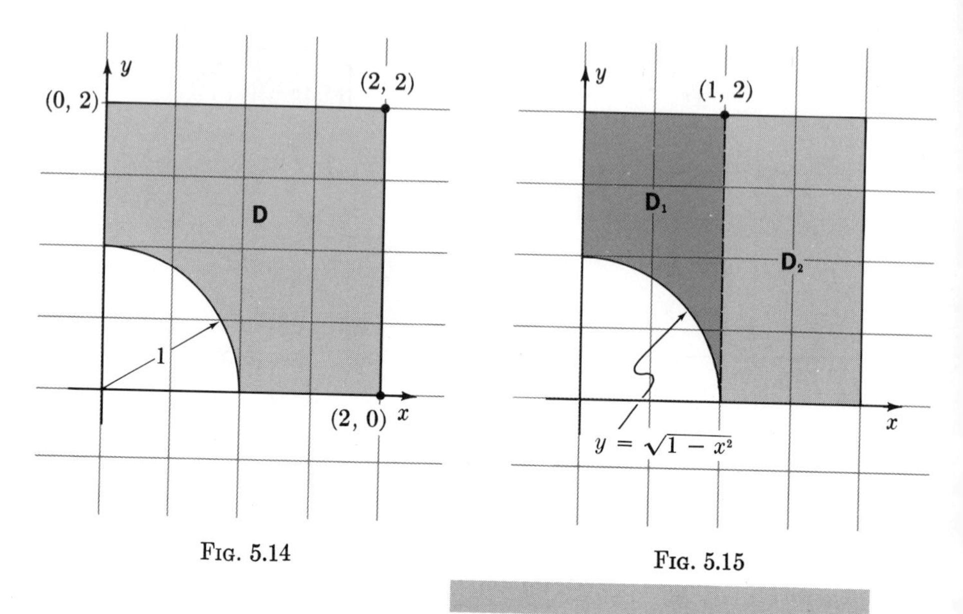

FIG. 5.14

FIG. 5.15

Area

It is often convenient to compute areas by double integrals, since

$$\iint_{\mathbf{D}} 1\, dx\, dy = \text{area}(\mathbf{D}).$$

EXAMPLE 5.6

Find the total area A bounded by $y = x^2$ and $y = x^4$.

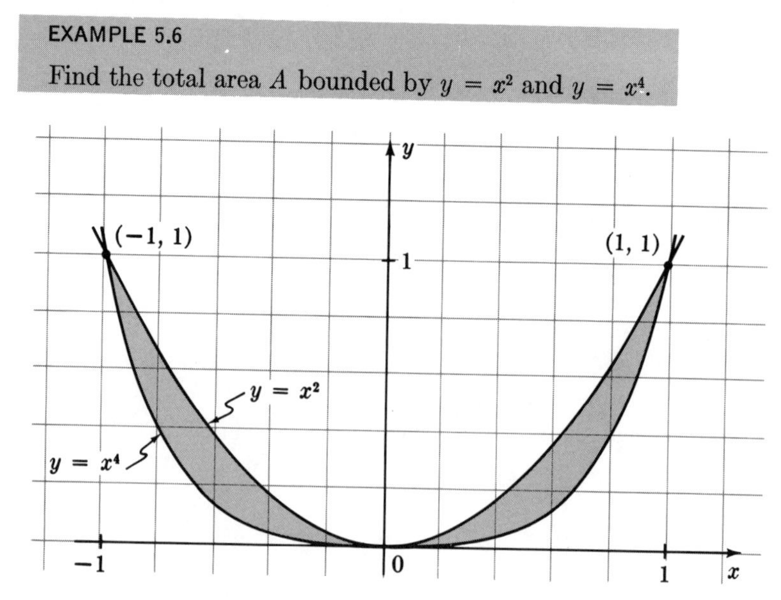

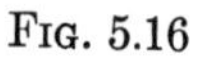

FIG. 5.16

Solution: Draw the domain (Fig. 5.16). Then

$$A = \iint dx\,dy = \int_{-1}^{1}\left(\int_{x^4}^{x^2} dy\right) dx = \int_{-1}^{1} (x^2 - x^4)\,dx = \frac{2}{3} - \frac{2}{5} = \frac{4}{15}.$$

Alternate Solution:

$$A = 2\int_0^1\left(\int_{\sqrt{y}}^{\sqrt[4]{y}} dx\right) dy = 2\int_0^1 (\sqrt[4]{y} - \sqrt{y})\,dy = 2\left(\frac{4}{5} - \frac{2}{3}\right) = \frac{4}{15}.$$

Answer: $A = \dfrac{4}{15}.$

EXERCISES

Find the volume under the surface $z = f(x, y)$, and over the indicated domain of the x, y-plane:

1. $z = 1$; Fig. 5.17
2. $z = y$; Fig. 5.17
3. $z = x$; Fig. 5.17
4. $z = 1 + x + y$; Fig. 5.17
5. $z = x^2$; Fig. 5.18
6. $z = xy$; Fig. 5.18
7. $z = y^2$; Fig. 5.18
8. $z = (x - y)^2$; Fig. 5.18

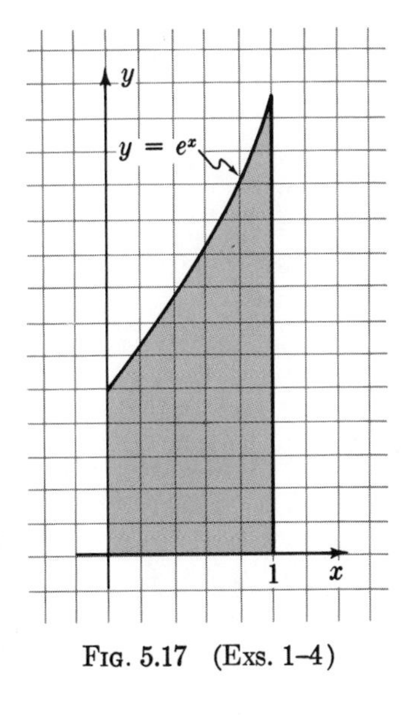

Fig. 5.17 (Exs. 1–4)

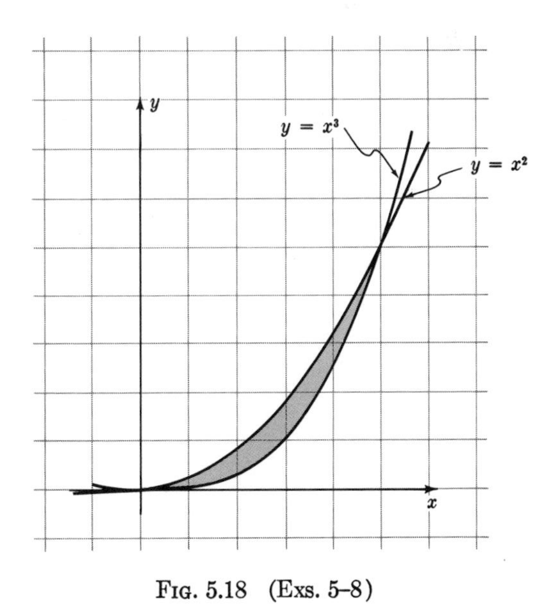

Fig. 5.18 (Exs. 5–8)

9. $z = 1$; Fig. 5.19
10. $z = 1 + x$; Fig. 5.19
11. $z = y$; Fig. 5.19

12. $z = 1$; Fig. 5.20
13. $z = 1 + x$; Fig. 5.20
14. $z = x^2y$; Fig. 5.20
15. $z = y^3$; Fig. 5.20
16. $z = x^2y^2$; Fig. 5.20

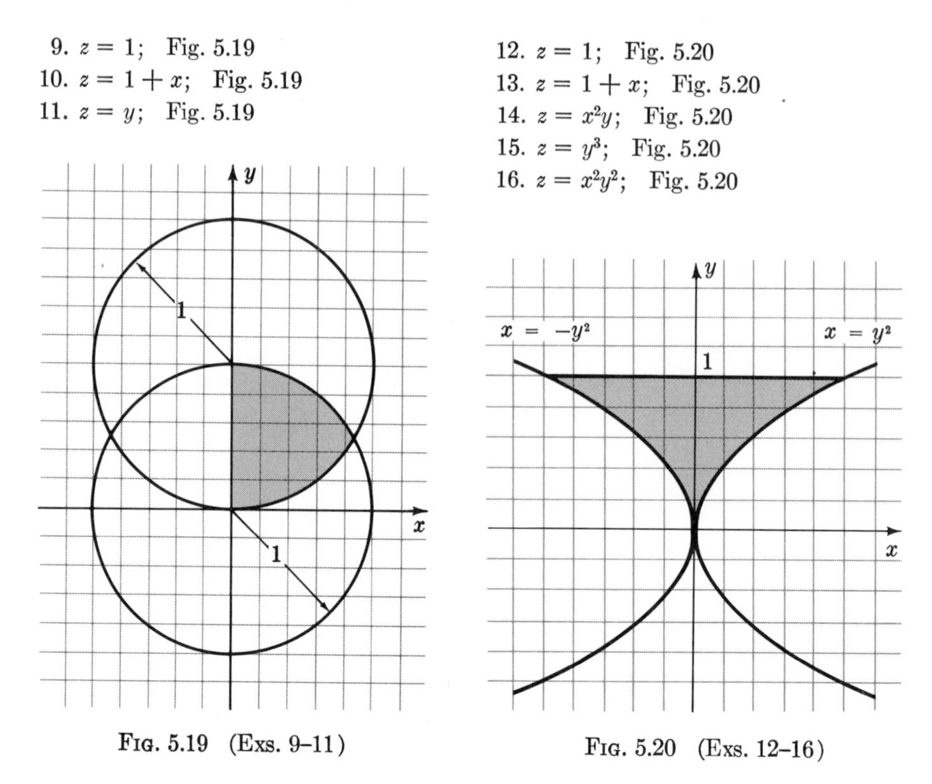

Fig. 5.19 (Exs. 9–11)

Fig. 5.20 (Exs. 12–16)

17. $z = 1$; Fig. 5.21

18. $z = 1 + x$; Fig. 5.21.

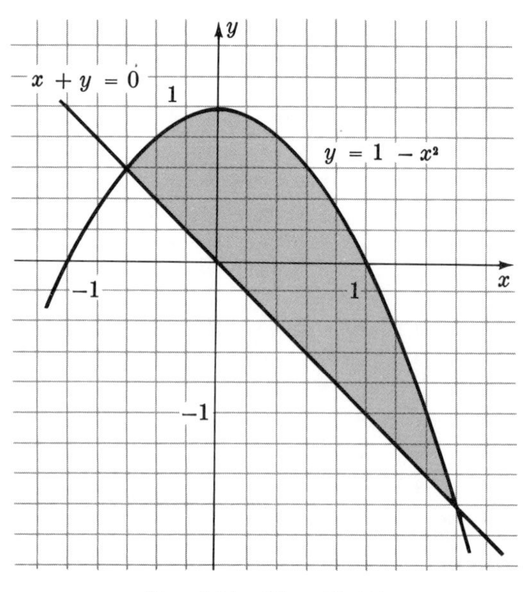

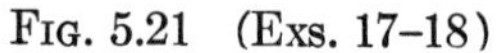

Fig. 5.21 (Exs. 17–18)

19. Describe each of the domains in Figs. 5.17–5.21 by inequalities between x and y.

Compute the double integral over the domain whose boundary curves are indicated; in each case draw a figure.

20. $\iint xe^{xy}\,dx\,dy; \quad x = 1, \quad x = 3, \quad xy = 1, \quad xy = 2$

21. $\iint x^2y\,dx\,dy; \quad y = 0, \quad x = 0, \quad x = (y-1)^2$

22. $\iint (x^3 + y^3)\,dx\,dy; \quad x^2 + y^2 = 1$

23. $\iint (x + y)^2\,dx\,dy; \quad x + y = 0, \quad y = x^2 + x$

24. $\iint (1 + xy)\,dx\,dy; \quad y = 0, \quad y = x, \quad y = 1 - x$

25. $\iint y^2\,dx\,dy; \quad y = \pm x, \quad y = \frac{1}{2}x + 3.$

26. Find $\iint_{\mathbf{D}} (1 + xy^2)\,dx\,dy$ over the domain bounded by the parabola $x = -y^2$ and the segments from $(2, 0)$ to $(-1, \pm 1)$.

6. POLAR COORDINATES

Suppose a domain (Fig. 6.1a) in the x, y-plane is described in polar coordinates by $r_0 \le r \le r_1$ and $\theta_0 \le \theta \le \theta_1$. Suppose a surface (Fig. 6.1b) is given by a function $z = f(r, \theta)$ over this domain. What is the volume of the solid between the surface and the domain?

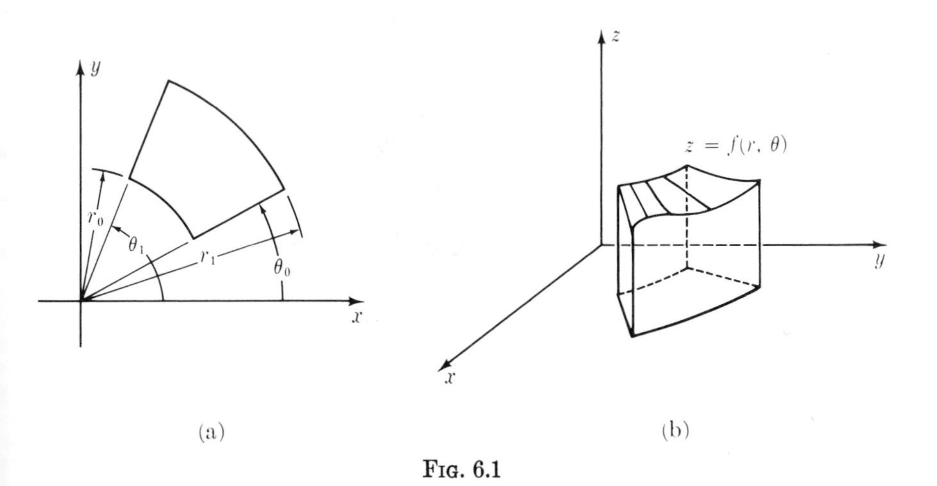

Fig. 6.1

Split the interval $r_0 \leq r \leq r_1$ into small pieces. Do the same for $\theta_0 \leq \theta \leq \theta_1$. The result is a decomposition of the plane domain into many small, almost rectangular, regions (Fig. 6.2). Each has dimensions dr and $r\,d\theta$, hence area $dA = r\,dr\,d\theta$. The portion of the solid lying over this elementary "rectangle" has height $z = f(r, \theta)$, hence its volume is $dV = f(r, \theta)\,r\,dr\,d\theta$. See Fig. 6.3.

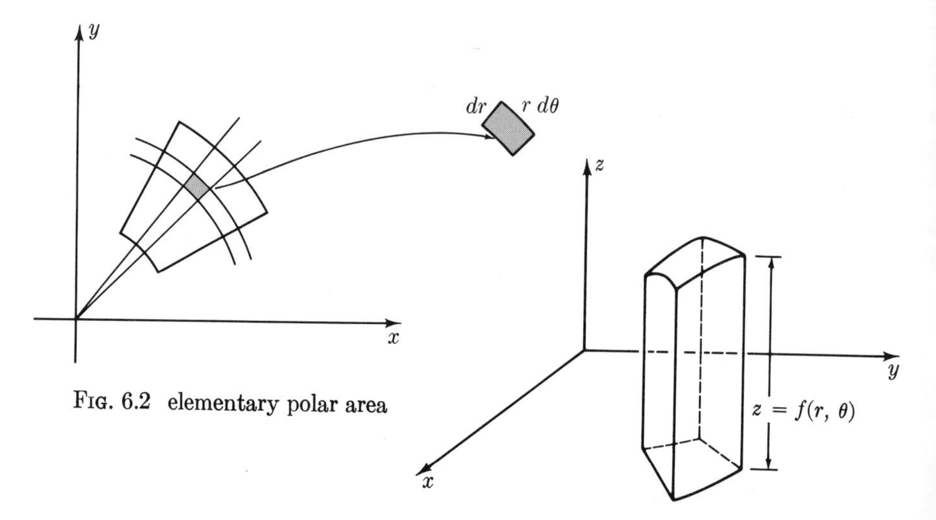

FIG. 6.2 elementary polar area

FIG. 6.3 "rectangular" column

The total volume is

$$V = \iint_{\mathbf{D}} f(r, \theta) r\,dr\,d\theta = \int_{\theta_0}^{\theta_1} \left(\int_{r_0}^{r_1} f(r, \theta)\,r\,dr \right) d\theta$$

$$= \int_{r_0}^{r_1} r \left(\int_{\theta_0}^{\theta_1} f(r, \theta)\,d\theta \right) dr.$$

EXAMPLE 6.1

Find the volume under the cone $z = r$, and over the domain $0 \leq r \leq a$ and $0 \leq \theta \leq \pi/2$.

Solution: The solid in question is shown in Fig. 6.4. Its volume is

$$V = \iint zr\,dr\,d\theta = \iint r^2\,dr\,d\theta = \int_0^a r^2\,dr \int_0^{\pi/2} d\theta = \frac{a^3}{3}\cdot\frac{\pi}{2}.$$

Answer: $V = \dfrac{\pi a^3}{6}$.

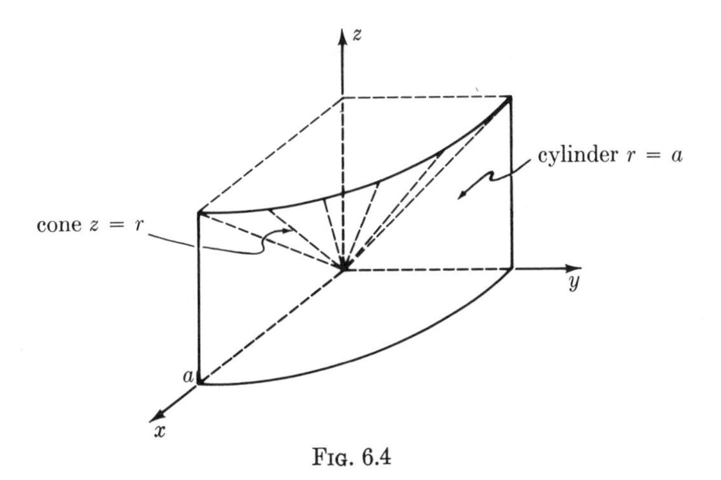

FIG. 6.4

Mass and Center of Gravity in Polar Coordinates

The mass and the center of gravity of a non-homogeneous sheet over the domain in Fig. 6.2 can be computed by double integrals. Since the element of area is

$$dA = r\,dr\,d\theta,$$

the elementary mass is

$$dM = \rho(r, \theta)\,r\,dr\,d\theta.$$

Note that ρ must be expressed in polar coordinates. Thus

$$\boxed{M = \iint_{\mathbf{D}} \rho r\,dr\,d\theta.}$$

To find the moment, express $\mathbf{x}$ in polar coordinates:

$$\mathbf{x} = (x, y) = (r\cos\theta, r\sin\theta).$$

Then $\bar{\mathbf{x}} = \mathbf{m}/M$, where

$$\boxed{\mathbf{m} = \iint_{\mathbf{D}} \mathbf{x}\,dM = \iint_{\mathbf{D}} \mathbf{x}\rho r\,dr\,d\theta = \iint_{\substack{r_0 \le r \le r_1 \\ \theta_0 \le \theta \le \theta_1}} (r\cos\theta, r\sin\theta)\rho r\,dr\,d\theta.}$$

EXAMPLE 6.2

Find the center of gravity of a homogeneous sheet in the shape of the semicircle $0 \le r \le a$ and $0 \le \theta \le \pi$.

Solution: Here ρ is constant. The mass is

$$M = \frac{1}{2}\pi a^2 \rho.$$

The moment is

$$\mathbf{m} = \iint \mathbf{x}\rho r\,dr\,d\theta = \iint (r\cos\theta,\, r\sin\theta)\rho r\,dr\,d\theta$$

$$= \rho\left(\iint r^2\cos\theta\,dr\,d\theta,\ \iint r^2\sin\theta\,dr\,d\theta\right)$$

$$= \rho\left(\int_0^a r^2\,dr\int_0^\pi \cos\theta\,d\theta,\ \int_0^a r^2\,dr\int_0^\pi \sin\theta\,d\theta\right) = \rho\left(0, \frac{2}{3}a^3\right).$$

Therefore

$$\bar{\mathbf{x}} = \frac{1}{M}\mathbf{m} = \frac{2}{\pi a^2}\left(0, \frac{2}{3}a^3\right).$$

Answer: $\bar{\mathbf{x}} = \left(0, \dfrac{4a}{3\pi}\right).$

EXAMPLE 6.3

Find the center of gravity of a sheet with density $\rho = r$ covering the quarter circle $0 \leq r \leq a$ and $0 \leq \theta \leq \pi/2$.

Solution: The mass is

$$M = \iint \rho r\,dr\,d\theta = \iint r^2\,dr\,d\theta = \int_0^a r^2\,dr\int_0^{\pi/2} d\theta = \frac{\pi a^3}{6}.$$

The moment is

$$\mathbf{m} = \iint \mathbf{x}\rho r\,dr\,d\theta = \iint (r\cos\theta,\, r\sin\theta)r^2\,dr\,d\theta$$

$$= \int_0^a r^3\,dr\int_0^{\pi/2} (\cos\theta, \sin\theta)\,d\theta = \frac{a^4}{4}(1, 1).$$

Therefore

$$\bar{\mathbf{x}} = \frac{1}{M}\mathbf{m} = \frac{6}{\pi a^3}\cdot\frac{a^4}{4}(1, 1).$$

Answer: $\bar{\mathbf{x}} = \left(\dfrac{3a}{2\pi}, \dfrac{3a}{2\pi}\right).$

Could you have predicted that $\bar{\mathbf{x}}$ lies on the line $y = x$ and $|\bar{\mathbf{x}}| > a/2$?

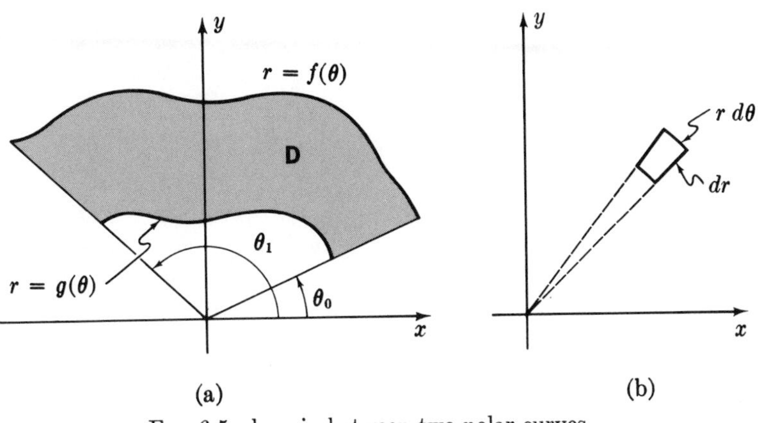

FIG. 6.5 domain between two polar curves

Other Domains

Now consider (Fig. 6.5a) a domain **D** defined by inequalities

$$g(\theta) \leq r \leq f(\theta), \qquad \theta_0 \leq \theta \leq \theta_1,$$

where $f(\theta)$ and $g(\theta)$ are continuous functions and $g(\theta) < f(\theta)$. The iteration formula for such a domain is

$$\iint\limits_{\mathbf{D}} H(x, y)\,dx\,dy = \iint\limits_{\mathbf{D}} H(r\cos\theta, r\sin\theta)\,r\,dr\,d\theta$$

$$= \int_{\theta_0}^{\theta_1}\left(\int_{r=g(\theta)}^{r=f(\theta)} H(r\cos\theta, r\sin\theta)\,r\,dr\right)d\theta.$$

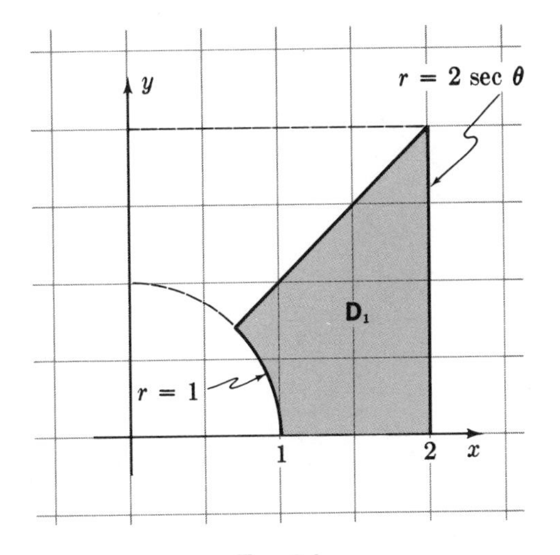

FIG. 6.6

EXAMPLE 6.4

Set up in polar coordinates the integral $\iint_{\mathbf{D}} \dfrac{dx\,dy}{x^2+y^2}$ over the domain **D** of Fig. 5.14. Do not evaluate.

Solution: The domain **D** splits naturally into two regions, one the reflection of the other in the line $y = x$; the function is symmetric in this line (Fig. 6.6). In polar coordinates, the line $x = 2$ is $r \cos\theta = 2$ or $r = 2 \sec\theta$. Consequently, the lower region $\mathbf{D}_1$ is determined by $0 \le \theta \le \pi/4$ and $1 \le r \le 2 \sec\theta$. The set-up is

$$\iint_{\mathbf{D}} \frac{dx\,dy}{x^2+y^2} = 2 \iint_{\mathbf{D}_1} \frac{r\,dr\,d\theta}{r^2}$$

$$= 2 \int_0^{\pi/4} \left(\int_1^{2\sec\theta} \frac{dr}{r} \right) d\theta = 2 \int_0^{\pi/4} \ln(2\sec\theta)\,d\theta.$$

EXAMPLE 6.5

Compute $\iint_{\mathbf{D}} xy\,dx\,dy$ over the domain **D** defined by $r = \sin 2\theta$, $0 \le \theta \le \pi/2$.

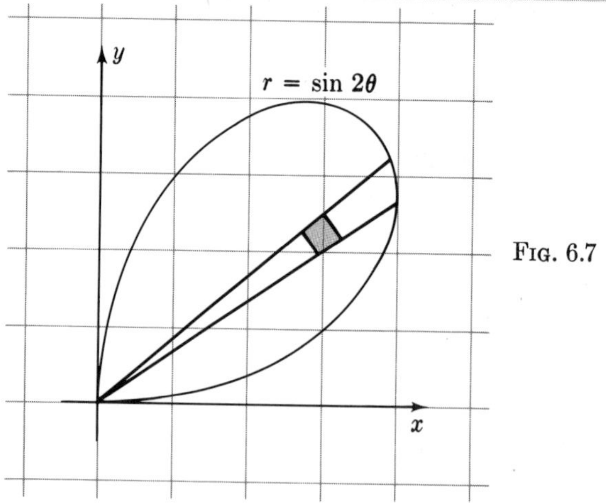

Fig. 6.7

Solution: Draw a figure (Fig. 6.7). In polar coordinates,

$$xy = (r\cos\theta)(r\sin\theta) = \frac{1}{2} r^2 \sin 2\theta,$$

hence

$$\iint_{\mathbf{D}} xy\,dx\,dy = \iint_{\mathbf{D}} \frac{1}{2} r^2 \sin 2\theta\, r\,dr\,d\theta = \frac{1}{2}\int_0^{\pi/2} (\sin 2\theta)\left(\int_0^{\sin 2\theta} r^3\,dr\right) d\theta$$

$$= \frac{1}{2}\int_0^{\pi/2} (\sin 2\theta)\left(\frac{1}{4}\sin^4 2\theta\right) d\theta = \frac{1}{8}\int_0^{\pi/2} \sin^5 2\theta\,d\theta = \frac{1}{16}\int_0^{\pi} \sin^5 \alpha\,d\alpha$$

$$= \frac{1}{8}\int_0^{\pi/2} \sin^5 \alpha\,d\alpha = \frac{1}{8}\cdot\frac{2\cdot 4}{1\cdot 3\cdot 5} = \frac{1}{15} \quad \text{(by tables).}$$

Answer: $\dfrac{1}{15}$.

EXAMPLE 6.6

Find the volume common to the three solid right circular cylinders bounded by $x^2 + y^2 = 1$, $y^2 + z^2 = 1$, $z^2 + x^2 = 1$.

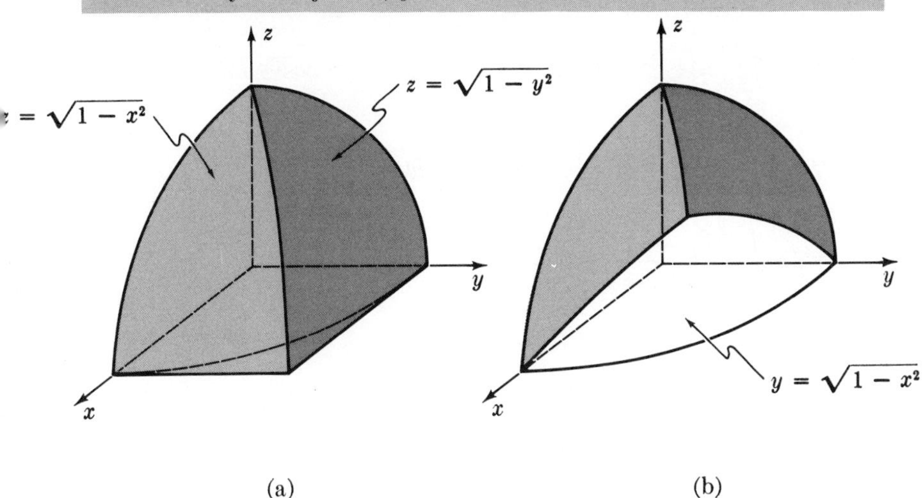

FIG. 6.8 three intersecting cylinders

Solution: The first problem is drawing the solid. By symmetry, only the portion in the first octant need be shown. In Fig. 6.8a the common part of $y^2 + z^2 = 1$ and $z^2 + x^2 = 1$ is shown. The base of the third cylinder $x^2 + y^2 = 1$ is dotted. In Fig. 6.8b this third cylinder is drawn, cutting off the desired solid. One sees that the solid is symmetric in the plane $x = y$, hence only the half to the left of this plane is needed (Fig. 6.9a). This is the solid under the surface $z = \sqrt{1 - x^2}$ and over the wedge-shaped domain **D** shown in Fig. 6.9b.

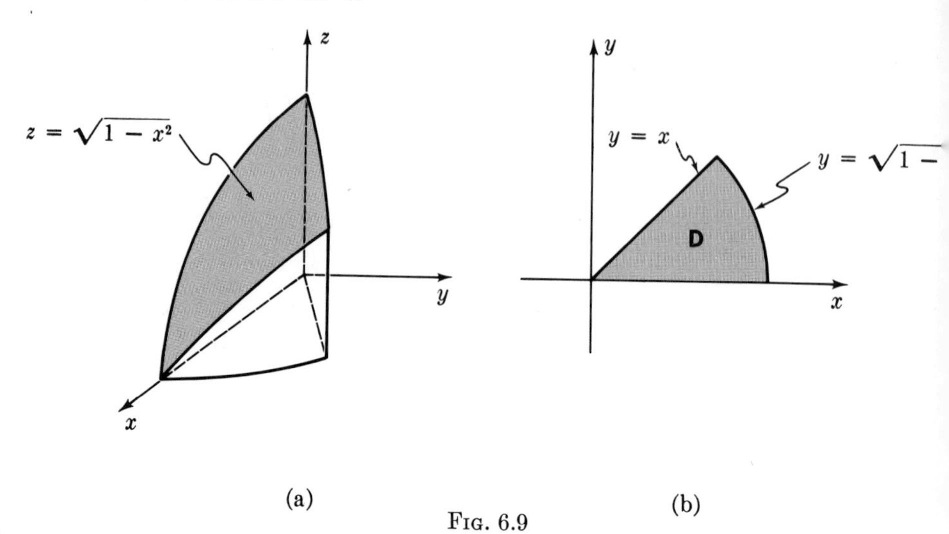

Fig. 6.9

Compute the volume using polar coordinates:

$$V = 16 \iint_{\mathbf{D}} \sqrt{1 - x^2}\, dx\, dy = 16 \iint_{\mathbf{D}} \sqrt{1 - r^2 \cos^2 \theta}\; r\, dr\, d\theta$$

$$= 16 \int_0^{\pi/4} \left[\int_0^1 r\sqrt{1 - r^2 \cos^2 \theta}\, dr \right] d\theta$$

$$= 16 \int_0^{\pi/4} \left[\left(\frac{-1}{3 \cos^2 \theta} \right) (1 - r^2 \cos^2 \theta)^{3/2} \Bigg|_{r=0}^{r=1} \right] d\theta = \frac{16}{3} \int_0^{\pi/4} \frac{1 - \sin^3 \theta}{\cos^2 \theta}\, d\theta.$$

Now

$$\frac{1 - \sin^3 \theta}{\cos^2 \theta} = \frac{1 - \sin^3 \theta}{1 - \sin^2 \theta} = \frac{1 + \sin \theta + \sin^2 \theta}{1 + \sin \theta} = \sin \theta + \frac{1}{1 + \sin \theta}.$$

From tables,

$$\int \frac{d\theta}{1 + \sin \theta} = -\tan\left(\frac{\pi}{4} - \frac{\theta}{2} \right) + C,$$

hence

$$V = \frac{16}{3} \left[-\cos \theta - \tan\left(\frac{\pi}{4} - \frac{\theta}{2} \right) \right] \Bigg|_0^{\pi/4} = \frac{16}{3} \left[1 - \frac{\sqrt{2}}{2} - \tan \frac{\pi}{8} + 1 \right]$$

$$= \frac{16}{3} \left[2 - \frac{\sqrt{2}}{2} - \frac{\sqrt{2}}{2 + \sqrt{2}} \right] = 8(2 - \sqrt{2}).$$

Answer: $8(2 - \sqrt{2})$.

NOTE: The expression for $\tan(\pi/8)$ comes from the half-angle formula $\tan(\theta/2) = \sin\theta/(1 + \cos\theta)$.

EXERCISES

Find the volume over the region given in polar coordinates and under the indicated surface. Sketch:

1. $z = x^2 + y^2$; $\quad 1 \le r \le 2, \quad 0 \le \theta \le \pi/2$
2. $z = x$; $\quad 0 \le r \le 1, \quad -\pi/2 \le \theta \le \pi/2$
3. $z = xy$; $\quad 1 \le r \le 2, \quad \pi/4 \le \theta \le \pi/2$.
4. Use polar coordinates to find the volume of a hemisphere of radius a.
5. Find the volume of the region bounded by the two paraboloids of revolution $z = x^2 + y^2$ and $z = 4 - 3(x^2 + y^2)$.
6. Find the volume of the lens-shaped region common to the sphere of radius 1 centered at $(0, 0, 0)$ and the sphere of radius 1 centered at $(0, 0, 1)$.
7. A drill of radius b bores on center through a sphere of radius a, where $a > b$. How much material is removed?
8. Find the center of gravity of a sheet with uniform density in the shape of a quarter circle $0 \le r \le a$ and $0 \le \theta \le \pi/2$.
9. (cont.) Now do it the easy way, using the result of Example 6.2.
10. Find the center of gravity of a homogeneous wedge of pie in the position $0 \le r \le a$ and $0 \le \theta \le \alpha$.
11. (cont.) Hold the radius a of Ex. 10 fixed, but let $\alpha \longrightarrow 0$. What is the limiting position of the center of gravity?
12. A quarter-circular sheet in the position $0 \le r \le a$ and $0 \le \theta \le \pi/2$ has density $\rho = a^2 - r^2$. Find its center of gravity.
13. Find the volume under the surface $z = x^4y^4$, and over the circle $x^2 + y^2 \le 1$.
14. Find the volume under the surface $z = r^3$, and over the quarter circle $0 \le r \le 1$, $0 \le \theta \le \pi/2$.
15. Find the volume under the cone $z = 5r$, and over the rose petal with boundary $r = \sin 3\theta, \quad 0 \le \theta \le \pi/3$.
16. Find the area bounded by the spiral $r = \theta, \quad 0 \le \theta \le 2\pi$, and the x-axis from 0 to 2π.
17. Evaluate over the unit disk $0 \le r \le 1$:
$$\iint x^4\,dx\,dy, \quad \iint x^2y^2\,dx\,dy, \quad \iint y^4\,dx\,dy.$$
18. Evaluate over the unit disk $0 \le r \le 1$:
$$\iint x^6\,dx\,dy, \quad \iint x^4y^2\,dx\,dy, \quad \iint x^2y^4\,dx\,dy, \quad \iint y^6\,dx\,dy.$$
19. Evaluate over the unit disk $0 \le r \le 1$:
$$\iint xy^2\,dx\,dy, \quad \iint x^4y^3\,dx\,dy, \quad \iint \sin(xy)\,dx\,dy.$$

20. Evaluate over the unit disk $0 \leq r \leq 1$:

$$\iint \frac{dx\,dy}{1 + x^2 + y^2}, \qquad \iint e^{x^2+y^2}\,dx\,dy.$$

21. Find the volume common to two right circular cylinders of radius 1 intersecting at right angles on center.

22. (cont.) Find the volume of the region in space consisting of all points (x, y, z), where one or more of the following three conditions hold:

(1) $x^2 + y^2 \leq 1, \; -2 \leq z \leq 2$

(2) $y^2 + z^2 \leq 1, \; -2 \leq x \leq 2$

(3) $z^2 + x^2 \leq 1, \; -2 \leq y \leq 2.$

[*Hint:* Use the results of Ex. 21 and Example 6.6; after that, its pure logic, not calculus.] The game of "jacks" has objects more or less in the given shape.

23. Evaluate the integral in Example 6.6 using rectangular coordinates.

11. Multiple Integrals

1. TRIPLE INTEGRALS

In this chapter we shall learn how to set up and apply triple integrals. As with double integrals, we shall work intuitively, and postpone the theory behind multiple integration until Chapter 12.

Let us consider the following problem. Suppose we have a bounded domain **D** in space filled by a non-homogeneous solid. At each point **x** the density of the solid is $\delta(\mathbf{x})$ gm/cm³. What is the total mass?

Our previous experience with problems of this type on the line and in the plane leads us to expect an answer of the form

$$M = \iiint_{\mathbf{D}} \delta(x)\, dx\, dy\, dz,$$

where the triple integral is defined by decomposing **D** into many small boxes, forming a suitable triple sum

$$\sum\sum\sum \delta(\bar{x}_i, \bar{y}_j, \bar{z}_k)(x_i - x_{i-1})(y_j - y_{j-1})(z_k - z_{k-1}),$$

and taking limits. The theory says that this process does indeed define the triple integral, and that in practice the triple integral can be evaluated by iteration.

Here is how iteration works when the domain is the part of a cylinder bounded between two surfaces, each the graph of a function. Precisely, suppose two surfaces $z = g(x, y)$ and $z = f(x, y)$ are defined over a domain **S** in the x, y-plane, and that $g(x, y) < f(x, y)$. See Fig. 1.1. These surfaces may be considered as the top and bottom of a domain **D** in the cylinder over **S**. Thus **D** consists of all points (x, y, z) where (x, y) is in **S** and

$$g(x, y) \leq z \leq f(x, y).$$

In this situation the iteration formula is

$$\iiint_{\mathbf{D}} \delta(\mathbf{x})\, dx\, dy\, dz = \iint_{\mathbf{S}} \left(\int_{z=g(x,y)}^{z=f(x,y)} \delta(x, y, z)\, dz \right) dx\, dy.$$

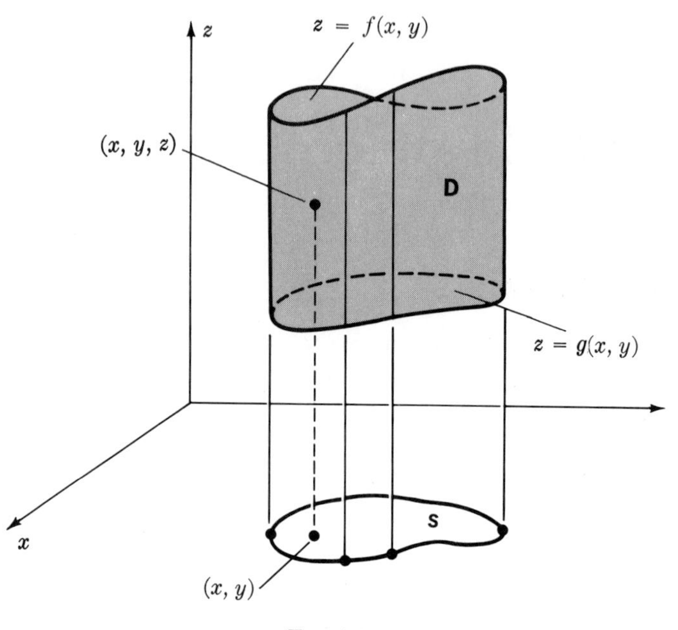

FIG. 1.1

REMARK: Some prefer the notation

$$\iint_{\mathbf{S}} dx\, dy \int_{g(x,y)}^{f(x,y)} \delta(x, y, z)\, dz.$$

The reason for the iteration formula is illustrated in Fig. 1.2. First, the little pieces of mass $\delta(x, y, z)\, dx\, dy\, dz$ in one column are added up by an integral in the vertical direction (x and y fixed, z variable). The result is

$$\left(\int_{g(x,y)}^{f(x,y)} \delta(x, y, z)\, dz\right) dx\, dy.$$

Then the masses of these individual columns are totaled by a double integral over **S**.

EXAMPLE 1.1

Find $\iiint (x^2 + y^2)\, dx\, dy\, dz$, taken over the block $1 \leq x \leq 2$, $0 \leq y \leq 1$, $2 \leq z \leq 5$.

Solution: The upper and lower boundaries are the planes $z = 5$ and

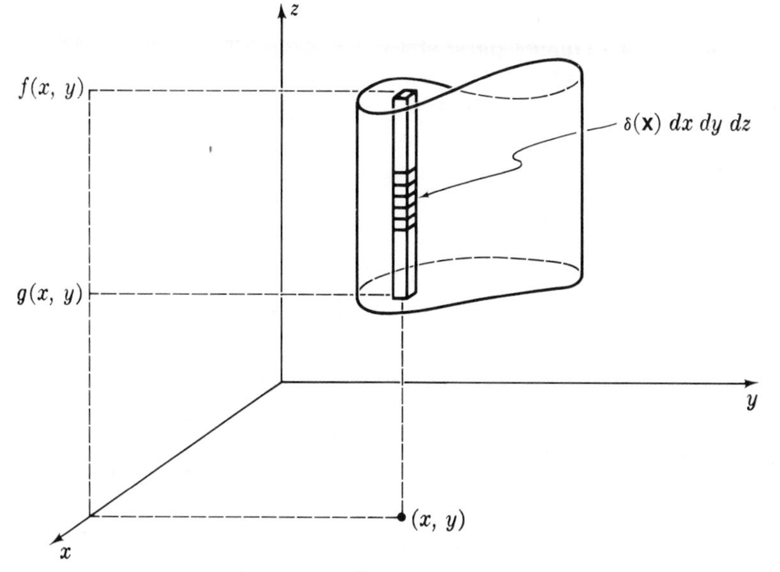

FIG. 1.2

$z = 2$. Therefore

$$\iiint (x^2 + y)\,dx\,dy\,dz = \iint_{\mathbf{S}} \left(\int_2^5 (x^2 + y)\,dz \right) dx\,dy,$$

where **S** is the rectangle $1 \le x \le 2$ and $0 \le y \le 1$. Now x and y are constant in the inner integral:

$$\int_2^5 (x^2 + y)\,dz = 3(x^2 + y),$$

hence

$$\iiint (x^2 + y)\,dx\,dy\,dz = \iint_{\mathbf{S}} 3(x^2 + y)\,dx\,dy$$

$$= \int_1^2 \left(\int_0^1 3(x^2 + y)\,dy \right) dx = \int_1^2 \left(3x^2 + \frac{3}{2} \right) dx = \frac{17}{2}.$$

Answer: $\dfrac{17}{2}$.

EXAMPLE 1.2

Compute $\iiint x^3y^2z\,dx\,dy\,dz$ over the domain **D** bounded by $x = 1, x = 2;\ \ y = 0, y = x^2;$ and $z = 0, z = 1/x$.

Solution: The domain **D** is the portion between the surfaces $z = 0$ and $z = 1/x$ of a solid cylinder parallel to the z-axis. The cylinder has base **S** in the x, y-plane, where **S** is shown in Fig. 1.3. The solid **D** itself is sketched in Fig. 1.4. (A rough sketch showing the general shape is quite satisfactory.)

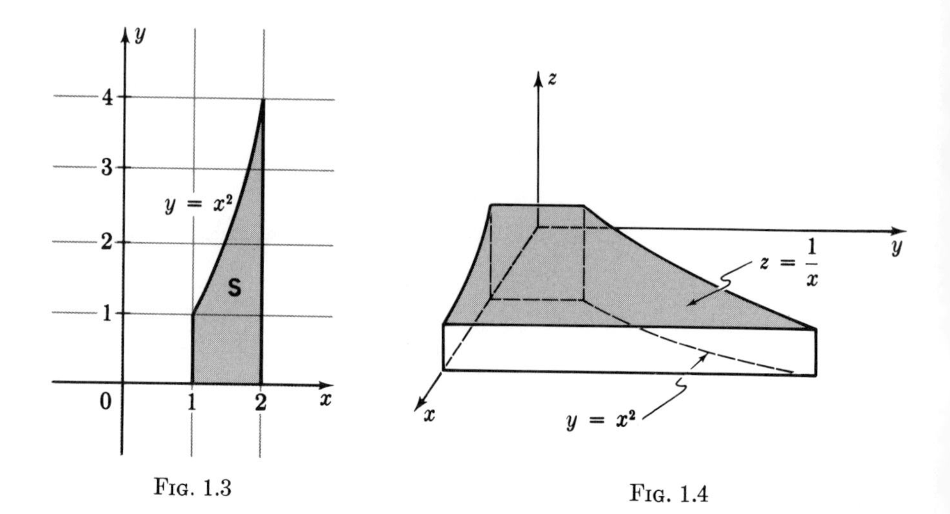

FIG. 1.3

FIG. 1.4

The iteration is

$$\iiint_{\mathbf{D}} x^3y^2z\,dx\,dy\,dz = \iint_{\mathbf{S}} \left(\int_0^{1/x} x^3y^2z\,dz\right) dx\,dy$$

$$= \iint_{\mathbf{S}} x^3y^2 \left(\frac{1}{2}z^2\bigg|_0^{1/x}\right) dx\,dy = \frac{1}{2}\iint_{\mathbf{S}} xy^2\,dx\,dy$$

$$= \frac{1}{2}\int_1^2 x\left(\int_0^{x^2} y^2\,dy\right) dx = \frac{1}{6}\int_1^2 x^7\,dx$$

$$= \frac{1}{48}(2^8 - 1) = \frac{255}{48} = \frac{85}{16}.$$

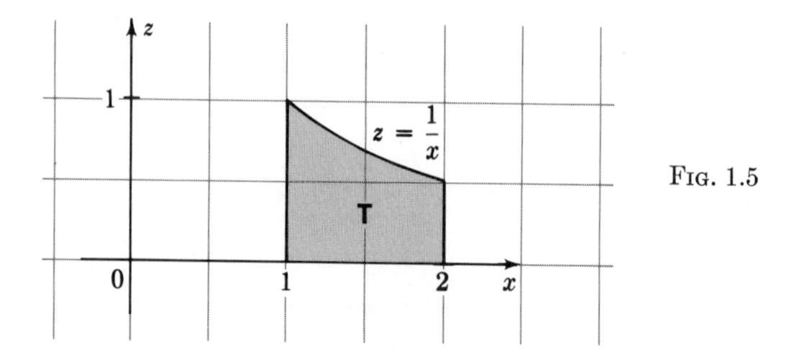

FIG. 1.5

Alternate Solution: The domain may be considered as the portion between the surfaces $y = 0$ and $y = x^2$ of a solid cylinder parallel to the y-axis. The cylinder has base **T** in the z, x-plane (Fig. 1.5). From this viewpoint, the first integration is with respect to y; the iteration is

$$\iiint\limits_{\mathbf{D}} x^3y^2z\,dx\,dy\,dz = \iint\limits_{\mathbf{T}} x^3z\left(\int_0^{x^2} y^2\,dy\right)dx\,dz = \iint\limits_{\mathbf{T}} \frac{1}{3}x^9z\,dx\,dz$$

$$= \frac{1}{3}\int_1^2 x^9\left(\int_0^{1/x} z\,dz\right)dx = \frac{1}{6}\int_1^2 x^7\,dx = \frac{1}{48}(2^8 - 1) = \frac{85}{16}.$$

Answer: $\dfrac{85}{16}$.

REMARK: It is bad technique to consider the region as a solid cylinder parallel to the x-axis, because the projection of the solid into the y, z-plane breaks into four parts. Therefore, the solid **D** itself must be decomposed into four parts, and the triple integral correspondingly expressed as a sum of four triple integrals (Fig. 1.6). The resulting computation is much longer than that in either of the previous solutions.

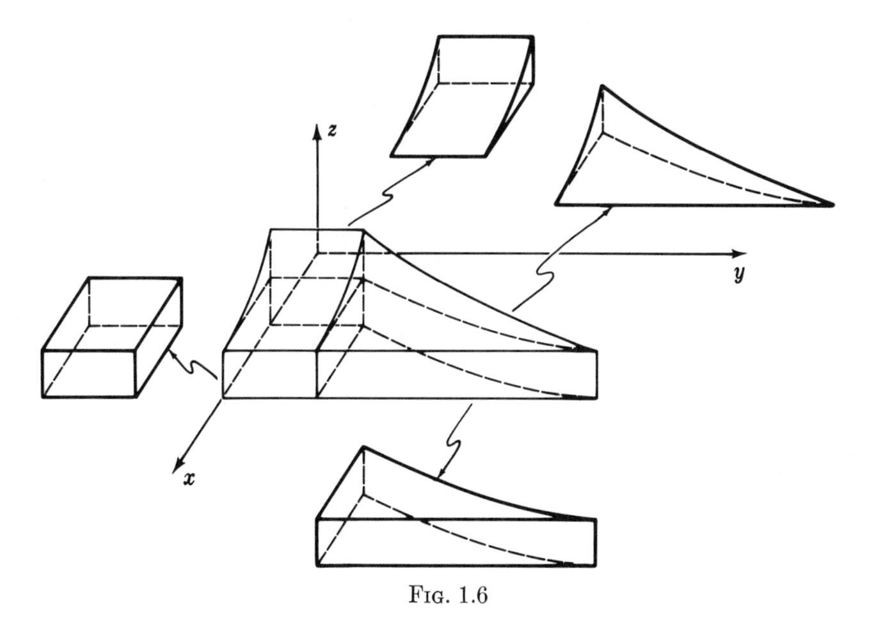

FIG. 1.6

In practice, try to pick an order of iteration which decomposes the required triple integral into as few summands as possible, hopefully only one. The typical summand has the form

$$\int_a^b\left[\int_{k(x)}^{h(x)}\left(\int_{g(x,y)}^{f(x,y)} \rho(x, y, z)\,dz\right)dy\right]dx.$$

(Possibly the variables are in some other order.) Once the integral

$$\int_{g(x,y)}^{f(x,y)} \rho(x, y, z)\, dz$$

is evaluated, the result is a function of x and y alone; z does not appear. Likewise, once the integral

$$\int_{k(x)}^{h(x)} \left(\int_{g(x,y)}^{f(x,y)} \rho(x, y, z)\, dz \right) dy$$

is evaluated, the result is a function of x alone; y does not appear.

Remember there are six possible orders of iteration for triple integrals. If you encounter an integrand you cannot find in tables, try a different order of iteration.

Domains and Inequalities

A domain in the plane or in space is frequently specified by a system of inequalities. A single inequality $f(x, y, z) \geq 0$ determines a domain whose boundary is $f(x, y, z) = 0$. To find the domain described by several such inequalities, draw the domain each describes, then form their intersection.

EXAMPLE 1.3

Draw the plane domain given by $x + y \leq 0, \quad y \geq x^2 + 2x$.

Solution: The first inequality determines the domain below (and on) the line $x + y = 0$. See Fig. 1.7a. The second inequality determines the domain above (and on) the parabola $y = x^2 + 2x$. See Fig. 1.7b. The line and parabola intersect at $(0, 0)$ and $(-3, 3)$. The domain satisfying both inequalities is shown in Fig. 1.7c.

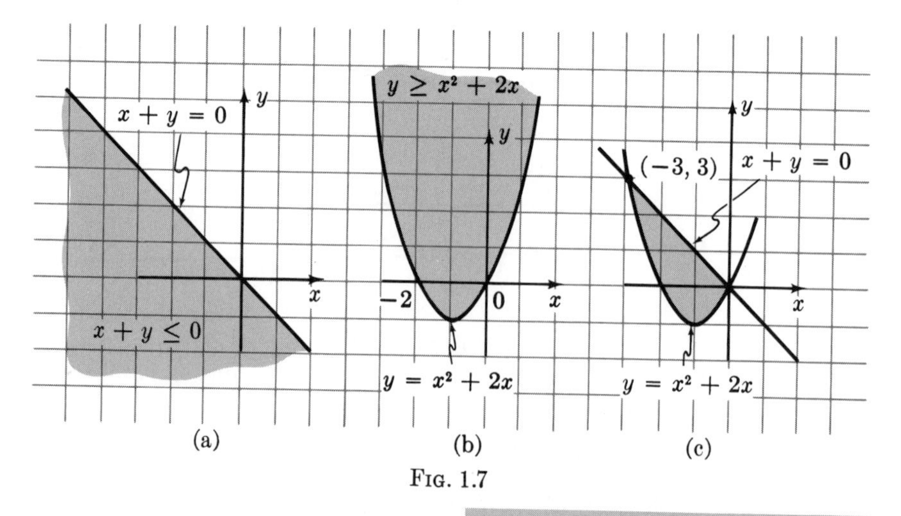

FIG. 1.7

EXAMPLE 1.4

Describe the plane domain **S** defined by $0 \leq x \leq y \leq a$. Write the integral $\iint_{\mathbf{S}} f(x, y)\,dx\,dy$ as an iterated integral in both orders.

Solution: The region is described by the three inequalities

$$x \geq 0, \qquad y \geq x, \qquad y \leq a.$$

Draw the corresponding domains and take their intersection, a triangle (Fig. 1.8).

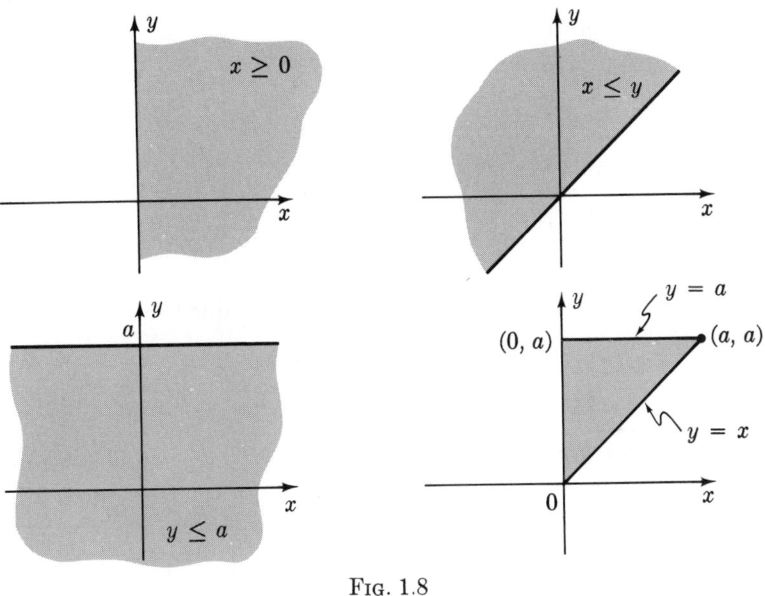

FIG. 1.8

To integrate first on x, arrange the inequalities describing the triangle in this order:

$$0 \leq y \leq a, \qquad 0 \leq x \leq y.$$

Think of y as fixed in the interval $[0, a]$. Then x runs from 0 to y, leading to

$$\int_0^y f(x, y)\,dx.$$

Now sum these quantities for y running from 0 to a. The result is

$$\iint_{\mathbf{S}} f(x, y)\,dx\,dy = \int_0^a \left(\int_0^y f(x, y)\,dx \right) dy.$$

To integrate first on y, describe the region in the other order:

$$0 \le x \le a, \qquad x \le y \le a.$$

Now the result is

$$\iint_{\mathbf{S}} f(x, y)\, dx\, dy = \int_0^a \left(\int_x^a f(x, y)\, dy \right) dx.$$

Answer: The triangle with vertices $(0, 0)$, $(0, a)$, (a, a);

$$\int_0^a \left(\int_0^y f(x, y)\, dx \right) dy = \int_0^a \left(\int_x^a f(x, y)\, dy \right) dx.$$

REMARK: A special case is interesting. Suppose $f(x, y) = g(x)$, a function of x alone. Then the right-hand side is

$$\int_0^a \left(\int_x^a g(x)\, dy \right) dx = \int_0^a g(x) \left(\int_x^a dy \right) dx = \int_0^a (a - x) g(x)\, dx.$$

The following formula is a consequence:

$$\int_0^a \left(\int_0^y g(x)\, dx \right) dy = \int_0^a (a - x) g(x)\, dx.$$

Thus the iterated integral of a function of one variable can be expressed as a simple integral.

Tetrahedra

If a domain **D** is specified by inequalities, it may be possible to arrange the inequalities so that limits of integration can be set up automatically. For example, suppose the inequalities can be arranged in this form:

$$a \le x \le b, \qquad h(x) \le y \le k(x), \qquad g(x, y) \le z \le f(x, y).$$

Then

$$\iiint_{\mathbf{D}} \delta(x, y, z)\, dx\, dy\, dz = \int_a^b \left[\int_{h(x)}^{k(x)} \left(\int_{g(x,y)}^{f(x,y)} \delta(x, y, z)\, dz \right) dy \right] dx.$$

Tetrahedral domains can be expressed by such inequalities, and they occur frequently enough that it is useful to practice setting up integrals over them.

EXAMPLE 1.5

A tetrahedron **T** has vertices at $(0, 0, 0)$, $(a, 0, 0)$, $(0, b, 0)$, $(0, 0, c)$, where $a, b, c > 0$. Set up $\iiint_{\mathbf{T}} \delta(x, y, z)\, dx\, dy\, dz$ as an iterated integral.

Solution: The slanted surface (Fig. 1.9) has equation

$$\frac{x}{a}+\frac{y}{b}+\frac{z}{c}=1.$$

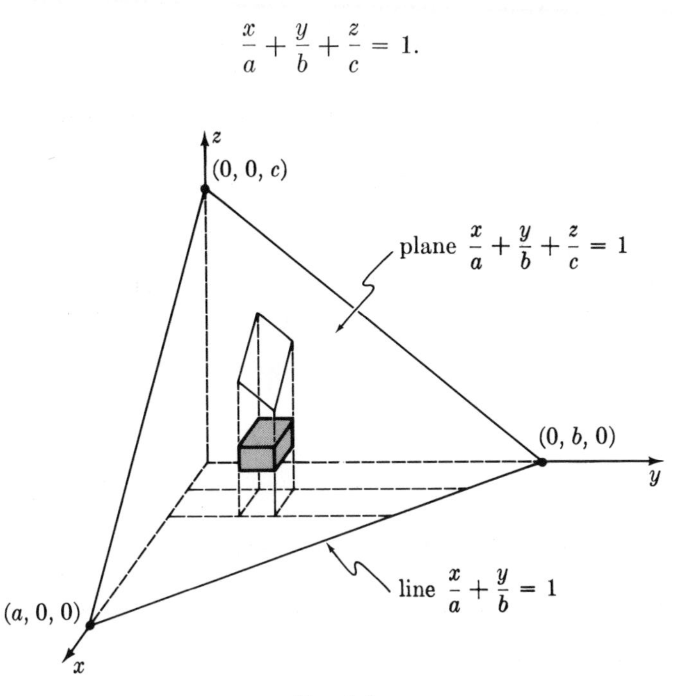

FIG. 1.9

The domain is defined by the inequalities

$$0 \le x, \qquad 0 \le y, \qquad 0 \le z, \qquad \frac{x}{a}+\frac{y}{b}+\frac{z}{c} \le 1.$$

Any order of iteration is satisfactory; for instance, choose the order of integration

$$\int \left[\int \left(\int \rho(x, y, z)\, dz\right) dy\right] dx.$$

To find the limits of integration, arrange the inequalities in this form:

$$a \le x \le b, \qquad h(x) \le y \le k(x), \qquad g(x, y) \le z \le f(x, y).$$

If x and y are fixed, then

$$0 \le z \le c\left(1-\frac{x}{a}-\frac{y}{b}\right).$$

If x is fixed, then

$$0 \le y \le b\left(1-\frac{x}{a}-\frac{z}{c}\right) \le b\left(1-\frac{x}{a}\right).$$

Finally,

$$0 \le x \le a\left(1 - \frac{y}{b} - \frac{z}{c}\right) \le a.$$

The resulting system of inequalities (equivalent to the original system) is

$$0 \le x \le a, \qquad 0 \le y \le b\left(1 - \frac{x}{a}\right), \qquad 0 \le z \le c\left(1 - \frac{x}{a} - \frac{y}{b}\right).$$

The corresponding iteration is

$$\iiint\limits_{\mathbf{T}} \rho(x, y, z)\, dx\, dy\, dz = \int_0^a \left[\int_0^{b(1 - \frac{x}{a})} \left(\int_0^{c(1 - \frac{x}{a} - \frac{y}{b})} \rho(x, y, z)\, dz\right) dy\right] dx.$$

EXERCISES

Evaluate the triple integral over the indicated domain:

1. $\iiint (2 - z)\, dx\, dy\, dz$; solid cylinder under $z = 2 - x^2$, based on the triangle with vertices $(0, 0, 0)$, $(1, 0, 0)$, $(0, 1, 0)$
2. $\iiint y\, dx\, dy\, dz$; solid cylinder under $z = x^2 + 2y^2$, based on the triangle in Ex. 1
3. $\iiint z^3\, dx\, dy\, dz$; pyramid with apex $(0, 0, 1)$, based on the square with vertices $(\pm 1, \pm 1)$
4. The same, except the square base has vertices $(\pm 1, 0)$, $(0, \pm 1)$
5. $\iiint x^3y^2z\, dx\, dy\, dz$; the domain lies in the first octant and is bounded by $x = 0$, $z = 0$, $y = x^2$, and $z = 1 - y^2$
6. $\iiint (3x^2 - z^2)\, dx\, dy\, dz$; the domain in the slab $0 \le y \le 1$ bounded by $z = y^2$, $z = -y^2$, and $y = x$, $y = -x$.
7. A solid cube has side a. Its density at each point is k times the product of the 6 distances of the point to the faces of the cube, where k is constant. Find the mass.
8. Charge is distributed over the tetrahedron with vertices **0**, **i**, **j**, **k**. The charge density at each point is a constant k times the product of the 4 distances from the point to the faces of the tetrahedron. Find the total charge.
9. Find $\iiint (x + y + z)^2\, dx\, dy\, dz$ over the domain in the first octant bounded by the coordinate planes, the plane $x + y + z = 2$, and the 3 planes $x = 1$, $y = 1$, $z = 1$.

10. Find $\iiint \frac{dx\,dy\,dz}{(x+y+z)^2}$ over the domain in the first octant between the planes $x+y+z=1$ and $x+y+z=4$. [*Hint:* Think!]

Sketch the domain:

11. $x^2+y^2 \le 1, \quad y+x^2 \ge 0$
12. $x^2+y^2 \le 1, \quad -x^2 \le y \le x^2$
13. $x^2+y^2 \ge 1, \quad (x-2)^2+y^2 \le 9$
14. $x \ge 3, \quad y \le -5, \quad y-x \ge -10$
15. $x+y \le 0, \quad xy \le 1, \quad (x-y)^2 \le 1$
16. $(x+y)^2 \le 1, \quad (x-y)^2 \le 1.$

Express the double integral $\iint f(x, y)\,dx\,dy$ over the specified domain as the sum of one or more iterated integrals in which y is the first variable integrated:

17. $x^2+y^2 \le 1, \quad x^2+(y-1)^2 \le 1$
18. $y \ge (x+1)^2, \quad y+2x \le 3$
19. $x \ge 0, \quad 0 \le y \le \pi, \quad x \le \sin y$
20. $x \ge 0, \quad x^2-y^2 \ge 1, \quad x^2+y^2 \le 9.$
21. Describe the domain $\mathbf{D}: 0 \le x \le y \le z \le 1$. Iterate $\iiint_{\mathbf{D}} \rho(x, y, z)\,dx\,dy\,dz$ in the orders x, y, z; z, y, x; and x, z, y.
22. Repeat Ex. 21 for $a \le x \le y \le z \le b$.
23. Repeat Ex. 21 for $0 \le x \le 2y \le 3z \le 6$.
24. Express a triple integral over the domain determined by $(x-1)^2+y^2 \le 4$, $0 \le z \le y$, $(x+1)^2+y^2 \le 4$ as one iterated integral (not a sum of two or more). Sketch the domain.
25. Find the integral of x over the tetrahedron with vertices $(5, 6, 3)$, $(4, 6, 3)$, $(5, 5, 3)$, $(5, 6, 2)$.
26. Set up the integral of $\delta(x, y, z)$ over the tetrahedron with vertices $(5, -5, 1)$, $(5, -5, -2)$, $(5, 1, 1)$, $(-2, -5, 1)$.
27. Express
$$\int_0^a \left[\int_0^z \left(\int_0^1 g(x)\,dx\right) dy\right] dz$$
as a simple integral.
28. Take four vertices of a unit cube, no two adjacent. Find the volume of the tetrahedron with these points as vertices.
29. (cont.) Now take the tetrahedron whose vertices are the remaining four vertices of the cube. The two tetrahedra intersect in a certain polyhedron. Describe it and find its volume.

2. CYLINDRICAL COORDINATES

Cylindrical coordinates are designed to fit situations with rotational (axial) symmetry about an axis.

The **cylindrical coordinates** of a point $\mathbf{x} = (x, y, z)$ are $\{r, \theta, z\}$, where

$\{r, \theta\}$ are the polar coordinates of (x, y) and z is the third rectangular coordinate (Fig. 2.1a). Each surface r = constant is a right circular cylinder, hence the name, cylindrical coordinates (Fig. 2.1b).

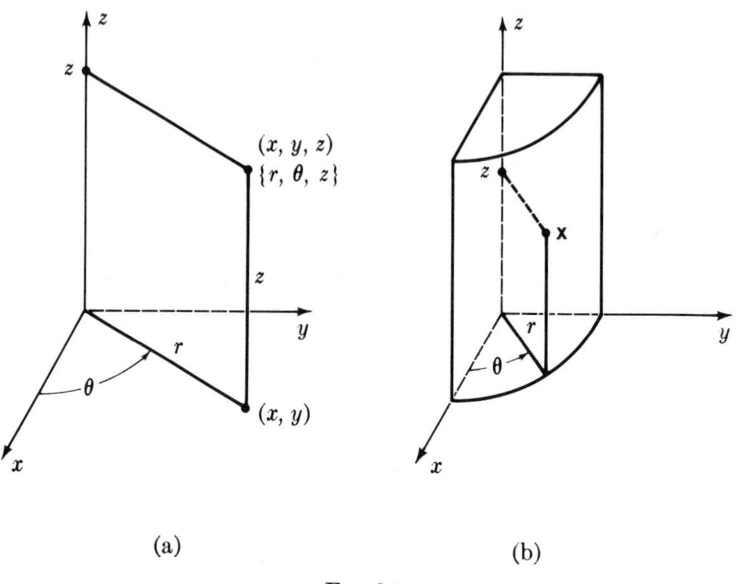

FIG. 2.1

Through each point **x** (not on the z-axis) pass three surfaces, r = constant, θ = constant, z = constant (Fig. 2.2). Each is orthogonal (perpendicular) to the other two at their common intersection **x**.

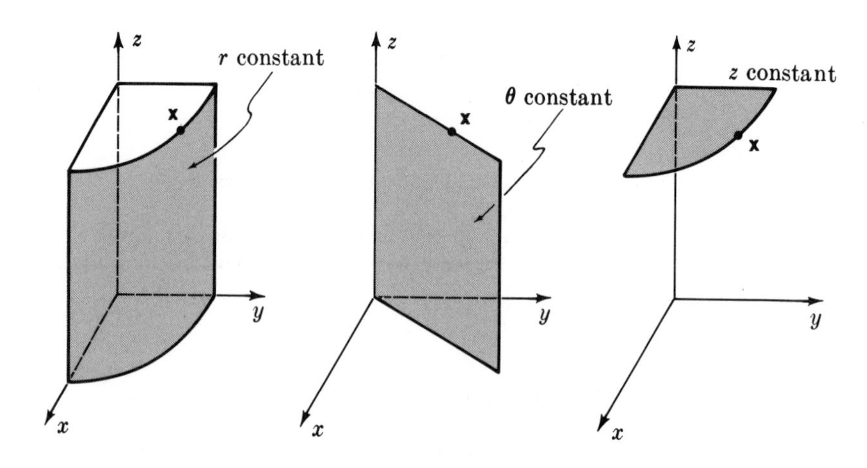

FIG. 2.2

The relations between the rectangular coordinates (x, y, z) and the cylindrical coordinates $\{r, \theta, z\}$ of a point are

$$\begin{cases} x = r\cos\theta \\ y = r\sin\theta \\ z = z \end{cases} \qquad \begin{cases} r^2 = x^2 + y^2 \\ \cos\theta = \dfrac{x}{r}, \quad \sin\theta = \dfrac{y}{r} \\ z = z. \end{cases}$$

The origin in the plane is given in polar coordinates by $r = 0$; the angle θ is undefined. Similarly, a point on the z-axis is given in cylindrical coordinates by $r = 0$, $z =$ constant; θ is undefined.

EXAMPLE 2.1

Graph the surfaces (i) $z = 2r$, (ii) $z = r^2$.

Solution: Both are surfaces of revolution about the z-axis, as is any surface $z = f(r)$. Since z depends only on r, not on θ, the height of the surface is constant above each circle $r = c$ in the x, y-plane. Thus the level curves are circles in the x, y-plane centered at the origin.

In (i), the surface meets the first quadrant of the y, z-plane in the line $z = 2y$. (In the first quadrant of the y, z-plane, $x = 0$ and $y \geq 0$. Since $r^2 = x^2 + y^2 = y^2$, it follows that $r = y$.) Rotated about the z-axis, this line spans a cone with apex at **0**. See Fig. 2.3a.

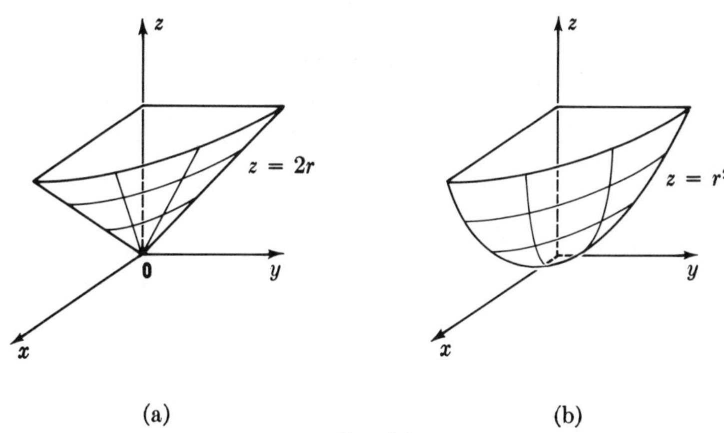

(a) (b)

FIG. 2.3

In (ii), the surface meets the y, z-plane in the parabola $z = y^2$. Rotated about the z-axis, this parabola generates a paraboloid of revolution (Fig. 2.3b).

EXAMPLE 2.2

Find the level surfaces of the function $f(r, \theta, z) = r$.

Solution: Each surface is defined by $r = c$. This is a right circular cylinder whose axis is the z-axis (Fig. 2.4).

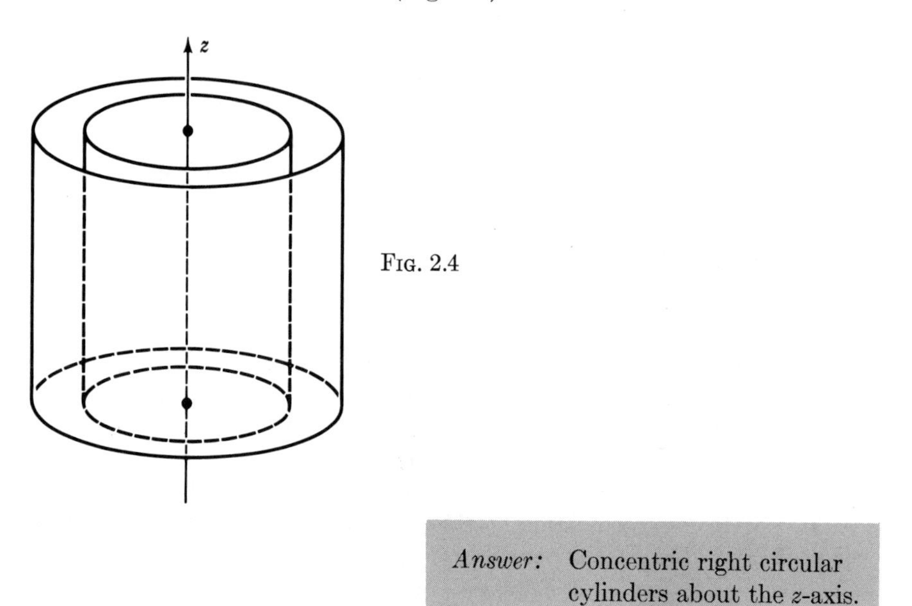

FIG. 2.4

Answer: Concentric right circular cylinders about the z-axis.

The Natural Frame

It is convenient to fit a frame of three mutually perpendicular vectors to cylindrical coordinates just as the frame **i**, **j**, **k** fits rectangular coordinates. At each point $\{r, \theta, z\}$ of space attach three mutually perpendicular unit

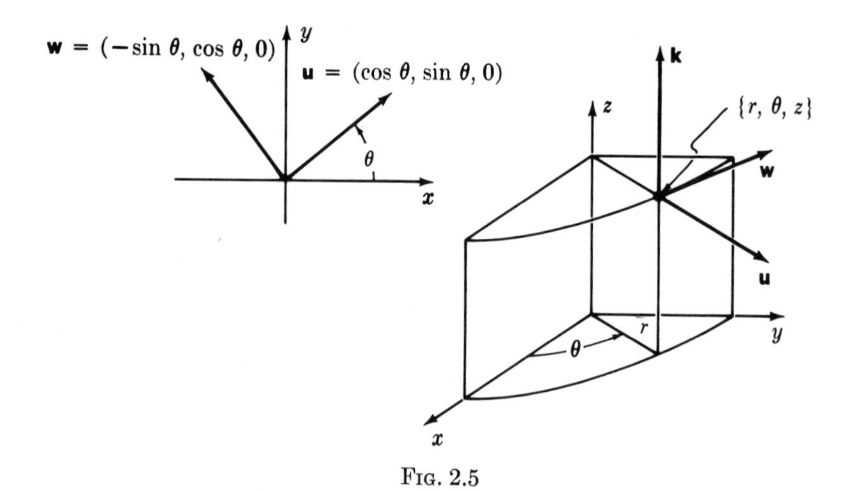

FIG. 2.5

vectors **u**, **w**, **k** chosen so

$$\begin{Bmatrix} \mathbf{u} \\ \mathbf{w} \\ \mathbf{k} \end{Bmatrix} \text{ points in the direction of increasing } \begin{Bmatrix} r \\ \theta \\ z \end{Bmatrix}.$$

Thus (Fig. 2.5)

$$\mathbf{u} = \frac{1}{r}(x, y, 0) = (\cos\theta, \sin\theta, 0),$$

$$\mathbf{w} = \frac{1}{r}(-y, x, 0) = (-\sin\theta, \cos\theta, 0),$$

$$\mathbf{k} = (0, 0, 1).$$

The vectors **u**, **w**, **k** form a right-hand system:

$$\mathbf{u} \times \mathbf{w} = \mathbf{k}, \qquad \mathbf{w} \times \mathbf{k} = \mathbf{u}, \qquad \mathbf{k} \times \mathbf{u} = \mathbf{w}.$$

Note that **u** and **w** depend on θ alone, while **k** is a constant vector, our old friend from the trio **i**, **j**, **k**. Note also

$$\frac{\partial \mathbf{u}}{\partial \theta} = \mathbf{w}, \qquad \frac{\partial \mathbf{w}}{\partial \theta} = -\mathbf{u}.$$

In situations with axial symmetry, it is frequently better to express vectors in terms of **u**, **w**, **k** rather than **i**, **j**, **k**.

Integrals

If a solid has axial symmetry, it is often convenient to place the z-axis on the axis of symmetry, and use cylindrical coordinates $\{r, \theta, z\}$ for the computation of integrals.

In polar coordinates $\{r, \theta\}$, the element of area is $r\,dr\,d\theta$. Hence the element of volume in cylindrical coordinates $\{r, \theta, z\}$ is

$$dV = r\,dr\,d\theta\,dz.$$

To gain further insight, let us argue intuitively for a moment. Given $\mathbf{x} = \{r, \theta, z\}$, how is this point displaced if we give small increments dr, $d\theta$, dz to its three coordinates? Figure 2.6 shows that the displacement in the direction of **u** is $dr\,\mathbf{u}$. The displacements in the directions of **w** and **k** are $r\,d\theta\,\mathbf{w}$ and $dz\,\mathbf{k}$. The displacement of **x** is approximately the sum of these three small vectors.

Accordingly we write

$$d\mathbf{x} = dr\,\mathbf{u} + r\,d\theta\,\mathbf{w} + dz\,\mathbf{k}.$$

Since the displacements are mutually perpendicular vectors, they span a small "rectangular box" of volume $(dr)(r\,d\theta)(dz)$. Thus we arrive again at the formula $dV = r\,dr\,d\theta\,dz$.

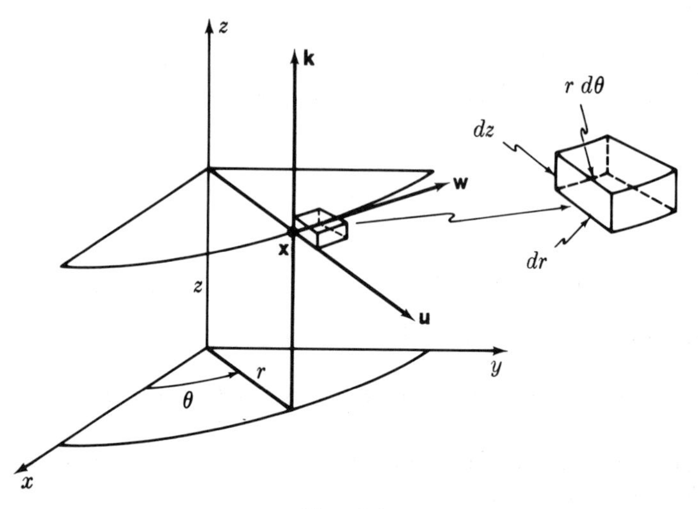

FIG. 2.6

EXAMPLE 2.3

Evaluate $\iiint (x^2 + y^2)^{1/2} z\,dx\,dy\,dz$, taken over the cone with apex $(0, 0, 1)$ and semicircular base bounded by the x-axis and $y = \sqrt{4 - x^2}$.

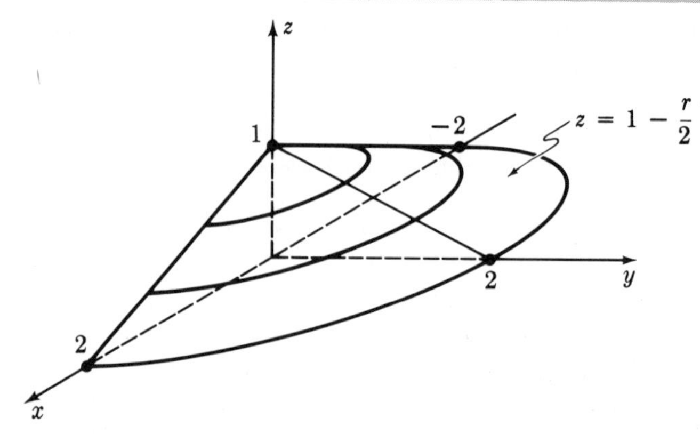

FIG. 2.7

Solution: Write the integral in cylindrical coordinates:

$$I = \iiint (rz)r\,dr\,d\theta\,dz.$$

The lateral surface of the cone (Fig. 2.7) is given by

$$z = 1 - \frac{1}{2}r,$$

so the solid is described by

$$0 \leq \theta \leq \pi, \qquad 0 \leq r \leq 2, \qquad 0 \leq z \leq 1 - \frac{1}{2}r.$$

Therefore

$$I = \left(\int_0^\pi d\theta\right)\int_0^2 r^2\left(\int_0^{1-r/2} z\,dz\right)dr = \pi\int_0^2 \frac{r^2}{2}\left(1 - \frac{1}{2}r\right)^2 dr$$

$$= \frac{\pi}{2}\int_0^2 r^2\left(1 - r + \frac{1}{4}r^2\right)dr = \frac{\pi}{8}\int_0^2 (r^4 - 4r^3 + 4r^2)\,dr = \frac{2\pi}{15}.$$

Answer: $\dfrac{2\pi}{15}$.

EXAMPLE 2.4

A region **D** in space is generated by revolving the plane region bounded by $z = 2x^2$, the x-axis, and $x = 1$ about the z-axis. Mass is distributed in **D** so that the density at each point is proportional to the distance of the point from the plane $z = -1$, and to the square of the distance of the point from the z-axis. Compute the total mass.

Solution: The density is

$$\delta = k(x^2 + y^2)(z + 1) = kr^2(z + 1),$$

where k is a constant. The portion of the solid in the first octant is shown in Fig. 2.8. In cylindrical coordinates, the solid is described by the inequalities

$$0 \leq \theta \leq 2\pi, \qquad 0 \leq r \leq 1, \qquad 0 \leq z \leq 2r^2.$$

Therefore its mass is

$$M = \iiint kr^2(z + 1)\,r\,dr\,d\theta\,dz = k\left(\int_0^{2\pi} d\theta\right)\int_0^1 r^3\left[\int_0^{2r^2} (z + 1)\,dz\right]dr$$

$$= 2\pi k\int_0^1 r^3\left[\frac{1}{2}(2r^2)^2 + (2r^2)\right]dr = \frac{7\pi}{6}k.$$

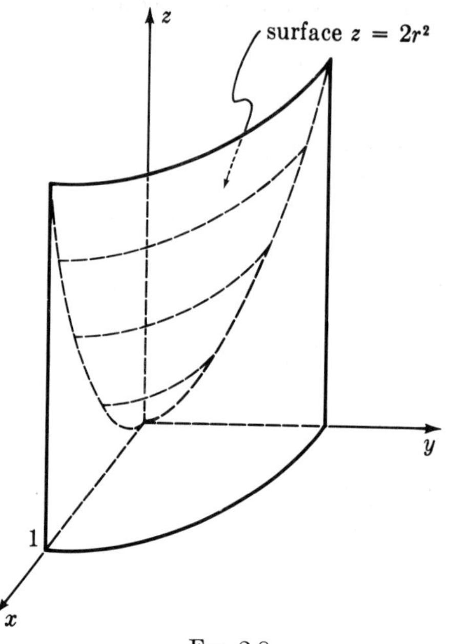

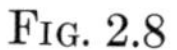
Fig. 2.8

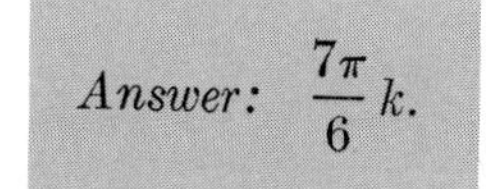
Answer: $\dfrac{7\pi}{6}k.$

EXERCISES

1. Find the rectangular coordinates of $\{2, 5\pi/4, -3\}$, $\{1, \pi, 2\}$.
2. Find the cylindrical coordinates of $(0, -2, -2)$, $(5, -5, 0)$, $(-1, -\sqrt{3}, -1)$.
3. A space curve is given by $r = r(t)$, $\theta = \theta(t)$, $z = z(t)$. Show that its arc length satisfies $\dot{s} = (\dot{r}^2 + r^2\dot{\theta}^2 + \dot{z}^2)^{1/2}$.
4. (cont.) Find the length of the spiral $r = a$, $\theta = bt$, $z = ct$ for $0 \leq t \leq 1$.
5. Prove analytically the formula

$$d\mathbf{x} = dr\,\mathbf{u} + r\,d\theta\,\mathbf{w} + dz\,\mathbf{k}.$$

6. Prove

$$\begin{bmatrix} d\mathbf{u} \\ d\mathbf{w} \\ d\mathbf{k} \end{bmatrix} = \begin{bmatrix} 0 & d\theta & 0 \\ -d\theta & 0 & 0 \\ 0 & 0 & 0 \end{bmatrix} \begin{bmatrix} \mathbf{u} \\ \mathbf{w} \\ \mathbf{k} \end{bmatrix}.$$

7*. Let $f = f(r, \theta, z)$. Prove

$$\operatorname{grad} f = \frac{\partial f}{\partial r}\mathbf{u} + \frac{1}{r}\frac{\partial f}{\partial \theta}\mathbf{w} + \frac{\partial f}{\partial z}\mathbf{k}.$$

[*Hint*: Use Ex. 5.]

8. Use cylindrical coordinates to find the volume of a right circular cone of base radius a and height h.

9. Find $\iiint xyz\,dx\,dy\,dz$ over the quarter cylinder $0 \leq r \leq a$, $0 \leq \theta \leq \frac{1}{2}\pi$, $0 \leq z \leq h$.

10. Find $\iiint z^2\,dx\,dy\,dz$ over the domain common to the sphere $x^2 + y^2 + z^2 \leq a^2$ and the cylinder $r \leq b$, where $b < a$.

11. Find $\iiint yz\,dx\,dy\,dz$ over the cylindrical wedge $0 \leq z \leq y$, $r \leq a$, $0 \leq \theta \leq \pi$.

12. Sketch the domain $r^2 \leq z \leq 2r^2$, $r \leq 1$. Now find $\iiint e^z\,dx\,dy\,dz$ over this domain.

13. Sketch the domain $|z| \leq r^2$, $r \leq a$, $x \geq 0$, $y \geq 0$. Now find $\iiint z^4\,dx\,dy\,dz$ over the domain.

14*. Sketch the domain $0 \leq r \leq \cos 2\theta$, $-\frac{1}{4}\pi \leq \theta \leq \frac{1}{4}\pi$, $0 \leq z \leq 1 - r^2$. Now find $\iiint z\,dx\,dy\,dz$. (Use the definite integral table inside the front cover to complete the solution.)

3. SPHERICAL COORDINATES

Spherical coordinates are designed to fit situations with central symmetry. The **spherical coordinates** $[\rho, \phi, \theta]$ of a point $\mathbf{x}$ are its distance $\rho = |\mathbf{x}|$ from the origin, its elevation angle ϕ, and its azimuth angle θ. (Often θ is called the longitude and ϕ the co-latitude.) See Fig. 3.1.

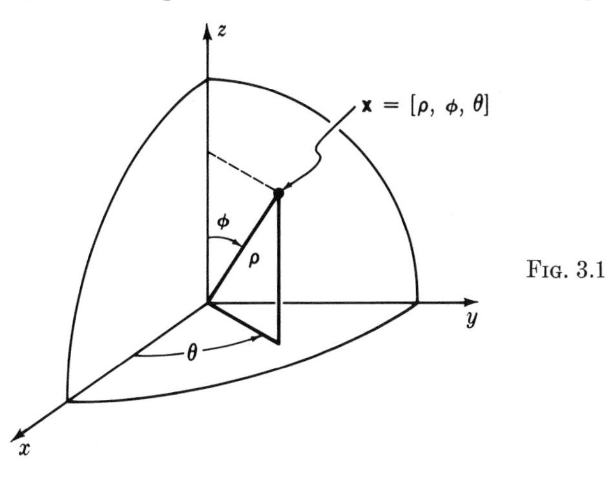

Fig. 3.1

Relations between the rectangular coordinates (x, y, z) of a point and its spherical coordinates may be read from Fig. 3.2. They are

$$\boxed{\begin{cases} x = \rho \sin\phi \cos\theta \\ y = \rho\sin\phi\sin\theta \\ z = \rho\cos\phi \end{cases} \qquad \begin{cases} \rho^2 = x^2 + y^2 + z^2 \\ \cos\phi = \dfrac{z}{\rho} \\ \tan\theta = \dfrac{y}{x}. \end{cases}}$$

Note that θ is not determined on the z-axis, so points of this axis are usually avoided. In general θ is determined up to a multiple of 2π, and $0 < \phi < \pi$.

The level surfaces

$$\left\{\begin{array}{l} \rho = \text{constant} \\ \phi = \text{constant} \\ \theta = \text{constant} \end{array}\right\} \quad \text{are} \quad \left\{\begin{array}{l} \text{concentric spheres about } \mathbf{0} \\ \text{right circular cones, apex } \mathbf{0} \\ \text{planes through the } z\text{-axis} \end{array}\right\}.$$

At each point $\mathbf{x}$ the three level surfaces intersect orthogonally (Fig. 3.3).

The Natural Frame

Select unit vectors $\boldsymbol{\lambda}, \boldsymbol{\mu}, \boldsymbol{\nu}$ at each point $\mathbf{x}$ of space (not on the z-axis) such that

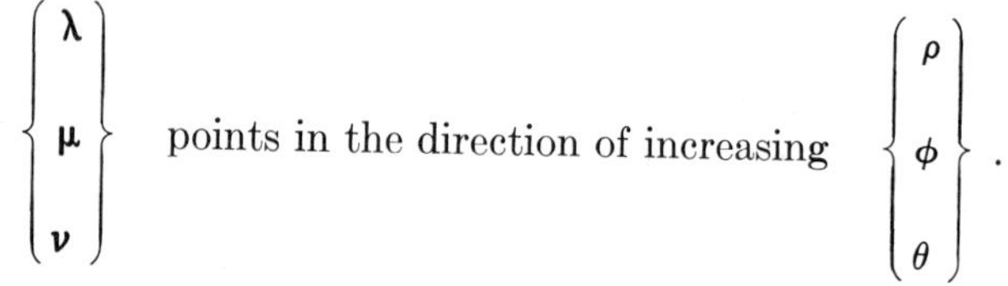

$$\left\{\begin{array}{c} \boldsymbol{\lambda} \\ \boldsymbol{\mu} \\ \boldsymbol{\nu} \end{array}\right\} \quad \text{points in the direction of increasing} \quad \left\{\begin{array}{c} \rho \\ \phi \\ \theta \end{array}\right\}.$$

See Fig. 3.4. Then $\boldsymbol{\lambda}, \boldsymbol{\mu}, \boldsymbol{\nu}$ is a right-hand system.

Our immediate problem is to express $\boldsymbol{\lambda}, \boldsymbol{\mu}, \boldsymbol{\nu}$ in terms of ρ, ϕ, θ. Here is a short cut for doing so:

$$\begin{aligned} \mathbf{x} &= \rho(\sin\phi\cos\theta, \sin\phi\sin\theta, \cos\phi), \\ d\mathbf{x} &= (\sin\phi\cos\theta, \sin\phi\sin\theta, \cos\phi)\,d\rho \\ &\quad + \rho(\cos\phi\cos\theta, \cos\phi\sin\theta, -\sin\phi)\,d\phi \\ &\quad + \rho(-\sin\phi\sin\theta, \sin\phi\cos\theta, 0)\,d\theta \\ &= \boldsymbol{\lambda}\,d\rho + \boldsymbol{\mu}\,\rho\,d\phi + \boldsymbol{\nu}\,\rho\sin\phi\,d\theta. \end{aligned}$$

It is easy to check that $\boldsymbol{\lambda}, \boldsymbol{\mu}, \boldsymbol{\nu}$ are unit vectors. Furthermore, they point in the directions of increasing ρ, ϕ, θ respectively. Precisely, if only ρ increases, then $d\phi = d\theta = 0$, hence $d\mathbf{x} = \boldsymbol{\lambda}\,d\rho$. Similarly, if only ϕ increases, then $d\mathbf{x} = \boldsymbol{\mu}\,\rho\,d\phi$,

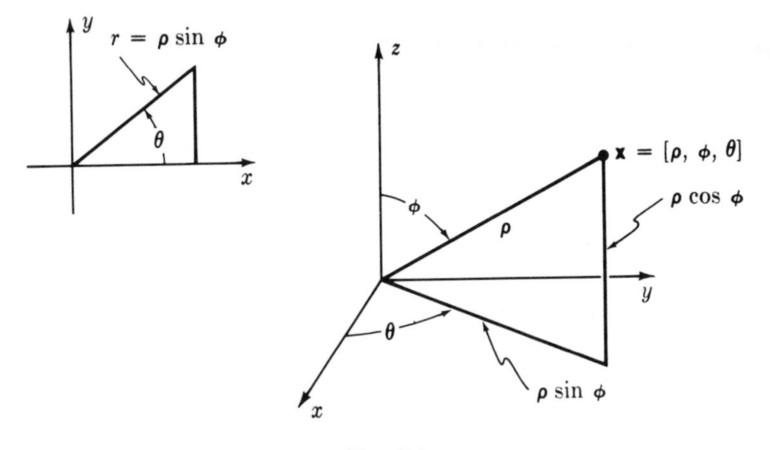
y
r = ρ sin φ
θ
x
z
x = [ρ, φ, θ]
ρ cos φ
φ
ρ
y
θ
ρ sin φ
x

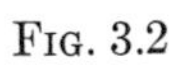
Fig. 3.2

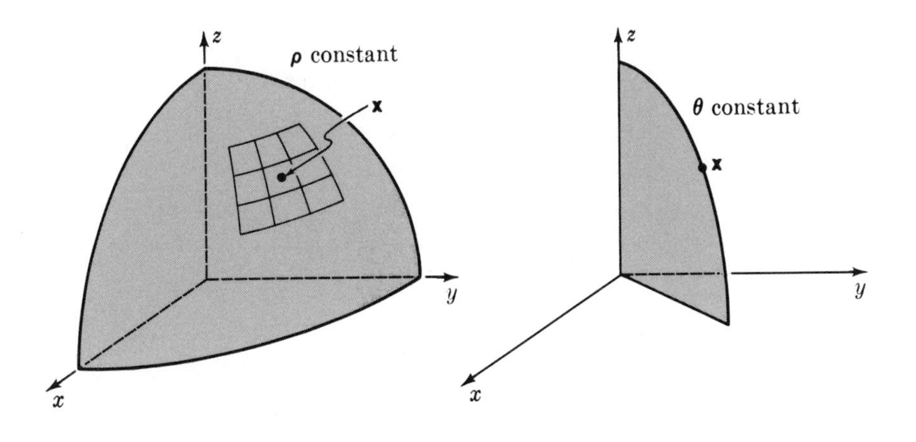
z
ρ constant
x
y
x
z
θ constant
x
y
x

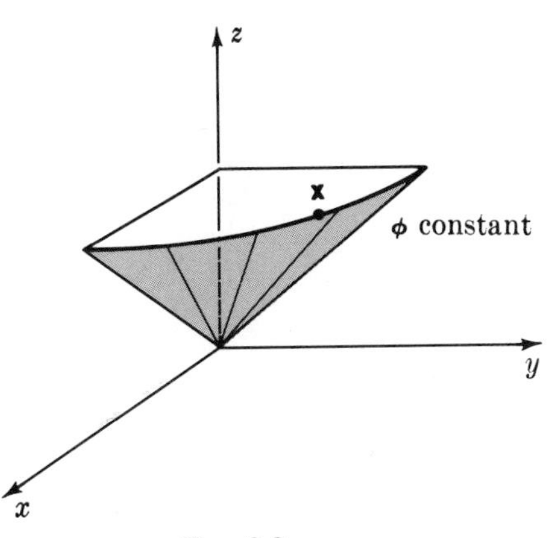
z
x
φ constant
y
x

Fig. 3.3

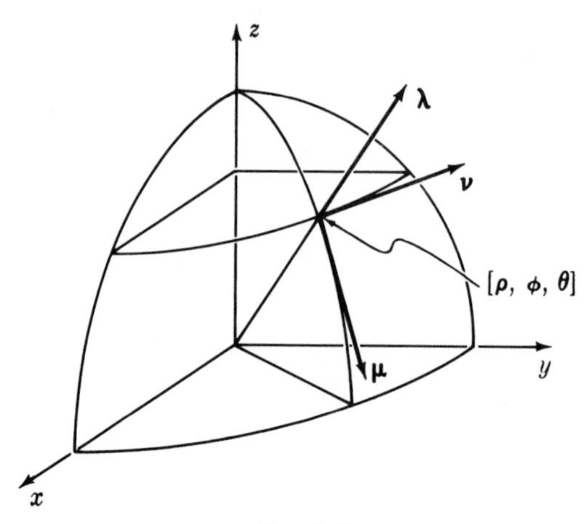

FIG. 3.4

and if only θ increases, then $d\mathbf{x} = \boldsymbol{\nu}\,\rho \sin\phi\, d\theta$. Conclusion:

$$\begin{cases} \boldsymbol{\lambda} = (\sin\phi\cos\theta,\ \sin\phi\sin\theta,\ \cos\phi) \\ \boldsymbol{\mu} = (\cos\phi\cos\theta,\ \cos\phi\sin\theta,\ -\sin\phi) \\ \boldsymbol{\nu} = (-\sin\theta,\ \cos\theta,\ 0), \end{cases}$$

$$d\mathbf{x} = \boldsymbol{\lambda}\, d\rho + \boldsymbol{\mu}\,\rho\, d\phi + \boldsymbol{\nu}\,\rho\sin\phi\, d\theta.$$

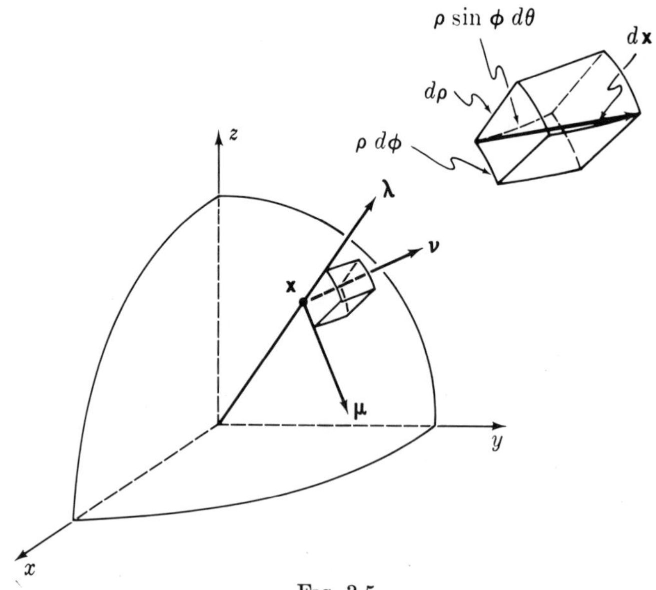

FIG. 3.5

Integrals

If a solid has central symmetry, it is often convenient to place the origin at the center of symmetry and use spherical coordinates $[\rho, \phi, \theta]$ for the computation of integrals.

The displacement of a point of space, in terms of the frame $\boldsymbol{\lambda}$, $\boldsymbol{\mu}$, $\boldsymbol{\nu}$ natural to spherical coordinates, is

$$\boxed{d\mathbf{x} = d\rho\, \boldsymbol{\lambda} + \rho\, d\phi\, \boldsymbol{\mu} + \rho \sin\phi\, d\theta\, \boldsymbol{\nu}.}$$

See Fig. 3.5. The element of volume of the small "rectangular" solid is the product of its sides:

$$\boxed{dV = (d\rho)(\rho\, d\phi)(\rho \sin\phi\, d\theta) = \rho^2 \sin\phi\, d\rho\, d\phi\, d\theta.}$$

EXAMPLE 3.1

Use spherical coordinates to find the volume of a sphere of radius a.

Solution:

$$V = \iiint \rho^2 \sin\phi\, d\rho\, d\phi\, d\theta$$

$$= \left(\int_0^a \rho^2\, d\rho\right)\left(\int_0^\pi \sin\phi\, d\phi\right)\left(\int_0^{2\pi} d\theta\right) = \left(\frac{a^3}{3}\right)(2)(2\pi).$$

Answer: $\dfrac{4}{3}\pi a^3.$

EXAMPLE 3.2

Find the volume of the portion of the unit sphere which lies in the right circular cone having its apex at the origin and making angle α with the positive z-axis.

Solution: The cone is specified by $0 \le \phi \le \alpha$, so the portion of the sphere is determined by $0 \le \theta \le 2\pi$, $0 \le \phi \le \alpha$, and $0 \le \rho \le 1$. See Fig. 3.6. Hence the volume is

$$\left(\int_0^{2\pi} d\theta\right)\left(\int_0^\alpha \sin\phi\, d\phi\right)\left(\int_0^1 \rho^2\, d\rho\right) = (2\pi)(1 - \cos\alpha)\left(\frac{1}{3}\right).$$

Answer: $\dfrac{2\pi}{3}(1 - \cos\alpha).$

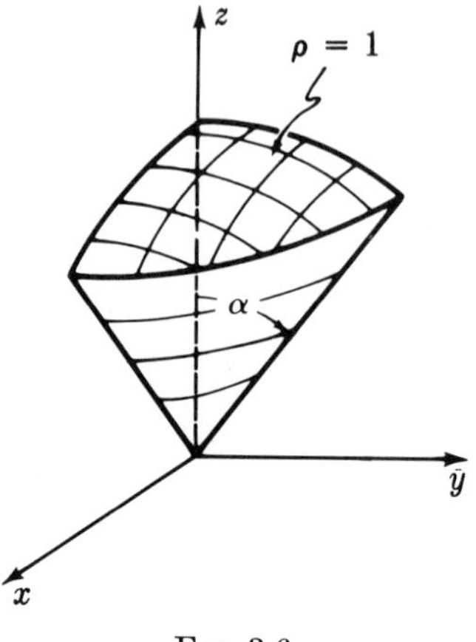

FIG. 3.6

REMARK: As a check, let $\alpha \longrightarrow \pi$. Then the volume should approach the volume of a sphere of radius 1. Does it?

EXAMPLE 3.3

A solid fills the region between concentric spheres of radii a and b, where $0 < a < b$. The density at each point is inversely proportional to its distance from the center. Find the total mass.

Solution: The solid is specified by $a \leq \rho \leq b$; the density is $\delta = k/\rho$. Hence

$$M = \iiint \frac{k}{\rho}\,\rho^2 \sin\phi \,d\rho\,d\phi\,d\theta$$

$$= k\left(\int_0^{2\pi} d\theta\right)\left(\int_0^{\pi} \sin\phi\,d\phi\right)\left(\int_a^b \rho\,d\rho\right) = (2\pi k)(2)\left(\frac{b^2 - a^2}{2}\right).$$

Answer: $2\pi k(b^2 - a^2)$.

REMARK: As $a \longrightarrow 0$, the solid tends to the whole sphere, with infinite density at the center. But $M \longrightarrow 2\pi k b^2$, which is finite.

EXAMPLE 3.4

A cylindrical hole of radius $\frac{1}{2}$ is bored through a sphere of radius 1. The surface of the hole passes through the center of the sphere. How much material is removed?

Solution: Center the sphere at $\mathbf{0}$ and let the cylinder (hole) be parallel to the z-axis, with axis through $(0, \frac{1}{2})$. See Fig. 3.7. By symmetry the volume

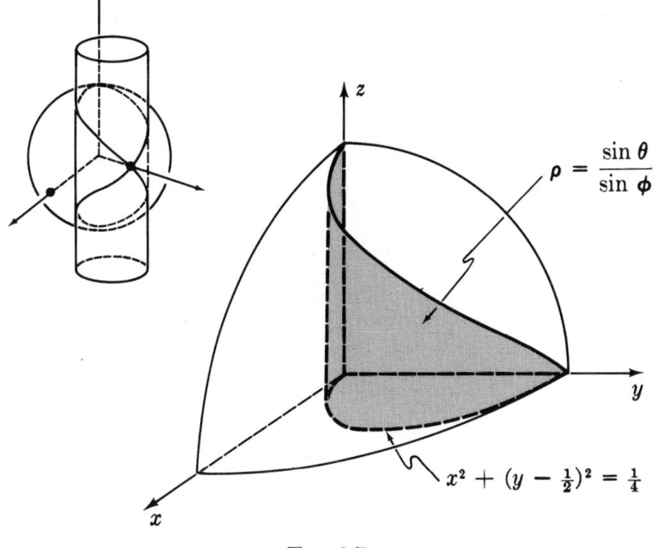

FIG. 3.7

is four times that in the first octant. The equation of the cylindrical surface is $x^2 + (y - \frac{1}{2})^2 = \frac{1}{4}$, or

$$x^2 + y^2 = y.$$

But in spherical coordinates, $x^2 + y^2 = \rho^2 \sin^2 \phi$ and $y = \rho \sin \phi \sin \theta$. Hence the equation of the cylinder is $\rho^2 \sin^2 \phi = \rho \sin \phi \sin \theta$, or

$$\rho = \frac{\sin \theta}{\sin \phi}.$$

The cylinder intersects the sphere in the curve (Fig. 3.8)

$$\rho = 1, \qquad \frac{\sin \theta}{\sin \phi} = 1.$$

In the first octant this is the curve

$$\rho = 1, \qquad \phi = \theta.$$

The first-octant portion of the volume naturally splits into two parts, separated by the cone $\phi = \theta$ of segments joining **0** to the curve of intersection. In the lower part, each radius from the origin ends at the cylinder, so the limits are

$$0 \le \theta \le \frac{\pi}{2}, \qquad \theta \le \phi \le \frac{\pi}{2}, \qquad 0 \le \rho \le \frac{\sin \theta}{\sin \phi}.$$

In the upper part, each radius from the origin ends on the sphere, so the

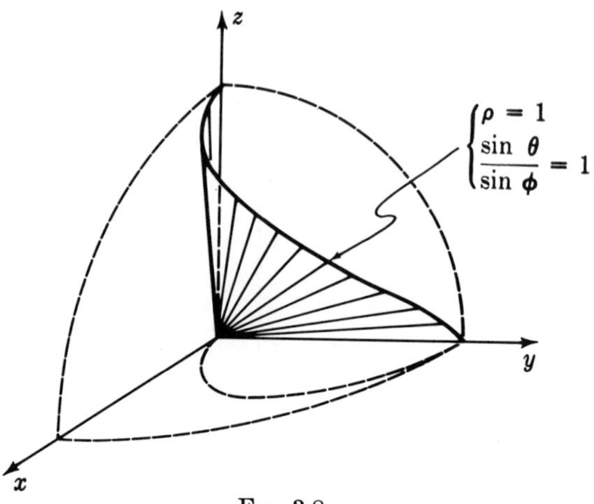

FIG. 3.8

limits are

$$0 \le \theta \le \frac{\pi}{2}, \qquad 0 \le \phi \le \theta, \qquad 0 \le \rho \le 1.$$

Thus $V = 4(V_1 + V_2)$, where V_1 is the volume of the lower part and V_2 that of the upper. Now

$$V_1 = \int_0^{\pi/2} d\theta \int_\theta^{\pi/2} \sin\phi\, d\phi \int_0^{\sin\theta/\sin\phi} \rho^2\, d\rho = \frac{1}{3}\int_0^{\pi/2} \sin^3\theta\, d\theta \int_\theta^{\pi/2} \frac{d\phi}{\sin^2\phi}$$

$$= \frac{1}{3}\int_0^{\pi/2} \sin^3\theta \cot\theta\, d\theta = \frac{1}{3}\int_0^{\pi/2} \sin^2\theta\cos\theta\, d\theta = \frac{1}{9},$$

and

$$V_2 = \int_0^{\pi/2} d\theta \int_0^{\theta} \sin\phi\, d\phi \int_0^1 \rho^2\, d\rho = \frac{1}{3}\int_0^{\pi/2} (1 - \cos\theta)\, d\theta = \frac{1}{3}\left(\frac{\pi}{2} - 1\right).$$

Therefore

$$V = 4\left(\frac{1}{9} + \frac{\pi}{6} - \frac{1}{3}\right) = \frac{2\pi}{3} - \frac{8}{9}.$$

Answer: $\frac{2\pi}{3} - \frac{8}{9}.$

Spherical Area

Suppose a point moves on the surface of the sphere $\rho = a$. Then $d\rho = 0$, so the formula for displacement specializes to

$$d\mathbf{x} = a\, d\phi\, \boldsymbol{\mu} + a\sin\phi\, d\theta\, \boldsymbol{\nu}.$$

The area of the small "rectangular" region on the surface of the sphere corresponding to changes $d\phi$ and $d\theta$ is

$$dA = (a\,d\phi)(a \sin\phi\,d\theta) = a^2 \sin\phi\,d\phi\,d\theta.$$

See Fig. 3.9. The integral of this expression over a region is the area of that region.

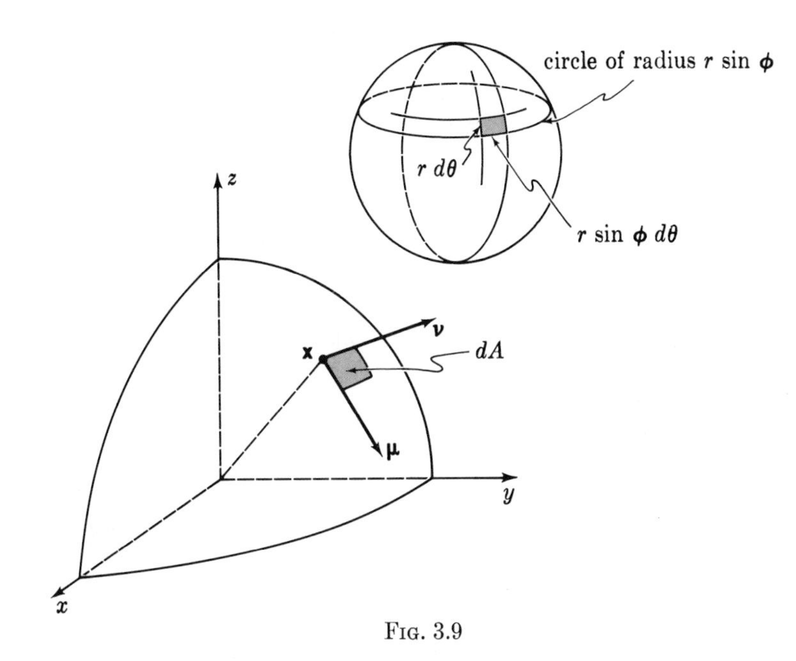

FIG. 3.9

EXAMPLE 3.5

Find the area of the polar cap, all points of co-latitude α or less on the unit sphere.

Solution: See Fig. 3.6. The region is defined on the sphere $\rho = 1$ by $0 \leq \phi \leq \alpha$, hence

$$A = \int_0^{2\pi} \left(\int_0^{\alpha} \sin\phi\,d\phi \right) d\theta = 2\pi(1 - \cos\alpha).$$

Answer: $2\pi(1 - \cos\alpha)$.

REMARK: Suppose **S** is a region on the unit sphere. The totality of infinite rays starting at **0** and passing through points of **S** is a cone which is called a **solid angle** (Fig. 3.10). A solid angle is measured by the area of the base region **S**. Since **S** lies on the unit sphere, its area is measured in square radians, a dimensionless unit. Thus the solid angle of the whole sphere has

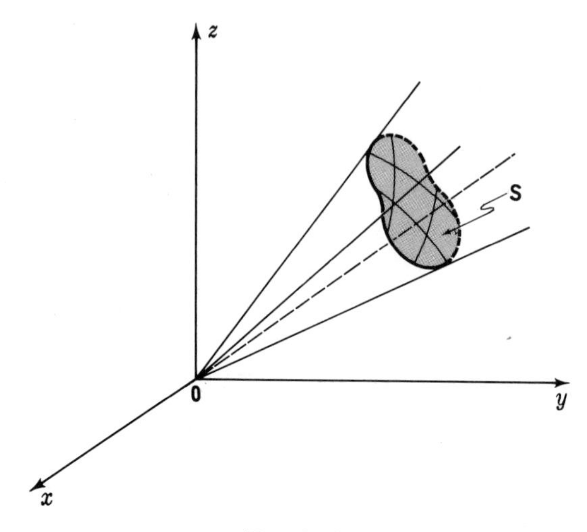

FIG. 3.10

measure 4π rad², or simply 4π. The solid angle determined by the first octant has measure $4\pi/8 = \pi/2$. The solid angle of the polar cap in Example 3.5 has measure $2\pi(1 - \cos\alpha)$. Incidentally, in the metric (SI) system, the unit for solid angles is the **steradian** (sr).

EXERCISES

1. Convert to rectangular coordinates: $[1, 3\pi/4, 3\pi/4]$, $[3, \pi/2, 5\pi/4]$, $[2, 2\pi/3, \pi/3]$
2. Convert to spherical coordinates:
$$(1, 1, 1),\ (1, -1, 1),\ (-1, -1, -1),\ (-1, 1, -1).$$
3. If $\mathbf{x}$ has spherical coordinates $[\rho, \phi, \theta]$, find the spherical coordinates of $-\mathbf{x}$ and of $3\mathbf{x}$.
4. Suppose a space curve is given by $\rho = \rho(t)$, $\phi = \phi(t)$, $\theta = \theta(t)$. Show that its arc length satisfies
$$\dot{s}^2 = \dot{\rho}^2 + \rho^2\dot{\phi}^2 + \rho^2(\sin^2\phi)\dot{\theta}^2.$$
5. (cont.) A **rhumb line** on a sphere of radius a is a curve that intersects each meridian at the same angle α. (Follow a constant compass setting.) Find the length of a rhumb line from the equator to the north pole.

6*. Let $f = f(r, \phi, \theta)$. Prove
$$\operatorname{grad} f = \frac{\partial f}{\partial \rho}\boldsymbol{\lambda} + \frac{1}{\rho}\frac{\partial f}{\partial \phi}\boldsymbol{\mu} + \frac{1}{\rho\sin\phi}\frac{\partial f}{\partial \theta}\boldsymbol{\nu}.$$

7. Prove
$$\begin{bmatrix} d\boldsymbol{\lambda} \\ d\boldsymbol{\mu} \\ d\boldsymbol{\nu} \end{bmatrix} = \begin{bmatrix} 0 & d\phi & \sin\phi\, d\theta \\ -d\phi & 0 & \cos\phi\, d\theta \\ -\sin\phi\, d\theta & -\cos\phi\, d\theta & 0 \end{bmatrix} \begin{bmatrix} \boldsymbol{\lambda} \\ \boldsymbol{\mu} \\ \boldsymbol{\nu} \end{bmatrix}.$$

8. Let $\mathbf{v}$ be a radial vector field of the form $\mathbf{v} = g(\rho)\boldsymbol{\lambda}$. Prove that $\mathbf{v}$ is a gradient field. [*Hint:* Use Ex. 6.]

9. Suppose $f[\rho, \phi, \theta]$ is homogeneous of degree n with respect to *rectangular* coordinates, that is, $f(t\mathbf{x}) = t^n f(\mathbf{x})$ for all $t > 0$. Prove $f[\rho, \phi, \theta] = \rho^n g(\phi, \theta)$, where $g(\phi, \theta) = f[1, \phi, \theta]$.

10. Find $\iiint \rho^n \, dx\, dy\, dz$ over the sphere $\rho \leq a$, where $n \geq 0$.

11. Find $\iiint (1 - \rho)^n \, dx\, dy\, dz$ over the sphere $\rho \leq 1$, where $n \geq 0$.

12. Find $\iiint z \, dx\, dy\, dz$ over the first octant portion of the sphere $\rho \leq a$.

13. Find $\iiint \rho^{-2} \, dx\, dy\, dz$ over the domain $z \geq 0$, $a \leq \rho \leq b$, where $0 < a < b$.

14. Find $\iiint z \, dx\, dy\, dz$ over the domain $1 \leq z$, $\rho \leq \frac{2}{3}\sqrt{3}$.

15. Two parallel planes at distance h intersect a sphere of radius a. Find the surface area of the spherical zone between the planes.

16. In Example 3.4, how much of the surface area of the sphere is removed?

Additional Volume Problems

Use any method you please to set up the integral. You may not be able to evaluate it, but at least try to reduce the answer to a simple integral, not a double or triple one.

17. Find the volume of the (solid) right circular torus obtained by revolving the circle $(y - A)^2 + z^2 = a^2$, $x = 0$ about the z-axis. Assume $0 < a \leq A$.

18. Suppose in Ex. 17 that $0 < A \leq a$. Let the portion of the circle to the right of the z-axis generate volume V_1 and the portion to the left generate volume V_2. Find $V_1 - V_2$.

19*. A sphere of radius a touches the sides of a cone of semi-apex angle α. See Fig. 3.11. Find the volume of the portion of the cone above the sphere.

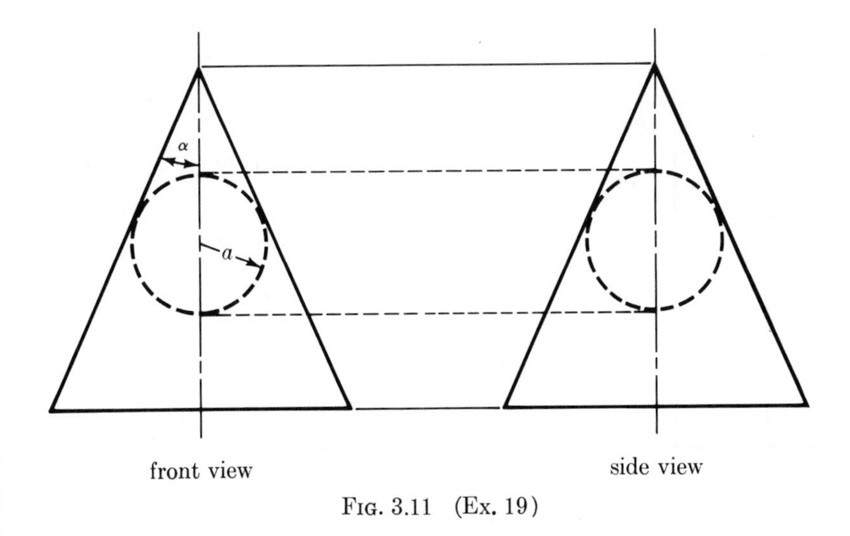

FIG. 3.11 (Ex. 19)

20*. A circular hole is bored through a right circular cone, dimensions as indicated in Fig. 3.12. The axes are perpendicular and the hole just fits. How much material is removed?

21*. (cont.) How much of the cone remains above the hole?

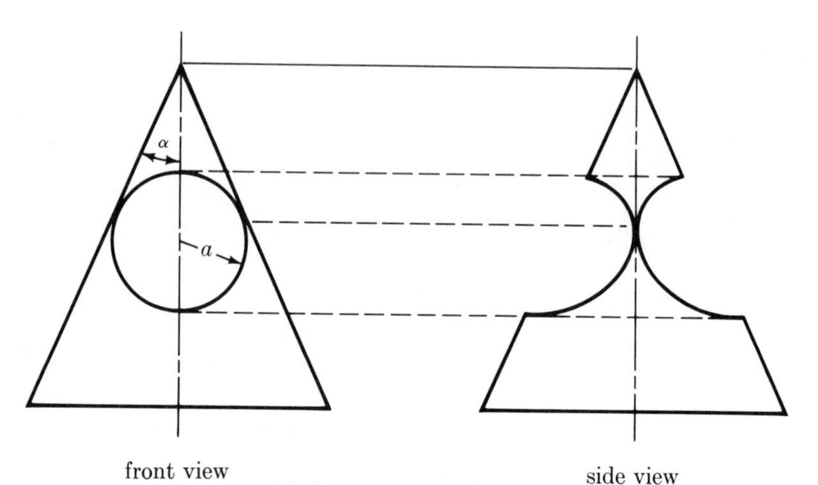

FIG. 3.12 (Exs. 20, 21)

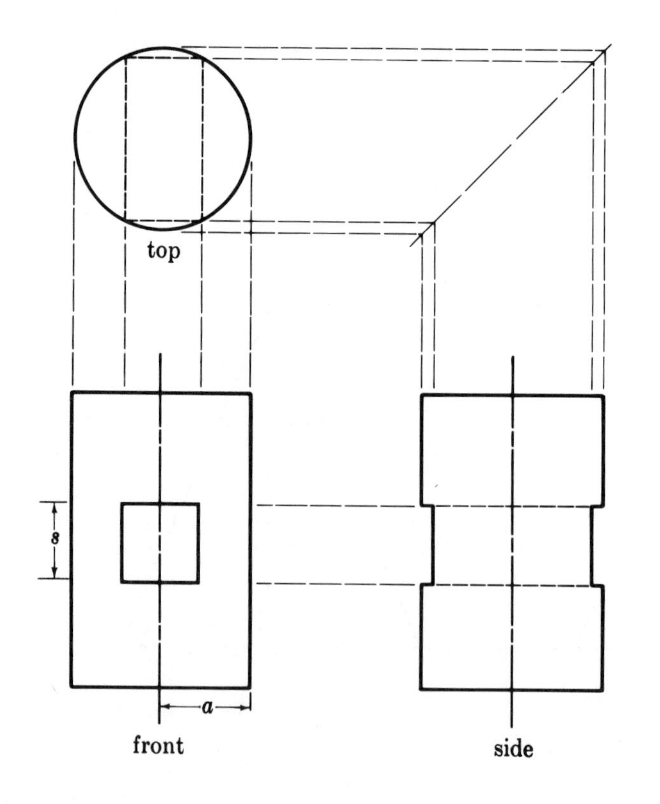

FIG. 3.13 (Ex. 25)

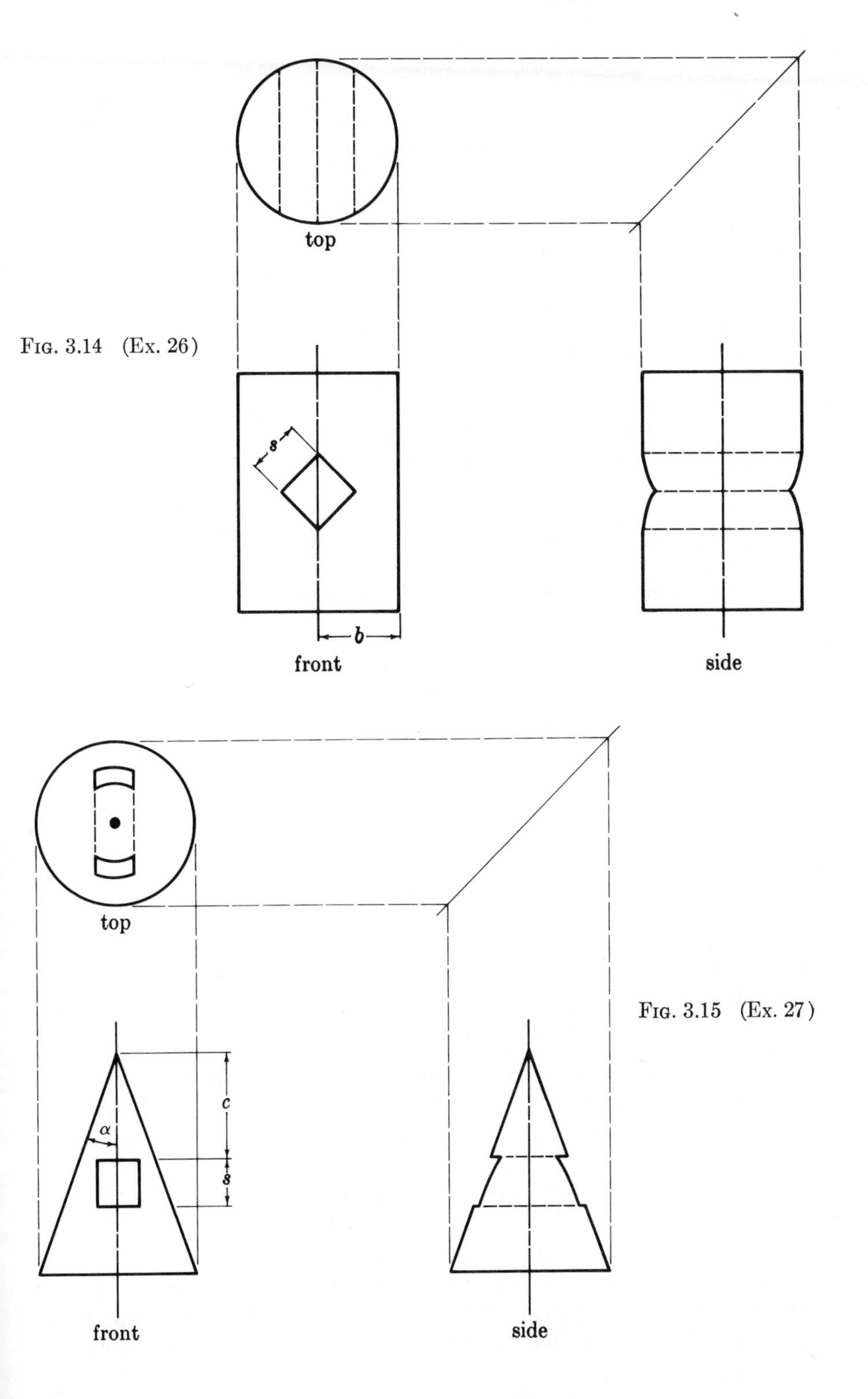

FIG. 3.14 (Ex. 26)

FIG. 3.15 (Ex. 27)

22*. A cylindrical hole of radius a is bored through a solid cylinder of radius $2a$; the hole is perpendicular to the solid cylinder and just touches a generator. Find the volume removed.

23. A circular hole of radius a is bored through a solid right circular cylinder of radius b. Assume the axis of the hole intersects the axis of the solid at a right angle and that $a \leq b$. Show that the volume of material removed is

$$8 \int_0^a \sqrt{a^2 - x^2}\sqrt{b^2 - x^2}\, dx = 8a^2 \int_0^{\pi/2} (\sin^2 \theta)\sqrt{b^2 - a^2 \cos^2 \theta}\, d\theta.$$

24. Do Ex. 23 assuming that the axes meet obliquely at angle α. Set up, but do not evaluate the integral. Can you evaluate it when $a = b$?

25. A square hole is bored through a right circular cylinder (Fig. 3.13). How much material is removed?

26*. A square hole is bored through a right circular cylinder (Fig. 3.14). How much material is removed?

27*. A square hole is bored through a right circular cone (Fig. 3.15). How much material is removed? (Do not evaluate the integral.)

4. CENTER OF GRAVITY

Suppose a solid $\mathbf{D}$ has density $\delta(\mathbf{x})$ at each point $\mathbf{x}$. Then its mass is

$$M = \iiint_{\mathbf{D}} \delta(\mathbf{x})\, dx\, dy\, dz.$$

Define its **moment about the origin**

$$\mathbf{m} = \iiint_{\mathbf{D}} \delta(\mathbf{x})\, \mathbf{x}\, dx\, dy\, dz,$$

and its **center of gravity**

$$\bar{\mathbf{x}} = \frac{1}{M}\mathbf{m}.$$

Note that $\mathbf{m}$ and $\bar{\mathbf{x}}$ are vectors. It is often convenient to express them in terms of components:

$$\mathbf{m} = (m_x, m_y, m_z), \qquad \bar{\mathbf{x}} = (\bar{x}, \bar{y}, \bar{z}).$$

The center of gravity may be considered as a sort of weighted average of the points of the solid. Recall in this connection that the center of gravity of a system of point-masses $M_1, \cdots, M_n$ located at $\mathbf{x}_1, \cdots, \mathbf{x}_n$ is

$$\bar{\mathbf{x}} = \frac{1}{M}(M_1\mathbf{x}_1 + M_2\mathbf{x}_2 + \cdots + M_n\mathbf{x}_n),$$

where $M = M_1 + \cdots + M_n$.

If a solid is symmetric in a coordinate plane, then the center of gravity lies on that coordinate plane. For example, suppose $\mathbf{D}$ is **symmetric** in the x, y-plane. This means that whenever a point (x, y, z) is in the solid, then

$(x, y, -z)$ is in the solid, *and* $\delta(x, y, z) = \delta(x, y, -z)$. The contribution to m_z at (x, y, z) is

$$\delta(x, y, z)\, z\, dx\, dy\, dz;$$

it is cancelled by the contribution

$$\delta(x, y, -z)(-z)\, dx\, dy\, dz = -\delta(x, y, z)\, z\, dx\, dy\, dz$$

at $(x, y, -z)$. Hence $m_z = 0$ and $\bar{z} = 0$.

Similarly, if **D** is symmetric in a coordinate axis, then the center of gravity lies on that axis. Finally, if **D** is symmetric in the origin, then $\bar{\mathbf{x}} = \mathbf{0}$.

To compute the center of gravity of a solid, exploit any symmetry it has by choosing an appropriate coordinate system. Of course, express the element of volume $dV = dx\, dy\, dz$ in the coordinate system chosen.

EXAMPLE 4.1

Find the center of gravity of a uniform hemisphere of radius a and mass M.

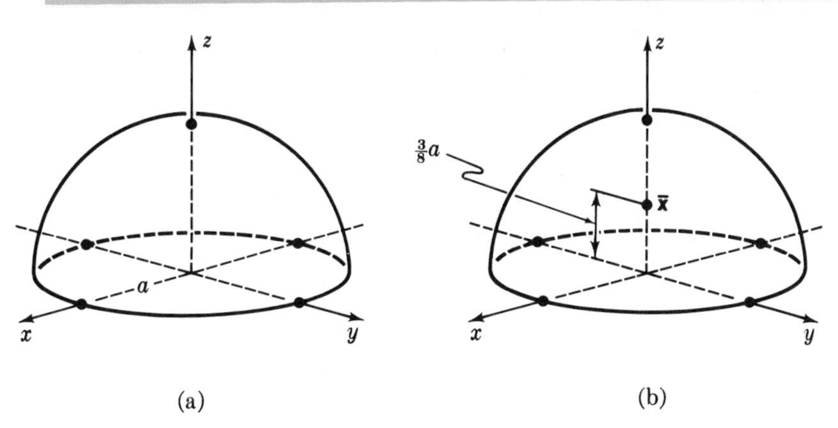

FIG. 4.1

Solution: To exploit symmetry, choose spherical coordinates, with the hemisphere defined by $\rho \leq a$ and $0 \leq \phi \leq \pi/2$. See Fig. 4.1a. The density δ is constant, so the mass is δ times the volume:

$$M = \frac{1}{2}\left(\frac{4}{3}\pi a^3\right)\delta = \frac{2}{3}\pi a^3 \delta.$$

Because the hemisphere is symmetric in the y, z-plane, $m_x = 0$. Likewise $m_y = 0$; only m_z requires computation:

$$m_z = \iiint \delta z\, dx\, dy\, dz = \iiint \delta(\rho\cos\phi)\rho^2 \sin\phi\, d\rho\, d\phi\, d\theta$$

$$= \delta \int_0^{2\pi} d\theta \int_0^{\pi/2} \cos\phi \sin\phi\, d\phi \int_0^a \rho^3\, d\rho = \delta(2\pi)\left(\frac{1}{2}\right)\left(\frac{a^4}{4}\right) = \frac{\pi}{4}\delta a^4.$$

Hence

$$\bar{z} = \frac{m_z}{M} = \frac{\pi\delta a^4/4}{2\pi\delta a^3/3} = \frac{3}{8}a.$$

Answer: The center of gravity lies on the axis of the hemisphere, $\frac{3}{8}$ of the distance from the equatorial plane to the pole (Fig. 4.1b).

REMARK: The answer is independent of δ. For uniform solids in general, the constant δ cancels when you divide $\mathbf{m}$ by M, so $\bar{\mathbf{x}}$ is a purely geometric quantity. It is then called the **center of gravity** or **centroid** of the geometric region (rather than the material solid). For uniform solids, from now on take $\delta = 1$ and $M = V$, the volume.

EXAMPLE 4.2

Find the center of gravity of a uniform right circular cone.

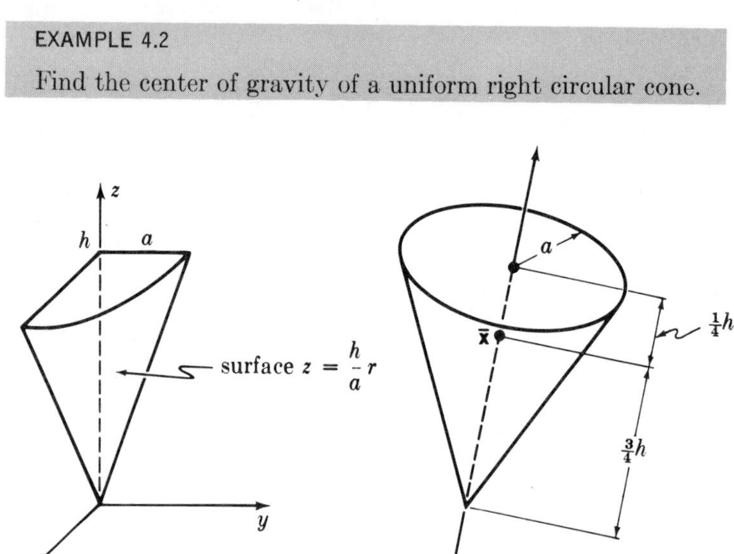

FIG. 4.2

Solution: Choose cylindrical coordinates with the apex of the cone at $\mathbf{0}$ and the base of radius a centered at $(0, 0, h)$. See Fig. 4.2a. The lateral surface of the cone is

$$z = \frac{h}{a}r, \qquad \text{that is,} \qquad r = \frac{a}{h}z,$$

and the volume of the cone is

$$V = \frac{1}{3}\pi a^2 h.$$

By symmetry $m_x = m_y = 0$. Compute m_z:

$$m_z = \iiint zr\,dr\,d\theta\,dz = \int_0^{2\pi}\left[\int_0^h \left(\int_0^{az/h} r\,dr\right) z\,dz\right] d\theta$$

$$= \int_0^{2\pi} d\theta \int_0^h \frac{1}{2}\left(\frac{a}{h}z\right)^2 z\,dz = (2\pi)\left(\frac{a^2}{2h^2}\right)\int_0^h z^3\,dz = \frac{\pi a^2}{h^2}\cdot\frac{h^4}{4} = \frac{\pi}{4}a^2h^2.$$

Hence

$$\bar{z} = \frac{m_z}{V} = \frac{\pi a^2h^2/4}{\pi a^2h/3} = \frac{3}{4}h.$$

Answer: The center of gravity is on the cone's axis, $\frac{1}{4}$ of the distance from the base to the apex (Fig. 4.2b).

EXAMPLE 4.3

The solid $0 \le x \le 1$, $0 \le y \le 2$, $0 \le z \le 3$ has density xyz gm/cm³. Find its center of gravity.

Solution:

$$M = \iiint xyz\,dx\,dy\,dz = \int_0^1 x\,dx \int_0^2 y\,dy \int_0^3 z\,dz = \frac{1}{2}\cdot\frac{4}{2}\cdot\frac{9}{2} = \frac{9}{2}\text{ gm}.$$

$$\mathbf{m} = \iiint (x, y, z)\,xyz\,dx\,dy\,dz$$

$$= \left(\iiint x^2yz\,dx\,dy\,dz,\quad \iiint xy^2z\,dx\,dy\,dz,\quad \iiint xyz^2\,dx\,dy\,dz\right)$$

$$= \left(\int_0^1 x^2\,dx \int_0^2 y\,dy \int_0^3 z\,dz,\quad \int_0^1 x\,dx \int_0^2 y^2\,dy \int_0^3 z\,dz,\quad \int_0^1 x\,dx \int_0^2 y\,dy \int_0^3 z^2\,dz\right)$$

$$= \left(\frac{1}{3}\cdot\frac{4}{2}\cdot\frac{9}{2},\quad \frac{1}{2}\cdot\frac{8}{3}\cdot\frac{9}{2},\quad \frac{1}{2}\cdot\frac{4}{2}\cdot\frac{27}{3}\right) = \frac{4\cdot 9}{3\cdot 2\cdot 2}(1, 2, 3) = 3(1, 2, 3)\text{ gm-cm}.$$

Hence

$$\bar{\mathbf{x}} = \frac{2}{9}(3)(1, 2, 3) = \left(\frac{2}{3}, \frac{4}{3}, 2\right)\text{ cm}.$$

Answer: $\bar{\mathbf{x}} = \left(\frac{2}{3}, \frac{4}{3}, 2\right)$ cm.

Plane Regions

The definitions of moment and center of gravity for solids can be modified in an obvious way to fit plane regions with a mass distribution.

EXAMPLE 4.4

Find the center of gravity of a uniform semicircular disk.

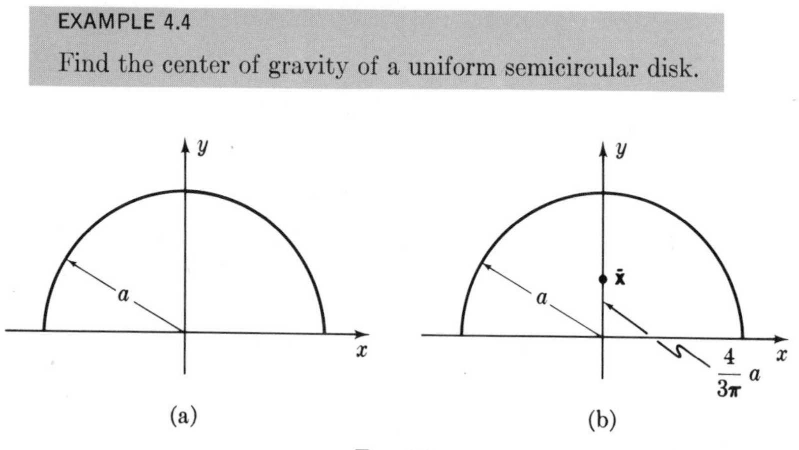

FIG. 4.3

Solution: Choose polar coordinates, taking the disk in the position $0 \le r \le a$, $0 \le \theta \le \pi$. See Fig. 4.3a. Take $\delta = 1$; the mass equals the area

$$A = \frac{1}{2}(\pi a^2).$$

By symmetry $m_x = 0$. Compute m_y:

$$m_y = \iint y\, r\, dr\, d\theta = \iint r^2 \sin\theta\, dr\, d\theta = \int_0^\pi \sin\theta\, d\theta \int_0^a r^2\, dr = \frac{2}{3}a^3.$$

Hence

$$\bar{\mathbf{x}} = \frac{1}{A}(m_x, m_y) = \frac{2}{\pi a^2}\left(0, \frac{2}{3}a^3\right) = \left(0, \frac{4}{3\pi}a\right).$$

Answer: $\bar{\mathbf{x}} = \left(0, \frac{4}{3\pi}a\right)$. See Fig. 4.3b.

There is a useful connection between the centers of gravity of plane regions and volumes of revolution.

First Pappus Theorem Suppose a region **D** in the x, y-plane, to the right of the y-axis, is revolved about the y-axis. Then the volume of the resulting solid is

$$V = 2\pi\bar{x}A,$$

where A is the area of the plane region **D** and $\bar{x}$ is the x-coordinate of its center of gravity (Fig. 4.4).

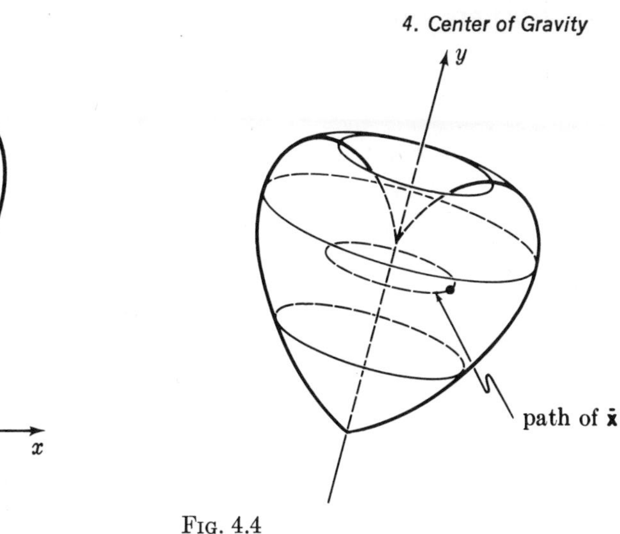

FIG. 4.4

In words, the volume is the area times the length of the circle traced by the center of gravity. Proof: A small portion $dx\,dy$ at $\mathbf{x}$ revolves into a thin ring of volume

$$dV = 2\pi x\,dx\,dy.$$

Hence

$$V = \iint_{\mathbf{D}} 2\pi x\,dx\,dy = 2\pi m_x.$$

But $m_x = \bar{x}A$, hence $V = 2\pi\bar{x}A$.

Wires

A non-homogeneous wire is described by its position, a space curve $\mathbf{x} = \mathbf{x}(s)$ where $a \leq s \leq b$, and its density $\delta = \delta(s)$. (Here s denotes arc length.) Its mass is

$$M = \int_a^b \delta(s)\,ds,$$

its moment is

$$\mathbf{m} = \int_a^b \mathbf{x}(s)\,\delta(s)\,ds,$$

and its center of gravity is

$$\bar{\mathbf{x}} = \frac{1}{M}\mathbf{m}.$$

If the wire is uniform, then $\delta(s)$ is a constant. In this case, the center of gravity is independent of δ, hence it is a property of the curve $\mathbf{x} = \mathbf{x}(s)$ alone; you can take $\delta = 1$ and replace M by L, the length.

EXAMPLE 4.5

Find the center of gravity of the uniform semicircle $r = a$, $y \geq 0$.

Solution: The length is $L = \pi a$. The moment is

$$\mathbf{m} = \int \mathbf{x}\, ds = \int_0^{\pi} (a\cos\theta,\ a\sin\theta)\, a\, d\theta$$

$$= a^2 \int_0^{\pi} (\cos\theta,\ \sin\theta)\, d\theta = a^2(0,\ 2).$$

Hence

$$\bar{\mathbf{x}} = \frac{1}{L}\mathbf{m} = \frac{1}{\pi a}a^2(0,\ 2) = \frac{a}{\pi}(0,\ 2).$$

Answer: $\left(0,\ \dfrac{2}{\pi}a\right).$

Suppose a plane curve is revolved about an axis in its plane, generating a surface of revolution. There is a useful relation between the center of gravity of the curve and the area of the surface.

Second Pappus Theorem Suppose a curve in the x, y-plane to the right of the y-axis is revolved about the y-axis. Then the area of the resulting surface is

$$A = 2\pi\bar{x}L,$$

where L is the length of the curve and $\bar{x}$ is the x-coordinate of the center of gravity (Fig. 4.5).

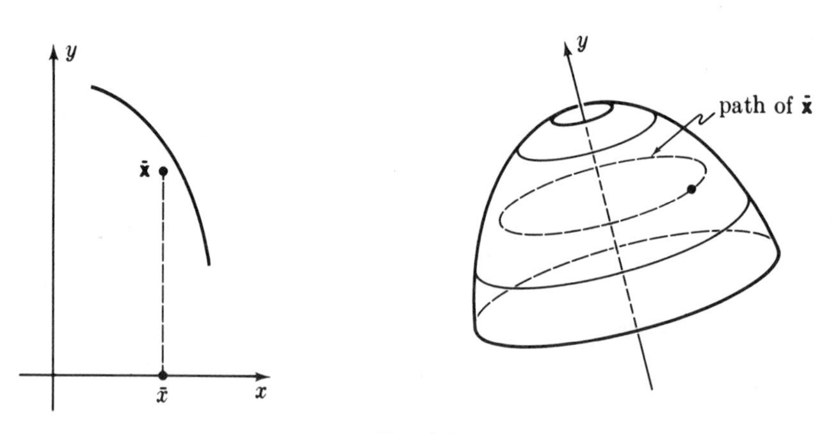

FIG. 4.5

In words, the area is the length of the curve times the length of the circle traced by the center of gravity. Proof: A short segment of length ds of the curve at the point $\mathbf{x}(s)$ revolves into the frustum of a cone with lateral area

$dA = 2\pi x\, ds$. See Fig. 4.6. Hence

$$A = \int_a^b 2\pi x\, ds = 2\pi \int_a^b x\, ds = 2\pi m_x = 2\pi \bar{x} L.$$

There is a useful aid to problem solving which we state for solids. With obvious modifications, it applies to wires or laminas.

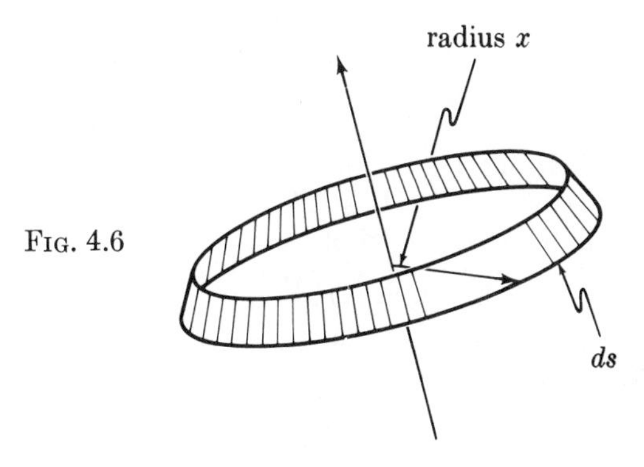

FIG. 4.6

> **Addition Law** Suppose a solid $\mathbf{D}$ of mass M and center of gravity $\bar{\mathbf{x}}$ is cut into two pieces $\mathbf{D}_0$ and $\mathbf{D}_1$, of masses M_0 and M_1, and centers of gravity $\bar{\mathbf{x}}_0$ and $\bar{\mathbf{x}}_1$. Then
>
> $$M = M_0 + M_1, \qquad \bar{\mathbf{x}} = \frac{1}{M}(M_0\bar{\mathbf{x}}_0 + M_1\bar{\mathbf{x}}_1).$$

The first formula is obvious. The second is simply a decomposition of the moment integral:

$$M\bar{\mathbf{x}} = \iiint_{\mathbf{D}} \delta(\mathbf{x})\,\mathbf{x}\, dV = \iiint_{\mathbf{D}_0} + \iiint_{\mathbf{D}_1} = M_0\,\bar{\mathbf{x}}_0 + M_1\,\bar{\mathbf{x}}_1.$$

EXERCISES

1. Find the center of gravity of the first octant portion of the uniform sphere $\rho \leq a$.
2. Suppose the density of the hemisphere $\rho \leq a$, $z \geq 0$ is $\delta = a - \rho$. Find the center of gravity.
3. Find the center of gravity of the uniform spherical cone $\rho \leq a$, $0 \leq \phi \leq \alpha$.
4. Find the center of gravity of the uniform hemispherical shell $a \leq \rho \leq b$, $z \geq 0$.
5. Find the center of gravity of a uniform circular wedge (sector).
6. The plane region bounded by $z = 1$ and the parabola $z = y^2$ is revolved about the z-axis. Find the center of gravity of the resulting uniform solid.
7. Find the center of gravity of a uniform wire in the shape of a quarter circle.

8. Find the center of gravity of the uniform spiral $\mathbf{x}(t) = (a\cos t, a\sin t, bt)$, $0 \leq t \leq t_0$.
9. A copper wire in the shape of a semicircle of radius 100 cm is steadily tapered from 0.1 to 0.5 cm in diameter. Find its center of gravity.
10. Verify the First Pappus Theorem for a semicircle revolved about its diameter.
11. Use the First Pappus Theorem to find the volume of a right circular torus.
12. Verify the First Pappus Theorem for a right triangle revolved about a leg.
13. Use the Second Pappus Theorem to find the surface area of a right circular torus.
14. Use the Second Pappus Theorem to obtain another solution of Example 4.5.
15. Find the center of gravity of the uniform spherical cap (surface) $\rho = a, \quad 0 \leq \phi \leq \alpha$.
16. Find the center of gravity of the uniform spherical cap (solid) $\rho \leq a, \quad a - h \leq z \leq a$.
17. Suppose a solid of density $\delta(\mathbf{x})$ is acted upon by a uniform (constant) gravitational field $\mathbf{f}$ so the force on a small portion at $\mathbf{x}$ is $[\delta(\mathbf{x})\, dV]\mathbf{f}$. Show that the solid is in equilibrium if a single force $-M\mathbf{f}$ is applied at $\bar{\mathbf{x}}$.
18. Find the center of gravity of the uniform triangle with vertices $\mathbf{a}$, $\mathbf{b}$, $\mathbf{c}$.
19. Find the center of gravity of the uniform tetrahedron with vertices $\mathbf{a}$, $\mathbf{b}$, $\mathbf{c}$, $\mathbf{d}$.

5. MOMENTS OF INERTIA

Recall that the kinetic energy of a moving particle of mass m and speed v is $K = \frac{1}{2}mv^2$. The kinetic energy of a system of moving particles is the sum of their individual kinetic energies. To define the kinetic energy of a moving solid body $\mathbf{D}$, decompose the body into elementary masses $\delta(\mathbf{x})\, dV$, where

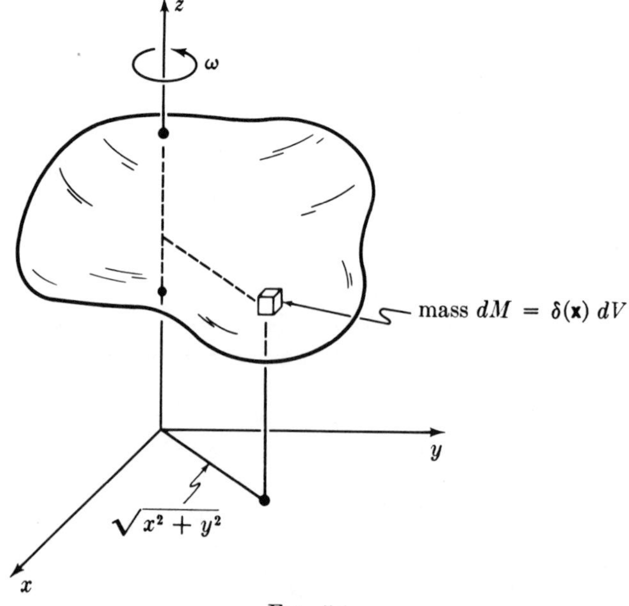

FIG. 5.1

$\delta(\mathbf{x})$ is the density at $\mathbf{x}$, and form the integral

$$K = \frac{1}{2} \iiint_{\mathbf{D}} \delta(\mathbf{x})\, |\mathbf{v}(\mathbf{x}, t)|^2\, dV.$$

Here $\mathbf{v}(\mathbf{x}, t)$ is the velocity of the point $\mathbf{x}$ of the body at time t. Thus K varies with time.

We shall compute the kinetic energy of a rigid body rotating with angular velocity $\boldsymbol{\omega}$ about an axis through the origin.

First consider rotation about the z-axis with angular speed ω, so the angular velocity is $\boldsymbol{\omega} = \omega\mathbf{k}$. See Fig. 5.1. An elementary volume $dV = dx\, dy\, dz$ at $\mathbf{x}$ has mass $dM = \delta(\mathbf{x})\, dV$ and speed $\omega\sqrt{x^2 + y^2}$, since $\sqrt{x^2 + y^2}$ is the distance from $\mathbf{x}$ to the z-axis. Hence its kinetic energy is

$$dK = \frac{1}{2}\,(\omega\sqrt{x^2 + y^2})^2\, \delta(\mathbf{x})\, dV = \frac{1}{2}\,\omega^2\, \delta(\mathbf{x})\,(x^2 + y^2)\, dV.$$

To find the kinetic energy of the entire solid $\mathbf{D}$, integrate, obtaining

$$K = \frac{1}{2}\, I_{zz}\,\omega^2,$$

where

$$I_{zz} = \iiint_{\mathbf{D}} \delta(\mathbf{x})\,(x^2 + y^2)\, dV$$

is the **moment of inertia** of $\mathbf{D}$ about the z-axis.

Similarly define

$$I_{xx} = \iiint_{\mathbf{D}} \delta(\mathbf{x})\,(y^2 + z^2)\, dV \qquad \text{and} \qquad I_{yy} = \iiint_{\mathbf{D}} \delta(\mathbf{x})\,(z^2 + x^2)\, dV.$$

From these moments of inertia one can compute the kinetic energy, provided the body rotates about one of the coordinate axes:

$$K = \frac{1}{2}\, I_{xx}\omega^2 \qquad \text{(rotation about the } x\text{-axis)},$$

$$K = \frac{1}{2}\, I_{yy}\omega^2 \qquad \text{(rotation about the } y\text{-axis)}.$$

Products of Inertia

For rotation about more general axes, however, three other quantities called **products of inertia** (also **mixed moments of inertia**) are needed:

$$I_{yz} = I_{zy} = -\iiint_{\mathbf{D}} \delta(\mathbf{x})yz\, dV, \qquad I_{zx} = I_{xz} = -\iint_{\mathbf{D}} \delta(\mathbf{x})zx\, dV,$$

$$I_{xy} = I_{yx} = -\iiint_{\mathbf{D}} \delta(\mathbf{x})xy\, dV.$$

The three moments and three products of inertia together form the symmetric **matrix of inertia**

$$I = \begin{bmatrix} I_{xx} & I_{xy} & I_{xz} \\ I_{yx} & I_{yy} & I_{yz} \\ I_{zx} & I_{zy} & I_{zz} \end{bmatrix}.$$

The quadratic form associated with this matrix gives the kinetic energy of a solid rotating about an axis through **0**.

Suppose **D** rotates about an axis through **0** with angular velocity $\boldsymbol{\omega} = (\omega_x, \omega_y, \omega_z)$. Then its kinetic energy is

$$K = \frac{1}{2}\boldsymbol{\omega} I \boldsymbol{\omega}' = \frac{1}{2}(\omega_x, \omega_y, \omega_z) \begin{bmatrix} I_{xx} & I_{xy} & I_{xz} \\ I_{yx} & I_{yy} & I_{yz} \\ I_{zx} & I_{zy} & I_{zz} \end{bmatrix} \begin{bmatrix} \omega_x \\ \omega_y \\ \omega_z \end{bmatrix}.$$

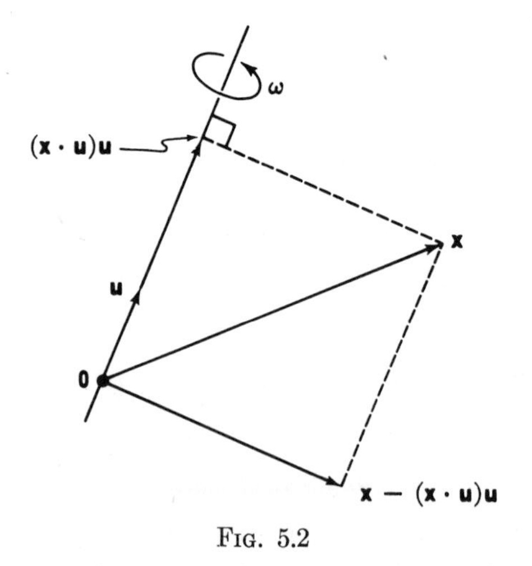

FIG. 5.2

Proof: Let **u** be a unit vector along the axis of rotation. Then $\boldsymbol{\omega} = \omega\mathbf{u}$, where $\omega^2 = \boldsymbol{\omega}\cdot\boldsymbol{\omega} = \omega_x^2 + \omega_y^2 + \omega_z^2$. If $\mathbf{x} \in \mathbf{D}$, the projection of **x** on the axis of rotation is $\mathbf{x}\cdot\mathbf{u}$. Therefore the distance of **x** from the axis is $[|\mathbf{x}|^2 - (\mathbf{x}\cdot\mathbf{u})^2]^{1/2}$. See Fig. 5.2. Hence the speed at **x** is $\omega[|\mathbf{x}|^2 - (\mathbf{x}\cdot\mathbf{u})^2]^{1/2}$, and the kinetic energy of an elementary mass at $\mathbf{x}$ is

$$dK = \frac{1}{2}\delta(\mathbf{x})\omega^2[|\mathbf{x}|^2 - (\mathbf{x}\cdot\mathbf{u})^2]\,dV = \frac{1}{2}\delta(\mathbf{x})[\omega^2\,|\mathbf{x}|^2 - (\mathbf{x}\cdot\boldsymbol{\omega})^2]\,dV.$$

But

$$\omega^2\,|\mathbf{x}|^2 - (\mathbf{x}\cdot\boldsymbol{\omega})^2 = \boldsymbol{\omega}\begin{bmatrix} |\mathbf{x}|^2 & 0 & 0 \\ 0 & |\mathbf{x}|^2 & 0 \\ 0 & 0 & |\mathbf{x}|^2 \end{bmatrix}\boldsymbol{\omega}' - \boldsymbol{\omega}\begin{bmatrix} x^2 & xy & xz \\ yx & y^2 & yz \\ zx & zy & z^2 \end{bmatrix}\boldsymbol{\omega}',$$

hence

$$dK = \frac{1}{2}\delta(\mathbf{x})\,\boldsymbol{\omega}\begin{bmatrix} y^2+z^2 & -xy & -xz \\ -yx & z^2+x^2 & -yz \\ -zx & -zy & x^2+y^2 \end{bmatrix}\boldsymbol{\omega}'\,dV.$$

The formula for K follows by integrating over $\mathbf{D}$.

Applications

Geometric symmetries simplify calculation of the moments and products of inertia. For example, suppose that a solid is symmetric in the x, y-plane. Recall this means that whenever a point (x, y, z) is in the solid, then so is $(x, y, -z)$, and $\delta(x, y, -z) = \delta(x, y, z)$. Then $I_{zx} = 0$, because the contribution

$$\delta(x, y, z)zx\,dV$$

at each point (x, y, z) above the x, y-plane is cancelled by the contribution

$$\delta(x, y, -z)(-z)(x)\,dV = -\delta(x, y, z)zx\,dV$$

at the symmetric point $(x, y, -z)$ below the x, y-plane. Likewise $I_{yz} = 0$ under the same symmetry condition. For example, a uniform right circular cone with axis the z-axis has all products of inertia 0, since it is symmetric in two coordinate planes.

EXAMPLE 5.1

Compute the moments and products of inertia for a uniform sphere of radius a and mass M with center at the origin. Measure length in cm and mass in gm.

Solution: Let δ denote the constant density; then $M = 4\pi a^3\delta/3$ gm. Because the sphere is symmetric in each coordinate plane, all products of inertia are 0.

By symmetry, $I_{xx} = I_{yy} = I_{zz}$. It appears most natural to use spherical coordinates and compute I_{zz}:

$$I_{zz} = \delta \iiint (x^2 + y^2)\, dV = \delta \iiint (\rho^2 \sin^2 \phi \cos^2 \theta + \rho^2 \sin^2 \phi \sin^2 \theta)\, dV$$

$$= \delta \iiint (\rho^2 \sin^2 \phi)\, \rho^2 \sin \phi \, d\rho \, d\phi \, d\theta$$

$$= \delta \int_0^a \rho^4 \, d\rho \int_0^\pi \sin^3 \phi \, d\phi \int_0^{2\pi} d\theta = \delta \left(\frac{a^5}{5}\right)\left(\frac{4}{3}\right)(2\pi) = \frac{8\pi a^5 \delta}{15} = \frac{2}{5} Ma^2.$$

Answer: $I_{xy} = I_{yz} = I_{zx} = 0,$

$I_{xx} = I_{yy} = I_{zz} = \frac{2}{5} Ma^2$ gm-cm^2.

EXAMPLE 5.2

Find the products of inertia of the first octant portion of the sphere of Example 5.1.

Solution: By symmetry $I_{xy} = I_{yz} = I_{zx}$. Choose spherical coordinates and compute that product of inertia whose formula seems the most symmetric; this is I_{xy}, since the z-axis is special in spherical coordinates:

$$I_{xy} = -\delta \iiint xy \, dx \, dy \, dz$$

$$= -\delta \iiint (\rho^2 \sin^2 \phi \cos \theta \sin \theta)\, \rho^2 \sin \phi \, d\rho \, d\phi \, d\theta$$

$$= -\delta \int_0^a \rho^4 \, d\rho \int_0^{\pi/2} \sin^3 \phi \, d\phi \int_0^{\pi/2} \cos \theta \sin \theta \, d\theta$$

$$= -\delta \left(\frac{a^5}{5}\right)\left(\frac{2}{3}\right)\left(\frac{1}{2}\right) = -\frac{a^5}{15}\delta.$$

But

$$M = \frac{1}{8}\left(\frac{4}{3}\pi a^3 \delta\right) = \frac{1}{6}\pi a^3 \delta.$$

Hence $I_{xy} = -2Ma^2/5\pi$.

Answer: $I_{xy} = I_{yz} = I_{zx} = -\frac{2}{5\pi} Ma^2$ gm-cm^2.

Note on units: The unit of work, or energy, in the CGS system is 1 erg = 1 dyne-cm. Remember 1 dyne = 1 gm-cm/sec^2.

EXAMPLE 5.3

The solid of Example 5.2 rotates with angular velocity $\boldsymbol{\omega} = (\omega/\sqrt{3})\,(1, 1, 1)$. Find its kinetic energy.

Solution: By symmetry, the moments of inertia are $\frac{1}{8}$ those of the full sphere, and the mass M is $\frac{1}{8}$ that of the full sphere. By Example 5.1

$$I_{xx} = I_{yy} = I_{zz} = \frac{2}{5}\,Ma^2,$$

where M denotes the mass of the *first octant portion* of the sphere. Since $\omega_x = \omega_y = \omega_z = \omega/\sqrt{3}$, the formula for kinetic energy yields

$$K = \frac{1}{2}\left(\frac{\omega^2}{3}\right)(3I_{xx} + 6I_{xy})$$

$$= \frac{1}{2}\left(\frac{\omega^2}{3}\right)\left(\frac{6}{5}\,Ma^2 - \frac{12}{5\pi}\,Ma^2\right) = \frac{1}{5}\left(1 - \frac{2}{\pi}\right)Ma^2\omega^2.$$

Answer: $K = \frac{1}{5}\left(1 - \frac{2}{\pi}\right)Ma^2\omega^2$ erg.

Parallel Axis Theorem

Take an axis β *anywhere* in space (not necessarily through the origin). Suppose a rigid body $\mathbf{D}$ rotates about this axis with angular speed ω. See Fig. 5.3. The speed at each point $\mathbf{x}$ is $\omega B_{\mathbf{x}}$, where $B_{\mathbf{x}}$ is the distance from $\mathbf{x}$ to the axis β. Hence the kinetic energy is

$$K = \frac{1}{2}\,I_\beta\,\omega^2, \qquad \text{where} \quad I_\beta = \iiint\limits_{\mathbf{D}} B_{\mathbf{x}}{}^2\,\delta(\mathbf{x})\,dV.$$

This defines I_β, the **moment of inertia of D about the axis** β. The Parallel Axis Theorem allows us to compute I_β in terms of the moment of inertia about a parallel axis through the center of gravity.

Parallel Axis Theorem If α is an axis through the center of gravity of $\mathbf{D}$, and β is an axis parallel to α, then

$$I_\beta = I_\alpha + Md^2,$$

where d is the distance between the axes and M is the mass of $\mathbf{D}$.

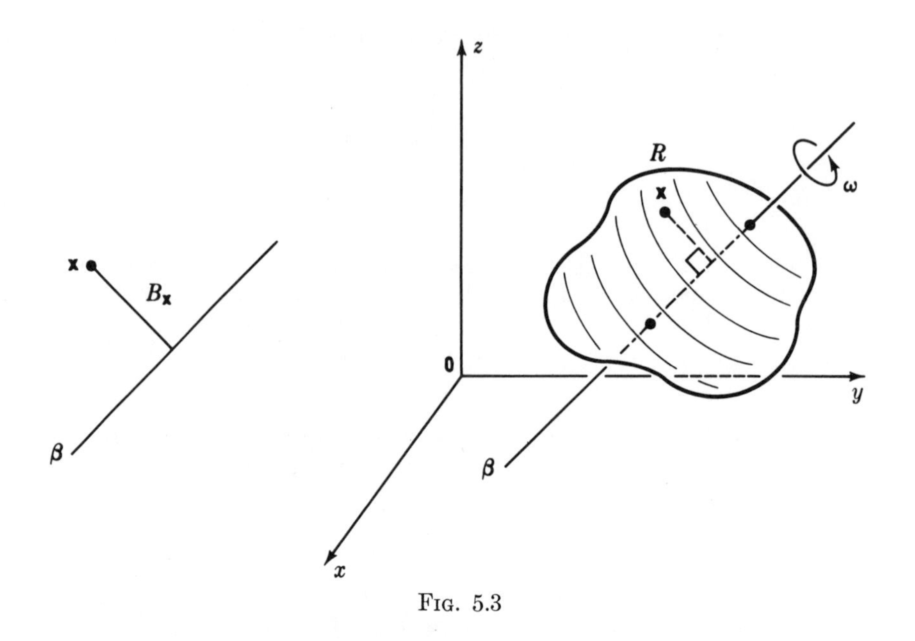

FIG. 5.3

The proof of this result in the general case is not hard, but involves some technicalities. So let us content ourselves with the special case in which $\bar{\mathbf{x}} = \mathbf{0}$ and α is the z-axis. Suppose β passes through $(a, b, 0)$. See Fig. 5.4.

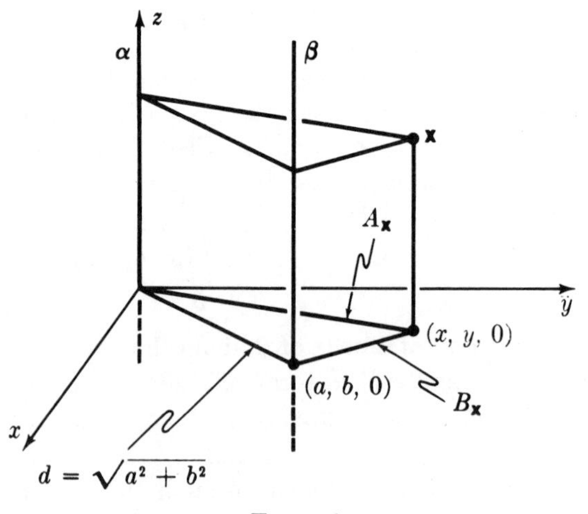

FIG. 5.4

For each point $\mathbf{x}$ of $\mathbf{D}$, the distance $A_{\mathbf{x}}$ of $\mathbf{x}$ from α is given by

$$A_{\mathbf{x}}^2 = x^2 + y^2 = |(x, y)|^2,$$

and the distance $B_{\mathbf{x}}$ of $\mathbf{x}$ from β is given by

$$\begin{aligned} B_{\mathbf{x}}{}^2 = (x-a)^2 + (y-b)^2 &= |(x, y) - (a, b)|^2 \\ &= |(x, y)|^2 - 2(a, b)\cdot(x, y) + |(a, b)|^2 \\ &= A_{\mathbf{x}}{}^2 - 2(a, b)\cdot(x, y) + d^2. \end{aligned}$$

Multiply by the element of mass, $\delta(\mathbf{x})\, dV$, and integrate. Result:

$$I_\beta = I_\alpha - 2(a, b)\cdot(m_x, m_y) + Md^2.$$

But $m_x = m_y = 0$ since $\bar{\mathbf{x}} = \mathbf{0}$. Hence $I_\beta = I_\alpha + Md^2$.

EXAMPLE 5.6

Find the moment of inertia of a uniform sphere of radius a and mass M about an axis tangent to the sphere.

Solution: From Example 5.1, the moment of inertia about any axis through the center (c.g.) is $2Ma^2/5$. The distance from a tangent axis β to the center is a, so the Parallel Axis Theorem implies

$$I_\beta = \frac{2}{5}Ma^2 + Ma^2 = \frac{7}{5}Ma^2.$$

Answer: $\dfrac{7}{5}Ma^2$.

EXERCISES

1. A uniform cylinder of mass M gm occupies the region $-h \leq z \leq h$, $x^2 + y^2 \leq a^2$, distance in cm. Find its moments and products of inertia. [*Hint:* Use cylindrical coordinates.]
2. (cont.) The cylinder rotates with angular speed $\boldsymbol{\omega}$ rad/sec about an axis through $\mathbf{0}$ and $(1, 1, 1)$. Find its kinetic energy.
3. Find the moments and products of inertia of the uniform hemisphere $\rho \leq a$, $z \geq 0$ of mass M.
4. The axis of a uniform right circular cone of mass M is the z-axis; its apex is at $\mathbf{0}$ and its base of radius a is at $z = h > 0$. Find its moments and products of inertia.
5. Find the moments and products of inertia of the uniform rectangular solid of mass M bounded by the planes $x = \pm a$, $y = \pm b$, $z = \pm c$.
6. Find the moments and products of inertia of the uniform rectangular solid of mass M bounded by the coordinate planes and the planes $x = a$, $y = b$, $z = c$.
7. The circular disk $y = 0$, $(x - A)^2 + z^2 \leq a^2$, $0 < a \leq A$ is revolved about the z-axis. Suppose the resulting anchor ring (solid torus) is a uniform solid of mass M. Find I_{zz}.
8. In Example 5.1, use symmetry to prove, without integrating, that

   $I_{zz} = \frac{2}{3}\delta \displaystyle\iiint \rho^2\, dV$. Then evaluate the integral.

9. Find I_{zz} for the uniform solid paraboloid of revolution of mass M bounded by $az = x^2 + y^2$ and $z = h$.

10*. Find the moments of inertia of the uniform solid ellipsoid of mass M,

$$\frac{x^2}{a^2} + \frac{y^2}{b^2} + \frac{z^2}{c^2} = 1.$$

[*Hint:* Stretch a suitable amount in each direction until the solid becomes a sphere; set $x = ua$, $y = vb$, and $z = wc$.]

11*. Find I_{zz} for the uniform solid elliptic paraboloid of mass M bounded by

$$z = \frac{x^2}{a} + \frac{y^2}{b}, \qquad z = h.$$

[*Hint:* Use the result of Ex. 9 and the hint of Ex. 10.]

12*. Find I_{zz} for the uniform solid of mass M bounded by the hyperboloid of revolution $\dfrac{z^2}{c^2} = \dfrac{1}{a^2}(x^2 + y^2) - 1$ and the planes $z = \pm h$.

13. Find the moments of inertia of the uniform solid of mass M in the region $-a \le x, y, z \le a$, $x^2 + y^2 + z^2 \ge a^2$.

14. Find the moment of inertia of a uniform solid right circular cylinder about a generator of its lateral surface. (This is important in problems concerning rolling.)

15*. Suppose a rigid body R is rotating about an axis through $\mathbf{0}$ with angular velocity $\boldsymbol{\omega}$. Show that the **angular momentum,** $\mathbf{J} = \iiint [\delta(\mathbf{x})\mathbf{x} \times \mathbf{v}]\, dV$ ($\mathbf{v}$ is velocity) is given by

$$\mathbf{J} = (\omega_x, \omega_y, \omega_z) \begin{bmatrix} I_{xx} & I_{xy} & I_{xz} \\ I_{yx} & I_{yy} & I_{yz} \\ I_{zx} & I_{zy} & I_{zz} \end{bmatrix}.$$

The definitions of moments and products of inertia can be easily modified to apply to wires and laminas rather than solids. Such moments can be computed directly, or sometimes by limit arguments.

16. Find the moments of inertia of the uniform circular disk $x^2 + y^2 \le a^2$, $z = 0$ of mass M. (The units are gm and cm.) [*Hint:* Let $h \longrightarrow 0$ in Ex. 1.]

17. A uniform wire of mass M lies along the z-axis from $z = -h$ to $z = h$. Find its moments of inertia. (The units are gm and cm.) [*Hint:* Let $a \longrightarrow 0$ in Ex. 1.]

18. A uniform rod of length L and mass M lies along the positive x-axis on the interval $a \le x \le a + L$. Find I_{yy}.

19. Find the moments of inertia of a uniform spherical shell (surface) of radius a and mass M about an axis through its center.

20. A uniform circular wire hoop $x^2 + y^2 = a^2$, $z = 0$ has mass M. Find its moments of inertia.

21. Find I_{zz} for the toroidal shell, the surface of the solid torus in Ex. 7.

22. Find the moments of inertia of the uniform cylindrical shell $x^2 + y^2 = a^2$, $-h \le z \le h$.

12. Integration Theory

1. INTRODUCTION

In this chapter we shall give the theoretical background for the techniques of multiple integration developed in the last two chapters. We shall concentrate mainly on the double integral, both because it is easier to visualize the plane than space, and because no really new ideas are involved in doing the three-dimensional theory once we master the two-dimensional case.

First we must give an accurate definition of the integral, one broad enough to apply at least to continuous functions. The idea will be to define the integral for step functions in an obvious way, then to obtain the integral of a more general function by squeezing the function between step functions.

Let us review briefly how we do this on the line. We work on a fixed closed interval $\mathbf{I} = [a, b]$. A **partition** of $\mathbf{I}$ consists of a finite increasing sequence of division points (not necessarily equally spaced):

$$a = x_0 < x_1 < x_2 < \cdots < x_n = b.$$

A **step function** is a function $s(x)$ on $[a, b]$ that is constant on the *open* intervals of a partition (Fig. 1.1). Thus

$$s(x) = B_i \qquad \text{for} \quad x_{i-1} < x < x_i.$$

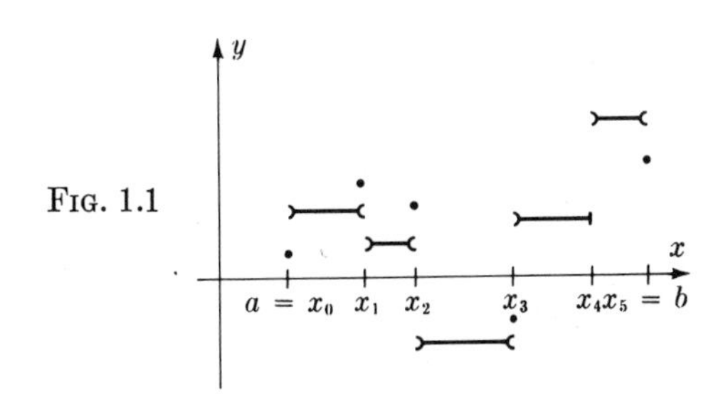

FIG. 1.1

The **integral** of $s(x)$ is defined by

$$\int_{\mathbf{I}} s(x)\, dx = \int_a^b s(x)\, dx = \sum_{i=1}^{n} B_i(x_i - x_{i-1}).$$

Now let $f(x)$ be any function with domain $\mathbf{I}$, arbitrary except we insist that f is bounded, $|f(x)| \leq M$. We call f **integrable** if for each $\epsilon > 0$ there exist step functions s and S on $\mathbf{I}$ such that

(1) $$s(x) \leq f(x) \leq S(x) \qquad \text{for all} \quad x \in \mathbf{I}$$

(2) $$\int_{\mathbf{I}} S(x)\,dx - \int_{\mathbf{I}} s(x)\,dx < \epsilon.$$

Next we show that if f is integrable, then there exists a *unique* number, called

$$\int_{\mathbf{I}} f(x)\,dx,$$

such that

$$\int_{\mathbf{I}} s(x)\,dx \leq \int_{\mathbf{I}} f(x)\,dx \leq \int_{\mathbf{I}} S(x)\,dx$$

for *all* step functions $s(x)$ and $S(x)$ that satisfy (1).

We prove that a continuous function is integrable, using its known *uniform* continuity to approximate it by step functions.

This then is the way integrals are developed on the line; we shall use the same approach in the plane. It is by no means the only way to achieve a theory of integration; we like it because it goes quickly, by easy natural steps. Also, the definitions, theorems, and proofs, with only minor modifications, cover the one-variable case. Therefore you do not have to know the theory of integration on the line to read this chapter.

2. STEP FUNCTIONS

We fix a closed rectangle $\mathbf{I}$ in the plane,

$$\mathbf{I} = \{(x, y) \mid a \leq x \leq b, \quad c \leq y \leq d\}$$

and denote its area by $|\mathbf{I}|$. Thus

$$|\mathbf{I}| = (b - a)(d - c).$$

A **partition** of $\mathbf{I}$ is a decomposition of $\mathbf{I}$ into a finite number of sub-rectangles by lines parallel to the axes. Thus we have partitions of $[a, b]$ and $[c, d]$,

$$a = x_0 < x_1 < x_2 < \cdots < x_m = b,$$

$$c = y_0 < y_1 < y_2 < \cdots < y_n = d,$$

defining the partition of $\mathbf{I}$. See Fig. 2.1. We denote the individual rectangles of the partition by

$$\mathbf{I}_{jk} = \{(x, y) \mid x_{j-1} \leq x \leq x_j, \quad y_{k-1} \leq y \leq y_k\},$$

$$j = 1, 2, \cdots, m, \qquad k = 1, 2, \cdots, n.$$

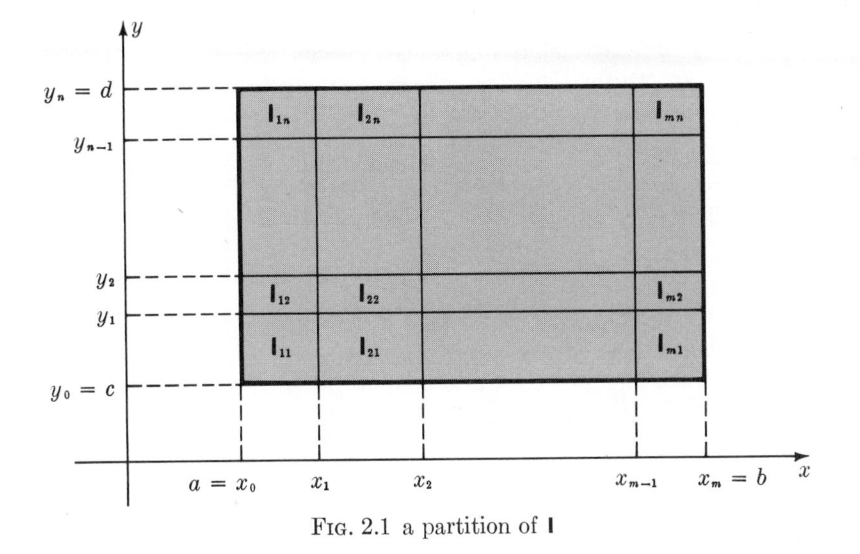

FIG. 2.1 a partition of **I**

> **Step Function** A **step function** on **I** is a real-valued function $s(x, y)$ with domain **I** such that
> (1) s is bounded;
> (2) there is a partition of **I** such that s is constant on the open interior of each rectangle $\mathbf{I}_{jk}$ of the partition. That is, $s(x, y) = B_{jk}$ for $x_{j-1} < x < x_j$ and $y_{k-1} < y < y_k$.

Note that s is unrestricted on the division *lines* of the partition (except that s must be bounded).

If $s(x, y)$ is a step function relative to some partition, then clearly it is a step function relative to any *finer* partition, a partition with more division lines (Fig. 2.2).

(a) partition

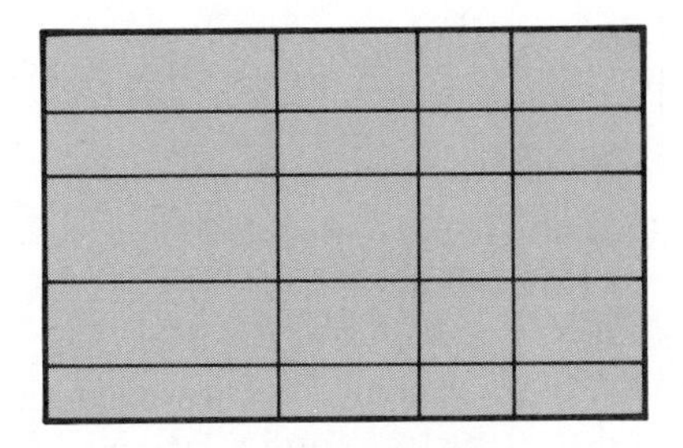

(b) finer partition (refinement)

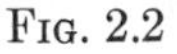

FIG. 2.2

We want to handle sums, products, and other combinations of step functions. But two different step functions may be associated with different partitions, so what do we do? We superpose the partitions (Fig. 2.3); then both functions are step functions relative to the new finer partition.

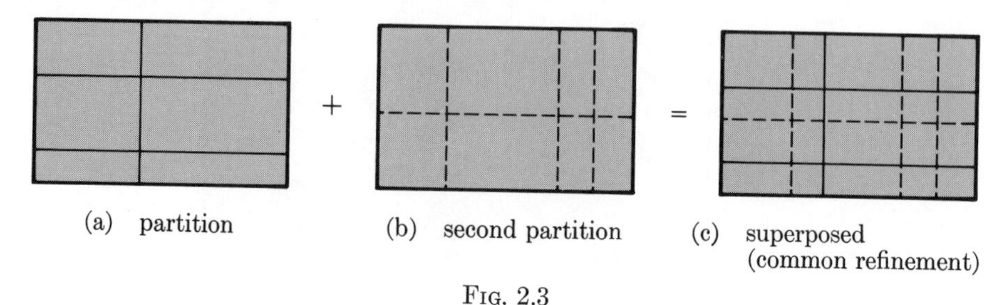

(a) partition (b) second partition (c) superposed (common refinement)

FIG. 2.3

We can use this construction to prove routinely the following result (see Exs. 1–4).

Theorem 2.1 Let s_1 and s_2 be step functions on $\mathbf{I}$ and k a constant. Then each of the following is a step function:

$$s_1 + s_2, \qquad ks_1, \qquad |s_1|, \qquad s_1 s_2$$

$$S(x, y) = \max\{s_1(x, y), s_2(x, y)\},$$

$$s(x, y) = \min\{s_1(x, y), s_2(x, y)\}.$$

We are now ready to define the integral of a step function.

Integrals of Step Functions

Integral of a Step Function Let s be a step function on $\mathbf{I}$ relative to a partition of $\mathbf{I}$ into rectangles $\mathbf{I}_{jk}$. Suppose $s(x, y) = B_{jk}$ for (x, y) in the interior of $\mathbf{I}_{jk}$. Define

$$\iint_{\mathbf{I}} s\,dx\,dy = \sum_{j,k} B_{jk}\,|\mathbf{I}_{jk}|.$$

(Recall that $|\mathbf{I}_{jk}|$ denotes the area of $\mathbf{I}_{jk}$.)

This definition looks innocent, yet it contains two subtle points. The first is that the *same* step function $s(x, y)$ may be associated with two *different* partitions, P_1 and P_2. Do we get the same answer when we compute the integral relative to P_1 and relative to P_2? Let us show that we do.

Suppose that $\mathbf{J}$ is a sub-rectangle of P_1 and that $s(x, y)$ takes the constant value $B_{\mathbf{J}}$ on $\mathbf{J}$. If P_3 is the superposed partition (common refinement) of P_1 and P_2, then $\mathbf{J}$ decomposes into a number of sub-rectangles $\mathbf{J}_{ij}$ of P_3, and their areas add up to the area of $\mathbf{J}$. On each, $s(x, y)$ has the same constant value $B_{\mathbf{J}}$.

Therefore,

$$\iint_{(P_1)} s(x, y)\, dx\, dy = \sum_{\mathbf{J}} B_{\mathbf{J}}\, |\mathbf{J}| = \sum_{\mathbf{J}} B_{\mathbf{J}} \sum |\mathbf{J}_{ij}|$$

$$= \sum_{\mathbf{J}} \sum_{i,j} B_{\mathbf{J}}\, |\mathbf{J}_{ij}| = \iint_{(P_3)} s(x, y)\, dx\, dy.$$

By the same reasoning,

$$\iint_{(P_2)} s(x, y)\, dx\, dy = \iint_{(P_3)} s(x, y)\, dx\, dy.$$

Hence,

$$\iint_{(P_1)} s(x, y)\, dx\, dy = \iint_{(P_2)} s(x, y)\, dx\, dy.$$

The second subtle point is this. Is there anything to be gained by using more general partitions of $\mathbf{I}$ into sub-rectangles than those arising from partitions of $[a, b]$ and $[c, d]$? The answer is no, as Fig. 2.4 indicates.

Now we state the main elementary properties of the integral of a step function. Except for the last two, their proofs are easy and are left as exercises.

Theorem 2.2 Let s_1, s_2, s be step functions on $\mathbf{I}$. Then:

(1) $$\iint_{\mathbf{I}} (s_1 + s_2)\, dx\, dy = \iint_{\mathbf{I}} s_1\, dx\, dy + \iint_{\mathbf{I}} s_2\, dx\, dy.$$

(2) $$\iint_{\mathbf{I}} ks\, dx\, dy = k \iint_{\mathbf{I}} s\, dx\, dy \qquad (k \text{ any constant}).$$

(3) If $s \geq 0$ on $\mathbf{I}$, then $$\iint_{\mathbf{I}} s\, dx\, dy \geq 0.$$

(4) If $s_1 \leq s_2$ on $\mathbf{I}$, then $$\iint_{\mathbf{I}} s_1\, dx\, dy \leq \iint_{\mathbf{I}} s_2\, dx\, dy.$$

(5) If $\mathbf{I}$ is partitioned into sub-rectangles $\mathbf{I}_{jk}$, then

$$\iint_{\mathbf{I}} s\, dx\, dy = \sum_{j,k} \iint_{\mathbf{I}_{jk}} s\, dx\, dy.$$

(6) $$\max\left\{\iint_{\mathbf{I}} s_1\, dx\, dy, \iint_{\mathbf{I}} s_2\, dx\, dy\right\} \leq \iint_{\mathbf{I}} \max\{s_1, s_2\}\, dx\, dy.$$

(7) $$\iint_{\mathbf{I}} \min\{s_1, s_2\}\, dx\, dy \leq \min\left\{\iint_{\mathbf{I}} s_1\, dx\, dy, \iint_{\mathbf{I}} s_2\, dx\, dy\right\}.$$

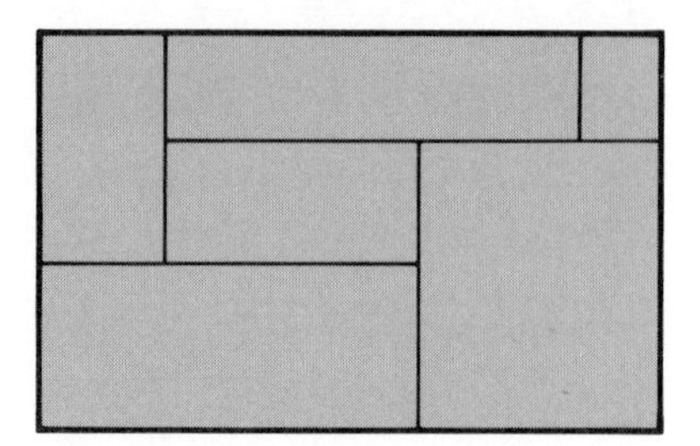

(a) more general partition

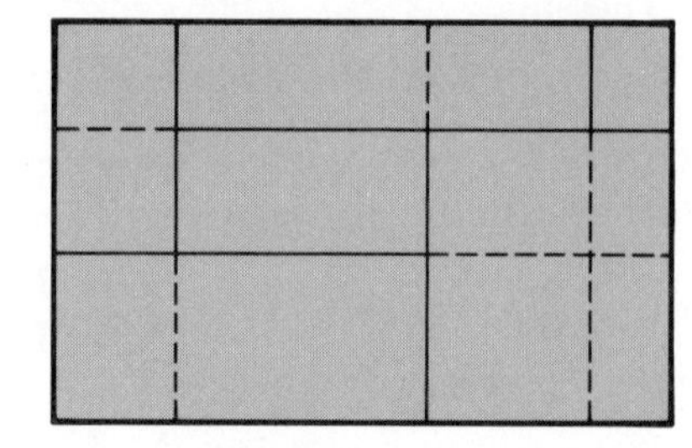

(b) the segments extended

FIG. 2.4

To prove (6), note that $s_1 \leq \max\{s_1, s_2\}$ on $\mathbf{I}$, hence

$$\iint s_1 \leq \iint \max\{s_1, s_2\}$$

by (4). Likewise

$$\iint s_2 \leq \iint \max\{s_1, s_2\};$$

consequently

$$\max\left\{\iint s_1, \iint s_2\right\} \leq \iint \max\{s_1, s_2\}.$$

The proof of (7) is similar.

REMARK: Note the meaning of (5). The given partition of $\mathbf{I}$ is *not* assumed to be related to a partition attached to the given step function.

Iteration

We complete the story of step functions with the **iteration formula**.

Theorem 2.3 Let $s(x, y)$ be a step function on $\mathbf{I}$, attached to the partition

$$a = x_0 < x_1 < \cdots < x_m = b, \qquad c = y_0 < y_1 < \cdots < y_n = d.$$

Define $t(x)$ by

$$t(x) = \begin{cases} \displaystyle\int_c^d s(x, y)\, dy & \text{for } x \neq x_j \\ \text{arbitrary} & \text{for } x = x_j. \end{cases}$$

Then $t(x)$ is a step function on $[a, b]$ and

$$\iint_{\mathbf{I}} s\, dx\, dy = \int_a^b t\, dx = \int_a^b \left(\int_c^d s(x, y)\, dy\right) dx.$$

Proof: As usual, let

$$\mathbf{I}_{jk} = \{(x, y) \mid x_{j-1} \le x \le x_j, \quad y_{k-1} \le y \le y_k\}$$

be the sub-rectangles of the partition and let $s(x, y) = B_{jk}$ for (x, y) in the open interior of $\mathbf{I}_{jk}$. For $x_{j-1} < x < x_j$ we have

$$t(x) = \int_c^d s(x, y)\, dy = \sum_{k=1}^{n} B_{jk}(y_k - y_{k-1}).$$

This proves that $t(x)$ is constant on the open interval (x_{j-1}, x_j), hence $t(x)$ is a step function on $[a, b]$. By definition,

$$\int_a^b t(x)\, dx = \sum_{j=1}^{m} \left(\sum_{k=1}^{n} B_{jk}(y_k - y_{k-1}) \right) (x_j - x_{j-1})$$

$$= \sum \sum B_{jk} |\mathbf{I}_{jk}| = \iint_{\mathbf{I}} s\, dx\, dy.$$

EXERCISES

Prove in detail Theorem 2.1 for

1. $s_1 + s_2$
2. $S(x, y)$
3. $|s_1|$
4. $s_1 s_2$

Prove Theorem 2.2:

5. part (1)
6. part (2)
7. part (3)
8. part (4)
9. part (5).

3. THE RIEMANN INTEGRAL

Let f be a bounded real-valued function with domain $\mathbf{I}$. We shall try to squeeze f between step functions whose integrals are arbitrarily close to each other. This is not possible for every function, but when it is, we are able to define the integral of f.

Since f is bounded, there exist step functions s and S such that $s \le f \le S$ on $\mathbf{I}$. Indeed, if $|f| \le B$ on $\mathbf{I}$, take for instance the constant functions $s = -B$ and $S = B$.

Integrable Function A bounded function f on $\mathbf{I}$ is called **integrable** if for each $\epsilon > 0$, there exist step functions s and S on $\mathbf{I}$ such that

(1) $s \le f \le S$ on $\mathbf{I}$,

(2) $\displaystyle\iint_{\mathbf{I}} S\, dx\, dy - \iint_{\mathbf{I}} s\, dx\, dy < \epsilon.$

The first thing we must do is show that if f is an integrable function, we can assign to f a unique number, its integral.

Theorem 3.1 Let f be integrable on $\mathbf{I}$. Consider *all* step functions s and S such that

$$s \leq f \leq S \quad \text{on} \quad \mathbf{I}.$$

Then

$$\sup_s \iint_{\mathbf{I}} s\,dx\,dy = \inf_S \iint_{\mathbf{I}} S\,dx\,dy.$$

The common value of the sup and inf is called the **integral** of f on $\mathbf{I}$, and is written

$$\iint_{\mathbf{I}} f\,dx\,dy.$$

Proof: If $s \leq f \leq S$ on $\mathbf{I}$, then

$$\iint_{\mathbf{I}} s\,dx\,dy \leq \iint_{\mathbf{I}} S\,dx\,dy$$

by Theorem 2.2, part (4). Hold S fixed and let s vary. It follows that

$$\sup_s \iint_{\mathbf{I}} s\,dx\,dy \leq \iint_{\mathbf{I}} S\,dx\,dy.$$

The left side of this inequality is independent of S. Now let S vary; it follows that

$$\sup_s \iint_{\mathbf{I}} s\,dx\,dy \leq \inf_S \iint_{\mathbf{I}} S\,dx\,dy. \tag{1}$$

This inequality holds for *any* bounded function. Next we show that if f is an *integrable* function, then the reverse inequality also holds, so (1) becomes an equality.

Let f be integrable. Given $\epsilon > 0$, there exist step functions s_0 and S_0 such that $s_0 \leq f \leq S_0$ and

$$\iint_{\mathbf{I}} S_0\,dx\,dy < \epsilon + \iint_{\mathbf{I}} s_0\,dx\,dy.$$

Therefore

$$\inf_S \iint_{\mathbf{I}} S\,dx\,dy \leq \iint_{\mathbf{I}} S_0\,dx\,dy < \epsilon + \iint_{\mathbf{I}} s_0\,dx\,dy$$

$$\leq \epsilon + \sup_s \iint_{\mathbf{I}} s\,dx\,dy.$$

This is true for all $\epsilon > 0$, hence

$$(2) \qquad \inf_S \iint_{\mathbf{I}} S\,dx\,dy \le \sup_s \iint_{\mathbf{I}} s\,dx\,dy.$$

By combining (1) and (2) we have the desired equality.

REMARK 1: The theorem implies that if f is integrable and $s \le f \le S$, then $\iint s \le \iint f \le \iint S$. This relation determines $\iint f$. In fact, $\iint f$ is the *only* number between all pairs $\iint s$ and $\iint S$.

REMARK 2: The integral we have defined is called the **Riemann (double) integral**, and integrable functions are also called **R-integrable**. It is possible to define integrals that apply to more functions—the most important is the Lebesgue integral—but their theories are more advanced.

Theorem 3.2 Let f be a step function on $\mathbf{I}$. Then f is integrable and the R-integral of f is the integral defined in Section 2.

In particular, if $f = B$ is constant on $\mathbf{I}$, then $\iint f = B\,|\mathbf{I}|$.

Proof: Given a step function f and $\epsilon > 0$, take $S = s = f$ in the definition of integrable function. Then $s \le f \le S$ and $\iint s = \iint S$, etc.

Continuous Functions

It is an important practical matter that all continuous functions are integrable. This is certainly not obvious from the definition.

Theorem 3.3 Each continuous function on $\mathbf{I}$ is integrable.

Proof: Let f be continuous on $\mathbf{I}$. By the theorems in Chapter 6, Section 3, f is bounded and is uniformly continuous on $\mathbf{I}$, since $\mathbf{I}$ is a bounded and closed set. Because f is bounded, we are allowed to ask whether or not it is integrable. Thus let $\epsilon > 0$. Since f is uniformly continuous, there is $\delta > 0$ such that $|f(\mathbf{x}) - f(\mathbf{z})| < \epsilon/|\mathbf{I}|$ whenever $|\mathbf{x} - \mathbf{z}| < \delta$.

Choose a partition with sub-rectangles $\mathbf{I}_{jk}$ so small that $|\mathbf{x} - \mathbf{z}| < \delta$ whenever $\mathbf{x}$ and $\mathbf{z}$ lie in the same $\mathbf{I}_{jk}$. Set

$$m_{jk} = \inf\{f(\mathbf{x}) \mid \mathbf{x} \in \mathbf{I}_{jk}\}, \qquad M_{jk} = \sup\{f(\mathbf{x}) \mid \mathbf{x} \in \mathbf{I}_{jk}\}.$$

Then $m_{jk} \le f(\mathbf{x}) \le M_{jk}$ for all $\mathbf{x} \in \mathbf{I}_{jk}$, and $M_{jk} - m_{jk} < \epsilon/|\mathbf{I}|$.

Now define step functions s and S by requiring that $s(\mathbf{x}) = m_{jk}$ and $S(\mathbf{x}) = M_{jk}$ for $\mathbf{x}$ in the open interior of $\mathbf{I}_{jk}$, and $s(\mathbf{x}) = S(\mathbf{x}) = f(\mathbf{x})$ for all points $\mathbf{x}$ on the division lines of the partition. Then $s \le f \le S$ on $\mathbf{I}$ and

$$\iint_{\mathbf{I}} S\,dx\,dy - \iint_{\mathbf{I}} s\,dx\,dy = \iint_{\mathbf{I}} (S - s)\,dx\,dy$$

$$< \iint_{\mathbf{I}} \frac{\epsilon}{|\mathbf{I}|}\,dx\,dy = \frac{\epsilon}{|\mathbf{I}|}\cdot|\mathbf{I}| = \epsilon.$$

Therefore f satisfies the definition of an integrable function.

Sums and Multiples

Theorem 3.4 Suppose f_1 and f_2 are integrable on $\mathbf{I}$ and c_1 and c_2 are constants. Then $c_1 f_1 + c_2 f_2$ is integrable and

$$\iint_{\mathbf{I}} (c_1 f_1 + c_2 f_2)\,dx\,dy = c_1 \iint_{\mathbf{I}} f_1\,dx\,dy + c_2 \iint_{\mathbf{I}} f_2\,dx\,dy.$$

Proof: We divide the proof into several steps.

(1) If f is integrable, then $-f$ is integrable and $\iint (-f) = -\iint f.$

For let $\epsilon > 0$. Choose s and S so $s \le f \le S$ and $\iint S - \iint s < \epsilon$. Then

$-S \le -f \le -s$ and $\iint (-s) - \iint (-S) = \iint S - \iint s < \epsilon$, etc.

(2) If f is integrable and $c > 0$, then cf is integrable and $\iint (cf) = c \iint f.$

This time choose s and S so $s \le f \le S$ and $\iint S - \iint s < \epsilon/c$. Then

$cs \le cf \le cS$ and $\iint cS - \iint cs = c\left(\iint S - \iint s\right) < \epsilon$, etc.

(3) If f is integrable and c is a constant, then cf is integrable and $\iint (cf) = c \iint f.$

We have just disposed of $c > 0$. If $c < 0$, then $cf = (-c)(-f)$ does it in two steps. The case $c = 0$ is obvious.

(4) If f_1 and f_2 are integrable, then $f_1 + f_2$ is integrable and

$$\iint (f_1 + f_2) = \iint f_1 + \iint f_2.$$

Given $\epsilon > 0$, choose step functions satisfying

$$s_1 \le f_1 \le S_1, \qquad s_2 \le f_2 \le S_2,$$

$$\iint S_1 - \iint s_1 < \frac{\epsilon}{2}, \qquad \iint S_2 - \iint s_2 < \frac{\epsilon}{2}.$$

Then $s_1 + s_2 \le f_1 + f_2 \le S_1 + S_2$ and

$$\iint (S_1 + S_2) - \iint (s_1 + s_2) = \left(\iint S_1 - \iint s_1\right) + \left(\iint S_2 - \iint s_2\right)$$

$$< \frac{\epsilon}{2} + \frac{\epsilon}{2} = \epsilon,$$

etc. Now given integrable functions f_1 and f_2, it follows from (3) that $c_1 f_1$ and $c_2 f_2$ are integrable, and from (4) that $c_1 f_1 + c_2 f_2$ is integrable. Again using (3) and (4) we have

$$\iint (c_1 f_1 + c_2 f_2) = c_1 \iint f_1 + c_2 \iint f_2.$$

Inequalities

Theorem 3.5 Suppose f and g are integrable on $\mathbf{I}$.

(1) If $f \ge 0$ on $\mathbf{I}$, then $\displaystyle\iint_{\mathbf{I}} f\,dx\,dy \ge 0$.

(2) If $f \le g$ on $\mathbf{I}$, then $\displaystyle\iint_{\mathbf{I}} f\,dx\,dy \le \iint_{\mathbf{I}} g\,dx\,dy$.

(3) If $m \le f \le M$ on $\mathbf{I}$, then $\displaystyle m\,|\mathbf{I}| \le \iint_{\mathbf{I}} f\,dx\,dy \le M\,|\mathbf{I}|$.

Proof: For (1), simply observe that $s_0 = 0$ is a step function and $s_0 \le f$ on $\mathbf{I}$. By definition, $\displaystyle\iint s \le \iint f$ for *any* step function s satisfying $s \le f$, in particular for s_0. Now (2) follows by applying (1) to $g - f$. Finally, (3) is immediate from the choice of $s = m$ and $S = M$ on $\mathbf{I}$.

For the next theorem we need a simple fact, which we state as a lemma.

Lemma Let A_1, A_2, B_1, B_2 be four real numbers. Then

$$\max\{B_1, B_2\} - \max\{A_1, A_2\} \leq \max\{B_1 - A_1, B_2 - A_2\}.$$

If in addition $B_1 \geq A_1$ and $B_2 \geq A_2$, then

$$\max\{B_1, B_2\} - \max\{A_1, A_2\} \leq (B_1 - A_1) + (B_2 - A_2).$$

Proof: For $i = 1, 2$,

$$B_i = A_i + (B_i - A_i) \leq \max\{A_1, A_2\} + \max\{B_1 - A_1, B_2 - A_2\}.$$

Here $\max\{B_1, B_2\} \leq$ (right-hand side), and the first inequality follows.

If $B_1 \geq A_1$ and $B_2 \geq A_2$, then $B_1 - A_1 \geq 0$, so $B_2 - A_2 \leq (B_1 - A_1) + (B_2 - A_2)$. Likewise $B_1 - A_1 \leq (B_1 - A_1) + (B_2 - A_2)$, so

$$\max\{B_1 - A_1, B_2 - A_2\} \leq (B_1 - A_1) + (B_2 - A_2).$$

The second inequality now follows from the first.

Theorem 3.6 Suppose f_1 and f_2 are integrable on $\mathbf{I}$. Then $\max\{f_1, f_2\}$ and $\min\{f_1, f_2\}$ are integrable and

(1) $$\max\left\{\iint_{\mathbf{I}} f_1\,dx\,dy, \iint_{\mathbf{I}} f_2\,dx\,dy\right\} \leq \iint_{\mathbf{I}} \max\{f_1, f_2\}\,dx\,dy,$$

(2) $$\iint_{\mathbf{I}} \min\{f_1, f_2\}\,dx\,dy \leq \min\left\{\iint_{\mathbf{I}} f_1\,dx\,dy, \iint_{\mathbf{I}} f_2\,dx\,dy\right\}.$$

Proof: Let $\epsilon > 0$ and choose step functions satisfying $s_1 \leq f_1 \leq S_1$, $s_2 \leq f_2 \leq S_2$, $\iint S_1 - \iint s_1 < \frac{1}{2}\epsilon$, and $\iint S_2 - \iint s_2 < \frac{1}{2}\epsilon$. Set $S = \max\{S_1, S_2\}$ and $s = \max\{s_1, s_2\}$, both step functions. Clearly $s \leq \max\{f_1, f_2\} \leq S$.

We shall prove first that $\iint S - \iint s < \epsilon$. This will imply that $\max\{f_1, f_2\}$ is integrable. By the lemma,

$$S - s \leq (S_1 - s_1) + (S_2 - s_2).$$

By (4) of Theorem 2.2,

$$\iint (S - s) \leq \iint (S_1 - s_1) + (S_2 - s_2) \leq \tfrac{1}{2}\epsilon + \tfrac{1}{2}\epsilon = \epsilon.$$

Hence $F = \max\{f_1, f_2\}$ is integrable. Since $f_1 \le F$ and $f_2 \le F$, we have $\iint f_1 \le \iint F$ and $\iint f_2 \le \iint F$ by (2) of Theorem 3.5. Relation (1) follows immediately.

Relation (2) is equivalent to (1) by the equation

$$\min\{f_1, f_2\} = -\max\{-f_1, -f_2\}.$$

Theorem 3.7 Suppose f is integrable on $\mathbf{I}$. Then $|f|$ is integrable and

$$\left|\iint_{\mathbf{I}} f\,dx\,dy\right| \le \iint_{\mathbf{I}} |f|\,dx\,dy.$$

Proof: To prove $|f|$ is integrable, we use the relation (easily verified)

$$|f| = \max\{f, 0\} - \min\{f, 0\}$$

and the last theorem. Since $f \le |f|$ and $-f \le |f|$, by Theorem 3.5 we have $\iint f \le \iint |f|$ and $-\iint f \le \iint |f|$. Hence $\left|\iint f\right| \le \iint |f|$.

REMARK: Why not prove this theorem directly, starting with $s \le f \le S$, etc.? Because even though we find easily that $|f| \le \max\{|s|, |S|\}$, it is hard to find a step function below $|f|$. Certainly $\min\{|s|, |S|\} \le |f|$ is wrong!

Additivity of the Integral

Theorem 3.8 Suppose f is a bounded function on $\mathbf{I}$, and suppose $\mathbf{I}$ is partitioned into sub-rectangles $\mathbf{I}_{jk}$. Then f is integrable on $\mathbf{I}$ if and only if f is integrable on all $\mathbf{I}_{jk}$. If so, then

$$\iint_{\mathbf{I}} f\,dx\,dy = \sum_{j,k} \iint_{\mathbf{I}_{jk}} f\,dx\,dy.$$

Proof: The theorem is true when f is a step function, by Theorem 2.2(5). We shall use this fact.

First suppose f is integrable on each of the mn sub-rectangles $\mathbf{I}_{jk}$. Let $\epsilon > 0$. Choose step functions satisfying $s_{jk} \le f \le S_{jk}$ on $\mathbf{I}_{jk}$ and

$$\iint_{\mathbf{I}_{jk}} S_{jk} - \iint_{\mathbf{I}_{jk}} s_{jk} < \frac{\epsilon}{mn}. \tag{1}$$

Define S and s on $\mathbf{I}$ by

$$\begin{cases} S = s = f & \text{on the dividing lines,} \\ S = S_{jk} \quad \text{and} \quad s = s_{jk} & \text{on the interior of } \mathbf{I}_{jk}. \end{cases}$$

Then $s \le f \le S$ on $\mathbf{I}$ and

$$(2)\qquad \iint_{\mathbf{I}} S - \iint_{\mathbf{I}} s = \sum_{j,k} \iint_{\mathbf{I}_{jk}} S - \sum_{j,k} \iint_{\mathbf{I}_{jk}} s$$

$$= \sum_{j,k} \iint_{\mathbf{I}_{jk}} S_{jk} - \sum_{j,k} \iint_{\mathbf{I}_{jk}} s_{jk}$$

$$= \sum_{j,k} \left(\iint_{\mathbf{I}_{jk}} S_{jk} - \iint_{\mathbf{I}_{jk}} s_{jk} \right) < \sum_{j,k} \frac{\epsilon}{mn} = \epsilon.$$

Therefore f is integrable on $\mathbf{I}$, and

$$(3)\qquad \iint_{\mathbf{I}} s \le \iint_{\mathbf{I}} f \le \iint_{\mathbf{I}} S.$$

But $s \le f \le S$ on each $\mathbf{I}_{jk}$, so

$$(4)\qquad \iint_{\mathbf{I}} s = \sum_{j,k} \iint_{\mathbf{I}_{jk}} s \le \sum_{j,k} \iint_{\mathbf{I}_{jk}} f \le \sum_{j,k} \iint_{\mathbf{I}_{jk}} S = \iint_{\mathbf{I}} S.$$

Compare the results of (2), (3), and (4):

$$(5)\qquad \iint_{\mathbf{I}} S - \iint_{\mathbf{I}} s < \epsilon, \qquad \iint_{\mathbf{I}} s \le \iint_{\mathbf{I}} f \le \iint_{\mathbf{I}} S,$$

$$\iint_{\mathbf{I}} s \le \sum \iint_{\mathbf{I}_{jk}} f \le \iint_{\mathbf{I}} S.$$

It follows readily that

$$(6)\qquad \left| \iint_{\mathbf{I}} f - \sum \iint_{\mathbf{I}_{jk}} f \right| < \epsilon.$$

Since ϵ is arbitrary, (6) implies

$$(7)\qquad \iint_{\mathbf{I}} f = \sum \iint_{\mathbf{I}_{jk}} f.$$

Now to prove the rest of the theorem, suppose f is integrable on $\mathbf{I}$. It is enough to prove f is integrable on each $\mathbf{I}_{jk}$, for then the first part of the proof will give the sum formula.

Let $\epsilon > 0$. Choose s and S so $s \le f \le S$ on $\mathbf{I}$ and $\displaystyle\iint S - \iint s < \epsilon$.

Then $s \leq f \leq S$ on $\mathbf{I}_{jk}$ and

$$\iint_{\mathbf{I}_{jk}} S - \iint_{\mathbf{I}_{jk}} s \leq \iint_{\mathbf{I}} (S - s) < \epsilon.$$

Hence f is integrable on $\mathbf{I}_{jk}$, which completes the proof.

Products

> **Theorem 3.9** If f and g are integrable on $\mathbf{I}$, then fg is integrable.

Proof: The identity

$$fg = \frac{1}{4}[(f + g)^2 - (f - g)^2]$$

shows that it is enough to prove that the square of an integrable function is integrable. Let f be integrable. Then $f^2 = |f|^2$ and $|f|$ is integrable. Hence we may assume $f \geq 0$. Also f is bounded, so $0 \leq f \leq B$.

Let $\epsilon > 0$. Choose step functions $0 \leq s \leq f \leq S \leq B$ and $\iint S - \iint s < \epsilon/2B$. Then $s^2 \leq f^2 \leq S^2$ and

$$S^2 - s^2 = (S + s)(S - s) \leq 2B(S - s),$$

hence $\iint S^2 - \iint s^2 \leq 2B \iint (S - s) < \epsilon$. This proves f^2 is integrable and completes the proof.

Remark: A much more general theorem is true, but its proof is very hard: If $H(z, w)$ is continuous for all z, w and if f and g are integrable on $\mathbf{I}$, then the composite function

$$h(x, y) = H[f(x, y), g(x, y)]$$

is integrable. The special case $H(z, w) = zw$ implies Theorem 3.9.

EXERCISES

1. Suppose a function f on $\mathbf{I}$ has the following property: if $\epsilon > 0$, there exists a step function s such that

$$|f(x, y) - s(x, y)| < \epsilon \qquad \text{for all} \quad (x, y) \in \mathbf{I}.$$

Prove f is integrable.

2. Suppose $g(x)$ is integrable on $[a, b]$. Set $f(x, y) = g(x)$ on $\mathbf{I}: a \leq x \leq b, c \leq y \leq d$. Prove f is integrable on $\mathbf{I}$.
3. Define f by $f(x, y) = 1$ if x and y are both rational, $f = 0$ otherwise. Prove that f is not integrable on any rectangle.
4. Suppose f is integrable on $a \leq x \leq b$, $c \leq y \leq d$. Define $g(x, y) = f(y, x)$.

   Prove that g is integrable on $c \leq x \leq d$, $a \leq y \leq b$. Find $\iint g$.
5. (cont.) Suppose f is integrable on $0 \leq x, y \leq 1$ and $f(x, y) + f(x, y) = 0$. Prove

   $\iint f = 0.$

6*. For each n, divide $\mathbf{I}$ into n^2 *equal* rectangles $\mathbf{I}_{jk}$. Choose *any* point $(x_{jk}, y_{jk}) \in \mathbf{I}_{jk}$ and set

$$A_n = \frac{|\mathbf{I}|}{n^2} \sum_{j,k} f(x_{jk}, y_{jk}).$$

Prove $A_n \longrightarrow \iint f$ if f is integrable. [*Hint:* First prove it for step functions.]

7. Let $f(x, y) = 0$ on a rectangle $\mathbf{I}$ except at a single point $\mathbf{p}$. Show that f is integrable and $\iint_{\mathbf{I}} f\,dx\,dy = 0$.

8*. Let $f(x, y)$ be defined in the square $\mathbf{I}$ with vertices $(0, 0)$, $(0, 1)$, $(1, 1)$, $(1, 0)$ by

$$f(x, y) = \begin{cases} 1 & \text{if } x = \frac{1}{2}, \frac{1}{3}, \frac{1}{4}, \cdots \\ 0 & \text{otherwise.} \end{cases}$$

Prove that f is integrable and $\iint_{\mathbf{I}} f\,dx\,dy = 0.$

9. Suppose $f(x, y) \geq 0$. Exercises 7 and 8 show that $\iint_{\mathbf{I}} f\,dx\,dy = 0$ does not imply that $f(x, y) = 0$ at each point of $\mathbf{I}$. Prove, however, that if f is *continuous*, then $\iint_{\mathbf{I}} f\,dx\,dy = 0$ does imply that $f(x, y) = 0$ at each point of $\mathbf{I}$.

4. ITERATION

Here is the main theorem on iteration of integrals over a rectangle $\mathbf{I} = \{(x, y) \mid a \leq x \leq b, \quad c \leq y \leq d\}$.

Theorem 4.1 Let f be a bounded function on the domain

$$\mathbf{I} = \{(x, y) \mid a \leq x \leq b, \quad c \leq y \leq d\}.$$

Suppose:

(1) f is integrable on $\mathbf{I}$.

(2) For each $x \in [a, b]$, with possibly a finite set of exceptions, the integral

$$g(x) = \int_c^d f(x, y)\, dy$$

exists.

Then g is integrable on $[a, b]$ and

$$\int_a^b g\, dx = \int_a^b \left(\int_c^d f(x, y)\, dy \right) dx = \iint_{\mathbf{I}} f\, dx\, dy.$$

Proof: Let $\epsilon > 0$. Choose step functions satisfying $s \leq f \leq S$ and $\iint S - \iint s < \epsilon$. Define step functions

$$t(x) = \int_c^d s(x, y)\, dy, \qquad T(x) = \int_c^d S(x, y)\, dy.$$

Then for each x, except perhaps the finitely many exceptions, the inequality $s \leq f \leq S$ implies

$$t(x) \leq g(x) \leq T(x).$$

By Theorem 2.3,

$$\int_a^b T(x)\, dx - \int_a^b t(x)\, dx = \iint_{\mathbf{I}} S\, dx\, dy - \iint_{\mathbf{I}} s\, dx\, dy < \epsilon.$$

This says that $g(x)$ is integrable on $[a, b]$. It also says that

$$\iint_{\mathbf{I}} s\, dx\, dy \leq \int_a^b g(x)\, dx \leq \iint_{\mathbf{I}} S\, dx\, dy$$

for each choice of s and S. But the only number that satisfies these inequalities for all s and S is $\iint_{\mathbf{I}} f$. It follows that

$$\int_a^b g(x)\, dx = \iint_{\mathbf{I}} f\, dx\, dy,$$

completing the proof.

REMARK: Hypothesis (1) is not sufficient for this theorem. There are examples of functions that satisfy (1) but not (2), so $g(x)$ is not even defined, let alone integrable.

Corollary If f is continuous on $\mathbf{I}$, then

$$\iint\limits_{\mathbf{I}} f\,dx\,dy = \int_a^b \left(\int_c^d f(x, y)\,dy\right) dx = \int_c^d \left(\int_a^b f(x, y)\,dx\right) dy.$$

Proof: By Theorem 3.3, f is integrable on $\mathbf{I}$. Because f is continuous on $\mathbf{I}$, for each x_0 the function $f(x_0, y)$ is continuous on $[c, d]$, hence integrable. Thus the hypotheses of Theorem 4.1 are satisfied, so the first equality follows. By symmetry so does the second.

Non-rectangular Domains

This completes the theory of the double integral over a rectangle. Our next job is to extend this theory to functions with non-rectangular domains.

The first problem is to decide which sets we shall allow as domains. This is not an easy question; the complete theory is beyond the scope of this course. We want a theory that includes at least the domains that arise in practice, but not so many more that the technical details become oppressive.

The kind of domain $\mathbf{D}$ we want can be described accurately this way. (1) It is a closed bounded subset of $\mathbf{R}^2$. (2) It is connected, i.e., any two points of $\mathbf{D}$ can be connected by a curve in $\mathbf{D}$. (3) It has many interior points (that is, each point of $\mathbf{D}$ is either an interior point or a limit of a sequence of interior points). (4) Its boundary consists of a finite number of arcs that have continuously turning tangents and a finite number of corners (Fig. 4.1a).

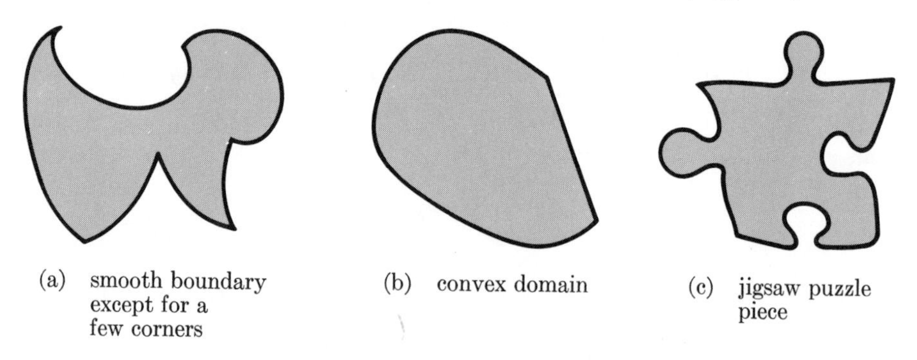

(a) smooth boundary except for a few corners

(b) convex domain

(c) jigsaw puzzle piece

FIG. 4.1 domains of integration

Many of the domains we deal with are convex (Fig. 4.1b). A domain is **convex** if it contains the line segment connecting any two of its points. Probably the most complicated domains arising in practice are shaped like the pieces of a jigsaw puzzle (Fig. 4.1c). Still, we can define the double integral of a continuous function on such a domain. We first cut the domain into a finite number of simple domains, then use iteration and the techniques developed in Chapter 10 to evaluate the integral on each of these.

Area

The double integral can be defined on domains that have *area* in a sense we define now.

Area Let $\mathbf{D}$ be a closed subset of the rectangle $\mathbf{I}$. Define the **characteristic function** $k_{\mathbf{D}}$ of $\mathbf{D}$ by

$$k_{\mathbf{D}}(x, y) = \begin{cases} 1 & \text{if} \quad (x, y) \in \mathbf{D} \\ 0 & \text{if} \quad (x, y) \notin \mathbf{D}. \end{cases}$$

The set $\mathbf{D}$ is **R-measurable** if $k_{\mathbf{D}}$ is integrable on $\mathbf{I}$. If so, the **area** of $\mathbf{D}$ is defined by

$$|\mathbf{D}| = \iint_{\mathbf{I}} k_{\mathbf{D}}\, dx\, dy.$$

Remark 1: Note that an R-measurable domain is automatically a *closed* set by this definition. Nevertheless, we shall stress this closure in the following theorems.

Remark 2: It will be left as an exercise to prove that if $\mathbf{I}$ and $\mathbf{J}$ are two rectangles, both containing $\mathbf{D}$, then $|\mathbf{D}|$ is the same whether computed in $\mathbf{I}$ or in $\mathbf{J}$. This is an easy, but necessary, result in order that $|\mathbf{D}|$ be properly defined.

The definition makes good sense. First it gives the right answer for any rectangle $\mathbf{J}$ in $\mathbf{I}$ (with sides parallel to the axes). Second, it is additive. This means that if two sets $\mathbf{D}$ and $\mathbf{E}$ are R-measurable and they do not intersect, then their union $\mathbf{D} \cup \mathbf{E}$ is R-measurable and $|\mathbf{D} \cup \mathbf{E}| = |\mathbf{D}| + |\mathbf{E}|$. This statement is contained in the next theorem.

Notation: At this point it is convenient to introduce the symbol $\varnothing$ for the empty set, the subset of $\mathbf{I}$ with no points. This set has the characteristic function $k_{\varnothing} = 0$, it is R-measurable, and $|\varnothing| = \iint 0 = 0$. The statement "$\mathbf{D}$ and $\mathbf{E}$ have no common points" is abbreviated simply "$\mathbf{D} \cap \mathbf{E} = \varnothing$".

Theorem 4.2 Suppose $\mathbf{D}$ and $\mathbf{E}$ are closed R-measurable subsets of $\mathbf{I}$.

(1) $\mathbf{D} \cap \mathbf{E}$ is R-measurable.
(2) $\mathbf{D} \cup \mathbf{E}$ is R-measurable.
(3) $|\mathbf{D} \cup \mathbf{E}| = |\mathbf{D}| + |\mathbf{E}| - |\mathbf{D} \cap \mathbf{E}|$.

Proof: First we observe that $k_{\mathbf{D}\cap\mathbf{E}} = k_{\mathbf{D}}k_{\mathbf{E}}$. For $k_{\mathbf{D}\cap\mathbf{E}}(\mathbf{x}) = 1$ precisely when $\mathbf{x} \in \mathbf{D} \cap \mathbf{E}$, that is, $\mathbf{x} \in \mathbf{D}$ and $\mathbf{x} \in \mathbf{E}$, which happens precisely when $k_{\mathbf{D}}(\mathbf{x}) = 1$ and $k_{\mathbf{E}}(\mathbf{x}) = 1$, that is, $k_{\mathbf{D}}(\mathbf{x})k_{\mathbf{E}}(\mathbf{x}) = 1$.

Now (1) follows because $k_{\mathbf{D}\cap\mathbf{E}}$ is the product of integrable functions, hence integrable.

For (2) we need the relation

$$k_{\mathbf{D}\cup\mathbf{E}} = k_{\mathbf{D}} + k_{\mathbf{E}} - k_{\mathbf{D}\cap\mathbf{E}}.$$

It is easily checked, because the right-hand side is 1 precisely when $\mathbf{x}$ is in $\mathbf{D}$ and not in $\mathbf{E}$ $(1 + 0 - 0)$, or $\mathbf{x}$ is in $\mathbf{E}$ and not in $\mathbf{D}$ $(0 + 1 - 0)$, or $\mathbf{x}$ is in both $(1 + 1 - 1)$, that is, when $\mathbf{x} \in \mathbf{D} \cup \mathbf{E}$.

Now (3) follows from

$$|\mathbf{D} \cup \mathbf{E}| = \iint_{\mathbf{I}} k_{\mathbf{D}\cup\mathbf{E}} = \iint_{\mathbf{I}} k_{\mathbf{D}} + \iint_{\mathbf{I}} k_{\mathbf{E}} - \iint_{\mathbf{I}} k_{\mathbf{D}\cap\mathbf{E}}$$
$$= |\mathbf{D}| + |\mathbf{E}| - |\mathbf{D} \cap \mathbf{E}|.$$

The next theorem includes a familiar situation, the area under a curve (the case $c = 0$).

Theorem 4.3 Let $h(x)$ be a continuous function on $[a, b]$ with $c \le h(x) \le d$. Define $\mathbf{D}$ by

$$\mathbf{D} = \{(x, y) \mid a \le x \le b, \quad c \le y \le h(x)\}.$$

Then $\mathbf{D}$ is R-measurable and

$$|\mathbf{D}| = \int_a^b [h(x) - c]\,dx.$$

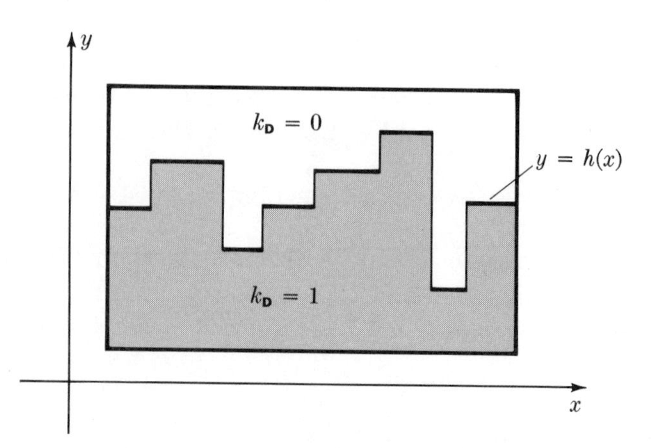

FIG. 4.2 The set (shaded) below the graph of h is $\mathbf{D}$.

Proof: First suppose h is a step function on $[a, b]$. Then $k_{\mathbf{D}}$ is a step function on $\mathbf{I}$. See Fig. 4.2. Hence $k_{\mathbf{D}}$ is integrable and

$$\iint\limits_{\mathbf{I}} k_{\mathbf{D}} = \int_a^b \left(\int_c^d k_{\mathbf{D}}(x, y)\, dy \right) dx.$$

For fixed x, we have $k_{\mathbf{D}}(x, y) = 1$ if $c \leq y \leq h(x)$ and $k_{\mathbf{D}}(x, y) = 0$ if $h(x) < y \leq d$. Hence

$$\int_c^d k_{\mathbf{D}}(x, y)\, dy = h(x) - c,$$

so

$$\iint\limits_{\mathbf{I}} k_{\mathbf{D}} = \int_a^b [h(x) - c]\, dx.$$

Now suppose h is continuous on $[a, b]$. Then $\mathbf{D}$ is closed (why?) and we must prove $\mathbf{D}$ is R-measurable. Given $\epsilon > 0$, there are step functions s and S on $[a, b]$ such that

$$c \leq s \leq h \leq S \leq d \qquad \text{and} \qquad \int_a^b (S - s)\, dx < \epsilon.$$

To estimate $k_{\mathbf{D}}$, we construct the sets

$$\mathbf{C} = \{(x, y) \mid a \leq x \leq b, \quad c \leq y \leq s(x)\}$$

and

$$\mathbf{E} = \{(x, y) \mid a \leq x \leq b, \quad c \leq y \leq S(x)\}.$$

Clearly $\mathbf{C} \subseteq \mathbf{D} \subseteq \mathbf{E}$, so $k_{\mathbf{C}} \leq k_{\mathbf{D}} \leq k_{\mathbf{E}}$. As we have seen, $k_{\mathbf{C}}$ and $k_{\mathbf{E}}$ are step functions, and

$$\iint\limits_{\mathbf{I}} k_{\mathbf{E}} - \iint\limits_{\mathbf{I}} k_{\mathbf{C}} = \int_a^b [S(x) - c]\, dx - \int_a^b [s(x) - c]\, dx$$

$$= \int_a^b [S(x) - s(x)]\, dx < \epsilon.$$

This proves that $k_{\mathbf{D}}$ is integrable on $\mathbf{I}$, hence $\mathbf{D}$ is R-measurable, and also that

$$\int_a^b [s(x) - c]\, dx \leq \iint\limits_{\mathbf{I}} k_{\mathbf{D}}\, dx\, dy \leq \int_a^b [S(x) - c]\, dx.$$

Since $\iint k_{\mathbf{D}} = |\mathbf{D}|$, this relation says that

$$\int (s - c)\, dx \leq |\mathbf{D}| \leq \iint (S - c)\, dx$$

for all s and S such that $s \le h \le S$. But the *only* number satisfying these inequantities is $\int (h - c)$, hence $|\mathbf{D}| = \int (h - c)$.

As a corollary, we obtain the area of a region bounded above and below by the graphs of two continuous functions.

Theorem 4.4 Suppose $g(x)$ and $h(x)$ are continuous functions on $[a, b]$ with $c \le g(x) \le h(x) \le d$. Define $\mathbf{D}$ by

$$\mathbf{D} = \{(x, y) \mid a \le x \le b, \quad g(x) \le y \le h(x)\}.$$

Then $\mathbf{D}$ is R-measurable and

$$|\mathbf{D}| = \int_a^b [h(x) - g(x)]\, dx.$$

Proof: This is a consequence of the last theorem, applied twice, and the relation (Fig. 4.3)

$$\mathbf{D} = \{(x, y) \mid c \le y \le h(x)\} \cap \{(x, y) \mid g(x) \le y \le d\}.$$

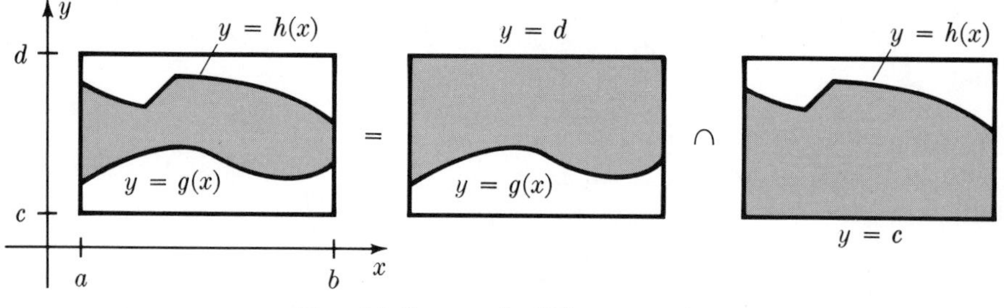

FIG. 4.3 See proof of Theorem 4.4.

REMARK: The preceding results are adequate for proving that standard geometric figures, like polygons and circles, have area. With a little patience one can now derive the usual area formulas for triangles, parallelograms, etc.

Integrals

Let $\mathbf{D}$ be a closed subset of a rectangle $\mathbf{I}$ and let f be a bounded real-valued function with domain $\mathbf{D}$. We seek conditions under which

$$\iint_{\mathbf{D}} f\, dx\, dy$$

can be defined. A first modest requirement is that $\mathbf{D}$ should be R-measurable.

Then at least

$$\iint_{\mathbf{D}} 1\,dx\,dy = \iint_{\mathbf{I}} k_{\mathbf{D}}\,dx\,dy$$

makes sense. We now need a reasonable definition of integrability on non-rectangular closed domains.

Integrable Function Let **D** be a closed R-measurable subset of **I**, and let f have domain **D**. Define F on **I** by

$$F = f \quad \text{on } \mathbf{D}, \qquad F = 0 \quad \text{elsewhere.}$$

We say f is **integrable** on **D** if F is integrable on **I**, and we define

$$\iint_{\mathbf{D}} f\,dx\,dy = \iint_{\mathbf{I}} F\,dx\,dy.$$

It can be shown that this definition is independent of the rectangle **I**.

Theorem 4.5 Suppose **D** is a closed R-measurable subset of **I**, and suppose f is integrable on the rectangle **I**. Then f is integrable on **D**.

Proof: The product $F = k_{\mathbf{D}}f$ of two functions integrable on **I** is itself integrable. But clearly $F = f$ on **D** and $F = 0$ outside **D**, so this integrable function F is exactly the F of the definition. Hence f is integrable on **D**.

The General Iteration Theorem

Theorem 4.6 Let $g(x)$ and $h(x)$ be continuous functions on $[a, b]$, with $c \leq g(x) \leq h(x) \leq d$. Define **D** as the subset of **I** bounded by the graphs of g and h:

$$\mathbf{D} = \{(x, y) \mid a \leq x \leq b, \quad g(x) \leq y \leq h(x)\}.$$

Let $f(x, y)$ be continuous on **D**. Then f is integrable on **D** and

$$\iint_{\mathbf{D}} f(x, y)\,dx\,dy = \int_a^b \left(\int_{g(x)}^{h(x)} f(x, y)\,dy\right) dx.$$

Proof: We begin by constructing a continuous function f^* on **I** such that $f^* = f$ on **D**. The idea is to make f^* constant on the vertical segments above and below **D**. Thus we set

$$f^*(x, y) = \begin{cases} f[x, g(x)] & \\ f(x, y) & \text{for} \\ f[x, h(x)] & \end{cases} \quad \begin{cases} c \leq y \leq g(x) \\ g(x) \leq y \leq h(x) \\ h(x) \leq y \leq d. \end{cases}$$

By definition, f^* agrees with f on **D**. The proof that f^* is continuous is routin and is left as an exercise. By Theorem 3.3, f^* is integrable on **I**, so Theorem 4. implies f is integrable on **D**.

Let $F = f$ on **D** and $F = 0$ outside **D**. By definition,

$$\iint_{\mathbf{D}} f\,dx\,dy = \iint_{\mathbf{I}} F\,dx\,dy.$$

For each fixed x,

$$\int_c^d F(x, y)\,dy$$

exists because for fixed x,

$$F(x, y) = \begin{cases} 0, & c \le y \le g(x) \\ f(x, y), & g(x) \le y \le h(x) \\ 0, & h(x) \le y \le d, \end{cases}$$

a step function on the two end intervals $[c, g(x)]$ and $[h(x), d]$ and a con tinuous function on the middle interval $[g(x), h(x)]$. Therefore the hypothese of Theorem 4.1 are both satisfied by f, so we conclude

$$\iint_{\mathbf{D}} f\,dx\,dy = \iint_{\mathbf{I}} F\,dx\,dy = \int_a^b \left(\int_c^d F\,dy \right) dx.$$

But

$$\int_c^d F\,dy = \int_c^{g(x)} F\,dy + \int_{g(x)}^{h(x)} F\,dy + \int_{h(x)}^d F\,dy = \int_{g(x)}^{h(x)} f(x, y)\,dy.$$

The theorem follows.

Theorem 4.7 Suppose **D** is a closed R-measurable subset of **I**, and suppose f is a continuous function with domain **D**. Then f is integrable on **D**.

A complete proof of this theorem, with the tools we have developed, is possible but arduous. Instead we shall give a short elegant proof, for which we borrow an important and plausible result from an advanced course on real functions, the Tietze Extension Theorem: There exists a continuous function f^* with domain **I** such that $f^* = f$ on **D**. With this assumed, the proof of Theorem 4.7 is very short indeed: Since f^* is continuous on **I**, it is integrable. Apply Theorem 4.5. Done!

REMARK: The Tietze Theorem stated precisely is this. Let **D** be a closed subset of $\mathbf{R}^2$ (also for $\mathbf{R}^3$, or $\mathbf{R}^n$ in general). Let f be a continuous real-valued function on **D**. Then there is a continuous function f^* on $\mathbf{R}^2$ such that $f^* = f$ on **D**.

Additivity

We close this section with a broad generalization of Theorem 3.8. First we require a more general definition of partition.

Partition Let $\mathbf{D}$ be R-measurable. A **partition** of $\mathbf{D}$ consists of R-measurable sets $\mathbf{D}_1, \cdots, \mathbf{D}_n$ such that

(1) $\mathbf{D} = \mathbf{D}_1 \cup \mathbf{D}_2 \cup \cdots \cup \mathbf{D}_n$,
(2) $|\mathbf{D}_i \cap \mathbf{D}_j| = 0 \quad \text{if} \quad i \neq j$.

Theorem 4.8 Suppose $\mathbf{D}$ is partitioned into $\mathbf{D}_1, \cdots, \mathbf{D}_n$, and suppose f is a bounded function on $\mathbf{D}$. Then f is integrable on $\mathbf{D}$ if and only if f is integrable on each $\mathbf{D}_j$. If so, then

$$\iint_{\mathbf{D}} f\,dx\,dy = \sum_{j=1}^{n} \iint_{\mathbf{D}_j} f\,dx\,dy.$$

Proof: Let the characteristic function of $\mathbf{D}_j$ be k_j, an integrable function by hypothesis. Suppose f is integrable on $\mathbf{D}$. Then fk_j is integrable on $\mathbf{D}$, which means f is integrable on $\mathbf{D}_j$ since $\mathbf{D}_j \subseteq \mathbf{D}$.

Now we compute $\sum k_j$. Clearly

$$\sum_{j=1}^{m} k_j = k_{\mathbf{D}} + e,$$

where the value of the error function $e = e(\mathbf{x})$ is one less than the number of sets $\mathbf{D}_j$ containing $\mathbf{x}$. Since $e(\mathbf{x}) > 0$ only if $\mathbf{x}$ belongs to two or more of the $\mathbf{D}_j$, and $|\mathbf{D}_i \cap \mathbf{D}_j| = 0$, we have

$$\iint_{\mathbf{D}} e \leq (m-1) \sum_{i \neq j} |\mathbf{D}_i \cap \mathbf{D}_j| = 0.$$

Therefore

$$\left| \iint_{\mathbf{D}} fe \right| \leq (\max_{\mathbf{D}} f) \iint e = 0, \qquad \text{hence} \qquad \iint_{\mathbf{D}} fe = 0.$$

But on $\mathbf{D}$,

$$f = fk_{\mathbf{D}} = \sum_{j=1}^{n} fk_j + fe,$$

so

$$\sum_{j=1}^{n} \iint_{\mathbf{D}_j} f = \sum_{j=1}^{n} \iint_{\mathbf{D}} fk_j = \iint_{\mathbf{D}} f - \iint_{\mathbf{D}} fe = \iint_{\mathbf{D}} f.$$

Now suppose f is integrable on each $\mathbf{D}_j$. Then the same relation, $f = \sum fk_j + fe$, implies f is integrable on $\mathbf{D}$. This completes the proof.

EXERCISES

1. In the proof of Theorem 4.1, we omitted to prove that $g(x)$ is *bounded*. Supply this missing step.
2. Prove the assertion about two rectangles in Remark 2 on p. 475.
3. Prove that the area of a line segment is 0.
4. Prove, on the basis of Theorem 4.4, that a triangle with vertices 0, (x_1, y_1), and (x_2, y_2), where $x_1 \leq 0 \leq x_2$, has area $\frac{1}{2}|x_1y_2 - x_2y_1|$.
5. (cont.) Prove that a triangle with vertices (x_1, y_1), (x_2, y_2), (x_3, y_3) has area

$$\frac{1}{2}\left\|\begin{matrix} x_1 & y_1 & 1 \\ x_2 & y_2 & 1 \\ x_3 & y_3 & 1 \end{matrix}\right\|.$$

6. (cont.) Prove $A = bh$ for a parallelogram.

7*. Prove that the function f^* in the proof of Theorem 4.6 is continuous.

5. CHANGE OF VARIABLES

Let us begin with a review of the one-variable situation.

Theorem 5.1 Suppose $\phi(u)$ is continuously differentiable on $[a, b]$ and suppose $\phi'(u) > 0$. Let $f(x)$ be integrable on $[\phi(a), \phi(b)]$. Then $f[\phi(u)]\phi'(u)$ is integrable on $[a, b]$ and

$$\int_{\phi(a)}^{\phi(b)} f(x)\,dx = \int_a^b f[\phi(u)]\phi'(u)\,du.$$

We shall not give a detailed proof of this theorem. We simply note that it is easily verified when f is a step function. The usual approximation—with some technical details—then proves it for f integrable.

Let us rather interpret Theorem 5.1 in such a way that its analogue in more dimensions, Theorem 5.2, will seem natural.

We write $x = \phi(u)$, where u runs over the interval $\mathbf{D} = [a, b]$ and x runs over the interval $\mathbf{E} = [\phi(a), \phi(b)]$. Since $\phi'(u) > 0$, it follows that x increases as u increases. Thus ϕ is a one-one mapping on the set $\mathbf{D}$ onto the set $\mathbf{E}$, and we may write $\mathbf{E} = \phi(\mathbf{D})$. With this notation, the formula of Theorem 5.1 can be written

$$\int_{\mathbf{E}} f(x)\,dx = \int_{\mathbf{D}} f[\phi(u)]\phi'(u)\,du,$$

or with $x = x(u)$ instead of $x = \phi(u)$,

$$\int_{\mathbf{E}} f(x)\,dx = \int_{\mathbf{D}} f[x(u)]\frac{dx}{du}\,du.$$

A similar formula holds in the plane and in space. There ϕ is a one-one mapping of some domain $\mathbf{D}$ of, say, $\mathbf{R}^2$ onto another domain $\mathbf{E}$. We shall need some information about such mappings, in particular, what replaces the derivative dx/du.

Jacobians

Let us note some facts about mappings. Let $\mathbf{S}$ be an open set in the u, v-plane and ϕ a continuously differentiable mapping of $\mathbf{S}$ into the x, y-plane. Write

$$\phi: \begin{cases} x = x(u, v) \\ y = y(u, v). \end{cases}$$

Thus the four partials $\partial x/\partial u, \cdots, \partial y/\partial v$ exist and are continuous on $\mathbf{S}$. We define the **Jacobian** of ϕ to be the determinant

$$\boxed{\frac{\partial(x, y)}{\partial(u, v)} = \begin{vmatrix} \dfrac{\partial x}{\partial u} & \dfrac{\partial x}{\partial v} \\ \dfrac{\partial y}{\partial u} & \dfrac{\partial y}{\partial v} \end{vmatrix}.}$$

It is a continuous function on $\mathbf{S}$.

Suppose ϕ maps $\mathbf{S}$ into an open set $\mathbf{T}$ and ψ in turn is a continuously differentiable mapping of $\mathbf{T}$ into the z, w-plane. The composite $\psi \circ \phi$ is a mapping of $\mathbf{S}$ into the z, w-plane. Thus

$$\begin{cases} z = z(x, y) = z[x(u, v), y(u, v)] \\ w = w(x, y) = w[x(u, v), y(u, v)]. \end{cases}$$

By the Chain Rule, we have the matrix relation

$$\begin{bmatrix} \dfrac{\partial z}{\partial u} & \dfrac{\partial z}{\partial v} \\ \dfrac{\partial w}{\partial u} & \dfrac{\partial w}{\partial v} \end{bmatrix} = \begin{bmatrix} \dfrac{\partial z}{\partial x} & \dfrac{\partial z}{\partial y} \\ \dfrac{\partial w}{\partial x} & \dfrac{\partial w}{\partial y} \end{bmatrix} \begin{bmatrix} \dfrac{\partial x}{\partial u} & \dfrac{\partial x}{\partial v} \\ \dfrac{\partial y}{\partial u} & \dfrac{\partial y}{\partial v} \end{bmatrix}.$$

Recall that the determinant of a product of matrices is the product of their determinants. Therefore

$$\boxed{\frac{\partial(z, w)}{\partial(u, v)} = \frac{\partial(z, w)}{\partial(x, y)} \frac{\partial(x, y)}{\partial(u, v)}.}$$

Suppose in particular that ϕ has an inverse and $\psi = \phi^{-1}$. Then ϕ is one-one on **S** onto **T**; also ψ is one-one on **T** onto **S**, and $\psi \circ \phi =$ identity. Then

$$\psi: \begin{cases} u = u(x, y) \\ v = v(x, y), \end{cases} \qquad \psi \circ \phi: \begin{cases} u = u[x(u, v), y(u, v)] \\ v = v[x(u, v), y(u, v)], \end{cases}$$

so

$$\frac{\partial(u, v)}{\partial(x, y)} \frac{\partial(x, y)}{\partial(u, v)} = \begin{vmatrix} 1 & 0 \\ 0 & 1 \end{vmatrix} = 1.$$

A continuously differentiable mapping ϕ that has a continuously differentiable inverse is called a **regular transformation**. We have proved:

> The Jacobian of a regular transformation is never 0.

Suppose the domain **S** of ϕ, in addition to being open, is connected. That is, any two of its points can be connected by a curve in **S**. Then the Jacobian of a regular transformation has constant sign in **S**, either always positive or always negative. For otherwise it would have to be zero at some point along a curve joining points where it had opposite signs. We shall concentrate on the case where the sign is always positive. We shall call a regular transformation with everywhere positive Jacobian a **proper transformation**.

Examples

Example 1. Translation:

$$\phi: \begin{cases} x = u + a \\ y = v + b \end{cases} \qquad \frac{\partial(x, y)}{\partial(u, v)} = \begin{vmatrix} 1 & 0 \\ 0 & 1 \end{vmatrix} = 1.$$

The transformation is proper on the whole plane.

Example 2. Linear transformation:

$$\phi: \begin{cases} x = a_{11}u + a_{12}v \\ y = a_{21}u + a_{22}v \end{cases} \qquad \frac{\partial(x, y)}{\partial(u, v)} = \begin{vmatrix} a_{11} & a_{12} \\ a_{21} & a_{22} \end{vmatrix}.$$

If we assume the determinant is positive, then ϕ is a proper transformation on the whole plane. If **D** is the unit square in the u, v-plane, then $\mathbf{E} = \phi(\mathbf{D})$ is a parallelogram (Fig. 5.1). More generally, any rectangle in the u, v-plane is mapped to a parallelogram.

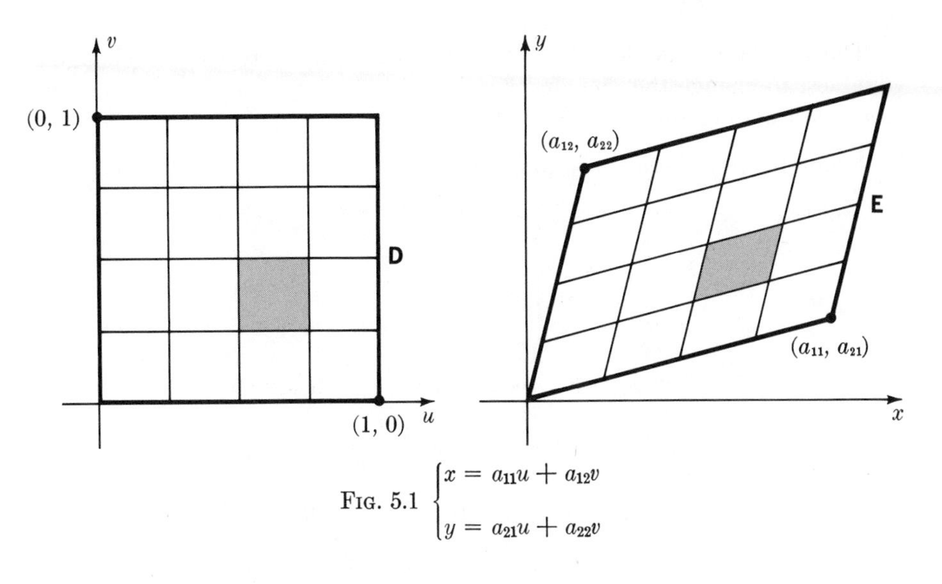

FIG. 5.1 $\begin{cases} x = a_{11}u + a_{12}v \\ y = a_{21}u + a_{22}v \end{cases}$

Example 3. Polar coordinates:

$$\phi: \quad \begin{cases} x = r\cos\theta \\ y = r\sin\theta \end{cases} \qquad \frac{\partial(x, y)}{\partial(r, \theta)} = \begin{vmatrix} \cos\theta & -r\sin\theta \\ \sin\theta & r\cos\theta \end{vmatrix} = r.$$

This is a proper transformation on any domain **D** of the r, θ-plane that avoids the θ-axis. For instance, the rectangle **D** in Fig. 5.2 is mapped to the circular region **E**.

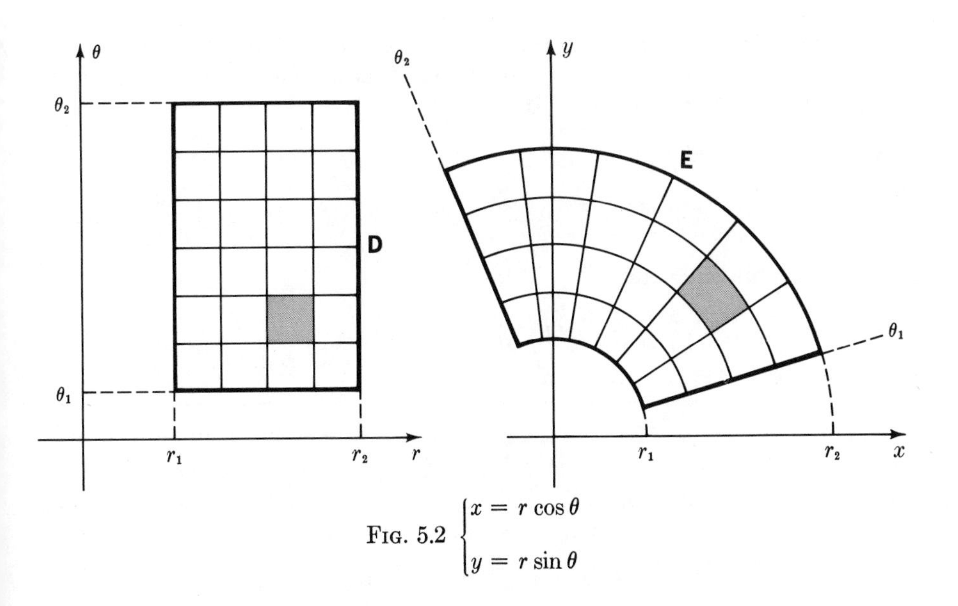

FIG. 5.2 $\begin{cases} x = r\cos\theta \\ y = r\sin\theta \end{cases}$

The last two examples illustrate the most common reason for changing variables, to change the domain of integration to a rectangle. Here is a further such example, but in 3-space.

Example 4. Spherical coordinates:

$$\phi: \begin{cases} x = \rho \sin \phi \cos \theta \\ y = \rho \sin \phi \sin \theta \\ z = \rho \cos \phi, \end{cases}$$

$$\frac{\partial(x, y, z)}{\partial(\rho, \phi, \theta)} = \begin{vmatrix} \sin \phi \cos \theta & \rho \cos \phi \cos \theta & -\rho \sin \phi \sin \theta \\ \sin \phi \sin \theta & \rho \cos \phi \sin \theta & \rho \sin \phi \cos \theta \\ \cos \phi & -\rho \sin \phi & 0 \end{vmatrix}$$

$$= \rho^2 \sin \phi \begin{vmatrix} \sin \phi \cos \theta & \cos \phi \cos \theta & -\sin \theta \\ \sin \phi \sin \theta & \cos \phi \sin \theta & \cos \theta \\ \cos \phi & -\sin \phi & 0 \end{vmatrix} = \rho^2 \sin \phi.$$

(The determinant can be expanded easily by minors of the third row.) A rectangular solid **D**: $0 < \rho_0 \leq \rho \leq \rho_1$, $0 < \phi_0 \leq \phi \leq \phi_1 < \pi$, $0 \leq \theta_0 \leq \theta \leq \theta_1 \leq 2\pi$ in ρ, ϕ, θ-space maps to a curved solid in x, y, z-space.

The Main Theorem

Theorem 5.2 Let **S** be an open subset of the u, v-plane and ϕ a proper transformation on **S** into the x, y-plane. Let **D** be an R-measurable subset of **S** and **E** $= \phi($**D**$)$ the image of **D** under ϕ. Then:

(1) **E** is R-measurable.

(2) If f is integrable on **E**, then the product $[(f \circ \phi)(u, v)] \dfrac{\partial(x, y)}{\partial(u, v)}$ is integrable on **D** and

$$\iint_{\mathbf{E}} f(x, y)\, dx\, dy = \iint_{\mathbf{D}} f[x(u, v), y(u, v)] \frac{\partial(x, y)}{\partial(u, v)}\, du\, dv.$$

A proof is hard and beyond the scope of this course. We shall note one case where the theorem is quite plausible, that of the linear transformation on a rectangle (Example 2 above). First suppose $f = 1$ on **E**, and look carefully again at Fig. 5.1. The formula in Theorem 5.2 boils down to

$$|\mathbf{E}| = |\mathbf{D}|\,(a_{11}a_{22} - a_{12}a_{21}) = a_{11}a_{22} - a_{12}a_{21}.$$

This is certainly correct, since the area of the parallelogram **E**, by vector algebra, is

$$|\mathbf{E}| = |(a_{11}, a_{21}) \times (a_{12}, a_{22})| = a_{11}a_{22} - a_{12}a_{21}.$$

In fact *any* rectangle $\mathbf{D}_1$ in the u, v-plane is mapped to a parallelogram $\mathbf{E}_1$ with $|\mathbf{E}_1| = |\mathbf{D}_1| (a_{11}a_{22} - a_{12}a_{21})$. Therefore if $f(x, y)$ is a *step function* on **E** relative to a division of **E** by lines parallel to the sides of **E**, then f is a step function on **D** and (2) in Theorem 5.2 is again true. This suggests an approximation procedure for proving (2) for any integrable function on **E**, but we must omit the details.

Once Theorem 5.2 is known for proper linear transformations on rectangles, it can be proved in general by another, deeper, approximation technique. This is to approximate the general proper transformation by a piecewise linear transformation; the Jacobian plays the role of an area distortion factor. The complete proof is quite formidable.

Polar Coordinates

The polar coordinate mapping

$$\begin{cases} x = r\cos\theta \\ y = r\sin\theta \end{cases} \qquad \frac{\partial(x, y)}{\partial(r, \theta)} = r$$

is a proper transformation in the open set

$$\mathbf{S} = \{(r, \theta) \mid 0 < r, \quad 0 < \theta < 2\pi\},$$

and (Fig. 5.3) it maps **S** onto

$$\mathbf{T} = \{(x, y) \mid y \neq 0 \quad \text{or} \quad y = 0 \text{ and } x < 0\}.$$

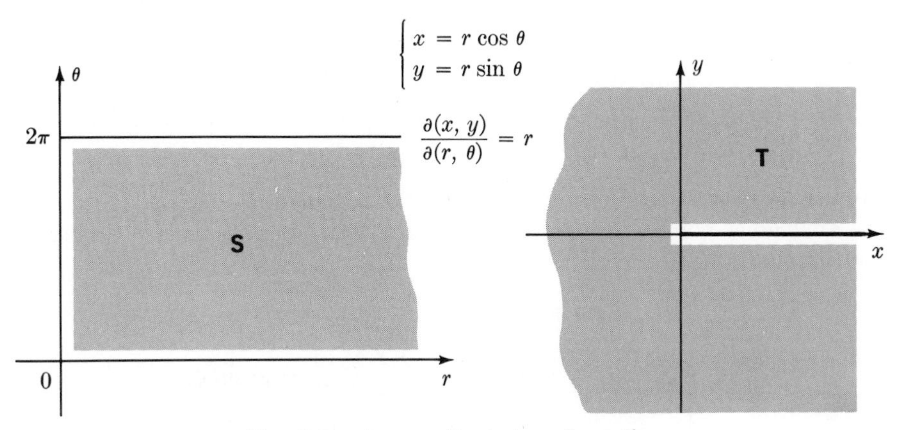

FIG. 5.3 polar coordinate transformation

There is a problem with $r = 0$ because many typical integration problems include $r = 0$. See Fig. 5.4.

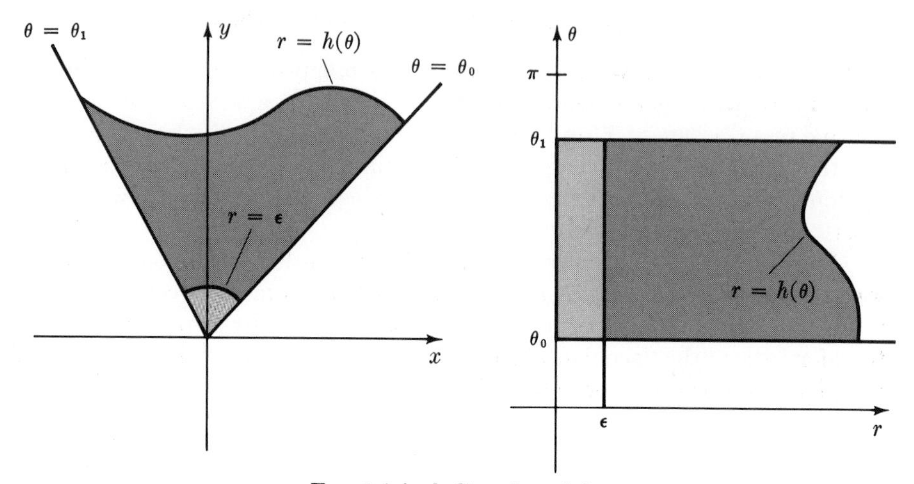

FIG. 5.4 including the origin

The problem is solved by deleting a small circular sector and taking limits:

$$\iint_{\mathbf{E}} f(x, y)\, dx\, dy = \lim_{\epsilon \to 0} \iint_{\mathbf{E},\, r \geq \epsilon} f\, dx\, dy$$

$$= \lim_{\epsilon \to 0} \iint_{\mathbf{D},\, r \geq \epsilon} f(r \cos \theta, r \sin \theta)\, r\, dr\, d\theta = \iint_{\mathbf{D}} f\, r\, dr\, d\theta.$$

The process works because the excluded region (circular sector in **E**, rectangular strip in **D**) has area that goes to 0 as $\epsilon \longrightarrow 0$.

Similarly we can take care of a domain **E** that goes completely around the origin, so that it hits both horizontal boundaries of **S** in Fig. 5.3. In summary, the formula

$$\iint_{\mathbf{E}} f\, dx\, dy = \iint_{\mathbf{D}} f \frac{\partial(x, y)}{\partial(r, \theta)}\, dr\, d\theta = \iint_{\mathbf{D}} f(r \cos \theta, r \sin \theta)\, r\, dr\, d\theta$$

holds without restriction on the R-measurable domain **D**.

Analogous considerations apply to the transformation from spherical to rectangular coordinates:

$$\begin{cases} x = \rho \sin \phi \cos \theta \\ y = \rho \sin \phi \sin \theta \\ z = \rho \cos \phi \end{cases} \qquad \frac{\partial(x, y, z)}{\partial(\rho, \phi, \theta)} = \rho^2 \sin \phi.$$

EXERCISES

Compute the Jacobian and prove that the mapping is a proper transformation:

1. $x = 2u + v, \quad y = u + 2v, \qquad$ all (u, v)
2. $x = u + v, \quad y = uv, \qquad u > v$
3. $x = u + v, \quad y = u^2 + v^2, \qquad v > u$
4. $x = u, \quad y = uv, \qquad u > 0$
5. $x = u, \quad y = uv, \quad z = uvw, \qquad u > 0, \quad v > 0$
6. $x = vw, \quad y = wu, \quad z = uv, \qquad u > 0, \quad v > 0, \quad w > 0.$
7. The transformation of **inversion** in the plane is defined by $\mathbf{x} \longrightarrow \mathbf{x}/|\mathbf{x}|^2$ for $\mathbf{x} \neq \mathbf{0}$. Find its Jacobian.
8. (cont.) Generalize to $\mathbf{R}^3$.
9. What domain in ρ, ϕ, θ-space corresponds to the solid sphere $|\mathbf{x}| \leq \epsilon$ in x, y, z-space? Show that its volume goes to 0 as $\epsilon \longrightarrow 0$.

6. APPLICATIONS OF INTEGRATION

Leibniz Rule

In this section we take up a number of important applications of multiple integration. The first concerns differentiation under the integral sign. Consider a function defined by a definite integral,

$$F(t) = \int_a^b f(x, t)\, dx.$$

Think of t as a parameter. When the variable x is "integrated out", there remains a function of t. Problem: find the derivative $F'(t)$. The answer is called the Leibniz Rule, or the rule for differentiating under the integral sign.

Leibniz Rule Suppose $f(x, t)$ and the partial derivative $f_t(x, t)$ are continuous on a rectangle

$$a \leq x \leq b, \qquad c \leq t \leq d.$$

Then

$$\frac{d}{dt}\int_a^b f(x, t)\, dx = \int_a^b f_t(x, t)\, dx$$

for $c \leq t \leq d$.

Proof: For each $t \in [c, d]$ let $\mathbf{D}_t$ denote the rectangle

$$\mathbf{D}_t = \{(x, s) \mid a \leq x \leq b, \quad c \leq s \leq t\}$$

in the x, s-plane. The idea is to evaluate

$$G(t) = \iint_{\mathbf{D}_t} f_s(x, s)\, dx\, ds$$

in two different ways, then to compute $G'(t)$. On the one hand,

$$(1) \qquad G(t) = \int_a^b \left(\int_c^t f_s(x, s)\, ds \right) dx = \int_a^b [f(x, t) - f(x, c)]\, dx,$$

where the inner integral is evaluated by the Fundamental Theorem of Calculus. On the other hand,

$$(2) \qquad G(t) = \int_c^t \left(\int_a^b f_s(x, s)\, dx \right) ds.$$

From (1),

$$\frac{d}{dt} G(t) = \frac{d}{dt} \int_a^b f(x, t)\, dx - \frac{d}{dt} \int_a^b f(x, c)\, dx = \frac{d}{dt} \int_a^b f(x, t)\, dx.$$

From (2),

$$\frac{d}{dt} G(t) = \frac{d}{dt} \int_c^t \left(\int_a^b f_s(x, s)\, dx \right) ds = \int_a^b f_t(x, t)\, dx.$$

The Leibniz Rule follows upon equating these expressions for $G'(t)$.

EXAMPLE 6.1

Find $\displaystyle \frac{d}{dt} \int_0^\pi \frac{\sin tx}{x}\, dx \quad$ at $\displaystyle t = \frac{1}{2}$.

Solution:

$$\frac{d}{dt} \int_0^\pi \frac{\sin tx}{x}\, dx = \int_0^\pi \frac{\partial}{\partial t} \left(\frac{\sin tx}{x} \right) dx = \int_0^\pi \cos tx\, dx = \frac{\sin \pi t}{t}.$$

When $t = \frac{1}{2}$, the value is $(\sin \frac{1}{2}\pi)/\frac{1}{2} = 2$.

Answer: 2.

REMARK: It is known that $F(t) = \displaystyle\int [(\sin tx)/x]\, dx$ cannot be expressed in terms of (a finite number of) the usual functions of calculus. In other words, you won't find it in a table of integrals, except as an infinite series. Nevertheless, $F(t)$ is a perfectly good differentiable function. But to compute its derivative, you need the Leibniz Rule.

Volume of an Ellipsoid

EXAMPLE 6.2

Find the volume enclosed by the ellipsoid

$$\frac{x^2}{a^2} + \frac{y^2}{b^2} + \frac{z^2}{c^2} = 1, \qquad a, b, c > 0.$$

Solution: Define sets

$$\mathbf{D} = \{(u, v, w) \mid u^2 + v^2 + w^2 \leq 1\},$$

$$\mathbf{E} = \left\{(x, y, z) \;\middle|\; \frac{x^2}{a^2} + \frac{y^2}{b^2} + \frac{z^2}{c^2} \leq 1\right\},$$

and the mapping

$$\phi: \quad x = au, \quad y = bv, \quad z = cw.$$

Then ϕ takes $\mathbf{D}$ onto $\mathbf{E}$, and ϕ is a proper transformation (non-singular linear actually), so

$$|\mathbf{E}| = \iiint_{\mathbf{E}} dx\,dy\,dz = \iiint_{\mathbf{D}} \frac{\partial(x, y, z)}{\partial(u, v, w)}\,du\,dv\,dw$$

$$= \iiint_{\mathbf{D}} abc\,du\,dv\,dw = abc\,|\mathbf{D}|.$$

But $\mathbf{D}$ is the solid unit sphere, so $|\mathbf{D}| = \frac{4}{3}\pi$.

Answer: $\frac{4}{3}\pi abc$.

Surface Area

Our aim is to give a reasonable definition of surface area in $\mathbf{R}^3$.

Up to now, a surface has always meant the graph of a function $z = f(x, y)$. A surface represented this way is called a **non-parametric surface**. There is, however, a more flexible concept called a **parametric surface**. Such a surface is the image under a nice mapping of a domain in the plane, just as a parametric curve is the image of an interval.

We start with a reasonable domain $\mathbf{D}$ in the u, v-plane (parameter plane) and a differentiable function $\mathbf{x} = \mathbf{x}(u, v)$ on $\mathbf{D}$ with values in $\mathbf{R}^3$. We assume that $\mathbf{x}$ maps $\mathbf{D}$ one-to-one on a set $\mathbf{S} \subset \mathbf{R}^3$. Then $\mathbf{S}$ is a parametric surface.

Each point of $\mathbf{S}$ is determined by the unique point of $\mathbf{D}$ from which it comes. Thus the point $\mathbf{x}(u, v)$ can be assigned the coordinates (u, v). In general, the grid of coordinate lines $u =$ const., $v =$ const. is mapped onto a grid of **coordinate curves** on $\mathbf{S}$. See Fig. 6.1.

The horizontal coordinate line through a point (a, b) of $\mathbf{D}$ can be written $u = a + t$, $v = b$. The corresponding curve on the surface is

$$\mathbf{x}(t) = \mathbf{x}(a + t, b).$$

The velocity vector of this curve is

$$\frac{d\mathbf{x}}{dt} = \mathbf{x}_u.$$

Thus $\mathbf{x}_u$ is tangent to the coordinate curve, hence tangent to the surface. A similar statement holds for $\mathbf{x}_v$.

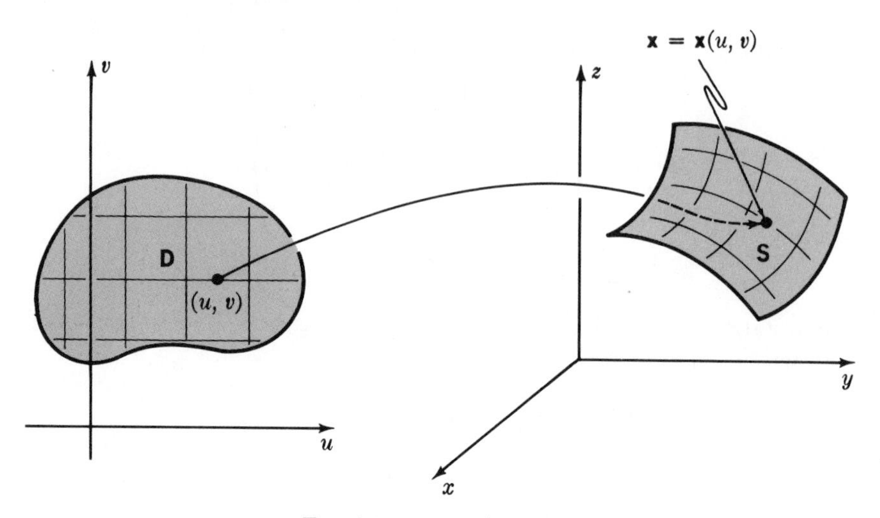

FIG. 6.1 parametric surface

We now make the assumption that $\mathbf{x}_u$ and $\mathbf{x}_v$ are non-collinear at each point of **S**. This guarantees that the two coordinate curves through each point of **S** are not tangent, but cross at a non-zero angle. In addition, it guarantees the existence of a well-defined tangent plane at each point of the surface. For the tangent plane is the plane determined by the non-collinear tangent vectors $\mathbf{x}_u$ and $\mathbf{x}_v$. Further evidence: the image of a curve $u = u(t)$, $v = v(t)$ in **D** is the curve $\mathbf{x}(t) = \mathbf{x}[u(t), v(t)]$. Its velocity vector (tangent to the surface) is

$$\frac{d\mathbf{x}}{dt} = \mathbf{x}_u \frac{du}{dt} + \mathbf{x}_v \frac{dv}{dt},$$

a vector which lies in the plane of $\mathbf{x}_u$ and $\mathbf{x}_v$.

Now the vector $\mathbf{x}_u \times \mathbf{x}_v$ plays an important role. It is not zero because $\mathbf{x}_u$ and $\mathbf{x}_v$ are non-collinear; it is perpendicular to the surface in the sense that it is perpendicular to $\mathbf{x}_u$ and to $\mathbf{x}_v$, hence to the tangent plane. Therefore

$$\mathbf{n} = \frac{\mathbf{x}_u \times \mathbf{x}_v}{|\mathbf{x}_u \times \mathbf{x}_v|}$$

is one of the two unit normals to the surface. The way it points defines for us the **top** of the parametric surface.

Let us return to our project of finding a reasonable definition for surface area. A small rectangle in **D** with sides du and dv maps to a small region on the surface. According to the formula $d\mathbf{x} = \mathbf{x}_u\, du + \mathbf{x}_v\, dv$, this region is closely approximated by the parallelogram (Fig. 6.2) in the tangent plane with sides $\mathbf{x}_u\, du$ and $\mathbf{x}_v\, dv$. Its area is

$$(*) \qquad dA = |(\mathbf{x}_u\, du) \times (\mathbf{x}_v\, dv)| = |\mathbf{x}_u \times \mathbf{x}_v|\, du\, dv.$$

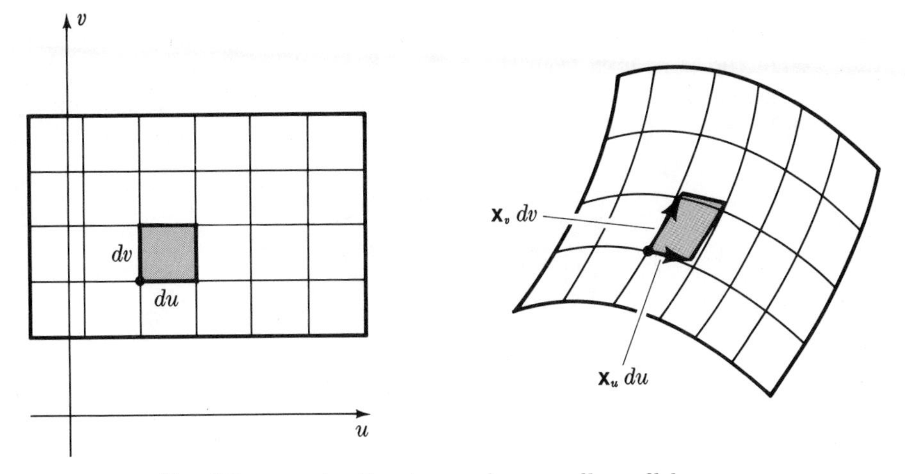

FIG. 6.2 approximation to area by a small parallelogram

Recall that

$$\mathbf{x}_u \times \mathbf{x}_v = \begin{vmatrix} \mathbf{i} & \mathbf{j} & \mathbf{k} \\ x_u & y_u & z_u \\ x_v & y_v & z_v \end{vmatrix} = \begin{vmatrix} y_u & z_u \\ y_v & z_v \end{vmatrix} \mathbf{i} + \begin{vmatrix} z_u & x_u \\ z_v & x_v \end{vmatrix} \mathbf{j} + \begin{vmatrix} x_u & y_u \\ x_v & y_v \end{vmatrix} \mathbf{k}$$

$$= \frac{\partial(y, z)}{\partial(u, v)} \mathbf{i} + \frac{\partial(z, x)}{\partial(u, v)} \mathbf{j} + \frac{\partial(x, y)}{\partial(u, v)} \mathbf{k}.$$

Consequently

$$|\mathbf{x}_u \times \mathbf{x}_v|^2 = \left(\frac{\partial(y, z)}{\partial(u, v)}\right)^2 + \left(\frac{\partial(z, x)}{\partial(u, v)}\right)^2 + \left(\frac{\partial(x, y)}{\partial(u, v)}\right)^2.$$

We substitute this into (*), then add up the small pieces of area by an integral. Thus we arrive at the following definition:

> **Surface Area** Let $\mathbf{x} = \mathbf{x}(u, v)$ define a parametric surface with domain $\mathbf{D}$. Its **surface area** is
>
> $$A = \iint_{\mathbf{D}} |\mathbf{x}_u \times \mathbf{x}_v| \, du \, dv$$
>
> $$= \iint_{\mathbf{D}} \sqrt{\left(\frac{\partial(y, z)}{\partial(u, v)}\right)^2 + \left(\frac{\partial(z, x)}{\partial(u, v)}\right)^2 + \left(\frac{\partial(x, y)}{\partial(u, v)}\right)^2} \, du \, dv.$$

There is one possible flaw in the definition. When we look at a surface, we see a definite area, quite independent of how we have *described* the surface. If the same *geometric* surface has two different parametrizations, how do we know that we get the same area from the two corresponding integrals?

This means that we have two plane domains **D** and **E**, each defining the surface **S** by differentiable functions $\mathbf{x}(u, v)$ and $\mathbf{y}(s, t)$ respectively (Fig. 6.3).

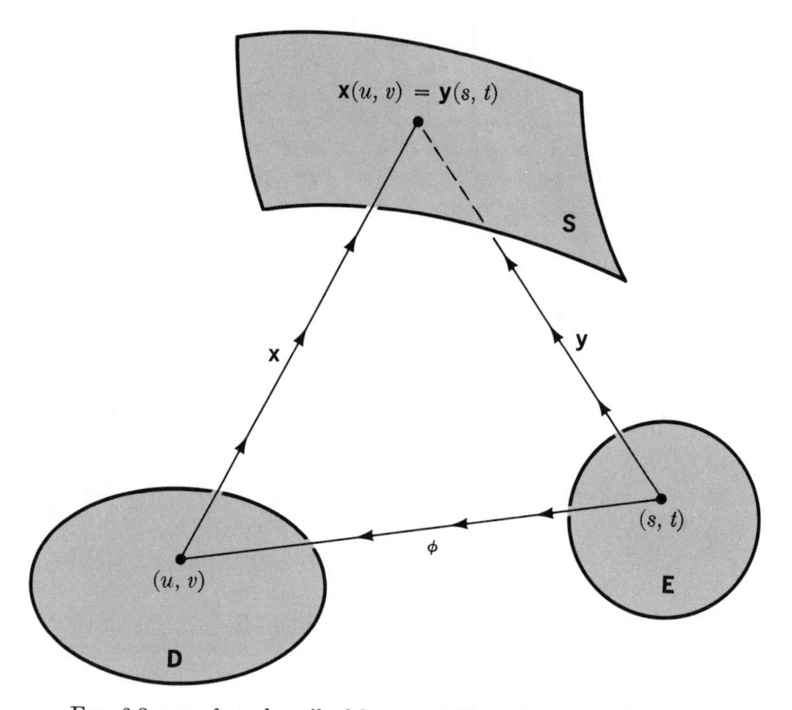

FIG. 6.3 a surface described by two different parametrizations

Since both **D** and **E** are in one-one correspondence with **S**, they are in one-one correspondence with each other. Thus there is a transformation

$$\phi: \begin{cases} u = u(s, t) \\ v = v(s, t) \end{cases}$$

from **E** onto **D** such that

$$\mathbf{y}(s, t) = \mathbf{x}[u(s, t), v(s, t)].$$

By the Chain Rule,

$$\mathbf{y}_s = \mathbf{x}_u u_s + \mathbf{x}_v v_s, \qquad \mathbf{y}_t = \mathbf{x}_u u_t + \mathbf{x}_v v_t,$$

hence

$$\mathbf{y}_s \times \mathbf{y}_t = (u_s v_t - v_s u_t)\mathbf{x}_u \times \mathbf{x}_v = \frac{\partial(u, v)}{\partial(s, t)} \mathbf{x}_u \times \mathbf{x}_v.$$

We also assume that there is a definite top side of the surface, the same whether computed by means of $\mathbf{x}(u, v)$ or $\mathbf{y}(s, t)$. This means that $\mathbf{x}_u \times \mathbf{x}_v$ and $\mathbf{y}_s \times \mathbf{y}_t$ point in the same direction, that is,

$$\frac{\partial(u, v)}{\partial(s, t)} > 0.$$

Hence ϕ is a *proper* transformation, and

$$|\mathbf{y}_s \times \mathbf{y}_t| = \frac{\partial(u, v)}{\partial(s, t)} |\mathbf{x}_u \times \mathbf{x}_v|.$$

Now we can compare the **D**-area and the **E**-area of the surface, using the change of variable formula:

$$A_{\mathbf{D}} = \iint_{\mathbf{D}} |\mathbf{x}_u \times \mathbf{x}_v| \, du \, dv$$

$$= \iint_{\mathbf{E}} |\mathbf{x}_u \times \mathbf{x}_v| \frac{\partial(u, v)}{\partial(s, t)} \, ds \, dt = \iint_{\mathbf{E}} |\mathbf{y}_s \times \mathbf{y}_t| \, ds \, dt = A_{\mathbf{E}}.$$

The areas are the same, computed either way. The definition is OK!

EXAMPLE 6.3

Find the area of the spiral ramp $\mathbf{x} = (u \cos v, u \sin v, bv)$ corresponding to the rectangle $\mathbf{D}$: $0 \leq u \leq a$, $0 \leq v \leq c$.

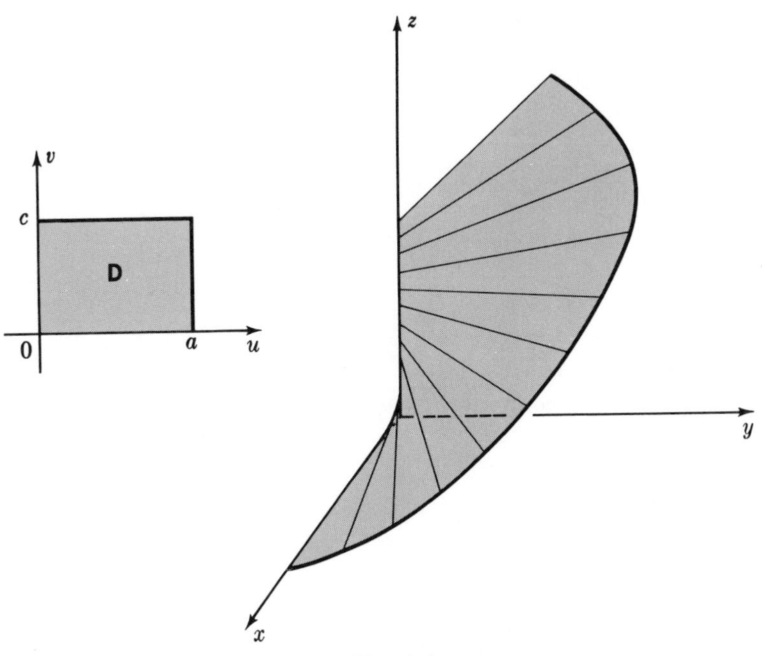

FIG. 6.4

Solution: Although not necessary, it is nice to sketch the surface (Fig. 6.4). Since

$$\mathbf{x}_u = (\cos v, \sin v, 0) \quad \text{and} \quad \mathbf{x}_v = (-u \sin v, u \cos v, b),$$

the element of area is

$$dA = \sqrt{\begin{vmatrix} \sin v & 0 \\ u\cos v & b \end{vmatrix}^2 + \begin{vmatrix} 0 & \cos v \\ b & -u\sin v \end{vmatrix}^2 + \begin{vmatrix} \cos v & \sin v \\ -u\sin v & u\cos v \end{vmatrix}^2}\, du\, dv$$

$$= \sqrt{b^2 \sin^2 v + b^2 \cos^2 v + u^2}\, du\, dv = \sqrt{b^2 + u^2}\, du\, dv.$$

As (u, v) ranges over the rectangle, the point $\mathbf{x}(u, v)$ runs over the spiral ramp. Hence

$$A = \iint_{\mathbf{D}} \sqrt{b^2 + u^2}\, du\, dv = \left(\int_0^a \sqrt{b^2 + u^2}\, du\right)\left(\int_0^c dv\right).$$

Answer: $\dfrac{c}{2}\left[a\sqrt{a^2 + b^2} + b^2 \ln\left(\dfrac{a + \sqrt{a^2 + b^2}}{b}\right)\right].$

A non-parametric surface, the graph of $z = f(x, y)$, is a special case of a parametric surface, provided we think of x and y as parameters. The variable point on the surface is $\mathbf{x} = (x, y, f(x, y))$. Then

$$\frac{\partial \mathbf{x}}{\partial x} = (1, 0, f_x), \qquad \frac{\partial \mathbf{x}}{\partial y} = (0, 1, f_y).$$

Consequently

$$\frac{\partial \mathbf{x}}{\partial x} \times \frac{\partial \mathbf{x}}{\partial y} = (1, 0, f_x) \times (0, 1, f_y) = (-f_x, -f_y, 1),$$

and the resulting formula for the element of area is

$$\boxed{dA = \sqrt{1 + f_x^2 + f_y^2}\, dx\, dy,}$$

a most useful expression.

The formula has a geometric interpretation. The unit normal to the surface is

$$\mathbf{N} = \frac{1}{\sqrt{1 + f_x^2 + f_y^2}}(-f_x, -f_y, 1).$$

Its third component (direction cosine) is

$$\cos \gamma = \frac{1}{\sqrt{1 + f_x^2 + f_y^2}},$$

where γ is the angle between the normal and the z-axis. Thus

$$(\cos \gamma)\, dA = dx\, dy,$$

which means that the small piece of surface of area dA projects onto a small portion of the x, y-plane of area $dx\, dy$.

EXAMPLE 6.4

Find the area of the portion of the hyperbolic paraboloid (saddle surface) $z = -x^2 + y^2$ defined on the domain $x^2 + y^2 \leq a^2$.

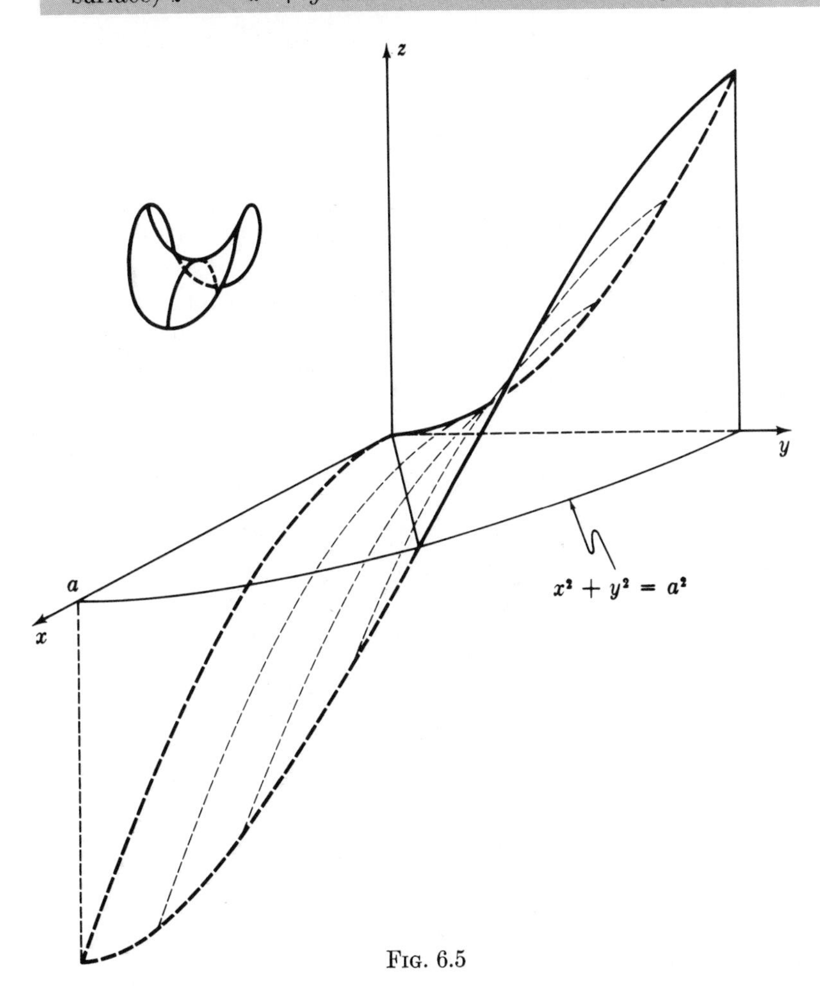

FIG. 6.5

Solution: First sketch the surface, then the portion corresponding to the range $x^2 + y^2 \leq a^2$. See Fig. 6.5. By the formula,

$$\begin{aligned} dA &= \sqrt{1 + f_x{}^2 + f_y{}^2}\, dx\, dy = \sqrt{1 + (-2x)^2 + (2y)^2}\, dx\, dy \\ &= \sqrt{1 + 4x^2 + 4y^2}\, dx\, dy. \end{aligned}$$

Use polar coordinates:

$$A = \iint \sqrt{1 + 4r^2}\, r\, dr\, d\theta$$

$$= \int_0^a r\sqrt{1 + 4r^2}\, dr \int_0^{2\pi} d\theta = \frac{1}{12}[(1 + 4a^2)^{3/2} - 1]\cdot 2\pi.$$

Answer: $\frac{\pi}{6}[(1 + 4a^2)^{3/2} - 1]$.

Green's Theorem

There is a useful connection between certain line integrals and double integrals. Suppose **D** is a domain in the x, y-plane whose boundary consists of one or several nice closed curves, curves consisting of arcs with continuously turning tangent vectors. The counterclockwise sense of rotation in the plane imparts a direction to each of the boundary curves (Fig. 6.6). We think of walking around the boundary curves so that **D** is always on our left.

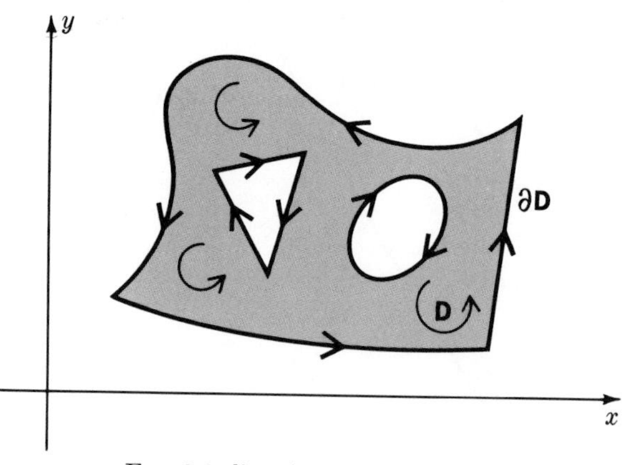

FIG. 6.6 direction on the boundary

We use the symbol $\partial\mathbf{D}$ to denote the whole boundary of **D**, with this direction imposed. Then we know the meaning of a line integral

$$\int_{\partial\mathbf{D}} P\, dx + Q\, dy.$$

The following theorem says that this *line* integral over $\partial\mathbf{D}$ is equal to a certain *double* integral over **D**.

Green's Theorem Suppose $P(x, y)$ and $Q(x, y)$ are continuously differentiable functions on the domain **D**. Then

$$\int_{\partial \mathbf{D}} P\,dx + Q\,dy = \iint_{\mathbf{D}} \left(\frac{\partial Q}{\partial x} - \frac{\partial P}{\partial y}\right) dx\,dy.$$

We shall not give a complete proof of the theorem but only prove one special case. The idea in general is to treat the theorem as two separate formulas, $\int P\,dx = -\iint P_y$ and $\int Q\,dy = \iint Q_x$. For the first, cut the domain by vertical lines into pieces, each of which is the region between the graphs of two functions. (This is the case we shall prove in a moment.) Then add up the results. The double integrals add up to the double integral over the whole domain. The line integrals add up to the line integral over the boundary, because the contributions from the vertical division lines cancel in pairs.

We shall prove

$$\int_{\partial \mathbf{D}} P\,dx = -\iint_{\mathbf{D}} \frac{\partial P}{\partial y}\,dx\,dy$$

for the domain **D** of Fig. 6.7.

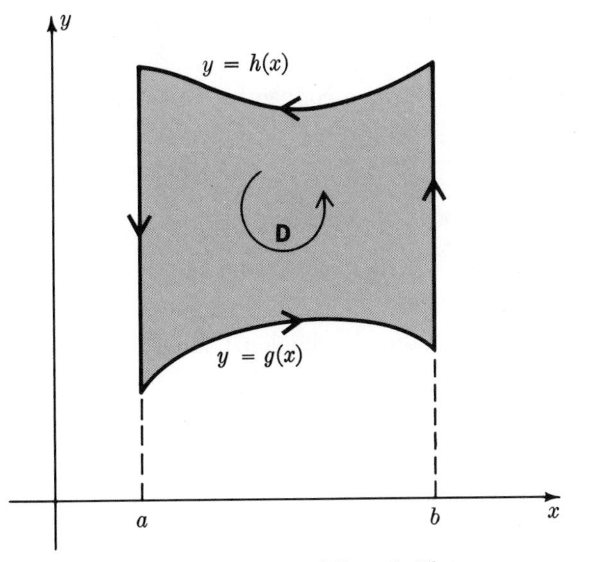

Fig. 6.7 special case of Green's Theorem

The boundary $\partial\mathbf{D}$ consists of four pieces, and accordingly the line integral decomposes into four summands. On the two vertical sides x is constant, hence

$dx = 0$, no contribution. Therefore

$$\int_{\partial \mathbf{D}} P\,dx = \int_b^a P[x, h(x)]\,dx + \int_a^b P[x, g(x)]\,dx$$

$$= \int_a^b \{-P[x, h(x)] + P[x, g(x)]\}\,dx.$$

$$= -\int_a^b \left(\int_{g(x)}^{h(x)} \frac{\partial P}{\partial y}\,dy\right) dx = -\iint_{\mathbf{D}} \frac{\partial P}{\partial y}\,dx\,dy.$$

This completes the proof.

Area Formula If **D** is a plane domain with a smooth boundary, then

$$|\mathbf{D}| = \frac{1}{2}\int_{\partial \mathbf{D}} -y\,dx + x\,dy.$$

Proof: Apply Green's Theorem with $P = -y$ and $Q = x$. Then $Q_x - P_y = 2$, so

$$\int_{\partial \mathbf{D}} -y\,dx + x\,dy = \iint_{\mathbf{D}} 2\,dx\,dy = 2\,|\mathbf{D}|.$$

REMARK: By similar applications of Green's Theorem, we also have

$$|\mathbf{D}| = \int_{\partial \mathbf{D}} x\,dy = -\int_{\partial \mathbf{D}} y\,dx.$$

The boxed formula is often more convenient than either of these because it has a certain amount of symmetry.

EXAMPLE 6.5

Find the area enclosed by the ellipse

$$\frac{x^2}{a^2} + \frac{y^2}{b^2} = 1.$$

Solution: Let **D** denote the domain bounded by the ellipse. Parameterize the ellipse as usual by

$$x = a\cos\theta, \quad y = b\sin\theta, \qquad 0 \le \theta \le 2\pi.$$

Then

$$|\mathbf{D}| = \frac{1}{2}\int_{\partial \mathbf{D}} -y\,dx + x\,dy$$

$$= \frac{1}{2}\int_0^{2\pi} -(b\sin\theta)(-a\sin\theta\,d\theta) + (a\cos\theta)(b\cos\theta\,d\theta)$$

$$= \frac{1}{2}\int_0^{2\pi} (ab\sin^2\theta + ab\cos^2\theta)\,d\theta = \pi ab.$$

Answer: πab.

EXERCISES

Evaluate $F(t)$ for $t > 0$ and then compute $F'(t)$. Now compute $F'(t)$ by the Leibniz Rule and compare your answers:

1. $F(t) = \int_0^1 e^{-tx}\,dx$
2. $F(t) = \int_0^1 \left(\arctan\frac{x}{t}\right) dx, \quad t > \frac{2}{\pi}$
3. $F(t) = \int_0^1 (t+x)^n\,dx$
4. $F(t) = \int_0^1 x^t\,dx, \quad t > -1.$

5*. Prove under suitable hypotheses:

$$\frac{d}{dt}\int_{g(t)}^{h(t)} f(x, t)\,dx = h'(t)f[h(t), t] - g'(t)f[g(t), t] + \int_{g(t)}^{h(t)} f_t(x, t)\,dx.$$

[*Hint:* Use the Chain Rule.]

Parametrize and set up an integral for surface area. Evaluate when you can:

6. sphere of radius a
7. lateral surface of a right circular cylinder of radius a and height h
8. lateral surface of a right circular cone of radius a and lateral height L
9. right circular torus obtained by revolving a circle of radius a about an axis in its plane at distance A from its center (where $a < A$)
10. ellipsoid $\frac{x^2}{a^2} + \frac{y^2}{b^2} + \frac{z^2}{c^2} = 1$

    [*Hint:* Use spherical coordinates and $\mathbf{x} = (a\sin\phi\cos\theta,\ b\sin\phi\sin\theta,\ c\cos\phi)$.]
11. (cont.) Reduce the double integral to a simple integral in the special case $a = b$.

Set up the area for the given non-parametric surface. Evaluate when you can:

12. $z = xy; \quad -1 \le x \le 1, \quad -1 \le y \le 1$ (do not evaluate)
13. $z = ax + by; \quad (x, y)$ in a domain $\mathbf{D}$
14. $z = x^2 + y^2; \quad x^2 + y^2 \le 1$
15. $z = \sqrt{1 - x^2 - y^2}; \quad x^2 + y^2 \le 1.$
16. Find the area of the triangle with vertices $(a, 0, 0)$, $(0, b, 0)$, and $(0, 0, c)$.
17. Prove that $\iint \mathbf{N}\,dA = \mathbf{0}$ for a closed surface.

    [*Hint:* Show that $\mathbf{i}\cdot\iint \mathbf{N}\,dA = 0$, etc.]
18. From each point of the space curve $\mathbf{x} = \mathbf{x}(s)$ draw a segment of length 1 in the direction of the unit tangent. These segments sweep out a surface. Show that its area is $\frac{1}{2}\int k(s)\,ds$, where $k(s)$ is the curvature and the integral is taken over the length of the curve.

19. (cont.) Interpret for a circle. Can you show that $\int k(s)\,ds = 2\pi$ for any closed oval (convex curve) in the plane? If so, interpret Ex. 18 for ovals.
20. Let $\mathbf{x} = \mathbf{x}(s)$ be a curve of length L on the unit sphere $\mathbf{x} = 1$. Connect each point of the curve to the origin, forming a surface. Show that the area of this surface is $\frac{1}{2}L$.

Let P and Q be continuously differentiable on a rectangle $\mathbf{I}$.

21. Suppose $(P, Q) = \operatorname{grad}\phi$. Prove

$$\int P\,dx + Q\,dy = 0$$

on every closed curve in $\mathbf{I}$.

22*. (cont.) Prove the converse.
23. Evaluate $\int 3y\,dx + 4x\,dy$ over the boundary of the rectangle with vertices at $(0, 1)$, $(6, 1)$, $(6, 3)$, $(0, 3)$ taken counterclockwise.
24. Evaluate $\int y^2e^x\,dx + 2ye^x\,dy$ over the circle $x^2 + y^2 = 1$.
25. Prove **Green's formula** under suitable hypotheses:

$$\int_{\partial\mathbf{D}} (-uv_y\,dx + uv_x\,dy) - \int_{\partial\mathbf{D}} (-vu_y\,dx + vu_x\,dy)$$
$$= \iint_{\mathbf{D}} [u(v_{xx} + v_{yy}) - v(u_{xx} + u_{yy})]\,dx\,dy.$$

26. (cont.) Test this formula when $\mathbf{D}$ is the unit disk, $u = 1$, and $v = \ln r$. Explain the result.
27*. Suppose P and Q are continuously differentiable functions on a rectangle $\mathbf{I}$, and $\partial P/\partial y = \partial Q/\partial x$. Prove there is a function $f(x, y)$ on $\mathbf{I}$ such that $\partial f/\partial x = P$ and $\partial f/\partial y = Q$.
28. (cont.) Let $\mathbf{a}$ and $\mathbf{b}$ be points of $\mathbf{I}$. Prove that the line integral $\int_{\mathbf{a}}^{\mathbf{b}} P\,dx + Q\,dy$ is the same for all paths in $\mathbf{I}$ from $\mathbf{a}$ to $\mathbf{b}$.

7. IMPROPER INTEGRALS [optional]

A multiple integral is improper if either the domain of integration is unbounded or the integrand has singularities or both. There are always two questions: Does the integral converge? If so, what is its value?

The Whole Plane

First we define

$$\iint_{\mathbf{R}^2} f(x, y)\,dx\,dy.$$

Once we have this, it is easy to define the integral over any unbounded domain $\mathbf{D}$. The integrand will be a real-valued function $f(x, y)$ defined on the whole plane $\mathbf{R}^2$. We can't expect to define the integral for all functions, so we shall restrict attention to functions having the following reasonable property:

> A function f on $\mathbf{R}^2$ is **locally integrable** if f is integrable on each *bounded* R-measurable set $\mathbf{D}$.

We shall define the integral by means of limits of the type

$$\lim_{n\to\infty} \iint_{\mathbf{D}_n} f\,dx\,dy,$$

where $\{\mathbf{D}_n\}$ is an expanding sequence of bounded R-measurable sets, $\mathbf{D}_1 \subseteq \mathbf{D}_2 \subseteq \mathbf{D}_3 \subseteq \cdots$, that in some sense fill out the plane.

> **Approximating Sequence** A sequence $\{\mathbf{D}_n\}$ of bounded R-measurable subsets of $\mathbf{R}^2$ is an **approximating sequence** provided
>
> (1) $\mathbf{D}_1 \subseteq \mathbf{D}_2 \subseteq \mathbf{D}_3 \subseteq \cdots$,
> (2) if $B > 0$, then there exists an n such that the disk of radius B and center $\mathbf{0}$ lies in $\mathbf{D}_n$, that is,
>
> $$\{\mathbf{x} \mid |\mathbf{x}| \le B\} \subseteq \mathbf{D}_n.$$

Thus each disk centered at $\mathbf{0}$, no matter how large, is eventually covered by the sets of any approximating sequence. In fact, any bounded set is eventually covered, because a bounded set is contained in some disk centered at $\mathbf{0}$.

Now we have enough preliminary definitions to define the integral.

> **Integral on $\mathbf{R}^2$** Let f be a locally integrable function on $\mathbf{R}^2$. Then f is called **integrable** on $\mathbf{R}^2$ provided:
>
> (1) For each approximating sequence $\{\mathbf{D}_n\}$, the limit
>
> $$\lim_{n\to\infty} \iint_{\mathbf{D}_n} f\,dx\,dy$$
>
> exists.
>
> (2) These limits all have the same value, which is then called
>
> $$\iint_{\mathbf{R}^2} f\,dx\,dy.$$

In every other case, we say

$$\iint_{\mathbf{R}^2} f\,dx\,dy$$

diverges. One common case of divergence occurs when

$$\lim_{n\to\infty} \iint_{\mathbf{D}_n} f\,dx\,dy = +\infty$$

for each approximating sequence $\{\mathbf{D}_n\}$. Then we write

$$\iint_{\mathbf{R}^2} f\,dx\,dy = +\infty.$$

The preceding definition has the drawback that it is impossible to test *every* approximating sequence. However, when the integrand has constant sign, it is enough to test any one approximating sequence.

> **Theorem 7.1** If $f \geq 0$ on $\mathbf{R}^2$ and
>
> $$\lim_{n\to\infty} \iint_{\mathbf{D}_n} f\,dx\,dy$$
>
> exists (finite or $+\infty$) for one single approximating sequence $\{\mathbf{D}_n\}$, then f is integrable on $\mathbf{R}^2$.

Proof: Take any other approximating sequence $\{\mathbf{E}_m\}$. For each m, the set $\mathbf{E}_m$ is bounded. Hence there exists an n such that $\mathbf{E}_m \subseteq \mathbf{D}_n$. Since $f \geq 0$, we conclude

$$\iint_{\mathbf{E}_m} f \leq \iint_{\mathbf{D}_n} f \leq \lim_{n\to\infty} \iint_{\mathbf{D}_n} f.$$

From this, we conclude that the *increasing* sequence

$$\left\{\iint_{\mathbf{E}_m} f\right\}$$

converges, possibly to $+\infty$, and

$$\lim_{m\to\infty} \iint_{\mathbf{E}_m} f \leq \lim_{n\to\infty} \iint_{\mathbf{D}_n} f.$$

Now we can apply the same argument with the $\mathbf{D}$'s and $\mathbf{E}$'s interchanged. The result is

$$\lim_{n\to\infty} \iint_{\mathbf{D}_n} f \leq \lim_{m\to\infty} \iint_{\mathbf{E}_m} f,$$

so equality follows. Since $\{\mathbf{E}_m\}$ is an arbitrary approximating sequence, the proof is complete.

There is a related result about absolute convergence. We shall state it, but save the (somewhat technical) proof for the exercises.

Theorem 7.2 Suppose

(1) f is locally integrable.
(2) $|f|$ is integrable on $\mathbf{R}^2$.

Then f is integrable on $\mathbf{R}^2$.

The point is that the test of Theorem 7.1 can be applied to $|f|$.

EXAMPLE 7.1

Prove that the integral

$$\iint_{\mathbf{R}^2} e^{-x^2-y^2}\,dx\,dy$$

is convergent and evaluate it.

Solution: The integrand is positive so we can use any convenient approximating sequence. We choose disks,

$$\mathbf{D}_n = \{\mathbf{x} \mid |\mathbf{x}| \leq n\},$$

and switch to polar coordinates:

$$\iint_{\mathbf{D}_n} e^{-x^2-y^2}\,dx\,dy = \iint_{\mathbf{D}_n} e^{-r^2} r\,dr\,d\theta = \left(\int_0^{2\pi} d\theta\right)\left(\int_0^n r\,e^{-r^2}\,dr\right)$$

$$= (2\pi)\left(\frac{1}{2}\right)(1 - e^{-n^2}) \longrightarrow \pi.$$

Answer: π.

This result has a surprising application.

EXAMPLE 7.2

Evaluate

$$\int_{-\infty}^{\infty} e^{-x^2}\,dx.$$

Solution: We solve Example 7.1 again, this time using a sequence of squares:

$$\mathbf{E}_n = \{(x, y) \mid |x| \leq n, \quad |y| \leq n\}.$$

Now

$$\iint_{\mathbf{E}_n} e^{-x^2-y^2}\,dx\,dy = \left(\int_{-n}^{n} e^{-x^2}\,dx\right)\left(\int_{-n}^{n} e^{-y^2}\,dy\right) = \left(\int_{-n}^{n} e^{-x^2}\,dx\right)^2.$$

We know from Theorem 7.1 and Example 7.1 that

$$\iint\limits_{\mathbf{E}_n} e^{-x^2-y^2}\,dx\,dy \longrightarrow \pi.$$

We conclude

$$\left(\int_{-\infty}^{\infty} e^{-x^2}\,dx\right)^2 = \pi.$$

Answer: $\sqrt{\pi}$.

REMARK: In probability theory, the function

$$\phi(x) = \frac{1}{\sqrt{2\pi}}\,e^{-x^2/2}$$

is important; it is the density function of the **normal distribution**. The simple change of variable $x = u/\sqrt{2}$ in Example 7.2 yields

$$\int_{-\infty}^{\infty} \phi(x)\,dx = 1.$$

The graph of ϕ is the familiar bell-shaped curve (Fig. 7.1); the area under the curve is 1.

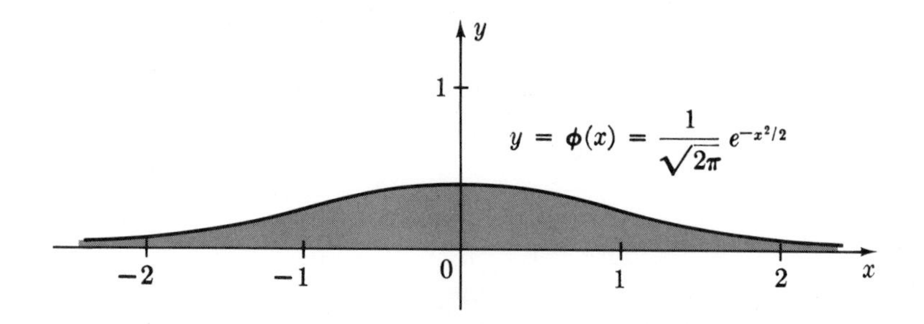

FIG. 7.1 normal distribution

Arbitrary Unbounded Domains

Now let $\mathbf{D}$ be an arbitrary unbounded subset of $\mathbf{R}^2$. We wish to define

$$\iint\limits_{\mathbf{D}} f(x, y)\,dx\,dy$$

To do so, we extend f to the whole plane by making it 0 outside of $\mathbf{D}$. Thus we define

$$F(x, y) = \begin{cases} f(x, y), & (x, y) \in \mathbf{D} \\ 0, & (x, y) \notin \mathbf{D}. \end{cases}$$

Then we set

$$\iint_{\mathbf{D}} f(x, y)\, dx\, dy = \iint_{\mathbf{R}^2} F(x, y)\, dx\, dy$$

if the right-hand side exists. For the domains that occur in practice, things go routinely from this point.

REMARK: Actually, it is reasonable to restrict attention to domains $\mathbf{D}$ for which the characteristic function $k_{\mathbf{D}}$ (one on $\mathbf{D}$, zero off) is locally integrable. This means that the intersection of $\mathbf{D}$ with any *bounded* R-measurable set is itself a (bounded) R-measurable set.

Before reading the next example, it is advisable to review Theorem 8.2, p. 110.

EXAMPLE 7.3

Evaluate

$$\int_0^{\infty} \frac{e^{-ax} - e^{-bx}}{x}\, dx, \qquad 0 < a < b.$$

Solution: The starting point is the observation

$$\frac{e^{-ax} - e^{-bx}}{x} = \int_a^b e^{-xy}\, dy.$$

Therefore it is natural to consider

$$I = \iint_{\mathbf{D}} e^{-xy}\, dx\, dy, \qquad \mathbf{D} = \{(x, y) \mid x \geq 0, \quad a \leq y \leq b\}.$$

To evaluate $\mathbf{I}$, take the rectangle

$$\mathbf{D}_n = \{(x, y) \mid 0 \leq x \leq n, \quad a \leq y \leq b\}.$$

On the one hand,

$$\iint_{\mathbf{D}_n} e^{-xy}\, dx\, dy = \int_0^n \left(\int_a^b e^{-xy}\, dy \right) dx$$

$$= \int_0^n \frac{e^{-ax} - e^{-bx}}{x}\, dx \longrightarrow \int_0^{\infty} \frac{e^{-ax} - e^{-bx}}{x}\, dx,$$

so

$$\iint_{\mathbf{D}} e^{-xy}\, dx\, dy = \int_0^{\infty} \frac{e^{-ax} - e^{-bx}}{x}\, dx.$$

On the other hand,

$$\iint\limits_{\mathbf{D}_n} e^{-xy}\,dx\,dy = \int_a^b \left(\int_0^n e^{-xy}\,dx\right) dy = \int_a^b \frac{1-e^{-ny}}{y}\,dy.$$

We infer from this that

$$\int_0^\infty \frac{e^{-ax}-e^{-bx}}{x}\,dx = \lim_{n\to\infty} \int_a^b \frac{1-e^{-ny}}{y}\,dy.$$

The trick now is to take the limit *under the integral sign* on the right. That is where Theorem 8.2, Chapter 3 comes in. We observe that

$$\frac{1-e^{-ny}}{y} \longrightarrow \frac{1}{y}$$

uniformly on $[a, b]$ because

$$\left|\frac{1-e^{-ny}}{y} - \frac{1}{y}\right| = \frac{e^{-ny}}{y} \leq \frac{e^{-na}}{a} \longrightarrow 0 \qquad \text{as} \quad n \longrightarrow \infty.$$

Therefore by Theorem 8.2,

$$\int_0^\infty \frac{e^{-ax}-e^{-bx}}{x}\,dx = \int_a^b \lim_{n\to\infty} \frac{1-e^{-ny}}{y}\,dy = \int_a^b \frac{dy}{y} = \ln\frac{b}{a}.$$

Answer: $\ln(b/a)$.

Singularities

Suppose there is a single point in the (bounded) domain of f at which f becomes infinite, or is indefined in some other way. We try to squeeze down on this isolated singularity of f by a growing sequence of subdomains, in the same way that we expanded to infinity in the case of the whole plane. If we limit ourselves to functions of constant sign, a result similar to Theorem 7.1 says that one such sequence suffices. The following example should illustrate the idea. Recall the integral

$$\int_0^1 \frac{dx}{x^p},$$

which is convergent for $p < 1$ and divergent for $p \geq 1$.

EXAMPLE 7.4

Find all p for which

$$\iint\limits_{r\leq 1} \frac{1}{r^p}\,dx\,dy \qquad (r^2 = x^2 + y^2)$$

is convergent.

Solution: The integrand is positive, except for a singularity at $r = 0$ when $p > 0$. In polar coordinates,

$$\iint_{r \le 1} \frac{1}{r^p}\,dx\,dy = \lim_{\epsilon \to 0+} \iint_{\epsilon \le r \le 1} \frac{1}{r^p}\,dx\,dy = \lim_{\epsilon \to 0+} \int_0^{2\pi} d\theta \int_\epsilon^1 \frac{r\,dr}{r^p} = 2\pi \int_0^1 \frac{dr}{r^{p-1}}.$$

Answer: For $p < 2$, it converges; for $p \ge 2$ it diverges.

Iteration

The main point about iteration of improper integrals is that it often doesn't work. *Be careful*; there are few useful results, and generally each case must be treated on its own merits. It is easy to get incorrect results.

EXAMPLE 7.5

Apply iteration blindly, both ways, to

$$I = \iint_{\substack{0 \le x \le 1 \\ 0 \le y \le 1}} \frac{x^2 - y^2}{(x^2 + y^2)^2}\,dx\,dy.$$

Solution: The formula

$$\int \frac{x^2 - a^2}{(x^2 + a^2)^2}\,dx = \frac{-x}{x^2 + a^2}$$

is valid, as is easily seen by differentiating the right-hand side. Therefore

$$I = \int_0^1 \left(\int_0^1 \frac{x^2 - y^2}{(x^2 + y^2)^2}\,dx \right) dy = \int_0^1 \left(\frac{-x}{x^2 + y^2} \right) \Bigg|_{x=0}^{x=1} dy$$

$$= \int_0^1 \frac{-1}{1 + y^2}\,dy = -\frac{\pi}{4}.$$

Now look carefully at the integrand. If x and y are interchanged, it changes sign. Therefore integration first on y, then on x, leads to $I = +\frac{1}{4}\pi$, an impossible situation.

What is wrong is that the improper double integral does not exist. This example is worth careful study.

To end on a happier note, we give an example where the result is correct, but it is beyond our technique to justify the formal steps. (We shall use integral 41 inside the back cover of the book.) Set

$$I = \iint_{\substack{0 \le x \\ 0 \le y}} e^{-xy} \sin x\,dx\,dy.$$

First,

$$I = \int_0^\infty \left(\int_0^\infty e^{-xy}\,dy \right) \sin x\,dx = \int_0^\infty \frac{\sin x}{x}\,dx.$$

Second,

$$I = \int_0^\infty \left(\int_0^\infty e^{-xy} \sin x\,dx \right) dy = \int_0^\infty \frac{-e^{-xy}}{1+y^2}(y \sin x + \cos x) \Bigg|_{x=0}^{x=\infty} dy$$

$$= \int_0^\infty \frac{dy}{1+y^2} = \frac{\pi}{2}.$$

Therefore

$$\int_0^\infty \frac{\sin x}{x}\,dx = \frac{\pi}{2}.$$

This result is correct, but this derivation is full of holes.

EXERCISES

Prove for integrals over $\mathbf{R}^2$:

1. $\displaystyle\iint \frac{dx\,dy}{1+x^2+y^2} = +\infty$

2. $\displaystyle\iint \frac{dx\,dy}{1+x^2+y^2+x^2y^2} = \pi^2.$

3. Suppose $0 \le g(x, y) \le f(x, y)$ on $\mathbf{R}^2$ and $\iint f < \infty$. Prove that $\iint g < \infty$.

4*. Prove Theorem 7.2. [*Hint:* Use a Cauchy Condition.]

5*. Prove divergent:

$$\iint_{\mathbf{R}^2} \frac{dx\,dy}{1+x^2y^2}.$$

6. Prove convergent:

$$\iint_{\mathbf{R}^2} \frac{dx\,dy}{1+x^4+y^4}.$$

Prove convergent and evaluate:

7. $\displaystyle\iint_{r\le 1} (\ln r)\,dx\,dy$

8. $\displaystyle\iint_{r\le 1} (1-r)^p\,dx\,dy, \quad p > -1$

9. $\displaystyle\iiint_{\rho\le 1} \rho^s\,dx\,dy\,dz, \qquad s > -3$

10. $\displaystyle\iiint_{\rho\le 1} (\ln \rho)\,dx\,dy\,dz$

11. $\displaystyle\iiint_{\rho\le 1} (1-\rho)^s\,dx\,dy\,dz, \qquad s > -1.$

12. Prove convergent for $p > 1$ and evaluate:

$$\iint_{r \geq 1} \frac{dx\,dy}{r^{2p}}.$$

13*. (cont.) Prove

$$\sum \frac{1}{(m^2 + n^2)^p} < \infty$$

for $p > 1$. The sum is taken over *all* integers m, n except $(0, 0)$.

Compute:

14. $\displaystyle\int_0^{\infty} e^{-tx^2}\,dx, \qquad t > 0$

15*. $\displaystyle\int_0^{\infty} x^2 e^{-x^2}\,dx.$

8. NUMERICAL INTEGRATION [optional]

In this section we discuss one method for approximating double integrals. It is an extension of Simpson's Rule.

Let us recall Simpson's Rule. To approximate an integral

$$\int_a^b f(x)\,dx,$$

we divide the interval $a \leq x \leq b$ into $2m$ equal parts of length h:

$$a = x_0 < x_1 < x_2 < \cdots < x_{2m} = b, \qquad h = \frac{b - a}{2m},$$

and use the formula

$$\int_a^b f(x)\,dx \approx \frac{h}{3} \sum_{i=0}^{2m} B_i f(x_i),$$

where the coefficients are 1, 4, 2, 4, 2, 4, 2, $\cdots$, 2, 4, 1.

We extend Simpson's Rule to double integrals in the following way. To approximate

$$\iint_{\mathbf{I}} f(x, y)\,dx\,dy,$$

where $\mathbf{I}$ denotes the rectangle $a \leq x \leq b$ and $c \leq y \leq d$, we divide the x-interval into $2m$ parts as before and also divide the y-interval into $2n$ equal parts of length k:

$$c = y_0 < y_1 < y_2 < \cdots < y_{2n} = d, \qquad k = \frac{d - c}{2n}.$$

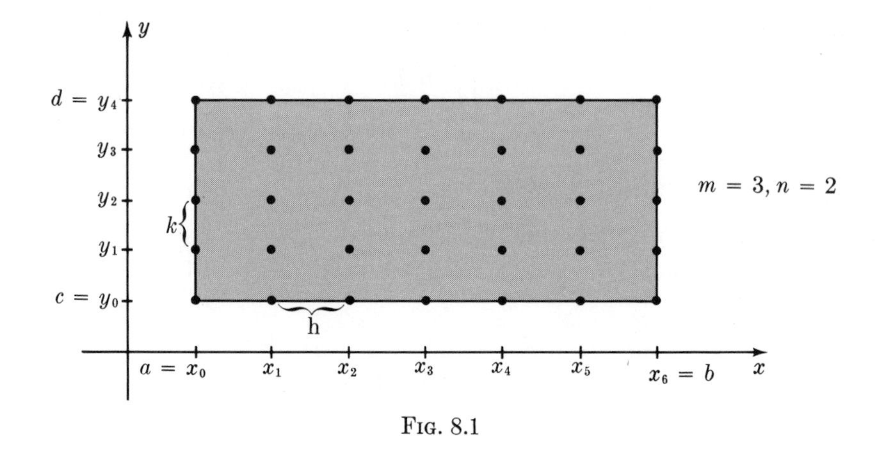

FIG. 8.1

We obtain $(2m + 1)(2n + 1)$ points of the rectangle $\mathbf{I}$, shown in Fig. 8.1. The Rule is

$$\iint_{\mathbf{I}} f(x, y)\, dx\, dy \approx \frac{hk}{9} \sum_{i=0}^{2m} \sum_{j=0}^{2n} A_{ij} f(x_i, y_j),$$

where the coefficients A_{ij} are certain products of the coefficients in the ordinary Simpson's Rule. Precisely,

$$A_{ij} = B_i C_j,$$

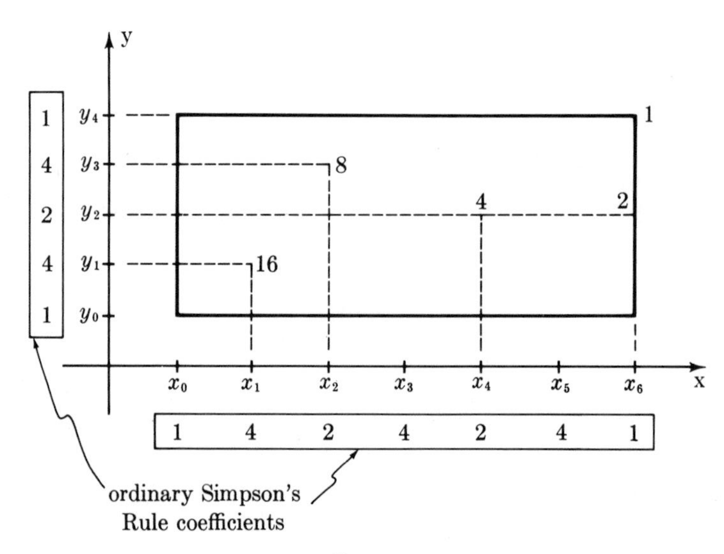

FIG. 8.2

where $B_0, B_1, \cdots, B_{2m}$ are the coefficients in the ordinary Simpson's Rule

$$\int_a^b p(x)\,dx \approx \frac{h}{3}\sum_{i=0}^{2m} B_i p(x_i),$$

and $C_0, C_1, \cdots, C_{2n}$ are the coefficients in the ordinary Simpson's Rule

$$\int_c^d q(y)\,dy \approx \frac{k}{3}\sum_{j=0}^{2n} C_j q(y_j).$$

In Fig. 8.2, several of these products are formed. Since B_i and C_j take values 1, 2, and 4, the coefficients A_{ij} take values 1, 2, 4, 8, and 16. The A_{ij} can be written in a matrix corresponding to the points (x_i, y_j) as in Fig. 8.2. For example, if $m = 3$ and $n = 2$, the matrix is

$$\begin{bmatrix} 1 & 4 & 2 & 4 & 2 & 4 & 1 \\ 4 & 16 & 8 & 16 & 8 & 16 & 4 \\ 2 & 8 & 4 & 8 & 4 & 8 & 2 \\ 4 & 16 & 8 & 16 & 8 & 16 & 4 \\ 1 & 4 & 2 & 4 & 2 & 4 & 1 \end{bmatrix}.$$

EXAMPLE 8.1

Estimate $\iint_{0 \le x,y \le 1} (x+y)^3\,dx\,dy$ by Simpson's Rule with $m = n = 1$.

Compare the result with the exact answer.

Solution: Here $h = k = \frac{1}{2}$. The coefficient matrix is

$$[A_{ij}] = \begin{bmatrix} 1 & 4 & 1 \\ 4 & 16 & 4 \\ 1 & 4 & 1 \end{bmatrix}.$$

Write the value $(x_i + y_j)^3$ in a matrix:

$$[(x_i+y_j)^3] = \begin{bmatrix} (1+0)^3 & \left(1+\frac{1}{2}\right)^3 & (1+1)^3 \\ \left(\frac{1}{2}+0\right)^3 & \left(\frac{1}{2}+\frac{1}{2}\right)^3 & \left(\frac{1}{2}+1\right)^3 \\ (0+0)^3 & \left(0+\frac{1}{2}\right)^3 & (0+1)^3 \end{bmatrix} = \begin{bmatrix} 1 & \frac{27}{8} & 8 \\ \frac{1}{8} & 1 & \frac{27}{8} \\ 0 & \frac{1}{8} & 1 \end{bmatrix}.$$

Now estimate the integral by

$$\frac{hk}{9}\sum_{i=0}^{2}\sum_{j=0}^{2} A_{ij}(x_i + y_j)^3.$$

To evaluate this sum, multiply corresponding terms of the two matrices and add the nine products:

$$\frac{1}{9}\left(\frac{1}{2}\right)\left(\frac{1}{2}\right)\left[1\cdot 1 + 4\cdot\frac{27}{8} + 1\cdot 8 + 4\cdot\frac{1}{8} + 16\cdot 1 + 4\cdot\frac{27}{8} + 1\cdot 0 + 4\cdot\frac{1}{8} + 1\cdot 1\right]$$

$$= \frac{1}{36}\left[1 + \frac{27}{2} + 8 + \frac{1}{2} + 16 + \frac{27}{2} + \frac{1}{2} + 1\right] = \frac{54}{36} = \frac{3}{2}.$$

The exact value is

$$\iint_{0\le x,y\le 1} (x+y)^3\,dx\,dy = \int_0^1\left(\int_0^1 (x+y)^3\,dx\right)dy$$

$$= \frac{1}{4}\int_0^1 [(y+1)^4 - y^4]\,dy = \frac{1}{20}[(y+1)^5 - y^5]\Big|_0^1 = \frac{30}{20} = \frac{3}{2}.$$

Answer: Simpson's Rule gives the estimate $\frac{3}{2}$, which is exact.

REMARK 1: Because Simpson's Rule is exact for cubics, the double integral rule is exact for cubics in two variables. (See exercises below.)

REMARK 2: The matrix of values $[f(x_i, y_j)]$ is arranged to conform to the layout of points (x_i, y_j) in the plane (Fig. 8.1).

The next example is an integral that cannot be evaluated exactly, only approximated.

EXAMPLE 8.2

Estimate $\displaystyle\iint_{0\le x,y\le \pi/2} \sin(xy)\,dx\,dy$, using $m = n = 1$.

Solution: Here $h = k = \pi/4$, and the coefficient matrix is

$$[A_{ij}] = \begin{bmatrix} 1 & 4 & 1 \\ 4 & 16 & 4 \\ 1 & 4 & 1 \end{bmatrix}.$$

The matrix of values of $\sin(xy)$ is

$$[\sin x_i y_j] = \begin{bmatrix} \sin 0 & \sin \frac{\pi^2}{8} & \sin \frac{\pi^2}{4} \\ \sin 0 & \sin \frac{\pi^2}{16} & \sin \frac{\pi^2}{8} \\ \sin 0 & \sin 0 & \sin 0 \end{bmatrix}.$$

The estimate is

$$\iint_{0 \le x, y \le \pi/2} \sin(xy)\, dx\, dy \approx \frac{\pi^2}{144}\left(16 \sin \frac{\pi^2}{16} + 8 \sin \frac{\pi^2}{8} + \sin \frac{\pi^2}{4}\right).$$

Error Estimate

The error estimate for Simpson's Rule in two variables is analogous to that in one variable:

$$|\text{error}| \le \frac{(b-a)(d-c)}{180}[h^4 M + k^4 N],$$

where

$$\left|\frac{\partial^4 f}{\partial x^4}\right| \le M \qquad \text{and} \qquad \left|\frac{\partial^4 f}{\partial y^4}\right| \le N.$$

We omit the proof.

EXAMPLE 8.3

Estimate the error in Example 8.2.

Solution:

$$\frac{\partial^4}{\partial x^4}(\sin xy) = y^4 \sin xy, \qquad \frac{\partial^4}{\partial y^4}(\sin xy) = x^4 \sin xy.$$

But $|\sin xy| \le 1$. Hence in the square $0 \le x,\ y \le \pi/2$,

$$\left|\frac{\partial^4 f}{\partial x^4}\right| = |x^4 \sin xy| \le \left(\frac{\pi}{2}\right)^4, \qquad \left|\frac{\partial^4 f}{\partial y^4}\right| = |y^4 \sin xy| \le \left(\frac{\pi}{2}\right)^4.$$

Apply the error estimate, with $m = n = 1$, $h = k = \pi/4$, and $M = N = (\pi/2)^4$:

$$|\text{error}| \le \frac{1}{45}\left(\frac{\pi}{4}\right)^2\left[2\left(\frac{\pi}{4}\right)^4\left(\frac{\pi}{2}\right)^4\right] = \frac{1}{45}\frac{\pi^{10}}{2^{15}} \approx 0.064.$$

EXERCISES

1. Use tables to complete Example 8.2.
2. Do Example 8.2 using $m = n = 2$, and compare your estimate to that of Ex. 1. Also estimate the error.
3. Suppose $f(x, y) = p(x)q(y)$. Show that the double integral Simpson's Rule estimate is just the product of the Simpson's Rule estimate for $\int p(x)\,dx$ by that for $\int q(y)\,dy$.
4. (cont.) Conclude that the rule is exact for polynomials involving only x^3y^3, x^3y^2, x^2y^3, x^3y, x^2y^2, xy^3, and lower degree terms.
5. The analogue of the Trapezoidal Rule is
$$\iint_{0 \leq x,y \leq 1} f(x, y)\,dx\,dy \approx \frac{1}{4}[f(0, 0) + f(0, 1) + f(1, 1) + f(1, 0)].$$
Show this rule is exact for polynomials $f(x, y) = A + Bx + Cy + Dxy$.
6. (cont.) Find the corresponding rule for a rectangle $a \leq x \leq b$, $c \leq x \leq d$, divided into rectangles of size h by k with $h = (b - a)/m$ and $k = (d - c)/n$.
7. (cont.) Let $\mathbf{I}$ denote the unit square $0 \leq x, y \leq 1$. Suppose $f(x, y) = 0$ at its four vertices. Prove that
$$\iint_{\mathbf{I}} f(x, y)\,dx\,dy = -\frac{1}{2}\iint_{\mathbf{I}} y(1 - y)f_{yy}(x, y)\,dx\,dy$$
$$= -\frac{1}{4}\int_0^1 x(1 - x)[f_{xx}(x, 1) + f_{xx}(x, 0)]\,dx.$$
8. (cont.) Suppose also that $|f_{xx}| \leq M$ and $|f_{yy}| \leq N$ on $\mathbf{I}$. Prove that
$$\left|\iint_{\mathbf{I}} f(x, y)\,dx\,dy\right| \leq \frac{1}{12}M + \frac{1}{12}N.$$
9. (cont.) Conclude that for any function $f(x, y)$, the error in the trapezoidal estimate (Ex. 5) is at most $(M + N)/12$.
[*Hint:* Use the result of Ex. 5 and interpolation.]
10. (cont.) Suppose $f(x, y)$ is defined on $0 \leq x \leq h$, $0 \leq y \leq k$, and $f = 0$ at the vertices of this rectangle $\mathbf{I}$. Suppose also $|f_{xx}| \leq M$ and $|f_{yy}| \leq N$. Deduce that
$$\left|\iint_{\mathbf{I}} f(x, y)\,dx\,dy\right| \leq \frac{hk}{12}(h^2M + k^2N).$$
11. Extend Simpson's Rule for double integrals to triple integrals.
12. (cont.) For which polynomials in 3 variables is the extended Simpson's Rule exact?
13. (cont.) Estimate the integral of $\sin(xyz)$ over the cube $0 \leq x, y, z \leq 1$, dividing the cube into 8 equal parts. Carry your work to 3 places.

13. Differential Equations

1. INTRODUCTION

We are familiar with differential equations of the type

$$\frac{dx}{dx} = f(x).$$

Each solution is an antiderivative of $f(x)$. We are also familiar with the differential equation

$$\frac{dy}{dx} = y,$$

each solution of which is a function $y = ce^x$.

These differential equations are instances of the most general **first order differential equation,**

$$\frac{dy}{dx} = q(x, y),$$

where $q(x, y)$ is a function of two variables. "First order" refers to the presence of the first derivative only, not the second or third, etc.

A differential equation $dy/dx = q(x, y)$ defines a **direction field**. At each point (x, y) of the domain of $q(x, y)$, imagine a short line segment of slope $q(x, y)$, as in Fig. 1.1. A **solution** of the differential equation is a function $y = y(x)$ whose graph has slope matching that of the direction field (Fig. 1.2).

The most important problem in this subject is the initial-value problem:

Initial-Value Problem Find the solution $y = y(x)$ of the differential equation

$$\frac{dy}{dx} = q(x, y)$$

whose graph passes through a given point (a, b), that is, which satisfies the **initial condition**

$$y(a) = b.$$

It is shown in advanced courses that if $q(x, y)$ is a reasonably behaved function, then the initial-value problem has a unique solution locally. That is, there is a neighborhood of a on which there exists one and only one differentiable function $y(x)$ satisfying $y'(x) = q[x, y(x)]$ and $y(a) = b$. This is true, for instance, if $q(x, y)$ has continuous partial derivatives in a neighborhood of (a, b). In Section 7, we shall discuss one method of proving such results, successive approximation.

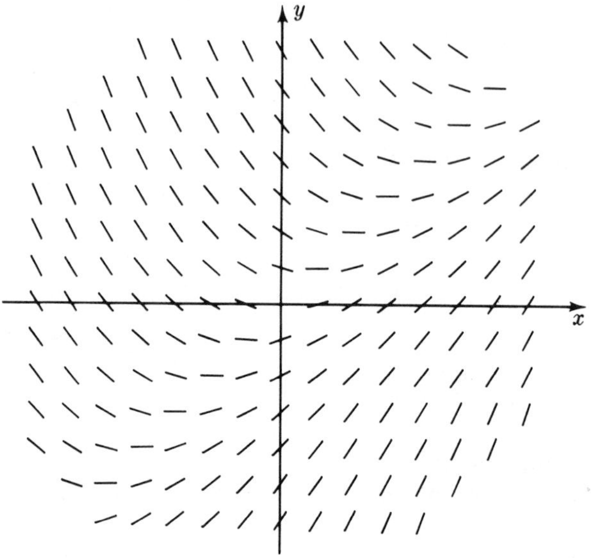

FIG. 1.1

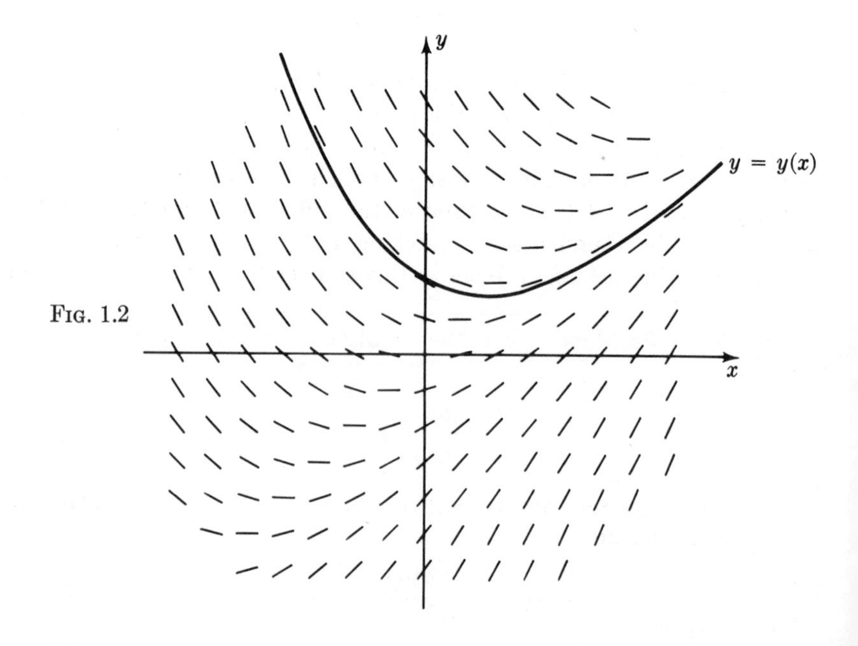

FIG. 1.2

Figure 1.3 shows two solutions of

$$\frac{dy}{dx} = q(x, y),$$

one satisfying the initial condition $y(0) = 0$, the other satisfying the initial condition $y(1) = -1$.

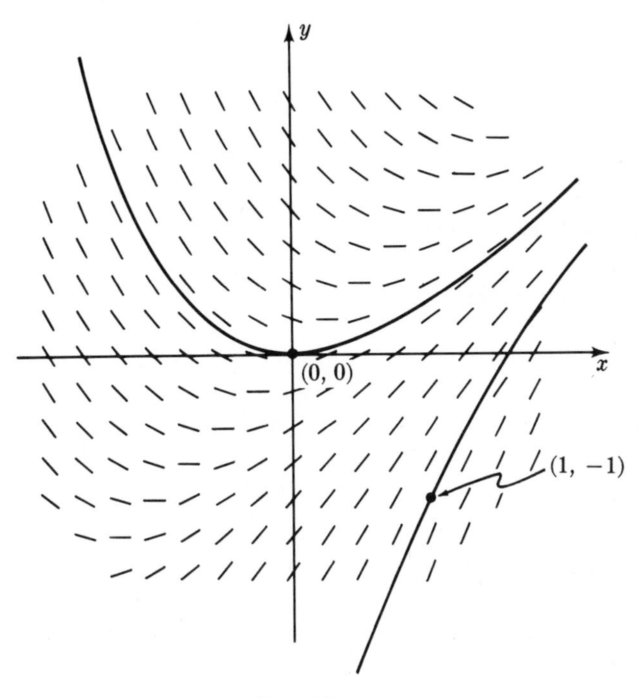

FIG. 1.3

Usually we first find the general solution to the differential equation, ignoring the initial condition. Somewhere along the line we are forced to integrate, thus introducing a constant of integration. We then select the constant so that the initial condition is satisfied.

EXAMPLE 1.1

Find a solution of the initial-value problem

$$\begin{cases} \dfrac{dy}{dx} = x^2 \\ y(-3) = 1. \end{cases}$$

Solution: Integrate:

$$y(x) = \frac{1}{3}x^3 + c.$$

This is the general solution of the differential equation. Now select the constant by substituting the initial condition:

$$1 = \frac{1}{3}(-3)^3 + c, \qquad c = 10.$$

Answer: $y(x) = \frac{1}{3}x^3 + 10.$

EXERCISES

Solve the initial-value problem:

1. $\dfrac{dy}{dx} = x^3, \quad y(0) = 0$
2. $\dfrac{dy}{dx} = x^2 + x - 5, \quad y(0) = -2$
3. $\dfrac{dy}{dx} = \dfrac{3}{x}, \quad y(1) = 1$
4. $\dfrac{dy}{dx} = \cos 2x, \quad y(\pi) = 0$
5. $\dfrac{dy}{dx} = e^{-x}, \quad y(0) = 5$
6. $\dfrac{dy}{dx} = xe^{x^2}, \quad y(0) = 1$
7. $\dfrac{dy}{dx} = y, \quad y(1) = 10$
8. $\dfrac{dy}{dx} = y, \quad y(1) = 0$
9. $\dfrac{dy}{dx} = x + \sin x, \quad y(0) = 0$
10. $\dfrac{dy}{dx} = \sqrt{x+1}, \quad y(0) = 2$
11. $\dfrac{dy}{dx} = \dfrac{x}{\sqrt{x^2+1}}, \quad y(0) = 0$
12. $\dfrac{dy}{dx} = xe^x, \quad y(1) = 5$
13. $\dfrac{dy}{dx} = (1 + x^2)(1 + x^3), \quad y(0) = 0.$
14. Sketch the direction field of $dy/dx = y^2$. Draw the solutions passing through $(0, 1)$; through $(0, -1)$. Rewrite the equation as $dx/dy = y^{-2}$ and solve.
15. Solve $xy' + y = 2x$.
    [*Hint:* Compute $(xy)'$.]
16. Solve $y' + y = e^{-x} \sin x$.
    [*Hint:* Compute $(e^x y)'$.]
17. Solve $x + yy' = x^2 + y^2$.
    [*Hint:* Compute the derivative of $\ln(x^2 + y^2)$.]
18. Solve $xy' - y = x^2 \sin x$.
    [Find the trick yourself this time.]

2. SEPARATION OF VARIABLES

The general first order differential equation is

$$\frac{dy}{dx} = q(x, y).$$

In this section, we study the special case in which $q(x, y) = f(x)g(y)$, the product of a function of x alone by a function of y alone. Examples are

$$\frac{dy}{dx} = x^2y^3, \qquad \frac{dy}{dx} = e^x \cos y, \qquad \frac{dy}{dx} = \frac{y}{\sqrt{1 + x^2}}.$$

Not included in this special case are

$$\frac{dy}{dx} = x + y, \qquad \frac{dy}{dx} = \sin(xy), \qquad \frac{dy}{dx} = \sqrt{1 + x^2 + y^2},$$

since in each of these equations the right-hand side is not of the form $f(x)g(y)$.

Solve

$$\frac{dy}{dx} = f(x)g(y)$$

by first "separating the variables," putting all the "y" on one side and all the "x" on the other. Write

$$\frac{dy}{g(y)} = f(x)\,dx.$$

Then integrate both sides (if you can):

$$G(y) = F(x) + c,$$

where $G(y)$ is an antiderivative of $1/g(y)$ and $F(x)$ is an antiderivative of $f(x)$. If this equation can be solved for y in terms of x, the resulting function is a solution of the differential equation. (This technique will be justified after several examples.)

Note that in separating the variables you must assume that $g(y) \neq 0$. If $g(y) = 0$ for $y = y_0$, it is simple to check that the constant function $y(x) = y_0$ is a solution of the equation.

EXAMPLE 2.1

Solve $\dfrac{dy}{dx} = xy.$

Solution: Obviously $y(x) = 0$ is a solution. Now assume $y(x) \neq 0$ and separate variables:

$$\frac{dy}{y} = x\,dx, \qquad \int \frac{dy}{y} = \int x\,dx,$$

$$\ln |y| = \frac{1}{2}x^2 + c,$$

$$|y| = ke^{x^2/2} \qquad (k = e^c),$$

where the constant $k = e^c$ is positive. If $y \geq 0$, then $|y| = y$; if $y < 0$, then $|y| = -y$. Hence

$$y = \begin{cases} ke^{x^2/2} & \text{if } y \geq 0 \\ -ke^{x^2/2} & \text{if } y < 0. \end{cases}$$

This is equivalent to

$$y = ae^{x^2/2},$$

where a is any non-zero constant.

Answer: $y = ae^{x^2/2}$, a any constant. (Note that the solution $y(x) = 0$ is included in this answer.)

Check:

$$\frac{dy}{dx} = \frac{d}{dx}(ae^{x^2/2}) = axe^{x^2/2} = xy.$$

EXAMPLE 2.2

Solve $\dfrac{dy}{dx} = \dfrac{2x(y-1)}{x^2+1}$.

Solution: Obviously $y(x) = 1$ is a solution. Now assume $y(x) \neq 1$ and separate variables:

$$\frac{dy}{y-1} = \frac{2x\,dx}{x^2+1}, \qquad \int \frac{dy}{y-1} = \int \frac{2x\,dx}{x^2+1},$$

$$\ln|y-1| = \ln(x^2+1) + c,$$

$$|y-1| = e^c(x^2+1) = k(x^2+1).$$

Therefore

$$y - 1 = \pm k(x^2+1),$$

where k is a positive constant. The answer may be written

$$y = 1 + a(x^2+1),$$

where a is an arbitrary constant. This includes the special solution $y(x) = 1$.

Answer: $y = 1 + a(x^2+1)$.

Check:

$$\frac{dy}{dx} = 2ax = 2x\frac{a(x^2+1)}{x^2+1} = \frac{2x(y-1)}{x^2+1}.$$

EXAMPLE 2.3

Solve $\dfrac{dy}{dx} = e^{x-y}$.

Solution:

$$\frac{dy}{dx} = e^x e^{-y}, \qquad e^y\,dy = e^x\,dx, \qquad e^y = e^x + c.$$

Answer: $y = \ln(e^x + c), \qquad e^x + c > 0.$

Check:

$$\frac{dy}{dx} = \frac{1}{e^x + c}\frac{d}{dx}(e^x + c) = \frac{e^x}{e^x + c} = \frac{e^x}{e^y} = e^{x-y}.$$

These examples show that the method works. Why does it work? Suppose $y = y(x)$ satisfies

$$\frac{dy}{dx} = f(x)g(y).$$

Let $G(y)$ be any antiderivative of $1/g(y)$. Note that $G(y)$ is indirectly a function of x. By the Chain Rule,

$$\frac{dG}{dx} = \frac{dG}{dy}\frac{dy}{dx} = \frac{1}{g(y)} f(x)g(y) = f(x).$$

Hence $G(y)$ is an antiderivative of $f(x)$. Therefore, if $F(x)$ is *any* antiderivative of $f(x)$,

$$G(y) = F(x) + c.$$

If you can solve this equation for y as a function of x, you have the general solution. If not, then at least you have a relation between x and y, often adequate for applications.

EXAMPLE 2.4

Solve the initial-value problem $\dfrac{dy}{dx} = \dfrac{2x}{5y^4 - 1}, \quad y(1) = 0.$

Solution:

$$(5y^4 - 1)\,dy = 2x\,dx, \qquad y^5 - y = x^2 + c.$$

We cannot solve this fifth degree equation for y; it is hopeless. Still we can substitute the initial data:

$$0 = 1 + c, \qquad c = -1;$$
$$y^5 - y = x^2 - 1.$$

This relation between x and y is as far as we shall get with the problem. But it is adequate for a graph and can be used to calculate y (given x) to any degree of accuracy.

Answer: $y^5 - y = x^2 - 1.$

EXAMPLE 2.5

Solve the initial-value problem $\dfrac{dy}{dx} = x^2y^2, \quad y(0) = b.$

Solution: Obviously $y(x) = 0$ is the solution if $b = 0$. Assume $y \neq 0$ and separate variables:

$$\frac{dy}{y^2} = x^2\,dx, \qquad -\frac{1}{y} = \frac{x^3}{3} + c.$$

Substitute the initial data:

$$-\frac{1}{b} = 0 + c = c.$$

Hence

$$-\frac{1}{y} = \frac{x^3}{3} - \frac{1}{b} = \frac{bx^3 - 3}{3b}.$$

Answer: $y = \dfrac{3b}{3 - bx^3}\cdot$ (The answer includes the solution $y(x) = 0$.)

EXAMPLE 2.6

Find all curves with the following geometric property: the slope of the curve at each point P is twice the slope of the line through P and the origin.

Solution: Translate the geometrical property into an equation. Suppose the graph of $y(x)$ is such a curve and $P = (x, y)$ a point on it. The slope at P is dy/dx. The slope of the line through P and $(0, 0)$ is y/x. Hence, the geometrical property is expressed by the differential equation

$$\frac{dy}{dx} = 2\frac{y}{x}\cdot$$

Obviously $y(x) = 0$ is a solution. Now assume $y(x) \neq 0$ and separate variables:

$$\frac{dy}{y} = 2\frac{dx}{x}, \qquad \ln|y| = 2\ln|x| + c = \ln x^2 + c.$$

Take exponentials, setting $k = e^c$:

$$|y| = kx^2.$$

As in Example 2.1, it follows that $y = ax^2$, where a is any constant.

Answer: The curves are the parabolas $y = ax^2$ and the line $y = 0$. See Fig. 2.1.

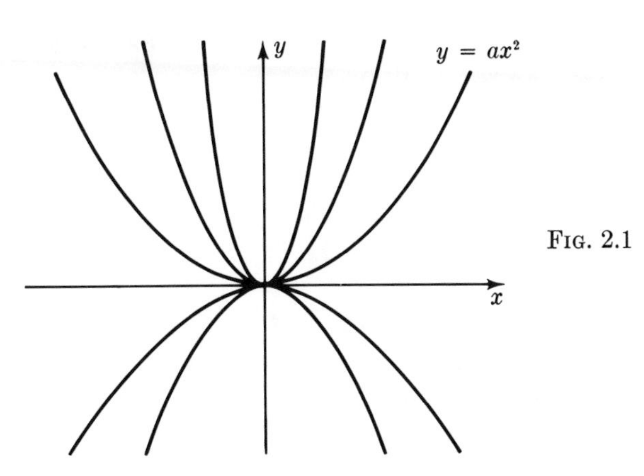

Fig. 2.1

EXAMPLE 2.7

Find all curves whose slope at each point P is the reciprocal of the slope of the line through P and the origin.

Solution: The problem calls for all functions $y(x)$ satisfying

$$\frac{dy}{dx} = \frac{x}{y}.$$

Separate variables:

$$y\,dy = x\,dx, \qquad \frac{1}{2}y^2 = \frac{1}{2}x^2 + c.$$

Answer: The curves are the rectangular hyperbolas $y^2 - x^2 = c$. See Fig. 2.2.

Note the special case $c = 0$, $y = \pm x$.

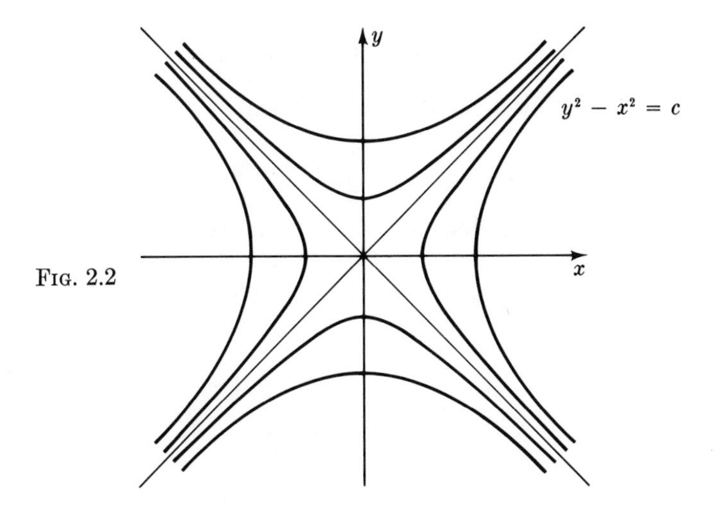

Fig. 2.2

EXERCISES

Solve:

1. $\dfrac{dy}{dx} = \dfrac{x^2}{y}$
2. $\dfrac{dy}{dx} = \dfrac{1}{xy}$
3. $\dfrac{dy}{dx} = \sqrt{\dfrac{y}{x}}$
4. $\dfrac{dy}{dx} = \dfrac{x+1}{xy}$
5. $\dfrac{dy}{dx} = xe^y$
6. $\dfrac{dy}{dx} = \left(\dfrac{1+y}{1+x}\right)^2.$

Show that the substitution $y = ux$ changes the differential equation into an equation for $u = u(x)$ whose variables separate, and solve:

7. $\dfrac{dy}{dx} = \dfrac{x^2 + y^2}{2x^2}$
8. $\dfrac{dy}{dx} = \dfrac{x+y}{x-y}.$

Solve the initial-value problem:

9. $\dfrac{dy}{dx} = -\dfrac{y}{2x}, \quad y(1) = 3$
10. $\dfrac{dy}{dx} = 2y \cot x, \quad y\left(\dfrac{\pi}{2}\right) = 5$
11. $\dfrac{dy}{dx} = \dfrac{(y-1)(y-2)}{x}, \quad y(4) = 0$
12. $\dfrac{dy}{dx} = y^3 \sin x, \quad y(0) = \dfrac{1}{2}.$

Solve the given differential equation; interpret the equation and its solution geometrically:

13. $\dfrac{dy}{dx} = -\dfrac{y}{x}$
14. $\dfrac{dy}{dx} = -\dfrac{x}{y}$
15. $\dfrac{dy}{dx} = \dfrac{y}{x}.$

3. LINEAR DIFFERENTIAL EQUATIONS

A differential equation of the form

$$\frac{dy}{dx} + p(x)y = q(x)$$

is called a **first order linear differential equation.**

The term "linear" can be described in the following way. Imagine a "black box" or "processor" that converts a function $y(x)$ into a function $w(x)$, as in Fig. 3.1.

FIG. 3.1

The black box is called **linear** if from a linear combination* of inputs, it produces the same linear combination of outputs (Fig. 3.2):

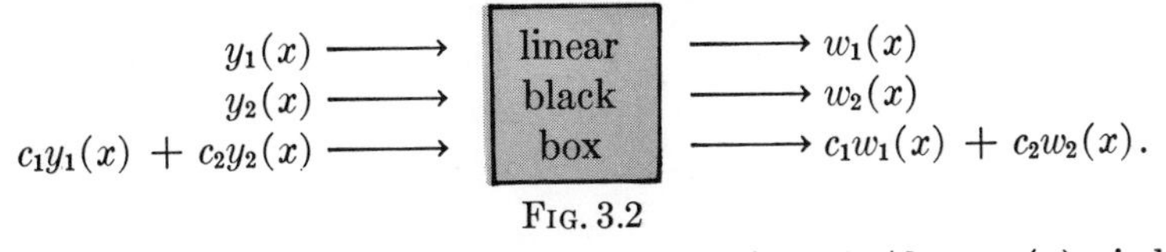

FIG. 3.2

In particular, the black box that converts y into $dy/dx + p(x)y$ is linear: when the input is $c_1y_1 + c_2y_2$, the output is

$$\frac{d}{dx}(c_1y_1 + c_2y_2) + p(x)(c_1y_1 + c_2y_2)$$

$$= c_1\left(\frac{dy_1}{dx} + p(x)y_1\right) + c_2\left(\frac{dy_2}{dx} + p(x)y_2\right) = c_1(\text{output}_1) + c_2(\text{output}_2).$$

This is why the differential equation $(dy/dx) + p(x)y = q(x)$ is called linear.

REMARK: The black boxes discussed here are nothing else than the linear functions of linear algebra, in an appropriate setting. (See Ch. 7, Sec. 1.) For example, let **D** be the set of all differentiable functions $y(x)$. Then the "black box" defined by $Ly = y' + p(x)y$ is a linear function defined on **D** since

$$L(ay_1 + by_2) = aLy_1 + bLy_2.$$

Not every black box is linear, however. For instance, consider the black box in Fig. 3.3.

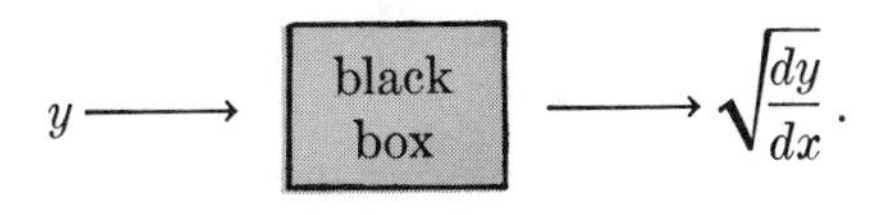

FIG. 3.3

This box is not linear since the output of a sum is not the sum of the outputs (Fig. 3.4).

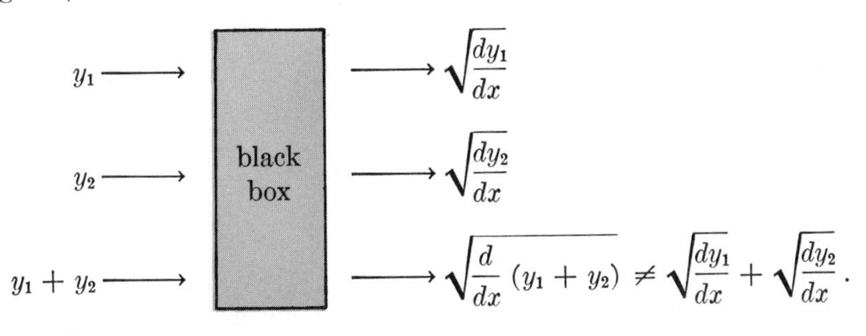

FIG. 3.4

* An expression $aX + bY + cZ + \cdots$, where a, b, c, $\cdots$ are numbers, is called a **linear combination** of X, Y, Z, $\cdots$.

Figure 3.5 shows another non-linear process.

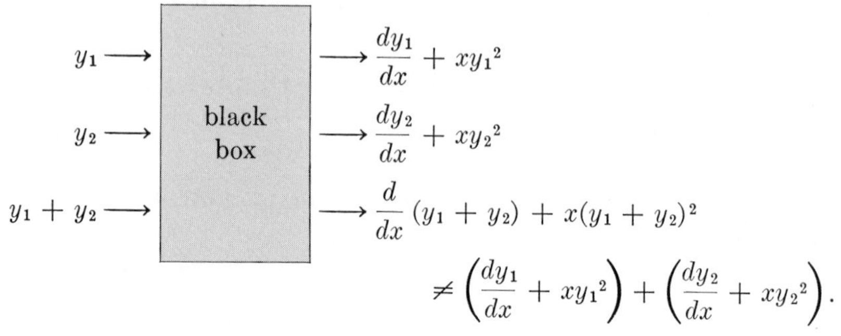

$$y_1 \longrightarrow \boxed{\text{black box}} \longrightarrow \frac{dy_1}{dx} + xy_1{}^2$$

$$y_2 \longrightarrow \boxed{\text{black box}} \longrightarrow \frac{dy_2}{dx} + xy_2{}^2$$

$$y_1 + y_2 \longrightarrow \boxed{\text{black box}} \longrightarrow \frac{d}{dx}(y_1 + y_2) + x(y_1 + y_2)^2$$

$$\neq \left(\frac{dy_1}{dx} + xy_1{}^2\right) + \left(\frac{dy_2}{dx} + xy_2{}^2\right).$$

FIG. 3.5

A Property of Linear Processes

An extremely useful property of linear black boxes is shown in Fig. 3.6.

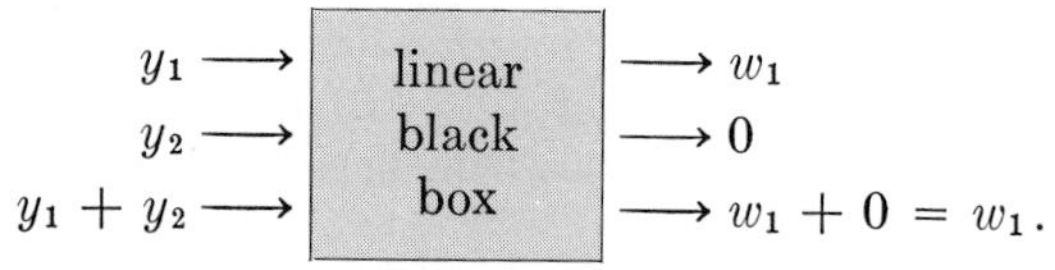

$$y_1 \longrightarrow \boxed{\text{linear black box}} \longrightarrow w_1$$

$$y_2 \longrightarrow \boxed{\text{linear black box}} \longrightarrow 0$$

$$y_1 + y_2 \longrightarrow \boxed{\text{linear black box}} \longrightarrow w_1 + 0 = w_1.$$

FIG. 3.6

If the output of y_2 is 0, then the output of $y_1 + y_2$ is the output of y_1. Hence output is unchanged when input is augmented by a function which is converted to zero.

This simple fact is the basis for a systematic study of linear differential equations. We return to the process that really interests us, the one that converts $y(x)$ into $dy/dx + p(x)y$. For brevity, we write

$$\frac{dy}{dx} + p(x)y = Ly,$$

indicating that Ly is the output corresponding to the input y. The differential equation in question is

$$\frac{dy}{dx} + p(x)y = q(x),$$

or

$$Ly = q.$$

A **solution** is an input $y(x)$ that results in the given output $q(x)$.

Suppose we can find a solution y_1. That means $Ly_1 = q$. If z is any function for which $Lz = 0$, then $L(y_1 + z) = q$. Thus $y_1 + z$ is also a solution. Furthermore, *every* solution must be of the form $y_1 + z$. To see why, assume y_1 and y_2 are solutions:

$$Ly_1 = q, \qquad Ly_2 = q.$$

Consider the difference $y_1 - y_2$. Because L is a *linear* process,

$$L(y_1 - y_2) = Ly_1 - Ly_2 = q - q = 0.$$

Therefore, when the input is $y_1 - y_2$, the output is 0; this says the difference of two solutions is one of the functions z.

Any two solutions of the differential equation

$$\frac{dy}{dx} + p(x)y = q(x)$$

differ by a solution of the **homogeneous equation**

$$\frac{dy}{dx} + p(x)y = 0.$$

Therefore, the general solution of

$$\frac{dy}{dx} + p(x)y = q(x)$$

is

$$u(x) + z(x),$$

where $u(x)$ is any solution of the equation and $z(x)$ is the general solution of the associated homogeneous equation.

Because of this analysis, the differential equation (often called the **non-homogeneous equation**)

$$\frac{dy}{dx} + p(x)y = q(x)$$

is solved in three steps:

(1) Find the general solution of the associated homogeneous equation

$$\frac{dy}{dx} + p(x)y = 0.$$

(2) Find any solution of

$$\frac{dy}{dx} + p(x)y = q(x).$$

(3) Add the results.

In the next two sections we shall treat steps (1) and (2) separately.

EXERCISES

Suppose the input to a black box is $y(x)$ and the output is one of the following. Determine in each case whether the black box is linear or non-linear:

1. $\int_0^1 y(x)\,dx$
2. $y(x) + x$
3. $\dfrac{d^2y}{dx^2} + 6\dfrac{dy}{dx} + y$
4. $3y(x)$
5. $\ln|y(x)|$
6. $y(x+1)$.
7. What is the general solution of the differential equation $dy/dx = x^2$? Is your answer a particular solution plus the general solution of the associated homogeneous equation?
8. A big black box consists of two linear black boxes in series:

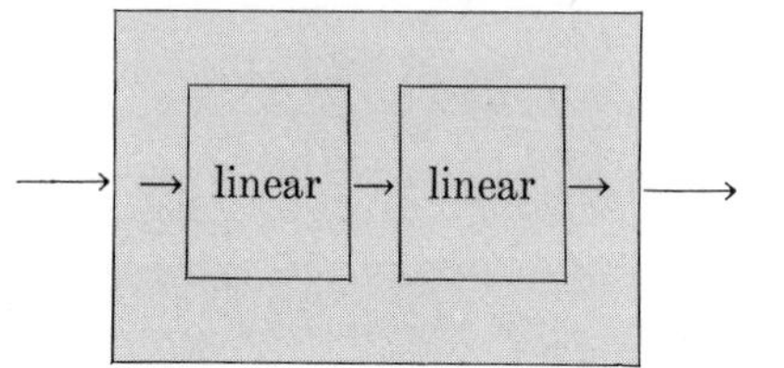

Is the big box linear?

4. HOMOGENEOUS EQUATIONS

Let us solve the homogeneous equation

$$\frac{dy}{dx} + p(x)y = 0.$$

Obviously $y(x) = 0$ is a solution. Assume $y(x) \neq 0$ and separate variables:

$$\frac{dy}{y} = -p(x)\,dx, \qquad \ln|y| = -P(x) + c,$$

where $P(x)$ is any antiderivative of $p(x)$. Now exponentiate both sides, setting $k = e^c$:

$$|y(x)| = ke^{-P(x)}.$$

Hence,

$$y(x) = \begin{cases} ke^{-P(x)} & \text{if } y \geq 0 \\ -ke^{-P(x)} & \text{if } y < 0. \end{cases}$$

These facts are summarized in the following statement.

The general solution of

$$\frac{dy}{dx} + p(x)y = 0$$

is

$$y(x) = ae^{-P(x)},$$

where $P(x)$ is any antiderivative of $p(x)$ and a is a constant.

EXAMPLE 4.1

Find the general solution of $\dfrac{dy}{dx} + (\sin x)y = 0.$

Solution: The equation is of the form

$$\frac{dy}{dx} + p(x)y = 0,$$

where $p(x) = \sin x$. An antiderivative of $p(x)$ is $P(x) = -\cos x$. By the formula, the general solution is

$$y(x) = ae^{-P(x)} = ae^{\cos x}.$$

Answer: $y(x) = ae^{\cos x}.$

EXAMPLE 4.2

Solve the initial-value problem $y' + 3y = 0, \quad y(1) = 4.$

Solution: In this case $p(x) = 3$. An antiderivative is $P(x) = 3x$ and the general solution is

$$y(x) = ae^{-3x}.$$

To choose the correct constant, substitute $x = 1$, $y = 4$:

$$4 = ae^{-3}, \qquad a = 4e^{3}.$$

Answer: $y(x) = 4e^{3(1-x)}.$

EXAMPLE 4.3

Solve the initial-value problem $\dfrac{dy}{dx} + \dfrac{1}{x}y = 0, \quad y(1) = 2.$

Solution: Since $p(x) = 1/x$ is not defined at $x = 0$, the solution may behave badly near $x = 0$. Therefore consider only $x > 0$.

An antiderivative of $1/x$ is $\ln x$; the general solution is

$$y(x) = ae^{-\ln x} = \frac{a}{x}$$

(which $\longrightarrow \infty$ as $x \longrightarrow 0$). To choose the correct constant, substitute $x = 1$, $y = 2$:

$$2 = \frac{a}{1}.$$

Answer: $y(x) = \dfrac{2}{x}, \quad x > 0.$

EXERCISES

Solve:

1. $\dfrac{dy}{dx} + 3\dfrac{y}{x} = 0$
2. $L\dfrac{di}{dt} + Ri = 0, \quad L, R$ constants
3. $\dfrac{dy}{d\theta} + y \tan\theta = 0$
4. $\dfrac{1}{2x}\dfrac{dy}{dx} + \dfrac{y}{x^2 + 1} = 0$
5. $x\dfrac{dy}{dx} + (1 + x)y = 0.$

Solve the initial-value problem:

6. $\dfrac{dy}{dx} + xy = 0, \quad y(0) = -1$
7. $(1 - x^2)\dfrac{dy}{dx} + xy = 0, \quad y(0) = 3$
8. $\dfrac{dy}{dx} + y\sqrt{x} = 0, \quad y(0) = 2$
9. $\dfrac{dy}{dx} + \dfrac{2y}{x} = 0, \quad y(2) = \dfrac{1}{5}.$
10. Solve $y'' - y' = 0$ by substituting $v = y'$.

5. NON-HOMOGENEOUS EQUATIONS

We now consider the non-homogeneous equation

$$\frac{dy}{dx} + p(x)y = q(x).$$

By our analysis of the problem, we need only one particular solution of this equation, to be added to the general solution of the associated homogeneous equation. We shall discuss two methods for finding a particular solution.

METHOD 1. GUESSING.

This works particularly well when

(a) $p(x)$ is a constant, and

(b) $q(x)$ and all of its derivatives can be expressed in terms of a few functions. For example, a quadratic polynomial and its derivatives can be expressed in terms of 1, x, and x^2. The function $\sin x$ and its derivatives can be expressed in terms of $\sin x$ and $\cos x$. Each of the functions

$$e^x, \qquad xe^x, \qquad x^2e^x, \qquad x\cos x, \qquad e^x \sin x$$

has a similar property; it and all its derivatives involve only a few functions. However, $1/x$ is not of this type; each of its derivatives involves a different function.

EXAMPLE 5.1

Find a solution of $\dfrac{dy}{dx} - y = x^2$.

Solution: The right side of the equation is a quadratic polynomial. Guess a quadratic polynomial

$$y(x) = Ax^2 + Bx + C,$$

and try to determine suitable coefficients A, B, C. Substitute $y(x)$ into the differential equation:

$$(2Ax + B) - (Ax^2 + Bx + C) = x^2,$$

$$-Ax^2 + (2A - B)x + (B - C) = x^2.$$

Equate coefficients:

$$-A = 1, \qquad 2A - B = 0, \qquad B - C = 0,$$

from which

$$A = -1, \qquad B = -2, \qquad C = -2.$$

Answer: $y(x) = -(x^2 + 2x + 2)$ is a solution.

EXAMPLE 5.2

Find a solution of $\dfrac{dy}{dx} - y = e^{3x}$.

Solution: Try $y = Ae^{3x}$. The equation becomes

$$3Ae^{3x} - Ae^{3x} = e^{3x}.$$

Hence,

$$2A = 1, \qquad A = \frac{1}{2}.$$

Answer: $y(x) = \dfrac{1}{2}e^{3x}$ is a solution.

EXAMPLE 5.3

Find a solution of $\dfrac{dy}{dx} - y = xe^{3x}$.

Solution: Notice that xe^{3x} and all its derivatives involve only the functions xe^{3x} and e^{3x}. Try

$$y(x) = Axe^{3x} + Be^{3x}.$$

The differential equation becomes

$$(Ae^{3x} + 3Axe^{3x} + 3Be^{3x}) - (Axe^{3x} + Be^{3x}) = xe^{3x},$$

$$2Axe^{3x} + (A + 2B)e^{3x} = xe^{3x}.$$

Thus, $2A = 1$, $A + 2B = 0$, from which $A = \frac{1}{2}$, $B = -\frac{1}{4}$.

Answer: $y(x) = \dfrac{1}{2}xe^{3x} - \dfrac{1}{4}e^{3x}$

is a solution.

EXAMPLE 5.4

Find a solution of $\dfrac{dy}{dx} - y = \sin x$.

Solution: All derivatives of $\sin x$ involve only $\sin x$ and $\cos x$. Try

$$y(x) = A\cos x + B\sin x.$$

The differential equation becomes

$$(-A\sin x + B\cos x) - (A\cos x + B\sin x) = \sin x,$$

$$-(A + B)\sin x + (B - A)\cos x = \sin x.$$

Therefore

$$-(A + B) = 1, \qquad B - A = 0.$$

Thus $A = B = -\frac{1}{2}$.

Answer: $y(x) = -\dfrac{1}{2}\cos x - \dfrac{1}{2}\sin x$

is a solution.

There is a special situation in which the method must be modified. Consider for instance the differential equation

$$\frac{dy}{dx} - y = e^x.$$

Try $y(x) = Ae^x$:

$$\frac{d}{dx}(Ae^x) - Ae^x = e^x,$$

$$Ae^x - Ae^x = e^x, \qquad 0 = e^x,$$

which is impossible. The guess fails because e^x is a solution of the homogeneous equation

$$\frac{dy}{dx} - y = 0.$$

Consequently, substituting $y = Ae^x$ makes the left side zero; there is no hope of equating it to the right side. The function xe^x is a solution, however, as is easily checked. This is an example of a general situation:

> If $q(x)$ is a solution of the homogeneous equation
>
> $$\frac{dy}{dx} + p(x)y = 0,$$
>
> then $y(x) = xq(x)$ is a solution of the non-homogeneous equation
>
> $$\frac{dy}{dx} + p(x)y = q(x).$$

This fact soon will seem more natural when we discuss Method 2. (See Example 5.8.) It is easily verified:

$$\begin{aligned}\frac{dy}{dx} + p(x)y &= \frac{d}{dx}[xq(x)] + p(x)[xq(x)] \\ &= q(x) + x\left(\frac{dq}{dx} + p(x)q(x)\right) \\ &= q(x) + x \cdot 0 = q(x).\end{aligned}$$

EXAMPLE 5.5

Find a solution of $\dfrac{dy}{dx} - \dfrac{1}{x}y = x.$

Solution: Notice that $y(x) = x$ is a solution of the homogeneous equation

$$\frac{dy}{dx} - \frac{1}{x}y = 0.$$

But x occurs also on the right-hand side of the differential equation. Therefore, according to the rule, a solution is $x \cdot x = x^2$. Check it!

Answer: $y(x) = x^2$ is a solution.

Summary

Here is a list of standard guesses for solving

$$\frac{dy}{dx} + p(x)y = q(x).$$

$q(x)$	Guess
a, constant	A, constant
$ax + b$	$Ax + B$
$ax^2 + bx + c$	$Ax^2 + Bx + C$
e^x	Ae^x
$\sin x$, $\cos x$	$A \cos x + B \sin x$
$e^x(ax + b)$	$e^x(Ax + B)$
$e^x(ax^2 + bx + c)$	$e^x(Ax^2 + Bx + C)$
$(ax + b) \sin x$	$(Ax + B) \cos x + (Cx + D) \sin x$
$e^x \sin x$, $e^x \cos x$	$e^x(A \cos x + B \sin x)$

METHOD 2.

Begin with any non-zero solution $u(x)$ of the homogeneous equation

$$\frac{du}{dx} + p(x)u = 0.$$

Look for a solution $y(x)$ of the non-homogeneous equation in the form

$$y(x) = u(x) \cdot v(x).$$

The unknown function is now $v(x)$. Substitute $y = uv$ and $y' = uv' + vu'$ into the differential equation $y' + py = q$:

$$uv' + vu' + puv = q,$$

$$uv' + v(u' + pu) = q.$$

But the expression in parentheses is 0. Hence,

$$uv' = q.$$

Since u is not 0,

$$\frac{dv}{dx} = \frac{q(x)}{u(x)}.$$

Antidifferentiate to find $v(x)$. (This may be difficult or impossible, a disadvantage of the method.)

If $V(x)$ is an antiderivative, then $V(x)u(x)$ is a solution of the differential equation. If you keep the constant of integration, then you get $[V(x) + c]u(x)$, which is the general solution of the differential equation. (Why?)

EXAMPLE 5.6

Find the general solution of $\dfrac{dy}{dx} - \dfrac{y}{x} = x^3 + 1.$

Solution: The associated homogeneous equation

$$\frac{du}{dx} - \frac{u}{x} = 0$$

has a solution $u(x) = x$. Set $y = xv$ and substitute:

$$x\frac{dv}{dx} + v - \frac{xv}{x} = x^3 + 1,$$

$$x\frac{dv}{dx} = x^3 + 1, \qquad \frac{dv}{dx} = x^2 + \frac{1}{x}.$$

Antidifferentiate:

$$v = \frac{1}{3}x^3 + \ln|x| + c.$$

The general solution is $xv(x)$.

Answer: $y(x) = \frac{1}{3}x^4 + x\ln|x| + cx.$

EXAMPLE 5.7

Solve $\frac{dy}{dx} + 2xy = x.$

Solution: Solve the associated homogeneous equation

$$\frac{du}{dx} + 2xu = 0:$$

$$\frac{du}{u} = -2x\,dx, \qquad \ln|u| = -x^2 + c.$$

One solution is

$$u(x) = e^{-x^2}.$$

Now set

$$y = e^{-x^2}v,$$

so

$$y' = -2xe^{-x^2}v + e^{-x^2}v'.$$

Substitute into the differential equation:

$$-2xe^{-x^2}v + e^{-x^2}v' + 2xe^{-x^2}v = x,$$

$$v' = xe^{x^2}.$$

Antidifferentiate:

$$v(x) = \frac{1}{2}e^{x^2} + c.$$

Hence the general solution is

$$y = e^{-x^2}v = \frac{1}{2} + ce^{-x^2}.$$

Hindsight: We should have guessed the particular solution $y = \frac{1}{2}$.

Alternate Solution: Separate variables:

$$\frac{dy}{dx} = x(1 - 2y), \qquad \frac{dy}{1 - 2y} = x\,dx.$$

$$-\frac{1}{2}\ln|1 - 2y| = \frac{1}{2}x^2 - \frac{1}{2}a,$$

$$\ln|1 - 2y| = a - x^2,$$

$$|1 - 2y| = 2ke^{-x^2} \qquad (2k = e^a).$$

It follows that

$$1 - 2y = \pm 2ke^{-x^2},$$

$$y = \frac{1}{2} + ce^{-x^2}, \qquad c \text{ any constant.}$$

Answer: $y(x) = \dfrac{1}{2} + ce^{-x^2}$.

Next we derive an assertion made in the discussion of Method 1 (rather than just pulling it out of a hat).

EXAMPLE 5.8

Find a solution of $\dfrac{dy}{dx} + p(x)y = q(x)$ if $q(x)$ is a solution of the associated homogeneous equation.

Solution: Try a solution $y(x) = v(x) \cdot q(x)$.

$$\frac{dy}{dx} + py = \frac{d}{dx}(vq) + pvq = q\frac{dv}{dx} + v\frac{dq}{dx} + pvq$$

$$= q\frac{dv}{dx} + v\left(\frac{dq}{dx} + pq\right) = q\frac{dv}{dx} + 0.$$

Hence, the differential equation becomes

$$q\frac{dv}{dx} = q, \qquad \frac{dv}{dx} = 1, \qquad v(x) = x.$$

Answer: $y(x) = xq(x)$ is a solution.

EXERCISES

Find a particular solution:

1. $\dfrac{dy}{dx} + 2y = x$

2. $\dfrac{dy}{dx} - y = 3x - 2$

3. $\dfrac{dy}{dx} + 4y = e^x$

4. $\dfrac{dy}{dx} - y = xe^x$

5. $\dfrac{dy}{dx} + y = x^2e^x$

6. $\dfrac{dy}{dx} + 2y = \cos x$

7. $2\dfrac{dy}{dx} + 5y = 3\sin x$

8. $6\dfrac{dy}{dx} + y = e^{3x} - x - 1$

9. $\dfrac{dy}{dx} + 2y = e^{-x}\cos x$

10. $L\dfrac{di}{dt} + Ri = e^{-kt}$

11. $L\dfrac{di}{dt} + Ri = E\cos t$

12. $\dfrac{dy}{d\theta} + ay = \cos 2\theta$

13. $\dfrac{dy}{dx} - y = x^5$

14. $\dfrac{dy}{dx} + 2y = e^{-2x}$

15. $\dfrac{dy}{dx} - y = e^x$

Use Method 2 of the text to find a particular solution:

16. $\dfrac{dy}{dx} + xy = x$

17. $\dfrac{dy}{dx} - \dfrac{2y}{x} = x^2 + 1$

18. $\dfrac{dy}{dx} - \dfrac{y}{x} = \dfrac{1}{x^2}$

19. $\dfrac{dy}{dx} - \dfrac{2xy}{x^2+1} = x.$

Find the general solution:

20. $\dfrac{dy}{dx} + 3y = 0$

21. $\dfrac{dy}{dx} + 3y = 1$

22. $\dfrac{dy}{dx} - 2y = 3x^2 + 2x + 1$

23. $\dfrac{dy}{dx} + y = xe^{2x} + 1$

24. $\dfrac{dy}{dx} - 2y = x^2e^x$

25. $L\dfrac{di}{dt} + Ri = E\cos\omega t$

26. $L\dfrac{di}{dt} + Ri = Ee^{kt}$

27. $\dfrac{dy}{dx} + 2xy = 2x$

28. $\dfrac{dy}{dx} + \dfrac{2xy}{x^2+1} = x$

29. $\dfrac{dy}{dx} + 3y = e^{-3x}.$

30. Show that the **Bernoulli Equation** $\dfrac{dy}{dx} + p(x)y = q(x)y^n$ can be reduced to a linear equation by the substitution $z = y^{1-n}$.

31. (cont.) Solve $y' + xy = \sqrt{y}$; find a formula for the answer, but do not try to evaluate it.

32.* Suppose $q(x)$ satisfies $|q(x)| \leq M$ and that $y(x)$ is a solution of the initial-value problem $\dfrac{dy}{dx} + y = q(x), \quad y(0) = 0$. Show that $|y(x)| \leq M$ for $x \geq 0$.

33. Sketch the direction field of $y' = x + y$. By inspection of the field find a particular solution of $y' - y = x$.

6. APPLICATIONS

We now consider some applications of the preceding material. In each case we must first interpret the geometrical or physical data in terms of a differential equation.

EXAMPLE 6.1

Find all curves $y = y(x)$ with the property that the tangent line at each point (x, y) on the curve meets the x-axis at $(x + 3, 0)$.

Solution: Sketch the data (Fig. 6.1). The slope of the tangent line is dy/dx on the one hand and $[y - 0]/[x - (x + 3)] = -y/3$ on the other.

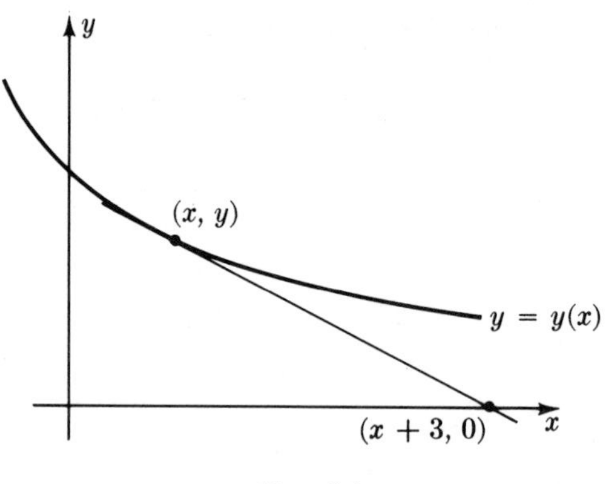

FIG. 6.1

Hence the differential equation is

$$\frac{dy}{dx} = -\frac{1}{3}y.$$

Answer: $y = ke^{-x/3}$.

EXAMPLE 6.2

Find all curves that intersect each of the rectangular hyperbolas $xy = c$ at right angles.

Solution: The family of hyperbolas is shown in Fig. 6.2. First construct the direction field which at each point (x, y) is tangent to the hyperbola $xy = c$ through the point.

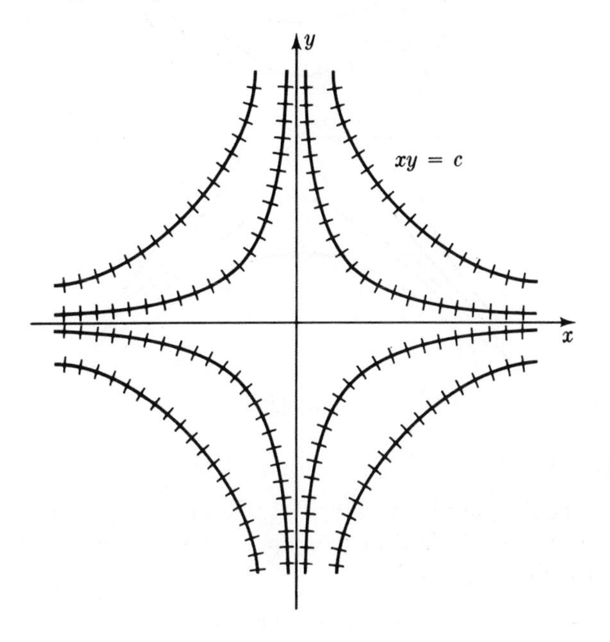

FIG. 6.2

Differentiate:

$$y + xy' = 0.$$

Hence

$$y' = -\frac{y}{x}.$$

The direction field perpendicular to the hyperbola through (x, y) is determined by the negative reciprocal of y'. In other words, the differential equation of the direction field is

$$\frac{dy}{dx} = \frac{x}{y}.$$

Separate variables:

$$y\,dy = x\,dx, \qquad y^2 = x^2 - k.$$

Answer: The perpendicular curves (Fig. 6.3) are the rectangular hyperbolas $x^2 - y^2 = k$.

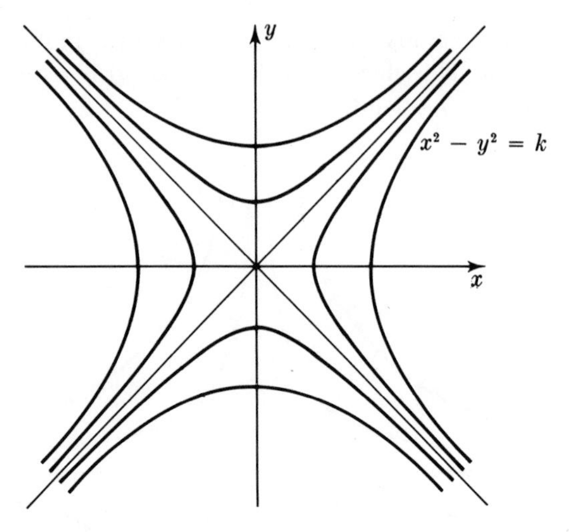

FIG. 6.3

EXAMPLE 6.3

A spherical mothball evaporates at a rate proportional to its surface area. If half of it evaporates in 3 weeks, when will the mothball disappear?

Solution: Let $r = r(t)$ denote the radius at time t in weeks, with $r(0) = r_0$. The volume at time t is $V(t) = \frac{4}{3}\pi r^3 = \frac{4}{3}\pi[r(t)]^3$, and the surface area is $A(t) = 4\pi r^2 = 4\pi[r(t)]^2$.

The data are

(i) $\dfrac{dV}{dt} = -kA,$

(ii) $V(3) = \dfrac{1}{2}\,V(0) = \dfrac{1}{2}\cdot\dfrac{4}{3}\,\pi r_0{}^3.$

By (i),

$$\frac{d}{dt}\left(\frac{4}{3}\,\pi r^3\right) = -4k\pi r^2,$$

that is,

$$4\pi r^2\,\frac{dr}{dt} = -4k\pi r^2,$$

from which

$$\frac{dr}{dt} = -k.$$

Hence

$$r = -kt + c.$$

Evaluate the constant c by substituting the initial condition, $r = r_0$ when $t = 0$:

$$r_0 = c.$$

Hence

$$r = r_0 - kt.$$

To find k, use condition (ii):

$$\frac{4}{3}\pi[r(3)]^3 = \frac{1}{2}\cdot\frac{4}{3}\pi r_0^3, \quad \text{hence} \quad r(3) = \frac{r_0}{\sqrt[3]{2}}.$$

But $r(3) = r_0 - 3k$, consequently

$$\frac{r_0}{\sqrt[3]{2}} = r_0 - 3k, \qquad k = \frac{r_0}{3}\left(1 - \frac{1}{\sqrt[3]{2}}\right).$$

Therefore, the formula for r is

$$r = r_0 - kt = r_0\left[1 - \frac{t}{3}\left(1 - \frac{1}{\sqrt[3]{2}}\right)\right].$$

From this it follows that the mothball disappears ($r = 0$) when

$$t = \frac{3}{\left(1 - \frac{1}{\sqrt[3]{2}}\right)} \approx 14.5.$$

Answer: The mothball disappears after approximately 14.5 weeks.

EXAMPLE 6.4

Let $T = T(t)$ denote the average temperature of a small piece of hot metal in a cooling bath of fixed temperature 50°. According to Newton's Law of Cooling, the metal cools at a rate proportional to the difference $T - 50$. Suppose the metal cools from 250° to 150° in 30 sec. How long would it take to cool from 450° to 60° ?

Solution: In mathematical terms, the Law of Cooling is a linear differential equation:

$$\frac{dT}{dt} = -k(T - 50),$$

that is,

$$\frac{dT}{dt} + kT = 50k.$$

The associated homogeneous equation

$$\frac{dT}{dt} + kT = 0,$$

has the general solution

$$T = ce^{-kt}.$$

It is easy to guess the particular solution $T = 50$. Conclusion: the general solution is

$$T = 50 + ce^{-kt}.$$

Setting $t = 0$, we find

$$T(0) = 50 + c, \qquad c = T(0) - 50.$$

Hence

$$T = 50 + [T(0) - 50]e^{-kt}.$$

We must determine the constant k. By hypothesis, if $T(0) = 250$, then $T(30) = 150$:

$$T(30) = 150 = 50 + [250 - 50]e^{-30k},$$

$$100 = 200e^{-30k}, \qquad k = -\frac{1}{30}\ln\frac{1}{2} = \frac{1}{30}\ln 2.$$

Therefore,

$$T = 50 + [T(0) - 50]e^{-kt}, \qquad \text{where} \quad k = \frac{1}{30}\ln 2.$$

In this formula we substitute $T = 60$ and $T(0) = 450$, then solve for t:

$$60 = 50 + [450 - 50]e^{-kt}, \qquad \frac{10}{400} = e^{-kt},$$

$$t = -\frac{1}{k}\ln\frac{1}{40} = \frac{1}{k}\ln 40 = \frac{30\ln 40}{\ln 2}.$$

Answer: $\dfrac{30\ln 40}{\ln 2} \approx 160$ sec.

EXAMPLE 6.5

Water flows from an open tank through a small hole of area A ft^2 at the base. When the depth in the tank is x ft, the rate of flow is $(0.60)A\sqrt{2gx}$, where $g = 32.17$ ft/sec^2. How long does it take to empty a full spherical tank of diameter 10 ft through a circular hole of diameter 2 in. at the bottom? (Assume a small hole at the top admits air.)

Solution: Let V denote the volume when the depth is x. Then

$$\frac{dV}{dt} = -k\sqrt{x}, \qquad \text{where} \quad k = (0.60)A\sqrt{2g}.$$

By the Chain Rule,

$$\frac{dV}{dx}\frac{dx}{dt} = -k\sqrt{x}.$$

Finding dV/dx is a problem in volume by slicing. Draw a cross section of the tank (Fig. 6.4). The change in volume is $dV \approx \pi y^2\,dx$. Hence

$$\frac{dV}{dx} = \pi y^2 = \pi[5^2 - (x-5)^2] = \pi(10x - x^2).$$

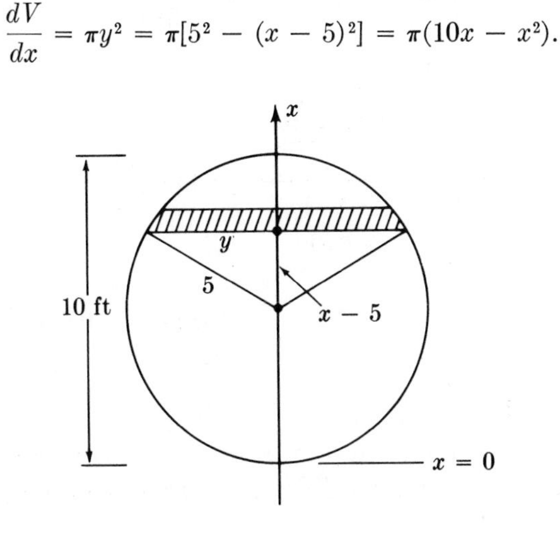

FIG. 6.4

The differential equation is

$$\pi(10x - x^2)\frac{dx}{dt} = -k\sqrt{x},$$

$$(10x^{1/2} - x^{3/2})\frac{dx}{dt} = \frac{-k}{\pi}, \qquad x(0) = 10.$$

Separate variables and integrate:

$$\frac{20}{3}x^{3/2} - \frac{2}{5}x^{5/2} = -\frac{k}{\pi}t + c.$$

Substitute the initial condition, $x = 10$ when $t = 0$:

$$\left(\frac{200}{3} - \frac{200}{5}\right)\sqrt{10} = c, \qquad c = \frac{80}{3}\sqrt{10}.$$

Solve for t when $x = 0$:

$$0 = -\frac{k}{\pi}t + c, \qquad t = \frac{\pi c}{k},$$

where $k = (0.60)A\sqrt{2g}$, and A is the area of a circle of 2 in. diameter,

$$A = \pi\left(\frac{1}{12}\right)^2 \text{ ft}^2.$$

Thus

$$k = (0.60)A\sqrt{2g} = \frac{\pi\sqrt{2g}}{240}.$$

The answer to the problem is

$$t = \frac{\pi c}{k} = \frac{\dfrac{80\pi\sqrt{10}}{3}}{\dfrac{\pi\sqrt{2g}}{240}} = 6400\sqrt{\frac{5}{g}} \quad \text{sec.}$$

Answer: $6400\sqrt{\dfrac{5}{g}} \approx 2520$ sec.

EXAMPLE 6.6

When a uniform steel rod of length L and cross-sectional area 1 is subjected to a tension T, it stretches to length $L + D$, where

$$\frac{D}{L} = kT, \qquad k \quad \text{a constant.}$$

(This is Hooke's Law, valid if T does not exceed a certain bound.)

Suppose the rod is suspended vertically by one end. How much does it stretch? Assume the weight of the rod is W.

Solution: First imagine the rod lying on a horizontal surface. Calibrate the rod by measuring distances from one end. In this way each point of the rod is identified with a number x between 0 and L.

Now suppose the rod is hung by the "zero" end. Let $y(x)$ denote the distance from the top to the point designated by x. Since the rod is stretched due to its own weight, $y(x) > x$ for $x > 0$. The amount the rod stretches is $y(L) - L$.

Consider a short portion of the rod from x to $x + h$. See Fig. 6.5. It is under tension $T(x)$, approximately the weight of the portion of the rod between the points marked x and L, i.e.,

$$T(x) \approx W\frac{L - x}{L}.$$

How much is this portion stretched by the tension $T(x)$? Its original length is h. When stretched, its ends are at $y(x)$ and $y(x + h)$, so its new length is $y(x + h) - y(x)$. The increase in length is $y(x + h) - y(x) - h$. By Hooke's Law,

$$\frac{y(x + h) - y(x) - h}{h} \approx kW\frac{L - x}{L}.$$

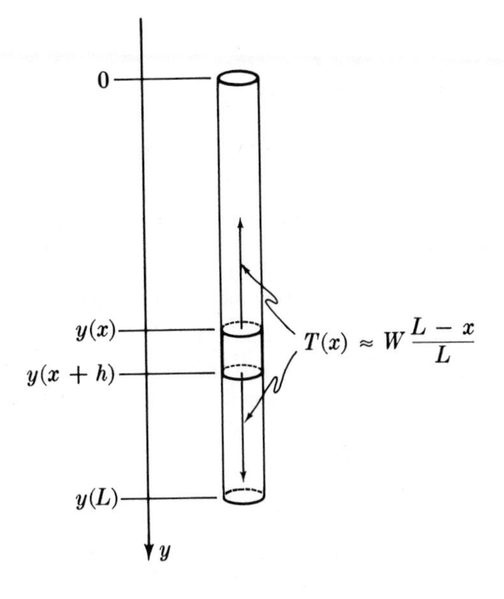

FIG. 6.5

Let $h \longrightarrow 0$:

$$\frac{dy}{dx} - 1 = kW\frac{L - x}{L}.$$

This is a differential equation for the position $y(x)$ of the point x. Solve it subject to the initial condition $y(0) = 0$. First integrate both sides:

$$y - x = -\frac{kW}{2}\frac{(L - x)^2}{L} + c.$$

By the initial condition,

$$0 = -\frac{kWL}{2} + c,$$

hence

$$y - x = \frac{1}{2}kWL - \frac{1}{2}\frac{kW}{L}(L - x)^2.$$

Set $x = L$:

$$y(L) - L = \frac{1}{2}kWL.$$

Answer: The hanging rod stretches $\frac{1}{2}kWL$ units of length.

EXERCISES

1. Find all curves that intersect each of the ellipses $x^2 + 3y^2 = c$ at right angles. Sketch the curves. (See Example 6.2.)
2. Find all curves that intersect each of the curves $y = ce^{-x^2}$ at right angles. Sketch them.
3. The **half-life** of a radioactive substance is the length of time in which a quantity of the substance is reduced to half its original mass. The rate of decay is proportional to the undecayed mass. If the half-life of a substance is 100 years, how long will it take a quantity of the substance to decay to $\frac{1}{10}$ its original mass?
4. (cont.) If $\frac{1}{4}$ of a radioactive substance decays in 2 hr, what is its half-life?
5. The population of a city is now 110,000. Ten years ago it was 100,000. If the rate of growth is proportional to the population, predict the population 20 years from now.
6. When quenched in oil at 50°C, a piece of hot metal cools from 250°C to 150°C in 30 sec. At what constant oil temperature will the metal cool from 250°C to 150°C in 45 sec? (See Example 6.4.)
7. (cont.) The same piece of metal is quenched in oil whose temperature is 50°C, but rising 0.5°C/sec. How long will it take for the metal to cool from 250°C to 150°C?
8. When a cool object is placed in hot gas it heats up at a rate proportional to the difference in temperature between the gas and the object. A cold metal bar at 0°C is placed in a tank of gas at constant temperature 200°C. In 40 sec its temperature rises to 50°C. How long would it take to heat up from 20°C to 100°C?
9. (cont.) How hot should the gas be so that the metal bar heats up from 0°C to 50°C in 20 sec?
10. Two substances, U and V, combine chemically to form a substance X; one gram of U combines with 2 gm of V to form 3 gm of X. Suppose 50 gm of U are allowed to react with 100 gm of V. According to the Law of Mass Action, the chemicals combine at a rate proportional to the product of the untransformed masses. Denote by $x(t)$ the mass of X at time t. (The reaction starts at $t = 0$.) Show that $x(t)$ satisfies the initial-value problem

$$\frac{dx}{dt} = k\left(50 - \frac{x}{3}\right)\left(100 - \frac{2x}{3}\right), \qquad x(0) = 0.$$

If 75 gm of X are formed in the first second, find a formula for $x(t)$.

11. (cont.) Suppose that 50 gm of U are allowed to react with 150 gm of V. (Not all of V will be transformed.) Set up an initial-value problem for $x(t)$ and solve. Verify that $x(t) \longrightarrow 150$ as $t \longrightarrow \infty$.
12. A cylindrical tank of diameter 10 ft contains water to a depth of 4 ft. How long will it take the tank to empty through a 2-in. hole in the bottom? (See Example 6.5.)
13. (cont.) Suppose the tank in Ex. 12 is a paraboloid of revolution opening upward. It contains water to a depth of 4 ft, and the diameter of the tank at the 4-ft level is 10 ft. How long will it take the tank to empty through a 2-in. hole in the bottom?
14. A falling body of mass m is subject to a downward force mg, where g is the gravitational constant. Due to air resistance, it is subject also to a retarding (up-

ward) force proportional to its velocity. Verify that Newton's Law, $F = ma$, asserts in this case that

$$\frac{dv}{dt} + \frac{k}{m} v = g.$$

Solve, and show that the velocity does not increase indefinitely but approaches a limiting value. (Assume an initial velocity v_0.)

15.* A thermometer is plunged into a hot liquid. A few seconds later it records T_0°C. Five seconds later it records T_1°C. Five seconds later still it records T_2°C. Show that the temperature of the liquid is

$$\frac{T_1^2 - T_0 T_2}{-T_0 + 2T_1 - T_2}.$$

(Assume the reading T changes at a rate proportional to the difference between T and the actual temperature.)

16. In Example 6.6, is the amount of stretching of the top half of the rod half the total stretching?

17.* Suppose the rod of Example 6.6 is a long right circular cone. Suspend it vertically by its base. How much does it stretch?

7. APPROXIMATE SOLUTIONS

Successive Approximations

An initial-value problem may be converted into an **integral equation**, an equation in which the unknown function occurs under the integral sign.

> Solving the initial-value problem
>
> $$\frac{dy}{dx} = f[x, y(x)], \qquad y(a) = b,$$
>
> is equivalent to solving the integral equation
>
> $$y(x) = b + \int_a^x f[t, y(t)]\, dt.$$

It is easy to verify this equivalence. On the one hand, if $y(x)$ satisfies the integral equation, then $y(a) = b$ and

$$\frac{dy}{dx} = \frac{d}{dx}\left[\int_a^x f[t, y(t)]\, dt\right] = f[x, y(x)].$$

On the other hand, if $y(x)$ satisfies the initial-value problem, then

$$y(x) - b = y(x) - y(a) = \int_a^x \frac{dy}{dt}\,dt = \int_a^x f[t, y(t)]\,dt.$$

One way to attack the integral equation is to compute a sequence

$$y_0(x), \quad y_1(x), \quad y_2(x), \quad \cdots$$

of approximate solutions. The following is an iteration procedure that generally produces better and better approximate solutions.

STEP (0): Choose an approximate solution $y_0(x)$. (The constant function $y_0(x) = b$ is a reasonable choice.)

STEP (1): Set

$$y_1(x) = b + \int_a^x f[t, y_0(t)]\,dt.$$

Usually, $y_1(x)$ is a better approximation than $y_0(x)$.

STEP (2): Set

$$y_2(x) = b + \int_a^x f[t, y_1(t)]\,dt,$$

a better approximation still.

. . . .

STEP (n): Set

$$y_n(x) = b + \int_a^x f[t, y_{n-1}(t)]\,dt.$$

EXAMPLE 7.1

Construct $y_4(x)$ for the initial-value problem

$$\frac{dy}{dx} = y \qquad y(0) = 1.$$

Start with $y_0(x) = 1$.

Solution: The integral equation is

$$y(x) = 1 + \int_0^x y(t)\,dt.$$

Apply the iteration procedure starting with $y_0(x) = 1$:

$$y_1(x) = 1 + \int_0^x y_0(t)\,dt = 1 + \int_0^x 1\cdot dt = 1 + x,$$

$$y_2(x) = 1 + \int_0^x (1+t)\,dt = 1 + x + \frac{1}{2}x^2,$$

$$y_3(x) = 1 + \int_0^x \left(1 + t + \frac{1}{2}t^2\right)dt = 1 + x + \frac{1}{2}x^2 + \frac{1}{6}x^3,$$

$$y_4(x) = 1 + \int_0^x \left(1 + t + \frac{1}{2}t^2 + \frac{1}{6}t^3\right)dt = 1 + x + \frac{1}{2}x^2 + \frac{1}{6}x^3 + \frac{1}{24}x^4.$$

Answer: $1 + x + \frac{1}{2}x^2 + \frac{1}{6}x^3 + \frac{1}{24}x^4.$

REMARK: The answer is the 4-th degree Taylor polynomial of e^x.

EXAMPLE 7.2

Start with the approximation $y_0(x) = 2$ and compute $y_3(x)$ for the initial-value problem

$$y' = x^2y - 1 \qquad y(0) = 2.$$

Solution: The corresponding integral equation is

$$y(x) = 2 + \int_0^x [t^2y(t) - 1]\,dt.$$

The successive approximations are

$$y_1(x) = 2 + \int_0^x [t^2y_0(t) - 1]\,dt$$

$$= 2 + \int_0^x (-1 + 2t^2)\,dt = 2 - x + \frac{2}{3}x^3.$$

$$y_2(x) = 2 + \int_0^x [t^2y_1(t) - 1]\,dt$$

$$= 2 + \int_0^x \left[-1 + t^2\left(2 - t + \frac{2}{3}t^3\right)\right]dt = 2 - x + \frac{2}{3}x^3 - \frac{1}{4}x^4 + \frac{1}{9}x^6.$$

$$y_3(x) = 2 + \int_0^x \left[-1 + t^2\left(2 - t + \frac{2}{3}t^3 - \frac{1}{4}t^4 + \frac{1}{9}t^6\right)\right]dt.$$

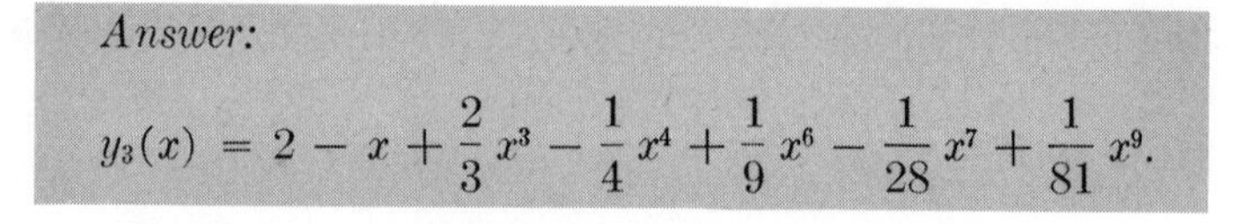

Answer:

$$y_3(x) = 2 - x + \frac{2}{3}x^3 - \frac{1}{4}x^4 + \frac{1}{9}x^6 - \frac{1}{28}x^7 + \frac{1}{81}x^9.$$

REMARK: The pattern of the iteration should be clear. If we want $y_4(x)$, we repeat each term in $y_3(x)$ and add the two new terms:

$$\int_0^x t^2\left(-\frac{1}{28}t^7 + \frac{1}{81}t^9\right) dt = \frac{-1}{280}x^{10} + \frac{1}{972}x^{12}.$$

This gives us $y(1) \approx y_4(1) \approx 1.50187$. Similarly the next two terms are

$$\int_0^x t^2\left(\frac{-1}{280}t^{10} + \frac{1}{972}t^{12}\right) dt = \frac{-1}{13\cdot 280}x^{13} + \frac{1}{15\cdot 972}x^{15},$$

which gives us $y(1) \approx y_5(1) \approx 1.50166$.

This **method of successive approximations,** also called the **Picard Method,** is often useful in finding approximate solutions. It is of great theoretical importance because it can be used to prove what we have taken for granted, that each initial-value problem for a sufficiently smooth direction field *has* a solution.

Taylor Polynomial Approximations

Consider an initial value problem

$$\frac{dy}{dx} = f[x, y(x)], \qquad y(a) = b,$$

where $f(x, y)$ is a polynomial, rational function, or more generally, a function that can be differentiated as often as we please. If we assume that $y(x)$ is differentiable (once) and $dy/dx = f[x, y(x)]$, then

$$\frac{d^2y(x)}{dx^2} = \frac{d}{dx}\{f[x, y(x)]\} = f_x[x, y(x)] + f_y[x, y(x)]y'(x)$$

exists by the Chain Rule. Hence $y(x)$ is differentiable twice. This implies the existence in turn of

$$\frac{d^3y(x)}{dx^3} = \frac{d}{dx}\{f_x[x, y(x)]\} + \frac{d}{dx}\{f_y[x, y(x)]\}y'(x) + f_y[x, y(x)]y''(x)$$
$$= f_{xx} + 2f_{xy}y' + f_{yy}(y')^2 + f_yy''.$$

By continuing this bootstrap operation, we find that $y^{(4)}, y^{(5)}, \cdots$ all exist.

What is more, we can calculate the values of these derivatives at $x = a$ in terms of an assumed initial value $y(a) = b$ of $y(x)$:

$$y'(a) = f(a, b), \qquad y''(a) = f_x(a, b) + f_y(a, b)y'(a),$$
$$y'''(a) = f_{xx}(a, b) + 2f_{xy}(a, b)y'(a) + f_{yy}(a, b)y'(a)^2 + f_y(a, b)y''(a),$$

etc. Hence we can compute the n-th degree Taylor polynomial of $y(x)$ at $x = a$:

$$p_n(x) = y(a) + \frac{y'(a)}{1!}(x-a) + \frac{y''(a)}{2!}(x-a)^2 + \cdots + \frac{y^{(n)}(a)}{n!}(x-a)^n.$$

It is proved in advanced courses that $p_n(x)$ is a good approximation to the solution $y(x)$ for n sufficiently large and x near a.

Let us try this technique on the same examples we used for successive approximations.

EXAMPLE 7.1

Compute the 5-th degree Taylor polynomial at $x = 0$ of the function $y(x)$ that satisfies the initial-value problem

$$\frac{dy}{dx} = y \qquad y(0) = 1.$$

Solution:

$$\begin{aligned} y' &= y & y'(0) &= 1 \\ y'' &= y' = y & y''(0) &= 1 \\ y''' &= y' = y & y'''(0) &= 1 \\ y^{(4)} &= y & y^{(4)}(0) &= 1 \\ y^{(5)} &= y & y^{(5)}(0) &= 1. \end{aligned}$$

Thus

$$y(x) \approx 1 + x + \frac{x^2}{2!} + \frac{x^3}{3!} + \frac{x^4}{4!} + \frac{x^5}{5!}.$$

Answer: $1 + x + \frac{x^2}{2!} + \frac{x^3}{3!} + \frac{x^4}{4!} + \frac{x^5}{5!}.$

REMARK: This simple problem has an explicit solution, $y = e^x$. The answer to the example is its Taylor polynomial $p_5(x)$ at $x = 0$. Observe that

$$p_5'(x) = 1 + x + \frac{x^2}{2!} + \frac{x^3}{3!} + \frac{x^4}{4!} = p_5(x) - \frac{x^5}{5!}.$$

Thus $p_5(x)$ satisfies the initial condition and, near $x = 0$, nearly satisfies the differential equation, $y' = y$.

EXAMPLE 7.2

Approximate the solution $y(x)$ of the initial-value problem

$$\frac{dy}{dx} = x^2y - 1 \qquad y(0) = 2$$

by its 9-th degree Taylor polynomial at $x = 0$.

Solution: The value $y(0) = 2$ is given; the values $y'(0)$, $y''(0)$, $\cdots$, $y^{(9)}(0)$ must be computed. Now

$$y' = x^2y - 1, \qquad y'(0) = -1.$$

Differentiate the expression for y':

$$y'' = 2xy + x^2y', \qquad y''(0) = 0.$$

Differentiate again:

$$y''' = 2y + 4xy' + x^2y''.$$

Substitute $x = 0$ and $y(0) = 2$ to obtain

$$y'''(0) = 4.$$

Continue in this way:

$$\begin{aligned}
y^{(4)} &= 6y' + 6xy'' + x^2y'' & y^{(4)}(0) &= 6y'(0) = -6\\
y^{(5)} &= 12y'' + 8xy''' + x^2y^{(4)} & y^{(5)}(0) &= 12y''(0) = 0\\
y^{(6)} &= 20y''' + 10xy^{(4)} + x^2y^{(5)} & y^{(6)}(0) &= 80\\
y^{(7)} &= 30y^{(4)} + 12xy^{(5)} + x^2y^{(6)} & y^{(7)}(0) &= -180\\
y^{(8)} &= 42y^{(5)} + 14xy^{(6)} + x^2y^{(7)} & y^{(8)}(0) &= 0\\
y^{(9)} &= 56y^{(6)} + 16xy^{(7)} + x^2y^{(8)} & y^{(9)}(0) &= 4480.
\end{aligned}$$

The coefficients of the Taylor approximation are

$$2, \quad -1, \quad 0, \quad \frac{4}{3!} = \frac{2}{3}, \quad \frac{-6}{4!} = \frac{-1}{4}, \quad 0,$$

$$\frac{80}{6!} = \frac{1}{9}, \quad \frac{-180}{7!} = \frac{-1}{28}, \quad 0, \quad \frac{4480}{9!} = \frac{1}{81}.$$

Answer:

$$y(x) \approx 2 - x + \frac{2}{3}x^3 - \frac{1}{4}x^4 + \frac{1}{9}x^6 - \frac{1}{28}x^7 + \frac{1}{81}x^9.$$

EXERCISES

Compute $y_3(x)$ by successive approximations for the initial-value problem; choose $y_0(x) = y(0)$:

1. $y' = xy + x$; $y(0) = 0$
2. $y' = xy + y$; $y(0) = 1$
3. $y' = xy + 1$; $y(0) = 2$
4. $y' = y + \sin x$; $y(0) = 1$
5. $y' = x^3y - x$; $y(0) = 1$
6. $y' = 1 + xy^2$; $y(0) = 0$
7. $y' = x + y^3$; $y(0) = 0$
8. $y' = -x^2 + y^2$; $y(0) = 0$
9. $y' = x^2 + y^2$; $y(0) = 0$
10. $y' = x^3$; $y(0) = 1$.

Without solving, compute the 5-th degree Taylor polynomial at $x = 0$ of the function satisfying the initial-value problem:

11. $y' = xy + x; \quad y(0) = 0$
12. $y' = xy + y; \quad y(0) = 1$
13. $y' = xy + 1; \quad y(0) = 2$
14. $y' = y + \sin x; \quad y(0) = 1$
15. $y' = x^3y - x; \quad y(0) = 1$
16. $y' = 1 + xy^2; \quad y(0) = 0$
17. $y' = x + y^3; \quad y(0) = 0$
18. $y' = -x^2 + y^2; \quad y(0) = 0.$

Compute the 4-th degree Taylor polynomial at $x = 0$ for the initial-value problem:

19. $y' = y \sin(x^2); \quad y(0) = 1$
20. $y' = 1 + x^2e^{-y}; \quad y(0) = 1$
21. $y' = e^x + xy^2; \quad y(0) = -1$
22. $y' = \cos(xy) - 1; \quad y(0) = -1$
23. $y' = x \sinh y; \quad y(0) = 0$
24. $y' = (1 + x^2)(1 + y^2); \quad y(0) = 2.$
25. The height $x(t)$ of a certain balloon satisfies $t\dot{x} = 0.5(t - x)^2$. When $t = 10$ sec, $x = 10$ ft. Estimate $x(12)$ to 0.1 ft accuracy.

14. Second Order Equations and Systems

1. LINEAR EQUATIONS

We shall study second order linear differential equations with constant coefficients:

$$\frac{d^2x}{dt^2} + p\,\frac{dx}{dt} + qx = r(t),$$

where p and q are constants. Equations of this type have many physical applications, particularly to elastic or electric phenomena.

First order linear equations are solved by finding one particular solution and adding it to the general solution of the associated homogeneous equation. The same method applies to second order linear equations for exactly the same reason. Regard the left-hand side of the equation as a "black box"; an input $x(t)$ leads to an output $\ddot{x} + p\dot{x} + qx$. The sum $x + y$ of two inputs leads to the sum of the corresponding outputs:

$$(x + y)^{\cdot\cdot} + p(x + y)^{\cdot} + q(x + y) = (\ddot{x} + p\dot{x} + qx) + (\ddot{y} + p\dot{y} + qy).$$

A constant multiple ax of an input leads to the same constant multiple of the output:

$$(ax)^{\cdot\cdot} + p(ax)^{\cdot} + q(ax) = a(\ddot{x} + p\dot{x} + qx).$$

Thus the "black box" is linear.

The general solution of

$$\frac{d^2x}{dt^2} + p\,\frac{dx}{dt} + qx = r(t)$$

is

$$x(t) + z(t),$$

where $x(t)$ is any solution of the differential equation and $z(t)$ is the general solution of the homogeneous equation

$$\frac{d^2x}{dt^2} + p\,\frac{dx}{dt} + qx = 0.$$

Therefore, as in the last chapter, we first treat homogeneous equations, then look for particular solutions of non-homogeneous equations.

2. HOMOGENEOUS EQUATIONS

In this section we study homogeneous equations

$$\ddot{x} + p\dot{x} + qx = 0,$$

where p and q are constants. There are three important special cases, each with $p = 0$, namely $q = 0$, $q = -k^2$, and $q = k^2$.

CASE 1: $\ddot{x} = 0$. This is easy. The solution is

$$x = at + b.$$

CASE 2: $\ddot{x} - k^2x = 0$. Here, the second derivative of x is proportional to x itself, which reminds us of the exponential function. It is not hard to guess two solutions: $x = e^{kt}$ and $x = e^{-kt}$. By linearity,

$$x = ae^{kt} + be^{-kt}$$

is also a solution. In Section 7 we shall prove this is the general solution.

CASE 3: $\ddot{x} + k^2x = 0$. Again it is easy to guess two solutions: $x = \cos kt$ and $x = \sin kt$. In Section 7 we shall prove the general solution is

$$x = a \cos kt + b \sin kt.$$

Summary

Differential equation	General solution
$\ddot{x} = 0$	$x = at + b$
$\ddot{x} - k^2x = 0$	$x = ae^{kt} + be^{-kt}$
$\ddot{x} + k^2x = 0$	$x = a \cos kt + b \sin kt$

These special cases are important because each homogeneous equation

$$\ddot{x} + p\dot{x} + qx = 0$$

can be reduced to one of them. The trick is analogous to reducing a quadratic equation $X^2 + pX + q = 0$ to the form $Y^2 \pm k^2 = 0$ by completing the square. Set

$$x(t) = e^{ht}y(t),$$

where $y(t)$ is a new unknown function and h is a constant. Compute derivatives by the Product Rule:

$$\begin{aligned}\dot{x} &= he^{ht}y + e^{ht}\dot{y}\\ &= e^{ht}(\dot{y} + hy);\\ \ddot{x} &= he^{ht}(\dot{y} + hy) - e^{ht}(\ddot{y} + h\dot{y})\\ &= e^{ht}(\ddot{y} + 2h\dot{y} + h^2y).\end{aligned}$$

Substitute these expressions into the differential equation:

$$e^{ht}(\ddot{y} + 2h\dot{y} + h^2y) + pe^{ht}(\dot{y} + hy) + qe^{ht}y = 0.$$

Cancel e^{ht} and group terms:

$$\ddot{y} + (2h + p)\dot{y} + (h^2 + ph + q)y = 0.$$

Now choose $h = -p/2$ to make the term in $\dot{y}$ drop out. The coefficient of y is then

$$h^2 + ph + q = \frac{p^2}{4} - \frac{p^2}{2} + q = \frac{-p^2 + 4q}{4},$$

and the differential equation becomes

$$\ddot{y} - \frac{p^2 - 4q}{4}y = 0.$$

This is one of the three special cases, depending on whether $p^2 - 4q$ is zero, positive, or negative.

EXAMPLE 2.1

Solve $\ddot{x} + 6\dot{x} + 9x = 0.$

Solution: Here

$$p = 6, \qquad q = 9, \qquad h = -\frac{p}{2} = -3, \qquad \frac{p^2 - 4q}{4} = 0.$$

The substitution $x = e^{-3t}y$ transforms the differential equation into

$$\ddot{y} = 0,$$

from which

$$y = at + b.$$

Answer: $x = e^{-3t}(at + b)$, where a and b are constants.

EXAMPLE 2.2

Solve $\ddot{x} - 3\dot{x} + 2x = 0.$

Solution: Here

$$p = -3, \qquad q = 2, \qquad h = -\frac{p}{2} = \frac{3}{2}, \qquad \frac{p^2 - 4q}{4} = \frac{1}{4} = \left(\frac{1}{2}\right)^2.$$

The substitution $x = e^{3t/2}y$ transforms the differential equation into

$$\ddot{y} - \left(\frac{1}{2}\right)^2 y = 0,$$

one of the standard forms. Conclusion:

$$y = ae^{t/2} + be^{-t/2}.$$

Answer: $x = e^{3t/2}(ae^{t/2} + be^{-t/2})$
$= ae^{2t} + be^{t}.$

EXAMPLE 2.3

Solve $\ddot{x} + 2\dot{x} + 5x = 0.$

Solution: In this example

$$p = 2, \qquad q = 5, \qquad h = -\frac{p}{2} = -1, \qquad \frac{p^2 - 4q}{4} = -4 = -2^2.$$

Set $x = e^{-t}y$. Then

$$\ddot{y} + 2^2 y = 0,$$

$$y = a \cos 2t + b \sin 2t.$$

Answer: $x = e^{-t}(a \cos 2t + b \sin 2t).$

Formula for Solving $\ddot{x} + p\dot{x} + qx = 0$

When you set

$$x = e^{-(p/2)t}y,$$

the differential equation

$$\ddot{x} + p\dot{x} + q = 0$$

is transformed into

$$\ddot{y} - \frac{\Delta}{4} y = 0, \qquad \text{where} \quad \Delta = p^2 - 4q.$$

Its solution depends on the sign of Δ:

CASE 1: $\Delta = 0$. Then

$$x(t) = e^{-(p/2)t}(at + b).$$

CASE 2: $\Delta > 0$. Write $\Delta = \sigma^2$. Then

$$x(t) = e^{-(p/2)t}(ae^{(\sigma/2)t} + be^{-(\sigma/2)t})$$
$$= ae^{[(-p+\sigma)/2]t} + be^{[(-p-\sigma)/2]t}.$$

CASE 3: $\Delta < 0$. Write $\Delta = -\sigma^2$. Then

$$x(t) = e^{-(p/2)t}\left[a\cos\left(\frac{\sigma}{2}t\right) + b\sin\left(\frac{\sigma}{2}t\right)\right].$$

Now compare these three cases with the three cases describing the roots of the quadratic equation

$$X^2 + pX + q = 0.$$

By the quadratic formula, the roots are

$$\alpha = -\frac{p}{2} + \frac{1}{2}\sqrt{\Delta}, \qquad \beta = -\frac{p}{2} - \frac{1}{2}\sqrt{\Delta}.$$

CASE 1: $\Delta = 0$. The roots are real and equal:

$$\alpha = \beta = -\frac{p}{2}.$$

CASE 2: $\Delta > 0$. Write $\Delta = \sigma^2$. The roots are real and distinct:

$$\alpha = \frac{-p+\sigma}{2}, \qquad \beta = \frac{-p-\sigma}{2}.$$

CASE 3: $\Delta < 0$. Write $\Delta = -\sigma^2$. The roots are complex:

$$\alpha = \frac{-p+\sigma i}{2}, \qquad \beta = \frac{-p-\sigma i}{2}, \qquad i = \sqrt{-1}.$$

To solve the differential equation

$$\ddot{x} + p\dot{x} + qx = 0,$$

find the roots α and β of the quadratic equation

$$X^2 + pX + q = 0.$$

1. If $\alpha = \beta$, then the solution is

$$x = e^{\alpha t}(at + b).$$

2. If α and β are real and distinct, then the solution is

$$x = ae^{\alpha t} + be^{\beta t}.$$

3. If α and β are complex,

$$\alpha = h + ki, \qquad \beta = h - ki,$$

then the solution is

$$x = e^{ht}(a\cos kt + b\sin kt).$$

Let us test these formulas on some numerical examples.

EXAMPLE 2.4

Solve each equation by the preceding rule:

(a) $\ddot{x} + 6\dot{x} + 9x = 0,$

(b) $\ddot{x} - 3\dot{x} + 2x = 0,$

(c) $\ddot{x} + 2\dot{x} + 5x = 0.$

Solution:

(a) $X^2 + 6X + 9 = 0,$

$(X + 3)^2 = 0, \quad \alpha = \beta = -3,$

$x = e^{-3t}(at + b).$

(b) $X^2 - 3X + 2 = 0,$

$(X - 2)(X - 1) = 0, \quad \alpha = 2, \quad \beta = 1,$

$x = ae^{2t} + be^{t}.$

(c) $X^2 + 2X + 5 = 0,$

$$\alpha = \frac{1}{2}(-2 + \sqrt{-16}) = -1 + 2i,$$

$$\beta = \frac{1}{2}(-2 - \sqrt{-16}) = -1 - 2i,$$

$x = e^{-t}(a \cos 2t + b \sin 2t).$

Answer:

(a) $x = e^{-3t}(at + b),$

(b) $x = ae^{2t} + be^{t},$

(c) $x = e^{-t}(a \cos 2t + b \sin 2t).$

EXERCISES

Find the general solution:

1. $\ddot{x} - 6\dot{x} + 5x = 0$
2. $\ddot{x} + 7\dot{x} + 10x = 0$
3. $\dfrac{d^2r}{d\theta^2} + 4r = 0$
4. $\dfrac{d^2y}{dx^2} + 6\dfrac{dy}{dx} + 13y = 0$
5. $2\dfrac{d^2y}{dx^2} - \dfrac{dy}{dx} + y = 0$
6. $\ddot{x} - 8\dot{x} + 16x = 0$

7. $\ddot{x} + 6\dot{x} = 0$

8. $4\dfrac{d^2y}{dx^2} + 4\dfrac{dy}{dx} + y = 0$

9. $\dfrac{d^2y}{dx^2} + 5\dfrac{dy}{dx} + 4y = 0$

10. $2\dfrac{d^2x}{dt^2} - 5\dfrac{dx}{dt} - 3x = 0$

11. $\dfrac{\ddot{x}}{a^3} - 4\dfrac{\dot{x}}{a} + 4ax = 0$

12. $\ddot{x} + \dot{x} + x = 0.$

Find the most general solution, subject to the indicated condition:

13. $\dfrac{d^2y}{dx^2} + 9y = 0, \quad y(0) = 0$

14. $\dfrac{d^2y}{dx^2} + 4y = 0, \quad y'(0) = 0$

15. $\ddot{x} - 4\dot{x} + 4x = 0, \quad x(0) = 1$

16. $\ddot{x} + 6\dot{x} + 8 = 0, \quad y(0) = 0$

17. $\dfrac{d^2r}{d\theta^2} - 2\dfrac{dr}{d\theta} + 2r = 0, \quad r(\pi) = 1$

18. $\dfrac{d^2r}{d\theta^2} - 2\dfrac{dr}{d\theta} + 2r = 0, \quad r'(\pi) = 1$

19. $\dfrac{d^2r}{d\theta^2} + r = 0, \quad r\left(\dfrac{\pi}{6}\right) = 0$

20. $\dfrac{d^2r}{d\theta^2} - 5\dfrac{dr}{d\theta} + 6r = 0, \quad r'(0) = 3.$

21. If a and b are positive constants, show that each solution of $\dfrac{d^2y}{dx^2} + a\dfrac{dy}{dx} + by = 0$ tends to zero as $x \longrightarrow \infty$.

22. Let $u(t)$ and $v(t)$ be solutions of $\ddot{x} + P(t)\dot{x} + Q(t)x = 0$. Set $w = \dot{u}v - u\dot{v}$. Show that $\dot{w} + Pw = 0$. Solve for w. (w is called the **Wronskian** of u and v.)

3. PARTICULAR SOLUTIONS

We return to the equation

$$\ddot{x} + p\dot{x} + qx = r(t).$$

The problem is to find any one solution. There are several ingenious methods for doing this, some quite complicated. We consider only the easiest one: guessing. This works just as in the last chapter.

EXAMPLE 3.1

Find a solution of $\ddot{x} + 3\dot{x} - x = t^2 - 1$.

Solution: Try $x = A + Bt + Ct^2$. Then $\dot{x} = B + 2Ct$ and $\ddot{x} = 2C$, so substitution in the equation yields

$$2C + 3(B + 2Ct) - (A + Bt + Ct^2) = t^2 - 1.$$

Equate coefficients:

$$\begin{aligned} t^2: &\quad -C = 1, \\ t: &\quad -B + 6C = 0, \\ 1: &\quad -A + 3B + 2C = -1. \end{aligned}$$

Therefore $C = -1$, $B = -6$, $A = -19$.

Answer: $x = -19 - 6t - t^2$ is a solution.

EXAMPLE 3.2

Find a solution of $\ddot{x} + 3\dot{x} - x = e^{-4t}$.

Solution: Try $x = Ae^{-4t}$. Then $\dot{x} = -4Ae^{-4t}$ and $\ddot{x} = 16Ae^{-4t}$. Substitute, and cancel e^{-4t}:

$$16A - 12A - A = 1, \quad 3A = 1.$$

Answer: $x = \frac{1}{3}e^{-4t}$ is a solution.

EXAMPLE 3.3

Find a solution of $\ddot{x} + 3\dot{x} - x = \cos t$.

Solution: Try $x = A\cos t + B\sin t$. Then $\dot{x} = -A\sin t + B\cos t$ and $\ddot{x} = -A\cos t - B\sin t$. Substitute:

$$(-A\cos t - B\sin t) + 3(-A\sin t + B\cos t) - (A\cos t + B\sin t) = \cos t.$$

Equate coefficients of $\cos t$ and $\sin t$:

$$\begin{cases} -2A + 3B = 1 \\ -3A - 2B = 0. \end{cases}$$

The solution of this system is $A = -\frac{2}{13}$ and $B = \frac{3}{13}$.

Answer: $x = \frac{1}{13}(-2\cos t + 3\sin t)$ is a solution.

EXAMPLE 3.4

Find the general solution of $\ddot{x} - x = t$.

Solution: The quadratic equation $X^2 - 1 = 0$ has roots 1 and -1. Hence the solution of the homogeneous equation $\ddot{x} - x = 0$ is $x = ae^t + be^{-t}$. We guess the particular solution $x = -t$.

Answer: $x(t) = ae^t + be^{-t} - t$, where a and b are constants.

Initial-Value Problems

The general solution of the differential equation

$$\ddot{x} + p\dot{x} + qx = r(t)$$

involves two arbitrary constants. By choosing them suitably you can usually find a solution satisfying two additional conditions. Most important is the case of initial conditions:

Find a solution $x = x(t)$ of the differential equation

$$\ddot{x} + p\dot{x} + qx = r(t)$$

such that

$$x(t_0) = a, \qquad \dot{x}(t_0) = b.$$

EXAMPLE 3.5

Solve the initial-value problem $\ddot{x} - x = t, \quad x(0) = 0, \quad \dot{x}(0) = 1.$

Solution: By the last example, the general solution of the differential equation is

$$x = ae^t + be^{-t} - t.$$

Hence

$$\dot{x} = ae^t - be^{-t} - 1.$$

Substitute the initial conditions:

$$\begin{cases} a + b = 0 \\ a - b - 1 = 1. \end{cases}$$

This system has the solution $a = 1$ and $b = -1$.

Answer: $x = e^t - e^{-t} - t.$

EXERCISES

Find a particular solution:

1. $\ddot{x} + 3\dot{x} = 1$
2. $2\ddot{x} + 3\dot{x} = 5$
3. $\ddot{x} - 4x = 2t + 1$
4. $\ddot{x} + x = t^2$
5. $\ddot{x} + 3\dot{x} - x = t^2 + 4t + 6$
6. $\ddot{x} + \dot{x} + x = t^3$
7. $\ddot{x} + \dot{x} + 2x = 2e^{3t}$
8. $\ddot{x} + x = \cosh t$
9. $3\dfrac{d^2y}{dx^2} + \dfrac{dy}{dx} - y = e^{-2x} + x$
10. $2\dfrac{d^2y}{dx^2} - 3y = \sin x$
11. $\dfrac{d^2y}{dx^2} + \dfrac{dy}{dx} - 3y = 4\sin 2x$
12. $\dfrac{d^2y}{dx^2} + 2\dfrac{dy}{dx} + y = 1 + x + \cos 3x$
13. $\dfrac{d^2y}{dx^2} + \dfrac{dy}{dx} + 2y = e^x \cos x$
14. $2\dfrac{d^2y}{dx^2} - y = x\cos x$
15. $\dfrac{d^2y}{dx^2} + y = e^x + e^{2x} + e^{3x} + e^{4x} + e^{5x}$
16. $\dfrac{d^2y}{dx^2} - \dfrac{dy}{dx} - y = xe^{3x}.$

Find the general solution:

17. $\ddot{x} + x = t^2$
18. $\ddot{x} - 7\dot{x} + 10x = 3e^t$

19. $\ddot{x} + \dot{x} - 6x = te^{-t}$

20. $\ddot{x} + \dot{x} - 5x = 2t + 3$

21. $\ddot{x} + 3\dot{x} = \cosh 2t$

22. $3\dfrac{d^2y}{dx^2} - \dfrac{dy}{dx} + y = x^2 + 5$

23. $3\dfrac{d^2i}{dt^2} + 4\dfrac{di}{dt} + 2i = 10\cos t$

24. $\dfrac{d^2r}{d\theta^2} - 2r = \sin\theta + \cos\theta.$

Solve the initial-value problem:

25. $\ddot{x} + x = 2t - 5, \quad x(0) = 0, \quad \dot{x}(0) = 1$

26. $\ddot{x} + 3x = e^t, \quad x(0) = 0, \quad \dot{x}(0) = 1$

27. $\ddot{x} - 4x = \sin 2t, \quad x(0) = 1, \quad \dot{x}(0) = 0$

28. $\ddot{x} - 2\dot{x} - 15x = 1, \quad x(1) = 0, \quad \dot{x}(1) = 0$

29. $4\ddot{x} - 7\dot{x} + 3x = e^{2t}, \quad x(0) = -1, \quad \dot{x}(0) = 2$

30. $\ddot{x} + 2\dot{x} + x = t^2, \quad x(0) = 0, \quad \dot{x}(0) = 5.$

31. The usual method for finding a particular solution of $\ddot{x} - 4x = e^{2t}$ fails. Why? Try $x = Ate^{2t}$.

32. (cont.) Find a solution of $\ddot{x} + x = \sin t$ by guessing.

33. Let z be a solution of the homogeneous equation $\ddot{x} + p\dot{x} + qx = 0$. Substitute $x(t) = z(t)u(t)$ in the equation $\ddot{x} + p\dot{x} + qx = r(t)$. Show that the resulting equation for u can be reduced to a first order equation by setting $w = \dot{u}$.

34. (cont.) Apply this technique to find a particular solution of $\ddot{x} + x = \sin t$.

35. Show that the equation $\ddot{x} + k^2x = r(t)$ has the particular solution

$$x(t) = \frac{1}{k}\int_0^t \sin k(t - s)r(s)\,ds.$$

You may presuppose the formula

$$\frac{d}{dt}\left(\int_0^t F(s, t)\,ds\right) = F(t, t) + \int_0^t \frac{\partial F}{\partial s}\,ds.$$

4. APPLICATIONS

In applications, the expression $a\cos kt + b\sin kt$ occurs frequently. There is a useful equivalent expression for it. Assume a and b are not both zero. If c denotes the positive square root of $a^2 + b^2$, then

$$\left(-\frac{a}{c}\right)^2 + \left(\frac{b}{c}\right)^2 = 1.$$

Thus the point $(b/c, -a/c)$ lies on the unit circle (Fig. 4.1). Hence there is an angle kt_0, determined up to an integer multiple of 2π, such that

$$\sin kt_0 = -\frac{a}{c}, \qquad \cos kt_0 = \frac{b}{c}.$$

By the addition law for the sine,

$$a \cos kt + b \sin kt = c\left(\frac{a}{c} \cos kt + \frac{b}{c} \sin kt\right)$$
$$= c(-\sin kt_0 \cos kt + \cos kt_0 \sin kt)$$
$$= c \sin k(t - t_0).$$

Thus the expression $a \cos kt + b \sin kt$ is nothing but a sine function (sinusoidal wave) in disguise.

[Alternatively, t_0 may be defined by $a/c = \cos kt_0$, $b/c = \sin kt_0$, from which $a \cos kt + b \sin kt = c \cos k(t - t_0)$.]

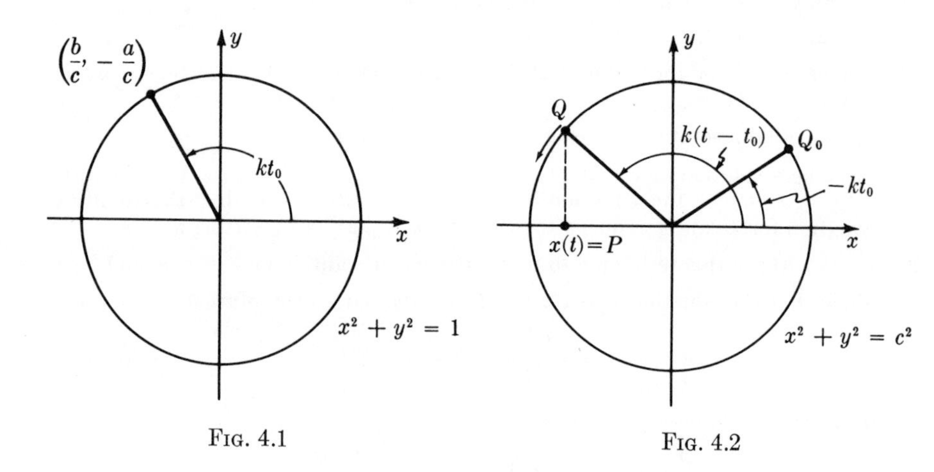

FIG. 4.1

FIG. 4.2

Simple Harmonic Motion

Imagine a point Q moving counterclockwise about the circle $x^2 + y^2 = c^2$. See Fig. 4.2. Starting at Q_0 when $t = 0$, the point Q moves with constant speed, making k revolutions in 2π sec. Its projection P on the x-axis has coordinate

$$x(t) = c \cos k(t - t_0).$$

As Q traverses its circle, P oscillates between $(c, 0)$ and $(-c, 0)$, making one complete oscillation in $2\pi/k$ sec. The motion of P is called **simple harmonic motion** with **amplitude** c, **period** $2\pi/k$, and **time lag** t_0 (**phase angle** $-kt_0$).

Note that the speed of P, unlike that of Q, is not constant. Why? When is it greatest?

REMARK: Since $\cos \alpha = \sin(\pi/2 - \alpha)$, harmonic motion may also be described in terms of a sine:

$$x(t) = c \cos k(t - t_0) = c \cos k(t_0 - t) = c \sin\left[\frac{\pi}{2} - k(t_0 - t)\right].$$

Conclusion:

$$x(t) = c \sin k(t - t_1), \qquad \text{where} \quad t_1 = t_0 - \frac{\pi}{2k}.$$

EXAMPLE 4.1 (Simple Harmonic Motion)

A particle moves along the x-axis. The only force acting on it is a force directed towards the origin and proportional to the displacement of the particle from the origin (a spring). Describe the motion. (Assume the spring has negligible length when unloaded.)

Solution: Call the mass m. The force is x times a negative constant, $-mk^2$, with $k > 0$. According to Newton's Law of Motion,

$$\text{mass} \times \text{acceleration} = \text{force}:$$

$$m\ddot{x} = -mk^2x, \qquad \ddot{x} + k^2x = 0.$$

The solution is

$$x = a \cos kt + b \sin kt = c \cos k(t - t_0),$$

where the constants c and t_0 depend on the initial position and velocity and $c > 0$. Hence the motion is simple harmonic.

Answer: $x = c \cos k(t - t_0)$.

EXAMPLE 4.2

Solve the same problem

(a) with initial conditions $x(0) = 0, \quad \dot{x}(0) = v_0 > 0$;

(b) with initial conditions $x(0) = x_0 > 0, \quad \dot{x}(0) = 0$.

Solution: By Example 4.1, the general solution is $x = c \cos k(t - t_0)$, where c and k are positive. Now determine c and t_0.

(a) Since $\dot{x} = -ck \sin k(t - t_0)$, the initial conditions are:

$$0 = c \cos(-kt_0), \qquad v_0 = -ck \sin(-kt_0),$$

that is,

$$\cos kt_0 = 0, \qquad \sin kt_0 = \frac{v_0}{ck}.$$

From the first equation, $kt_0 = \pi/2$ or $kt_0 = -\pi/2$; from the second, $\sin kt_0 > 0$. Hence the only possibility is $kt_0 = \pi/2$, that is, $t_0 = \pi/2k$. It follows that $v_0 = ck$, so $c = v_0/k$. Therefore the desired solution is

$$x = \frac{v_0}{k} \cos\left(kt - \frac{\pi}{2}\right) = \frac{v_0}{k} \sin kt.$$

(b) In this case the initial conditions are:

$$x_0 = c\cos(-kt_0), \qquad 0 = -ck\sin(-kt_0),$$

that is,

$$\cos kt_0 = \frac{x_0}{c}, \qquad \sin kt_0 = 0.$$

From the second equation, $kt_0 = 0$ or π; from the first, $\cos kt_0 > 0$. Hence $kt_0 = 0$. It follows that $c = x_0$, and the desired solution is $x = x_0 \cos kt$.

Answer: (a) $x = \dfrac{v_0}{k}\sin kt;$

(b) $x = x_0 \cos kt.$

REMARK: As this example shows, it is simpler to describe harmonic motion by the sine form when the initial position is $x_0 = 0$ and by the cosine form when the initial velocity is $v_0 = 0$.

EXAMPLE 4.3

A 16-lb weight hangs at rest on the end of an 8-ft spring attached to a ceiling. When the spring is stretched, it exerts a restoring force proportional to displacement from equilibrium position. Suppose the weight is pulled down 6 in. and released. Describe its motion.

Solution: Let x denote the distance from the ceiling in feet (Fig. 4.3) The force due to the spring is $F(x) = -cx$. When the spring is at rest $F = -16$ and $x = 8$. Hence $c = 2$ and $F(x) = -2x$ lb.

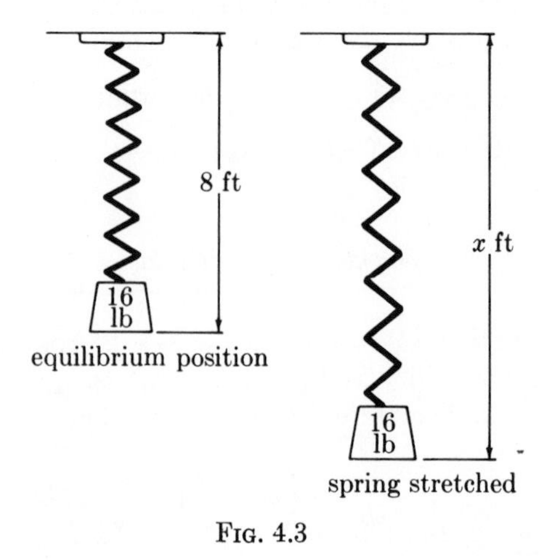

FIG. 4.3

When the weight is at position x, the forces acting on it are its weight, 16 lb, and the spring, $-2x$ lb. Since the units of force and length are pounds and feet, mass must be in slugs: $m = 16/g = \frac{16}{32}$ slugs. Thus mass × acceleration = force,

$$\frac{16}{32}\ddot{x} = 16 - 2x, \qquad \ddot{x} + 4x = 32.$$

$$\text{Initial conditions:} \quad x(0) = 8.5, \qquad \dot{x}(0) = 0.$$

The associated homogeneous equation has solution

$$x = c\cos 2(t - t_0),$$

and it is easy to guess the particular solution, $x = 8$. Consequently,

$$x = c\cos 2(t - t_0) + 8$$

is the general solution. It describes simple harmonic motion centered about $x = 8$ rather than $x = 0$. That is reasonable since you expect the weight to oscillate about its equilibrium point at $x = 8$.

The initial conditions are

$$x(0) = 8.5, \qquad \dot{x}(0) = 0.$$

From these you can solve for c and t_0. But good scientists are lazy; they refuse to do the same work twice. Use the answer to part (b), of Example 4.2. It describes harmonic motion with 0 initial velocity by $x = x_0 \cos kt$, where x_0 is initial displacement. In this problem, initial displacement is $\frac{1}{2}$. Hence, you can write down the answer.

Answer: $x = \dfrac{1}{2}\cos 2t + 8.$

Other Applications

> EXAMPLE 4.4
>
> A particle of mass 1 attached to a spring slides along a straight surface. It is subject to a restoring force proportional to displacement and a retarding force (due to friction) assumed proportional to velocity. Describe the motion of the particle assuming it starts at the origin (equilibrium point) with initial velocity v_0.

Solution: The forces are $-k^2x$ (spring) and $-p\dot{x}$ (friction), where k and p are positive constants. The equation of motion is

$$\text{mass} \times \text{acceleration} = \text{force},$$

$$\ddot{x} = -p\dot{x} - k^2x, \qquad \ddot{x} + p\dot{x} + k^2x = 0.$$

$$\text{Initial conditions:} \quad x(0) = 0, \qquad \dot{x}(0) = v_0.$$

Before solving, think about the problem. If there is no friction, that is if p is zero, then the motion is simple harmonic. If the friction is small, then the motion should be nearly simple harmonic motion, except for a gradual slowing down due to friction. If the friction is large, the motion should be considerably inhibited, in fact there might not be oscillations.

From this physical reasoning, it seems clear that the relative size of the constants k and p is crucial.

To solve the differential equation, examine the roots of the corresponding quadratic equation:

$$X^2 + pX + k^2 = 0.$$

They are

$$\alpha = -\frac{p}{2} + \frac{1}{2}\sqrt{\Delta}, \qquad \beta = -\frac{p}{2} - \frac{1}{2}\sqrt{\Delta}, \qquad \text{where} \quad \Delta = p^2 - 4k^2.$$

The nature of the solutions depends on the sign of Δ. In terms of k and p, the crucial question is whether $p > 2k$, $p < 2k$, or $p = 2k$.

CASE 1: $p < 2k$ (underdamped case: friction small compared to spring force). In this case $\Delta < 0$; set $\Delta = -4q^2$. Then the general solution of the differential equation is

$$x = e^{-(p/2)t}(a \cos qt + b \sin qt).$$

Use the first initial condition, $x(0) = 0$; it implies $a = 0$. Hence

$$x = be^{-(p/2)t} \sin qt.$$

Now use the second initial condition, $\dot{x}(0) = v_0$:

$$\dot{x}(t) = b\left(-\frac{p}{2} e^{-(p/2)t} \sin qt + qe^{-(p/2)t} \cos qt\right),$$

$$v_0 = \dot{x}(0) = bq, \qquad b = \frac{v_0}{q}.$$

Therefore, the solution of the initial-value problem is:

$$x = \frac{v_0}{q} e^{-(p/2)t} \sin qt.$$

This is a damped oscillatory motion (Fig. 4.4). The particle oscillates with constant period $2\pi/q$, but the amplitude $\longrightarrow 0$ as $t \longrightarrow \infty$.

CASE 2: $p > 2k$ (overdamped case: friction large compared to spring force). Now $\Delta > 0$. The roots of the quadratic are

$$\alpha = \frac{1}{2}(-p + \sqrt{\Delta}), \qquad \beta = \frac{1}{2}(-p - \sqrt{\Delta}).$$

Both roots are negative ($\sqrt{\Delta} = \sqrt{p^2 - 4k^2} < p$). Let us call the roots $-r$

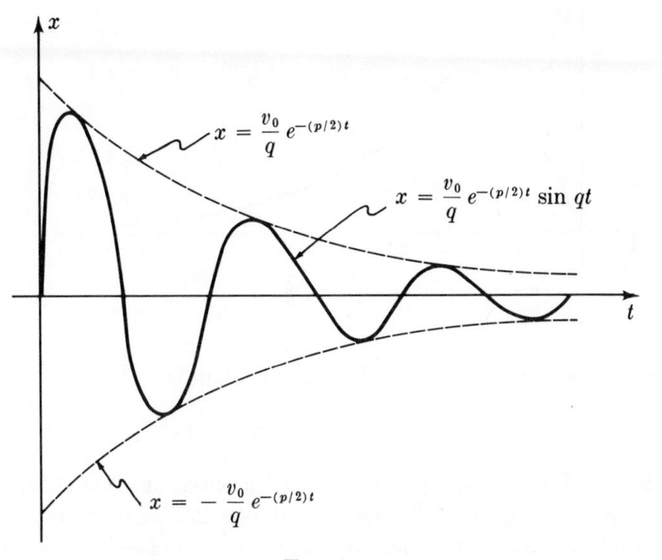

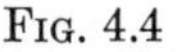
FIG. 4.4

and $-s$, where $r > s > 0$. The general solution is

$$x = ae^{-rt} + be^{-st}.$$

Since

$$\dot{x} = -are^{-rt} - bse^{-st},$$

the initial conditions become

$$0 = a + b, \qquad v_0 = -ar - bs.$$

It follows that

$$-a = b = \frac{v_0}{r - s}.$$

Therefore, the solution of the initial-value problem is

$$x = c(e^{-st} - e^{-rt}), \qquad \text{where} \quad c = \frac{v_0}{r - s} > 0.$$

The graph of x as a function of t is shown in Fig. 4.5. The particle moves away from the origin at first, then reverses direction and approaches the origin, never quite reaching it again.

CASE 3: $p = 2k$ (borderline case, critical damping). Now $\Delta = 0$, and the quadratic equation has two equal roots, $\alpha = \beta = -p/2$. The general solution is

$$x = e^{-(p/2)t}(a + bt).$$

Thus $x(0) = a$ and $\dot{x}(0) = b - \frac{1}{2}pa$, so the initial conditions reduce to

$$0 = a, \qquad v_0 = b.$$

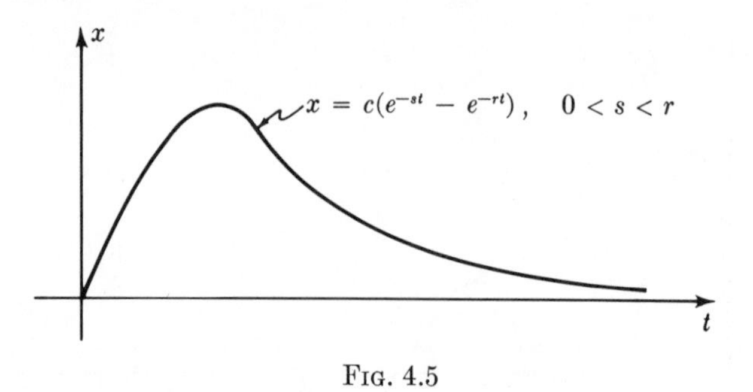

FIG. 4.5

Therefore the solution of the initial-value problem is

$$x = v_0 t e^{-(p/2)t} = v_0 t e^{-kt}.$$

By solving $\dot{x} = 0$, it is seen that x reaches its maximum value $2v_0/pe = v_0/ke$ when $t = 2/p$. See Fig. 4.6. The particle moves $2v_0/pe$ units from the origin, reverses direction and approaches the origin again as $t \longrightarrow \infty$. As is physically plausible, the maximum distance from the origin is proportional to the initial velocity, inversely proportional to the coefficient of friction.

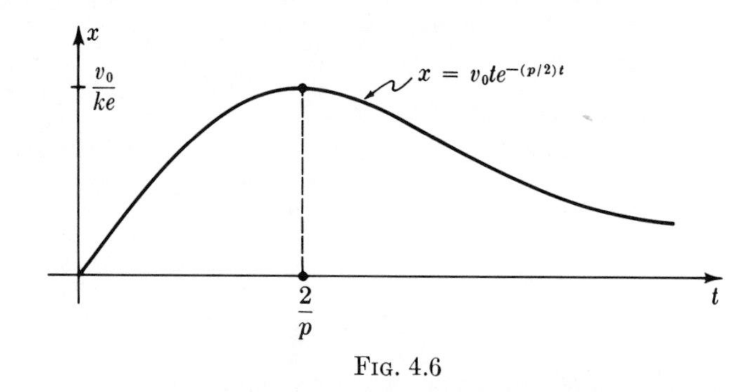

FIG. 4.6

Answer: Let $\Delta = p^2 - 4k^2$.

If $\Delta < 0$, then $x = \dfrac{v_0}{q} e^{-(p/2)t} \sin qt$, where $q = \dfrac{1}{2}\sqrt{-\Delta}$.

If $\Delta > 0$, then $x = \dfrac{v_0}{r - s}(e^{-st} - e^{-rt})$, where

$$r = \frac{1}{2}(p + \sqrt{\Delta}), \qquad s = \frac{1}{2}(p - \sqrt{\Delta}).$$

If $\Delta = 0$, then $x = v_0 t e^{-(p/2)t} = v_0 t e^{-kt}$.

The next example provides a simple mechanical model for a variety of physical phenomena.

EXAMPLE 4.5

Suppose in Example 4.4 an external force $F = A \sin \omega t$ is imposed. What is the nature of the motion when t is large? (Describe the solution $x(t)$ as $t \longrightarrow \infty$.) Assume small friction.

Solution: The differential equation is

$$\ddot{x} + p\dot{x} + k^2 x = A \sin \omega t.$$

Guess a particular solution of the form

$$x = a \cos \omega t + b \sin \omega t.$$

Substitute; after some computation the differential equation becomes

$$[(k^2 - \omega^2)a + p\omega b] \cos \omega t + [(k^2 - \omega^2)b - p\omega a] \sin \omega t = A \sin \omega t.$$

Equation coefficients of $\cos \omega t$ and $\sin \omega t$:

$$\begin{cases} ra + sb = 0 \\ -sa + rb = A \end{cases} \qquad \text{where} \quad r = k^2 - \omega^2 \quad \text{and} \quad s = p\omega.$$

Solve:

$$a = \frac{-As}{r^2 + s^2}, \qquad b = \frac{Ar}{r^2 + s^2}.$$

Hence, a particular solution is

$$x = \frac{A}{r^2 + s^2}(-s \cos \omega t + r \sin \omega t).$$

By the usual method, convert it to the form

$$x = \frac{A}{\sqrt{r^2 + s^2}} \sin \omega (t - t_0),$$

where

$$\sin \omega t_0 = \frac{s}{\sqrt{r^2 + s^2}} \quad \text{and} \quad \cos \omega t_0 = \frac{r}{\sqrt{r^2 + s^2}}.$$

If friction is small, the general solution of the homogeneous equation

$$\ddot{x} + p\dot{x} + k^2 x = 0$$

is

$$x = ce^{-(p/2)t} \sin q(t - t_1), \qquad q = \frac{1}{2}\sqrt{4k^2 - p^2}.$$

(See the answer to Example 4.4, first case.)

Combine results: the general solution of

$$\ddot{x} + p\dot{x} + k^2 x = A \sin \omega t$$

is

$$x = ce^{-(p/2)t} \sin q(t - t_1) + \frac{A}{\sqrt{(k^2 - \omega^2)^2 + (p\omega)^2}} \sin \omega(t - t_0).$$

The constants c and t_1 can be determined from initial conditions, but they are not needed to predict the behavior of x for large values of t. As $t \longrightarrow \infty$, the term $ce^{-(p/2)t} \sin q(t - t_1) \longrightarrow 0$. (It is called a **transient.**) Thus after sufficient time has elapsed,

$$x(t) \approx C \sin \omega(t - t_0);$$

all that is visible is simple harmonic motion. The amplitude of this motion is

$$C = \frac{A}{\sqrt{(k^2 - \omega^2)^2 + (p\omega)^2}}.$$

In case p is small and $\omega \approx k$, the denominator is small, hence the amplitude C is large. This is the phenomenon of **resonance.** It occurs when there is a periodic force with frequency near the "natural frequency" of the system. Destructive vibrations in machinery and vibrations caused by soldiers marching in step on a bridge are results of resonance.

The approximation

$$x(t) \approx C \sin \omega(t - t_0),$$

valid for large values of t, is called the **steady state solution.** Notice that it is independent of the initial conditions; these affect only the constants in the transient term.

Answer: For large t, $x(t) \approx C \sin \omega(t - t_0)$, where

$$C = \frac{A}{\sqrt{(k^2 - \omega^2)^2 + (p\omega)^2}} = \frac{A}{\sqrt{r^2 + s^2}},$$

and

$$\begin{cases} \sin \omega t_0 = \dfrac{s}{\sqrt{r^2 + s^2}} \\ \\ \cos \omega t_0 = \dfrac{r}{\sqrt{r^2 + s^2}} \end{cases} \qquad \begin{cases} r = k^2 - \omega^2 \\ \\ s = p\omega. \end{cases}$$

The standard electric model for this situation is a simple circuit with resistance R, inductance L, capacitance C, current I, and with voltage $E = -A \cos \omega t$. The equation of the circuit is

$$\frac{d^2I}{dt^2} + \frac{R}{L}\frac{dI}{dt} + \frac{1}{LC} I = \frac{1}{L}\frac{dE}{dt} = \frac{A\omega}{L} \sin \omega t.$$

The steady state solution is

$$I(t) \approx \frac{A\omega}{L\sqrt{\left(\frac{1}{LC} - \omega^2\right)^2 + \left(\frac{R}{L}\omega\right)^2}} \sin \omega (t - t_0).$$

Resonance (tuned circuit) occurs when $\omega^2 = 1/LC$. Then $\sin \omega t_0 = 1$ and $\cos \omega t_0 = 0$. Hence $\omega t_0 = \pi/2$ and

$$I(t) \approx \frac{A}{R} \sin\left(\omega t - \frac{\pi}{2}\right).$$

EXERCISES

Express as $c \cos k(t - t_0)$ and as $c \sin k(t - t_1)$. Use trigonometric tables if necessary:

1. $\cos 3t + \sin 3t$
2. $\sin t - \cos t$
3. $\sqrt{3} \cos 2t - \sin 2t$
4. $\cos t + \sqrt{3} \sin t$
5. $4 \sin t - 3 \cos t$
6. $\sin \frac{t}{2} + 5 \cos \frac{t}{2}$.

The next four problems concern a particle in simple harmonic motion about the origin. Its position at time t is $x(t)$ and its velocity is $v(t)$.

7. Find $x(t)$ if the period is 2 sec, the amplitude is 6, and $x(0) = 3$.
8. Find $x(t)$ if $x(0) = 0$, $v(0) = 5$, and the period is 2 sec.
9. Find the amplitude if $x(0) = 0$, $v(0) = 10$, and $v(3) = 5$.
10. Estimate the amplitude to 3 significant digits if $x(0) = 0$, $v(0) = 10$, and $x(1) = 6$.
11. A 16-lb weight hangs at rest on the end of an 8-ft spring attached to a ceiling. (See Example 4.3.) The weight is pulled down k in. and released. As it passes through its original (equilibrium) position, its speed is 1.5 ft/sec. Find k.
12. (cont.) Suppose the weight slides along a wall. Due to friction there is a retarding force proportional to the velocity. The motion is a damped vibration with period $2\pi/\sqrt{3}$ sec. After how long will the amplitude of the oscillation be reduced to half its original magnitude?
13. Suppose the weight in Example 4.3 is w lb, not 16 lb. What is the period of its motion?
14. A pendulum is made of a small weight at the end of a long wire. Its motion is described by the differential equation

$$\frac{d^2\theta}{dt^2} + \frac{g}{L} \sin \theta = 0,$$

where θ is the angle between the wire and the vertical. If the pendulum swings only through a small arc, then $\sin \theta$ can be approximated by θ, thus simplifying the differential equation. Do so and find the approximate period of the pendulum.

15. A cylindrical buoy floats vertically in the water. Its weight is 100 lb and its diameter is 2 ft. When depressed slightly and released, it oscillates with simple harmonic motion. Find the period of the oscillation.
[*Hint:* This is just a spring problem in disguise. Use Archimedes' Law: A body

in water is subjected to an upward buoyant force equal to the weight of the water displaced. Take the density of water to be 62.4 lb/ft^3.]

16. In the overdamped motion of Fig. 4.5, find the maximum value of $x(t)$. Show that $x(t) > 0$ for $t > 0$ and $x(t) < 0$ for $t < 0$.
17. Consider the underdamped motion of Fig. 4.4. Show that the graph crosses the t-axis infinitely often for $t > 0$.
18. A rocket sled is subjected to $6g$ acceleration for 5 sec. After the engine shuts down, the sled undergoes a deceleration (ft/sec^2) equal to 0.05 times its velocity (ft/sec). What is the speed of the sled 10 sec after engine shutdown? How far has it traveled?

The external forces acting on a projectile are gravity and air resistance. At low altitude and low speed, it may be assumed that air resistance is proportional to speed.

If a projectile is shot straight up, it rises to its maximum height and then falls to the ground. Whether it takes longer to rise or to fall, or equal times, is not obvious. However, as we shall see, in falling there is a terminal speed. Hence a projectile shot up with initial speed faster than the terminal speed necessarily takes longer in falling than in rising. The next five exercises show this is *always* so.

19. A projectile is shot straight up with initial velocity v_0. Show that its height satisfies the initial-value problem $\ddot{y} + k\dot{y} = -g$, $y(0) = 0$, $\dot{y}(0) = v_0$. Derive the solution

$$y(t) = A(1 - e^{-kt}) - \frac{g}{k}t, \qquad \text{where} \quad A = \frac{g + kv_0}{k^2}.$$

20. (cont.) Show that v approaches a terminal velocity as $t \longrightarrow \infty$. Find it.
21. (cont.) Show that the projectile reaches its maximum height at time

$$t_1 = \frac{1}{k}\ln\left(\frac{g + kv_0}{g}\right).$$

Show that the projectile returns to ground at time $t_2 > 0$, where

$$\left(\frac{g + kv_0}{g}\right)(1 - e^{-kt_2}) = kt_2.$$

22. (cont.) Show that $t_1 \longrightarrow v_0/g$ as $k \longrightarrow 0$ and interpret.
[*Hint:* Express the derivative of $\ln(1 + cx)$ at $x = 0$ as a limit; alternatively use the first order Taylor Approximation.]
Guess what t_2 approaches as $k \longrightarrow 0$.
23*. (cont.) Prove $(t_2 - t_1) > t_1$. Begin by showing that $\dot{y}(t_1 + \tau) + \dot{y}(t_1 - \tau) > 0$ for $\tau > 0$. Interpret physically. Then integrate the inequality over the interval $0 \leq \tau \leq t_1$ to obtain $y(t_1 + \tau) > y(t_1 - \tau)$ for $\tau > 0$. Deduce that $2t_1 < t_2$.
24*. (cont.) Now try a concrete problem. Suppose the initial velocity is 500 ft/sec and the maximum height is 3000 ft. Assume $g = 32.2$ ft/sec^2. Estimate k to 3 significant digits. Then estimate t_1 and t_2 to 2 significant digits.

5. POWER SERIES SOLUTIONS

In the theory of differential equations, there is an important result concerning initial-value problems of the type

$$\frac{dy}{dx} = f(x, y), \qquad y(a) = b,$$

where $f(x, y)$ has a convergent power series expansion near (a, b). It says that the solution $y = y(x)$ has a convergent power series expansion near $x = a$. Therefore we may assume a solution of the form

$$y(x) = \sum_{n=0}^{\infty} a_n (x - a)^n$$

and then try to find suitable coefficients $a_0, a_1, a_2, \cdots$. This technique often leads to an exact power series solution, rather than to a Taylor polynomial approximation as in the last chapter.

Similar remarks apply to initial-value problems of the form

$$\frac{d^2y}{dx^2} = f\left(x, y, \frac{dy}{dx}\right), \qquad y(a) = b_0, \quad y'(a) = b_1.$$

EXAMPLE 5.1

Find a power series solution to

$y' = xy, \qquad y(0) = a_0.$

Solution: Try a solution of the form

$$y(x) = a_0 + a_1x + a_2x^2 + \cdots,$$

where the coefficients must be determined. Substitute this power series into the differential equation:

$$\frac{d}{dx}(a_0 + a_1x + a_2x^2 + \cdots) = x(a_0 + a_1x + a_2x^2 + \cdots),$$

$$a_1 + 2a_2x + 3a_3x^2 + \cdots + (k+1)a_{k+1}x^k + \cdots$$
$$= a_0x + a_1x^2 + \cdots + a_{k-1}x^k + \cdots.$$

Equate coefficients:

$$a_1 = 0, \quad 2a_2 = a_0, \quad 3a_3 = a_1, \quad \cdots, \quad (k+1)a_{k+1} = a_{k-1}, \quad \cdots.$$

Hence

$$a_{k+1} = \frac{a_{k-1}}{k+1}.$$

This **recurrence relation** expresses each coefficient in terms of the next to last coefficient. From it, all coefficients can be computed successively. For instance, all odd coefficients are zero; since $a_1 = 0$, it follows that $a_3 = 0$, from which it follows that $a_5 = 0$, and so on. The even coefficients can be expressed in terms of a_0. Apply the recurrence relation with $k = 1, 3, 5$:

$$a_2 = \frac{1}{2}a_0, \qquad a_4 = \frac{1}{4}a_2 = \frac{1}{4}\cdot\frac{1}{2}a_0, \qquad a_6 = \frac{1}{6}a_4 = \frac{1}{6}\cdot\frac{1}{4}\cdot\frac{1}{2}a_0.$$

The pattern is clear:

$$a_{2n} = \frac{1}{2n}\cdot\frac{1}{2n-2}\cdot\frac{1}{2n-4}\cdots\frac{1}{6}\cdot\frac{1}{4}\cdot\frac{1}{2}\,a_0 = \frac{1}{2^n\cdot n!}\,a_0.$$

Therefore the desired power series is

$$y(x) = a_0\left[1 + \frac{x^2}{2} + \frac{x^4}{2^2\cdot 2!} + \frac{x^6}{2^3\cdot 3!} + \cdots + \frac{x^{2n}}{2^n\cdot n!} + \cdots\right] = a_0 \sum_{n=0}^{\infty} \frac{x^{2n}}{2^n\cdot n!}.$$

When written in a slightly different form, this series is a familiar one:

$$y(x) = a_0 \sum_{n=0}^{\infty} \frac{1}{n!}\left(\frac{x^2}{2}\right)^n = a_0 e^{x^2/2}.$$

Answer: $y = a_0 e^{x^2/2}$.

REMARK: This differential equation can be solved also by separation of variables. The next example, however, is not as simple.

EXAMPLE 5.2

Obtain a power series solution of the initial-value problem

$$\begin{cases} y'' + xy' + y = 0 \\ y(0) = 0, \qquad y'(0) = 3. \end{cases}$$

Solution: Try a power series

$$y(x) = a_0 + a_1 x + a_2 x^2 + \cdots + a_n x^n + \cdots.$$

From the initial conditions, $a_0 = 0$ and $a_1 = 3$. Now substitute the series into the differential equation:

$$\begin{aligned} y'' + xy' + y = {} & (2a_2 + 3\cdot 2a_3 x + 4\cdot 3a_4 x^2 + \cdots + n(n-1)a_n x^{n-2} + \cdots) \\ & + x(a_1 + 2a_2 x + 3a_3 x^2 + \cdots + na_n x^{n-1} + \cdots) \\ & + (a_0 + a_1 x + a_2 x^2 + a_3 x^3 + \cdots + a_n x^n + \cdots) = 0. \end{aligned}$$

Collect powers of x:

$$\begin{aligned} & (a_0 + 2a_2) + (2a_1 + 3\cdot 2a_3)x + (3a_2 + 4\cdot 3a_4)x^2 + \cdots \\ & + [a_n + na_n + (n+2)(n+1)a_{n+2}]x^n + \cdots = 0. \end{aligned}$$

All coefficients on the left-hand side are zero, in particular the coefficient of x^n. It follows that

$$a_{n+2} = \left(-\frac{1}{n+2}\right)a_n.$$

From this recurrence relation, the even coefficients can be expressed as multiples of a_0 and the odd coefficients as multiples of a_1. Since $a_0 = 0$, all even coefficients equal 0. Apply the recurrence relation with $n = 1$, then $n = 3$, then $n = 5$, and so on:

$$a_3 = -\frac{1}{3}a_1, \qquad a_5 = -\frac{1}{5}a_3 = \left(-\frac{1}{5}\right)\left(-\frac{1}{3}\right)a_1 = \frac{1}{3\cdot 5}a_1,$$

$$a_7 = -\frac{1}{7}a_5 = -\frac{1}{3\cdot 5\cdot 7}a_1.$$

In general,

$$a_{2n+1} = (-1)^n \frac{1}{1\cdot 3\cdot 5\cdot 7\cdots(2n+1)}a_1$$

$$= (-1)^n \frac{2\cdot 4\cdot 6\cdots(2n)}{1\cdot 2\cdot 3\cdots(2n+1)}a_1 = (-1)^n \frac{2^n\cdot n!}{(2n+1)!}a_1.$$

Since $a_1 = 3$, all coefficients are now determined.

Answer: $y = 3\sum_{n=0}^{\infty}(-1)^n \frac{2^n\cdot n!}{(2n+1)!}x^{2n+1}.$

EXERCISES

Solve by a power series at $x = 0$; check your answer by solving the differential equation exactly:

1. $\dfrac{dy}{dx} = x + y$

2. $\dfrac{dy}{dx} + 3y = e^x$

   [*Hint:* Expand e^x in a power series.]

3. $\dfrac{d^2y}{dx^2} + 4y = 0$

4. $\dfrac{dy}{dx} - y = x^2.$

Obtain a power series solution at $x = 0$:

5. $x\dfrac{d^2y}{dx^2} = y$

6. $x\dfrac{d^2y}{dx^2} + \dfrac{dy}{dx} + xy = 0$

(Exercise 6 is a special case of Bessel's Equation, which has important applications in fluid flow, electric fields, aerodynamics, and other physical problems.)

7. $x^2\dfrac{d^2y}{dx^2} + (x^2 + x)\dfrac{dy}{dx} - y = 0$

8. $\dfrac{d^2y}{dx^2} + x\dfrac{dy}{dx} - y = 0.$

For the initial-value problem, find a power series solution at $x = 0$ up to and including the term in x^4:

9. $\dfrac{dy}{dx} = 1 - x^2 - y^2; \quad y(0) = 2$

10. $\dfrac{dy}{dx} = \dfrac{x}{x + y + 1}; \quad y(0) = 0$

11. $\dfrac{dy}{dx} = 1 - y + x^3y^2; \quad y(0) = -1$

12. $\dfrac{d^2y}{dx^2} + y = \dfrac{e^x}{1-x}; \quad y(0) = 1.$

Find a power series solution at $x = 2$ up to and including the term in $(x-2)^4$:

13. $x\dfrac{dy}{dx} = y^2; \quad y(2) = 1$

14. $\dfrac{d^2y}{dx^2} + (x-2)\dfrac{dy}{dx} + y = x; \quad y(2) = 1.$

[*Hint:* Write $x = (x-2) + 2$.]

6. MATRIX POWER SERIES

Power series whose variable takes matrix values are a useful tool in the study of linear systems of differential equations (next section). Given a power series

$$f(z) = c_0 + c_1z + c_2z^2 + \cdots + c_nz^n + \cdots$$

and a matrix

$$A = \begin{bmatrix} a_{11} & a_{12} \\ a_{21} & a_{22} \end{bmatrix},$$

we boldly substitute A for z:

$$f(A) = c_0I + c_1A + c_2A^2 + \cdots + c_nA^n + \cdots.$$

(Note that the constant term is replaced by c_0 times the identity matrix I.)

Is this substitution meaningful? Each power A^n can be written

$$A^n = \begin{bmatrix} a_{11}^{(n)} & a_{12}^{(n)} \\ a_{21}^{(n)} & a_{22}^{(n)} \end{bmatrix},$$

where $^{(n)}$ is a superscript, not an exponent. The sum $f(A)$ is a matrix computed component-wise:

$$f(A) = c_0\begin{bmatrix} 1 & 0 \\ 0 & 1 \end{bmatrix} + \sum_{n=1}^{\infty} c_n \begin{bmatrix} a_{11}^{(n)} & a_{12}^{(n)} \\ a_{21}^{(n)} & a_{22}^{(n)} \end{bmatrix}$$

$$= \begin{bmatrix} c_0 + \sum_{n=1}^{\infty} c_n a_{11}^{(n)} & \sum_{n=1}^{\infty} c_n a_{12}^{(n)} \\ \sum_{n=1}^{\infty} c_n a_{21}^{(n)} & c_0 + \sum_{n=1}^{\infty} c_n a_{22}^{(n)} \end{bmatrix}.$$

If the series for $f(z)$ converges for $|z| < R$, it can be shown that the four series in $f(A)$ converge, provided the characteristic roots λ and μ of A satisfy

$|\lambda| < R$ and $|\mu| < R$. In particular, if $R = \infty$, then $f(A)$ converges for *all* matrices A. We shall sketch a proof of this fact for the case of real characteristic roots. The complex case is proved similarly, once the proper groundwork is laid (next chapter and an extension of Chapter 7, Section 8). We begin with two important examples.

EXAMPLE 6.1

Let $f(z) = c_0 + c_1 z + c_1 z^2 + \cdots$ converge for $|z| < R$ and let

$$A = \begin{bmatrix} \lambda & 0 \\ 0 & \mu \end{bmatrix}, \qquad \text{where} \quad |\lambda| < R, \quad |\mu| < R.$$

Prove that the power series for $f(A)$ converges and find its sum.

Solution:

$$A^n = \begin{bmatrix} \lambda^n & 0 \\ 0 & \mu^n \end{bmatrix},$$

hence

$$f(A) = \begin{bmatrix} c_0 + \sum_{n=1}^{\infty} c_n \lambda^n & 0 \\ 0 & c_0 + \sum_{n=1}^{\infty} c_n \mu^n \end{bmatrix} = \begin{bmatrix} f(\lambda) & 0 \\ 0 & f(\mu) \end{bmatrix}.$$

Answer: $\begin{bmatrix} f(\lambda) & 0 \\ 0 & f(\mu) \end{bmatrix}.$

EXAMPLE 6.2

Let $f(z) = c_0 + c_1 z + c_2 z^2 + \cdots$ converge for $|z| < R$, and let

$$A = \begin{bmatrix} \lambda & 1 \\ 0 & \lambda \end{bmatrix}, \qquad \text{where} \quad |\lambda| < R.$$

Prove that the power series for $f(A)$ converges and find its sum.

Solution:

$$A = \lambda I + N,$$

where

$$N = \begin{bmatrix} 0 & 1 \\ 0 & 0 \end{bmatrix}, \qquad N^2 = 0.$$

Hence by the Binomial Theorem,

$$A^n = (\lambda I + N)^n = \lambda^n I + n\lambda^{n-1}N.$$

(Higher powers N^2, N^3, $\cdots$ are all 0.) Thus

$$f(A) = c_0 I + \sum_{n=1}^{\infty} c_n A^n = c_0 I + \sum_{n=1}^{\infty} c_n(\lambda^n I + n\lambda^{n-1}N)$$

$$= \left(c_0 + \sum_{n=1}^{\infty} c_n\lambda^n\right) I + \left(\sum_{n=1}^{\infty} nc_n\lambda^{n-1}\right) N = f(\lambda)I + f'(\lambda)N.$$

We know that $f(\lambda)$ converges because $|\lambda| < R$. We also know that $f'(\lambda)$ converges by Theorem 8.8(3), p. 114.

Answer: $\begin{bmatrix} f(\lambda) & f'(\lambda) \\ 0 & f(\lambda) \end{bmatrix}$.

In general, if A has real characteristic roots, then there is a non-singular matrix P such that either

$$A = P\begin{bmatrix} \lambda & 0 \\ 0 & \mu \end{bmatrix}P^{-1} \quad \text{or} \quad A = P\begin{bmatrix} \lambda & 1 \\ 0 & \lambda \end{bmatrix}P^{-1}.$$

(See Theorem 8.1, p. 272 and Theorem 8.2, p. 273.) An easy computation as in Examples 6.1 and 6.2 shows that

$$A^n = P\begin{bmatrix} \lambda^n & 0 \\ 0 & \mu^n \end{bmatrix}P^{-1} \quad \text{or} \quad A^n = P\begin{bmatrix} \lambda^n & n\lambda^{n-1} \\ 0 & \lambda^n \end{bmatrix}P^{-1},$$

hence either

$$f(A) = P\begin{bmatrix} f(\lambda) & 0 \\ 0 & f(\mu) \end{bmatrix}P^{-1} \quad \text{or} \quad f(A) = P\begin{bmatrix} f(\lambda) & f'(\lambda) \\ 0 & f(\lambda) \end{bmatrix}P^{-1}.$$

This establishes the convergence of $f(A)$ provided the characteristic roots satisfy $|\lambda| < R$ and $|\mu| < R$, where R is the radius of convergence of $f(z)$.

The Exponential Function

The matrix exponential is particularly useful in the study of linear systems. It is given by

$$e^A = I + \sum_{n=1}^{\infty} \frac{1}{n!} A^n,$$

a formula valid for all A since $e^z = 1 + \sum z^n/n!$ converges for all z. Let us evaluate e^A in two important instances.

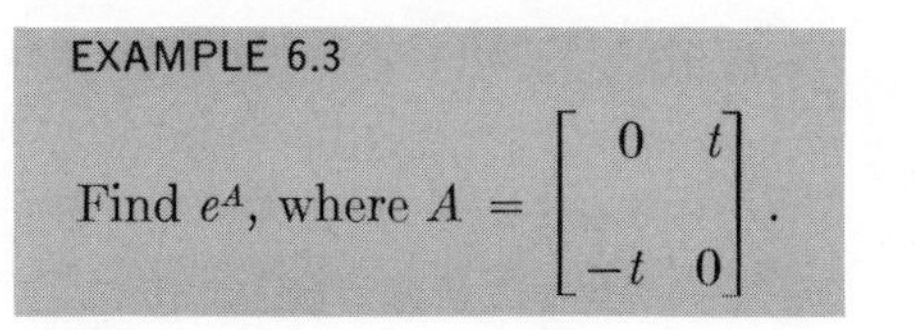

EXAMPLE 6.3

Find e^A, where $A = \begin{bmatrix} 0 & t \\ -t & 0 \end{bmatrix}$.

Solution: Set

$$C = \begin{bmatrix} 0 & 1 \\ -1 & 0 \end{bmatrix} \quad \text{so} \quad A = tC.$$

Note that

$$C^2 = \begin{bmatrix} 0 & 1 \\ -1 & 0 \end{bmatrix}\begin{bmatrix} 0 & 1 \\ -1 & 0 \end{bmatrix} = \begin{bmatrix} -1 & 0 \\ 0 & -1 \end{bmatrix} = -I.$$

It follows readily that successive powers of C are

$$I,\ C,\ -I,\ -C,\ I,\ C,\ -I,\ -C,\ I,\ \cdots,$$

repeating in groups of four (compare to the powers of i). Since $A^n = t^nC^n$,

$$\begin{aligned} e^A &= I + \frac{t}{1!}C - \frac{t^2}{2!}I - \frac{t^3}{3!}C + \frac{t^4}{4!}I + \frac{t^5}{5!}C - \frac{t^6}{6!}I - \frac{t^7}{7!}C + \cdots \\ &= \left(1 - \frac{t^2}{2!} + \frac{t^4}{4!} - \frac{t^6}{6!} + \cdots\right)I + \left(\frac{t}{1!} - \frac{t^3}{3!} + \frac{t^5}{5!} - \frac{t^7}{7!} + \cdots\right)C \\ &= (\cos t)I + (\sin t)C. \end{aligned}$$

Answer: $\begin{bmatrix} \cos t & \sin t \\ -\sin t & \cos t \end{bmatrix}$.

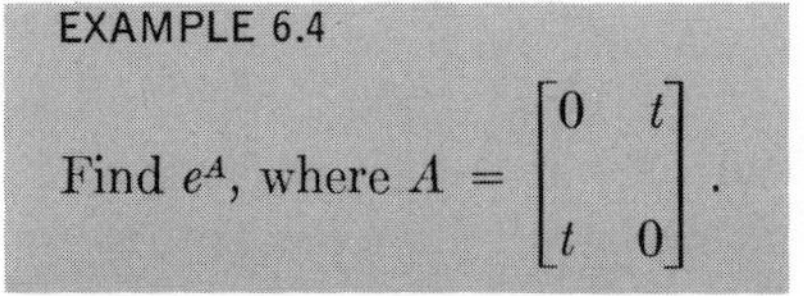

EXAMPLE 6.4

Find e^A, where $A = \begin{bmatrix} 0 & t \\ t & 0 \end{bmatrix}$.

Solution: Proceed as in the last example:

$$A = tC, \quad C = \begin{bmatrix} 0 & 1 \\ 1 & 0 \end{bmatrix}, \quad C^2 = I.$$

Successive powers of C are

$$I,\ C,\ I,\ C,\ \cdots,$$

hence

$$e^A = I + \frac{t}{1!}C + \frac{t^2}{2!}I + \frac{t^3}{3!}C + \cdots$$

$$= \left(1 + \frac{t^2}{2!} + \frac{t^4}{4!} + \cdots\right)I + \left(\frac{t}{1!} + \frac{t^3}{3!} + \cdots\right)C = (\cosh t)I + (\sinh t)C.$$

Answer: $\begin{bmatrix} \cosh t & \sinh t \\ \sinh t & \cosh t \end{bmatrix}$.

We close with two properties of the matrix exponential, whose proofs will be given in Section 8:

(1) $e^A e^B = e^{A+B}$ if $AB = BA$.
(2) $(e^A)^{-1} = e^{-A}$.

EXERCISES

Set $A = \begin{bmatrix} -3 & 9 \\ -1 & 3 \end{bmatrix}$. Find:

1. e^A
2. $\sin A$
3. $(I - A)^{-1}$
4. $\cos A$.

Set $A = \begin{bmatrix} \lambda & 0 \\ 1 & \lambda \end{bmatrix}$. Find:

5. $f(A)$ for $f(z) = c_0 + c_1 z + \cdots$
6. e^{tA}.

Compute e^A:

7. $A = \begin{bmatrix} 0 & 1 \\ 0 & 1 \end{bmatrix}$
8. $A = \begin{bmatrix} \lambda & 1 \\ 0 & 0 \end{bmatrix}$.

9. Set

$$P = \begin{bmatrix} 5 & 2 \\ 2 & 1 \end{bmatrix} \quad \text{and} \quad A = \begin{bmatrix} 0 & 1 \\ 1 & 0 \end{bmatrix}.$$

Verify that $e^{PAP^{-1}} = Pe^A P^{-1}$.

10. Set $B = \ln(I - A)$. Test the formula $e^B = I - A$ in two cases (where $|\lambda| < 1$ and $|\mu| < 1$):

$$A = \begin{bmatrix} \lambda & 0 \\ 0 & \mu \end{bmatrix}; \quad A = \begin{bmatrix} \lambda & 1 \\ 0 & \lambda \end{bmatrix}.$$

11. If

$$A = \begin{bmatrix} a & b \\ -b & a \end{bmatrix},$$

show that

$$e^A = e^a \begin{bmatrix} \cos b & \sin b \\ -\sin b & \cos b \end{bmatrix}.$$

12. Compute e^A for $A = \begin{bmatrix} a & b \\ b & a \end{bmatrix}$.

7. SYSTEMS

We consider **first order systems** of the form

$$\begin{cases} \dot{x} = f(t, x, y) \\ \dot{y} = g(t, x, y). \end{cases}$$

Here $f(t, x, y)$ and $g(t, x, y)$ are given continuously differentiable functions of three variables. In the corresponding initial-value problem, an initial point (t_0, x_0, y_0) is given and we seek two functions, $x(t)$ and $y(t)$, satisfying

$$\begin{cases} \dfrac{dx(t)}{dt} = f[t, x(t), y(t)] \\ \dfrac{dy(t)}{dt} = g[t, x(t), y(t)] \end{cases} \quad \text{and} \quad \begin{cases} x(t_0) = x_0 \\ y(t_0) = y_0. \end{cases}$$

Such systems include the most general second order equation

$$\ddot{x} = f(t, x, \dot{x})$$

in one unknown $x(t)$. For simply set $y = \dot{x}$ and a system results:

$$\begin{cases} \dot{x} = y \\ \dot{y} = f(t, x, y). \end{cases}$$

A **linear system** has the form

$$\begin{cases} \dot{x} = a_{11}(t)x + a_{12}(t)y + b_1(t) \\ \dot{y} = a_{21}(t)x + a_{22}(t)y + b_2(t). \end{cases}$$

It can be written in matrix form as

$$\dot{\mathbf{x}} = A\mathbf{x} + \mathbf{b},$$

where

$$\mathbf{x} = \begin{bmatrix} x \\ y \end{bmatrix}, \quad \dot{\mathbf{x}} = \begin{bmatrix} \dot{x} \\ \dot{y} \end{bmatrix}, \quad A = A(t) = \begin{bmatrix} a_{11} & a_{12} \\ a_{21} & a_{22} \end{bmatrix}, \quad \mathbf{b} = \mathbf{b}(t) = \begin{bmatrix} b_1 \\ b_2 \end{bmatrix}.$$

(We do not use our usual notation $\mathbf{x}'$ for column vectors because of a possible confusion with derivatives.)

We shall emphasize linear systems with constant coefficient matrix A.

Homogeneous Systems

Consider a homogeneous linear system with constant coefficients

$$\dot{\mathbf{x}} = A\mathbf{x}, \qquad \mathbf{x} = \begin{bmatrix} x \\ y \end{bmatrix}, \qquad A = \begin{bmatrix} a_{11} & a_{12} \\ a_{21} & a_{22} \end{bmatrix}.$$

> The solution of the initial-value problem
> $$\dot{\mathbf{x}} = A\mathbf{x}, \qquad \mathbf{x}(0) = \mathbf{x}_0$$
> is
> $$\mathbf{x} = e^{tA}\mathbf{x}_0.$$

We shall prove here only that $\mathbf{x} = e^{tA}\mathbf{x}_0$ is a solution, not that it is the *only* solution. This fact is harder and will be proved in the next section. We have

$$\mathbf{x}(t) = \mathbf{x}_0 + \sum_{n=1}^{\infty} \frac{t^n}{n!} A^n \mathbf{x}_0,$$

hence $\mathbf{x}(0) = \mathbf{x}_0$ and

$$\dot{\mathbf{x}} = \sum_{n=1}^{\infty} \frac{t^{n-1}}{(n-1)!} A^n \mathbf{x}_0 = A \sum_{n=1}^{\infty} \frac{t^{n-1}}{(n-1)!} A^{n-1}\mathbf{x}_0 = Ae^{tA}\mathbf{x}_0 = A\mathbf{x}.$$

REMARK: The solution is formally the same as that of the initial-value problem for one differential equation

$$\dot{\mathbf{x}} = a\mathbf{x}, \qquad \mathbf{x}(0) = \mathbf{x}_0.$$

Here the constant a can be thought of as a 1 by 1 matrix, and the solution is

$$\mathbf{x} = \mathbf{x}_0 e^{at}.$$

EXAMPLE 7.1

Solve the initial-value problem

$$\begin{cases} \dot{x} = \lambda x \\ \dot{y} = \mu y \end{cases} \qquad \begin{cases} x(0) = x_0 \\ y(0) = y_0. \end{cases}$$

Solution: In matrix form, the system is $\dot{\mathbf{x}} = A\mathbf{x}$ and $\mathbf{x}(0) = \mathbf{x}_0$, where

$$\mathbf{x} = \begin{bmatrix} x \\ y \end{bmatrix}, \qquad A = \begin{bmatrix} \lambda & 0 \\ 0 & \mu \end{bmatrix}, \qquad \mathbf{x}_0 = \begin{bmatrix} x_0 \\ y_0 \end{bmatrix}.$$

By Example 6.1,

$$e^{tA} = \begin{bmatrix} e^{\lambda t} & 0 \\ 0 & e^{\mu t} \end{bmatrix},$$

hence the solution is

$$\mathbf{x} = e^{tA}\mathbf{x}_0 = \begin{bmatrix} e^{\lambda t} & 0 \\ 0 & e^{\mu t} \end{bmatrix} \begin{bmatrix} x_0 \\ y_0 \end{bmatrix} = \begin{bmatrix} x_0 e^{\lambda t} \\ y_0 e^{\mu t} \end{bmatrix}.$$

Answer: $\begin{cases} x = x_0 e^{\lambda t} \\ y = y_0 e^{\mu t}. \end{cases}$

REMARK: The system really consists of two entirely unrelated equations. Such a system is called **uncoupled**. The next example is not so simple.

EXAMPLE 7.2

Solve the initial-value problem

$$\begin{cases} \dot{x} = \lambda x + y \\ \dot{y} = \lambda y \end{cases} \qquad \begin{cases} x(0) = x_0 \\ y(0) = y_0. \end{cases}$$

Solution: In matrix form the system is $\dot{\mathbf{x}} = A\mathbf{x}$, where

$$\mathbf{x} = \begin{bmatrix} x \\ y \end{bmatrix}, \qquad A = \begin{bmatrix} \lambda & 1 \\ 0 & \lambda \end{bmatrix}, \qquad \mathbf{x}_0 = \begin{bmatrix} x_0 \\ y_0 \end{bmatrix}.$$

By the calculation in Example 6.2,

$$(tA)^n = t^n A^n = \begin{bmatrix} \lambda^n t^n & n\lambda^{n-1} t^n \\ 0 & \lambda^n t^n \end{bmatrix},$$

hence

$$e^{tA} = \begin{bmatrix} 1 & 0 \\ 0 & 1 \end{bmatrix} + \sum_{n=1}^{\infty} \begin{bmatrix} \lambda^n t^n & n\lambda^{n-1} t^n \\ 0 & \lambda^n t^n \end{bmatrix} = \begin{bmatrix} e^{\lambda t} & te^{\lambda t} \\ 0 & e^{\lambda t} \end{bmatrix}.$$

Therefore the solution in matrix form is

$$\mathbf{x} = \begin{bmatrix} e^{\lambda t} & te^{\lambda t} \\ 0 & e^{\lambda t} \end{bmatrix} \begin{bmatrix} x_0 \\ y_0 \end{bmatrix}.$$

Answer: $x = (x_0 + y_0 t)e^{\lambda t}$

$y = y_0 e^{\lambda t}.$

Second Order Equations

We already noted that the general second order equation $\ddot{x} = f(t, x, \dot{x})$ can be reduced to a first order system. One technique is to set $y = \dot{x}$ and then the system is $\dot{x} = y$, $\dot{y} = f(t, x, y)$. But there are other techniques; for instance set $y = \dot{x} + tx$. Then $\dot{y} = \ddot{x} + x + t\dot{x} = f(t, x, y) + x + t(y - tx)$, so the system is

$$\dot{x} = -tx + y, \qquad \dot{y} = (1 - t^2)x + ty + f(t, x, y).$$

This particular reduction is not especially recommended, but it indicates the variety possible. Let us try one easy example.

EXAMPLE 7.3

Find the general solution of $\ddot{x} + k^2x = 0, \quad k > 0.$

Solution: Set

$$\dot{x} = ky.$$

Then

$$k\dot{y} + k^2x = \ddot{x} + k^2x = 0,$$

$$\dot{y} = -kx.$$

Thus the system of first order equations

$$\begin{cases} \dot{x} = ky \\ \dot{y} = -kx, \end{cases}$$

is equivalent to the given second order equation. The system may be written

$$\dot{\mathbf{x}} = A\mathbf{x}, \qquad \text{where} \quad A = \begin{bmatrix} 0 & k \\ -k & 0 \end{bmatrix}.$$

By Example 6.3,

$$e^{tA} = \begin{bmatrix} \cos kt & \sin kt \\ -\sin kt & \cos kt \end{bmatrix},$$

hence the solution to the system is

$$\begin{bmatrix} x \\ y \end{bmatrix} = \begin{bmatrix} \cos kt & \sin kt \\ -\sin kt & \cos kt \end{bmatrix} \begin{bmatrix} x_0 \\ y_0 \end{bmatrix}.$$

Answer: $x = x_0 \cos kt + y_0 \sin kt$, x_0 and y_0 constants.

Non-homogeneous Systems

Suppose we wish to solve a non-homogeneous system with constant coefficients:

$$\dot{\mathbf{x}} = A\mathbf{x} + \mathbf{b}(t), \qquad A = \begin{bmatrix} a_{11} & a_{12} \\ a_{21} & a_{22} \end{bmatrix}, \qquad \mathbf{b} = \begin{bmatrix} b_1(t) \\ b_2(t) \end{bmatrix}.$$

Here is a direct and effective method. Define $\mathbf{y}(t)$ by $\mathbf{x} = e^{tA}\mathbf{y}$. Then

$$(e^{tA}\mathbf{y})^{\boldsymbol{\cdot}} = Ae^{tA}\mathbf{y} + \mathbf{b},$$

hence the given differential equation is equivalent to

$$Ae^{tA}\mathbf{y} + e^{tA}\dot{\mathbf{y}} = Ae^{tA}\mathbf{y} + \mathbf{b}, \qquad e^{tA}\dot{\mathbf{y}} = \mathbf{b}, \qquad \dot{\mathbf{y}} = e^{-tA}\mathbf{b}.$$

Now integrate:

$$\mathbf{y}(t) = \mathbf{y}_0 + \int_0^t e^{-sA}\,\mathbf{b}(s)\,ds,$$

where the integral of a vector function is taken componentwise. Therefore we have the solution formula

$$\mathbf{x}(t) = e^{tA}\left(\mathbf{x}_0 + \int_0^t e^{-sA}\,\mathbf{b}(s)\,ds\right).$$

EXAMPLE 7.4

Solve the system

$$\begin{cases} \dfrac{dx}{dt} = x + y + t^2 \\[2ex] \dfrac{dy}{dt} = y + t, \end{cases} \qquad x(0) = y(0) = 0.$$

Solution: Write the system as $\dot{\mathbf{x}} = A\mathbf{x} + \mathbf{b}$, where

$$\mathbf{x} = \begin{bmatrix} x(t) \\ y(t) \end{bmatrix}, \qquad A = \begin{bmatrix} 1 & 1 \\ 0 & 1 \end{bmatrix}, \qquad \mathbf{b}(t) = \begin{bmatrix} t^2 \\ t \end{bmatrix}.$$

In Example 7.2 it was shown that

$$e^{tA} = \begin{bmatrix} e^t & te^t \\ 0 & e^t \end{bmatrix}, \qquad e^{-sA} = \begin{bmatrix} e^{-s} & -se^{-s} \\ 0 & e^{-s} \end{bmatrix}.$$

Hence

$$\int_0^t e^{-sA}\mathbf{b}(s)\,ds = \int_0^t \begin{bmatrix} e^{-s} & -se^{-s} \\ 0 & e^{-s} \end{bmatrix}\begin{bmatrix} s^2 \\ s \end{bmatrix} ds = \int_0^t \begin{bmatrix} 0 \\ se^{-s} \end{bmatrix} ds$$

$$= \begin{bmatrix} \int_0^t 0\,ds \\ \int_0^t se^{-s}\,ds \end{bmatrix} = \begin{bmatrix} 0 \\ -e^{-t}(t+1)+1 \end{bmatrix}.$$

The initial condition is $\mathbf{x}(0) = \mathbf{0}$, so the solution formula yields

$$\mathbf{x}(t) = e^{tA}\left[\mathbf{0} + \int_0^t e^{-sA}\mathbf{b}(s)\,ds\right]$$

$$= \begin{bmatrix} e^t & te^t \\ 0 & e^t \end{bmatrix}\begin{bmatrix} 0 \\ 1 - e^{-t}(t+1) \end{bmatrix} = \begin{bmatrix} te^t - t(t+1) \\ e^t - (t+1) \end{bmatrix}.$$

Answer: $x = te^t - t^2 - t, \quad y = e^t - t - 1.$

EXERCISES

1. Show that the substitution $\dot{x} = ky$ changes $\ddot{x} - k^2x = 0$ to the system

$$\begin{cases} \dot{x} = ky \\ \dot{y} = kx, \end{cases}$$

   and solve.

2. Show that the substitution $y = \dot{x} - \lambda x$ changes $\ddot{x} - 2\lambda\dot{x} + \lambda^2 x = 0$ to the system

$$\begin{cases} \dot{x} = \lambda x + y \\ \dot{y} = \lambda y, \end{cases}$$

   and solve.

3. Let $\mathbf{x}$ be a solution of $\dot{\mathbf{x}} = A\mathbf{x}$. Show that $\mathbf{y} = P^{-1}\mathbf{x}$ is a solution of $\dot{\mathbf{y}} = (P^{-1}AP)\mathbf{y}$. (Here P is a non-singular constant matrix.)

4. Solve

$$\begin{cases} \dot{x} = ky + \cos kt \\ \dot{y} = -kx + \sin kt. \end{cases}$$

5. Solve

$$\begin{cases} \dot{x} = \lambda x + y + te^{\lambda t} \\ \dot{y} = \lambda y + e^{\lambda t}. \end{cases}$$

8. UNIQUENESS OF SOLUTIONS [optional]

We differentiate vector functions and matrix functions componentwise. Let us note a few useful formulas:

If $\mathbf{x} = \mathbf{x}(t)$, $B = B(t)$, and $C = C(t)$, then
$(B\mathbf{x})^{\cdot} = B^{\cdot}\mathbf{x} + B\dot{\mathbf{x}}, \qquad (BC)^{\cdot} = B^{\cdot}C + BC^{\cdot},$
$(B^{-1})^{\cdot} = -B^{-1}B^{\cdot}B^{-1}$ if B is non-singular.

The first two formulas are easily proved. For instance,

$$\Big(\sum_j b_{ij}c_{jk}\Big)^{\cdot} = \sum_j (\dot{b}_{ij}c_{jk} + b_{ij}\dot{c}_{jk})$$

implies the second. Note the order of multiplication in the products. The third formula follows from the second:

$$BB^{-1} = I, \qquad B^{\cdot}B^{-1} + B(B^{-1})^{\cdot} = I^{\cdot} = 0, \qquad B^{-1}B^{\cdot}B^{-1} + (B^{-1})^{\cdot} = 0.$$

Homogeneous Systems

We claimed (p. 557) that the general solution of $\ddot{x} - k^2x = 0$ is

$$x = ae^{kt} + be^{-kt},$$

assuming $k \neq 0$. An equivalent form of this solution is

$$x = a\cosh kt + b\sinh kt.$$

We also claimed that the general solution of $\ddot{x} + k^2x = 0$ is

$$x = a\cos kt + b\sin kt.$$

Both are reasonable from the system point of view. If we set $\dot{x} = ky$ and

$$\mathbf{x} = \begin{bmatrix} x \\ y \end{bmatrix},$$

the two systems are $\dot{\mathbf{x}} = A\mathbf{x}$, where either

$$A = \begin{bmatrix} 0 & k \\ k & 0 \end{bmatrix} \qquad \text{or} \qquad A = \begin{bmatrix} 0 & k \\ -k & 0 \end{bmatrix}.$$

We saw earlier that for these matrices either

$$e^{tA} = \begin{bmatrix} \cosh kt & \sinh kt \\ \sinh kt & \cosh kt \end{bmatrix} \qquad \text{or} \qquad e^{tA} = \begin{bmatrix} \cos kt & \sin kt \\ -\sin kt & \cos kt \end{bmatrix}.$$

The claims about $\ddot{x} \pm k^2 x = 0$ follow from the assertion that $\mathbf{x} = e^{tA}\mathbf{x}_0$. We shall prove this in general for *any* homogeneous system.

A Uniqueness Theorem

> **Theorem 8.1** The general solution of
> $$\dot{\mathbf{x}} = A\mathbf{x},$$
> where A is any constant matrix, is
> $$\mathbf{x} = e^{tA}\mathbf{x}_0, \qquad \mathbf{x}_0 = \mathbf{x}(0).$$

Proof: Suppose $\mathbf{x}$ is any solution of $\dot{\mathbf{x}} = A\mathbf{x}$. Set $B(t) = e^{tA}$. Then

$$\dot{B} = \frac{d}{dt}\left(I + \sum_{n=1}^{\infty} \frac{t^n}{n!} A^n\right) = \sum_{n=1}^{\infty} \frac{t^{n-1}}{(n-1)!} A^n = AB.$$

Therefore

$$\begin{aligned}(B^{-1}\mathbf{x})^{\cdot} = (B^{-1})^{\cdot}\mathbf{x} + B^{-1}\dot{\mathbf{x}} &= -B^{-1}(AB)B^{-1}\mathbf{x} + B^{-1}(A\mathbf{x}) \\ &= -B^{-1}A\mathbf{x} + B^{-1}A\mathbf{x} = \mathbf{0}.\end{aligned}$$

It follows that $B^{-1}\mathbf{x} = \mathbf{x}_0$, a constant vector, hence $\mathbf{x} = B\mathbf{x}_0$. Obviously $\mathbf{x}(0) = \mathbf{x}_0$ since $B(0) = I$. This completes the proof.

Another Uniqueness Proof

The theorem we just proved shows that the initial value problem $\dot{\mathbf{x}} = A\mathbf{x}$, $\mathbf{x}(0) = \mathbf{x}_0$ has *only one solution*, namely $\mathbf{x} = e^{tA}\mathbf{x}_0$. There is another way to prove there is at most one solution without actually knowing there is any solution. The method involves estimating successive approximations, and it is important because it applies with suitable modifications to many systems besides homogeneous ones with constant coefficients.

Suppose $\mathbf{x}_1(t)$ and $\mathbf{x}_2(t)$ are two solutions of the initial value problem $\dot{\mathbf{x}} = A\mathbf{x}$, $\mathbf{x}(0) = \mathbf{x}_0$. Their difference $\mathbf{x}(t) = \mathbf{x}_1(t) - \mathbf{x}_2(t)$ is a solution of the initial-value problem $\dot{\mathbf{x}} = A\mathbf{x}$, $\mathbf{x}(0) = \mathbf{0}$. If we can conclude that $\mathbf{x}(t) = \mathbf{0}$, then $\mathbf{x}_1 = \mathbf{x}_2$.

> **Theorem 8.2** Let $\mathbf{x}(t)$ be a solution of the differential equation $\dot{\mathbf{x}} = A\mathbf{x}$ with $\mathbf{x}(0) = \mathbf{0}$. Then $\mathbf{x}(t) = \mathbf{0}$.

Proof: We may write

$$\mathbf{x}(t) = \int_0^t \dot{\mathbf{x}}(s)\, ds = \int_0^t A\mathbf{x}(s)\, ds = A\int_0^t \mathbf{x}(s)\, ds.$$

Let us work on a fixed interval $-b \le t \le b$. We know (p. 240) that there is a constant c such that $|A\mathbf{v}| \le c\,|\mathbf{v}|$ for all $\mathbf{v}$. Also, since $\mathbf{x}(s)$ is continuous on $[-b, b]$, there is a constant M such that $|\mathbf{x}(s)| \le M$ for $-b \le s \le b$. Now we are set up for a remarkable bootstrap operation. We start:

$$|\mathbf{x}(t)| \le c\left|\int_0^t \mathbf{x}(s)\,ds\right| \le c\left|\int_0^t |\mathbf{x}(s)|\,ds\right| = c\left|\int_0^t M\,ds\right| = Mc\,|t|.$$

Next step:

$$|\mathbf{x}(t)| \le c\left|\int_0^t |\mathbf{x}(s)|\,ds\right| \le c\left|\int_0^t Mc\,|s|\,ds\right| = \frac{1}{2}Mc^2\,|t|^2.$$

Once again:

$$|\mathbf{x}(t)| \le c\left|\int_0^t |\mathbf{x}(s)|\,ds\right| \le c\left|\int_0^t \frac{1}{2}Mc^2\,|s|^2\,ds\right| = \frac{1}{3!}Mc^3\,|t|^3.$$

Continuing in this way, we prove

$$|\mathbf{x}(t)| \le \frac{1}{n!}Mc^n\,|t|^n,$$

for all n. Hence

$$|\mathbf{x}(t)| \le \frac{1}{n!}Mc^nb^n = \frac{(cb)^n}{n!}.$$

As $n \longrightarrow \infty$, we have $(cb)^n/(n!) \longrightarrow 0$. Hence $\mathbf{x}(t) = \mathbf{0}$ for $|t| \le b$. But b is arbitrary, so $\mathbf{x}(t) = \mathbf{0}$ for all t. End of the proof.

Here is a simple, but useful, corollary of the uniqueness theorem.

Theorem 8.3 The only *matrix* solution of

$$X^{\cdot} = AX, \qquad X(0) = X_0$$

is

$$X = e^{tA}X_0.$$

Proof: Write $X = [\mathbf{x}, \mathbf{y}]$ in columns. Then

$$\dot{\mathbf{x}} = A\mathbf{x} \qquad \mathbf{x}(0) = \mathbf{x}_0$$

$$\dot{\mathbf{y}} = A\mathbf{y} \qquad \mathbf{y}(0) = \mathbf{y}_0,$$

so the uniqueness theorem for *vector* systems implies

$$\mathbf{x} = e^{tA}\mathbf{x}_0, \qquad \mathbf{y} = e^{tA}\mathbf{y}_0.$$

Hence $X = e^{tA}X_0$. (See Ex. 1 for another proof.)

In Section 6, we stated without proof the following two properties of the matrix exponential:

(1) $e^A e^B = e^{A+B}$ if $AB = BA$.

(2) $(e^A)^{-1} = e^{-A}$.

Property (1) can be proved by multiplying power series, but there is a tricky convergence problem. A better way is to use differential equations. We know that $e^{t(A+B)}$ is the *only* solution of $X^{\cdot} = (A + B)X$, $X(0) = I$.

Set $Y(t) = e^{tA}e^{tB}$. From $AB = BA$ we deduce $e^{tA}B = Be^{tA}$:

$$e^{tA}B = \left(\sum_0^\infty \frac{t^n}{n!} A^n\right) B = \sum_0^\infty \frac{t^n}{n!} (A^n B)$$

$$= \sum_0^\infty \frac{t^n}{n!} (BA^n) = B \sum_0^\infty \frac{t^n}{n!} A^n = Be^{tA}.$$

Therefore

$$Y^{\cdot} = Ae^{tA}e^{tB} + e^{tA}Be^{tB} = Ae^{tA}e^{tB} + Be^{tA}e^{tB} = (A + B)Y.$$

Since $Y(0) = I$, uniqueness implies $Y(t) = e^{t(A+B)}$. Set $t = 1$ to complete the proof of (1).

Certainly $A(-A) = (-A)A$. Hence

$$I = e^0 = e^{A+(-A)} = e^A e^{-A}, \qquad (e^A)^{-1} = e^{-A}.$$

This proves (2).

EXERCISES

1. Prove the Theorem 8.3 by considering $e^{-tA}X$.
2*. Show that Theorem 8.2 is true when $A = A(t)$ is a matrix of continuous functions.
3. Set

$$A = \begin{bmatrix} 0 & 1 \\ 0 & 0 \end{bmatrix} \quad \text{and} \quad B = \begin{bmatrix} 0 & 0 \\ 1 & 0 \end{bmatrix}.$$

Compute e^{A+B} and $e^A e^B$. Explain the discrepancy.

4. Interpret as trigonometric identities ($\exp A$ is another notation for e^A):

$$\exp\begin{bmatrix} 0 & \alpha+\beta \\ -(\alpha+\beta) & 0 \end{bmatrix} = \exp\begin{bmatrix} 0 & \alpha \\ -\alpha & 0 \end{bmatrix} \exp\begin{bmatrix} 0 & \beta \\ -\beta & 0 \end{bmatrix}.$$

The trace of a 2×2 matrix A is $\operatorname{tr} A = a_{11} + a_{22}$.

5. Show that $|B(t)|^{\cdot} = \operatorname{tr}(B^{\cdot} \operatorname{cof} B)$.
6. (cont.) Set $g(t) = |e^{tA}|$. Prove $\dot{g} = (\operatorname{tr} A)g$.
7. (cont.) Prove $|e^A| = e^{\operatorname{tr} A}$.

8. (cont.) Let $X(t)$ satisfy $X^{\cdot} = AX$. The **Wronskian** of X is $w(t) = |X(t)|$. Show that $w(t) = w(0)e^{(\operatorname{tr} A)t}$.
9. Show that
$$x(t) = \int_0^t (t - s)g(s)\,ds$$
is the solution of $\ddot{x} = g$, $x(0) = \dot{x}(0) = 0$.
10. Suppose $\mathbf{x}$ is the solution of $\dot{\mathbf{x}} = A\mathbf{x} + \mathbf{b}(t)$ with $\mathbf{x}(0) = \mathbf{0}$ and $\mathbf{y}$ is the solution of $\dot{\mathbf{y}} = A\mathbf{y} + \mathbf{c}(t)$ with $\mathbf{y}(0) = \mathbf{0}$. Show that $\mathbf{x} + \mathbf{y}$ is the solution of $\dot{\mathbf{z}} = A\mathbf{z} + (\mathbf{b} + \mathbf{c})$ with $\mathbf{z}(0) = \mathbf{0}$. (This is the **principle of superposition.**)

15. Complex Analysis

1. INTRODUCTION

The simple equation

$$x^2 + 1 = 0$$

has no solution in terms of *real numbers*, the numbers of ordinary experience; neither do the equations

$$x^2 + 3 = 0, \qquad x^2 + x + 1 = 0, \qquad x^2 - 4x + 10 = 0.$$

Yet such equations arise in scientific computations. For this reason, the real number system is enlarged by introducing a new number i satisfying

$$\boxed{i^2 = -1.}$$

The result is the system of **complex numbers.** It consists of all expressions $a + bi$, where a and b are real numbers. These expressions are treated by the usual rules of algebra, except that i^2 is replaced by -1. Thus

$$\begin{aligned} (a + bi) + (c + di) &= (a + c) + (b + d)i, \\ (a + bi)\,(c + di) &= ac + bdi^2 + adi + bci \\ &= (ac - bd) + (ad + bc)i. \end{aligned}$$

Two complex numbers $a + bi$ and $c + di$ are **equal** if and only if $a = c$ and $b = d$. A real number a is considered as a special type of complex number: $a = a + 0 \cdot i$.

REMARKS ON NOTATION: Sometimes $a + ib$ is the preferred notation. For example, $-1 + i\sqrt{3}$ looks better than $-1 + \sqrt{3}\,i$ because of the possible confusion with $-1 + \sqrt{3i}$. Similarly $\cos\theta + i\sin\theta$ is better than $\cos\theta + \sin\theta\, i$. In engineering, j is often used instead of i.

In terms of complex numbers, the quadratic equation

$$x^2 + 1 = 0$$

has two roots, $\pm i$. Furthermore, every quadratic equation

$$ax^2 + bx + c = 0$$

has complex roots. By the quadratic formula, the roots are

$$x = -\frac{b}{2a} \pm \frac{\sqrt{D}}{2a},$$

where $D = b^2 - 4ac$. If D is non-negative, then $\sqrt{D}$ is a real number, so the roots are real. If D is negative, write

$$\sqrt{D} = \sqrt{(-1)(-D)} = \sqrt{-1}\sqrt{-D} = i\sqrt{-D};$$

then the equation has complex roots

$$x = \frac{-b}{2a} \pm \frac{\sqrt{-D}}{2a}\, i.$$

EXAMPLE 1.1

Find the roots of $x^2 + x + 1 = 0$ and of $x^2 - 4x + 10 = 0$.

Solution: Apply the quadratic formula. For the first equation $a = b = c = 1$, $D = -3$. For the second equation $a = 1$, $b = -4$, $c = 10$, $D = -24$.

Answer: $-\frac{1}{2} \pm \frac{\sqrt{3}}{2}\, i;\quad 2 \pm i\sqrt{6}.$

All quadratic equations with real coefficients can be solved in terms of complex numbers. What about cubic equations

$$ax^3 + bx^2 + cx + d = 0\ ?$$

If a new quantity i is needed to solve quadratics, is another new quantity needed to solve cubics, still another to solve quartics, and so on?

This question was answered in 1799 by C. F. Gauss in the Fundamental Theorem of Algebra. It asserts that each polynomial equation of any degree with real coefficients, has a complex root. Thus, the complex number system is rich enough to contain solutions for all polynomial equations. Once the number i is adjoined to the real number system, that is enough.

REMARK: Later we shall learn that each complex number has two complex square roots. Hence, the quadratic formula provides complex roots for any quadratic equation with complex coefficients. The Fundamental Theorem of Algebra also includes the statement that each polynomial equation with *complex* coefficients has complex roots.

EXERCISES

Solve and check your answers by direct substitution:

1. $x^2 - 8x + 25 = 0$
2. $x^2 + 25 = 0$
3. $x^2 + x + 2 = 0$
4. $x^2 - 6x + 9 = 0$
5. $3x^2 - 2x + 3 = 0$
6. $2x^2 + x + 2 = 0$
7. $225x^2 + 15x + 61 = 0$
8. $5x^2 - 4x + 1 = 0.$

Find all roots:

9. $x^3 - 1 = 0$
10. $x^3 + 1 = 0$
11. $x^3 - x^2 + x - 1 = 0$
12. $x^4 - 2x^2 + 1 = 0$
13. $x^4 - 1 = 0$
14. $x^4 + 5x^2 + 4 = 0.$
15. Compute $(1 + i)^2$. Use the result to solve the equation $x^2 = i$.
16. Compute $[(-1 + i\sqrt{3})/2]^3$.
17. Compute $i + i^2 + i^3 + \cdots + i^{1492}$.

2. COMPLEX ARITHMETIC

Complex numbers can be pictured as vectors in a plane. Think of the number 1 as a unit horizontal vector and the number i as a unit vertical vector (Fig. 2.1a). Then a complex number $a + bi$ is a linear combination of these two vectors (Fig. 2.1b).

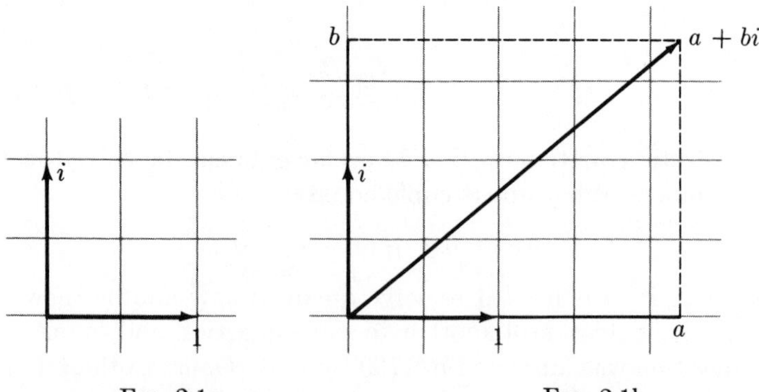

FIG. 2.1a FIG. 2.1b

The horizontal component of $a + bi$ is called its **real part,** written $\text{Re}(a + bi)$; the vertical component is called its **imaginary part,** written $\text{Im}(a + bi)$. Thus

$$\boxed{\text{Re}(a + bi) = a, \qquad \text{Im}(a + bi) = b.}$$

Note that $\text{Im}(a + bi)$ is the real number b, *not* bi.

Associated with each complex number $a + bi$ is the number $a - bi$ called its (**complex**) **conjugate** (Fig. 2.2). Conjugates are denoted by bars:

$$\boxed{\overline{a + bi} = a - bi.}$$

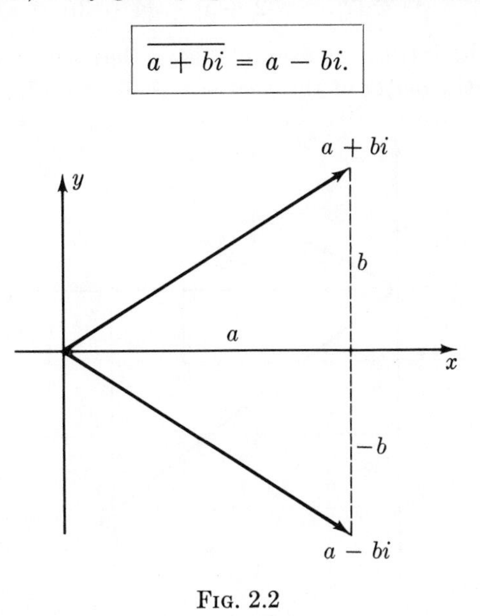

FIG. 2.2

Set $z = a + bi$; the following basic relations hold:

$$\boxed{\begin{array}{ll} z + \bar{z} = 2a = 2\,\mathrm{Re}(z), & z - \bar{z} = 2bi = 2i\,\mathrm{Im}(z), \\ \bar{\bar{z}} = z, & z\bar{z} = a^2 + b^2. \end{array}}$$

By the quadratic formula, if the roots of a quadratic equation with real coefficients are not real, then they are conjugate complex numbers.

The **modulus** or **absolute value** of a complex number $a + bi$ is

$$\boxed{|a + bi| = \sqrt{a^2 + b^2}.}$$

It is the length of the vector $a + bi$. Notice that

$$|z|^2 = z\bar{z}.$$

If $z \neq 0$, then $|z| > 0$.

The absolute value of a complex number is a measure of its size. Since complex numbers fill the plane, you cannot say that one complex number is greater than or less than another; the statements

$$i < 2, \qquad 2 + i < 4 - 3i$$

make no sense. Yet you can compare absolute values; the statements

$$|i| < |2|, \qquad |2 + i| < |4 - 3i|$$

do make sense. The latter, for example, says that the *length* of the vector $2 + i$ is less than the *length* of the vector $4 - 3i$. See Fig. 2.3.

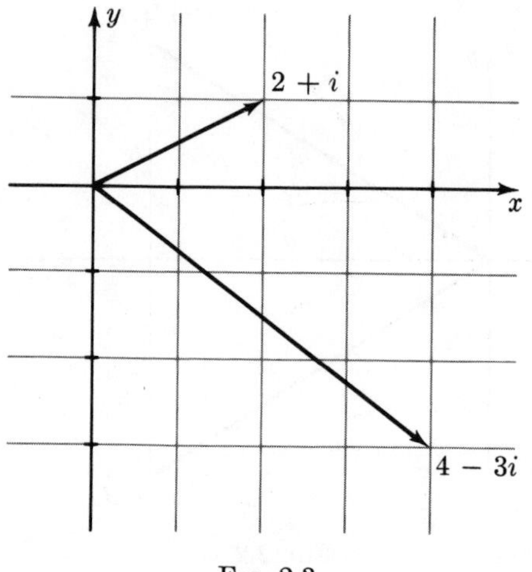

FIG. 2.3

All complex numbers of the same absolute value determine a circle (Fig. 2.4).

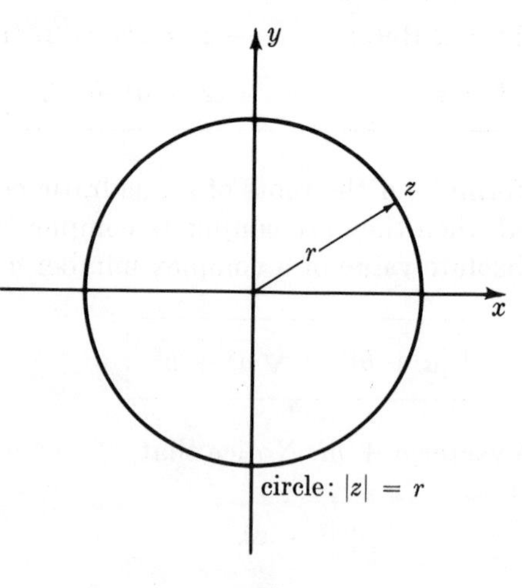

FIG. 2.4

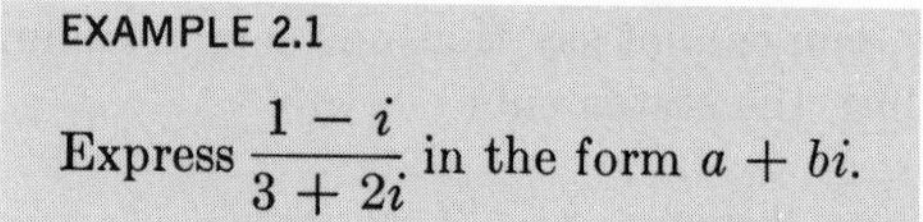

EXAMPLE 2.1

Express $\dfrac{1 - i}{3 + 2i}$ in the form $a + bi$.

Solution: Here is a standard trick. Multiply numerator and denominator by $3 - 2i$, the conjugate of the denominator:

$$\frac{1-i}{3+2i} = \frac{1-i}{3+2i} \cdot \frac{3-2i}{3-2i} = \frac{(3-2)-(3+2)i}{3^2+2^2} = \frac{1-5i}{13}.$$

Answer: $\frac{1}{13} - \frac{5}{13}i.$

EXAMPLE 2.2

If $|z| = 1$, show that $\frac{1}{z} = \bar{z}$.

Solution: Let $z = a + bi$. Then by the trick of the last example,

$$\frac{1}{z} = \frac{1}{a+bi} = \frac{1}{a+bi} \cdot \frac{a-bi}{a-bi} = \frac{a-bi}{a^2+b^2}.$$

But $a^2 + b^2 = |z|^2 = 1$. Hence,

$$\frac{1}{z} = a - bi = \bar{z}.$$

Shorter Solution:

$$z\bar{z} = |z|^2 = 1, \quad \bar{z} = \frac{1}{z}.$$

EXERCISES

Express $\bar{z}$ in the form $a + bi$:

1. $z = \frac{1}{2} - \frac{1}{3}i$
2. $z = 0.4 + 1.7i$
3. $z = \overline{(2-i)}$
4. $z = (2+i)^{-1}$
5. $z = (-1-i)(-2+3i)$
6. $z = (1+i)(2+i)(3+i)$
7. $z = \dfrac{i}{1-i}$
8. $z = \dfrac{1-i}{i}$
9. $z = \dfrac{2+i}{2-i}$
10. $z = \dfrac{2+3i}{-1+i}$.
11. Find $|z|$ in Ex. 1–10.
12. Show that $\overline{zw} = \bar{z}\bar{w}$.
13. Show that $|zw| = |z|\,|w|$.
14. Show that $|\bar{z}| = |z|$.
15. Show that the equation $\bar{z} = z$ is satisfied only by real numbers.
16. Let $f(x) = a_0x^3 + a_1x^2 + a_2x + a_3$, where the coefficients are real numbers. Let z be a complex zero of $f(x)$. Prove that $\bar{z}$ is also a zero. (*Hint:* $\bar{0} = 0$.)

17. Do the same for an n-th degree polynomial with real coefficients.
18. Show that if $(a + bi)^4 = 1$, then $a + bi$ is one of the numbers ± 1, $\pm i$.

3. POLAR FORM

When a complex number z is written $z = a + bi$, it is said to be in **rectangular form.** If $z \neq 0$, it is sometimes convenient to express z in **polar form:**

$$z = r(\cos\theta + i\sin\theta), \qquad r > 0.$$

As is seen in Fig. 3.1,

$$r = |z| = \sqrt{a^2 + b^2}; \qquad \cos\theta = \frac{a}{r}, \qquad \sin\theta = \frac{b}{r}.$$

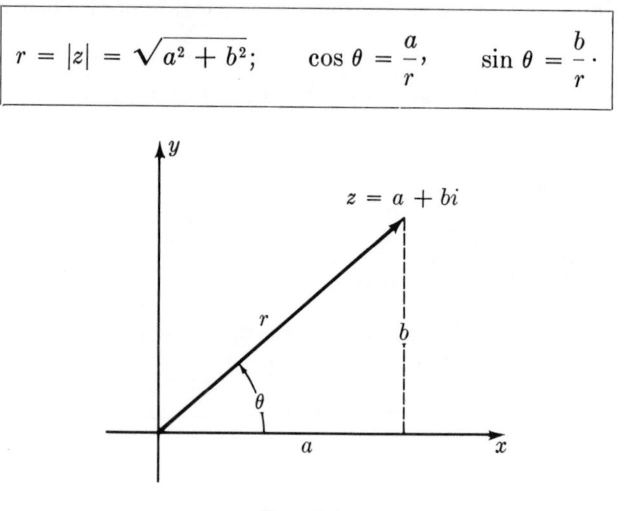

FIG. 3.1

The complex number $\cos\theta + i\sin\theta$ is a unit vector making angle θ with the positive x-axis. Hence, the polar form $z = r(\cos\theta + i\sin\theta)$ expresses the vector z as a unit vector in the same direction stretched by a factor $|z|$. The angle θ is called the **argument** of z and is written arg z. It is determined only up to a multiple of 2π. For example, $\arg(1 + i)$ can be taken to be $\pi/4$ or $9\pi/4$ or $-15\pi/4$, etc. If $z = r(\cos\theta + i\sin\theta)$, then

$$\bar{z} = r[\cos(-\theta) + i\sin(-\theta)] = r(\cos\theta - i\sin\theta).$$

See Fig. 3.2.

Polar form is particularly useful in situations involving multiplication (or division) of complex numbers; it makes multiplication easy. Suppose

$$z_1 = r_1(\cos\theta_1 + i\sin\theta_1), \qquad z_2 = r_2(\cos\theta_2 + i\sin\theta_2).$$

Then

$$\begin{aligned} z_1z_2 &= r_1r_2[(\cos\theta_1\cos\theta_2 - \sin\theta_1\sin\theta_2) + i(\sin\theta_1\cos\theta_2 + \cos\theta_1\sin\theta_2)] \\ &= r_1r_2[\cos(\theta_1 + \theta_2) + i\sin(\theta_1 + \theta_2)]. \end{aligned}$$

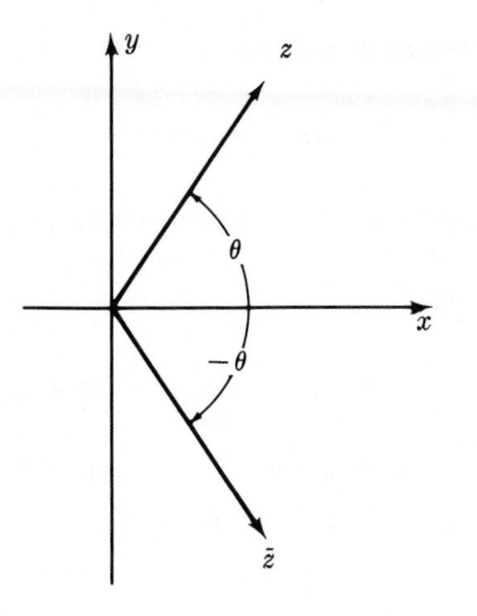

FIG. 3.2

The product is again in polar form; its modulus is r_1r_2 and its argument is $\theta_1 + \theta_2$.

To multiply two complex numbers, multiply their absolute values and add their arguments:

$$|z_1z_2| = |z_1|\,|z_2|, \qquad \arg z_1z_2 = \arg z_1 + \arg z_2 .$$

Similarly, to divide two complex numbers, divide their absolute values and subtract their arguments:

$$\left|\frac{z_1}{z_2}\right| = \frac{|z_1|}{|z_2|}, \qquad \arg\frac{z_1}{z_2} = \arg z_1 - \arg z_2 .$$

From this rule follows De Moivre's Theorem, a formula for powers of a complex number.

De Moivre's Theorem For each positive integer n,

$$[r(\cos\theta + i\sin\theta)]^n = r^n(\cos n\theta + i\sin n\theta).$$

EXAMPLE 3.1

Describe geometrically the effect of multiplying a complex number by i, by $1 - i$.

Solution: Write i and $1 - i$ in polar form:

$$i = 1\left[\cos\frac{\pi}{2} + i\sin\frac{\pi}{2}\right], \qquad 1 - i = \sqrt{2}\left[\cos\left(-\frac{\pi}{4}\right) + i\sin\left(-\frac{\pi}{4}\right)\right].$$

According to the rule for multiplication,

$$|iz| = |i|\,|z| = |z|, \qquad \arg iz = \arg z + \arg i = \arg z + \frac{\pi}{2}\cdot$$

Therefore multiplying the vector z by i simply rotates the vector 90° counterclockwise (Fig. 3.3).

According to the rule,

$$|(1 - i)z| = \sqrt{2}\,|z|, \qquad \arg(1 - i)z = \arg z - \frac{\pi}{4}\cdot$$

Therefore multiplying the vector z by $1 - i$ rotates the vector 45° clockwise and stretches it by a factor $\sqrt{2}$. See Fig. 3.3.

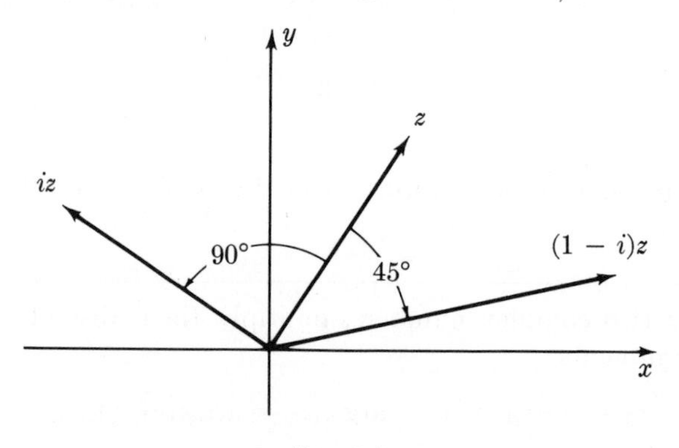

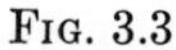
FIG. 3.3

Answer: rotation by 90°;
rotation by −45° and stretching by a factor $\sqrt{2}$.

EXAMPLE 3.2

Compute $(1 + i)^8$.

Solution: By the Binomial Theorem,

$$\begin{aligned}(1 + i)^8 &= 1 + 8i + 28i^2 + 56i^3 + 70i^4 + 56i^5 + 28i^6 + 8i^7 + i^8\\ &= 1 + 8i - 28 - 56i + 70 + 56i - 28 - 8i + 1 = 16.\end{aligned}$$

Alternate Solution: Write $1 + i$ in polar form, then use De Moivre's Theorem:

$$1 + i = \sqrt{2}\left(\cos\frac{\pi}{4} + i\sin\frac{\pi}{4}\right),$$

$$(1 + i)^8 = (\sqrt{2})^8\left(\cos\frac{8\pi}{4} + i\sin\frac{8\pi}{4}\right) = 16(1 + 0\cdot i) = 16.$$

Answer: 16.

The following example illustrates a powerful technique for deriving trigonometric identities.

EXAMPLE 3.3

Derive the identities

$$\cos 3\theta = \cos^3 \theta - 3 \cos \theta \sin^2 \theta,$$

$$\sin 3\theta = 3 \cos^2 \theta \sin \theta - \sin^3 \theta.$$

Solution: Compute $(\cos \theta + i \sin \theta)^3$ two ways, by De Moivre's Theorem and by the Binomial Theorem. The results must be equal:

$$\begin{aligned}\cos 3\theta + i \sin 3\theta &= \cos^3 \theta + 3 \cos^2 \theta \,(i \sin \theta) + 3 \cos \theta \,(i \sin \theta)^2 + (i \sin \theta)^3 \\ &= (\cos^3 \theta - 3 \cos \theta \sin^2 \theta) + i(3 \cos^2 \theta \sin \theta - \sin^3 \theta).\end{aligned}$$

Now equate real and imaginary parts on both sides of this equation.

Roots of Unity

The equation

$$z^4 = 1$$

has four complex roots, ± 1 and $\pm i$. Thus the number 1 has four 4-th roots, which are complex numbers equally spaced around the circle $|z| = 1$.

The situation for n-th roots is similar. Write the equation $z^n = 1$ in polar form, setting $z = r(\cos \theta + i \sin \theta)$:

$$[r(\cos \theta + i \sin \theta)]^n = 1(\cos 0 + i \sin 0), \qquad r > 0.$$

By De Moivre's Theorem,

$$r^n(\cos n\theta + i \sin n\theta) = 1(\cos 0 + i \sin 0).$$

Consequently,

$$r^n = 1, \qquad \cos n\theta = 1, \qquad \sin n\theta = 0.$$

It follows that

$$r = 1, \qquad n\theta = 2\pi k, \qquad \theta = \frac{2\pi k}{n}, \qquad k \text{ an integer.}$$

Thus

$$z = \cos \frac{2\pi k}{n} + i \sin \frac{2\pi k}{n},$$

where k is an integer. This formula yields exactly n distinct values: for $k = 0, 1, 2, \cdots, n - 1$. (Why?) For $k = 1$, call the root

$$\omega = \cos \frac{2\pi}{n} + i \sin \frac{2\pi}{n}.$$

Since (De Moivre again)

$$\omega^k = \cos\frac{2\pi k}{n} + i\sin\frac{2\pi k}{n},$$

the other roots are the powers of ω, namely $1 = \omega^0, \omega, \omega^2, \cdots, \omega^{n-1}$.

The equation

$$z^n = 1$$

has exactly n complex roots $1, \omega, \omega^2, \cdots, \omega^{n-1}$, where

$$\omega = \cos\frac{2\pi}{n} + i\sin\frac{2\pi}{n}\cdot$$

These numbers are called n-th **roots of unity.** They lie equally spaced around the circle $|z| = 1$. See Fig. 3.4.

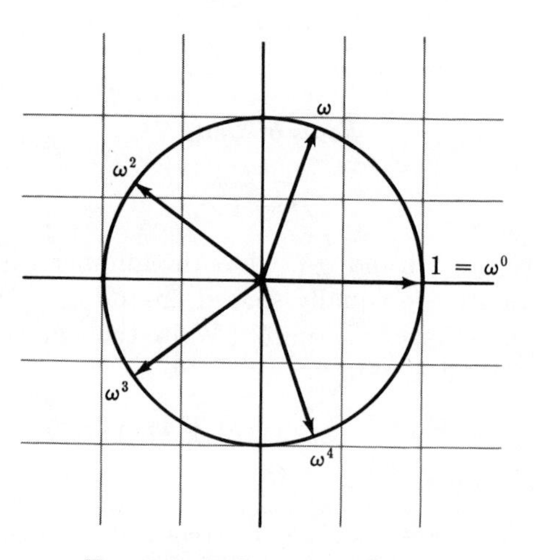

FIG. 3.4 5-th roots of unity.

EXAMPLE 3.4

Find the cube roots of 1.

Solution: Use the formulas given above with $n = 3$. The cube roots are

$$\omega^0 = 1,$$

$$\omega = \cos\frac{2\pi}{3} + i\sin\frac{2\pi}{3},$$

$$\omega^2 = \cos\frac{4\pi}{3} + i\sin\frac{4\pi}{3}\cdot$$

Answer: $1, -\frac{1}{2}+\frac{\sqrt{3}}{2}i, -\frac{1}{2}-\frac{\sqrt{3}}{2}i.$

As a check, cube the complex numbers in the answer by the Binomial Theorem.

Not only the number 1, but any (non-zero) complex number α has complex n-th roots, exactly n of them.

Each non-zero complex number α has exactly n complex n-th roots. If

$$\alpha = r(\cos\theta + i\sin\theta),$$

set

$$\beta = r^{1/n}\left(\cos\frac{\theta}{n} + i\sin\frac{\theta}{n}\right).$$

Then the n-th roots of α are

$$\beta, \beta\omega, \beta\omega^2, \cdots, \beta\omega^{n-1},$$

where $1, \omega, \cdots, \omega^{n-1}$ are the n-th roots of unity. In polar form,

$$\beta\omega^k = r^{1/n}\left[\cos\left(\frac{\theta}{n}+\frac{2\pi k}{n}\right) + i\sin\left(\frac{\theta}{n}+\frac{2\pi k}{n}\right)\right].$$

It follows that the n-th roots of α are equally spaced points on the circle of radius $|\alpha|^{1/n}$ centered at the origin.

The assertion is easily verified. By De Moivre's Theorem

$$\beta^n = r(\cos\theta + i\sin\theta) = \alpha.$$

Hence β is an n-th root of α. Furthermore,

$$(\beta\omega^k)^n = \beta^n \cdot \omega^{kn} = \alpha \cdot 1,$$

so $\beta, \beta\omega, \beta\omega^2, \cdots, \beta\omega^{n-1}$ are n-th roots of α. There are no others; for if γ is an n-th root of α, then

$$\left(\frac{\gamma}{\beta}\right)^n = \frac{\alpha}{\alpha} = 1.$$

Hence γ/β is an n-th root of unity; $\gamma/\beta = \omega^k$ for some k. Therefore $\gamma = \beta\omega^k$.

In practice, we compute n-th roots from the above formula for $\beta\omega^k$, not by actually multiplying β and ω^k.

EXAMPLE 3.5

Find the cube roots of $\frac{27}{2}(1 + i\sqrt{3})$.

Solution: The polar form of this number is

$$27(\cos 60° + i \sin 60°).$$

Apply the above formula for n-th roots with $n = 3$ and $\theta = 60°$. The three cube roots are

$$\begin{aligned} \beta &= 3(\cos 20° + i \sin 20°), \\ \beta\omega &= 3[\cos(20 + 120)° + i \sin(20 + 120)°], \\ \beta\omega^2 &= 3[\cos(20 + 240)° + i \sin(20 + 240)°]. \end{aligned}$$

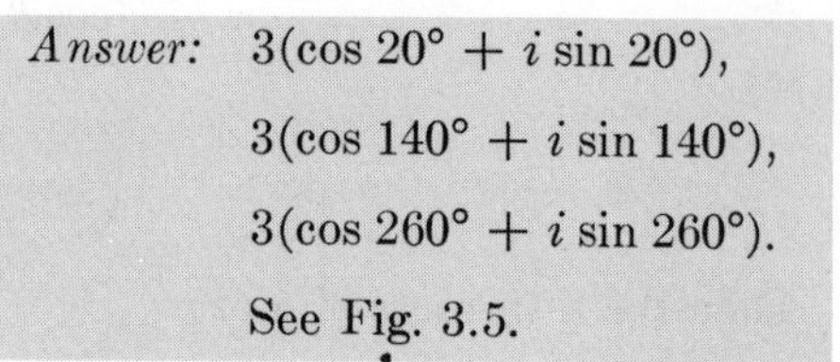

Answer: $3(\cos 20° + i \sin 20°)$,

$3(\cos 140° + i \sin 140°)$,

$3(\cos 260° + i \sin 260°)$.

See Fig. 3.5.

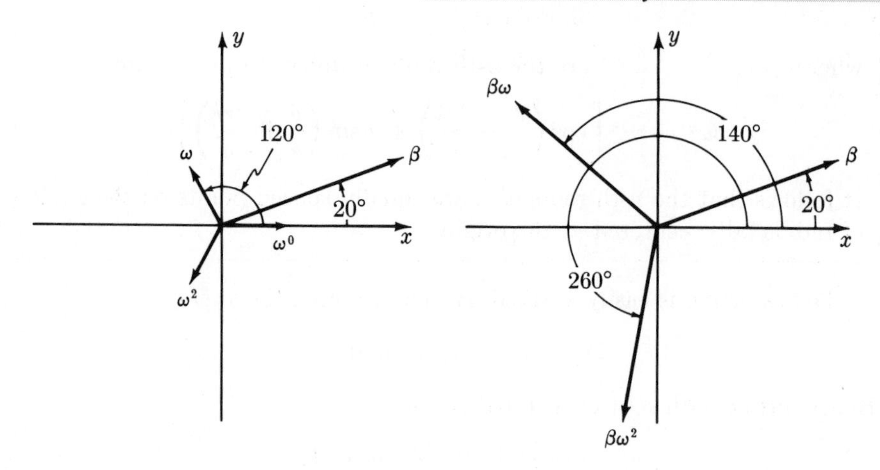

FIG. 3.5

EXERCISES

Express in polar form:

1. $-1 + i$
2. $-1 - i$
3. $1 + i\sqrt{3}$
4. $3 - 2i$
5. -4
6. $-3i$
7. $1 + 4i$
8. $4 - 3i$.

Multiply and express the answer in rectangular form:

9. $\left(\cos \frac{\pi}{6} + i \sin \frac{\pi}{6}\right)\left(\cos \frac{\pi}{3} + i \sin \frac{\pi}{3}\right)$

10. $(\cos 17° + i \sin 17°)(\cos 208° + i \sin 208°)$
11. $(\cos 135° + i \sin 135°)(1 + i)$
12. $(\cos 17° - i \sin 17°)(\cos 197° + i \sin 197°)$.

Express in polar form:

13. $\dfrac{1+i}{1-i}$

14. $\dfrac{1-i}{-1+i\sqrt{3}}$

15. $\dfrac{i\sqrt{3}}{-1-i\sqrt{3}}$

16. $\dfrac{5}{\cos\frac{\pi}{5} + i \sin\frac{\pi}{5}}$.

Compute by De Moivre's Theorem:

17. $(1-i)^4$

18. $(-1+i)^6$

19. $(\sqrt{3}-i)^7$

20. $\left[\dfrac{1+i\sqrt{3}}{2}\right]^{11}$

21. $\dfrac{1}{(1+i)^3}$

22. $\left[\dfrac{1+i}{\sqrt{2}}\right]^{1000}$.

23. Verify De Moivre's Theorem for $n = 2, 3, 4, 5$.
24. Find the 5-th roots of unity.
25. Find the 8-th roots of unity.
26. Find the 5-th roots of $(1 + i)$.
27. Find the 6-th roots of -1.
28. Find the 7-th roots of $-i$.
29. Find the 10-th roots of i.
30. Examine the 8-th roots of unity, $1, \omega, \omega^2, \cdots$. One root has the value 1, another (which?) is a square root of unity, and two others (which?) are 4-th roots of unity. The remaining four (which?) are called **primitive** 8-th roots of unity.
31. (cont.) Similarly analyze 6-th roots, 9-th roots, and 10-th roots of unity. Which are the primitive ones?
32. Show that each 5-th root of unity other than 1 satisfies $x^4 + x^3 + x^2 + x + 1 = 0$. (*Hint:* Factor $x^5 - 1$.)
33. Factor $x^6 - 1$. Show that each primitive 6-th root of unity satisfies $x^2 - x + 1 = 0$.
34. Set $\alpha = \dfrac{-1 \pm \sqrt{5}}{2}$ and $\omega = \dfrac{\alpha \pm \sqrt{\alpha^2 - 4}}{2}$. Show that $\omega^5 = 1$. (*Hint:* Use Ex. 32.)
35. Verify that De Moivre's Theorem holds also for exponents which are negative integers.
36. Let $1, \omega, \omega^2, \cdots, \omega^{n-1}$ be the n-th roots of unity. Show that $1 + \omega + \omega^2 + \cdots + \omega^{n-1} = 0$.

37. (cont.) Show that

$$1 + \cos\frac{2\pi}{n} + \cos\frac{4\pi}{n} + \cdots + \cos\frac{2(n-1)\pi}{n} = 0,$$

$$\sin\frac{2\pi}{n} + \sin\frac{4\pi}{n} + \cdots + \sin\frac{2(n-1)\pi}{n} = 0.$$

38. Prove that $\cos 3\theta = 4\cos^3\theta - 3\cos\theta$. (*Hint:* See Example 3.3.)

39. Express $\cos 4\theta$ in terms of $\cos\theta$.

4. COMPLEX EXPONENTIALS

Let us try to give a meaning to e^z for complex numbers z. The exponential series

$$e^x = 1 + x + \frac{x^2}{2!} + \cdots + \frac{x^n}{n!} + \cdots$$

converges for all real values of x. We now boldly substitute iy for x:

$$\begin{aligned} e^{iy} &= 1 + iy + \frac{(iy)^2}{2!} + \frac{(iy)^3}{3!} + \cdots \\ &= 1 + iy - \frac{y^2}{2!} - \frac{iy^3}{3!} + \frac{y^4}{4!} + \frac{iy^5}{5!} - \cdots \\ &= \left(1 - \frac{y^2}{2!} + \frac{y^4}{4!} - \frac{y^6}{6!} + \cdots\right) + i\left(y - \frac{y^3}{3!} + \frac{y^5}{5!} - \cdots\right). \end{aligned}$$

The quantities in parentheses are familiar; they are the series for $\cos y$ and $\sin y$. These observations suggest the following definition:

For any complex number of the form iy with y real, define

$$e^{iy} = \cos y + i \sin y.$$

For a general complex number $z = x + iy$, define

$$e^z = e^x \cdot e^{iy} = e^x(\cos y + i \sin y).$$

If this is to be a reasonable definition of e^z, the usual rules of exponents should hold. Let us verify that $e^{z_1} \cdot e^{z_2} = e^{z_1+z_2}$. Suppose $z_1 = x_1 + iy_1$ and $z_2 = x_2 + iy_2$. Then

$$\begin{aligned} e^{z_1} \cdot e^{z_2} &= e^{x_1}(\cos y_1 + i \sin y_1) \cdot e^{x_2}(\cos y_2 + i \sin y_2) \\ &= e^{x_1} \cdot e^{x_2}(\cos y_1 + i \sin y_1)(\cos y_2 + i \sin y_2) \\ &= e^{x_1+x_2}[\cos(y_1 + y_2) + i \sin(y_1 + y_2)] \\ &= e^{x_1+x_2+i(y_1+y_2)} = e^{(x_1+iy_1)+(x_2+iy_2)} = e^{z_1+z_2}, \end{aligned}$$

which completes the verification.

We observe that

$$e^{-iy} = \cos(-y) + i\sin(-y) = \cos y - i\sin y.$$

Hence e^{-iy} is both the reciprocal and the conjugate of e^{iy}. This is not surprising since $|e^{iy}| = 1$, and, as shown in Example 2.2, if $|z| = 1$, then $1/z = \bar{z}$.

EXAMPLE 4.1

Find the value of:

(1) e^{i}, (2) $e^{2\pi i}$, (3) $e^{\pi i}$, (4) $e^{\ln 2+(\pi i/4)}$, (5) $e^{2\pi i/n}$.

Solution: In each case, use the definition

$$e^{x+iy} = e^{x}(\cos y + i\sin y).$$

(1) $e^{i} = \cos 1 + i\sin 1 \quad (1 \text{ rad}).$

(2) $e^{2\pi i} = \cos 2\pi + i\sin 2\pi = 1.$

(3) $e^{\pi i} = \cos \pi + i\sin \pi = -1.$

(4) $e^{\ln 2+\pi i/4} = e^{\ln 2}\left(\cos\frac{\pi}{4} + i\sin\frac{\pi}{4}\right) = 2\left(\frac{\sqrt{2}}{2} + i\frac{\sqrt{2}}{2}\right) = \sqrt{2}\,(1+i).$

(5) $e^{2\pi i/n} = \cos\frac{2\pi}{n} + i\sin\frac{2\pi}{n}\cdot$

REMARK: From the answer to (5) we recognize that $e^{2\pi i/n}$ is an n-th root of unity. This agrees with the answer to (2) since

$$(e^{2\pi i/n})^{n} = e^{2\pi i} = 1.$$

Trigonometric Functions

From the basic relations

$$e^{i\theta} = \cos\theta + i\sin\theta,$$
$$e^{-i\theta} = \cos\theta - i\sin\theta,$$

follow two important formulas:

$$\cos\theta = \frac{e^{i\theta} + e^{-i\theta}}{2} = \operatorname{Re}(e^{i\theta}),$$

$$\sin\theta = \frac{e^{i\theta} - e^{-i\theta}}{2i} = \operatorname{Im}(e^{i\theta}).$$

These formulas are extremely useful and are worth memorizing. They convert problems about trigonometric functions into problems about exponentials which are often simpler to handle.

EXAMPLE 4.2

Derive the identity

$$1 + 2\cos\theta + 2\cos 2\theta + \cdots + 2\cos n\theta = \frac{\sin\left(n + \frac{1}{2}\right)\theta}{\sin\frac{1}{2}\theta}.$$

Solution: Let C_n denote the left-hand side of the identity,

$$\begin{aligned} C_n &= 1 + 2\cos\theta + 2\cos 2\theta + \cdots + 2\cos n\theta \\ &= 1 + (e^{i\theta} + e^{-i\theta}) + (e^{2i\theta} + e^{-2i\theta}) + \cdots + (e^{ni\theta} + e^{-ni\theta}). \end{aligned}$$

Rearrange:

$$\begin{aligned} C_n &= e^{-ni\theta} + e^{-(n-1)i\theta} + \cdots + e^{-i\theta} + 1 + e^{i\theta} + \cdots + e^{ni\theta} \\ &= e^{-ni\theta}(1 + e^{i\theta} + e^{2i\theta} + \cdots + e^{2ni\theta}). \end{aligned}$$

The expression in parentheses is a finite geometric series whose sum is known:

$$C_n = e^{-ni\theta} \cdot \frac{1 - e^{(2n+1)i\theta}}{1 - e^{i\theta}} = \frac{e^{-ni\theta} - e^{(n+1)i\theta}}{1 - e^{i\theta}}.$$

Here is an important trick. Remembering that $e^{i\alpha} - e^{-i\alpha} = 2i\sin\alpha$, multiply numerator and denominator by $e^{-i\theta/2}$:

$$C_n = \frac{e^{-(n+\frac{1}{2})i\theta} - e^{(n+\frac{1}{2})i\theta}}{e^{-i\theta/2} - e^{i\theta/2}} = \frac{-2i\sin\left(n + \frac{1}{2}\right)\theta}{-2i\sin\frac{\theta}{2}} = \frac{\sin\left(n + \frac{1}{2}\right)\theta}{\sin\frac{\theta}{2}}.$$

Hyperbolic Functions

There is a close connection between trigonometric and hyperbolic functions. Recall the definitions

$$\cosh x = \frac{e^x + e^{-x}}{2}, \qquad \sinh x = \frac{e^x - e^{-x}}{2}.$$

Formally replacing x by iy, we find

$$\cosh iy = \cos y, \qquad \sinh iy = i\sin y.$$

In order to give meaning to these identities, we define $\sin z$, $\cos z$, $\sinh z$, and $\cosh z$ for complex numbers z.

If z is a complex number,

$$\cos z = \frac{e^{iz} + e^{-iz}}{2}, \qquad \sin z = \frac{e^{iz} - e^{-iz}}{2i}.$$

Similarly,

$$\cosh z = \frac{e^{z} + e^{-z}}{2}, \qquad \sinh z = \frac{e^{z} - e^{-z}}{2}.$$

Thus $\cos z$ and $\cosh z$ are closely related functions, as are $\sin z$ and $\sinh z$. Directly from the definitions, one obtains the basic relations:

$$\cos z = \cosh iz, \qquad \sin z = \frac{1}{i}\sinh iz = -i\sinh iz,$$

valid for all complex numbers z. Equivalent relations are obtained on replacing z by iw:

$$\cosh w = \cos iw, \qquad i\sinh w = \sin iw,$$

valid for all complex numbers w.

It is proved in more advanced courses that each identity involving sines and cosines, valid for all real values of the variables, is valid also for all complex values of the variables. (For example, the identity

$$\cos(z + w) = \cos z \cos w - \sin z \sin w$$

holds for all complex values of z and w as well as for all real values.) In particular, in any identity involving sines and cosines of variables $z, w, \cdots$, the variables may be replaced by $iz, iw, \cdots$. It follows that each identity involving sines and cosines has a counterpart involving hyperbolic functions.

EXAMPLE 4.3

Show that $\cosh(z + w) = \cosh z \cosh w + \sinh z \sinh w$.

Solution:

$$\begin{aligned}\cosh(z + w) &= \cos i(z + w) = \cos(iz + iw)\\ &= \cos iz \cos iw - \sin iz \sin iw\\ &= \cosh z \cosh w - (i\sinh z)(i\sinh w)\\ &= \cosh z \cosh w + \sinh z \sinh w.\end{aligned}$$

EXAMPLE 4.4

Find a formula for $1 + 2\cosh z + 2\cosh 2z + \cdots + 2\cosh nz$.

Solution: Use the corresponding formula for cosines found in Example 4.2:

$$1 + 2\cosh z + \cdots + 2\cosh nz = 1 + 2\cos iz + \cdots + 2\cos niz$$

$$= \frac{\sin i\left(n + \frac{1}{2}\right)z}{\sin \frac{iz}{2}}.$$

But

$$\frac{\sin i\left(n + \frac{1}{2}\right)z}{\sin \frac{iz}{2}} = \frac{i \sinh\left(n + \frac{1}{2}\right)z}{i \sinh \frac{z}{2}}.$$

Answer:

$$1 + 2\cosh z + 2\cosh 2z + \cdots + 2\cosh nz$$

$$= \frac{\sinh\left(n + \frac{1}{2}\right)z}{\sinh \frac{z}{2}}.$$

EXERCISES

1. Evaluate
   (a) $e^{\pi i/3}$
   (b) e^{1-i}
   (c) $e^{1/(1-2i)}$
   (d) $\sin\left(\frac{\pi}{4} + 2i\right)$
   (e) $\cos 3i$
   (f) $\cosh \pi i$.
2. Show that

$$\sin(x + iy) = \sin x \cosh y + i \sinh y \cos x,$$
$$\cos(x + iy) = \cos x \cosh y - i \sin x \sinh y.$$

3. Find all complex numbers z for which
   (a) $\sin z$ is real
   (b) $\cos z = 0$.
4. Show that for all complex numbers z,
   (a) $e^{z+2\pi i} = e^z$
   (b) $\sin(z + 2\pi) = \sin z$
   (c) $\cos(z + 2\pi) = \cos z$
   (d) $\sinh(z + 2\pi i) = \sinh z$
   (e) $\cosh(z + 2\pi i) = \cosh z$.
5. Is $|\cos z| \le 1$ for all complex numbers z?

6. Derive the trigonometric identities:

(a) $\cos^4 x = \dfrac{1}{2^4}[\cos 4x + 4\cos 2x + 6 + 4\cos(-2x) + \cos(-4x)]$

(b) $\cos^6 x = \dfrac{1}{2^6}[\cos 6x + 6\cos 4x + 15\cos 2x + 20$
$+ 15\cos(-2x) + 6\cos(-4x) + \cos(-6x)].$

(*Hint:* Write $\cos x = \frac{1}{2}(e^{ix} + e^{-ix})$ and use the Binomial Theorem.) Guess a general formula for $\cos^{2n} x$.

7. Derive the identity

$$1 + 2\cosh z + 2\cosh 2z + \cdots + 2\cosh nz = \frac{\sinh\left(n + \frac{1}{2}\right)z}{\sinh \frac{z}{2}}$$

by converting the left-hand side into a sum of exponentials.

8. Define $\tanh z = \dfrac{\sinh z}{\cosh z}$, $\operatorname{sech} z = \dfrac{1}{\cosh z}\cdot$ Find a relation between $\tanh^2 z$ and $\operatorname{sech}^2 z$.

9. Express $\cosh 4z$ in terms of $\cosh z$.
(*Hint:* Express $\cos 4z$ in terms of $\cos z$.)

10. Derive the hyperbolic identities:
(a) $\sinh 2z = 2\sinh z \cosh z$ (b) $\cosh 2z = 2\cosh^2 z - 1$
(c) $\sinh 3z = 4\sinh^3 z + 3\sinh z$.

5. INTEGRATION AND DIFFERENTIATION

In this section we deal with functions having complex values, for example,

$$f(x) = e^{ix} = \cos x + i \sin x.$$

Such a function can be written as

$$f(x) = u(x) + i\, v(x),$$

where $u(x)$ and $v(x)$ are real-valued functions.

If $f(x) = u(x) + i\, v(x)$, then the derivative of $f(x)$ is defined by

$$f'(x) = u'(x) + i\, v'(x).$$

Similarly,

$$\int f(x)\, dx = \int u(x)\, dx + i \int v(x)\, dx.$$

Many formulas for differentiation and integration extend to complex-valued functions. We shall consider just one case, complex exponentials.

If α is a complex number, then

$$\frac{d}{dx}(e^{\alpha x}) = \alpha e^{\alpha x}$$

and

$$\int e^{\alpha x}\,dx = \frac{1}{\alpha}e^{\alpha x} + C, \qquad \alpha \neq 0.$$

Let us verify the first formula; the second formula follows from it. Suppose $\alpha = a + bi$. Then

$$\frac{d}{dx}(e^{\alpha x}) = \frac{d}{dx}[e^{(a+bi)x}] = \frac{d}{dx}[e^{ax}(\cos bx + i\sin bx)]$$

$$= \frac{d}{dx}(e^{ax}\cos bx) + i\frac{d}{dx}(e^{ax}\sin bx).$$

By ordinary differentiation,

$$\frac{d}{dx}(e^{\alpha x}) = (ae^{ax}\cos bx - be^{ax}\sin bx) + i(ae^{ax}\sin bx + be^{ax}\cos bx)$$

$$= ae^{ax}(\cos bx + i\sin bx) + ibe^{ax}(\cos bx + i\sin bx)$$

$$= (a + bi)e^{ax}(\cos bx + i\sin bx) = \alpha e^{\alpha x}.$$

EXAMPLE 5.1

Evaluate $\int e^{ax}\cos bx\,dx$.

Solution: This can be done using integration by parts twice. It is easier, however, using complex exponentials. From

$$e^{(a+bi)x} = e^{ax}\cos bx + ie^{ax}\sin bx$$

follows

$$\int e^{(a+bi)x}\,dx = \int e^{ax}\cos bx\,dx + i\int e^{ax}\sin bx\,dx.$$

Therefore the desired integral is the real part of the integral on the left. By the preceding rule,

$$\int e^{(a+bi)x}\,dx = \frac{e^{(a+bi)x}}{a+bi} + C.$$

To find the real part, write

$$\frac{e^{(a+bi)x}}{a+bi} = \frac{e^{ax}(\cos bx + i\sin bx)}{a+bi}\cdot\frac{a-bi}{a-bi}$$

$$= \frac{e^{ax}[(a\cos bx + b\sin bx) + i(a\sin bx - b\cos bx)]}{a^2+b^2}.$$

Answer: $\dfrac{e^{ax}(a\cos bx + b\sin bx)}{a^2+b^2} + C.$

REMARK: By comparing imaginary parts, we get free of charge the formula

$$\int e^{ax}\sin bx\,dx = \frac{e^{ax}(a\sin bx - b\cos bx)}{a^2+b^2} + C.$$

EXAMPLE 5.2

Compute $\displaystyle\int_0^{2\pi} e^{inx}\,dx$, where n is a non-zero integer.

Solution:

$$\int_0^{2\pi} e^{inx}\,dx = \int_0^{2\pi}\cos nx\,dx + i\int_0^{2\pi}\sin nx\,dx = \frac{\sin nx}{n}\bigg|_0^{2\pi} - i\,\frac{\cos nx}{n}\bigg|_0^{2\pi} = 0.$$

Alternate Solution: An antiderivative of e^{inx} is $e^{inx}/(in)$. Hence

$$\int_0^{2\pi} e^{inx}\,dx = \frac{1}{in}\,e^{inx}\bigg|_0^{2\pi} = \frac{1}{in}\,(e^{2\pi in} - e^0) = \frac{1}{in}\,(1-1) = 0.$$

(Remember, $e^{2\pi i} = 1$.)

Answer: 0.

EXAMPLE 5.3

Compute $\displaystyle\int_0^{2\pi}\sin kx\cos nx\,dx$, $\quad k, n$ positive integers.

Solution: Write

$$\int_0^{2\pi}\sin kx\cos nx\,dx = \int_0^{2\pi}\left(\frac{e^{ikx}-e^{-ikx}}{2i}\right)\left(\frac{e^{inx}+e^{-inx}}{2}\right)dx$$

$$= \frac{1}{4i}\left[\int_0^{2\pi} e^{i(k+n)x}\,dx - \int_0^{2\pi} e^{-i(k+n)x}\,dx\right.$$

$$\left. + \int_0^{2\pi} e^{i(k-n)x}\,dx - \int_0^{2\pi} e^{-i(k-n)x}\,dx\right].$$

Now use the result of the last example. If $k \neq n$, then all four integrals on the right are 0. If $k = n$, the first two integrals are 0, the third and fourth cancel.

Answer: 0.

EXAMPLE 5.4

Evaluate $\int x \cos^3 x \, dx$.

Solution: Write

$$x \cos^3 x = x \left(\frac{e^{ix} + e^{-ix}}{2} \right)^3 = \frac{x}{8} (e^{3ix} + 3e^{2ix}e^{-ix} + 3e^{ix}e^{-2ix} + e^{-3ix})$$

$$= \frac{x}{8} (e^{3ix} + 3e^{ix} + 3e^{-ix} + e^{-3ix}).$$

Hence,

$$\int x \cos^3 x \, dx$$

$$= \frac{1}{8} \left(\int xe^{3ix} \, dx + 3 \int xe^{ix} \, dx + 3 \int xe^{-ix} \, dx + \int xe^{-3ix} \, dx \right)$$

$$= \frac{1}{8} (I_3 + 3I_1 + 3I_{-1} + I_{-3}),$$

where

$$I_n = \int xe^{inx} \, dx.$$

To evaluate I_n, integrate by parts:

$$I_n = \int xe^{inx} \, dx = \int x \, d\left(\frac{e^{inx}}{in} \right) = \frac{xe^{inx}}{in} - \int \frac{e^{inx}}{in} \, dx = \frac{xe^{inx}}{in} - \frac{e^{inx}}{(in)^2}$$

$$= e^{inx} \left(\frac{1}{n^2} + \frac{x}{in} \right) = \frac{e^{inx}}{n^2} (1 - inx).$$

Notice that I_n and I_{-n} are conjugates. Hence

$$I_n + I_{-n} = 2 \operatorname{Re}(I_n) = 2 \operatorname{Re} \left[\frac{(\cos nx + i \sin nx)(1 - inx)}{n^2} \right]$$

$$= \frac{2}{n^2} (\cos nx + nx \sin nx).$$

Now the answer follows easily:

$$\int x \cos^3 x \, dx = \frac{1}{8} (I_3 + I_{-3}) + \frac{3}{8} (I_1 + I_{-1}) + C$$

$$= \frac{1}{8} \cdot \frac{2}{3^2} (\cos 3x + 3x \sin 3x) + \frac{3}{8} \cdot 2(\cos x + x \sin x) + C.$$

Answer:

$$\frac{\cos 3x}{36} + \frac{x \sin 3x}{12} + \frac{3 \cos x}{4} + \frac{3x \sin x}{4} + C.$$

EXERCISES

1. Compute the 12-th derivative of $e^{x\sqrt{3}/2}\left(\cos\frac{x}{2} + i\sin\frac{x}{2}\right)$.

Evaluate the integral using complex exponentials:

2. $\displaystyle\int \cos x \cosh x\, dx$

3. $\displaystyle\int e^x \cos^4 x\, dx$

4. $\displaystyle\int_0^{\pi} \cos^{2n} x\, dx$

5. $\displaystyle\int_0^{2\pi} \sin kx \sin nx\, dx$, k, n positive integers

6. $\displaystyle\int_0^{2\pi} \cos kx \cos nx\, dx$, k, n positive integers.

7. $\displaystyle\int x \sin^3 x\, dx$

8. $\displaystyle\int x \cos^2 x\, dx$

9. Compute $\displaystyle\int_0^{2\pi} (a_0 + a_1 \cos x + a_2 \cos 2x + \cdots + a_n \cos nx)^2\, dx$.

   (*Hint:* Use Ex. 6.)

6. APPLICATIONS TO DIFFERENTIAL EQUATIONS

In Chapter 14 we developed a systematic method for solving the linear differential equation

$$ay'' + by' + cy = 0.$$

Now we present another approach, via complex exponentials. The basic idea is this:

> If $y(x) = u(x) + iv(x)$ is a complex-valued function that satisfies the differential equation (with real coefficients)
>
> $$ay'' + by' + cy = 0,$$
>
> then $u(x)$ and $v(x)$ also satisfy the equation. In other words, the real and the imaginary parts of a solution are also solutions.

Let us verify this statement. We are given

$$ay'' + by' + cy = 0.$$

Hence

$$a(u'' + iv'') + b(u' + iv') + c(u + iv) = 0,$$

that is,

$$(au'' + bu' + cu) + i(av'' + bv' + cv) = 0.$$

Since the left-hand side equals 0, so do its real and imaginary parts:

$$au'' + bu' + cu = 0 \qquad \text{and} \qquad av'' + bv' + cv = 0.$$

Thus $u(x)$ and $v(x)$ are solutions.

EXAMPLE 6.1

Solve $y'' + 9y = 0$.

Solution: Try $y = e^{px}$, where p is allowed to be complex:

$$y'' + 9y = p^2e^{px} + 9e^{px} = 0, \qquad p^2 + 9 = 0, \qquad p = \pm 3i.$$

Thus e^{3ix} is a complex solution. Therefore, its real and imaginary parts, $\cos 3x$ and $\sin 3x$, are also solutions. Hence, the general solution, which involves two arbitrary constants, is $c_1 \cos 3x + c_2 \sin 3x$. Note that e^{-3ix} is another complex solution, but its real and imaginary parts are $\cos 3x$ and $-\sin 3x$, nothing new. (The general solution could be written $ae^{3ix} + be^{-3ix}$.)

Answer: $y = c_1 \cos 3x + c_2 \sin 3x.$

EXAMPLE 6.2

Solve $y'' - 5y' + 6y = 0$.

Solution: Try $y = e^{px}$:

$$y'' - 5y' + 6y = p^2e^{px} - 5pe^{px} + 6e^{px} = 0,$$

$$p^2 - 5p + 6 = 0, \qquad p = 2, 3.$$

Hence e^{2x} and e^{3x} are solutions.

Answer: $y = c_1e^{2x} + c_2e^{3x}.$

EXAMPLE 6.3

Solve $y'' + 4y' + 13y = 0$.

Solution: Try $y = e^{px}$:

$$y'' + 4y' + 13y = (p^2 + 4p + 13)e^{px} = 0,$$

$$p^2 + 4p + 13 = 0, \qquad p = -2 \pm 3i.$$

Hence $y = e^{(-2+3i)x}$ is a complex solution. Its real and imaginary parts are $e^{-2x} \cos 3x$ and $e^{-2x} \sin 3x$, also solutions.

Answer: $y = e^{-2x}(c_1 \cos 3x + c_2 \sin 3x).$

Particular Solutions

Complex exponentials can be applied also to non-homogeneous equations

$$ay'' + by' + cy = f(x),$$

where the function $f(x)$ can be expressed in terms of complex exponentials.

EXAMPLE 6.4

Find a particular solution of the differential equation $y'' + 9y = \cos 2x$.

Solution: Since $\cos 2x = \mathrm{Re}(e^{2ix})$, find a complex solution of the differential equation

$$y'' + 9y = e^{2ix},$$

and take its real part. Try $y = ae^{2ix}$:

$$y'' + 9y = (2i)^2 ae^{2ix} + 9ae^{2ix} = -4ae^{2ix} + 9ae^{2ix} = e^{2ix},$$

$$5a = 1.$$

Thus $\frac{1}{5}e^{2ix}$ is a complex solution.

Answer: $\mathrm{Re}\left(\frac{1}{5}e^{2ix}\right) = \frac{1}{5}\cos 2x.$

EXAMPLE 6.5

Find a particular solution of $y'' + 4y' + 13y = \cos x$.

Solution: Since $\cos x = \mathrm{Re}(e^{ix})$, find a complex solution of

$$y'' + 4y' + 13y = e^{ix},$$

and take its real part. Try $y = ae^{ix}$:

$$y'' + 4y' + 13y = -ae^{ix} + 4aie^{ix} + 13ae^{ix} = e^{ix}.$$

Hence

$$-a + 4ai + 13a = 1, \qquad a = \frac{1}{12 + 4i} = \frac{12 - 4i}{12^2 + 4^2} = \frac{3 - i}{40}.$$

Therefore a complex solution is

$$y = \frac{1}{40}(3 - i)e^{ix} = \frac{1}{40}(3 - i)(\cos x + i \sin x).$$

Its real part is

$$\mathrm{Re}(y) = \frac{1}{40}(3 \cos x + \sin x).$$

Answer: $\frac{1}{40}(3 \cos x + \sin x).$

EXAMPLE 6.6

Find a particular solution of $y'' + 2y' + 3y = e^x \sin 2x$.

Solution: Since $e^x \sin 2x = \text{Im}(e^{(1+2i)x})$, find a complex solution of

$$y'' + 2y' + 3y = e^{(1+2i)x},$$

and take its imaginary part. Try $y = ae^{(1+2i)x}$:

$$y'' + 2y' + 3y = a[(1 + 2i)^2 + 2(1 + 2i) + 3]e^{(1+2i)x} = e^{(1+2i)x},$$

$$a[(1 + 4i + 4i^2) + (2 + 4i) + 3] = 1,$$

$$a = \frac{1}{2 + 8i} = \frac{2 - 8i}{2^2 + 8^2} = \frac{1 - 4i}{34}.$$

Thus a complex solution is

$$y = \frac{1}{34}(1 - 4i)e^{(1+2i)x} = \frac{1}{34}(1 - 4i)e^x(\cos 2x + i \sin 2x),$$

and its imaginary part is

$$\text{Im}(y) = \frac{1}{34}e^x(\sin 2x - 4 \cos 2x).$$

Answer: $\frac{1}{34}e^x(\sin 2x - 4 \cos 2x)$.

EXERCISES

Using complex exponentials, find the general real solution:

1. $y''' - y = 0$
2. $y''' + y'' + y' + y = 0$
3. $y^{(4)} - y'' - 6y = 0$
4. $y^{(4)} - y = 0$.
5. Find the general solution of $y'' - 6y' + 10y = \cos 2x$ by the method of this section.

Find a particular solution by the method of this section:

6. $y'' - a^2y = \cos bx$
7. $y'' + a^2y = x \cos bx$
8. $y'' + a^2y = e^{-bx} \cos ax$
9. $y'' + y' - 6y = 3 \sin x$
10. $L\frac{d^2I}{dt^2} + R\frac{dI}{dt} + \frac{1}{C}I = E \cos \omega t$
11. $\frac{d^2r}{d\theta^2} - \frac{dr}{d\theta} = a \sin \theta$.

7. APPLICATIONS TO POWER SERIES [optional]

Here is a puzzling fact about power series. We know the geometric series $1 + x + x^2 + \cdots$ diverges for $|x| \geq 1$. This is not surprising because it represents $1/(1 - x)$, a function which "blows up" at $x = 1$. But the power series for $1/(1 + x^2)$ also diverges for $|x| \geq 1$ even though $1/(1 + x^2)$ is perfectly well-behaved for all x. What goes wrong?

The answer to this question requires a broader view of power series. When investigating a real series

$$a_0 + a_1(x - x_0) + a_2(x - x_0)^2 + \cdots + a_n(x - x_0)^n + \cdots,$$

it often helps to regard the series as a special case of the complex series

$$a_0 + a_1(z - x_0) + a_2(z - x_0)^2 + \cdots + a_n(z - x_0)^n + \cdots.$$

(The coefficients are the same, but z is allowed to be complex.) For example, the real series

$$\frac{1}{1 + x^2} = 1 - x^2 + x^4 - x^6 + \cdots$$

is a special case of the complex series

$$\frac{1}{1 + z^2} = 1 - z^2 + z^4 - z^6 + \cdots.$$

The *complex* function $1/(1 + z^2)$ is well-behaved for all *real* values of z but "blows up" at $z = i$ and $z = -i$. Hence, its power series cannot be expected to converge for all z. In fact, the series converges only in the disk $|z| < 1$. The same is true of

$$\frac{1}{1 - z} = 1 + z + z^2 + z^3 + \cdots,$$

whose convergence is limited to the disk $|z| < 1$ because of the "bad point" at $z = 1$. In both examples, the series fail to converge for $|z| > 1$ due to troublesome points on the circle $|z| = 1$.

In general, the complex power series

$$a_0 + a_1(z - z_0) + a_2(z - z_0)^2 + \cdots$$

converges to the function $f(z)$ in a region S of the complex plane if

$$|f(z) - [a_0 + a_1(z - z_0) + \cdots a_n(z - z_0)^n]| \longrightarrow 0$$

as $n \longrightarrow \infty$ for each point z of S. Formally, this definition is the same as the corresponding definition for real power series.

The basic fact about convergence of complex power series is this:

> Given a power series
>
> $$a_0 + a_1(z - z_0) + a_2(z - z_0)^2 + \cdots + a_n(z - z_0)^n + \cdots,$$
>
> precisely one of three cases holds:
>
> (i) The series converges only for $z = z_0$.
>
> (ii) The series converges for all values of z.
>
> (iii) There is a positive number R such that the series converges for each z satisfying $|z - z_0| < R$ and diverges for each z satisfying $|z - z_0| > R$.

The proof of the corresponding fact for real power series (p. 75) applies almost without change.

In case (iii), the series converges in the circle (Fig. 7.1) with center at z_0 and radius R. The number R is called the **radius of convergence.**

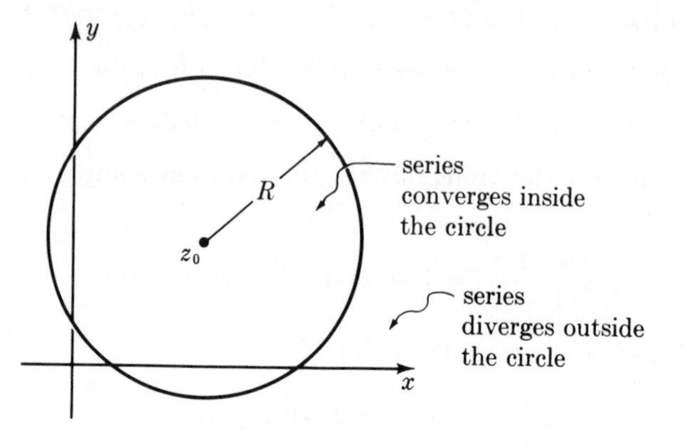

FIG. 7.1

The real power series

$$a_0 + a_1(x - x_0) + \cdots + a_n(x - x_0)^n + \cdots$$

is a special case of the complex power series

$$a_0 + a_1(z - x_0) + \cdots + a_n(z - x_0)^n + \cdots .$$

The latter converges in a circle of radius R centered at x_0 on the real axis (Fig. 7.2). The real series converges on that part of the real axis contained in this circle, namely, an interval (diameter of the circle of convergence). Half of this interval is a radius of the circle; hence the term "radius of convergence" as applied to real series.

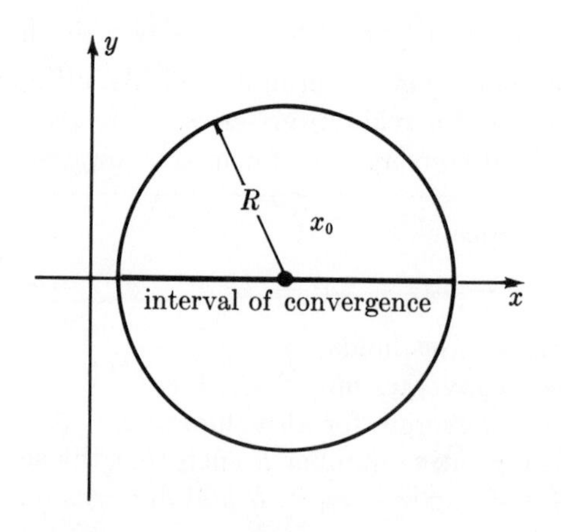

FIG. 7.2

It is shown in more advanced courses that the power series for $f(z)$ at $z = z_0$ converges inside the largest circle centered at z_0 in which $f(z)$ and all its derivatives are well-behaved. For example, take $f(z) = 1/(1 - z)$, which has one bad point, at $z = 1$. Its power series at $z = i$ converges in the circle shown in Fig. 7.3.

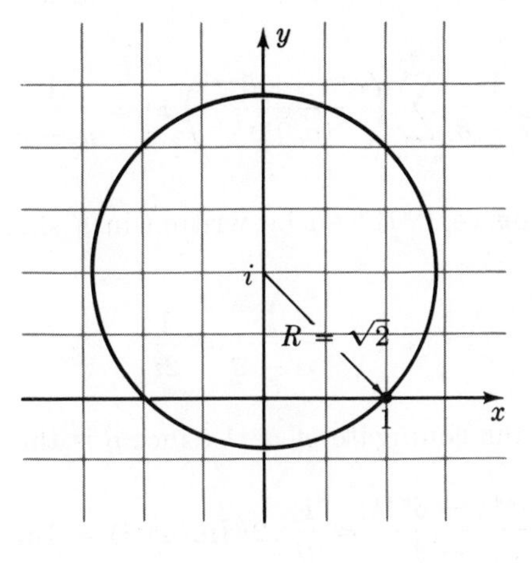

FIG. 7.3

It is known that the ratio test applies to complex power series as well as to real ones.

Partial Fractions

Complex power series have applications to problems involving real functions. An example is the power series expansion by partial fractions, where the denominator has complex zeros.

EXAMPLE 7.1

Find the power series for $\dfrac{1}{x^2 - 6x + 10}$ at $x = 0$. What is its radius of convergence?

Solution: First, factor the denominator. Since the equation

$$x^2 - 6x + 10 = 0$$

has roots $3 + i$ and $3 - i$,

$$x^2 - 6x + 10 = [x - (3 + i)]\,[x - (3 - i)].$$

Now use partial fractions. (Use abbreviations $\alpha = 3 + i$ and $\beta = 3 - i$. Note that $\alpha\beta = 10$.)

$$\frac{1}{x^2 - 6x + 10} = \frac{1}{(x-\alpha)(x-\beta)} = \frac{1}{\alpha-\beta}\left(-\frac{1}{x-\beta} + \frac{1}{x-\alpha}\right)$$

$$= \frac{1}{\alpha-\beta}\left(\sum_{n=0}^{\infty}\frac{x^n}{\beta^{n+1}} - \sum_{n=0}^{\infty}\frac{x^n}{\alpha^{n+1}}\right) = \frac{1}{\alpha-\beta}\sum_{n=0}^{\infty}\left(\frac{1}{\beta^{n+1}} - \frac{1}{\alpha^{n+1}}\right)x^n$$

$$= \frac{1}{\alpha-\beta}\sum_{n=0}^{\infty}\left(\frac{\alpha^{n+1}-\beta^{n+1}}{(\alpha\beta)^{n+1}}\right)x^n = \frac{1}{\alpha-\beta}\sum_{n=0}^{\infty}\left(\frac{\alpha^{n+1}-\beta^{n+1}}{10^{n+1}}\right)x^n.$$

This is the answer. It can be written in a slightly different form by observing that

$$\frac{1}{\alpha-\beta} = \frac{1}{2i},$$

and that β^{n+1} is the conjugate of α^{n+1} (since β is the conjugate of α). Thus

$$\frac{\alpha^{n+1}-\beta^{n+1}}{\alpha-\beta} = \frac{1}{2i}\cdot 2i\,\mathrm{Im}(\alpha^{n+1}) = \mathrm{Im}(\alpha^{n+1}).$$

The series converges inside the largest circle centered at the origin in which $1/(x^2 - 6x + 10)$ is well-behaved. This circle passes through $3 + i$ and $3 - i$, the two points where the function "blows up." Its radius is $|3 + i| = |3 - i| = \sqrt{10}$.

Answer: $$\frac{1}{x^2 - 6x + 10} = \sum_{n=0}^{\infty}\frac{1}{10^{n+1}}\mathrm{Im}(\alpha^{n+1})x^n,$$

$\alpha = 3 + i; \quad R = \sqrt{10}.$

REMARK: Let us write out a few terms of the preceding series:

$\alpha = 3 + i \qquad \mathrm{Im}(\alpha) = 1$

$\alpha^2 = (3+i)^2 = 3^2 + 6i + i^2 \qquad \mathrm{Im}(\alpha^2) = 6$

$\alpha^3 = (3+i)^3 = 3^3 + 3\cdot 3^2 i + 3\cdot 3i^2 + i^3 \qquad \mathrm{Im}(\alpha^3) = 3\cdot 3^2 - 1 = 26$

$\alpha^4 = (3+i)^4 = 3^4 + 4\cdot 3^3 i + 6\cdot 3^2 i^2 + 4\cdot 3i^3 + i^4 \qquad \mathrm{Im}(\alpha^4) = 4\cdot 3^3 - 4\cdot 3 = 96.$

Hence

$$\frac{1}{x^2 - 6x + 10} = \frac{1}{10} + \frac{6}{10^2}x + \frac{26}{10^3}x^2 + \frac{96}{10^4}x^3 + \cdots.$$

As a quick check, let us try $x = 1$ (permissible since the series converges for $|x| < \sqrt{10}$). The left side is then $\frac{1}{5} = 0.2000$; the sum of the first four terms on the right side is

$$0.1000 + 0.0600 + 0.0260 + 0.0096 = 0.1956.$$

EXERCISES

1. Expand $1/(1 - z)$ in powers of $z - i$.
[*Hint:* Write $(1 - z) = (1 - i) - (z - i)$.]

Find the power series at $z = 0$ and write out the first four terms explicitly. What is the radius of convergence?

2. $\dfrac{1}{z^2 + z + 1}$

3. $\dfrac{1}{z^2 + 4z + 13}$

4. $\dfrac{z + 1}{z^2 - z + 1}$

5. $\dfrac{3z + 5z^2}{1 - 2z + 5z^2}$.

6. Find the sum of the series $1 + r \cos \theta + r^2 \cos 2\theta + \cdots + r^n \cos n\theta + \cdots$, $-1 < r < 1$.
[*Hint:* $r^n \cos n\theta = \operatorname{Re}(z^n)$, where $z = re^{i\theta}$.]

7. If $z = x + iy$, prove $\dfrac{\partial}{\partial x}(z^n) = nz^{n-1}$, $\quad \dfrac{\partial}{\partial y}(z^n) = niz^{n-1}$.

8. Write $z^n = u(x, y) + i\, v(x, y)$, in real and imaginary parts. Compute u and v for $n = \pm 1, \pm 2$, and 3.

9. (cont.) Show, using Ex. 7, that $\dfrac{\partial u}{\partial x} = \dfrac{\partial v}{\partial y}$, $\quad \dfrac{\partial u}{\partial y} = -\dfrac{\partial v}{\partial x}$.

10. (cont.) Show that $\dfrac{\partial^2 u}{\partial x^2} + \dfrac{\partial^2 u}{\partial y^2} = 0$, $\quad \dfrac{\partial^2 v}{\partial x^2} + \dfrac{\partial^2 v}{\partial y^2} = 0$.

11. (cont.) Let $f(z) = a_0 + a_1 z + a_2 z^2 + \cdots$ be the sum of a complex power series. Separate real and imaginary parts: $f(z) = u(x, y) + i\, v(x, y)$.
Show that

$$\frac{\partial u}{\partial x} = \frac{\partial v}{\partial y}, \qquad \frac{\partial u}{\partial y} = -\frac{\partial v}{\partial x},$$

$$\frac{\partial^2 u}{\partial x^2} + \frac{\partial^2 u}{\partial y^2} = 0, \qquad \frac{\partial^2 v}{\partial x^2} + \frac{\partial^2 v}{\partial y^2} = 0.$$

TABLE 1 Trigonometric Functions

Degrees	sin	cos	tan	cot	
0°	.0000	1.000	.0000	—	90°
1°	.0175	.9998	.0175	57.29	89°
2°	.0349	.9994	.0349	28.64	88°
3°	.0523	.9986	.0524	19.08	87°
4°	.0698	.9976	.0699	14.30	86°
5°	.0872	.9962	.0875	11.43	85°
6°	.1045	.9945	.1051	9.514	84°
7°	.1219	.9925	.1228	8.144	83°
8°	.1392	.9903	.1405	7.115	82°
9°	.1564	.9877	.1584	6.314	81°
10°	.1736	.9848	.1763	5.671	80°
11°	.1908	.9816	.1944	5.145	79°
12°	.2079	.9781	.2126	4.705	78°
13°	.2250	.9744	.2309	4.331	77°
14°	.2419	.9703	.2493	4.011	76°
15°	.2588	.9659	.2679	3.732	75°
16°	.2756	.9613	.2867	3.487	74°
17°	.2924	.9563	.3057	3.271	73°
18°	.3090	.9511	.3249	3.078	72°
19°	.3256	.9455	.3443	2.904	71°
20°	.3420	.9397	.3640	2.747	70°
21°	.3584	.9336	.3839	2.605	69°
22°	.3746	.9272	.4040	2.475	68°
23°	.3907	.9205	.4245	2.356	67°
24°	.4067	.9135	.4452	2.246	66°
25°	.4226	.9063	.4663	2.145	65°
26°	.4384	.8988	.4877	2.050	64°
27°	.4540	.8910	.5095	1.963	63°
28°	.4695	.8829	.5317	1.881	62°
29°	.4848	.8746	.5543	1.804	61°
30°	.5000	.8660	.5774	1.732	60°
31°	.5150	.8572	.6009	1.664	59°
32°	.5299	.8480	.6249	1.600	58°
33°	.5446	.8387	.6494	1.540	57°
34°	.5592	.8290	.6745	1.483	56°
35°	.5736	.8192	.7002	1.428	55°
36°	.5878	.8090	.7265	1.376	54°
37°	.6018	.7986	.7536	1.327	53°
38°	.6157	.7880	.7813	1.280	52°
39°	.6293	.7771	.8098	1.235	51°
40°	.6428	.7660	.8391	1.192	50°
41°	.6561	.7547	.8693	1.150	49°
42°	.6691	.7431	.9004	1.111	48°
43°	.6820	.7314	.9325	1.072	47°
44°	.6947	.7193	.9657	1.036	46°
45°	.7071	.7071	1.000	1.000	45°
	cos	sin	ctn	tan	Degrees

TABLE 2 Trigonometric Functions for Angles in Radians*

Rad.	Sin	Tan	Cot	Cos	Rad.	Sin	Tan	Cot	Cos
.00	.00000	.00000	∞	1.00000	**.50**	.47943	.54630	1.8305	.87758
.01	.01000	.01000	99.997	0.99995	.51	.48818	.55936	1.7878	.87274
.02	.02000	.02000	49.993	.99980	.52	.49688	.57256	1.7465	.86782
.03	.03000	.03001	33.323	.99955	.53	.50553	.58592	1.7067	.86281
.04	.03999	.04002	24.987	.99920	.54	.51414	.59943	1.6683	.85771
.05	.04998	.05004	19.983	.99875	.55	.52269	.61311	1.6310	.85252
.06	.05996	.06007	16.647	.99820	.56	.53119	.62695	1.5950	.84726
.07	.06994	.07011	14.262	.99755	.57	.53963	.64097	1.5601	.84190
.08	.07991	.08017	12.473	.99680	.58	.54802	.65517	1.5263	.83646
.09	.08988	.09024	11.081	.99595	.59	.55636	.66956	1.4935	.83094
.10	.09983	.10033	9.9666	.99500	**.60**	.56464	.68414	1.4617	.82534
.11	.10978	.11045	9.0542	.99396	.61	.57287	.69892	1.4308	.81965
.12	.11971	.12058	8.2933	.99281	.62	.58104	.71391	1.4007	.81388
.13	.12963	.13074	7.6489	.99156	.63	.58914	.72911	1.3715	.80803
.14	.13954	.14092	7.0961	.99022	.64	.59720	.74454	1.3431	.80210
.15	.14944	.15114	6.6166	.98877	.65	.60519	.76020	1.3154	.79608
.16	.15932	.16138	6.1966	.98723	.66	.61312	.77610	1.2885	.78999
.17	.16918	.17166	5.8256	.98558	.67	.62099	.79225	1.2622	.78382
.18	.17903	.18197	5.4954	.98384	.68	.62879	.80866	1.2366	.77757
.19	.18886	.19232	5.1997	.98200	.69	.63654	.82534	1.2116	.77125
.20	.19867	.20271	4.9332	.98007	**.70**	.64422	.84229	1.1872	.76484
.21	.20846	.21314	4.6917	.97803	.71	.65183	.85953	1.1634	.75836
.22	.21823	.22362	4.4719	.97590	.72	.65938	.87707	1.1402	.75181
.23	.22798	.23414	4.2709	.97367	.73	.66687	.89492	1.1174	.74517
.24	.23770	.24472	4.0864	.97134	.74	.67429	.91309	1.0952	.73847
.25	.24740	.25534	3.9163	.96891	.75	.68164	.93160	1.0734	.73169
.26	.25708	.26602	3.7591	.96639	.76	.68892	.95045	1.0521	.72484
.27	.26673	.27676	3.6133	.96377	.77	.69614	.96967	1.0313	.71791
.28	.27636	.28755	3.4776	.96106	.78	.70328	.98926	1.0109	.71091
.29	.28595	.29841	3.3511	.95824	.79	.71035	1.0092	.99084	.70385
.30	.29552	.30934	3.2327	.95534	**.80**	.71736	1.0296	.97121	.69671
.31	.30506	.32033	3.1218	.95233	.81	.72429	1.0505	.95197	.68950
.32	.31457	.33139	3.0176	.94924	.82	.73115	1.0717	.93309	.68222
.33	.32404	.34252	2.9195	.94604	.83	.73793	1.0934	.91455	.67488
.34	.33349	.35374	2.8270	.94275	.84	.74464	1.1156	.89635	.66746
.35	.34290	.36503	2.7395	.93937	.85	.75128	1.1383	.87848	.65998
.36	.35227	.37640	2.6567	.93590	.86	.75784	1.1616	.86091	.65244
.37	.36162	.38786	2.5782	.93233	.87	.76433	1.1853	.84365	.64483
.38	.37092	.39941	2.5037	.92866	.88	.77074	1.2097	.82668	.63715
.39	.38019	.41105	2.4328	.92491	.89	.77707	1.2346	.80998	.62941
.40	.38942	.42279	2.3652	.92106	**.90**	.78333	1.2602	.79355	.62161
.41	.39861	.43463	2.3008	.91712	.91	.78950	1.2864	.77738	.61375
.42	.40776	.44657	2.2393	.91309	.92	.79560	1.3133	.76146	.60582
.43	.41687	.45862	2.1804	.90897	.93	.80162	1.3409	.74578	.59783
.44	.42594	.47078	2.1241	.90475	.94	.80756	1.3692	.73034	.58979
.45	.43497	.48306	2.0702	.90045	.95	.81342	1.3984	.71511	.58168
.46	.44395	.49545	2.0184	.89605	.96	.81919	1.4284	.70010	.57352
.47	.45289	.50797	1.9686	.89157	.97	.82489	1.4592	.68531	.56530
.48	.46178	.52061	1.9208	.88699	.98	.83050	1.4910	.67071	.55702
.49	.47063	.53339	1.8748	.88233	.99	.83603	1.5237	.65631	.54869
.50	.47943	.54630	1.8305	.87758	**1.00**	.84147	1.5574	.64209	.54030
Rad.	Sin	Tan	Cot	Cos	Rad.	Sin	Tan	Cot	Cos

* Tables 2–5 are reproduced from "Handbook of Tables for Mathematics" (R. C. Weast and S. M. Selby, eds.), 3rd ed., Chemical Rubber Co., Cleveland, Ohio, and used by permission.

TABLE 2 Trigonometric Functions for Angles in Radians (Continued)

Rad.	Sin	Tan	Cot	Cos	Rad.	Sin	Tan	Cot	Cos
1.00	.84147	1.5574	.64209	.54030	**1.50**	.99749	14.101	.07091	.07074
1.01	.84683	1.5922	.62806	.53186	1.51	.99815	16.428	.06087	.06076
1.02	.85211	1.6281	.61420	.52337	1.52	.99871	19.670	.05084	.05077
1.03	.85730	1.6652	.60051	.51482	1.53	.99917	24.498	.04082	.04079
1.04	.86240	1.7036	.58699	.50622	1.54	.99953	32.461	.03081	.03079
1.05	.86742	1.7433	.57362	.49757	1.55	.99978	48.078	.02080	.02079
1.06	.87236	1.7844	.56040	.48887	1.56	.99994	92.621	.01080	.01080
1.07	.87720	1.8270	.54734	.48012	1.57	1.00000	1255.8	.00080	.00080
1.08	.88196	1.8712	.53441	.47133	1.58	.99996	−108.65	−.00920	−.00920
1.09	.88663	1.9171	.52162	.46249	1.59	.99982	−52.067	−.01921	−.01920
1.10	.89121	1.9648	.50897	.45360	**1.60**	.99957	−34.233	−.02921	−.02920
1.11	.89570	2.0143	.49644	.44466	1.61	.99923	−25.495	−.03922	−.03919
1.12	.90010	2.0660	.48404	.43568	1.62	.99879	−20.307	−.04924	−.04918
1.13	.90441	2.1198	.47175	.42666	1.63	.99825	−16.871	−.05927	−.05917
1.14	.90863	2.1759	.45959	.41759	1.64	.99761	−14.427	−.06931	−.06915
1.15	.91276	2.2345	.44753	.40849	1.65	.99687	−12.599	−.07937	−.07912
1.16	.91680	2.2958	.43558	.39934	1.66	.99602	−11.181	−.08944	−.08909
1.17	.92075	2.3600	.42373	.39015	1.67	.99508	−10.047	−.09953	−.09904
1.18	.92461	2.4273	.41199	.38092	1.68	.99404	− 9.1208	−.10964	−.10899
1.19	.92837	2.4979	.40034	.37166	1.69	.99290	− 8.3492	−.11977	−.11892
1.20	.93204	2.5722	.38878	.36236	**1.70**	99166	− 7.6966	−.12993	−.12884
1.21	.93562	2.6503	.37731	.35302	1.71	.99033	− 7.1373	−.14011	−.13875
1.22	.93910	2.7328	.36593	.34365	1.72	.98889	− 6.6524	−.15032	−.14865
1.23	.94249	2.8198	.35463	.33424	1.73	.98735	− 6.2281	−.16056	−.15853
1.24	.94578	2.9119	.34341	.32480	1.74	.98572	− 5.8535	−.17084	−.16840
1.25	.94898	3.0096	.33227	.31532	1.75	.98399	− 5.5204	−.18115	−.17825
1.26	.95209	3.1133	.32121	.30582	1.76	.98215	− 5.2221	−.19149	−.18808
1.27	.95510	3.2236	.31021	.29628	1.77	.98022	− 4.9534	−.20188	−.19789
1.28	.95802	3.3413	.29928	.28672	1.78	.97820	− 4.7101	−.21231	−.20768
1.29	.96084	3.4672	.28842	.27712	1.79	.97607	− 4.4887	−.22278	−.21745
1.30	.96356	3.6021	.27762	.26750	**1.80**	.97385	− 4.2863	−.23330	−.22720
1.31	.96618	3.7471	.26687	.25785	1.81	.97153	− 4.1005	−.24387	−.23693
1.32	.96872	3.9033	.25619	.24818	1.82	.96911	− 3.9294	−.25449	−.24663
1.33	.97115	4.0723	.24556	.23848	1.83	.96659	− 3.7712	−.26517	−.25631
1.34	.97348	4.2556	.23498	.22875	1.84	.96398	− 3.6245	−.27590	−.26596
1.35	.97572	4.4552	.22446	.21901	1.85	.96128	− 3.4881	−.28669	−.27559
1.36	.97786	4.6734	.21398	.20924	1.86	.95847	− 3.3608	−.29755	−.28519
1.37	.97991	4.9131	.20354	.19945	1.87	.95557	− 2.2419	−.30846	−.29476
1.38	.98185	5.1774	.19315	.18964	1.88	.95258	− 3.1304	−.31945	−.30430
1.39	.98370	5.4707	.18279	.17981	1.89	.94949	− 3.0257	−.33051	−.31381
1.40	.98545	5.7979	.17248	.16997	**1.90**	.94630	− 2.9271	−.34164	−.32329
1.41	.98710	6.1654	.16220	.16010	1.91	.94302	− 2.8341	−.35284	−.33274
1.42	.98865	6.5811	.15195	.15023	1.92	.93965	− 2.7463	−.36413	−.34215
1.43	.99010	7.0555	.14173	.14033	1.93	.93618	− 2.6632	−.37549	−.35153
1.44	.99146	7.6018	.13155	.13042	1.94	.93262	− 2.5843	−.38695	−.36087
1.45	.99271	8.2381	.12139	.12050	1.95	.92896	− 2.5095	−.39849	−.37018
1.46	.99387	8.9886	.11125	.11057	1.96	.92521	− 2.4383	−.41012	−.37945
1.47	.99492	9.8874	.10114	.10063	1.97	.92137	− 2.3705	−.42185	−.38868
1.48	.99588	10.983	.09105	.09067	1.98	.91744	− 2.3058	−.43368	−.39788
1.49	.99674	12.350	.08097	.08071	1.99	.91341	− 2.2441	−.44562	−.40703
1.50	.99749	14.101	.07091	.07074	**2.00**	.90930	− 2.1850	−.45766	−.41615
Rad.	Sin	Tan	Cot	Cos	Rad.	Sin	Tan	Cot	Cos

TABLE 3 Four-Place Mantissas for Common Logarithms

N	0	1	2	3	4	5	6	7	8	9	Proportional Parts								
											1	2	3	4	5	6	7	8	9
10	0000	0043	0086	0128	0170	0212	0253	0294	0334	0374	*4	8	12	17	21	25	29	33	37
11	0414	0453	0492	0531	0569	0607	0645	0682	0719	0755	4	8	11	15	19	23	26	30	34
12	0792	0828	0864	0899	0934	0969	1004	1038	1072	1106	3	7	10	14	17	21	24	28	31
13	1139	1173	1206	1239	1271	1303	1335	1367	1399	1430	3	6	10	13	16	19	23	26	29
14	1461	1492	1523	1553	1584	1614	1644	1673	1703	1732	3	6	9	12	15	18	21	24	27
15	1761	1790	1818	1847	1875	1903	1931	1959	1987	2014	*3	6	8	11	14	17	20	22	25
16	2041	2068	2095	2122	2148	2175	2201	2227	2253	2279	3	5	8	11	13	16	18	21	24
17	2304	2330	2355	2380	2405	2430	2455	2480	2504	2529	2	5	7	10	12	15	17	20	22
18	2553	2577	2601	2625	2648	2672	2695	2718	2742	2765	2	5	7	9	12	14	16	19	21
19	2788	2810	2833	2856	2878	2900	2923	2945	2967	2989	2	4	7	9	11	13	16	18	20
20	3010	3032	3054	3075	3096	3118	3139	3160	3181	3201	2	4	6	8	11	13	15	17	19
21	3222	3243	3263	3284	3304	3324	3345	3365	3385	3404	2	4	6	8	10	12	14	16	18
22	3424	3444	3464	3483	3502	3522	3541	3560	3579	3598	2	4	6	8	10	12	14	15	17
23	3617	3636	3655	3674	3692	3711	3729	3747	3766	3784	2	4	6	7	9	11	13	15	17
24	3802	3820	3838	3856	3874	3892	3909	3927	3945	3962	2	4	5	7	9	11	12	14	16
25	3979	3997	4014	4031	4048	4065	4082	4099	4116	4133	2	3	5	7	9	10	12	14	15
26	4150	4166	4183	4200	4216	4232	4249	4265	4281	4298	2	3	5	7	8	10	11	13	15
27	4314	4330	4346	4362	4378	4393	4409	4425	4440	4456	2	3	5	6	8	9	11	13	14
28	4472	4487	4502	4518	4533	4548	4564	4579	4594	4609	2	3	5	6	8	9	11	12	14
29	4624	4639	4654	4669	4683	4698	4713	4728	4742	4757	1	3	4	6	7	9	10	12	13
30	4771	4786	4800	4814	4829	4843	4857	4871	4886	4900	1	3	4	6	7	9	10	11	13
31	4914	4928	4942	4955	4969	4983	4997	5011	5024	5038	1	3	4	6	7	8	10	11	12
32	5051	5065	5079	5092	5105	5119	5132	5145	5159	5172	1	3	4	5	7	8	9	11	12
33	5185	5198	5211	5224	5237	5250	5263	5276	5289	5302	1	3	4	5	6	8	9	10	12
34	5315	5328	5340	5353	5366	5378	5391	5403	5416	5428	1	3	4	5	6	8	9	10	11
35	5441	5453	5465	5478	5490	5502	5514	5527	5539	5551	1	2	4	5	6	7	9	10	11
36	5563	5575	5587	5599	5611	5623	5635	5647	5658	5670	1	2	4	5	6	7	8	10	11
37	5682	5694	5705	5717	5729	5740	5752	5763	5775	5786	1	2	3	5	6	7	8	9	10
38	5798	5809	5821	5832	5843	5855	5866	5877	5888	5899	1	2	3	5	6	7	8	9	10
39	5911	5922	5933	5944	5955	5966	5977	5988	5999	6010	1	2	3	4	5	7	8	9	10
40	6021	6031	6042	6053	6064	6075	6085	6096	6107	6117	1	2	3	4	5	6	8	9	10
41	6128	6138	6149	6160	6170	6180	6191	6201	6212	6222	1	2	3	4	5	6	7	8	9
42	6232	6243	6253	6263	6274	6284	6294	6304	6314	6325	1	2	3	4	5	6	7	8	9
43	6335	6345	6355	6365	6375	6385	6395	6405	6415	6425	1	2	3	4	5	6	7	8	9
44	6435	6444	6454	6464	6474	6484	6493	6503	6513	6522	1	2	3	4	5	6	7	8	9
45	6532	6542	6551	6561	6571	6580	6590	6599	6609	6618	1	2	3	4	5	6	7	8	9
46	6628	6637	6646	6656	6665	6675	6684	6693	6702	6712	1	2	3	4	5	6	7	7	8
47	6721	6730	6739	6749	6758	6767	6776	6785	6794	6803	1	2	3	4	5	5	6	7	8
48	6812	6821	6830	6839	6848	6857	6866	6875	6884	6893	1	2	3	4	4	5	6	7	8
49	6902	6911	6920	6928	6937	6946	6955	6964	6972	6981	1	2	3	4	4	5	6	7	8
50	6990	6998	7007	7016	7024	7033	7042	7050	7059	7067	1	2	3	3	4	5	6	7	8
51	7076	7084	7093	7101	7110	7118	7126	7135	7143	7152	1	2	3	3	4	5	6	7	8
52	7160	7168	7177	7185	7193	7202	7210	7218	7226	7235	1	2	2	3	4	5	6	7	7
53	7243	7251	7259	7267	7275	7284	7292	7300	7308	7316	1	2	2	3	4	5	6	6	7
54	7324	7332	7340	7348	7356	7364	7372	7380	7388	7396	1	2	2	3	4	5	6	6	7
N	0	1	2	3	4	5	6	7	8	9	1	2	3	4	5	6	7	8	9

* Interpolation in this section of the table is inaccurate.

TABLE 3 Four-Place Mantissas for Common Logarithms (Continued)

N	0	1	2	3	4	5	6	7	8	9	Proportional Parts 1	2	3	4	5	6	7	8	9
55	7404	7412	7419	7427	7435	7443	7451	7459	7466	7474	1	2	2	3	4	5	5	6	7
56	7482	7490	7497	7505	7513	7520	7528	7536	7543	7551	1	2	2	3	4	5	5	6	7
57	7559	7566	7574	7582	7589	7597	7604	7612	7619	7627	1	2	2	3	4	5	5	6	7
58	7634	7642	7649	7657	7664	7672	7679	7686	7694	7701	1	1	2	3	4	4	5	6	7
59	7709	7716	7723	7731	7738	7745	7752	7760	7767	7774	1	1	2	3	4	4	5	6	7
60	7782	7789	7796	7803	7810	7818	7825	7832	7839	7846	1	1	2	3	4	4	5	6	6
61	7853	7860	7868	7875	7882	7889	7896	7903	7910	7917	1	1	2	3	4	4	5	6	6
62	7924	7931	7938	7945	7952	7959	7966	7973	7980	7987	1	1	2	3	3	4	5	6	6
63	7993	8000	8007	8014	8021	8028	8035	8041	8048	8055	1	1	2	3	3	4	5	5	6
64	8062	8069	8075	8082	8089	8096	8102	8109	8116	8122	1	1	2	3	3	4	5	5	6
65	8129	8136	8142	8149	8156	8162	8169	8176	8182	8189	1	1	2	3	3	4	5	5	6
66	8195	8202	8209	8215	8222	8228	8235	8241	8248	8254	1	1	2	3	3	4	5	5	6
67	8261	8267	8274	8280	8287	8293	8299	8306	8312	8319	1	1	2	3	3	4	5	5	6
68	8325	8331	8338	8344	8351	8357	8363	8370	8376	8382	1	1	2	3	3	4	4	5	6
69	8388	8395	8401	8407	8414	8420	8426	8432	8439	8445	1	1	2	2	3	4	4	5	6
70	8451	8457	8463	8470	8476	8482	8488	8494	8500	8506	1	1	2	2	3	4	4	5	6
71	8513	8519	8525	8531	8537	8543	8549	8555	8561	8567	1	1	2	2	3	4	4	5	5
72	8573	8579	8585	8591	8597	8603	8609	8615	8621	8627	1	1	2	2	3	4	4	5	5
73	8633	8639	8645	8651	8657	8663	8669	8675	8681	8686	1	1	2	2	3	4	4	5	5
74	8692	8698	8704	8710	8716	8722	8727	8733	8739	8745	1	1	2	2	3	4	4	5	5
75	8751	8756	8762	8768	8774	8779	8785	8791	8797	8802	1	1	2	2	3	3	4	5	5
76	8808	8814	8820	8825	8831	8837	8842	8848	8854	8859	1	1	2	2	3	3	4	5	5
77	8865	8871	8876	8882	8887	8893	8899	8904	8910	8915	1	1	2	2	3	3	4	4	5
78	8921	8927	8932	8938	8943	8949	8954	8960	8965	8971	1	1	2	2	3	3	4	4	5
79	8976	8982	8987	8993	8998	9004	9009	9015	9020	9025	1	1	2	2	3	3	4	4	5
80	9031	9036	9042	9047	9053	9058	9063	9069	9074	9079	1	1	2	2	3	3	4	4	5
81	9085	9090	9096	9101	9106	9112	9117	9122	9128	9133	1	1	2	2	3	3	4	4	5
82	9138	9143	9149	9154	9159	9165	9170	9175	9180	9186	1	1	2	2	3	3	4	4	5
83	9191	9196	9201	9206	9212	9217	9222	9227	9232	9238	1	1	2	2	3	3	4	4	5
84	9243	9248	9253	9258	9263	9269	9274	9279	9284	9289	1	1	2	2	3	3	4	4	5
85	9294	9299	9304	9309	9315	9320	9325	9330	9335	9340	1	1	2	2	3	3	4	4	5
86	9345	9350	9355	9360	9365	9370	9375	9380	9385	9390	1	1	2	2	3	3	4	4	5
87	9395	9400	9405	9410	9415	9420	9425	9430	9435	9440	0	1	1	2	2	3	3	4	4
88	9445	9450	9455	9460	9465	9469	9474	9479	9484	9489	0	1	1	2	2	3	3	4	4
89	9494	9499	9504	9509	9513	9518	9523	9528	9533	9538	0	1	1	2	2	3	3	4	4
90	9542	9547	9552	9557	9562	9566	9571	9576	9581	9586	0	1	1	2	2	3	3	4	4
91	9590	9595	9600	9605	9609	9614	9619	9624	9628	9633	0	1	1	2	2	3	3	4	4
92	9638	9643	9647	9652	9657	9661	9666	9671	9675	9680	0	1	1	2	2	3	3	4	4
93	9685	9689	9694	9699	9703	9708	9713	9717	9722	9727	0	1	1	2	2	3	3	4	4
94	9731	9736	9741	9745	9750	9754	9759	9763	9768	9773	0	1	1	2	2	3	3	4	4
95	9777	9782	9786	9791	9795	9800	9805	9809	9814	9818	0	1	1	2	2	3	3	4	4
96	9823	9827	9832	9836	9841	9845	9850	9854	9859	9863	0	1	1	2	2	3	3	4	4
97	9868	9872	9877	9881	9886	9890	9894	9899	9903	9908	0	1	1	2	2	3	3	4	4
98	9912	9917	9921	9926	9930	9934	9939	9943	9948	9952	0	1	1	2	2	3	3	4	4
99	9956	9961	9965	9969	9974	9978	9983	9987	9991	9996	0	1	1	2	2	3	3	3	4
N	0	1	2	3	4	5	6	7	8	9	1	2	3	4	5	6	7	8	9

TABLE 4 Antilogarithms

	0	1	2	3	4	5	6	7	8	9	Proportional Parts 1	2	3	4	5	6	7	8	9
.00	1000	1002	1005	1007	1009	1012	1014	1016	1019	1021	0	0	1	1	1	1	2	2	2
.01	1023	1026	1028	1030	1033	1035	1038	1040	1042	1045	0	0	1	1	1	1	2	2	2
.02	1047	1050	1052	1054	1057	1059	1062	1064	1067	1069	0	0	1	1	1	1	2	2	2
.03	1072	1074	1076	1079	1081	1084	1086	1089	1091	1094	0	0	1	1	1	1	2	2	2
.04	1096	1099	1102	1104	1107	1109	1112	1114	1117	1119	0	1	1	1	1	2	2	2	2
.05	1122	1125	1127	1130	1132	1135	1138	1140	1143	1146	0	1	1	1	1	2	2	2	2
.06	1148	1151	1153	1156	1159	1161	1164	1167	1169	1172	0	1	1	1	1	2	2	2	2
.07	1175	1178	1180	1183	1186	1189	1191	1194	1197	1199	0	1	1	1	1	2	2	2	2
.08	1202	1205	1208	1211	1213	1216	1219	1222	1225	1227	0	1	1	1	1	2	2	2	3
.09	1230	1233	1236	1239	1242	1245	1247	1250	1253	1256	0	1	1	1	1	2	2	2	3
.10	1259	1262	1265	1268	1271	1274	1276	1279	1282	1285	0	1	1	1	1	2	2	2	3
.11	1288	1291	1294	1297	1300	1303	1306	1309	1312	1315	0	1	1	1	2	2	2	3	3
.12	1318	1321	1324	1327	1330	1334	1337	1340	1343	1346	0	1	1	1	2	2	2	2	3
.13	1349	1352	1355	1358	1361	1365	1368	1371	1374	1377	0	1	1	1	2	2	2	3	3
.14	1380	1384	1387	1390	1393	1396	1400	1403	1406	1409	0	1	1	1	2	2	2	3	3
.15	1413	1416	1419	1422	1426	1429	1432	1435	1439	1442	0	1	1	1	2	2	2	3	3
.16	1445	1449	1452	1455	1459	1462	1466	1469	1472	1476	0	1	1	1	2	2	2	3	3
.17	1479	1483	1486	1489	1493	1496	1500	1503	1507	1510	0	1	1	1	2	2	2	3	3
.18	1514	1517	1521	1524	1528	1531	1535	1538	1542	1545	0	1	1	1	2	2	2	3	3
.19	1549	1552	1556	1560	1563	1567	1570	1574	1578	1581	0	1	1	1	2	2	3	3	3
.20	1585	1589	1592	1596	1600	1603	1607	1611	1614	1618	0	1	1	1	2	2	3	3	3
.21	1622	1626	1629	1633	1637	1641	1644	1648	1652	1656	0	1	1	2	2	2	3	3	3
.22	1660	1663	1667	1671	1675	1679	1683	1687	1690	1694	0	1	1	2	2	2	3	3	3
.23	1698	1702	1706	1710	1714	1718	1722	1726	1730	1734	0	1	1	2	2	2	3	3	4
.24	1738	1742	1746	1750	1754	1758	1762	1766	1770	1774	0	1	1	2	2	2	3	3	4
.25	1778	1782	1786	1791	1795	1799	1803	1807	1811	1816	0	1	1	2	2	2	3	3	4
.26	1820	1824	1828	1832	1837	1841	1845	1849	1854	1858	0	1	1	2	2	3	3	3	4
.27	1862	1866	1871	1875	1879	1884	1888	1892	1897	1901	0	1	1	2	2	3	3	3	4
.28	1905	1910	1914	1919	1923	1928	1932	1936	1941	1945	0	1	1	2	2	3	3	4	4
.29	1950	1954	1959	1963	1968	1972	1977	1982	1986	1991	0	1	1	2	2	3	3	4	4
.30	1995	2000	2004	2009	2014	2018	2023	2028	2032	2037	0	1	1	2	2	3	3	4	4
.31	2042	2046	2051	2056	2061	2065	2070	2075	2080	2084	0	1	1	2	2	3	3	4	4
.32	2089	2094	2099	2104	2109	2113	2118	2123	2128	2133	0	1	1	2	2	3	3	4	4
.33	2138	2143	2148	2153	2158	2163	2168	2173	2178	2183	0	1	1	2	2	3	3	4	4
.34	2188	2193	2198	2203	2208	2213	2218	2223	2228	2234	1	1	2	2	3	3	4	4	5
.35	2239	2244	2249	2254	2259	2265	2270	2275	2280	2286	1	1	2	2	3	3	4	4	5
.36	2291	2296	2301	2307	2312	2317	2323	2328	2333	2339	1	1	2	2	3	3	4	4	5
.37	2344	2350	2355	2360	2366	2371	2377	2382	2388	2393	1	1	2	2	3	3	4	4	5
.38	2399	2404	2410	2415	2421	2427	2432	2438	2443	2449	1	1	2	2	3	3	4	4	5
.39	2455	2460	2466	2472	2477	2483	2489	2495	2500	2506	1	1	2	2	3	3	4	5	5
.40	2512	2518	2523	2529	2535	2541	2547	2553	2559	2564	1	1	2	2	3	4	4	5	5
.41	2570	2576	2582	2588	2594	2600	2606	2612	2618	2624	1	1	2	2	3	4	4	5	5
.42	2630	2636	2642	2649	2655	2661	2667	2673	2679	2685	1	1	2	2	3	4	4	5	6
.43	2692	2698	2704	2710	2716	2723	2729	2735	2742	2748	1	1	2	3	3	4	4	5	6
.44	2754	2761	2767	2773	2780	2786	2793	2799	2805	2812	1	1	2	3	3	4	4	5	6
.45	2818	2825	2831	2838	2844	2851	2858	2864	2871	2877	1	1	2	3	3	4	5	5	6
.46	2884	2891	2897	2904	2911	2917	2924	2931	2938	2944	1	1	2	3	3	4	5	5	6
.47	2951	2958	2965	2972	2979	2985	2992	2999	3006	3013	1	1	2	3	3	4	5	5	6
.48	3020	3027	3034	3041	3048	3055	3062	3069	3076	3083	1	1	2	3	4	4	5	6	6
.49	3090	3097	3105	3112	3119	3126	3133	3141	3148	3155	1	1	2	3	4	4	5	6	6
	0	1	2	3	4	5	6	7	8	9	1	2	3	4	5	6	7	8	9

TABLE 4 Antilogarithms (Continued)

	0	1	2	3	4	5	6	7	8	9	Proportional Parts								
											1	2	3	4	5	6	7	8	9
.50	3162	3170	3177	3184	3192	3199	3206	3214	3221	3228	1	1	2	3	4	4	5	6	7
.51	3236	3243	3251	3258	3266	3273	3281	3289	3296	3304	1	2	2	3	4	5	5	6	7
.52	3311	3319	3327	3334	3342	3350	3357	3365	3373	3381	1	2	2	3	4	5	5	6	7
.53	3388	3396	3404	3412	3420	3428	3436	3443	3451	3459	1	2	2	3	4	5	6	6	7
.54	3467	3475	3483	3491	3499	3508	3516	3524	3532	3540	1	2	2	3	4	5	6	6	7
.55	3548	3556	3565	3573	3581	3589	3597	3606	3614	3622	1	2	2	3	4	5	6	7	7
.56	3631	3639	3648	3656	3664	3673	3681	3690	3698	3707	1	2	3	3	4	5	6	7	8
.57	3715	3724	3733	3741	3750	3758	3767	3776	3784	3793	1	2	3	3	4	5	6	7	8
.58	3802	3811	3819	3828	3837	3846	3855	3864	3873	3882	1	2	3	4	4	5	6	7	8
.59	3890	3899	3908	3917	3926	3936	3945	3954	3963	3972	1	2	3	4	5	5	6	7	8
.60	3981	3990	3999	4009	4018	4027	4036	4046	4055	4064	1	2	3	4	5	6	6	7	8
.61	4074	4083	4093	4102	4111	4121	4130	4140	4150	4159	1	2	3	4	5	6	7	8	9
.62	4169	4178	4188	4198	4207	4217	4227	4236	4246	4256	1	2	3	4	5	6	7	8	9
.63	4266	4276	4285	4295	4305	4315	4325	4335	4345	4355	1	2	3	4	5	6	7	8	9
.64	4365	4375	4385	4395	4406	4416	4426	4436	4446	4457	1	2	3	4	5	6	7	8	9
.65	4467	4477	4487	4498	4508	4519	4529	4539	4550	4560	1	2	3	4	5	6	7	8	9
.66	4571	4581	4592	4603	4613	4624	4634	4645	4656	4667	1	2	3	4	5	6	7	9	10
.67	4677	4688	4699	4710	4721	4732	4742	4753	4764	4775	1	2	3	4	5	7	8	9	10
.68	4786	4797	4808	4819	4831	4842	4853	4864	4875	4887	1	2	3	4	6	7	8	9	10
.69	4898	4909	4920	4932	4943	4955	4966	4977	4989	5000	1	2	3	5	6	7	8	9	10
.70	5012	5023	5035	5047	5058	5070	5082	5093	5105	5117	1	2	4	5	6	7	8	9	11
.71	5129	5140	5152	5164	5176	5188	5200	5212	5224	5236	1	2	4	5	6	7	8	10	11
.72	5248	5260	5272	5284	5297	5309	5321	5333	5346	5358	1	2	4	5	6	7	9	10	11
.73	5370	5383	5395	5408	5420	5433	5445	5458	5470	5483	1	3	4	5	6	8	9	10	11
.74	5495	5508	5521	5534	5546	5559	5572	5585	5598	5610	1	3	4	5	6	8	9	10	12
.75	5623	5636	5649	5662	5675	5689	5702	5715	5728	5741	1	3	4	5	7	8	9	10	12
.76	5754	5768	5781	5794	5808	5821	5834	5848	5861	5875	1	3	4	5	7	8	9	11	12
.77	5888	5902	5916	5929	5943	5957	5970	5984	5998	6012	1	3	4	5	7	8	10	11	12
.78	6026	6039	6053	6067	6081	6095	6109	6124	6138	6152	1	3	4	6	7	8	10	11	13
.79	6166	6180	6194	6209	6223	6237	6252	6266	6281	6295	1	3	4	6	7	9	10	11	13
.80	6310	6324	6339	6353	6368	6383	6397	6412	6427	6442	1	3	4	6	7	9	10	12	13
.81	6457	6471	6486	6501	6516	6531	6546	6561	6577	6592	2	3	5	6	8	9	11	12	14
.82	6607	6622	6637	6653	6668	6683	6699	6714	6730	6745	2	3	5	6	8	9	11	12	14
.83	6761	6776	6792	6808	6823	6839	6855	6871	6887	6902	2	3	5	6	8	9	11	13	14
.84	6918	6934	6950	6966	6982	6998	7015	7031	7047	7063	2	3	5	6	8	10	11	13	15
.85	7079	7096	7112	7129	7145	7161	7178	7194	7211	7228	2	3	5	7	8	10	12	13	15
.86	7244	7261	7278	7295	7311	7328	7345	7362	7379	7396	2	3	5	7	8	10	12	13	15
.87	7413	7430	7447	7464	7482	7499	7516	7534	7551	7568	2	3	5	7	9	10	12	14	16
.88	7586	7603	7621	7638	7656	7674	7691	7709	7727	7745	2	4	5	7	9	11	12	14	16
.89	7762	7780	7798	7816	7834	7852	7870	7889	7907	7925	2	4	5	7	9	11	13	14	16
.90	7943	7962	7980	7998	8017	8035	8054	8072	8091	8110	2	4	6	7	9	11	13	15	17
.91	8128	8147	8166	8185	8204	8222	8241	8260	8279	8299	2	4	6	8	9	11	13	15	17
.92	8318	8337	8356	8375	8395	8414	8433	8453	8472	8492	2	4	6	8	10	12	14	15	17
.93	8511	8531	8551	8570	8590	8610	8630	8650	8670	8690	2	4	6	8	10	12	14	16	18
.94	8710	8730	8750	8770	8790	8810	8831	8851	8872	8892	2	4	6	8	10	12	14	16	18
.95	8913	8933	8954	8974	8995	9016	9036	9057	9078	9099	2	4	6	8	10	12	15	17	19
.96	9120	9141	9162	9183	9204	9226	9247	9268	9290	9311	2	4	6	8	11	13	15	17	19
.97	9333	9354	9376	9397	9419	9441	9462	9484	9506	9528	2	4	7	9	11	13	15	17	20
.98	9550	9572	9594	9616	9638	9661	9683	9705	9727	9750	2	4	7	9	11	13	16	18	20
.99	9772	9795	9817	9840	9863	9886	9908	9931	9954	9977	2	5	7	9	11	14	16	18	20
	0	1	2	3	4	5	6	7	8	9	1	2	3	4	5	6	7	8	9

TABLE 5 Exponential Functions

x	e^x	$\text{Log}_{10}(e^x)$	e^{-x}	x	e^x	$\text{Log}_{10}(e^x)$	e^{-x}
0.00	1.0000	0.00000	1.000000	**0.50**	1.6487	0.21715	0.606531
0.01	1.0101	.00434	0.990050	0.51	1.6653	.22149	.600496
0.02	1.0202	.00869	.980199	0.52	1.6820	.22583	.594521
0.03	1.0305	.01303	.970446	0.53	1.6989	.23018	.588605
0.04	1.0408	.01737	.960789	0.54	1.7160	.23452	.582748
0.05	1.0513	0.02171	0.951229	**0.55**	1.7333	0.23886	0.576950
0.06	1.0618	.02606	.941765	0.56	1.7507	.24320	.571209
0.07	1.0725	.03040	.932394	0.57	1.7683	.24755	.565525
0.08	1.0833	.03474	.923116	0.58	1.7860	.25189	.559898
0.09	1.0942	.03909	.913931	0.59	1.8040	.25623	.554327
0.10	1.1052	0.04343	0.904837	**0.60**	1.8221	0.26058	0.548812
0.11	1.1163	.04777	.895834	0.61	1.8404	.26492	.543351
0.12	1.1275	.05212	.886920	0.62	1.8589	.26926	.537944
0.13	1.1388	.05646	.878095	0.63	1.8776	.27361	.532592
0.14	1.1503	.06080	.869358	0.64	1.8965	.27795	.527292
0.15	1.1618	0.06514	0.860708	**0.65**	1.9155	0.28229	0.522046
0.16	1.1735	.06949	.852144	0.66	1.9348	.28663	.516851
0.17	1.1853	.07383	.843665	0.67	1.9542	.29098	.511709
0.18	1.1972	.07817	.835270	0.68	1.9739	.29532	.506617
0.19	1.2092	.08252	.826959	0.69	1.9937	.29966	.501576
0.20	1.2214	0.08686	0.818731	**0.70**	2.0138	0.30401	0.496585
0.21	1.2337	.09120	.810584	0.71	2.0340	.30835	.491644
0.22	1.2461	.09554	.802519	0.72	2.0544	.31269	.486752
0.23	1.2586	.09989	.794534	0.73	2.0751	.31703	.481909
0.24	1.2712	.10423	.786628	0.74	2.0959	.32138	.477114
0.25	1.2840	0.10857	0.778801	**0.75**	2.1170	0.32572	0.472367
0.26	1.2969	.11292	.771052	0.76	2.1383	.33006	.467666
0.27	1.3100	.11726	.763379	0.77	2.1598	.33441	.463013
0.28	1.3231	.12160	.755784	0.78	2.1815	.33875	.458406
0.29	1.3364	.12595	.748264	0.79	2.2034	.34309	.453845
0.30	1.3499	0.13029	0.740818	**0.80**	2.2255	0.34744	0.449329
0.31	1.3634	.13463	.733447	0.81	2.2479	.35178	.444858
0.32	1.3771	.13897	.726149	0.82	2.2705	.35612	.440432
0.33	1.3910	.14332	.718924	0.83	2.2933	.36046	.436049
0.34	1.4049	.14766	.711770	0.84	2.3164	.36481	.431711
0.35	1.4191	0.15200	0.704688	**0.85**	2.3396	0.36915	0.427415
0.36	1.4333	.15635	.697676	0.86	2.3632	.37349	.423162
0.37	1.4477	.16069	.690734	0.87	2.3869	.37784	.418952
0.38	1.4623	.16503	.683861	0.88	2.4109	.38218	.414783
0.39	1.4770	.16937	.677057	0.89	2.4351	.38652	.410656
0.40	1.4918	0.17372	0.670320	**0.90**	2.4596	0.39087	0.406570
0.41	1.5068	.17806	.663650	0.91	2.4843	.39521	.402524
0.42	1.5220	.18240	.657047	0.92	2.5093	.39955	.398519
0.43	1.5373	.18675	.650509	0.93	2.5345	.40389	.394554
0.44	1.5527	.19109	.644036	0.94	2.5600	.40824	.390628
0.45	1.5683	0.19543	0.637628	**0.95**	2.5857	0.41258	0.386741
0.46	1.5841	.19978	.631284	0.96	2.6117	.41692	.382893
0.47	1.6000	.20412	.625002	0.97	2.6379	.42127	.379083
0.48	1.6161	.20846	.618783	0.98	2.6645	.42561	.375311
0.49	1.6323	.21280	.612626	0.99	2.6912	.42995	.371577
0.50	1.6487	0.21715	0.606531	**1.00**	2.7183	0.43429	0.367879

TABLE 5 Exponential Functions (Continued)

x	e^x	$\text{Log}_{10}(e^x)$	e^{-x}
1.00	2.7183	0.43429	0.367879
1.01	2.7456	.43864	.364219
1.02	2.7732	.44298	.360595
1.03	2.8011	.44732	.357007
1.04	2.8292	.45167	.353455
1.05	2.8577	0.45601	0.349938
1.06	2.8864	.46035	.346456
1.07	2.9154	.46470	.343009
1.08	2.9447	.46904	.339596
1.09	2.9743	.47338	.336216
1.10	3.0042	0.47772	0.332871
1.11	3.0344	.48207	.329559
1.12	3.0649	.48641	.326280
1.13	3.0957	.49075	.323033
1.14	3.1268	.49510	.319819
1.15	3.1582	0.49944	0.316637
1.16	3.1899	.50378	.313486
1.17	3.2220	.50812	.310367
1.18	3.2544	.51247	.307279
1.19	3.2871	.51681	.304221
1.20	3.3201	0.52115	0.301194
1.21	3.3535	.52550	.298197
1.22	3.3872	.52984	.295230
1.23	3.4212	.53418	.292293
1.24	3.4556	.53853	.289384
1.25	3.4903	0.54287	0.286505
1.26	3.5254	.54721	.283654
1.27	3.5609	.55155	.280832
1.28	3.5966	.55590	.278037
1.29	3.6328	.56024	.275271
1.30	3.6693	0.56458	0.272532
1.31	3.7062	.56893	.269820
1.32	3.7434	.57327	.267135
1.33	3.7810	.57761	.264477
1.34	3.8190	.58195	.261846
1.35	3.8574	0.58630	0.259240
1.36	3.8962	.59064	.256661
1.37	3.9354	.59498	.254107
1.38	3.9749	.59933	.251579
1.39	4.0149	.60367	.249075
1.40	4.0552	0.60801	0.246597
1.41	4.0960	.61236	.244143
1.42	4.1371	.61670	.241714
1.43	4.1787	.62104	.239309
1.44	4.2207	.62538	.236928
1.45	4.2631	0.62973	0.234570
1.46	4.3060	.63407	.232236
1.47	4.3492	.63841	.229925
1.48	4.3929	.64276	.227638
1.49	4.4371	.64710	.225373
1.50	4.4817	0.65144	0.223130
1.50	4.4817	0.65144	0.223130
1.51	4.5267	.65578	.220910
1.52	4.5722	.66013	.218712
1.53	4.6182	.66447	.216536
1.54	4.6646	.66881	.214381
1.55	4.7115	0.67316	0.212248
1.56	4.7588	.67750	.210136
1.57	4.8066	.68184	.208045
1.58	4.8550	.68619	.205975
1.59	4.9037	.69053	.203926
1.60	4.9530	0.69487	0.201897
1.61	5.0028	.69921	.199888
1.62	5.0531	.70356	.197899
1.63	5.1039	.70790	.195930
1.64	5.1552	.71224	.193980
1.65	5.2070	0.71659	0.192050
1.66	5.2593	.72093	.190139
1.67	5.3122	.72527	.188247
1.68	5.3656	.72961	.186374
1.69	5.4195	.73396	.184520
1.70	5.4739	0.73830	0.182684
1.71	5.5290	.74264	.180866
1.72	5.5845	.74699	.179066
1.73	5.6407	.75133	.177284
1.74	5.6973	.75567	.175520
1.75	5.7546	0.76002	0.173774
1.76	5.8124	.76436	.172045
1.77	5.8709	.76870	.170333
1.78	5.9299	.77304	.168638
1.79	5.9895	.77739	.166960
1.80	6.0496	0.78173	0.165299
1.81	6.1104	.78607	.163654
1.82	6.1719	.79042	.162026
1.83	6.2339	.79476	.160414
1.84	6.2965	.79910	.158817
1.85	6.3598	0.80344	0.157237
1.86	6.4237	.80779	.155673
1.87	6.4883	.81213	.154124
1.88	6.5535	.81647	.152590
1.89	6.6194	.82082	.151072
1.90	6.6859	0.82516	0.149569
1.91	6.7531	.82950	.148080
1.92	6.8210	.83385	.146607
1.93	6.8895	.83819	.145148
1.94	6.9588	.84253	.143704
1.95	7.0287	0.84687	0.142274
1.96	7.0993	.85122	.140858
1.97	7.1707	.85556	.139457
1.98	7.2427	.85990	.138069
1.99	7.3155	.86425	.136695
2.00	7.3891	0.86859	0.135335

TABLE 5 Exponential Functions (Continued)

x	e^x	$\text{Log}_{10}(e^x)$	e^{-x}
2.00	7.3891	0.86859	0.135335
2.01	7.4633	.87293	.133989
2.02	7.5383	.87727	.132655
2.03	7.6141	.88162	.131336
2.04	7.6906	.88596	.130029
2.05	7.7679	0.89030	0.128735
2.06	7.8460	.89465	.127454
2.07	7.9248	.89899	.126186
2.08	8.0045	.90333	.124930
2.09	8.0849	.90768	.123687
2.10	8.1662	0.91202	0.122456
2.11	8.2482	.91636	.121238
2.12	8.3311	.92070	.120032
2.13	8.4149	.92505	.118837
2.14	8.4994	.92939	.117655
2.15	8.5849	0.93373	0.116484
2.16	8.6711	.93808	.115325
2.17	8.7583	.94242	.114178
2.18	8.8463	.94676	.113042
2.19	8.9352	.95110	.111917
2.20	9.0250	0.95545	0.110803
2.21	9.1157	.95979	.109701
2.22	9.2073	.96413	.108609
2.23	9.2999	.96848	.107528
2.24	9.3933	.97282	.106459
2.25	9.4877	0.97716	0.105399
2.26	9.5831	.98151	.104350
2.27	9.6794	.98585	.103312
2.28	9.7767	.99019	.102284
2.29	9.8749	.99453	.101266
2.30	9.9742	0.99888	0.100259
2.31	10.074	1.00322	.099261
2.32	10.176	1.00756	.098274
2.33	10.278	1.01191	.097296
2.34	10.381	1.01625	.096328
2.35	10.486	1.02059	0.095369
2.36	10.591	1.02493	.094420
2.37	10.697	1.02928	.093481
2.38	10.805	1.03362	.092551
2.39	10.913	1.03796	.091630
2.40	11.023	1.04231	0.090718
2.41	11.134	1.04665	.089815
2.42	11.246	1.05099	.088922
2.43	11.359	1.05534	.088037
2.44	11.473	1.05968	.087161
2.45	11.588	1.06402	0.086294
2.46	11.705	1.06836	.085435
2.47	11.822	1.07271	.084585
2.48	11.941	1.07705	.083743
2.49	12.061	1.08139	.082910
2.50	12.182	1.08574	0.082085

x	e^x	$\text{Log}_{10}(e^x)$	e^{-x}
2.50	12.182	1.08574	0.082085
2.51	12.305	1.09008	.081268
2.52	12.429	1.09442	.080460
2.53	12.554	1.09877	.079659
2.54	12.680	1.10311	.078866
2.55	12.807	1.10745	0.078082
2.56	12.936	1.11179	.077305
2.57	13.066	1.11614	.076536
2.58	13.197	1.12048	.075774
2.59	13.330	1.12482	.075020
2.60	13.464	1.12917	0.074274
2.61	13.599	1.13351	.073535
2.62	13.736	1.13785	.072803
2.63	13.874	1.14219	.072078
2.64	14.013	1.14654	.071361
2.65	14.154	1.15088	0.070651
2.66	14.296	1.15522	.069948
2.67	14.440	1.15957	.069252
2.68	14.585	1.16391	.068563
2.69	14.732	1.16825	.067881
2.70	14.880	1.17260	0.067206
2.71	15.029	1.17694	.066537
2.72	15.180	1.18128	.065875
2.73	15.333	1.18562	.065219
2.74	15.487	1.18997	.064570
2.75	15.643	1.19431	0.063928
2.76	15.800	1.19865	.063292
2.77	15.959	1.20300	.062662
2.78	16.119	1.20734	.062039
2.79	16.281	1.21168	.061421
2.80	16.445	1.21602	0.060810
2.81	16.610	1.22037	.060205
2.82	16.777	1.22471	.059606
2.83	16.945	1.22905	.059013
2.84	17.116	1.23340	.058426
2.85	17.288	1.23774	0.057844
2.86	17.462	1.24208	.057269
2.87	17.637	1.24643	.056699
2.88	17.814	1.25077	.056135
2.89	17.993	1.25511	.055576
2.90	18.174	1.25945	0.055023
2.91	18.357	1.26380	.054476
2.92	18.541	1.26814	.053934
2.93	18.728	1.27248	.053397
2.94	18.916	1.27683	.052866
2.95	19.106	1.28117	0.052340
2.96	19.298	1.28551	.051819
2.97	19.492	1.28985	.051303
2.98	19.688	1.29420	.050793
2.99	19.886	1.29854	.050287
3.00	20.086	1.30288	0.049787

TABLE 5 Exponential Functions (Continued)

x	e^x	$\text{Log}_{10}(e^x)$	e^{-x}
3.00	20.086	1.30288	0.049787
3.01	20.287	1.30723	.049292
3.02	20.491	1.31157	.048801
3.03	20.697	1.31591	.048316
3.04	20.905	1.32026	.047835
3.05	21.115	1.32460	0.047359
3.06	21.328	1.32894	.046888
3.07	21.542	1.33328	.046421
3.08	21.758	1.33763	.045959
3.09	21.977	1.34197	.045502
3.10	22.198	1.34631	0.045049
3.11	22.421	1.35066	.044601
3.12	22.646	1.35500	.044157
3.13	22.874	1.35934	.043718
3.14	23.104	1.36368	.043283
3.15	23.336	1.36803	0.042852
3.16	23.571	1.37237	.042426
3.17	23.807	1.37671	.042004
3.18	24.047	1.38106	.041586
3.19	24.288	1.38540	.041172
3.20	24.533	1.38974	0.040762
3.21	24.779	1.39409	.040357
3.22	25.028	1.39843	.039955
3.23	25.280	1.40277	.039557
3.24	25.534	1.40711	.039164
3.25	25.790	1.41146	0.038774
3.26	26.050	1.41580	.038388
3.27	26.311	1.42014	.038006
3.28	26.576	1.42449	.037628
3.29	26.843	1.42883	.037254
3.30	27.113	1.43317	0.036883
3.31	27.385	1.43751	.036516
3.32	27.660	1.44186	.036153
3.33	27.938	1.44620	.035793
3.34	28.219	1.45054	.035437
3.35	28.503	1.45489	0.035084
3.36	28.789	1.45923	.034735
3.37	29.079	1.46357	.034390
3.38	29.371	1.46792	.034047
3.39	29.666	1.47226	.033709
3.40	29.964	1.47660	0.033373
3.41	30.265	1.48094	.033041
3.42	30.569	1.48529	.032712
3.43	30.877	1.48963	.032387
3.44	31.187	1.49397	.032065
3.45	31.500	1.49832	0.031746
3.46	31.817	1.50266	.031430
3.47	32.137	1.50700	.031117
3.48	32.460	1.51134	.030807
3.49	32.786	1.51569	.030501
3.50	33.115	1.52003	0.030197

x	e^x	$\text{Log}_{10}(e^x)$	e^{-x}
3.50	33.115	1.52003	0.030197
3.51	33.448	1.52437	.029897
3.52	33.784	1.52872	.029599
3.53	34.124	1.53306	.029305
3.54	34.467	1.53740	.029013
3.55	34.813	1.54175	0.028725
3.56	35.163	1.54609	.028439
3.57	35.517	1.55043	.028156
3.58	35.874	1.55477	.027876
3.59	36.234	1.55912	.027598
3.60	36.598	1.56346	0.027324
3.61	36.966	1.56780	.027052
3.62	37.338	1.57215	.026783
3.63	37.713	1.57649	.026516
3.64	38.092	1.58083	.026252
3.65	38.475	1.58517	0.025991
3.66	38.861	1.58952	.025733
3.67	39.252	1.59386	.025476
3.68	39.646	1.59820	.025223
3.69	40.045	1.60255	.024972
3.70	40.447	1.60689	0.024724
3.71	40.854	1.61123	.024478
3.72	41.264	1.61558	.024234
3.73	41.679	1.61992	.023993
3.74	42.098	1.62426	.023754
3.75	42.521	1.62860	0.023518
3.76	42.948	1.63295	.023284
3.77	43.380	1.63729	.023052
3.78	43.816	1.64163	.022823
3.79	44.256	1.64598	.022596
3.80	44.701	1.65032	0.022371
3.81	45.150	1.65466	.022148
3.82	45.604	1.65900	.021928
3.83	46.063	1.66335	.021710
3.84	46.525	1.66769	.021494
3.85	46.993	1.67203	0.021280
3.86	47.465	1.67638	.021068
3.87	47.942	1.68072	.020858
3.88	48.424	1.68506	.020651
3.89	48.911	1.68941	.020445
3.90	49.402	1.69375	0.020242
3.91	49.899	1.69809	.020041
3.92	50.400	1.70243	.019841
3.93	50.907	1.70678	.019644
3.94	51.419	1.71112	.019448
3.95	51.935	1.71546	0.019255
3.96	52.457	1.71981	.019063
3.97	52.985	1.72415	.018873
3.98	53.517	1.72849	.018686
3.99	54.055	1.73283	.018500
4.00	54.598	1.73718	0.018316

TABLE 5 Exponential Functions (Continued)

x	e^x	$\text{Log}_{10}(e^x)$	e^{-x}
4.00	54.598	1.73718	0.018316
4.01	55.147	1.74152	.018133
4.02	55.701	1.74586	.017953
4.03	56.261	1.75021	.017774
4.04	56.826	1.75455	.017597
4.05	57.397	1.75889	0.017422
4.06	57.974	1.76324	.017249
4.07	58.557	1.76758	.017077
4.08	59.145	1.77192	.016907
4.09	59.740	1.77626	.016739
4.10	60.340	1.78061	0.016573
4.11	60.947	1.78495	.016408
4.12	61.559	1.78929	.016245
4.13	62.178	1.79364	.016083
4.14	62.803	1.79798	.015923
4.15	63.434	1.80232	0.015764
4.16	64.072	1.80667	.015608
4.17	64.715	1.81101	.015452
4.18	65.366	1.81535	.015299
4.19	66.023	1.81969	.015146
4.20	66.686	1.82404	0.014996
4.21	67.357	1.82838	.014846
4.22	68.033	1.83272	.014699
4.23	68.717	1.83707	.014552
4.24	69.408	1.84141	.014408
4.25	70.105	1.84575	0.014264
4.26	70.810	1.85009	.014122
4.27	71.522	1.85444	.013982
4.28	72.240	1.85878	.013843
4.29	72.966	1.86312	.013705
4.30	73.700	1.86747	0.013569
4.31	74.440	1.87181	.013434
4.32	75.189	1.87615	.013300
4.33	75.944	1.88050	.013168
4.34	76.708	1.88484	.013037
4.35	77.478	1.88918	0.012907
4.36	78.257	1.89352	.012778
4.37	79.044	1.89787	.012651
4.38	79.838	1.90221	.012525
4.39	80.640	1.90655	.012401
4.40	81.451	1.91090	0.012277
4.41	82.269	1.91524	.012155
4.42	83.096	1.91958	.012034
4.43	83.931	1.92392	.011914
4.44	84.775	1.92827	.011796
4.45	85.627	1.93261	0.011679
4.46	86.488	1.93695	.011562
4.47	87.357	1.94130	.011447
4.48	88.235	1.94564	.011333
4.49	89.121	1.94998	.011221
4.50	90.017	1.95433	0.011109

x	e^x	$\text{Log}_{10}(e^x)$	e^{-x}
4.50	90.017	1.95433	0.011109
4.51	90.922	1.95867	.010998
4.52	91.836	1.96301	.010889
4.53	92.759	1.96735	.010781
4.54	93.691	1.97170	.010673
4.55	94.632	1.97604	0.010567
4.56	95.583	1.98038	.010462
4.57	96.544	1.98473	.010358
4.58	97.514	1.98907	.010255
4.59	98.494	1.99341	.010153
4.60	99.484	1.99775	0.010052
4.61	100.48	2.00210	.009952
4.62	101.49	2.00644	.009853
4.63	102.51	2.01078	.009755
4.64	103.54	2.01513	.009658
4.65	104.58	2.01947	0.009562
4.66	105.64	2.02381	.009466
4.67	106.70	2.02816	.009372
4.68	107.77	2.03250	.009279
4.69	108.85	2.03684	.009187
4.70	109.95	2.04118	0.009095
4.71	111.05	2.04553	.009005
4.72	112.17	2.04987	.008915
4.73	113.30	2.05421	.008826
4.74	114.43	2.05856	.008739
4.75	115.58	2.06290	0.008652
4.76	116.75	2.06724	.008566
4.77	117.92	2.07158	.008480
4.78	119.10	2.07593	.008396
4.79	120.30	2.08027	.008312
4.80	121.51	2.08461	0.008230
4.81	122.73	2.08896	.008148
4.82	123.97	2.09330	.008067
4.83	125.21	2.09764	.007987
4.84	126.47	2.10199	.007907
4.85	127.74	2.10633	0.007828
4.86	129.02	2.11067	.007750
4.87	130.32	2.11501	.007673
4.88	131.63	2.11936	.007597
4.89	132.95	2.12370	.007521
4.90	134.29	2.12804	0.007447
4.91	135.64	2.13239	.007372
4.92	137.00	2.13673	.007299
4.93	138.38	2.14107	.007227
4.94	139.77	2.14541	.007155
4.95	141.17	2.14976	0.007083
4.96	142.59	2.15410	.007013
4.97	144.03	2.15844	.006943
4.98	145.47	2.16279	.006874
4.99	146.94	2.16713	.006806
5.00	148.41	2.17147	0.006738

TABLE 5 Exponential Functions (Continued)

x	e^x	$\text{Log}_{10}(e^x)$	e^{-x}
5.00	148.41	2.17147	0.006738
5.01	149.90	2.17582	.006671
5.02	151.41	2.18016	.006605
5.03	152.93	2.18450	.006539
5.04	154.47	2.18884	.006474
5.05	156.02	2.19319	0.006409
5.06	157.59	2.19753	.006346
5.07	159.17	2.20187	.006282
5.08	160.77	2.20622	.006220
5.09	162.39	2.21056	.006158
5.10	164.02	2.21490	0.006097
5.11	165.67	2.21924	.006036
5.12	167.34	2.22359	.005976
5.13	169.02	2.22793	.005917
5.14	170.72	2.23227	.005858
5.15	172.43	2.23662	0.005799
5.16	174.16	2.24096	.005742
5.17	175.91	2.24530	.005685
5.18	177.68	2.24965	.005628
5.19	179.47	2.25399	.005572
5.20	181.27	2.25833	0.005517
5.21	183.09	2.26267	.005462
5.22	184.93	2.26702	.005407
5.23	186.79	2.27136	.005354
5.24	188.67	2.27570	.005300
5.25	190.57	2.28005	0.005248
5.26	192.48	2.28439	.005195
5.27	194.42	2.28873	.005144
5.28	196.37	2.29307	.005092
5.29	198.34	2.29742	.005042
5.30	200.34	2.30176	0.004992
5.31	202.35	2.30610	.004942
5.32	204.38	2.31045	.004893
5.33	206.44	2.31479	.004844
5.34	208.51	2.31913	.004796
5.35	210.61	2.32348	0.004748
5.36	212.72	2.32782	.004701
5.37	214.86	2.33216	.004654
5.38	217.02	2.33650	.004608
5.39	219.20	2.34085	.004562
5.40	221.41	2.34519	0.004517
5.41	223.63	2.34953	.004472
5.42	225.88	2.35388	.004427
5.43	228.15	2.35822	.004383
5.44	230.44	2.36256	.004339
5.45	232.76	2.36690	0.004296
5.46	235.10	2.37125	.004254
5.47	237.46	2.37559	.004211
5.48	239.85	2.37993	.004169
5.49	242.26	2.38428	.004128
5.50	244.69	2.38862	0.004087

x	e^x	$\text{Log}_{10}(e^x)$	e^{-x}
5.50	244.69	2.38862	0.0040868
5.55	257.24	2.41033	.0038875
5.60	270.43	2.43205	.0036979
5.65	284.29	2.45376	.0035175
5.70	298.87	2.47548	.0033460
5.75	314.19	2.49719	0.0031828
5.80	330.30	2.51891	.0030276
5.85	347.23	2.54062	.0028799
5.90	365.04	2.56234	.0027394
5.95	383.75	2.58405	.0026058
6.00	403.43	2.60577	0.0024788
6.05	424.11	2.62748	.0023579
6.10	445.86	2.64920	.0022429
6.15	468.72	2.67091	.0021335
6.20	492.75	2.69263	.0020294
6.25	518.01	2.71434	0.0019305
6.30	544.57	2.73606	.0018363
6.35	572.49	2.75777	.0017467
6.40	601.85	2.77948	.0016616
6.45	632.70	2.80120	.0015805
6.50	665.14	2.82291	0.0015034
6.55	699.24	2.84463	.0014301
6.60	735.10	2.86634	.0013604
6.65	772.78	2.88806	.0012940
6.70	812.41	2.90977	.0012309
6.75	854.06	2.93149	0.0011709
6.80	897.85	2.95320	.0011138
6.85	943.88	2.97492	.0010595
6.90	992.27	2.99663	.0010078
6.95	1043.1	3.01835	.0009586
7.00	1096.6	3.04006	0.0009119
7.05	1152.9	3.06178	.0008674
7.10	1212.0	3.08349	.0008251
7.15	1274.1	3.10521	.0007849
7.20	1339.4	3.12692	.0007466
7.25	1408.1	3.14863	0.0007102
7.30	1480.3	3.17035	.0006755
7.35	1556.2	3.19206	.0006426
7.40	1636.0	3.21378	.0006113
7.45	1719.9	3.23549	.0005814
7.50	1808.0	3.25721	0.0005531
7.55	1900.7	3.27892	.0005261
7.60	1998.2	3.30064	.0005005
7.65	2100.6	3.32235	.0004760
7.70	2208.3	3.34407	.0004528
7.75	2321.6	3.36578	0.0004307
7.80	2440.6	3.38750	.0004097
7.85	2565.7	3.40921	.0003898
7.90	2697.3	3.43093	.0003707
7.95	2835.6	3.45264	.0003527
8.00	2981.0	3.47436	0.0003355

TABLE 5 Exponential Functions (Continued)

x	e^x	$\text{Log}_{10}(e^x)$	e^{-x}
8.00	2981.0	3.47436	0.0003355
8.05	3133.8	3.49607	.0003191
8.10	3294.5	3.51779	.0003035
8.15	3463.4	3.53950	.0002887
8.20	3641.0	3.56121	.0002747
8.25	3827.6	3.58293	0.0002613
8.30	4023.9	3.60464	.0002485
8.35	4230.2	3.62636	.0002364
8.40	4447.1	3.64807	.0002249
8.45	4675.1	3.66979	.0002139
8.50	4914.8	3.69150	0.0002035
8.55	5166.8	3.71322	.0001935
8.60	5431.7	3.73493	.0001841
8.65	5710.1	3.75665	.0001751
8.70	6002.9	3.77836	.0001666
8.75	6310.7	3.80008	0.0001585
8.80	6634.2	3.82179	.0001507
8.85	6974.4	3.84351	.0001434
8.90	7332.0	3.86522	.0001364
8.95	7707.9	3.88694	.0001297
9.00	8103.1	3.90865	0.0001234
9.05	8518.5	3.93037	.0001174
9.10	8955.3	3.95208	.0001117
9.15	9414.4	3.97379	.0001062
9.20	9897.1	3.99551	.0001010
9.25	10405	4.01722	0.0000961
9.30	10938	4.03894	.0000914
9.35	11499	4.06065	.0000870
9.40	12088	4.08237	.0000827
9.45	12708	4.10408	.0000787
9.50	13360	4.12580	0.0000749
9.55	14045	4.14751	.0000712
9.60	14765	4.16923	.0000677
9.65	15522	4.19094	.0000644
9.70	16318	4.21266	.0000613
9.75	17154	4.23437	0.0000583
9.80	18034	4.25609	.0000555
9.85	18958	4.27780	.0000527
9.90	19930	4.29952	.0000502
9.95	20952	4.32123	0.0000477
10.00	22026	4.34294	0.0000454

Answers To Selected Exercises

CHAPTER 1

Section 1, page 2

1. $\frac{3}{2}(1 - 3^{-10})$ **3.** $255/256$
5. $3(x^{n+1} - 3^{n+1})/x^n(x - 3)$
7. $(\sqrt{r} + r)(1 - r^4)/(1 - r)$ **9.** $\frac{5}{7}$ **11.** 2^{-9}
13. $1/(1 + x^2)$ **15.** 9 ft
17. 60 miles (one hour at 60 mph) **19.** $\frac{1}{9}$
21. $43/99$ **23.** $\frac{1}{2} + \frac{1}{4} + \cdots + 1/2n =$
$\frac{1}{2}(1 + \frac{1}{2} + \cdots + 1/n) \longrightarrow \infty.$ **25.** 1600 because

$$\frac{1}{101} + \cdots + \frac{1}{1600} = \left(\frac{1}{101} + \cdots + \frac{1}{200}\right) + \left(\frac{1}{201} + \cdots + \frac{1}{400}\right)$$

$$+ \left(\frac{1}{401} + \cdots + \frac{1}{800}\right) + \left(\frac{1}{801} + \cdots + \frac{1}{1600}\right)$$

$$> \frac{1}{2} + \cdots + \frac{1}{2} = 2.$$

Section 2, page 8

1. C **3.** D **5.** C **7.** C **9.** C
11. $s_n < 1 + 1/(p - 1)$ for $p = 2$.
13. $\sum_{j=1}^{n}(a_j + b_j) = (\sum_{j=1}^{n} a_j) + (\sum_{j=1}^{n} b_j) \longrightarrow A + B.$
15. $\sum a_n$ converges, so $a_n \longrightarrow 0$. Therefore $0 \leq a_n < 1$ for $n \geq N$, so $0 \leq a_n{}^2 < a_n$. Then $\sum a_n{}^2$ converges by comparison with $\sum a_n$.

Section 3, page 11

1. C **3.** D **5.** C **7.** C **9.** C **11.** C
13. All real x. Given x, set $a_n = x^{2n}/n!$. Then $a_{n+1}/a_n = x^2/(n + 1) \longrightarrow 0$, so the series converges by the ratio test.
15. $-\frac{1}{3} < x < \frac{1}{3}$

17. $a_n < r^n$ and $r < 1$; convergence by comparison with the geometric series.
19. Take $a_n = 1/n$ and $b_n = 1/n^2$. There is no contradiction; the test in the text requires $b_n/a_n \longrightarrow L > 0$.

Section 4, page 15

1. C, not abs **3.** D **5.** C abs
7. C, not abs **9.** C abs **11.** D

13. $1 - \dfrac{1}{2} + \dfrac{1}{4\cdot 2!} - \dfrac{1}{8\cdot 3!} + \dfrac{1}{16\cdot 4!} - \dfrac{1}{32\cdot 5!} \approx 0.60651$

15. $s_{2n} = s_{2n-2} + (a_{2n-1} + a_{2n}) > s_{2n-2}$ and
$s_{2n} = a_1 + [(a_2 + a_3) + (a_4 + a_5) + \cdots$
$+ (a_{2n-2} + a_{2n-1}) + (a_{2n})] < a_1$ since each term in parentheses is negative. Similarly $s_{2n+1} < s_{2n-1}$ and $s_{2n+1} > 0$.

Section 5, page 22

1. $\frac{1}{8}$ **3.** 1 **5.** $\pi/4$ **7.** 2 **9.** $\frac{1}{3}\ln(3 + \sqrt{10})$
11. $\pi/4$ **13.** $2/s^3$ **15.** $1/(s - a)$ **17.** $s/(s^2 - 1)$
19. infinite **21.** infinite **23.** $\ln 2 \approx 0.69315$
25. $\ln 100 \approx 4.60517$

Section 6, page 29

1. D **3.** C **5.** D **7.** C **9.** C **11.** D

15. $V = \pi \int_1^\infty dx/x^2 = \pi$, $A \geq \int_1^\infty dx/x = \infty$. The word "paint" implies a layer of uniform thickness. **17.** $s \geq 0$ **19.** $s > \frac{1}{2}$

Section 7, page 34

1. D **3.** C **5.** D **7.** C **9.** $\sum_1^\infty 1/n^2 < 1 + \int_1^\infty dx/x^2 = 2$

11. about $e^{1000} \approx 1.97 \times 10^{434}$ **13.** C **15.** D
17. The substitution $x^2 = u$ reduces it to Ex. 16.
19. By the Mean Value Theorem,

$$f(x) = \frac{\arctan bx - \arctan ax}{x} = \frac{bx - ax}{x}\,\frac{1}{1 + \theta^2},$$

where $ax \leq \theta < bx$. Hence $f(x) \leq (b - a)/(1 + a^2x^2)$, etc.

Section 8, page 39

1. C **3.** D **5.** D **7.** C **9.** C **11.** D **13.** C
15. C

Section 9, page 43

1. $I = \int_0^\pi = \int_0^{\pi/2} + \int_{\pi/2}^\pi = \int_0^{\pi/2} x \ln \sin x$

$+ \int_0^{\pi/2} (x + \frac{1}{2}\pi) \ln \cos x = \frac{1}{2}\pi \int_0^{\pi/2} \ln \cos x$

$+ \int_0^{\pi/2} x(\ln \sin x + \ln \cos x)$. Now $\int_0^{\pi/2} \ln \cos x$

$= -\frac{1}{2}\pi \ln 2$ by Example 9.1, and

$\int_0^{\pi/2} x(\ln \sin x + \ln \cos x) = \int_0^{\pi/2} x \ln(\frac{1}{2} \sin 2x)$

$= \int_0^{\pi/2} x \ln \sin 2x - (\ln 2) \int_0^{\pi/2} x = \frac{1}{4} \int_0^\pi x \ln \sin x$

$- \frac{1}{8}\pi^2 \ln 2$. Hence $I = -\frac{1}{4}\pi^2 \ln 2 + \frac{1}{4}I - \frac{1}{8}\pi^2 \ln 2$, $\frac{3}{4}I = -\frac{3}{8}\pi^2 \ln 2$.

3. Set $\cos x = t$. Then $I = \int_0^1 \ln \sqrt{1 - t^2}\, dt$

$= \frac{1}{2} \int_0^1 \ln(1 - t^2)\, dt = \frac{1}{2} \int_0^1 \ln(1 + t)\, dt + \frac{1}{2} \int_0^1 \ln(1 - t)\, dt$

$= \frac{1}{2} \int_1^2 \ln u\, du + \frac{1}{2} \int_0^1 \ln u\, du = \frac{1}{2} \int_0^2 \ln u\, du = \frac{1}{2}(u \ln u - u) \Big|_0^2$

$= \ln 2 - 1$.

5. By parts, $I_n = \int_0^1 (\ln x)^n\, dx = x(\ln x)^n \Big|_0^1 - \int_0^1 nx(\ln x)^{n-1}(dx/x)$

$= -nI_{n-1}$, etc.

7. $I = \int_0^\infty \operatorname{sech} x\, dx = 2 \arctan e^x \Big|_0^\infty = 2(\frac{1}{2}\pi - \frac{1}{4}\pi) = \frac{1}{2}\pi.$

9. Set

$$\frac{1}{x^4 + 1} = \frac{Ax + B}{x^2 + \sqrt{2}x + 1} + \frac{Cx + D}{x^2 - \sqrt{2}x + 1},$$

clear of fractions, and equate powers of x.

Section 10, page 51

1. $100! = \sqrt{200\pi}(100/e)^{100}$. By 4-place tables, $\log(100!) \approx 158.0$. Answer: 158. **3.** $\ln 2$ **5.** $\ln 3$

7. From a figure like 10.1, with $y = 1/\sqrt{x}$,

$$\sum_{1}^{n-1} \frac{1}{\sqrt{k}} = \int_1^n \frac{dx}{\sqrt{x}} + a_n = 2\sqrt{n} - 2 + a_n,$$

where $a_1 < a_2 < a_3 < \cdots < 1$. Hence $a_n \longrightarrow L$,

$$\sum_{1}^{n} \frac{1}{\sqrt{k}} - 2\sqrt{n} = \frac{1}{\sqrt{n}} - 2 + a_n \longrightarrow L - 2.$$

CHAPTER 2

Section 2, page 55

1. $(x-1)^2 + 7(x-1) + 8$ **3.** $2(x+1)^3 - (x+1)^2 + 9(x+1)$
5. $2(x+2)^4 - 11(x+2)^3 + 18(x+2)^2$
7. $5(x+1)^5 - 21(x+1)^4 + 31(x+1)^3 - 19(x+1)^2 + 5(x+1)$
9. $x^4 - 7x^3 + 5x^2 + 3x - 6$ **11.** 62.23 **13.** -157.2

Section 3, page 62

1. $2x - \dfrac{8}{3!}x^3 + \dfrac{2^5}{5!}x^5 - \cdots + (-1)^{n-1}\dfrac{2^{2n-1}}{(2n-1)!}x^{2n-1},$

$$|r_{2n-1}(x)| \le \frac{2^{2n+1}}{(2n+1)!}|x|^{2n+1}$$

3. $x + x^2 + \dfrac{x^3}{2!} + \dfrac{x^4}{3!} + \cdots + \dfrac{x^n}{(n-1)!},$

$$|r_n(x)| \le \frac{(x+n+1)e^x}{(n+1)!}x^{n+1} \text{ if } x \ge 0, \qquad |r_n(x)| \le |x|^{n+1}/n! \text{ if } x < 0$$

5. $(x-1) + 3(x-1)^2/2! + 2(x-1)^3/3! - 2(x-1)^4/4! +$
$4(x-1)^5/5! + \cdots + (-1)^n 2(x-1)^n/n(n-1)(n-2);$
if $n \ge 4$: $|r_n(x)| \le 2(n-2)!(x-1)^{n+1}/(n+1)!$ for $x \ge 1$,
$|r_n(x)| \le 2(n-2)!(1-x)^{n+1}/(n+1)!x^{n-1}$ for $0 < x \le 1$

7. $x^2 - x^3 + \frac{1}{2}x^4 - \frac{1}{6}x^5 + \cdots + (-1)^n x^n/(n-2)!;$
$|r_n(x)| \le x^{n+1}/(n-1)!$ for $x \ge 0$,
$|r_n(x)| \le [x^2 - 2(n+1)x + n(n+1)]\,|x|^{n+1}/e^x(n+1)!$ for $x \le 0$

9. $x^2 - x^4/3! + x^6/5! + \cdots + (-1)^{n-1}x^{2n}/(2n-1)!;$
$|r_{2n}(x)| \le [|x| + 2n + 2]\,|x|^{2n+2}/(2n+2)!$

11. $1 + x - \frac{1}{2}x^2 - \frac{1}{6}x^3 + x^4/4! + x^5/5! + \cdots + \sigma_n x^n/n!$, where
$\sigma_{4k} = \sigma_{4k+1} = 1,\ \sigma_{4k+2} = \sigma_{4k+3} = -1;\ |r_n(x)| \le |x|^{n+1}/(n+1)!$

13. $2x + 2x^5/5! + 2x^9/9! + \cdots + 2x^{4n+1}/(4n+1)!$;
$|r_{4n+1}(x)| \leq (1 + \cosh x)\,|x|^{4n+3}/(4n+3)!$
15. $|\text{error}| \leq \frac{3}{128} < 0.024$

Section 4, page 66

1. $1 - x^2 + \frac{1}{2}x^4 - \frac{1}{6}x^6 + x^8/4! - x^{10}/5! + x^{12}/6!$
3. $1/(11 \times 2^{11}) < 5 \times 10^{-5}$
5. $p_4(x) = x^2 - \frac{1}{3}x^4$, $|\text{error}| \leq \frac{1}{2}(2x)^6/6! < 5 \times 10^{-8}$
7. $|\sin x - p_9(x)| < (\pi/2)^{11}/11! < 3.6 \times 10^{-6}$
9. $\sin(5\pi/8) \approx 1 - (\frac{1}{8}\pi)^2/2! + (\frac{1}{8}\pi)^4/4! - (\frac{1}{8}\pi)^6/6! \approx 0.92388$,
$|\text{error}| < 2 \times 10^{-8}$

Section 5, page 68

1. $\sum_1^\infty (-1)^{i-1}3^{2i-1}x^{2i-1}/(2i-1)!$ 3. $\sum_1^\infty (-1)^{i-1}2^{2i-1}x^{2i}/(2i)!$
5. $\sum_0^\infty (-1)^i 2^i x^i/i!$ 7. $\sum_0^\infty 3^i(x + \frac{2}{3})^i/i!$ 9. $\sum_0^\infty 3^i x^i$
11. $\sum_0^\infty (-1)^i (x-2)^i/2^{i+1}$

CHAPTER 3

Section 1, page 77

1. $1/(4-x)$, 1 3. $1/(1+x^2)$, 1 5. e^{x+1}, ∞
7. $1/(1-125x^3)$, $\frac{1}{5}$ 9. $e^{-(x-1)^4}$, ∞ 11. $1/(1-e^x)$, $x < 0$
13. $e^{-\sin x}$, all x 15. $1/(1-2\sqrt{x})$, $0 \leq x < \frac{1}{4}$
17. $1/(1-a^2-x^2)$, $|x| < (1-a^2)^{1/2}$ if $|a| < 1$, no x if $|a| \geq 1$
19. $\cos x^{1/3}$, all x

Section 2, page 81

1. 1 3. 1 5. 2 7. 1 9. $1/e$ 11. $\frac{4}{3}$ 13. 1
15. 5 17. 1 19. ∞

Section 3, page 89

1. $\sum_0^\infty 5^n x^n$ 3. $\sum_0^\infty x^{3n}/n!$ 5. $\sum_0^\infty x^n/(2n)!$
7. $-1 + \sum_2^\infty (n-1)x^n/n!$ 9. $\sum_0^\infty x^{n+2}$
11. $\sum_1^\infty (-1)^{n+1}2^{2n-1}x^{2n}/(2n)!$
13. $1 + 2x + 7x^2 + 14x^3 + 37x^4 + 74x^5 + 175x^6$
15. $1 + x^2 + x^3 + \frac{3}{2}x^4 + \frac{13}{6}x^5 + \frac{73}{24}x^6$ 17. $x^3 - \frac{1}{2}x^5$
19. $8!$ 21. $8!/4! = 1680$

Section 4, page 95

1. $\sum_0^\infty (3/2^{n+1} - 4/3^{n+1})x^n$, $|x| < 2$
3. $\frac{1}{2}\sum_0^\infty (1/2^n - 1 - 1/3^{n+1})x^n$, $|x| < 1$

5. $\sum_0^\infty (-1)^n(x^{3n} + x^{3n+1})$, $|x| < 1$ **7.** $(4 - 3x)/(1 - x)^2$, $|x| < 1$
9. $-\frac{1}{4}\ln(1 - x^4)$, $|x| < 1$ **11.** $(2 - x)/(1 - x)^2$, $|x| < 1$

Section 5, page 101

1. $\sum_0^\infty \frac{1}{2}(-1)^n(n + 1)(n + 2)x^n$ **3.** $1 + \sum_1^\infty (n + 1)4^n x^{2n}$
5. $1 + \frac{1}{2}x^3 + \sum_2^\infty (-1)^{n-1}[3\cdot 7\cdot 11\cdots(4n - 5)]x^{3n}/2^n\cdot n!$
7. $\sqrt{2}\{1 + \frac{1}{4}(x - 1) + \sum_2^\infty (-1)^n[3\cdot 5\cdot 7\cdots(2n - 3)](x - 1)^n/4^n\cdot n!\}$
9. $1 + \frac{1}{2}x^2 + \frac{1}{2}x^3 + \frac{1}{8}x^4$ **11.** $\sqrt{3}(2x + \frac{1}{3}x^2 - \frac{49}{36}x^3 - \frac{47}{216}x^4)$
13. $1 - \frac{4}{3}x - \frac{8}{9}x^2 + \frac{52}{27}x^3 - \frac{91}{81}x^4$
15. $3(1 + \frac{1}{81})^{1/4} \approx 3 + \frac{1}{108} - 2^{-5} \times 3^{-6} \approx$
$3.00000 + 0.00926 - 0.00004 \approx 3.0092$

17. $$\sum_0^\infty \frac{(2n)!}{2^{2n}(n!)^2}\frac{x^{2n+1}}{2n + 1}$$

Section 6, page 104

1. The alternating series $1 - \frac{1}{2} + \frac{1}{3} - \cdots$ has $1/n \longrightarrow 0$.
3. $(\frac{1}{5})^5/5! < 4 \times 10^{-6}$, hence $e^{-1/5} \approx 1 - \frac{1}{5} + \frac{1}{2}(\frac{1}{5})^2 - \frac{1}{6}(\frac{1}{5})^3 + \frac{1}{24}(\frac{1}{5})^4 \approx 0.81874$ **5.** 4, including the constant

Section 7, page 108

1. $\sum_0^\infty (-1)^n x^{2n+1}/(2n + 1)(2n + 1)!$ **3.** $\sum_0^\infty (-1)^n x^{4n+2}/(4n + 2)$
5. 0.10003 **7.** 0.25049 **9.** $k = \frac{9}{41}$, length $= (2\pi)(41)(1 - S)$, $S \approx (\frac{1}{2})^2(\frac{9}{41})^2 + (\frac{1}{2}\cdot\frac{3}{4})^2(\frac{1}{3})(\frac{9}{41})^4 \approx 0.01216$; length ≈ 254.5

Section 8, page 115

1. Let $\epsilon > 0$. Then there exists N such that $(\frac{9}{10})^n < \epsilon$ for all $n \geq N$ since $(\frac{9}{10})^n \longrightarrow 0$. Therefore $|f_n(x) - 0| \leq (\frac{9}{10})^n < \epsilon$ for $n \geq N$ and all x such that $0 \leq x \leq \frac{9}{10}$, so $f_n \longrightarrow 0$ uniformly on $[0, \frac{9}{10}]$. Suppose $f_n \longrightarrow F$ uniformly on $[0, 1]$. Then F would be continuous on $[0, 1]$ by Theorem 8.1 since each f_n is continuous. But $F(x) = 0$ for $0 \leq x < 1$ and $F(1) = 1$, so F is not continuous.
3. By elementary calculus, $0 \leq xe^{-nx} \leq 1/ne$ for $0 \leq x < \infty$. Therefore $xe^{-nx} \longrightarrow 0$ uniformly on $[0, \infty)$.
5. $|(\sin nx)/n^2| \leq 1/n^2$; use the M-test.
7. Let $0 < a$. Then $|e^{-nx}\sin nx| \leq e^{-na}$ for $a \leq x < \infty$. By the M-test, the series converges uniformly on $[a, \infty)$. By Theorem 8.1, the sum is continuous there. This is true for each $a > 0$, hence the sum is continuous for $0 < x < \infty$.
9. Both series converge uniformly by the M-test. Apply Theorem 8.4 (3).
11. $|a_n \sin nx| \leq |a_n|$. Apply the M-test with $M_n = |a_n|$.

CHAPTER 4

Section 1, page 122

9. yes **11.** no **13.** $(5, 2, 4)$
15. $(3, -2, 10)$ **17.** $(1, -16, 9)$
19. $(-1, 10, 4)$ **21.** $u_1 + v_1 = v_1 + u_1$, etc.
23. $(a + b)v_1 = av_1 + bv_1$, etc.
25. The diagonals of a parallelogram bisect each other. But $\frac{1}{2}(\mathbf{v} + \mathbf{w})$ is the midpoint of the diagonal from $\mathbf{0}$ to $\mathbf{v} + \mathbf{w}$. Hence it is the midpoint of the diagonal from $\mathbf{v}$ to $\mathbf{w}$. *Alternate solution*: The vector $\mathbf{w} - \mathbf{v}$ is parallel to the segment from $\mathbf{v}$ to $\mathbf{w}$, and has the same length. Hence the midpoint is $\frac{1}{2}(\mathbf{w} - \mathbf{v}) + \mathbf{v} = \frac{1}{2}(\mathbf{v} + \mathbf{w})$.
27. $\frac{1}{3}(\mathbf{u} + \mathbf{v} + \mathbf{w}) = \frac{2}{3}[\frac{1}{2}(\mathbf{u} + \mathbf{v})] + \frac{1}{3}\mathbf{w}$. Hence $\frac{1}{3}(\mathbf{u} + \mathbf{v} + \mathbf{w})$ is on the segment joining $\mathbf{w}$ to the midpoint of $\mathbf{uv}$ (a median). Likewise it is on the other two medians, so all three medians intersect.

Section 2, page 129

1. 29 **3.** -1 **5.** $\sqrt{11}$ **7.** 1
9. $\arccos(-12/25)$ **11.** $\frac{1}{2}\pi$
13. $\arccos(-1/\sqrt{15})$ **15.** $\sqrt{42}$ **17.** $5\sqrt{2}$
19. $\frac{1}{2}\sqrt{2}, 0, \frac{1}{2}\sqrt{2}$ **21.** $\frac{1}{7}\sqrt{14}, \frac{1}{14}\sqrt{14}, -\frac{3}{14}\sqrt{14}$
23. $(1, 1, 0), (0, 2, 1)$ for instance
25. $|\mathbf{v}\cdot\mathbf{w}| = |\mathbf{v}|\cdot|\mathbf{w}|\,|\cos\theta| \le |\mathbf{v}|\cdot|\mathbf{w}|$. Equality if $\mathbf{v} = \mathbf{0}$, $\mathbf{w} = \mathbf{0}$, or $\theta = 0, \pi$.
27. $|\mathbf{v} + \mathbf{w}|^2 - |\mathbf{v} - \mathbf{w}|^2 = (\mathbf{v} + \mathbf{w})\cdot(\mathbf{v} + \mathbf{w}) - (\mathbf{v} - \mathbf{w})\cdot(\mathbf{v} - \mathbf{w}) = (|\mathbf{v}|^2 + 2\mathbf{v}\cdot\mathbf{w} + |\mathbf{w}|^2) - (|\mathbf{v}|^2 - 2\mathbf{v}\cdot\mathbf{w} + |\mathbf{w}|^2) = 4\mathbf{v}\cdot\mathbf{w}$.
29. $\mathbf{u}\cdot\mathbf{v} = |\mathbf{u}|\cdot|\mathbf{v}|\cos\theta = \cos\theta$, etc. When $\mathbf{u} = (\cos\alpha, \sin\alpha, 0)$ and $\mathbf{v} = (\cos\beta, \sin\beta, 0)$ are plane vectors, the formula reduces to the addition law $\cos(\alpha - \beta) = \cos\alpha\cos\beta + \sin\alpha\sin\beta$.

Section 3, page 132

1. $\mathbf{x}\cdot\mathbf{n} = \frac{1}{3}$, $\mathbf{n} = (\frac{1}{3}, -\frac{2}{3}, \frac{2}{3})$
3. $\mathbf{x}\cdot\mathbf{n} = 3$, $\mathbf{n} = (-\frac{8}{9}, \frac{1}{9}, -\frac{4}{9})$
5. $\mathbf{x}\cdot\mathbf{n} = \sqrt{3}$, $\mathbf{n} = \frac{1}{3}\sqrt{3}(1, 1, 1)$
7. $\mathbf{x} = \mathbf{x}_0 + t(\mathbf{x}_1 - \mathbf{x}_0)$ for instance
9. The plane through $\mathbf{x}_0$ parallel to the given plane is $\mathbf{x}\cdot\mathbf{n} = \mathbf{x}_0\cdot\mathbf{n}$. The line $\mathbf{x} = t\mathbf{n}$, perpendicular to these parallel planes, meets them in $p\mathbf{n}$ and $(\mathbf{x}_0\cdot\mathbf{n})\mathbf{n}$ respectively. Hence the distance between the planes is $|p\mathbf{n} - (\mathbf{x}_0\cdot\mathbf{n})\mathbf{n}| = |p - \mathbf{x}_0\cdot\mathbf{n}|$.
11. They are parallel if and only if the direction of the line is perpendicular to the normal to the plane, that is, $\mathbf{v}\cdot\mathbf{n} = 0$.
13. Substitute $\mathbf{x} = \mathbf{x}_0 + t\mathbf{v}$ into $\mathbf{x}\cdot\mathbf{n} = p$ and solve for t. The result is $t = (p - \mathbf{x}_0\cdot\mathbf{n})/(\mathbf{v}\cdot\mathbf{n})$. Now substitute this t into $\mathbf{x} = \mathbf{x}_0 + t\mathbf{v}$.

15. Let $\mathbf{x} = \mathbf{x}_0 + t\mathbf{u}$ be the required point. Then $(\mathbf{x} - \mathbf{y}_0)\cdot\mathbf{u} = 0$, so
$$t = (\mathbf{x} - \mathbf{x}_0)\cdot\mathbf{u} = (\mathbf{x} - \mathbf{y}_0 + \mathbf{y}_0 - \mathbf{x}_0)\cdot\mathbf{u} = (\mathbf{y}_0 - \mathbf{x}_0)\cdot\mathbf{u}.$$
Therefore $\mathbf{x} = \mathbf{x}_0 + [(\mathbf{y}_0 - \mathbf{x}_0)\cdot\mathbf{u}]\mathbf{u}$.

Section 4, page 140

1. $(-\frac{1}{3}, \frac{2}{3})$ **3.** $(-1, 1)$ **5.** $(1/19, 7/19)$
7. $(-8/19, 1/19, 7/19)$ **9.** $(0, 0, 0)$
11. $(6/28, -7/28, -3/28)$ **13.** two parallel planes
15. The first and third planes are parallel. **17.** $(t, \frac{2}{3}t - \frac{1}{3})$
19. $(t, -\frac{1}{2}t - \frac{1}{2}, -\frac{5}{2}t + \frac{1}{2})$
21. Subtract a times the second equation from the third; then subtract a times the first equation from the second. The results are $(b - a)y + (c - a)z = d_2'$ and $b(b - a)y + c(c - a)z = d_3'$. Subtract b times the first of *these* from the second: $(c - b)(c - a)z = d_3''$, etc.

Section 5, page 147

1. $(-5, 2, -14)$ **3.** $(-4, 8, -4)$ **5.** $(2, -2, 0)$ **7.** $(0, 0, 0)$
9. $(-7, 8, 2)$ **11.** $(0, -10, 10)$ **13.** $(3, -1, 4)$
15. A determinant changes sign when two rows are interchanged, hence $\mathbf{u}\cdot(\mathbf{v}\times\mathbf{w}) = -\mathbf{v}\cdot(\mathbf{u}\times\mathbf{w}) = +\mathbf{v}\cdot(\mathbf{w}\times\mathbf{u})$.
17. $(av_2)w_3 - (av_3)w_2 = a(v_2w_3 - v_3w_2)$, etc.
19. Express both sides in terms of components and compare.
23. By the second formula, $(\mathbf{a}\times\mathbf{b})\times(\mathbf{a}\times\mathbf{c}) = [(\mathbf{a}\times\mathbf{c})\cdot\mathbf{a}]\mathbf{b} - [(\mathbf{a}\times\mathbf{c})\cdot\mathbf{b}]\mathbf{a} = -[(\mathbf{a}\times\mathbf{c})\cdot\mathbf{b}]\mathbf{a}$.

Section 6, page 155

1. 2 **3.** $\mathbf{x} = (-2, 1, 0) + t(-5, 1, 1)$
5. $\mathbf{x} = (\frac{8}{5}, \frac{1}{5}, 1) + t(3, 1, -5)$
7. $(3, 1)$ **9.** $(1, 4, -7)$ **11.** $(6, -5, -19)$
13. $s(5, 0, -4) + t(0, 5, -3)$ **15.** $x = 1$
17. $2x - y - z = 3$ **19.** $x/a + y/b + z/c = 1$
21. Up to sign, the determinant is the volume V of a parallelepiped with sides $a = (a_1^2 + a_2^2 + a_3^2)^{1/2}$, $b = (b_1^2 + b_2^2 + b_3^2)^{1/2}$, $c = (c_1^2 + c_2^2 + c_3^2)^{1/2}$. The base area is $A = bc \sin\alpha \leq bc$. The height is $h = a \sin\gamma$, where γ is the angle between c and the base. Hence $V = Ah \leq abc$.
23. $\mathbf{F} = (0, 0, n)$, $\mathbf{p} = (\sum_1^n \mathbf{p}_j)/n$
25. $(\mathbf{p} + \mathbf{c})\times\mathbf{F} - (\mathbf{q} + \mathbf{c})\times\mathbf{F} = (\mathbf{p} - \mathbf{q})\times\mathbf{F}$

CHAPTER 5

Section 1, page 161

1. $\dot{\mathbf{x}} = (e^t, 2e^{2t}, 3e^{3t})$ **3.** $(1, 3, 4)$
5. $\mathbf{v} = (2t, 3t^2 + 4t^3, 0)$, $|\mathbf{v}| = (4t^2 + 9t^4 + 24t^5 + 16t^6)^{1/2}$

7. $\mathbf{v} = (-A\omega \sin \omega t, A\omega \cos \omega t, B)$, $|\mathbf{v}| = (A^2\omega^2 + B^2)^{1/2}$
9. $d\,|\mathbf{x}|^2/dt = d(\mathbf{x}\cdot\mathbf{x})/dt = 2\mathbf{x}\cdot\dot{\mathbf{x}} = 0$, hence $|\mathbf{x}|^2 = \text{const}$
11. $(x_1 + y_1)^\cdot = \dot{x}_1 + \dot{y}_1$, etc. 13. $(x_2y_3 - x_3y_2)^\cdot = (\dot{x}_2y_3 - \dot{x}_3y_2) + (x_2\dot{y}_3 - x_3\dot{y}_2)$, etc.
15. $[\mathbf{x}(t) - \mathbf{y}(\tau_0)]\cdot[\mathbf{x}(t) - \mathbf{y}(\tau_0)]$ is minimum at $t = t_0$, hence its derivative is 0 there: $[\mathbf{x}(t_0) - \mathbf{y}(\tau_0)]\cdot\dot{\mathbf{x}}(t_0) = 0$, etc.

Section 2, page 167

1. $(a_1^2 + a_2^2 + a_3^2)^{1/2}$ 3. $2\pi\sqrt{2}$ 5. $\int_{x_0}^{x_1} (1 + a^2n^2x^{2n-2})^{1/2}\,dx$
7. $\sqrt{17} - \frac{1}{4}\ln(\sqrt{17} - 4)$ 9. $(\sqrt{2}/2)(1, -\sin t, \cos t)$
11. $(a_1^2 + a_2^2 + a_3^2)^{-1/2}(a_1, a_2, a_3)$

Section 3, page 173

1. $2\sqrt{5}/25$ 3. $\sqrt{19}/7\sqrt{14}$
5. $\dot{\mathbf{N}}\cdot\mathbf{N} = \frac{1}{2}d(\mathbf{N}\cdot\mathbf{N})/dt = 0$; $\dot{\mathbf{T}}\cdot\mathbf{N} + \mathbf{T}\cdot\dot{\mathbf{N}} = 0$, $\dot{\mathbf{T}} = k\mathbf{N}$, hence $\dot{\mathbf{N}}\cdot\mathbf{T} = -k$. Thus $\dot{\mathbf{N}} = -k\mathbf{T} + 0\mathbf{N} = -k\mathbf{T}$.
7. $(\pi/2, 1)$ 9. $k = |y''|/(\cdots)$. But $y'' = 0$ at an inflection.
11. $d\alpha/dt = (d\alpha/ds)(ds/dt) = 3e^2/(1+e^4)^{3/2}$ rad/sec
13. $k = (2/x^3)/[1 + (-1/x^2)^2]^{3/2} \longrightarrow 0$ as $x \longrightarrow 0$ or $x \longrightarrow \infty$

Section 4, page 180

1. Use the solution of Example 4.3. The shell strikes the hill when $y/x = \tan\beta$, that is, when $t = (2v_0/g)(\sin\alpha - \tan\beta\cos\alpha)$.
3. $\mathbf{v} = (1, 2t)$, $\mathbf{a} = (0, 2)$
5. $\mathbf{v} = \mathbf{T}$ and $\mathbf{a} = d\mathbf{T}/ds = k\mathbf{N}$ since $ds/dt = 1$. Hence the tangential component of $\mathbf{a}$ is always $\mathbf{0}$ and the normal component is $\mathbf{a}$ itself. Since $\mathbf{x} = (x, \sin x)$, $\mathbf{T}(ds/dx) = (1, \cos x)$ and $\mathbf{a}(ds/dx) + \mathbf{T}(d^2s/dx^2) = (0, -\sin x)$. Hence, $(ds/dx)^2 = 1 + \cos^2 x$, $(ds/dx)(d^2s/dx^2) = -\sin x \cos x$. Evaluate at $x = 0$: $\mathbf{a} = \mathbf{0}$; at $x = \pi/2$: $\mathbf{a} = (0, -\sqrt{2}/2)$.
7. $\mathbf{v} = (-a\omega\sin\omega t, a\omega\cos\omega t, b)$, $\mathbf{a} = -a\omega^2(\cos\omega t, \sin\omega t, 0)$. Since these are perpendicular, the tangential component of $\mathbf{a}$ is $\mathbf{0}$; the normal component is $\mathbf{a}$.
11. $8000\pi/24 \approx 1047$ mph, $8000\pi(\cos 40°)/24 \approx 802$ mph, 0

Section 5, page 186

1. 3 3. $6\pi^2$ 5. $\frac{2}{3}$ 7. 1 9. g 11. $(\frac{3}{2}, 2, \frac{5}{2})$
13. $(0, \frac{2}{5}, 0)$
15. $\mathbf{T}\,ds = d\mathbf{x} = (dx, dy)$. Rotate $\pi/2$ clockwise. Then $\mathbf{T}$ rotates into the outward normal $\mathbf{N}_0$: $\mathbf{N}_0\,ds = (dy, -dx)$. The area of the small triangle formed by the two nearby radii and $d\mathbf{x}$ is $\frac{1}{2}\mathbf{x}\cdot\mathbf{N}_0\,ds = \frac{1}{2}(x\,dy - y\,dx)$.

17. The center is $(a\theta, a)$ and $\mathbf{x}(\theta) =$ (moving point) $-$ (center) $= -a(\sin\theta, \cos\theta)$; hence, the moving point is $\mathbf{x} = (a\theta, a) - a(\sin\theta, \cos\theta)$.
19. $8a$ **23.** $(1 - e^{-1}, \frac{1}{2}(1 - e^{-2}), \frac{1}{3}(1 - e^{-3}))$
25. Since $\mathbf{x}$ and $\dot{\mathbf{x}}$ are perpendicular unit vectors, $\mathbf{x} \times \dot{\mathbf{x}}$ is a unit vector.

Section 6, page 192

1. $(0, 1)$ **3.** $(\sqrt{2}, \sqrt{2})$ **5.** $(\frac{1}{2}\sqrt{3}, \frac{1}{2})$ **7.** $\{\sqrt{2}, 7\pi/4\}$
9. $\{2, 5\pi/6\}$ **11.** $\frac{1}{2}a[\sqrt{2} + \ln(1 + \sqrt{2})]$

13. $6a \displaystyle\int_0^{\pi/6} (8\sin^2 3\theta + 1)^{1/2}\, d\theta$ **15.** $\pi a^2/4$ **17.** $\pi a^2/2$

19. $3\pi a^2/2$ **21.** a^2
23. $(2b^2 + a^2)(\pi/2 - \text{arc}\cos b/a) + 3b\sqrt{a^2 - b^2}$

CHAPTER 6

Section 2, page 204

1. Componentwise: $x_{nj} \longrightarrow a_j$ and $y_{nj} \longrightarrow b_j$, hence $x_{nj} + y_{nj} \longrightarrow a_j + b_j$.
3. $\mathbf{x}_n \cdot \mathbf{y}_n = x_{n1}y_{n1} + x_{n2}y_{n2} + x_{n3}y_{n3} \longrightarrow x_1y_1 + x_2y_2 + x_3y_3 = \mathbf{x} \cdot \mathbf{y}$.
5. $x_{n2}y_{n3} - x_{n3}y_{n2} \longrightarrow x_2y_3 - x_3y_2$, etc.
7. Let $(x_n, y_n) \longrightarrow (x, y)$ and $y_n \geq x_n^2$. Then $y_n - x_n^2 \geq 0$ and $y_n - x_n^2 \longrightarrow y - x^2$, so $y - x^2 \geq 0$, $y \geq x^2$.
9. If $(x_n, y_n) \longrightarrow (x, y)$ and $x_n^2/a^2 - y_n^2/b^2 \leq 1$, then $x^2/a^2 - y^2/b^2 = \lim(x_n^2/a^2 - y_n^2/b^2) \leq 1$.
11. If $\mathbf{x}_n \in \mathbf{R}^3$ and $\mathbf{x}_n \longrightarrow \mathbf{x}$, then $\mathbf{x} \in \mathbf{R}^3$.
13. Let $\{x_0, y_0\} \in \mathbf{S}$ so $0 < x_0 < 1$. Set $\delta = \min\{x_0, 1 - x_0\} > 0$. Then $\{\mathbf{x} \mid |\mathbf{x} - \mathbf{x}_0| < \delta\} \subseteq \mathbf{S}$.
15. Let $\mathbf{x}_0 \in \mathbf{S} \cap \mathbf{T}$. Since $\mathbf{x}_0 \in \mathbf{S}$, there is $\delta_1 > 0$ such that $\{\mathbf{x} \mid |\mathbf{x} - \mathbf{x}_0| < \delta_1\} \subseteq \mathbf{S}$. Likewise there is $\delta_2 > 0$ for $\mathbf{T}$. Set $\delta = \min\{\delta_1, \delta_2\} > 0$. Then $|\mathbf{x} - \mathbf{x}_0| < \delta$ implies $\mathbf{x} \in \mathbf{S}$ and $\mathbf{x} \in \mathbf{T}$, hence $\mathbf{x} \in \mathbf{S} \cap \mathbf{T}$.
17. open, not closed **19.** neither **21.** closed, not open

Section 3, page 208

1. Let f_1 and f_2 be continuous at $\mathbf{a}$. If $\epsilon > 0$, there exist $\delta_1 > 0$ and $\delta_2 > 0$ such that $|\mathbf{x} - \mathbf{a}| < \delta_j$ implies $|f_j(\mathbf{x}) - f_j(\mathbf{a})| < \frac{1}{2}\epsilon$. Set $\delta = \min\{\delta_1, \delta_2\}$. Then $|\mathbf{x} - \mathbf{a}| < \delta$ implies

$$|[f_1(\mathbf{x}) + f_2(\mathbf{x})] - [f_1(\mathbf{a}) + f_2(\mathbf{a})]| \leq |f_1(\mathbf{x}) - f_1(\mathbf{a})| + |f_2(\mathbf{x}) - f_2(\mathbf{a})| < \tfrac{1}{2}\epsilon + \tfrac{1}{2}\epsilon = \epsilon.$$

3. By the triangle inequality, $||\mathbf{x}| - |\mathbf{a}|| \leq |\mathbf{x} - \mathbf{a}|$, etc.

5. $g(x, y) = x + y$ is continuous (sum of continuous functions), so $f(x, y) = 1/g(x, y)$ is continuous (quotient of continuous functions, non-zero denominator). Now set $\mathbf{x}_n = (1/n, 1/n)$. Then $|\mathbf{x}_{n+1} - \mathbf{x}_n| \longrightarrow 0$, but $|f(\mathbf{x}_{n+1}) - f(\mathbf{x}_n)| = \frac{1}{2} \not\longrightarrow 0$. Hence f is not uniformly continuous.
7. Let $f(\mathbf{a}) > c$. Set $\epsilon = f(\mathbf{a}) - c > 0$. Then there exists $\delta > 0$ such that $|\mathbf{x} - \mathbf{a}| < \delta$ implies $|f(\mathbf{x}) - f(\mathbf{a})| < \epsilon$, in particular $f(\mathbf{x}) > f(\mathbf{a}) - \epsilon = c$. Hence $\mathbf{a}$ is an interior point; the set is open.
9. A continuous function has a minimum on a bounded closed set.
11. *Every* point on the surface of a sphere is closest to the center.
13. Since f is continuous on a bounded closed set, the circle, f has a minimum there. But $f(x, y) = (x - 3y)^2 + y^2 > 0$ on the circle, hence the minimum is positive.
15. 1, at $\mathbf{0}$ 17. yes, composite of e^t and $t = -x^2 - y^2$
19. $x \longrightarrow (u, v) = (x, b)$ is continuous and $(u, v) \longrightarrow f(u, v)$ is continuous. The composite is $x \longrightarrow f(x, b)$.
21. At each point of the set $\{(x, y) \mid x \neq y\}$, the function f is continuous, a quotient of continuous functions. Next, if $x \neq y$, the Mean Value Theorem says there is a z between x and y such that $f(x, y) = g'(z)$. To prove f is continuous at (a, a), let $\epsilon > 0$. Choose δ so $|z - a| < \delta$ implies $|g'(z) - g'(a)| < \epsilon$. If $|(x, y) - (a, a)| < \delta$, then either (1) $x = y$ and $|f(x, y) - f(a, a)| = |g'(x) - g'(a)| < \epsilon$, or (2) $x \neq y$ and $f(x, y) = g'(z)$ where z is between x and y and hence $|z - a| < \delta$. Then $|f(x, y) - f(a, a)| = |g'(z) - g'(a)| < \epsilon$. Thus f is continuous at (a, a).

Section 4, page 212

15. $z = 2x^2 + 3c^2$, $y = c$
17. Let $(x_n, y_n, z_n) \longrightarrow (x, y, z)$ and $f(x_n, y_n) = z_n$. Then $(x_n, y_n) \longrightarrow (x, y) \in \mathbf{D}$ since $\mathbf{D}$ is closed, and $z_n \to z$. But $z_n = f(x_n, y_n) \longrightarrow f(x, y)$ by continuity. Hence $f(x, y) = z$, so (x, y, z) is also on the graph.

Section 5, page 216

1. 1, 2 3. $3y$, $3x$ 5. $4x/(y+1)$, $-2x^2/(y+1)^2$
7. $\sin y$, $x \cos y$ 9. $2 \cos 2x$, $-3 \sin 3y$
11. $2y \cos 2xy$, $2x \cos 2xy$ 13. $1/y - y/x^2$, $-x/y^2 + 1/x$
15. e^y, xe^y 17. ye^{xy}, xe^{xy} 19. $2e^{2x} \sin y$, $e^{2x} \cos y$
21. $4x$, 1 23. 32, 0, $3x^3z^2$
25. $z_x + 3z_y = 6(3x - y) + 3[-2(3x - y)] = 0$
27. $z_x^2 - z_y^2 = (2x)^2 - (-2y)^2 = 4(x^2 - y^2) = 4z$

Section 6, page 223

1. max = 4 at (0, 0) 3. min = 0 at (2, −3)
5. min = 4 at (0, 0)

7. max $= 2\sqrt{3}/9$ at $(\sqrt{3}/3, 0)$, neither max nor min at $(-\sqrt{3}/3, 0)$
9. cube of side 2 ft **11.** $x = y = (2V/3)^{1/3}$, $z = (9V/4)^{1/3}$

CHAPTER 7

Section 1, page 226

1. $3x + 2y - 5z$
3. $F(\mathbf{x}) = \mathbf{p}\cdot\mathbf{x}$ for some vector $\mathbf{p} \neq \mathbf{0}$. Hence $F(\mathbf{x}) = 0$ defines the plane through $\mathbf{0}$ perpendicular to $\mathbf{p}$.
5. $F(x, y, z) = xF(\mathbf{i}) + yF(\mathbf{j}) + zF(\mathbf{k})$. Given $F(\mathbf{i})$ and $F(\mathbf{j})$ the value $F(\mathbf{k})$ may still be any constant, so F is not determined.
7. Let F and G be linear and suppose $H = F + G$. Then
$$\begin{aligned} H(a\mathbf{u} + b\mathbf{v}) &= [F + G](a\mathbf{u} + b\mathbf{v}) = F(a\mathbf{u} + b\mathbf{v}) + G(a\mathbf{u} + b\mathbf{v}) \\ &= aF(\mathbf{u}) + bF(\mathbf{v}) + aG(\mathbf{u}) + bG(\mathbf{v}) \\ &= a[F(\mathbf{u}) + G(\mathbf{u})] + b[F(\mathbf{v}) + G(\mathbf{v})] \\ &= a[F + G](\mathbf{u}) + b[F + G](\mathbf{v}) = aH(\mathbf{u}) + bH(\mathbf{v}). \end{aligned}$$
9.
$$\begin{aligned} L(af + bg) &= \int_0^1 [af(t) + bg(t)]\,dt = a\int_0^1 f(t)\,dt + b\int_0^1 g(t)\,dt \\ &= aL(f) + bL(g). \end{aligned}$$
11. Expand the squares on the left side. The result is $\sum_{i=1}^3 \sum_{j=1}^3 a_i^2 x_j^2$, all mixed terms canceling out. This is precisely the same as the right side.
13. Let $F(\mathbf{x}) = \mathbf{p}\cdot\mathbf{x} + d$, where $d \neq 0$. If $a + b = 1$, then
$$\begin{aligned} F(a\mathbf{u} + b\mathbf{v}) &= \mathbf{p}\cdot(a\mathbf{u} + b\mathbf{v}) + d = a\mathbf{p}\cdot\mathbf{u} + b\mathbf{p}\cdot\mathbf{v} + (a + b)d \\ &= a(\mathbf{p}\cdot\mathbf{u} + d) + b(\mathbf{p}\cdot\mathbf{v} + d) = aF(\mathbf{u}) + bF(\mathbf{v}). \end{aligned}$$
15. It is enough to show that $G(\mathbf{x}) = F(\mathbf{x}) - F(\mathbf{0})$ is a homogeneous linear function.
$$\begin{aligned} G(a\mathbf{u} + b\mathbf{v}) &= F(a\mathbf{u} + b\mathbf{v}) - F(\mathbf{0}) \\ &= F(a\mathbf{u} + b\mathbf{v} + [1 - a - b]\mathbf{0}) - F(\mathbf{0}) \\ &= aF(\mathbf{u}) + bF(\mathbf{v}) + (1 - a - b)F(\mathbf{0}) - F(\mathbf{0}) \\ &= a[F(\mathbf{u}) - F(\mathbf{0})] + b[F(\mathbf{v}) - F(\mathbf{0})] = aG(\mathbf{u}) + bG(\mathbf{v}). \end{aligned}$$

Section 2, page 231

1. According to the text, we may take $c^2 = |\mathbf{p}_1|^2 + |\mathbf{p}_2|^2 + |\mathbf{p}_3|^2$, where $\mathbf{p}_1$, $\mathbf{p}_2$, $\mathbf{p}_3$ are the rows of A. Thus $|\mathbf{p}_1|^2 = a_{11}^2 + a_{12}^2 + a_{13}^2$, etc.
3.
$$\begin{aligned} F(b\mathbf{x} + c\mathbf{y}) &= \mathbf{a}\times(b\mathbf{x} + c\mathbf{y}) = \mathbf{a}\times(b\mathbf{x}) + \mathbf{a}\times(c\mathbf{y}) \\ &= b(\mathbf{a}\times\mathbf{x}) + c(\mathbf{a}\times\mathbf{y}) = bF(\mathbf{x}) + cF(\mathbf{y}). \end{aligned}$$
5.
$$\begin{aligned} F(x_1, y_1, z_1) + F(x_2, y_2, z_2) &= (x_1 + x_2, y_1 + y_2, 2) \\ &\neq F(x_1 + x_2, y_1 + y_2, z_1 + z_2) \end{aligned}$$
7. $\begin{bmatrix} 1 & 0 & 0 \\ 0 & 1 & 0 \\ 1 & 1 & 1 \end{bmatrix}$ **9.** $\begin{bmatrix} 0 & 1 & 0 \\ 0 & 1 & 0 \\ 0 & 2 & 4 \end{bmatrix}$

11. Identity transformation; each vector is fixed
13. Projection onto the x_3, x_1-plane
15. Each vector is transformed into its negative; reflection through **0**.
17. Reflection in the x_1, x_2-plane
19. Stretching by a factor of 2 in the x_1-direction, and by a factor of 3 in the x_2-direction, followed by a reflection in the x_1, x_2-plane
21. $F(\mathbf{x}) = F(x\mathbf{i} + y\mathbf{j} + z\mathbf{k}) = xF(\mathbf{i}) + yF(\mathbf{j}) + zF(\mathbf{k})$, again a vector in the plane of $F(\mathbf{i})$, $F(\mathbf{j})$, $F(\mathbf{k})$.
23. Let $G(\mathbf{x}) = \mathbf{a}\cdot F(\mathbf{x})$. Then $G(b\mathbf{x} + c\mathbf{y}) = \mathbf{a}\cdot F(b\mathbf{x} + c\mathbf{y}) = \mathbf{a}\cdot[bF(\mathbf{x}) + cF(\mathbf{y})] = b[\mathbf{a}\cdot F(\mathbf{x})] + c[\mathbf{a}\cdot F(\mathbf{y})] = bG(\mathbf{x}) + cG(\mathbf{y})$.

Section 3, page 238

1. 1 **3.** 10

5. $\begin{bmatrix} 17 \\ 39 \end{bmatrix}$ **7.** $\begin{bmatrix} 2 \\ 2 \\ 2 \end{bmatrix}$ **9.** 3 **11.** $\begin{bmatrix} 9 & 0 \\ 9 & 0 \end{bmatrix}$ **13.** $\begin{bmatrix} -7 & 20 & 26 \\ -1 & 3 & 2 \\ -2 & 4 & 5 \end{bmatrix}$

15. $\begin{bmatrix} -6 & -6 & -6 \\ 3 & 3 & 3 \\ 6 & 6 & 6 \end{bmatrix}$ **17.** $\begin{bmatrix} a_1 & a_2 & a_3 \\ 0 & 0 & 0 \\ 0 & 0 & 0 \end{bmatrix}$

19. $\begin{bmatrix} 0 & 0 & 0 \\ 0 & 0 & 0 \\ 0 & 0 & 0 \end{bmatrix}$ **21.** $x^2 + 2y^2 + 3z^2$

23. $\begin{bmatrix} 0 & 0 \\ 0 & 0 \end{bmatrix}$ **25.** No. For example, $\begin{bmatrix} a & b & 0 \\ 0 & 0 & 0 \\ 0 & 0 & 0 \end{bmatrix}\begin{bmatrix} 0 & 0 & 0 \\ 0 & 0 & 0 \\ c & d & e \end{bmatrix} = 0.$

27. $$\mathbf{x}A = (x, y, z)\begin{bmatrix} a_1 & a_2 & a_3 \\ b_1 & b_2 & b_3 \\ c_1 & c_2 & c_3 \end{bmatrix}$$

$$= (a_1x + b_1y + c_1z,\ a_2x + b_2y + c_2z,\ a_3x + b_3y + c_3z), \text{ hence}$$

$$(\mathbf{x}A)' = \begin{bmatrix} a_1x + b_1y + c_1z \\ a_2x + b_2y + c_2z \\ a_3x + b_3y + c_3z \end{bmatrix} = \begin{bmatrix} a_1 & b_1 & c_1 \\ a_2 & b_2 & c_2 \\ a_3 & b_3 & c_3 \end{bmatrix}\begin{bmatrix} x \\ y \\ z \end{bmatrix} = A'\mathbf{x}'.$$

29. $$\mathbf{x}A\mathbf{y}' = (x_1, x_2, x_3)\begin{bmatrix} a_{11} & a_{12} & a_{13} \\ a_{21} & a_{22} & a_{23} \\ a_{31} & a_{32} & a_{33} \end{bmatrix}\begin{bmatrix} y_1 \\ y_2 \\ y_3 \end{bmatrix}$$

$$= (x_1, x_2, x_3)\begin{bmatrix} a_{11}y_1 + a_{12}y_2 + a_{13}y_3 \\ a_{21}y_1 + a_{22}y_2 + a_{23}y_3 \\ a_{31}y_1 + a_{32}y_2 + a_{33}y_3 \end{bmatrix}$$

$$= x_1(a_{11}y_1 + a_{12}y_2 + a_{13}y_3) + x_2(a_{21}y_1 + a_{22}y_2 + a_{23}y_3) + x_3(a_{31}y_1 + a_{32}y_2 + a_{33}y_3)$$

$$= \sum_{i,j} x_i a_{ij} y_j = \sum_{i,j} y_j a_{ij} x_i = \sum_{i,j} y_i a_{ji} x_j.$$

Similarly

$$\mathbf{y}A'\mathbf{x}' = \sum_{i,j} y_i a_{ji} x_j.$$

31. $(A + B)' = [a_{ij} + b_{ij}]' = [a_{ji} + b_{ji}] = [a_{ji}] + [b_{ji}] = A' + B'$

Section 4, page 243

1. $(98, -1, 11)'$ **3.** $F \circ G$ has the matrix

$$AB = \begin{bmatrix} 5 & 1 & 0 \\ 8 & 1 & 0 \\ 0 & 0 & 0 \end{bmatrix}$$

5. $$\begin{bmatrix} a_1 & b_1 & c_1 \\ a_2 & b_2 & c_2 \\ a_3 & b_3 & c_3 \end{bmatrix}\begin{bmatrix} x \\ y \\ z \end{bmatrix} = \begin{bmatrix} d_1 \\ d_2 \\ d_3 \end{bmatrix}$$

7. $x = 26r + 11s - 26t,\ y = 5r - 11s + 8t,\ z = 3r + 12s - 17t$

9. $$\mathbf{a}' \times \mathbf{x}' = \begin{bmatrix} a_2x_3 - a_3x_2 \\ a_3x_1 - a_1x_3 \\ a_1x_2 - a_2x_1 \end{bmatrix} = \begin{bmatrix} 0 & -a_3 & a_2 \\ a_3 & 0 & -a_1 \\ -a_2 & a_1 & 0 \end{bmatrix}\begin{bmatrix} x_1 \\ x_2 \\ x_3 \end{bmatrix}$$

11. $\mathbf{a}' \times (\mathbf{b}' \times \mathbf{x}') = A(\mathbf{b}' \times \mathbf{x}') = A(B\mathbf{x}') = (AB)\mathbf{x}'$. However,

$$AB = \begin{bmatrix} 0 & -a_3 & a_2 \\ a_3 & 0 & -a_1 \\ -a_2 & a_1 & 0 \end{bmatrix}\begin{bmatrix} 0 & -b_3 & b_2 \\ b_3 & 0 & -b_1 \\ -b_2 & b_1 & 0 \end{bmatrix}$$

$$= \begin{bmatrix} -a_2b_2 - a_3b_3 & a_2b_1 & a_3b_1 \\ a_1b_2 & -a_3b_3 - a_1b_1 & a_3b_2 \\ a_1b_3 & a_2b_3 & -a_1b_1 - a_2b_2 \end{bmatrix}$$

$$= -(a_1b_1 + a_2b_2 + a_3b_3)I + \begin{bmatrix} a_1b_1 & a_2b_1 & a_3b_1 \\ a_1b_2 & a_2b_2 & a_3b_2 \\ a_1b_3 & a_2b_3 & a_3b_3 \end{bmatrix}$$

$$= -(\mathbf{a}' \cdot \mathbf{b}')I + \mathbf{b}'\mathbf{a}.$$

13. By induction: suppose $|A^n\mathbf{x}'| \le c^n\,|\mathbf{x}'|$. Then
$|A^{n+1}\mathbf{x}'| = |A(A^n\mathbf{x}')| = c\,|A^n\mathbf{x}'| \le c\cdot c^n\,|\mathbf{x}'| = c^{n+1}\,|\mathbf{x}'|$, etc.

15.
$$\sum_{n=0}^{\infty}\frac{1}{n!}A^n(b\mathbf{x}' + c\mathbf{y}') = \lim_{m\to\infty}\sum_{n=0}^{m}\frac{1}{n!}A^n(b\mathbf{x}' + c\mathbf{y}')$$
$$= \lim_{m\to\infty}\left[b\sum_{n=0}^{m}\frac{1}{n!}A^n\mathbf{x}' + c\sum_{n=0}^{m}\frac{1}{n!}A^n\mathbf{y}'\right]$$
$$= b\lim_{m\to\infty}\sum_{n=0}^{m}\frac{1}{n!}A^n\mathbf{x}' + c\lim_{m\to\infty}\sum_{n=0}^{m}\frac{1}{n!}A^n\mathbf{y}'$$
$$= b\sum_{n=0}^{\infty}\frac{1}{n!}A^n\mathbf{x}' + c\sum_{n=0}^{\infty}\frac{1}{n!}A^n\mathbf{y}'.$$

Section 5, page 252

1. $4x^2 - 2xy + 3y^2$ **3.** $x^2 + 2y^2 + 8yz + 3y^2$

5. $\begin{bmatrix}1 & 0\\ 0 & -1\end{bmatrix}$ **7.** $\begin{bmatrix}1 & 1\\ 1 & 1\end{bmatrix}$ **9.** $\begin{bmatrix}a^2 & ab\\ ab & b^2\end{bmatrix}$

11. $\begin{bmatrix}0 & \frac12 & \frac12\\ \frac12 & 1 & \frac12\\ \frac12 & \frac12 & 0\end{bmatrix}$ **13.** $\begin{bmatrix}a^2 & ab & ac\\ ab & b^2 & bc\\ ac & bc & c^2\end{bmatrix}$

15. No **17.** Yes **19.** $\frac12(A + A')$

21. Let $f(\mathbf{x}) = \mathbf{x}A\mathbf{x}'$ and $g(\mathbf{x}) = \mathbf{x}B\mathbf{x}'$ be positive definite. Then $f(\mathbf{x}) + g(\mathbf{x}) \ge 0 + 0 = 0$ with equality if and only if $\mathbf{x}' = \mathbf{0}$. Since $f(\mathbf{x}) + g(\mathbf{x}) = \mathbf{x}(A + B)\mathbf{x}'$, the matrix of the sum is $A + B$.

23. Let $B = \frac12(A + A')$. Then $B' = (\frac12A + \frac12A')' = (\frac12A)' + (\frac12A')' = \frac12A' + \frac12A = \frac12(A + A') = B$.

25. $(AB)' = (BA)' = A'B' = AB$.

27. We seek the smallest positive k such that $x^2 + 2xy + 2y^2 - k(x^2 + y^2) = (1 - k)x^2 + 2xy + (2 - k)y^2$ is not positive definite. It *is* positive definite if $1 - k > 0$, $2 - k > 0$, and $(1 - k)(2 - k) - 1 > 0$. The zeros of $(1 - k)(2 - k) - 1 = k^2 - 3k + 1$ are $k = \frac12(3 \pm \sqrt5)$. The answer is $k = \frac12(3 - \sqrt5)$, which satisfies $0 < k < 1$.

29. $f(\mathbf{x})$ is negative semi-definite if and only if $-f(\mathbf{x})$ is positive semi-definite if and only if $-a \ge 0$, $-c \ge 0$, and $(-a)(-c) - (-b)^2 \ge 0$ if and only if $a \le 0$, $c \le 0$, $ac - b^2 \ge 0$.

31. There exists $\mathbf{x}_0 \ne \mathbf{0}$ such that $f(\mathbf{x}_0) = 0$. By rotating coordinates, we may assume $\mathbf{x}_0 = (0, 1)$. Then $c = 0$, so $f = x(ax + 2by)$. If $b = 0$, then $f = ax^2$ is not indefinite, hence $b \ne 0$. The factors x, $ax + 2by$ are not proportional, and remain that way when we rotate back.

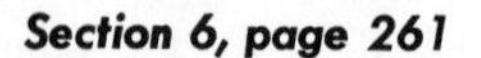

Section 6, page 261

1.

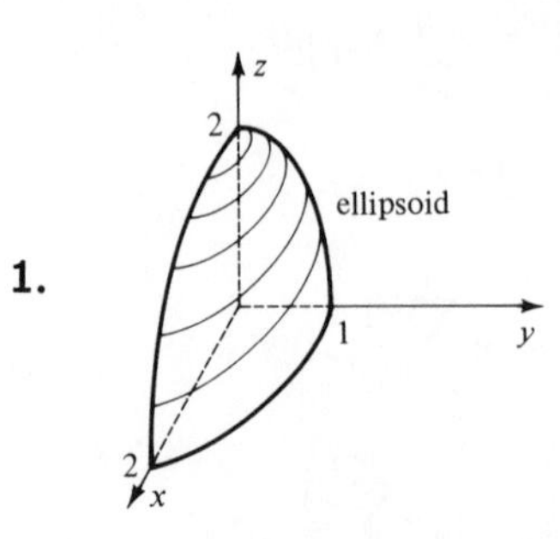

3.

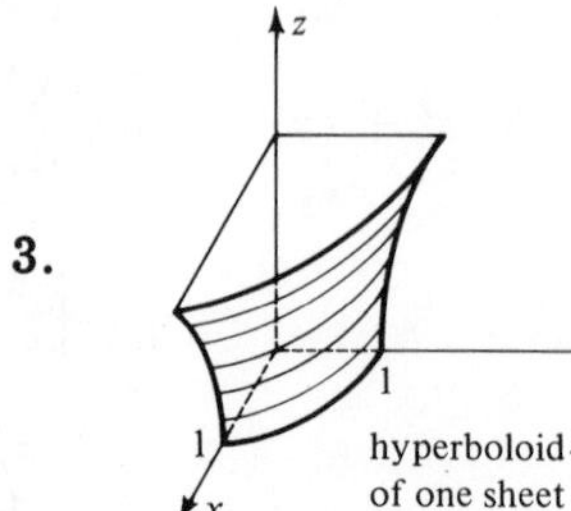

5.

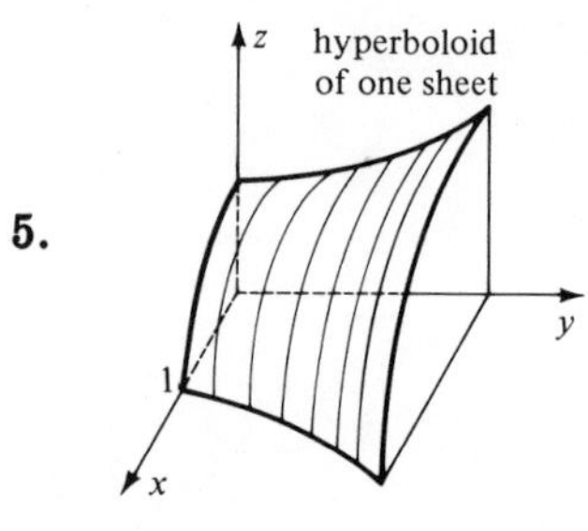

7.

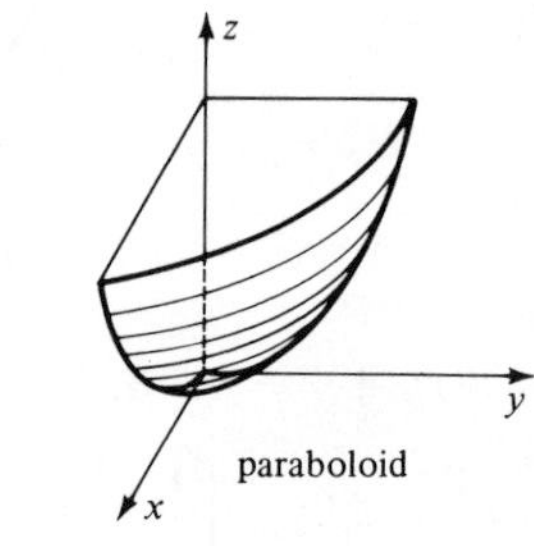

9.

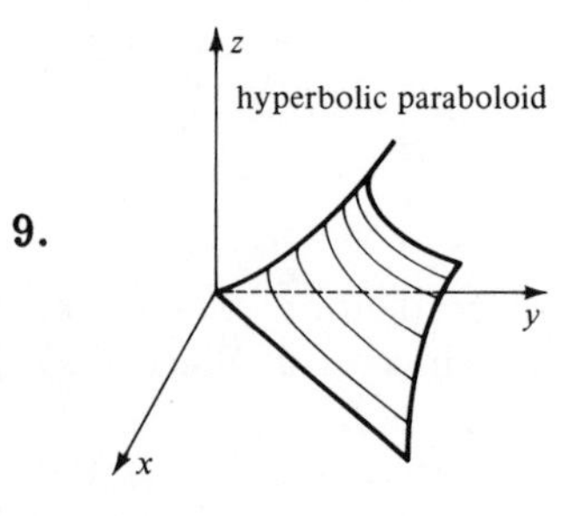

11. $z = (x + 1)^2 + y^2 - 1$; paraboloid with axis $x = -1$, $y = 0$

13. **15.**

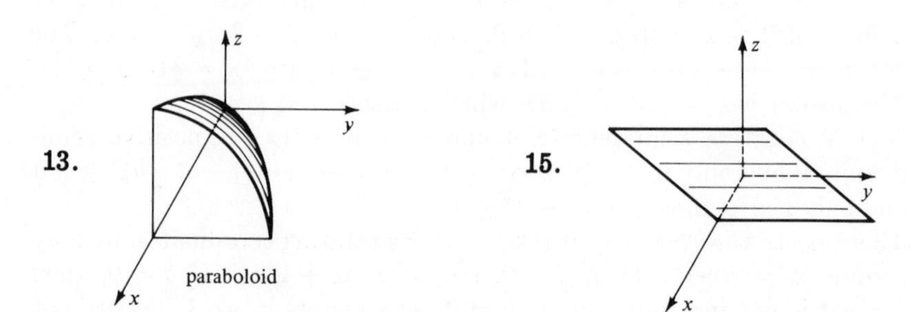

17.

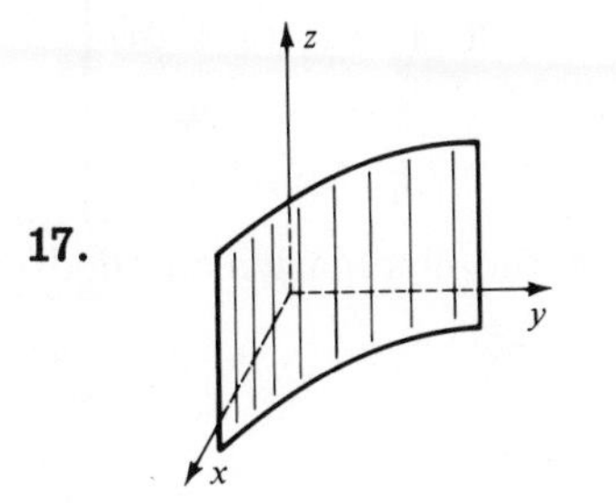

19.

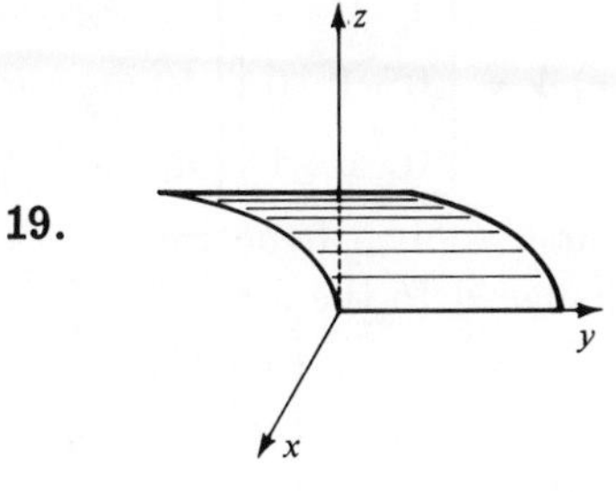

21.

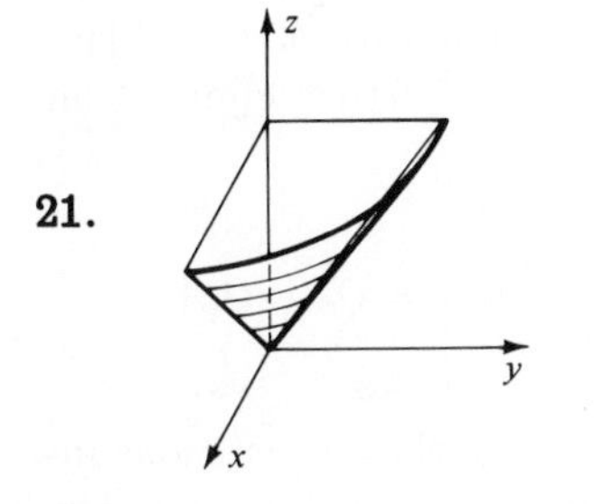

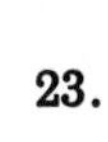

23.

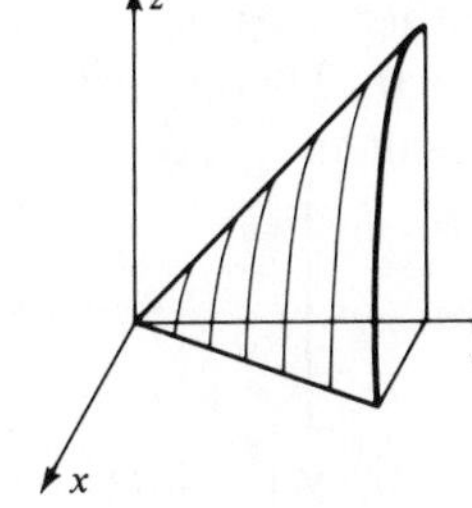

25. $f(x/z,\ y/z) = 0$ **27.** $f(y,\ \sqrt{x^2+z^2}) = 0$

Section 7, page 268

1. $\dfrac{1}{7}\begin{bmatrix} 3 & -1 \\ 1 & 2 \end{bmatrix}$ **3.** $-\dfrac{1}{17}\begin{bmatrix} -2 & 3 \\ 3 & 4 \end{bmatrix}$ **5.** $\dfrac{1}{a^2}\begin{bmatrix} a & -1 \\ 0 & a \end{bmatrix}$

7. $\begin{bmatrix} 1 & 0 \\ -b & 1 \end{bmatrix}$ **9.** $\dfrac{1}{a}\begin{bmatrix} 1 & -b \\ 0 & a \end{bmatrix}$ **11.** $-\dfrac{1}{6}\begin{bmatrix} -6 & -3 & 3 \\ -2 & 3 & -1 \\ 2 & 0 & -2 \end{bmatrix}$

13. $-\dfrac{1}{6}\begin{bmatrix} 23 & 22 & -30 \\ -11 & -10 & 12 \\ -5 & -4 & 6 \end{bmatrix}$ **15.** $\dfrac{1}{5}\begin{bmatrix} -1 & -3 & 5 \\ -4 & -7 & 10 \\ -3 & -9 & 10 \end{bmatrix}$

17. $[\frac{5}{2},\ 2,\ \frac{3}{2}]'$ **19.** $[-\frac{3}{7},\ -\frac{1}{2},\ \frac{11}{14}]'$

21. $(AB)(B^{-1}A^{-1}) = A(BB^{-1})A^{-1} = AIA^{-1} = AA^{-1} = I$, so AB has an inverse and it is $B^{-1}A^{-1}$

23. $A'(A^{-1})' = (A^{-1}A)' = I' = I$, so A' has an inverse and it is $(A^{-1})'$.

25. $$P_{12}A = \begin{bmatrix} 0 & 1 & 0 \\ 1 & 0 & 0 \\ 0 & 0 & 1 \end{bmatrix}\begin{bmatrix} a_1 & a_2 & a_3 \\ b_1 & b_2 & b_3 \\ c_1 & c_2 & c_3 \end{bmatrix} = \begin{bmatrix} b_1 & b_2 & b_3 \\ a_1 & a_2 & a_3 \\ c_1 & c_2 & c_3 \end{bmatrix}$$

27. $$R_{31}(c)A = \begin{bmatrix} 1 & 0 & c \\ 0 & 1 & 0 \\ 0 & 0 & 1 \end{bmatrix}\begin{bmatrix} a_1 & a_2 & a_3 \\ b_1 & b_2 & b_3 \\ d_1 & d_2 & d_3 \end{bmatrix} = \begin{bmatrix} a_1 + cd_1 & a_2 + cd_2 & a_3 + cd_3 \\ b_1 & b_2 & b_3 \\ d_1 & d_2 & d_3 \end{bmatrix}$$

29. We first apply to the rows of A the steps of the elimination method to change A to the form

$$B = \begin{bmatrix} b_{11} & b_{12} & b_{13} \\ 0 & b_{22} & b_{23} \\ 0 & b_{32} & b_{33} \end{bmatrix} = Q_r Q_{r-1} \cdots Q_2 Q_1 A.$$

Each Q in these steps is a P_{ij} or $R_{ij}(c)$ by Exs. 25, 27. Also each of these steps either does not change $|A|$ or changes its sign only, hence $|B| = \pm |A| \neq 0$. Therefore b_{22} and b_{32} are not both 0. Further elimination steps result in

$$\begin{bmatrix} c_{11} & c_{12} & c_{13} \\ 0 & c_{22} & c_{23} \\ 0 & 0 & c_{33} \end{bmatrix} = Q_s \cdots Q_1 A, \qquad \text{then} \qquad \begin{bmatrix} c_{11} & 0 & 0 \\ 0 & c_{22} & 0 \\ 0 & 0 & c_{33} \end{bmatrix} = Q_t \cdots Q_1 A.$$

Now multiply the first row by c_{11}^{-1}, etc. By Ex. 26, these operations involve further multiplications by Q's, now of the form $D_1(c_{11}^{-1})$, etc. The final result is $Q_n \cdots Q_1 A = I$.

31. Each row operation is applied simultaneously to A and to I:

$$A = \begin{bmatrix} 2 & 1 \\ -1 & 3 \end{bmatrix} \longrightarrow \begin{bmatrix} 1 & -3 \\ 2 & 1 \end{bmatrix} \longrightarrow \begin{bmatrix} 1 & -3 \\ 0 & 7 \end{bmatrix} \longrightarrow \begin{bmatrix} 1 & -3 \\ 0 & 1 \end{bmatrix}$$

$$\longrightarrow \begin{bmatrix} 1 & 0 \\ 0 & 1 \end{bmatrix} = I$$

$$I = \begin{bmatrix} 1 & 0 \\ 0 & 1 \end{bmatrix} \longrightarrow \begin{bmatrix} 0 & -1 \\ 1 & 0 \end{bmatrix} \longrightarrow \begin{bmatrix} 0 & -1 \\ 1 & 2 \end{bmatrix} \longrightarrow \begin{bmatrix} 0 & -1 \\ \frac{1}{7} & \frac{2}{7} \end{bmatrix}$$

$$\longrightarrow \begin{bmatrix} \frac{3}{7} & -\frac{1}{7} \\ \frac{1}{7} & \frac{2}{7} \end{bmatrix} = A^{-1}$$

The explicit sequence of steps is explained by

$$A^{-1} = R_{21}(3)D_2(\tfrac{1}{7})R_{12}(-2)D_1(-1)P_{12}.$$

33.

$$A = \begin{bmatrix} 1 & 1 & 1 \\ 1 & -1 & 2 \\ 1 & 1 & 4 \end{bmatrix} \longrightarrow \begin{bmatrix} 1 & 1 & 1 \\ 0 & -2 & 1 \\ 0 & 0 & 3 \end{bmatrix} \longrightarrow \begin{bmatrix} 1 & 1 & 1 \\ 0 & 1 & -\frac{1}{2} \\ 0 & 0 & 1 \end{bmatrix}$$

$$\longrightarrow \begin{bmatrix} 1 & 0 & \frac{3}{2} \\ 0 & 1 & -\frac{1}{2} \\ 0 & 0 & 1 \end{bmatrix} \longrightarrow \begin{bmatrix} 1 & 0 & 0 \\ 0 & 1 & 0 \\ 0 & 0 & 1 \end{bmatrix} = I$$

$$I = \begin{bmatrix} 1 & 0 & 0 \\ 0 & 1 & 0 \\ 0 & 0 & 1 \end{bmatrix} \longrightarrow \begin{bmatrix} 1 & 0 & 0 \\ -1 & 1 & 0 \\ -1 & 0 & 1 \end{bmatrix} \longrightarrow \begin{bmatrix} 1 & 0 & 0 \\ \frac{1}{2} & -\frac{1}{2} & 0 \\ -\frac{1}{3} & 0 & \frac{1}{3} \end{bmatrix}$$

$$\longrightarrow \begin{bmatrix} \frac{1}{2} & \frac{1}{2} & 0 \\ \frac{1}{2} & -\frac{1}{2} & 0 \\ -\frac{1}{3} & 0 & \frac{1}{3} \end{bmatrix} \longrightarrow \begin{bmatrix} 1 & \frac{1}{2} & -\frac{1}{2} \\ \frac{1}{3} & -\frac{1}{2} & \frac{1}{6} \\ -\frac{1}{3} & 0 & \frac{1}{3} \end{bmatrix} = A^{-1}$$

35.

$$A = \begin{bmatrix} 2 & 2 & 6 \\ -1 & 2 & -9 \\ 1 & 3 & -2 \end{bmatrix} \longrightarrow \begin{bmatrix} 1 & 3 & -2 \\ -1 & 2 & -9 \\ 2 & 2 & 6 \end{bmatrix} \longrightarrow \begin{bmatrix} 1 & 3 & -2 \\ 0 & 5 & -11 \\ 0 & -4 & 10 \end{bmatrix}$$

$$\longrightarrow \begin{bmatrix} 1 & 3 & -2 \\ 0 & 1 & -1 \\ 0 & -4 & 10 \end{bmatrix} \longrightarrow \begin{bmatrix} 1 & 0 & 1 \\ 0 & 1 & -1 \\ 0 & 0 & 6 \end{bmatrix}$$

$$\longrightarrow \begin{bmatrix} 1 & 0 & 1 \\ 0 & 1 & -1 \\ 0 & 0 & 1 \end{bmatrix} \longrightarrow \begin{bmatrix} 1 & 0 & 0 \\ 0 & 1 & 0 \\ 0 & 0 & 1 \end{bmatrix} = I$$

$$I = \begin{bmatrix} 1 & 0 & 0 \\ 0 & 1 & 0 \\ 0 & 0 & 1 \end{bmatrix} \longrightarrow \begin{bmatrix} 0 & 0 & 1 \\ 0 & 1 & 0 \\ 1 & 0 & 0 \end{bmatrix} \longrightarrow \begin{bmatrix} 0 & 0 & 1 \\ 0 & 1 & 1 \\ 1 & 0 & -2 \end{bmatrix}$$

$$\longrightarrow \begin{bmatrix} 0 & 0 & 1 \\ 1 & 1 & -1 \\ 1 & 0 & -2 \end{bmatrix} \longrightarrow \begin{bmatrix} -3 & -3 & 4 \\ 1 & 1 & -1 \\ 5 & 4 & -6 \end{bmatrix}$$

$$\longrightarrow \begin{bmatrix} -3 & -3 & 4 \\ 1 & 1 & -1 \\ \frac{5}{6} & \frac{4}{6} & -\frac{6}{6} \end{bmatrix}$$

$$\longrightarrow \frac{1}{6}\begin{bmatrix} -23 & -22 & 30 \\ 11 & 10 & -12 \\ 5 & 4 & -6 \end{bmatrix} = A^{-1}$$

37. $(A + B)^{\cdot} = [a_{ij} + b_{ij}]^{\cdot} = [\dot{a}_{ij} + \dot{b}_{ij}] = [\dot{a}_{ij}] + [\dot{b}_{ij}] = A^{\cdot} + B^{\cdot}$.

39. $(AB)^{\cdot} = [\sum a_{ij}b_{jk}]^{\cdot} = [\sum \dot{a}_{ij}b_{jk} + \sum a_{ij}\dot{b}_{jk}] = [\sum \dot{a}_{ij}b_{jk}] + [\sum a_{ij}\dot{b}_{jk}] = A^{\cdot}B + AB^{\cdot}$

Section 8, page 278

1. $2, a\begin{bmatrix}1\\0\end{bmatrix}$; $-3, b\begin{bmatrix}0\\1\end{bmatrix}$ **3.** $3, a\begin{bmatrix}1\\0\end{bmatrix}$; $2, b\begin{bmatrix}1\\-1\end{bmatrix}$

5. $-1, a\begin{bmatrix}0\\1\end{bmatrix}$ **7.** $0, a\begin{bmatrix}2\\-1\end{bmatrix}$; $3, b\begin{bmatrix}1\\1\end{bmatrix}$

9. $1, a\begin{bmatrix}1\\-2\end{bmatrix}$; $5, b\begin{bmatrix}1\\2\end{bmatrix}$ **11.** $-2, a\begin{bmatrix}1\\1\end{bmatrix}$

13. $3, a\begin{bmatrix}2\\-3\end{bmatrix}$; $4, b\begin{bmatrix}1\\-1\end{bmatrix}$

15. $\sqrt{3}, a\begin{bmatrix}2\\-1+\sqrt{3}\end{bmatrix}$; $-\sqrt{3}, b\begin{bmatrix}1-\sqrt{3}\\1\end{bmatrix}$

17. $\Delta = (a-d)^2 + 4bc \geq 0$ if $bc \geq 0$.

19. Equate coefficients in
$$f(t) = (t-\lambda)(t-\mu) = t^2 - (a+d)t + (ad-bc).$$

21. $\begin{bmatrix}1 & 1\\-2 & 2\end{bmatrix}$ (See **9**.) **23.** $\begin{bmatrix}2 & 1\\-1 & 1\end{bmatrix}$ (See **7**.)

25. $\begin{bmatrix}0 & 1\\1 & 0\end{bmatrix}$ **27.** $3, a\begin{bmatrix}1\\0\\0\end{bmatrix}$; $1, b\begin{bmatrix}0\\1\\0\end{bmatrix}$; $2, c\begin{bmatrix}0\\0\\1\end{bmatrix}$

29. $-2, a\begin{bmatrix}0\\0\\1\end{bmatrix}$ **31.** $0, a\begin{bmatrix}0\\1\\0\end{bmatrix}$; $1, b\begin{bmatrix}3\\0\\1\end{bmatrix}$

33. $f(t) = (t-\lambda)(t-\mu) = t^2 - (a+c)t + (ac-b^2)$, hence $(a-\lambda)(a-\mu) = f(a) = a^2 - a(a+c) + (ac-b^2) = -b^2$. Therefore $(a-\lambda)(a-\mu) \leq 0$, so $\lambda \leq a \leq \mu$

35. Let $A = \begin{bmatrix}a & b\\b & c\end{bmatrix}$. By Theorem 8.5, there are *real* characteristic roots λ and μ. If $\lambda \neq \mu$, use Theorem 8.1. If $\lambda = \mu$, the discriminant of $f(t)$ is 0, so $\Delta = (a-c)^2 + 4b^2 = 0$, $a = c$, $b = 0$. Hence A is already diagonal.

37. Equate coefficients of t^2 in the expansion of $f(t)$. Only the term $(t-a_{11})(t-a_{22})(t-a_{33})$ contributes to t^2, and the contribution is $-(a_{11}+a_{22}+a_{33})$.

39. Try

$$A = \begin{bmatrix} 0 & a & b \\ 0 & 0 & c \\ 0 & 0 & 0 \end{bmatrix}$$

CHAPTER 8

Section 1, page 287

1. Take $hx + ky = 0$; $|f(\mathbf{x})/|\mathbf{x}|| = |xy^2/|\mathbf{x}|| \le |\mathbf{x}|^3/|\mathbf{x}| = |\mathbf{x}|^2 \longrightarrow 0$.
3. The rational function f is continuous where the denominator is non-zero, that is, $\mathbf{x} \neq \mathbf{0}$. It is also continuous at $\mathbf{0}$ because if $\mathbf{x} \neq \mathbf{0}$, then

$$|f(\mathbf{x})| = \frac{|xy|\,|x+y|}{|\mathbf{x}|^2} \le \frac{|\mathbf{x}|\,|\mathbf{x}|\,(|\mathbf{x}| + |\mathbf{x}|)}{|\mathbf{x}|^2} \le 2\,|\mathbf{x}|.$$

Hence $f(\mathbf{x}) \longrightarrow 0 = f(\mathbf{0})$ as $\mathbf{x} \longrightarrow \mathbf{0}$.
5. $f(\mathbf{0}) = f_x(\mathbf{0}) = f_y(\mathbf{0}) = 0$, so if f is differentiable at $\mathbf{0}$, then $f(\mathbf{x})/|\mathbf{x}| \longrightarrow 0$ as $\mathbf{x} \longrightarrow \mathbf{0}$. But $f(x, x)/|(x, x)| = \pm\frac{1}{2}\sqrt{2} \not\longrightarrow 0$.
7. Let $f = f_1 f_2$, where $f_i(\mathbf{c} + \mathbf{x}) = f_i(\mathbf{c}) + \mathbf{k}_i \cdot \mathbf{x} + e_i(\mathbf{x})$ and $e_i(\mathbf{x})/|\mathbf{x}| \longrightarrow 0$ as $\mathbf{x} \longrightarrow \mathbf{0}$. Then $f(\mathbf{c} + \mathbf{x}) = f(\mathbf{c}) + \mathbf{k} \cdot \mathbf{x} + e(\mathbf{x})$, where $\mathbf{k}$ is constant and

$$e(\mathbf{x}) = f_1(\mathbf{c})e_2(\mathbf{x}) + f_2(\mathbf{c})e_1(\mathbf{x}) + (\mathbf{k}_1 \cdot \mathbf{x})(\mathbf{k}_2 \cdot \mathbf{x}) + (\mathbf{k}_1 \cdot \mathbf{x})e_2(\mathbf{x}) + (\mathbf{k}_2 \cdot \mathbf{x})e_1(\mathbf{x}) + e_1(\mathbf{x})e_2(\mathbf{x}).$$

It follows, one term at a time, that $e(\mathbf{x})/|\mathbf{x}| \longrightarrow 0$ as $|\mathbf{x}| \longrightarrow 0$. The only questionable term is $(\mathbf{k}_1 \cdot \mathbf{x})(\mathbf{k}_2 \cdot \mathbf{x})$. But $|\mathbf{k}_i \cdot \mathbf{x}| \le |\mathbf{k}_i| \cdot |\mathbf{x}|$, so

$$\frac{|(\mathbf{k}_1 \cdot \mathbf{x})(\mathbf{k}_2 \cdot \mathbf{x})|}{|\mathbf{x}|} \le \frac{|\mathbf{k}_1| \cdot |\mathbf{k}_2| \cdot |\mathbf{x}|^2}{|\mathbf{x}|} \longrightarrow 0.$$

9. $f/g = f \cdot (1/g)$; use **7** and **8**.

Section 2, page 291

1. $(9t^2 + 2t)e^{t^2(3t+1)}$ 3. $4t^3 \cos(1/t) + t^2 \sin(1/t) - 2t$ 5. $9t^8$
7. $4te^{-t}[(2 - t) \sin 4t + 4t \cos 4t]$
9. $(s^2 - t)^2(2s^2 + t)/2s^3t^2$, $-(s^2 - t)^2(2s^2 + t)/4s^2t^3$
11. $(st^4 + st^2 + t)/z$, $(s + s^2t + 2s^2t^3)/z$, where $z^2 = 1 + s^2t^4 + (1 + st)^2$
13. 3.5π ft^3/hr
15. $(\partial/\partial u)F(u + v, u - v) + (\partial/\partial v)F(u + v, u - v) = (F_x + F_y) + (F_x - F_y) = 2F_x$
17. 1 19. 2 21. -1
23. Apply $\partial/\partial x$ to $f(t\mathbf{x}) = t^n f(\mathbf{x})$ by the Chain Rule: $tf_x(t\mathbf{x}) = t^n f_x(\mathbf{x})$.

Section 3, page 296

1. $z = 0$ **3.** $4x + 8y - z = 8$ **5.** $4x - 13y + z = -20$
7. $(-1, -1, 1)/\sqrt{3}$ **9.** $(0, -3, 1)/\sqrt{10}$ **11.** $y = \pm x, z = 0$

Section 4, page 301

1. lines; grad $z = (1, -2)$ **3.** parabolas; grad $z = (2x, 1)$
5. planes normal to $(1, 1, 1)$
7. Consider $xyz = 1$, a typical level surface. Because of symmetry in the coordinate axes, there are four sheets to the surface. The first octant sheet looks something like a paraboloid of revolution with axis (a, a, a), $a > 0$, but it flattens out towards the coordinate planes as you move away from the origin.
9. $(1, 1, 1)$, $(2x, 8y, 18z)$, (yz, zx, xy), $(2x, 2y, -2z)$
11. $-2r^{-3}(\cos 3\theta, \sin 3\theta)$
13. $x - y = 0$ **15.** $2x - z = 1$ **17.** $\mathbf{a}$
19.
$$\operatorname{div}(f\mathbf{u}) = \operatorname{div}(fu, fv, fw) = \frac{\partial(fu)}{\partial x} + \frac{\partial(fv)}{\partial y} + \frac{\partial(fw)}{\partial z}$$
$$= \left(\frac{\partial f}{\partial x}u + f\frac{\partial u}{\partial x}\right) + \left(\frac{\partial f}{\partial y}v + f\frac{\partial v}{\partial y}\right) + \left(\frac{\partial f}{\partial z}w + f\frac{\partial w}{\partial z}\right)$$
$$= \left(\frac{\partial f}{\partial x}u + \frac{\partial f}{\partial y}v + \frac{\partial f}{\partial z}w\right) + f\left(\frac{\partial u}{\partial x} + \frac{\partial v}{\partial y} + \frac{\partial w}{\partial z}\right)$$
$$= (\operatorname{grad} f)\cdot\mathbf{u} + f \operatorname{div}\mathbf{u}.$$
21.
$$\begin{aligned}\operatorname{curl}(f\mathbf{u}) &= \operatorname{curl}(fu, fv, fw)\\ &= ((fw)_y - (fv)_z, (fu)_z - (fw)_x, (fv)_x - (fu)_y)\\ &= ([fw_y + f_yw - fv_z - f_zv], [fu_z + f_zu - fw_x - f_xw],\\ &\qquad [fv_x + f_xv - fu_y - f_yu])\\ &= f\cdot(w_y - v_z, u_z - w_x, v_x - u_y)\\ &\qquad + (f_yw - f_zv, f_zu - f_xw, f_xv - f_yu)\\ &= f \operatorname{curl}\mathbf{u} + (f_x, f_y, f_z) \times (u, v, w) = f\operatorname{curl}\mathbf{u} + (\operatorname{grad} f) \times \mathbf{u}.\end{aligned}$$
23. By **21**, $\operatorname{curl}[f(\rho)\mathbf{x}] = [\operatorname{grad} f(\rho)] \times \mathbf{x} + f(\rho) \operatorname{curl}\mathbf{x} = [(f'(\rho)/\rho)\mathbf{x}] \times \mathbf{x} + \mathbf{0} = \mathbf{0}$.
25. $\operatorname{curl}(\mathbf{a} \times \mathbf{x}) = \operatorname{curl}(bz - cy, cx - az, ay - bx) = ((ay - bx)_y - (cx - az)_z, (bz - cy)_z - (ay - bx)_x, (cx - az)_x - (bz - cy)_y) = (2a, 2b, 2c) = 2\mathbf{a}$

Section 5, page 306

1. 1, 1, 1 **3.** 0, -3, 1 **5.** 1 **7.** 6

Section 6, page 309

1. $-2, -\frac{4}{3}$
3. $F = \operatorname{grad}[\frac{1}{3}(x^3 + y^3 + z^3) + xyz]$, $\frac{1}{3}(a^3 + b^3 + c^3) + abc$.

5. Take $x = a\cos\theta$, $y = a\sin\theta$, $0 \le \theta \le 2\pi$. Then

$$\left(-y\frac{dx}{d\theta} + x\frac{dy}{d\theta}\right)\Big/ (x^2 + y^2) = a^2/a^2 = 1, \text{ so } \int = \int_0^{2\pi} d\theta = 2\pi.$$

7. $(n-2)^{-1}a^{2-n}$

Section 7, page 315

1. $(1 - \sin y)/(x\cos y - 1)$ 3. $y(e^{xy} - 3y)/x(6y - e^{xy})$
5. $(e^x \sin y + e^y \sin x)/(e^y \cos x - e^x \cos y)$ 7. $-2x^3/9y^5$
9. $1, \frac{1}{3}$
11. $$\frac{1}{3^{(n-2)/2}}(x^2 + y^2 + z^2)^{n/2} \le x^n + y^n + z^n \le (x^2 + y^2 + z^2)^{n/2}$$
13. $F[x, y, z(x, y)] = 0$. Compute $\partial/\partial x$ by the Chain Rule: $F_x + F_z z_x = 0$, etc.

Section 8, page 318

1. $-(z/x)\,dx - (z/y)\,dy$ 3. $(1/2z)\,dx + (y/z)\,dy$
5. $r^2 = x^2 + y^2$, hence $2r\,dr = 2x\,dx + 2y\,dy$. Also $x\,dy - y\,dx =$ $(r\cos\theta)(dr\sin\theta + r\,d\theta\cos\theta) - (r\sin\theta)(dr\cos\theta - r\,d\theta\sin\theta) =$ $r^2(\cos^2\theta + \sin^2\theta)\,d\theta = r^2\,d\theta$.
7. $f_x(t\,dx + x\,dt) + f_y(t\,dy + y\,dt) + f_z(t\,dz + z\,dt) =$ $nt^{n-1}f\,dt + t^n(f_x\,dx + f_y\,dy + f_z\,dz)$. Hence $xf_x(t\mathbf{x}) + yf_y(t\mathbf{x}) + zf_z(t\mathbf{x}) = nt^{n-1}f(\mathbf{x})$ and $f_x(t\mathbf{x}) = t^{n-1}f_x(\mathbf{x})$, $f_y(t\mathbf{x}) = t^{n-1}f_y(\mathbf{x})$, $f_z(t\mathbf{x}) = t^{n-1}f_z(\mathbf{x})$.
9. $d\mathbf{u} = (-\sin\theta\,d\theta, \cos\theta\,d\theta) = \mathbf{w}\,d\theta$, $d\mathbf{w} = (-\cos\theta\,d\theta, -\sin\theta\,d\theta) =$ $-\mathbf{u}\,d\theta$, $d\mathbf{x} = d(r\mathbf{u}) = \mathbf{u}\,dr + r\,d\mathbf{u} = \mathbf{u}\,dr + r\,\mathbf{w}\,d\theta$.
11. Set $z = (x^2 - y^3)^{1/2}$ so $dz = \frac{1}{2}(2x\,dx - 3y^2\,dy)(x^2 - y^3)^{-1/2}$. Set $x = 6$, $y = 3$, $dx = -0.01$, $dy = 0.02$. Then $dz = -0.11$ so $z + dz \approx 3.00 - 0.11 = 2.89$. (By 5-place tables: 2.8895.)

CHAPTER 9

Section 1, page 328

1. $20x^3y^4$ 3. $-2/y^3$ 5. $-\sin(x+y)$ 7. $-e^{x/y}(x+y)/y^3$
9. $mnx^{m-1}y^{n-1}$ 11. $x^{y-1}(1 + y\ln x)$ 13. $-2(x+y)/(x-y)^3$
15. $2(y-x)/(1+xy)^3$ 17. $(-9x^2 + 25xy - 18y^2)/(x-y)^3(x-2y)^3$
19. $2b$ 21. $f_{xx} = f_{yy} = g''(x+y) + h''(x-y)$

Section 2, page 333

1. $18xy^2$, $18x^2y$ 3. $8y^3$, $24xy^2$
5. $-2y\cos(xy) + xy^2\sin(xy)$, $-2x\cos(xy) + x^2y\sin(xy)$
7. $e^{xy}(\sin x)(2y + xy^2 - x) + 2e^{xy}(\cos x)(1 + xy)$, $x^2e^{xy}\cos x + (x^2y + 2x)e^{xy}\sin x$

9. $(x^{1/y}/x^2y^4)[(y-1)(\ln x)+y^2-2y]$, $(x^{1/y}/xy^5)[2y^2+4y(\ln x)+\ln^2 x]$
11. $8y(y^2-5x^2)/(x^2+y^2)^4$, $8x(x^2-5y^2)/(x^2+y^2)^4$
13. $f(x, y) = g(x) + xh(y) + k(y)$ **15.** cubic polynomials
17. $ax + by + c$ **19.** none

21.
$$\begin{bmatrix} m(m-1)x^{m-2}y^nz^p & mnx^{m-1}y^{n-1}z^p & mpx^{m-1}y^nz^{p-1} \\ mnx^{m-1}y^{n-1}z^p & n(n-1)x^my^{n-2}z^p & npx^my^{n-1}z^{p-1} \\ mpx^{m-1}y^nz^{p-1} & npx^my^{n-1}z^{p-1} & p(p-1)x^my^nz^{p-2} \end{bmatrix}$$

23.
$$\begin{bmatrix} -\sin w & -2\sin w & -3\sin w \\ -2\sin w & -4\sin w & -6\sin w \\ -3\sin w & -6\sin w & -9\sin w \end{bmatrix}, \qquad w = x + 2y + 3z$$

25. 10 **27.** $g(x, y) + h(y, z) + k(z, x)$

Section 3, page 338

1. $p_2(x, y) = 1 + 2(x-1) + 2(y-1) + (x-1)^2 + 4(x-1)(y-1) + (y-1)^2$, $p_1(x, y) = 1 + 2(x-1) + 2(y-1)$, the linear part of $p_2(x, y)$
3. $p_1(x, y) = 0$, $p_2(x, y) = xy$ **5.** $p_1(x, y) = 1$, $p_2(x, y) = 1 + (x-1)y$
7. $p_1(x, y) = p_2(x, y) = -x - (y - \frac{1}{2}\pi)$
9. $p_1(x, y) = (x - \frac{1}{2}) + 2(y - \frac{1}{4})$,
$p_2(x, y) = p_1(x, y) - \frac{1}{2}[(x-\frac{1}{2})^2 + 4(x-\frac{1}{2})(y-\frac{1}{4}) + 4(y-\frac{1}{4})^2]$
11. 1.1200; exact: $(1.1)^{1.2} \approx 1.12117$
13. 3.9993; exact: 3.99929499
15. It is the tangent plane at $(a, b, f(a, b))$.
17. $p_1(x, y) = 1 + \frac{1}{2}x + y$ at $(0, 0)$. Also $f_{xx} = -1/4(1 + x + 2y)^{3/2}$, $f_{xy} = -1/2(\cdots)^{3/2}$, $f_{yy} = -1/(\cdots)^{3/2}$. Hence $|f_{xx}|$, $|f_{xy}|$, $|f_{yy}|$ are bounded by

$$\frac{1}{(1+x+2y)^{3/2}} \le \frac{1}{(1-0.1-0.2)^{3/2}} = \frac{1}{(0.7)^{3/2}} < \frac{1}{(0.64)^{3/2}}$$

$$= \frac{1}{(0.8)^3} = \frac{1}{0.512} < 2.$$

Take $M_2 = 2$: $|r_1(x, y)| < M_2|(x, y)|^2 \le 2(0.02) = 0.04$.

Section 4, page 344

1. min **3.** neither **5.** max **7.** min **9.** neither

Section 5, page 348

1. $(1, 1, 1)$ **3.** $1, -1$
5. $P = (\sum_1^n x_k/n, \sum_1^n y_k/n)$, where $P_k = (x_k, y_k)$ **7.** $\sqrt{2}$
9. distance ≈ 1.207 at $x \approx 0.7533$, $y \approx 0.5674$, $z \approx 0.7527$

Section 6, page 351

1. min **3.** min **5.** min **7.** neither **9.** min
11. neither **13.** max 0 **15.** min $-\frac{5}{2}$ at $(-\frac{1}{2}, -1, \frac{3}{2})$
17. min **19.** max **21.** neither

Section 7, page 359

1. $\sqrt{13}, -\sqrt{13}$
3. None; no level curve $x - y = c$ is tangent to the hyperbola.
5. $\frac{1}{2}, -\frac{1}{2}$ **7.** square **9.** height $= \sqrt{2}$ (radius) **11.** none
13. max $2^{1-p/q}$, min 1 **15.** $\frac{2}{9}\sqrt{3}, \frac{1}{4}\sqrt{2}$

Section 8, page 367

1. $8abc/3\sqrt{3}$ **3.** $\sqrt{14}, -\sqrt{14}$ **5.** cube
7. $\frac{1}{27}$, at $(\frac{1}{3}, \frac{1}{3}, \frac{1}{3})$ **9.** max $3^{1-p/q}$, min 1
11. Maximize $xy + yz + zx$ subject to $x^2 + y^2 + z^2 = 1$.
13. max: $\frac{2}{9}(27 + \sqrt{3})$ ft³; min: $\frac{2}{9}(27 - \sqrt{3})$ ft³; sides: $\mu, \mu, 6 - 2\mu$, where $\mu = \frac{1}{3}(6 \mp \sqrt{3})$ ft
15. The conditions for an extremum at $\mathbf{x}$ are $2\mathbf{x}A - 2\lambda\mathbf{x}B = 0$, $\mathbf{x}B\mathbf{x}' = 1$. Hence $\mathbf{x} \neq \mathbf{0}$ and $\mathbf{x}A = \lambda\mathbf{x}B$, that is, $\mathbf{x}AB^{-1} = \lambda\mathbf{x}$. (Note that $\mathbf{x}A\mathbf{x}' = \lambda$.)
17. $$H_f - \lambda H_g = \begin{bmatrix} 2 & 0 & 0 \\ 0 & 2 & 0 \\ 0 & 0 & -2 \end{bmatrix} - \lambda \begin{bmatrix} 0 & 0 & 0 \\ 0 & 0 & 0 \\ 0 & 0 & 0 \end{bmatrix}$$
is not positive definite.

CHAPTER 10

Section 2, page 376

1. $\frac{15}{2}$ **3.** $\frac{1}{9}$ **5.** $\frac{1}{16}$ **7.** 0 **9.** 0 **11.** $\frac{8}{3}$ **13.** 0
15. 0 **17.** $\pi/2$ **19.** $\frac{9}{2}\ln 3 + 3\ln 2 - \frac{15}{4}$

Section 3, page 380

1. $\ln\frac{4}{3}$ **3.** $\frac{4}{9}$ **5.** $\frac{1}{2}$ **7.** 0 **9.** 18
11. $(3^{n+2} - 2^{n+3} + 1)/(n + 1)(n + 2)$ **15.** $A = \iint f(x, y)\, dx\, dy$

Section 4, page 386

1. 16/3 **3.** 16/3 **5.** 46/3 **7.** 27/4 gm **9.** 6.5 gm
11. (7/15, 7/15) **13.** $(\pi - 1, (\frac{3}{8}\pi^2 - \pi + 1)/(\frac{1}{2}\pi - 1))$
15. (0, 14/5)

Section 5, page 397

1. $e - 1$ **3.** 1 **5.** $\frac{1}{30}$ **7.** $\frac{1}{70}$ **9.** $\frac{2}{3}\pi - \frac{1}{2}\sqrt{3}$
11. $\frac{1}{3}\pi - \frac{1}{4}\sqrt{3}$ **13.** $\frac{2}{3}$ **15.** $\frac{1}{3}$ **17.** $\frac{5}{6}\sqrt{5}$
19. Fig. 5.17: $0 \le x \le 1$, $0 \le y \le e^x$; Fig. 5.18: $0 \le x \le 1$, $x^3 \le y \le x^2$; Fig. 5.19: $-\frac{1}{2}\sqrt{3} \le x \le \frac{1}{2}\sqrt{3}$, $1 - (1 - x^2)^{1/2} \le y \le (1 - x^2)^{1/2}$; Fig. 5.20: $0 \le y \le 1$, $-y^2 \le x \le y^2$; Fig. 5.21: $\frac{1}{2}(1 - \sqrt{5}) \le x \le \frac{1}{2}(1 + \sqrt{5})$, $-x \le y \le 1 - x^2$ **21.** $\frac{1}{168}$ **23.** $\frac{32}{105}$ **25.** 104

Section 6, page 407

1. $15\pi/8$ **3.** $15/16$ **5.** 2π **7.** $\frac{4}{3}\pi[a^3 - (a^2 - b^2)^{3/2}]$
11. $(\frac{2}{3}a, 0)$ **13.** $3\pi/640$ **15.** $20/27$
17. $\pi/8$, $\pi/24$, $\pi/8$ **19.** 0, 0, 0 by odd symmetry **21.** $16/3$

CHAPTER 11

Section 1, page 418

1. $\frac{59}{60}$ **3.** $\frac{1}{15}$ **5.** $\frac{1}{315}$ **7.** $ka^9/216$ **9.** $\frac{16}{5}$

17. $\displaystyle\int_{-\sqrt{3}/2}^{\sqrt{3}/2} \left(\int_{1-\sqrt{1-x^2}}^{\sqrt{1-x^2}} f\,dy \right) dx$ **19.** $\displaystyle\int_0^1 \left(\int_{\arcsin x}^{\pi - \arcsin x} f\,dy \right) dx$

21. tetrahedron with vertices at (0, 0, 0), (0, 0, 1), (0, 1, 1), (1, 1, 1);

$$\int_0^1 \left[\int_0^z \left(\int_0^y \rho\,dx \right) dy \right] dz, \quad \int_0^1 \left[\int_x^1 \left(\int_y^1 \rho\,dz \right) dy \right] dx,$$

$$\int_0^1 \left[\int_y^1 \left(\int_0^y \rho\,dx \right) dz \right] dy$$

23. tetrahedron with vertices (0, 0, 0), (0, 0, 2), (0, 3, 2), (6, 3, 2);

$$\int_0^2 \left[\int_0^{3z/2} \left(\int_0^{2y} \rho\,dx \right) dy \right] dz, \quad \int_0^6 \left[\int_{x/2}^3 \left(\int_{2y/3}^2 \rho\,dz \right) dy \right] dx,$$

$$\int_0^3 \left[\int_{2y/3}^2 \left(\int_0^{2y} \rho\,dx \right) dz \right] dy$$

25. $\displaystyle\int_4^5 \left[\int_{10-x}^6 \left(\int_{13-x-y}^3 x\,dz \right) dy \right] dx = \frac{19}{24}$

27. $\displaystyle\frac{1}{2} \int_0^a (a - x)^2 g(x)\,dx$

Section 2, page 426

1. $(-\sqrt{2}, -\sqrt{2}, -3)$, $(-1, 0, 2)$
3. By direct calculation, $\dot{x}^2 + \dot{y}^2 + \dot{z}^2 = \dot{r}^2 + r^2\dot{\theta}^2 + \dot{z}^2$.
5. $d\mathbf{x} = d(r\cos\theta, r\sin\theta, z)$
$= (dr\cos\theta - r\,d\theta\sin\theta, dr\sin\theta + r\,d\theta\cos\theta, dz)$
$= dr(\cos\theta, \sin\theta, 0) + r\,d\theta(-\sin\theta, \cos\theta, 0) + dz(0, 0, 1)$.
7. $df = (\operatorname{grad} f)\cdot d\mathbf{x}$, hence $f_r\,dr + f_\theta\,d\theta + f_z\,dz = (\operatorname{grad} f)\cdot\mathbf{u}\,dr + (\operatorname{grad} f)\cdot\mathbf{w}\,r\,d\theta + (\operatorname{grad} f)\cdot\mathbf{k}\,dz$. Therefore $(\operatorname{grad} f)\cdot\mathbf{u} = f_r$, etc. by equating coefficients of dr, $d\theta$, dz, and the formula follows.
9. $a^4h^2/16$ **11.** $2a^5/15$ **13.** $\pi a^{12}/60$

Section 3, page 436

1. $(-\frac{1}{2}, \frac{1}{2}, -\frac{1}{2}\sqrt{2})$, $(-\frac{3}{2}\sqrt{2}, -\frac{3}{2}\sqrt{2}, 0)$, $(\frac{1}{2}\sqrt{3}, \frac{3}{2}, -1)$
3. $[\rho, \pi - \phi, \theta + \pi]$, $[3\rho, \phi, \theta]$ **5.** $\frac{1}{2}\pi a\sec\alpha$
7. $d\boldsymbol{\lambda} = d(\sin\phi\cos\theta, \sin\phi\sin\theta, \cos\phi) =$
$d\phi\,(\cos\phi\cos\theta, \cos\phi\sin\theta, -\sin\phi) +$
$d\theta\,(-\sin\phi\sin\theta, \sin\phi\cos\theta, 0) = d\phi\,\boldsymbol{\mu} + \sin\phi\,d\theta\,\boldsymbol{\nu}$, etc.
9. In *spherical* coordinates, $[\rho, \phi, \theta] = \rho[1, \phi, \theta]$, hence $f[\rho, \phi, \theta] = \rho^n f[1, \phi, \theta] = \rho^n g(\phi, \theta)$
11. $8\pi/(n+1)(n+2)(n+3)$ **13.** $2\pi(b-a)$
15. $2\pi ah$ **17.** $2\pi^2Aa^2$
19. Center the sphere at the origin and take its apex at $(a\csc\alpha, 0, 0)$. Treat like a volume of revolution problem, using polar coordinates in the x, y-plane:

$$V = \int_0^{(\pi/2)-\alpha}\left(\int_0^{a\csc(\theta+\alpha)}(2\pi r\sin\theta)r\,dr\right)d\theta = \frac{\pi a^3(1-\sin\alpha)^2}{3\sin\alpha}.$$

Check: if $\alpha \longrightarrow \pi/2$, then $V \longrightarrow 0$; if $\alpha \longrightarrow 0$, then $V \approx \frac{1}{3}\pi a^3\csc\alpha = \frac{1}{3}(\pi a^2)(a\csc\alpha)$, the volume of a cone of radius a and height $a\csc\alpha$.
23. Take the axis of the solid cylinder along the z-axis and bore along the y-axis. Section by planes perpendicular to the x-axis; the cross-section is a rectangle of dimensions $2\sqrt{a^2 - x^2}$, $2\sqrt{b^2 - x^2}$. This yields the first integral. Substitute $x = a\cos\theta$ for the second.
25. $2a^2s\arcsin(s/2a) + \frac{1}{2}s^2\sqrt{4a^2 - s^2}$
27. $V = \dfrac{s^2}{2b}\displaystyle\int_c^{c+s}\left[\sqrt{x^2 - b^2} + \frac{x^2}{b}\operatorname{arc\,csc}\left(\frac{x}{b}\right)\right]dx$, where $b = s/2\tan\alpha$

Section 4, page 447

1. $(\frac{3}{8}a, \frac{3}{8}a, \frac{3}{8}a)$ **3.** $(0, 0, \frac{3}{8}a(1 + \cos\alpha))$
5. For the wedge $0 \le r \le a$ and $-\alpha \le \theta \le \alpha$, $\bar{\mathbf{x}} = (\frac{2}{3}(\sin\alpha)\alpha^{-1}a, 0)$.
7. For the wire $r = a$ and $0 \le \theta \le \pi/2$, $\bar{\mathbf{x}} = (2a/\pi, 2a/\pi)$.
9. Let the wire be $\mathbf{x}(\theta) = 100(\cos\theta, \sin\theta)$, where $0 \le \theta \le \pi$ and

$\delta(\theta) = 0.01 + 0.24\theta/\pi$. Then $\bar{x} = -4800/13\pi^2 \approx -37.4$ cm, $\bar{y} = 200/\pi \approx 63.7$ cm

11. $V = 2\pi^2 A a^2$ for a circle of radius a revolved about an axis in its plane at distance A from the center of the circle. (See Ex. **17**, p. 437)

13. Data is in Ex. 11, area $= 4\pi^2 Aa$. **15.** $(0, 0, \frac{1}{2}a(1 + \cos\alpha))$

19. $\frac{1}{4}(\mathbf{a} + \mathbf{b} + \mathbf{c} + \mathbf{d})$

Section 5, page 455

1. products: 0, $I_{xx} = I_{yy} = M(3a^2 + 4h^2)/12$ gm-cm², $I_{zz} = \frac{1}{2}Ma^2$ gm-cm²

3. products: 0, $I_{xx} = I_{yy} = I_{zz} = \frac{2}{5}Ma^2$

5. products: 0, $I_{xx} = \frac{1}{3}M(b^2 + c^2)$, etc.

7. $M = 2\pi^2 A a^2 \delta$. Let $u = x - A$. Then by slicing into cylindrical shells,

$$I_{zz} = \delta \int_{-a}^{a} (A + u)^2[2\pi(A + u)(2\sqrt{a^2 - u^2})\,du] = \tfrac{1}{4}M(4A^2 + 3a^2).$$

9. $I_{zz} = \frac{1}{3}Mha$ **11.** $\frac{1}{6}Mh(a + b)$

13. $\frac{2}{5}Ma^2(10 - \pi)/(6 - \pi)$

15. $\mathbf{v} = \boldsymbol{\omega} \times \mathbf{x}$, hence

$$\begin{aligned}
\mathbf{x} \times \mathbf{v} &= \mathbf{x} \times (\boldsymbol{\omega} \times \mathbf{x}) = |\mathbf{x}|^2 \boldsymbol{\omega} - (\mathbf{x} \cdot \boldsymbol{\omega})\mathbf{x} \\
&= (x^2 + y^2 + z^2)(\omega_x, \omega_y, \omega_z) - (x\omega_x + y\omega_y + z\omega_z)(x, y, z) \\
&= ((y^2 + z^2)\omega_x - xy\omega_y - xz\omega_z, \cdots) \\
&= (\omega_x, \omega_y, \omega_z)\begin{bmatrix} y^2 + z^2 & -xy & -xz \\ -yx & z^2 + x^2 & -yz \\ -zx & -zy & x^2 + y^2 \end{bmatrix}.
\end{aligned}$$

Multiply by δ and integrate.

17. $I_{xx} = I_{yy} = \frac{1}{3}Mh^2$ gm-cm², $I_{zz} = 0$

19. $I_{xx} = I_{yy} = I_{zz} = \frac{2}{3}Ma^2$ **21.** $\frac{1}{2}M(2A^2 + 3a^2)$

CHAPTER 12

Section 2, Page 463

1. s_1 and s_2 are constant on each rectangle of the common refinement; so is $s_1 + s_2$.

3. $|s_1|$ is constant wherever s_1 is.

5. Take a partition with respect to which *both* s_1 and s_2 are step functions. Say $s_1(x, y) = B_{jk}$ and $s_2(x, y) = C_{jk}$ on the interior of $\mathbf{I}_{jk}$. Then

$$\begin{aligned}
\iint (s_1 + s_2) &= \sum (B_{jk} + C_{jk})\,|\mathbf{I}_{jk}| \\
&= \sum B_{jk}\,|\mathbf{I}_{jk}| + \sum C_{jk}\,|\mathbf{I}_{jk}| = \iint s_1 + \iint s_2.
\end{aligned}$$

7. Each $B_{jk} \geq 0$, hence $\sum B_{jk} |\mathbf{I}_{jk}| \geq 0$.
9. Given s, choose a defining partition for s that is *finer* than the given fixed partition of $\mathbf{I}$. Then each $\mathbf{I}_{jk}$ is partitioned into rectangles $\mathbf{I}_{jk,rs}$ inside each of which s is constant, say $s = B_{jk,rs}$. Hence

$$\iint_{\mathbf{I}} s = \sum_{j,k,r,s} B_{jk,rs} |\mathbf{I}_{jk,rs}| = \sum_{j,k} \sum_{r,s} B_{jk,rs} |\mathbf{I}_{jk,rs}| = \sum_{j,k} \iint_{\mathbf{I}_{jk}} s$$

Section 3, page 471

1. Given $\epsilon > 0$, choose s so $|f - s| < \epsilon_1$, where $\epsilon_1 = \epsilon/3|\mathbf{I}|$. Then $s - \epsilon_1 \leq f \leq s + \epsilon_1$, the functions $s \pm \epsilon_1$ are step functions, and $\iint(s + \epsilon_1) - \iint(s - \epsilon_1) = 2 \iint \epsilon_1 = \frac{2}{3}\epsilon < \epsilon$.
3. Suppose $s \leq f \leq S$. Let $\mathbf{J}$ be any rectangle on which s is constant. Since $\mathbf{J}$ contains points with irrational coordinates, $s \leq 0$ on $\mathbf{J}$. Similarly, $S \geq 1$ on each rectangle where S is constant. Hence $\iint S - \iint s \geq \iint 1 - \iint 0 = |\mathbf{I}|$, so $\iint S - \iint s < \epsilon$ cannot be satisfied if $\epsilon \leq |\mathbf{I}|$.
5. From Ex. 4, $\iint f(y, x) = \iint f(x, y)$, hence $2 \iint f(x, y) = 0$.
7. Assume $f(\mathbf{p}) > 0$. Given $\epsilon > 0$, take a partition $\mathbf{I}_{jk}$ so fine that each $\mathbf{I}_{jk}$ has area less than $\epsilon/f(\mathbf{p})$. Suppose $\mathbf{p} \in \mathbf{I}_{rs}$. Define $s(\mathbf{x}) = 0$; define $S(\mathbf{x}) = f(\mathbf{p})$ on $\mathbf{I}_{rs}$ and $S(\mathbf{x}) = 0$ otherwise. Then $s \leq f \leq S$ and

$$\iint_{\mathbf{I}} (S - s) = \iint_{\mathbf{I}} S < f(\mathbf{p})[\epsilon/f(\mathbf{p})] = \epsilon.$$

Thus f is integrable and $0 \leq \iint_{\mathbf{I}} f < \epsilon$. Hence $\iint_{\mathbf{I}} f = 0$. A similar argument holds if $f(\mathbf{p}) < 0$.
9. Suppose $f(\mathbf{p}) > 0$ for some $\mathbf{p}$ in $\mathbf{I}$. By continuity there exists a small square $\mathbf{S}$ centered at p on which $f(\mathbf{x}) > \frac{1}{2}f(\mathbf{p})$. Then $\iint_{\mathbf{I}} f > \frac{1}{2}f(\mathbf{p}) \cdot |\mathbf{S}| > 0$, a contradiction.

Section 4, page 482

1. Say $|f| \leq M$ on $\mathbf{I}$. If $g(x)$ is defined, then

$$|g(x)| = \left| \int_c^d f(x, y)\, dy \right| \leq \int_c^d |f(x, y)|\, dy \leq M(d - c).$$

3. First step: the segment from $\mathbf{x}_0$ to $\mathbf{x}_1$ obviously can be enclosed in the rectangle with $\mathbf{x}_0$ and $\mathbf{x}_1$ as opposite vertices, sides parallel to the axes. Its area is $|(x_1 - x_0)(y_1 - y_0)| \leq \frac{1}{2}|\mathbf{x}_1 - \mathbf{x}_0|^2$. Second step: divide the segment into n equal parts. The division points are $\mathbf{z}_j = \mathbf{x}_0 + (j/n)(\mathbf{x}_1 - \mathbf{x}_0)$. By the first step, the segment from $\mathbf{z}_j$ to $\mathbf{z}_{j-1}$ is enclosed in a rectangle of area $\leq \frac{1}{2}|\mathbf{z}_j - \mathbf{z}_{j-1}| = |\mathbf{x}_1 - \mathbf{x}_0|^2/n^2$. The whole segment is covered by these n rectangles; their *total* area is at most $|\mathbf{x}_1 - \mathbf{x}_0|^2/n$. This is an upper bound for the area of the segment. Let $n \longrightarrow \infty$.
5. The triangle with vertices $\mathbf{0}$, $\mathbf{x}_2 - \mathbf{x}_1$, $\mathbf{x}_3 - \mathbf{x}_1$ is a translate of the given

triangle, so it has the same area. But by subtracting rows,

$$\frac{1}{2}\left|\begin{vmatrix} x_1 & y_1 & 1 \\ x_2 & y_2 & 1 \\ x_3 & y_3 & 1 \end{vmatrix}\right| = \frac{1}{2}\left|\begin{vmatrix} x_1 & y_1 & 1 \\ x_2 - x_1 & y_2 - y_1 & 0 \\ x_3 - x_1 & y_3 - y_1 & 0 \end{vmatrix}\right|$$

$$= \frac{1}{2}\left|(x_2 - x_1)(y_3 - y_1) - (x_3 - x_1)(y_2 - y_1)\right|,$$

which is its area by Ex. 4.

7. Because g and h are continuous, the following three sets are closed:

$$\mathbf{S}_1 = \{(x, y) \mid c \le y \le g(x), \quad a \le x \le b\}$$

$$\mathbf{S}_2 = \{(x, y) \mid g(x) \le y \le h(x), \quad a \le x \le b\}$$

$$\mathbf{S}_3 = \{(x, y) \mid h(x) \le y \le d, \quad a \le x \le b\}.$$

On $\mathbf{S}_1$, $f^*(x, y) = f[x, g(x)]$ is continuous because it is a composite function of continuous functions. Similarly, f^* is continuous on $\mathbf{S}_3$. Also f^* is continuous on $\mathbf{S}_2$ because it equals f, a continuous function. The assertion follows directly from the following lemma. *Lemma:* If g is continuous on closed sets $\mathbf{S}$ and $\mathbf{T}$, then g is continuous on $\mathbf{S} \cup \mathbf{T}$. For let $\mathbf{x}_0 \in \mathbf{S} \cup \mathbf{T}$. To prove g is continuous at $\mathbf{x}_0$. *Case 1:* $\mathbf{x}_0 \in \mathbf{S} \cap \mathbf{T}$. Then if $\epsilon > 0$, there is $\delta_1 > 0$ such that $|g(\mathbf{x}) - g(\mathbf{x}_0)| < \epsilon$ for all $\mathbf{x}$ such that $\mathbf{x} \in \mathbf{S}$ and $|\mathbf{x} - \mathbf{x}_0| < \delta_1$, and there is $\delta_2 > 0$ such that $|g(\mathbf{x}) - g(\mathbf{x}_0)| < \epsilon$ for all $\mathbf{x}$ such that $\mathbf{x} \in \mathbf{T}$ and $|\mathbf{x} - \mathbf{x}_0| < \delta_2$. Choose $\delta = \min\{\delta_1, \delta_2\}$, etc. *Case 2:* $\mathbf{x}_0$ is in one set, not in the other. Say $\mathbf{x}_0 \in \mathbf{S}$ and $\mathbf{x}_0 \notin \mathbf{T}$. Since $\mathbf{T}$ is closed, there is $\delta > 0$ such that $\mathbf{x} \notin \mathbf{T}$ whenever $|\mathbf{x} - \mathbf{x}_0| < \delta$. The continuity of g at $\mathbf{x}_0$ in $\mathbf{S} \cup \mathbf{T}$ now follows from the continuity of g at $\mathbf{x}_0$ in $\mathbf{S}$.

Section 5, page 489

1. 3; inverse: $u = \frac{2}{3}x - \frac{1}{3}y$, $v = -\frac{1}{3}x + \frac{2}{3}y$.
3. $2(v - u)$; inverse: $u = \frac{1}{2}x - \frac{1}{2}\sqrt{2y - x^2}$, $v = \frac{1}{2}x + \frac{1}{2}\sqrt{2y - x^2}$ on the domain $y > \frac{1}{2}x^2$.
5. u^2v; inverse: $u = x$, $v = y/x$, $w = z/y$ on $x > 0$, $y > 0$, z arbitrary.
7. $-1/(x^2 + y^2)^2$. **9.** The cube $0 \le \rho \le \epsilon$, $0 \le \phi \le \pi$, $0 \le \theta \le 2\pi$; $2\pi^2\epsilon \longrightarrow 0$.

Section 6, page 501

1. $F = (1 - e^{-t})/t$, $F' = e^{-t}/t - 1/t^2 + e^{-t}/t^2$
3. $F = [(1 + t)^{n+1} - t^{n+1}]/(n + 1)$, $F' = (1 + t)^n - t^n$
5. Define

$$G(u, v, t) = \int_u^v f(x, t)\, dx.$$

By the Chain Rule,

$$\frac{d}{dt} G[g(t), h(t), t] = G_u\dot{g} + G_v\dot{h} + G_t$$

$$= -f[g(t), t]\dot{g} + f[h(t), t]\dot{h} + \int_g^h f_t(x, t)\, dx.$$

7. $\displaystyle\int_0^{2\pi} \left(\int_0^h a\, dz \right) d\theta = 2\pi a h$

9. $\mathbf{x} = ((A + a \cos \alpha) \cos \theta,\ (A + a \cos \alpha) \sin \theta,\ a \sin \alpha),\ 0 \le \theta,\ \alpha \le 2\pi$

$$\iint a(A + a \cos \alpha)\, d\theta\, d\alpha = 4\pi^2 A a$$

11. $\displaystyle 2\pi a \int_0^{\pi} (c^2 \sin^2 \phi + a^2 \cos^2 \phi)^{1/2} \sin \phi\, d\phi$

13. $\displaystyle\iint_{\mathbf{D}} (1 + a^2 + b^2)^{1/2}\, dx\, dy = (1 + a^2 + b^2)^{1/2} \cdot |\mathbf{D}|$

15. $\displaystyle\int_0^{2\pi} \left(\int_0^1 \frac{r\, dr}{\sqrt{1 - r^2}} \right) d\theta = 2\pi$

17. Set $\mathbf{A} = \displaystyle\iint \mathbf{N}\, dA$. It is enough to show that $\mathbf{A} \cdot \mathbf{i} = \mathbf{A} \cdot \mathbf{j} = \mathbf{A} \cdot \mathbf{k} = 0.$

For example, $\mathbf{A} \cdot \mathbf{k} = \displaystyle\iint \mathbf{N} \cdot \mathbf{k}\, dA = \iint \cos \gamma\, dA = \iint dx\, dy$. Each point of the projection of the closed surface on the x, y-plane is covered an even number of times, half positively, half negatively, hence

$$\iint dx\, dy = 0.$$

19. $A = \pi = \pi[(\sqrt{1 + a^2})^2 - a^2]$. For any oval, $k\, ds = d\alpha$, where α is the angle between the tangent and the positive x-axis. Hence $\displaystyle\int k\, ds = 2\pi$.

Thus the area swept out by the unit tangent is π, regardless of the oval.

21. $\displaystyle\int_{\partial \mathbf{D}} P\, dx + Q\, dy = \iint_{\mathbf{D}} (Q_x - P_y)\, dx\, dy$

$$= \iint_{\mathbf{D}} [(\phi_y)_x - (\phi_x)_y]\, dx\, dy = 0.$$

Alternate solution, not assuming second derivatives exist:

$$\int_{\mathbf{x}_0}^{\mathbf{x}_1} P\,dx + Q\,dy = \int_{\mathbf{x}_0}^{\mathbf{x}_1} \phi_x\,dx + \phi_y\,dy = \int_{\mathbf{x}_0}^{\mathbf{x}_1} d\phi = \phi(\mathbf{x}_1) - \phi(\mathbf{x}_0).$$

This is 0 if $\mathbf{x}_0 = \mathbf{x}_1$ (closed curve).

23. 12

25. $$\int_{\partial\mathbf{D}} (-uv_y + vu_y)\,dx + (uv_x - vu_x)\,dy$$

$$= \iint_{\mathbf{D}} [(uv_x - vu_x)_x - (-uv_y + vu_y)_y]\,dx\,dy$$

$$= \iint_{\mathbf{D}} [u(v_{xx} + v_{yy}) - v(u_{xx} + u_{yy})]\,dx\,dy.$$

The other terms cancel each other. The proof requires u and v to have continuous second partials on $\mathbf{D}$.

27. Let (a, b) be the lower left-hand corner of $\mathbf{I}$ and let (x, y) be any point of $\mathbf{I}$. These points are opposite vertices of a sub-rectangle $\mathbf{J}$, and

$$\int_{\partial\mathbf{J}} P\,dx + Q\,dy = \iint_{\mathbf{J}} (Q_x - P_y)\,dx\,dy = 0.$$

This implies

$$\int_a^x P(s, b)\,ds + \int_b^y Q(x, t)\,dt - \int_a^x P(s, y)\,ds - \int_b^y Q(a, t)\,dt = 0.$$

Hence we can define

$$f(x, y) = \int_a^x P(s, b)\,ds + \int_b^y Q(x, t)\,dt$$

$$= \int_a^x P(s, y)\,ds + \int_b^y Q(a, t)\,dt.$$

By the first form of f, we have $\partial f/\partial y = Q(x, y)$. By the second, $\partial f/\partial x = P(x, y)$.

Section 7, page 510

1. $$\iint_{r \le a} dx\,dy/(1 + r^2) = \int_0^{2\pi} d\theta \int_0^a \frac{r\,dr}{1 + r^2} = \pi \ln(1 + a^2) \longrightarrow \infty$$

as $a \longrightarrow \infty$.

3. $$\iint_{\mathbf{D}_n} g \le \iint_{\mathbf{D}_n} f \le \iint_{\mathbf{R}^2} f.$$

Hence $\{\iint_{\mathbf{D}_n} g\}$ is a bounded increasing sequence (since $g \geq 0$), so it converges.

5. Let **D** be the domain $0 \leq y \leq 1/x$, $1 \leq x \leq b$. Then $1 + x^2y^2 \leq 2$ on **D**, so

$$\iint_{\mathbf{D}} \frac{dx\,dy}{1 + x^2y^2} \geq \frac{1}{2}|\mathbf{D}| = \frac{1}{2}\ln b \longrightarrow \infty.$$

Alternate solution:

$$\int_0^a dy \int_0^a \frac{dx}{1 + x^2y^2} = \int_0^{a^2} \frac{\arctan u}{u}\,du > \int_1^{a^2} \frac{\arctan u}{u}\,du \geq \frac{\pi}{4}\int_1^{a^2} \frac{du}{u}$$

$$= \frac{\pi}{2}\ln a \longrightarrow \infty.$$

7. $$\iint_{\epsilon \leq r \leq 1} (\ln r)\,dx\,dy = 2\pi \int_\epsilon^1 r(\ln r)\,dr$$

$$= (2\pi)\left(-\tfrac{1}{4} - \tfrac{1}{2}\epsilon^2 \ln \epsilon + \tfrac{1}{4}\epsilon^2\right) \longrightarrow -\tfrac{1}{2}\pi$$

as $\epsilon \longrightarrow 0$.

9. $$\iiint_{\epsilon \leq \rho \leq 1} \rho^s\,dx\,dy\,dz = (2\pi)(2)\int_\epsilon^1 \rho^{s+2}\,d\rho, \text{ etc.}$$

11. Let $0 < a < 1$. Then

$$\iiint_{\rho \leq a} (1 - \rho)^s\,dx\,dy\,dz = 4\pi \int_0^a \rho^2(1 - \rho)^s\,d\rho$$

$$= 4\pi \int_{1-a}^1 \rho^s(1 - \rho)^2\,d\rho$$

$$= 4\pi\left[\frac{1 - (1 - a)^{s+1}}{s + 1} - 2\,\frac{1 - (1 - a)^{s+2}}{s + 2} + \frac{1 - (1 - a)^{s+3}}{s + 3}\right]$$

$$\longrightarrow 4\pi\left[\frac{1}{s + 1} - \frac{2}{s + 2} + \frac{1}{s + 3}\right].$$

13. Drop the 4 points $(\pm 1, \pm 1)$ and drop all terms $(m, 0)$, $(0, n)$; that part converges by the integral test. For each remaining point (m, n), let $S_{m,n}$ be the square of sides 1 for which (m, n) is the farthest point from the origin. Then these squares do not overlap and on $S_{m,n}$, $(m^2 + n^2)^{-p} \leq (x^2 + y^2)^{-p}$, hence

$$\frac{1}{(m^2 + n^2)^p} \leq \iint_{S_{m,n}} \frac{dx\,dy}{(x^2 + y^2)^p}.$$

Now sum.

15. $\frac{1}{4}\sqrt{\pi}$

Section 8, page 516

1. 1.1946

3. $(\frac{1}{3}h \sum B_i p_i)(\frac{1}{3}k \sum C_j q_j) = \frac{1}{9}hk \sum B_i C_j p_i q_j = \frac{1}{9}hk \sum A_{ij} f_{ij}$

5. Both expressions are $A + \frac{1}{2}B + \frac{1}{2}C + \frac{1}{4}D$

7. Integrate by parts twice:

$$\int_0^1 y(1-y) f_{yy}(x, y)\, dy = \int_0^1 (2y-1) f_y(x, y)\, dy$$

$$= f(x, 1) + f(x, 0) - 2\int_0^1 f(x, y)\, dy.$$

Now integrate on x:

$$\iint y(1-y) f_{yy}(x, y)\, dx\, dy = \int_0^1 [f(x, 1) + f(x, 0)]\, dx$$

$$- 2\iint f(x, y)\, dx\, dy.$$

Integrate by parts twice:

$$\int_0^1 x(1-x)[f_{xx}(x, 1) + f_{xx}(x, 0)]\, dx$$

$$= \int_0^1 (2x-1)[f_x(x, 1) + f_x(x, 0)]\, dx = -2\int_0^1 [f(x, 1) + f(x, 0)]\, dx.$$

The last step uses the hypothesis $f(0, 0) = f(0, 1) = f(1, 0) = f(1, 1) = 0$.

9. Define $p(x, y) = A + Bx + Cy + Dxy$ by $A = f(0, 0)$, $A + B = f(1, 0)$, $A + C = f(0, 1)$, $A + B + C + D = f(1, 1)$. Then $g(x, y) = f(x, y) - p(x, y)$ satisfies the hypotheses of Ex. 8, etc.

11.

$$\int_a^b \left[\int_c^d \left(\int_e^f F(x, y, z)\, dz\right) dy\right] dx$$

$$\approx \frac{hjk}{27} \sum_{r=0}^{2m} \sum_{s=0}^{2n} \sum_{t=0}^{2p} A_r B_s C_t F(x_r, y_s, z_t),$$

$$h = \frac{b-a}{2m},\ j = \frac{d-c}{2n},\ k = \frac{f-e}{2p},$$

$(A_0, A_1, A_2, \cdots) = (1, 4, 2, 4, 2, \cdots)$, etc.

13.

$$\iiint \approx \tfrac{1}{27}(\tfrac{1}{2})^3[64 \sin \tfrac{1}{8} + 3(16) \sin \tfrac{1}{4} + 3(4) \sin \tfrac{1}{2} + \sin 1] \approx 0.122$$

CHAPTER 13

Section 1, page 520

1. $y = \frac{1}{4}x^4$ **3.** $y = 1 + 3 \ln x$ **5.** $y = 6 - e^{-x}$ **7.** $y = 10e^{x-1}$
9. $y = \frac{1}{2}x^2 - \cos x + 1$ **11.** $y = (x^2 + 1)^{1/2} - 1$
13. $y = x + \frac{1}{3}x^3 + \frac{1}{4}x^4 + \frac{1}{6}x^6$ **15.** $y = x + c/x$
17. $\ln(x^2 + y^2) = 2x + c$

Section 2, page 526

1. $y^2 = \frac{2}{3}x^3 + c$ **3.** $y = (\sqrt{x} + c)^2$ **5.** $y = \ln[2/(c - x^2)]$
7. $y = x[1 + 2/(c - \ln x)]$ **9.** $y = 3/\sqrt{x}$
11. $y = (4 - x)/(2 - x)$ **13.** (Slope at P) $= -$(slope $\overline{OP}$); $xy = c$; rectangular hyperbolas with axes $y = \pm x$
15. (Slope at P) $=$ (slope $\overline{OP}$); $y = cx$; straight lines through (0, 0)

Section 3, page 530

1. linear **3.** linear **5.** non-linear
7. $y = \frac{1}{3}x^3 + c$; yes, $y = \frac{1}{3}x^3$ is a particular solution, and $y = c$ is the general solution of $y' = 0$.

Section 4, page 532

1. $y = c/x^3$ **3.** $y = c \cos \theta$ **5.** $y = ce^{-x}/x$
7. $y = 3(1 - x^2)^{1/2}$ **9.** $y = \frac{4}{5}x^{-2}$

Section 5, page 539

1. $y = \frac{1}{2}x - \frac{1}{4}$ **3.** $y = \frac{1}{5}e^x$ **5.** $y = \frac{1}{4}e^x(2x^2 - 2x + 1)$
7. $\frac{1}{29}(15 \sin x - 6 \cos x)$ **9.** $y = \frac{1}{2}e^{-x}(\sin x + \cos x)$
11. $i = E(R \cos t + L \sin t)/(R^2 + L^2)$
13. $y = -(x^5 + 5x^4 + 20x^3 + 60x^2 + 120x + 120)$
15. $y = xe^x$ **17.** $y = x^3 - x$ **19.** $y = \frac{1}{2}(x^2 + 1) \ln(x^2 + 1)$
21. $y = ce^{-3x} + \frac{1}{3}$ **23.** $y = ce^{-x} + \frac{1}{3}xe^{2x} - \frac{1}{9}e^{2x} + 1$
25. $i = ce^{-Rt/L} + E(R \cos \omega t + L\omega \sin \omega t)/(R^2 + L^2\omega^2)$
27. $y = ce^{-x^2} + 1$ **29.** $y = ce^{-3x} + xe^{-3x}$
33. The slope is constant c along each line $x + y = c$. But each such line has slope -1, so $c = -1$ gives the solution $x + y = -1$.

Section 6, page 548

1. $y = cx^3$ **3.** $100(\log 10)/(\log 2) \approx 332$ years
5. $11^3 \times 10^2 \approx 133000$ **7.** about 32.1 sec
9. $100(2 + \sqrt{3}) \approx 373.2°C$ **13.** $8000/\sqrt{2g} \approx 997$ sec

Section 7, page 554

1. $\frac{1}{2}x^2 + \frac{1}{8}x^4 + \frac{1}{48}x^6$ **3.** $2 + x + x^2 + \frac{1}{3}x^3 + \frac{1}{4}x^4 + \frac{1}{15}x^5 + \frac{1}{24}x^6$
5. $1 - \frac{1}{2}x^2 + \frac{1}{4}x^4 - \frac{1}{12}x^6 + \frac{1}{32}x^8 - \frac{1}{120}x^{10} + \frac{1}{384}x^{12}$

7. $\frac{1}{2}x^2 + \frac{1}{56}x^7 + \dfrac{1}{16\cdot 56}x^{12} + \dfrac{3}{34\cdot(56)^2}x^{17} + \dfrac{1}{22\cdot(56)^3}x^{22}$

9. $\frac{1}{3}x^3 + \frac{1}{63}x^7 + \dfrac{2}{33\cdot 63}x^{11} + \dfrac{1}{15\cdot(63)^2}x^{15}$

11. $\frac{1}{2}x^2 + \frac{1}{8}x^4$ **13.** $2 + x + x^2 + \frac{1}{3}x^3 + \frac{1}{4}x^4 + \frac{1}{15}x^5$
15. $1 - \frac{1}{2}x^2 + \frac{1}{4}x^4$ **17.** $\frac{1}{2}x^2$ **19.** $1 + \frac{1}{3}x^3$
21. $-1 + x + x^2 - \frac{1}{2}x^3 - \frac{5}{24}x^4$ **23.** 0
25. $10 + \frac{1}{60}(t - 10)^3 - \frac{1}{800}(t - 10)^4$, $x(12) \approx 10.1$ ft

CHAPTER 14

Section 2, page 561

1. $x = ae^t + be^{5t}$ **3.** $r = a\cos 2\theta + b\sin 2\theta$
5. $y = e^{x/4}[a\cos(x\sqrt{7}/4) + b\sin(x\sqrt{7}/4)]$ **7.** $x = a + be^{-6t}$
9. $y = ae^{-x} + be^{-4x}$ **11.** $x = e^{2a^2t}(bt + c)$ **13.** $y = a\sin 3x$
15. $x = e^{2t}(at + 1)$ **17.** $r = e^{\theta}(-e^{-\pi}\cos\theta + b\sin\theta)$
19. $r = a(\cos\theta - \sqrt{3}\sin\theta)$

Section 3, page 564

1. $x = \frac{1}{3}t$ **3.** $x = -\frac{1}{2}t - \frac{1}{4}$ **5.** $x = -t^2 - 10t - 38$
7. $x = \frac{1}{7}e^{3t}$ **9.** $y = \frac{1}{9}e^{-2x} - x - 1$
11. $y = \frac{4}{53}(-2\cos 2x - 7\sin 2x)$ **13.** $y = \frac{1}{6}e^x(\cos x + \sin x)$

15. $y = \dfrac{e^x}{1^2+1} + \dfrac{e^{2x}}{2^2+1} + \dfrac{e^{3x}}{3^2+1} + \dfrac{e^{4x}}{4^2+1} + \dfrac{e^{5x}}{5^2+1}$

17. $x = a\cos t + b\sin t + t^2 - 2$
19. $x = ae^{2t} + be^{-3t} + \frac{1}{36}e^{-t}(1 - 6t)$
21. $x = a + be^{-3t} + \frac{1}{10}(-2\cosh 2t + 3\sinh 2t)$
23. $i = e^{-2t/3}(a\cos t\sqrt{2}/3 + b\sin t\sqrt{2}/3) + \frac{10}{17}(4\sin t - \cos t)$
25. $x = 5\cos t - \sin t + 2t - 5$ **27.** $x = \frac{1}{16}(9e^{2t} + 7e^{-2t} - 2\sin 2t)$
29. $x = 10e^t - \frac{56}{5}e^{3t/4} + \frac{1}{5}e^{2t}$
31. Because e^{2t} is a solution of the homogeneous equation; $x = \frac{1}{4}te^{2t}$
33. $z\dot{w} + (2\dot{z} + pz)w = r$

Section 4, page 575

1. $\sqrt{2}\cos 3(t - \frac{1}{12}\pi) = \sqrt{2}\sin 3(t + \frac{1}{12}\pi)$
3. $2\cos 2(t + \frac{1}{12}\pi) = 2\sin 2(t + \frac{1}{3}\pi)$

5. $5\cos(t-\alpha) = 5\sin(t-\beta)$, where $\alpha = \arccos(-\frac{3}{5}) \approx 126° 52'$ and $\beta = \arcsin\frac{3}{5} \approx 36° 52'$

7. $x = 6\cos\pi(t \pm \frac{1}{3})$ **9.** $90/\pi$ **11.** 9

13. π, assuming the equilibrium position is 8 ft from the ceiling; $2\pi\sqrt{L/g}$ if it is L ft from the ceiling

15. $2\pi/[(62.4)\pi g/100]^{1/2} \approx 0.791$ sec

Section 5, page 579

1. $y = ce^x - x - 1$ **3.** $y = a\cos 2x + b\sin 2x$

5. $y = a_1\sum_1^\infty nx^n/(n!)^2$ **7.** $(2a_1/x)(e^{-x} + x - 1)$

9. $y = 2 - 3x + 6x^2 - \frac{34}{3}x^3 + \frac{61}{3}x^4 + \cdots$

11. $y = -1 + 2x - x^2 + \frac{1}{3}x^3 + \frac{1}{6}x^4 + \cdots$

13. $y = 1 + \frac{1}{2}(x-2) + \frac{1}{8}(x-2)^2 + \frac{1}{24}(x-2)^3 + \frac{1}{96}(x-2)^4 + \cdots$

Section 6, page 584

1. $e^A = I + A = \begin{bmatrix} -2 & 9 \\ -1 & 4 \end{bmatrix}$ **3.** $(I - A)^{-1} = I + A = \begin{bmatrix} -2 & 9 \\ -1 & 4 \end{bmatrix}$

5. $f(A) = \begin{bmatrix} f(\lambda) & 0 \\ f'(\lambda) & f(\lambda) \end{bmatrix}$ **7.** $I + (e-1)A = \begin{bmatrix} 1 & e-1 \\ 0 & e \end{bmatrix}$

9. $P^{-1} = \begin{bmatrix} 1 & -2 \\ -2 & 5 \end{bmatrix}, \quad e^A = \begin{bmatrix} 1 & 0 \\ 1 & 1 \end{bmatrix}, \quad e^{PAP^{-1}} = \begin{bmatrix} 3 & -4 \\ 1 & -1 \end{bmatrix}$

11. $A = aI + \begin{bmatrix} 0 & b \\ -b & 0 \end{bmatrix}$, so $e^A = e^{aI}e^{\begin{bmatrix} 0 & b \\ -b & 0 \end{bmatrix}}$, etc.

Section 7, page 590

1. $x = x_0\cosh kt + y_0\sinh kt$ **3.** $\dot{\mathbf{y}} = P^{-1}\dot{\mathbf{x}} = P^{-1}A\mathbf{x} = P^{-1}AP\mathbf{y}$

5. $x = (a + bt)e^{\lambda t} + t^2e^{\lambda t}$, $y = be^{\lambda t} + te^{\lambda t}$

Section 8, page 594

1. $(e^{-tA}X)^\cdot = -e^{-tA}AX + e^{-tA}\dot{X} = e^{-tA}(\dot{X} - AX) = 0$, so $e^{-tA}X = X_0$, etc.

3. $e^A = I + A, \quad e^B = I + B,$

$$e^Ae^B = \begin{bmatrix} 2 & 1 \\ 1 & 1 \end{bmatrix}, \qquad e^{A+B} = \begin{bmatrix} \cosh 1 & \sinh 1 \\ \sinh 1 & \cosh 1 \end{bmatrix}.$$

Since $AB \neq BA$, there is no reason for equality of e^Ae^B and e^{A+B}.

5. Just compute:

$$B = \begin{bmatrix} a & b \\ c & d \end{bmatrix}, \qquad \operatorname{cof} B = \begin{bmatrix} d & -b \\ -c & a \end{bmatrix},$$

$$\operatorname{tr}(\dot{B} \operatorname{cof} B) = \operatorname{tr}\left(\begin{bmatrix} \dot{a} & \dot{b} \\ \dot{c} & \dot{d} \end{bmatrix}\begin{bmatrix} d & -b \\ -c & a \end{bmatrix}\right)$$

$$= \operatorname{tr}\begin{bmatrix} \dot{a}d - \dot{b}c & \cdot \\ \cdot & a\dot{d} - b\dot{c} \end{bmatrix} = (ad - bc)^{\boldsymbol{\cdot}}.$$

7. Solve for g: $g(t) = g(0)\, e^{t(\operatorname{tr} A)} = e^{t(\operatorname{tr} A)}$ since $g(0) = |e^0| = 1$. Set $t = 1$.

9. Integrate twice and change the order of iteration:

$$\dot{x}(t) = \int_0^t g(s)\, ds, \qquad x(t) = \int_0^t \dot{x}(u)\, du$$

$$= \int_0^t \left(\int_0^u g(s)\, ds\right) du = \int_0^t \left(\int_s^t du\right) g(s)\, ds$$

$$= \int_0^t (t - s) g(s)\, ds.$$

CHAPTER 15

Section 1, page 598

1. $4 \pm 3i$ **3.** $\frac{1}{2}(-1 \pm i\sqrt{7})$ **5.** $\frac{1}{3}(1 \pm 2i\sqrt{2})$
7. $\frac{1}{30}(-1 \pm 9i\sqrt{3})$ **9.** $1, \frac{1}{2}(-1 \pm i\sqrt{3})$ **11.** $1, \pm i$
13. $\pm 1, \pm i$ **15.** $(1 + i)^2 = 2i$; $\pm(1 + i)/\sqrt{2}$ **17.** 0

Section 2, page 601

1. $\frac{1}{2} + \frac{1}{3}i$ **3.** $2 - i$ **5.** $5 + i$ **7.** $-\frac{1}{2} - \frac{1}{2}i$ **9.** $\frac{3}{5} - \frac{4}{5}i$
11. [1] $\frac{1}{6}\sqrt{13}$, [3] $\sqrt{5}$, [5] $\sqrt{26}$, [7] $\frac{1}{2}\sqrt{2}$, [9] 1

Section 3, page 608

1. $\sqrt{2}(\cos \frac{3}{4}\pi + i \sin \frac{3}{4}\pi)$ **3.** $2(\cos \frac{1}{3}\pi + i \sin \frac{1}{3}\pi)$
5. $4(\cos \pi + i \sin \pi)$
7. $\sqrt{17}(\cos \theta + i \sin \theta)$, where $\tan \theta = 4$, $0 < \theta < \frac{1}{2}\pi$ **9.** i
11. $-\sqrt{2}$ **13.** $\cos \frac{1}{2}\pi + i \sin \frac{1}{2}\pi$
15. $\frac{1}{2}\sqrt{3}(\cos 7\pi/6 + i \sin 7\pi/6)$ **17.** -4 **19.** $-64\sqrt{3} + 64i$
21. $-\frac{1}{4} - \frac{1}{4}i$ **25.** $\pm 1, \pm i, (\pm 1 \pm i)/\sqrt{2}$, all signs

27. $\pm i$, $\frac{1}{2}(\pm\sqrt{3} \pm i)$, all signs
29. $\cos\theta + i\sin\theta$, $\theta = \frac{1}{20}\pi \pm \frac{1}{5}\pi k$, $0 \le k \le 9$
39. $\cos 4\theta = 8\cos^4\theta - 8\cos^2\theta + 1$

Section 4, page 614

1. $\frac{1}{2}(1 + i\sqrt{3})$, $e(\cos 1 - i\sin 1)$, $e^{1/5}(\cos\frac{2}{5} + i\sin\frac{2}{5})$, $(\cosh 2 + i\sinh 2)/\sqrt{2}$, $\cosh 3$, -1
3. Set $z = x + iy$. By **2**, $\sin z$ is real only if $\sinh y\cos x = 0$, i.e., $y = 0$ or $x = \frac{1}{2}\pi \pm k\pi$, $k = 0, 1, 2, \cdots$. Likewise $\cos z$ is real only if $\cos x\cosh y = 0$ and $\sin x\sinh y = 0$, that is, $\cos x = 0$ and $y = 0$, $z = \frac{1}{2}\pi \pm k\pi$.
5. no; for example, $\cos iy = \cosh y \longrightarrow \infty$ as $y \longrightarrow \infty$
9. $\cosh 4x = 8\cosh^4 x - 8\cosh^2 x + 1$

Section 5, page 619

1. $e^{x\sqrt{3}/2}(\cos\frac{1}{2}x + i\sin\frac{1}{2}x)$
3. $\frac{1}{8}e^x[\frac{1}{17}(\cos 4x + 4\sin 4x) + \frac{4}{5}(\cos 2x + 2\sin 2x) + 3] + c$
5. 0 if $k \ne n$; π if $k = n$
7. $\frac{1}{36}(-\sin 3x + 3x\cos 3x + 27\sin x - 27x\cos x) + c$
9. $\pi(2a_0^2 + a_1^2 + a_2^2 + \cdots + a_n^2)$

Section 6, page 622

1. $y = ae^x + e^{-x/2}(b\cos\frac{1}{2}x\sqrt{3} + c\sin\frac{1}{2}x\sqrt{3})$
3. $y = ae^{x\sqrt{3}} + be^{-x\sqrt{3}} + c\cos x\sqrt{2} + d\sin x\sqrt{2}$
5. $y = e^{3x}(a\cos x + b\sin x) + \frac{1}{30}(\cos 2x - 2\sin 2x)$
7. $y = (a^2 - b^2)^{-2}[(a^2 - b^2)x\cos bx + 2b\sin bx]$ if $a \ne b$; $y = \frac{1}{4}a^{-2}(ax^2\sin ax + x\cos ax + a\sin ax)$ if $a = b$
9. $y = \frac{3}{50}(-\cos x - 7\sin x)$
11. $\frac{1}{2}a(\cos\theta - \sin\theta)$

Section 7, page 627

1. $\sum_0^\infty (\frac{1}{2})^{n+1}(1 + i)^{n+1}(z - i)^n$
3. $\frac{1}{3}\sum_0^\infty (\frac{1}{13})^{n+1}\operatorname{Im}(-2 + 3i)^{n+1}z^n$, $R = \sqrt{13}$
5. $1 - \sum_0^\infty \operatorname{Re}(1 + 2i)^{n+1}z^n$, $R = \sqrt{5}/5$
7. $z^n = \sum_0^n \binom{n}{k} x^{n-k}(iy)^k$, $\partial(z^n)/\partial x = \sum_0^{n-1} (n - k)\binom{n}{k} x^{n-k-1}(iy)^k$. But $(n - k)\binom{n}{k} = n\binom{n-1}{k}$, hence $\partial(z^n)/\partial x = n\sum_0^{n-1} \binom{n-1}{k} x^{n-k-1}(iy)^k = n(x + iy)^{n-1} = nz^{n-1}$; a similar argument for $\partial z^n/\partial y$.
9. $\partial(z^n)/\partial y = i\,\partial(z^n)/\partial x$, $u_y + iv_y = i[u_x + iv_x] = -v_x + iu_x$. Equate real and imaginary parts.
11. Apply Exs. 9 and 10 term-by-term.

Index

A

Absolute convergence, 14
  value, 599
Acceleration, 174
  polar form, 194
Addition law, for centers of gravity, 447
  of matrices, 237
  of vectors, 119
Additivity, 469, 481
Adjoint (cofactor), 263
Affine function, 227(Ex. 12)
Alternating series, 12, 102
Amplitude, 566
Angular momentum, 185, 456(Ex. 15)
  speed, 180
  velocity, 180
Approximating sequence, 503
Approximation
  by Taylor polynomials, 56
  to solutions of differential equations, 549
Arc length, 163, 189
Area, 191, 396, 475, 500
Argument, 602
Aristotle, 3(Ex. 26)
Associative law, 236, 242

B

Bernoulli's equation, 539(Ex. 30)
Bessel's equation, 579(Ex. 6)
Binomial series, 97
Bolzano–Weierstrass theorem, 205
Bounded set, 207

C

Cardioid, 193(Ex. 19)
Cauchy criterion, 5, 17, 205
  test, 5
Cauchy–Schwarz inequality, 129(Ex. 25)
Center of gravity, 383, 401, 440, 442
Central force, 194
Centripetal acceleration, 179
Centroid, 442
Chain rule, 159, 288
Change of variable, 482, 486
Characteristic function, 475
  polynomial, 270
  roots, 270
  vector, 269
Cissoid, 193(Ex. 20)
Closed set, 200
Cofactor matrix, 263
Comparison of series and integrals, 31
  test for series, 6
Completeness, 74
Complex exponential, 610
  numbers, 596
Components of acceleration, 178
  of a vector, 118
Composite functions, 206
Composition of linear functions, 241
Cone, 257
Conjugate, 599
Constrained maxima, 352
Continuity, 205
Contour line, 297
Convergence, 4, 199
Convex, 474
Coordinate curves, 491
  functions, 227
  planes, 116
Coordinates, space, 117
Couple, 156(Ex. 25)
Cramer's rule, 140
Cross product, 141
Curl, 301(Ex. 19)
Curvature, 168
Curve length, 163
Cusp, 166
Cycloid, 188(Ex. 17)
Cylinder, 260
Cylindrical coordinates, 419

D

Definite integrals, 105
De Moivre's theorem, 603
Density, 382
Derivative of a matrix, 269(Exs. 37–40), 591
  of a vector, 158
Determinants, 139
Difference of vectors, 121
Differentiable function, 284, 286
Differential, 316
  equation, 517
Direction cosines, 128
  field, 517
Directional derivative, 302
Distance formula, 124
Divergence, 301(Ex. 19)
Divergent series, 4
Domain, 197
Dot product, 124
Double integral, 372

E

Eccentricity, 195(Ex. 7)
Electric current, 574
Elimination, 134
Ellipsoid, 255, 490
Elliptic integrals, 106
  paraboloid, 259
Equilibrium, 154
Euler's constant, 44
  relation, 291(Ex. 24)
Even function, 86
Exponential function, 582
  of a matrix, 244(Ex. 15), 582

F

First order differential equation, 517
  system, 585
Folium, 188(Ex. 21)
Functions of several variables, 197

G

Gauss, C. F., 597
Generating function, 96(Ex. 17)
Generator of a cone, 257
  cylinder, 260
Geometric series, 1
Gradient, 297
Graph of a function, 209
Green's formula, 502(Ex. 25)
  theorem, 499

H

Half-life, 548(Ex. 3)
Harmonic motion, 566
  series, 2
Hessian matrix, 340
Homogeneous differential equation, 529, 557
  equations, 150
  function, 291(Exs. 16–26)
  linear function, 226
  system, 586, 591
Hooke's law, 546
Hyperbola, 187(Ex. 16)
Hyperbolic functions, 612
  paraboloid, 259
Hyperboloid, 256, 257

I

Identity matrix, 261
Imaginary part, 598
Implicit differentiation, 311
  function, 310
Improper double integral, 502
  integral, 17, 18, 35
Inconsistent system, 135
Indefinite form, 251
Inequalities, 414
Infinite series, 4
Infinitely differentiable, 68
Initial condition, 517
  -value problem, 517, 563
Inner product, 124
Integrable function, 457, 463, 479
  on $\mathbf{R}^2$, 503
Integral, 464
  equation, 549
  of step function, 457, 460
  of vector function, 184
Interior point, 202
Intersection of planes, 137, 148
  of sets, 198
Interval, 198
  of convergence, 76
Inverse of a matrix, 262
Inversion, 489(Ex. 7)
Iterated integral, 379

Iteration, 409, 462, 473, 479
  formula, 378, 393

J

Jacobi identity, 147(Ex. 24)
Jacobian, 483
  matrix, 321
Jordan form, 271
  for symmetric matrices, 276

K

Kepler's laws, 194–196
Kinetic energy, 184

L

Lagrange multiplier, 354
Laplace transform, 21
Leibniz rule, 489
Lemniscate, 193(Ex. 21)
Length of a curve, 163
  of a vector, 123
Level curve, 297
  surface, 299
Limaçon, 193(Ex. 23)
Line integral, 182
Linear differential equation, 526
  function, 224
  system, 133, 266
  transformation, 227
Locally integrable, 503

M

Mass, 382, 401
Matrix, 229
  exponential, 582
  of a linear transformation, 228
  of inertia, 450
  multiplication, 235
Maxima and minima, 207, 217
  with constraints, 352
Mean value theorem, 285
Measurable, 475
Minor, 139
Mixed moments of inertia, 450
  partials, 327
Modulus, 599
Moment, 383, 440
  of inertia, 449, 454
Momentum, 185
Mothball problem, 542
M-test, 112
Multiplication by a scalar, 120
Multiplier, 354

N

Natural frame, 422, 428
Negative definite matrix, 248
Negative semi-definite, 249
Newton's law of cooling, 543
  of motion, 175
Non-homogeneous differential equation, 529
  systems, 589
Non-parametric surface, 491
Non-singular matrix, 265
Normal distribution, 506
  form, 131
  to a surface, 295
  vector, 172
Numerical integration, 511

O

Odd function, 86
Open set, 203
Orientation, 145
Orthogonal, 298
  projection, 230

P

Pappus theorems, 444, 446
Paraboloids, 258
Parallel-axis theorem, 454
Parameters, 153
Parametric equation, 130
  form of a plane, 152
  surface, 491
Partial derivative, 213
  fractions, 93, 625
  sum, 4
Particular solution, 562
Partition, 457, 458, 481
Period, 566
  of orbit, 196
Phase angle, 566
Picard method, 552

Plane, 131
  through three points, 154
Polar angle, 189
  coordinates, 188, 400
Positive definite, 246
  semi-definite, 249
Potential, 308
Power series, 72
Principal minor, 250
Product of inertia, 450
  of matrices, 235
Proper transformation, 484
p-series, 7
Pure partials, 326

Q

Quadratic form, 244, 366
Quadric surface, 253

R

Radius, 189
  of convergence, 76, 624
Ratio test, 10, 78
Real part, 598
Recurrence relation, 577
Regular transformation, 484
Resonance, 574
Riemann integral, 465
Right-hand rule, 116
R-integrable, 465
R-measurable, 475
Root test, 12(Ex. 17)
Roots of unity, 605
Rose curves, 193(Exs. 12–17)
Row-by-column multiplication, 232
Row-by-matrix-by-column, 234
Rulings on a quadric, 297(Exs. 11–13)

S

Saddle point, 344
Scalar multiple, 120
Second derivative test, 340, 367
  order differential equation, 588
Semi-definite, 249
Separation of variables, 520
Set notation, 198
Simple harmonic motion, 566
  integral, 369
Simpson's rule, 512
Singular matrix, 265
Singularity, 35
Skew lines, 151
Solid angle, 435
Solution of differential equation, 517
Speed, 158
Spherical area, 434
  coordinates, 427
Spiral of Archimedes, 193(Ex. 11)
Steady state, 574
Step function, 457, 459
Steradian, 436
Stirling's formula, 47
Strophoid, 193(Ex. 22)
Successive approximations, 549
Sum of vectors, 119
Summation notation, 5
Superposition, 595
Surface area, 493
Symmetric matrix, 245
System determinant, 140
System of differential equations, 585

T

Tangent plane, 292
  vector, 167
Taylor formula, 58
  polynomials, 56, 334
  series, 68
Term by term operations on series, 90
Tetrahedra, 416
Torque, 146, 156(Ex. 24)
Transformation, linear, 227
  of variables, 482
Transient, 574
Transpose, 236
Triangle inequality, 123
Trigonometric functions, 611
Triple cross product, 147(Ex. 21), 244(Ex. 12)
  integral, 409
  scalar product, 145

U

Uncoupled system, 587
Underdetermined system, 136

Uniform continuity, 208
convergence, 109
Union of sets, 197
Uniqueness of power series, 82
theorem, 592
Unit normal, 172
tangent, 167
vector, 127

V

Vector, 118
algebra, 122
field, 306
Velocity, 158, 194
Voltage, 309
Volume of parallepiped, 147

W

Wallis's product, 45
Weierstrass M-test, 112
–Bolzano theorem, 205
Witch of Agnesi, 188(Ex. 20)
Wronskian, 562(Ex. 22), 595(Ex. 8)

Z

Zeno's paradoxes, 3(Ex. 26)

4
5
6
7
8
9
0
1
2
3